Moscow Lectures

Volume 4

More information about this series at http://www.springer.com/series/15875

Vladimir I. Bogachev • Oleg G. Smolyanov

Real and Functional Analysis

NATIONAL RESEARCH
UNIVERSITY

Skoltech
Skolkovo Institute of Science and Technology

 Springer

Vladimir I. Bogachev
Department of Mechanics
and Mathematics
Moscow State University
Moscow, Russia

Higher School of Economics
National Research University
Moscow, Russia

Oleg G. Smolyanov
Department of Mechanics
and Mathematics
Moscow State University
Moscow, Russia

Moscow Institute of Physics
and Technology
Dolgoprudnyi, Russia

This book is an expanded and revised version of the work first published in Russian in 2009 (1st edition) and 2011 (2nd edition) with the publisher *Regular and Chaotic Dynamics*, Moscow - Izhevsk, under the title Действительный и функциональный анализ: университетский курс.

ISSN 2522-0314   ISSN 2522-0322   (electronic)
Moscow Lectures
ISBN 978-3-030-38221-6   ISBN 978-3-030-38219-3   (eBook)
https://doi.org/10.1007/978-3-030-38219-3

Mathematics Subject Classification (2010): 28-01, 46-01, 46A03, 46Bxx, 46Fxx, 47-01

This Springer imprint is published by the registered company Springer Nature Switzerland AG
The registered company address is: Gewerbestrasse 11, 6330 Cham, Switzerland

Preface to the Book Series *Moscow Lectures*

You hold a volume in a textbook series of Springer Nature dedicated to the Moscow mathematical tradition. Moscow mathematics has very strong and distinctive features. There are several reasons for this, all of which go back to good and bad aspects of Soviet organization of science. In the twentieth century, there was a veritable galaxy of great mathematicians in Russia, while it so happened that there were only few mathematical centers in which these experts clustered. A major one of these, and perhaps the most influential, was Moscow.

There are three major reasons for the spectacular success of Soviet mathematics:

1. Significant support from the government and the high prestige of science as a profession. Both factors were related to the process of rapid industrialization in the USSR.
2. Doing research in mathematics or physics was one of very few intellectual activities that had no mandatory ideological content. Many would-be computer scientists, historians, philosophers, or economists (and even artists or musicians) became mathematicians or physicists.
3. The Iron Curtain prevented international mobility.

These are specific factors that shaped the structure of Soviet science. Certainly, factors (2) and (3) are more on the negative side and cannot really be called favorable but they essentially came together in combination with the totalitarian system. Nowadays, it would be impossible to find a scientist who would want all of the three factors to be back in their totality. On the other hand, these factors left some positive and long lasting results.

An unprecedented concentration of many bright scientists in few places led eventually to the development of a unique "Soviet school". Of course, mathematical schools in a similar sense were formed in other countries too. An example is the French mathematical school, which has consistently produced first-rate results over a long period of time and where an extensive degree of collaboration takes place. On the other hand, the British mathematical community gave rise to many prominent successes but failed to form a "school" due to a lack of collaborations. Indeed, a school as such is not only a large group of closely collaborating individuals but also a group knit tightly together through student-advisor relationships. In the USA, which is currently the world leader in terms of the level and volume of mathematical research, the level of mobility is very high, and for this reason there are no US mathematical schools in the Soviet or French sense of the term. One can talk not only about the Soviet school of mathematics but also, more specifically, of the Moscow, Leningrad, Kiev, Novosibirsk, Kharkov, and other schools. In all these places, there were constellations of distinguished scientists with large numbers of students, conducting regular seminars. These distinguished scientists were often not merely advisors and leaders, but often they effectively became spiritual leaders in a very general sense.

A characteristic feature of the Moscow mathematical school is that it stresses the necessity for mathematicians to learn mathematics as broadly as they can, rather than focusing on a narrow field in order to get important results as soon as possible.

The Moscow mathematical school is particularly strong in the areas of algebra/algebraic geometry, analysis, geometry and topology, probability, mathematical physics and dynamical systems. The scenarios in which these areas were able to develop in Moscow have passed into history. However, it is possible to maintain and develop the Moscow mathematical tradition in new formats, taking into account modern realities such as globalization and mobility of science. There are three recently created centers—the Independent University of Moscow, the Faculty of Mathematics at the National Research University Higher School of Economics (HSE) and the Center for Advanced Studies at Skolkovo Institute of Science and Technology (SkolTech)—whose mission is to strengthen the Moscow mathematical tradition in new ways. HSE and SkolTech are universities offering officially licensed full-time educational programs. Mathematical curricula at these universities follow not only the Russian and Moscow tradition but also new global developments in mathematics. Mathematical programs at the HSE are influenced by those of the Independent University of Moscow (IUM). The IUM is not a formal university; it is rather a place where mathematics students of different universities can attend special topics courses as well as courses elaborating the core curriculum. The IUM was the main initiator of the HSE Faculty of Mathematics. Nowadays, there is a close collaboration between the two institutions.

While attempting to further elevate traditionally strong aspects of Moscow mathematics, we do not reproduce the former conditions. Instead of isolation and academic inbreeding, we foster global sharing of ideas and international cooperation. An important part of our mission is to make the Moscow tradition of mathematics at a university level a part of global culture and knowledge.

The "Moscow Lectures" series serves this goal. Our authors are mathematicians of different generations. All follow the Moscow mathematical tradition, and all teach or have taught university courses in Moscow. The authors may have taught mathematics at HSE, SkolTech, IUM, the Science and Education Center of the Steklov Institute, as well as traditional schools like MechMath in MGU or MIPT. Teaching and writing styles may be very different. However, all lecture notes are supposed to convey a live dialog between the instructor and the students. Not only personalities of the lecturers are imprinted in these notes, but also those of students.

We hope that expositions published within the "Moscow lectures" series will provide clear understanding of mathematical subjects, useful intuition, and a feeling of life in the Moscow mathematical school.

Moscow, Russia   Igor M. Krichever
                                                              Vladlen A. Timorin
                                                             Michael A. Tsfasman
                                                              Victor A. Vassiliev

Preface

Wherefore I perceive that there is nothing better,
than that a man should rejoice in his own works; for
that is his portion: for who shall bring him to see what
shall be after him?

Ecclesiastes.

This book is the result of the substantial reworking and enlarging of lectures on real and functional analysis given over the years by the authors at the Department of Mechanics and Mathematics of Moscow State University and also at the Faculty of Mathematics of the Higher School of Economics in Moscow.

A course on the theory of functions and functional analysis, with the title "Analysis-III", was first introduced in the university curriculum (for all students of the third year, not as a 8000-course) at MekhMat in the 1940s on the initiative of A. N. Kolmogorov, who became the first lecturer together with S. V. Fomin, who gave this course at the Department of Physics. Several years later the classical textbook [331] by Kolmogorov and Fomin was published, which to the present day remains among the best university courses. Later this course was given at Mekhmat by I. M. Gelfand, G. E. Shilov and other eminent mathematicians. Analysis-III gathered several previously existing courses of the theory of functions of a real variable, integral equations, and also elements of variational calculus. At present at Mekhmat and at other top mathematical departments, the course of real analysis (more precisely, the theory of the Lebesgue integral) and variational calculus (optimal control) have again become separate courses (though, there exist university programmes of functional analysis that include the Lebesgue integral). On the other hand, advanced courses of functional analysis now include, unlike in Kolmorogov's time, the spectral theory of selfadjoint operators and elements of the theory of Sobolev spaces.

Several excellent more recent books on functional analysis are suitable as textbooks, such as Reed, Simon [502], Rudin [522], Werner [628], and many others mentioned in the bibliographic comments (which list several hundred titles that we looked through over several decades in the libraries of dozens of universities and mathematical institutes all over the world), as well as fundamental treatises of a reference nature, such as Dunford, Schwartz [**164**], Edwards [**171**], and Kantorovich, Akilov [**312**]. However, most of them either do not cover all the material included in advanced courses of real and functional analysis (in particular, needed for PhD students) or, at the opposite extreme, they contain too much additional material not separated from the necessary minimum and by their organization and style are not textbooks, but rather advanced courses for researchers and readers already familiar with the subject. This book aims at giving a modern exposition of all the material from functional analysis traditionally presented in a one-year or one-and-a-half year course (including real analysis) and necessary for the study of partial differential equations, mathematical physics, optimal control, and the theory of stochastic processes.

There are three levels of exposition in this book oriented towards different categories of readers: 1) a relatively standard course for mathematical departments of universities; this part occupies less than one half of the whole book; the corresponding material is distributed over the main sections of Chapters 1–9 and includes approximately one half of all the exercises; 2) complementary material for advanced students and PhD students which includes the main sections of Chapters 10–12 and a small number of subsections from the additional sections (called "Complements and exercises") in other chapters as well as some exercises; 3) finally, more special information deserving the attention of more professional readers (this information is presented in the sections "Complements and exercises" in all chapters).

We emphasize that the core material connected with levels 1) and 2) occupies less than 300 pages, i.e., somewhat less than half of the book (this material corresponds to approximately 100 academic lecturing hours, and a lecture hour enables one to cover on average 2–3 pages of book text).

This structure makes the present text different from many existing books on the subject, although we have been obviously influenced by many of these books. Our goal is to offer a modern textbook for a broad readership covering all necessary material in real and functional analysis for advanced students and PhD students. Of course, as mentioned in 3), we also step out of any formal curriculum and present additional useful and interesting facts which either demonstrate the connections of this area with other areas and applications, or belong to the classical foundations of functional analysis and admit a relatively simple presentation (for example, we give with complete proofs the Eberlein–Shmulian and Krein–Milman Theorems). However, all such facts are placed in complementary sections and can be omitted on the first reading. In addition, these complements are not used in the main part of the book and are independent of each other. Naturally, certain questions are presented in the aforementioned treatises in more detail and in greater depth, and there are interesting applications discussed in other textbooks but not mentioned at all here. As follows from what has been said, we have not aimed at creating a universal encyclopedia of functional analysis. The bibliographic comments give further references for reading in many classical and modern directions, which along with complementary sections can make this book useful also for a broader audience of researchers in the most diverse fields.

Every chapter is accompanied by a collection of relatively simple exercises (marked by °), which can be used in class or for self-study, and also by a number of more difficult problems oriented towards more advanced students. Problems of this kind can be offered at PhD examinations in the case when the demonstrated level of training gives good grounds to hope for success. We tried to avoid too difficult exercises, preferring to mention some useful facts without proof with reference to the literature (a limited number of very difficult problems are marked by *). Some difficult exercises are given with solutions or references to the literature, but also many routine exercises are provided with hints.

The book ends with a detailed subject index at the beginning of which we placed the list of notation ordered according to the first letter of the name, with the exception of symbols beginning with mathematical symbols. The terms containing several words are usually indicated in the index with all significant words. For example, the term "Banach space" is also given as "space – Banach".

The sections are numbered consecutively within each chapter, and the section number is preceded by the chapter number; for example, the second section of Chapter 3 is numbered as §3.2. For assertions and formulas we use the triple numeration: the chapter number, the section number, and the assertion number (all assertions are numbered consecutively within the section independently of their type, and their numbers stand in front of their names). For example, a reference to Definition 1.2.5 refers the reader to Section §1.2 in Chapter 1, where one finds "1.2.5. Definition". The example following this definition is presented as "1.2.6. Example". In such a system, the number 6 does not mean, of course, that this is the sixth example in the given section, but only indicates the general number in the series of the numbered assertions in this section. Similarly arranged numbers of formulas are included in parentheses.

Let us make some more specific comments on all chapters. Chapter 1 gives a concise introduction to metric and topological spaces including Baire's Theorem and compactness. Although this material is often presented in courses of advanced calculus, it is reasonable to recall basic concepts and facts in a course on functional analysis. Along with some standard topological concepts we also discuss convergence of nets, which is useful for the study of nonmetrizable functional spaces. Of course, the main material of this book can be read without reference to topologies and nets. Chapters 2–4 contain Lebesgue's theory of measure and integral along with connections between differentiation and integration. This material can be included also in a course of advanced calculus. It would be possible to avoid using Lebesgue's integral in a course of functional analysis, at least in an introductory course. However, a serious discussion of functional spaces, operators, distributions and Fourier transforms obviously requires inclusion of Lebesgue's integral as a prerequisite. As a kind of irreducible minimum we would mention the classical theorems about passage to the limit in the integral, Fubini's theorem, the spaces L^1 and L^2, and Hölder's inequality. Spaces of integrable functions and operators on them belong to the classics of functional analysis and were widely employed by Banach in his famous book that laid the foundations of this subject. The proper course on functional analysis begins in Chapter 5 devoted to normed spaces; in particular to Banach and Hilbert spaces. In Chapter 6 we study linear functionals and operators on Banach spaces, which is the heart of linear functional analysis. This is the largest chapter in the book. It contains the three main results about linear operators: the Hahn–Banach Theorem, the Banach–Steinhaus Theorem, and the Banach Inverse Mapping Theorem (along with its equivalent: the Closed Graph Theorem). Among many other important things discussed here are weak topologies, weak compactness and compact operators. Chapter 7 continues the study of linear operators, but focuses on the spectral theory of linear operators. Most of the chapter deals with Hilbert spaces and selfadjoint operators, but some fundamental facts are established for general Banach spaces. The most important result of this chapter is the Spectral Theorem for bounded selfadjoint operators on Hilbert spaces, according to which a general selfadjoint operator is the operator of multiplication by a real measurable function on some L^2 space. Chapter 8 gives an introduction to locally convex spaces, which is motivated by the theory of distributions (generalized functions), the fundamentals of which are presented in this chapter, as well as by the discussion of weak topologies in the previous chapter. Chapter 9 introduces two other very important objects of modern analysis: Fourier transforms of functions and distributions and Sobolev classes of functions with generalized derivatives. In Chapter 10 we turn to unbounded linear operators including unbounded selfadjoint operators. The main motivation for the study of unbounded selfadjoint operators is their application in quantum mechanics (this chapter also contains a brief discussion of mathematical models of quantum mechanics). Of course, there are other important motivations such as applications to differential operators and the study of generators of operator semigroups, which are also discussed here. The material of Chapters 7 and 10 leads naturally to the subject of Chapter 11: Banach algebras, a far reaching generalization of the spaces of bounded operators and bounded continuous functions. The key results of Chapters 7–11 were discovered shortly after the appearance of Banach's book and represent very important new steps in the development of the linear theory. Nonlinear functional analysis is beyond the scope of our book, but nevertheless we could not stop without giving a short introduction to this actively developing area in the final Chapter 12, where we discuss differentiability in infinite-dimensional spaces and some classical results about nonlinear mappings.

We are grateful to our colleagues and students for suggested remarks on the text, we are particularly indebted to E. A. Alekhno for his substantial editing work and also to A. V. Arutyunov, M. D. Avdeeva, N. H. Bingham, P. A. Borodin, D. Elworthy, A. V. Gorshkov, A. Ya. Helemskii, A. V. Korolev, E. D. Kosov, E. P. Krugova, A. A. Lipchius, I. Marshall, S. N. Popova, O. V. Pugachev, A. V. Shaposhnikov, S. V. Shaposhnikov, and N. A. Tolmachev.

Contents

CHAPTER 1

Metric and Topological Spaces

In this chapter we consider the most important concepts of modern analysis — metric and topological spaces. Large parts of the chapter consist of definitions and examples (probably, already familiar to the reader from a course of calculus), moreover, some additional information about compact spaces needed for the sequel is given.

1.1. Elements of Set Theory

We assume some knowledge of fundamentals of set theory, but here we recall some concepts and facts for convenience of further references. An important role in functional analysis is played by the non-defined notions of set and mapping. A mapping f from a set X to a set Y (notation: $f\colon X \to Y$) is called *injective* or an *injection* if $f(x) \neq f(y)$ whenever $x \neq y$. A mapping f is called *surjective* or a *surjection* if $f(X) = Y$, i.e., for every $y \in Y$ there is $x \in X$ with $f(x) = y$. If f is injective and surjective, then f is called a *bijection* or a *one-to-one mapping*. Let $f^{-1}(y)$ denote the *preimage* of y, i.e., $f^{-1}(y) := \{x \in X\colon f(x) = y\}$. The preimage of a set $E \subset Y$ is $f^{-1}(E) := \{x \in X\colon f(x) \in E\}$. For every $A \subset X$, the set $f(A) := \{f(x)\colon x \in A\}$ is called the *image* of A under the mapping f. The restriction of a mapping f to a subset $E \subset X$ is denoted by $f|_E$.

The product of nonempty sets X and Y is the set $X \times Y$ of all pairs of the form (x, y), where $x \in X$ and $y \in Y$. The product of an arbitrary family of nonempty sets X_t, where the index t runs through some nonempty set T, is defined as the family of all collections $\{x_t\}_{t \in T}$ with $x_t \in X_t$. It is denoted by $\prod_{t \in T} X_t$.

The sets of natural, integer, rational, real, and complex numbers are denoted, respectively, by $\mathbb{N}$, $\mathbb{Z}$, $\mathbb{Q}$, $\mathbb{R}$, and $\mathbb{C}$. The symbol $\mathbb{R}^n$ denotes the n-dimensional coordinate space.

Throughout we employ the following two set-theoretic concepts: an equivalence relation and a relation of partial order.

Suppose that we are given some set $\mathcal{R}$ of pairs of elements of a set X, i.e., a subset $\mathcal{R} \subset X \times X$. We shall say that $\mathcal{R}$ defines on the set X an *equivalence relation* and write $x \sim y$ whenever $(x, y) \in \mathcal{R}$, provided that the following conditions are fulfilled:

(i) $x \sim x$ for all $x \in X$,
(ii) if $x \sim y$, then $y \sim x$,
(iii) if $x \sim y$ and $y \sim z$, then $x \sim z$.

V. I. Bogachev, O. G. Smolyanov, *Real and Functional Analysis*,
Moscow Lectures 4, https://doi.org/10.1007/978-3-030-38219-3_1

By simple examples one can show that these three conditions are independent.

Any equivalence relation splits X into disjoint *equivalence classes* consisting of pairwise equivalent elements. For example, if $x \sim y$ only for $x = y$, then every class consists of precisely one element. If, conversely, all elements are equivalent, then we obtain only one equivalence class. Yet another example: let $x \sim y$ for $x, y \in \mathbb{R}^1$ if $x - y \in \mathbb{Q}$. Then equivalence classes are countable. It is often useful to choose a representative in every equivalence class. It turns out that for accomplishing this desire, perfectly innocent at the first glance, one needs a special axiom.

Axiom of choice. *For every collection of nonempty pairwise disjoint sets, there exists a set containing precisely one element from each of these sets.*

Using this axiom is important for many constructions of functional analysis, and without this axiom at least for countable collections very little remains from "continuous mathematics". Nevertheless, it is useful to remember that this is indeed an axiom that does not follow from the principles of the so-called naive set theory.

We say that a set X is equipped with a relation of *partial order* or is *partially ordered* if a certain collection $\mathcal{P}$ of pairs $(x, y) \in X \times X$ is distinguished, for which we write $x \leqslant y$, such that

(i) $x \leqslant x$, (ii) if $x \leqslant y$ and $y \leqslant z$, then $x \leqslant z$.

If $x \leqslant y$, we also write $y \geqslant x$. Note that we do not include the equality $x = y$ in the case $x \leqslant y$ and $y \leqslant x$ unlike a number of other authors.

It is important that there is no requirement that all elements be pairwise comparable. For example, the plane $\mathbb{R}^2$ can be partially ordered as follows: $x = (x_1, x_2) \leqslant y = (y_1, y_2)$ if $x_1 \leqslant y_1$ and $x_2 \leqslant y_2$.

If all elements in X are pairwise comparable, then X is called *linearly ordered*. For example, the real line with its usual order is linearly ordered, but the aforementioned coordinate-wise order on the plane is not. However, the plane can be equipped with a natural linear order: the so-called *lexicographic order*, for which $x \leqslant y$ if either $x_1 < y_1$ or $x_1 = y_1$ and $x_2 \leqslant y_2$.

In a partially ordered set some parts can be linearly ordered. Such parts are called *chains*. For example, the real line as a part of the plane with the coordinate-wise order is a chain.

If X is a partially ordered set and $M \subset X$, an element $\mu \in X$ is called a *majorant* or *upper bound* of the set M if $m \leqslant \mu$ for all $m \in M$. If m is a majorant of M such that $m \leqslant \widehat{m}$ for every other majorant $\widehat{m}$ of the set M, then m is called the *least upper bound* or *supremum* of M. An element $m \in X$ is called *maximal* if we have $m' \leqslant m$ whenever $m \leqslant m'$. Note that there is no requirement that all elements of X be smaller than m. For example, if $x \leqslant y$ only in case $x = y$, every element is maximal. Similarly one defines a *minorant* (or *lower bound*), the *infimum the greatest lower bound*, and a *minimal* element.

A linearly ordered set X is called *well-ordered* if every nonempty part of X has a minimal element.

For example, the set of natural numbers with its natural order is well-ordered, and the sets of rational and real numbers are not.

The axiom of choice is equivalent to the following assertion (if we accept it as an axiom, then the axiom of choice becomes a theorem); for a proof, see [271], [331], and [364].

The Zermelo theorem. *Every nonempty set can be well-ordered.*

Let us give yet another corollary of the axiom of choice (which is also equivalent to it).

Zorn's lemma (or the Kuratowski–Zorn lemma). *If every chain in a partially ordered set X has a majorant, then X has a maximal element.*

We recall once again that a maximal element need not be unique.

A typical example of using Zorn's lemma is the proof of existence of algebraic bases in linear spaces. A system of distinct vectors $\{v_\alpha\}$ in a linear space X is *linearly independent* (or *algebraically independent*) if $c_1 v_{\alpha_1} + \cdots + c_n v_{\alpha_n} \neq 0$ for all numbers c_i not vanishing simultaneously and all distinct indices $\alpha_1, \ldots, \alpha_n$. Such a system is called an *algebraic basis* (a *Hamel basis*) in the space X if every vector in X is a finite linear combination of vectors v_α. In the zero space the zero vector is regarded as a basis.

1.1.1. Proposition. *Every real or complex linear space possesses an algebraic basis. Any two such bases have the same cardinality. In addition, an algebraic basis of a linear subspace can be complemented to an algebraic basis of the whole space.*

PROOF. We can assume that the given space X contains nonzero vectors. Then there are systems of algebraically independent vectors in X. Let Λ denote the family of all such systems. We can introduce in Λ the following order: $\lambda_1 \leqslant \lambda_2$ if $\lambda_1 \subset \lambda_2$. Clearly, this is a partial order. We have to show that the set Λ contains a maximal element, i.e., a system λ of algebraically independent vectors that is not a proper subset of any other system of independent vectors. Such maximal system will be a basis, since the existence of a vector v that is not a linear combination of vectors in λ would mean that the system $\lambda \cup v$ is also independent contrary to the maximality of λ. The existence of a maximal element follows from Zorn's lemma. In order to apply it we have to verify that every chain Λ_0 in Λ has a majorant. In other words, given a set Λ_0 of independent collections of vectors such that any two collections are comparable (one of the two is contained in the other), we have to find an independent system of vectors containing all systems in Λ_0. For such a system we must take the union λ_0 of all systems in Λ_0. The fact that the obtained system is independent is clear from the following reasoning. If vectors $v_1, \ldots, v_n$ belong to λ_0, then there exist systems $\lambda_1, \ldots, \lambda_n \in \Lambda_0$ such that $v_i \in \lambda_i$ for all $i = 1, \ldots, n$. Since the systems λ_i are pairwise comparable, there is the largest one λ_{i_0} among them. Then all vectors v_i belong to λ_{i_0} and hence are linearly independent.

A minor modification of this reasoning enables us to complement an algebraic basis of a subspace to a basis of the whole space: it suffices to take for elements Λ

independent systems containing a fixed basis of the given subspace. This reasoning applies to any field.

Finally, the assertion about the equicardinality of algebraic bases of the space X in case of a finite-dimensional space is known from linear algebra. If X is infinite-dimensional and γ_1 and γ_2 are two algebraic bases, then the cardinality of γ_2 does not exceed the cardinality of γ_1. Indeed, to every element $v \in \gamma_2$ we associate a finite set of elements $S \subset \gamma_1$ through which it is linearly represented. This finite set S is associated to at most finitely many elements in γ_2 (not exceeding the cardinality of S, since by means of k vectors one cannot linearly express more than k linearly independent vectors). Hence the cardinality of γ_2 does not exceed the cardinality of the set of finite subsets of γ_1, which has the same cardinality as γ_1 (Exercise 1.9.27). Thus, the cardinality of γ_2 does not exceed the cardinality of γ_1, and the opposite inequality is true as well. □

1.2. Metric Spaces

One of the most fundamental concepts of modern analysis is the concept of metric space.

1.2.1. Definition. *A set X equipped with a function $d\colon X \times X \to [0, +\infty)$ is called a metric space with a metric d if*
1) $d(x, y) = 0$ *precisely when $x = y$,*
2) $d(x, y) = d(y, x)$ *for all $x, y \in X$,*
3) $d(x, z) \leqslant d(x, y) + d(y, z)$ *for all $x, y, z \in X$.*

Condition 3) is called the *triangle inequality*. A metric space X with a metric d is denoted by (X, d). The same set can be equipped with different metrics. When speaking of a metric space X without indicating the metric it is meant that some metric is fixed. If only conditions 2) and 3) are fulfilled, then d is called a *semimetric*.

Any subset of a metric space becomes a metric space with respect to the metric of the original space.

1.2.2. Example. (i) Any set X is a metric space with the discrete metric $d(x, y) = 1$ whenever $x \neq y$, $d(x, x) = 0$.

(ii) The set of all bounded functions $B(\Omega)$ on a nonempty set Ω is a metric space with the metric

$$d(f, g) = \sup_{x \in \Omega} |f(x) - g(x)|. \qquad (1.2.1)$$

A special case is the space l^∞ of all bounded sequences on $\mathbb{N}$ with the metric

$$d(x, y) = \sup_n |x_n - y_n|, \quad x = (x_n), y = (y_n).$$

(iii) The set $C[a, b]$ of all continuous functions on the interval $[a, b]$ is a metric space with the metric (1.2.1).

(iv) The set l^2 of all infinite real sequences $x = (x_n)$ such that $\sum_{n=1}^{\infty} x_n^2 < \infty$ is equipped with the metric

$$d(x, y) = \left(\sum_{n=1}^{\infty} (x_n - y_n)^2 \right)^{1/2}.$$

(v) Let X be a nonempty set, let (Y, d_Y) be a nonempty metric space, and let $B(X, Y)$ be the set of all mappings $f \colon X \to Y$ bounded in the following sense: there exists $y_0 \in Y$ and $r > 0$ (depending on f) such that $d_Y\big(f(x), y_0\big) \leqslant r$ for all $x \in X$, i.e., $f(X)$ is contained in a ball. The function
$$d(f, g) := \sup_{x \in X} d_Y\big(f(x), g(x)\big)$$
makes $B(X, Y)$ a metric space.

A verification of the fact that in these examples we obtain metric spaces is a simple exercise. The coordinate space $\mathbb{R}^n$ possesses many metrics in addition to the usual Euclidean metric (for example, the metric $d(x, y) = \sup_i |x_i - y_i|$).

For a general metric space (X, d), one can naturally introduce many concepts familiar for the real line or the n-dimensional space. For example, the *open ball* of radius $r > 0$ centered at a point x_0 is the set
$$B(x_0, r) := \{x \in X \colon d(x, x_0) < r\}.$$
The term "open" will be soon clear. The *closed ball* of radius $r > 0$ centered at x_0 is the set
$$\{x \in X \colon d(x, x_0) \leqslant r\}.$$
A set A is called *bounded* if it is contained in a ball. Its *diameter* is the number $\operatorname{diam} A := \sup\{d(x, y) \colon x, y \in A\}$.

1.2.3. Definition. *A sequence of points x_n in X is called convergent to a point x if, for every $\varepsilon > 0$, there exists a number n_ε such that $d(x_n, x) < \varepsilon$ for all $n \geqslant n_\varepsilon$. Then x is called the limit of $\{x_n\}$ and is denoted by the symbol $\lim\limits_{n \to \infty} x_n$.*

A sequence $\{x_n\} \subset X$ is called Cauchy or fundamental if, for every $\varepsilon > 0$, there exists a number n_ε such that $d(x_n, x_k) < \varepsilon$ for all $n, k \geqslant n_\varepsilon$.

A convergent sequence is Cauchy, since $d(x_n, x_k) \leqslant d(x_n, x) + d(x, x_k)$. Clearly, there is only one limit. Any Cauchy sequence is bounded, since the union of a ball and a finite set is bounded.

A point x is called a *limit point* for a set M in a metric space X if every open ball of a positive radius centered at x contains a point from M distinct from x. The point x is called an *adherent point* of M if every open ball of a positive radius centered at x contains a point from M (possibly, the point x itself). The following two cases are possible:

1) $x \in M$ and some open ball $B(x, r)$ of a positive radius has no other points of M (then x is called an *isolated point* of M), 2) there exists a sequence of distinct points of M converging to x.

With this terminology isolated points are adherent points, but not limit points. Though, this terminological distinction, which we follow just to be consistent with some other textbooks (which, however, are not all consistent), is not very important; sometimes it is more convenient to include isolated point to limit points. Some nuances also arise when speaking of limit points of sequences: for example, if we have a sequence $\{x_n\}$ in which all elements are the same point x, then definitely x is the limit of this sequence, but not the limit point (in the above terminology) for the set of points x_n (since the latter set is a singleton).

If every ball of positive radius centered at x contains an uncountable set of points from M, then x is called an *accumulation* or *condensation* point of the set M.

1.2.4. Definition. *A set U in a metric space X is called open if, for every point x in U, there exists $r(x) > 0$ such that $B(x, r(x)) \subset U$. The empty set is open by definition. A set F in X is called closed if its complement $X \backslash F$ is open.*

A set F is closed if and only if F contains all its limit points. Indeed, a limit point of a closed set cannot belong to the open complement. On the other hand, if F contains all its limit points and $X \backslash F$ is not open, then, for some $x \notin F$, each ball $B(x, 1/n)$ contains a point x_n from F, i.e., x is a limit point of F and hence $x \in F$, which is impossible.

It is readily seen that the open ball $B(x_0, r)$ of radius $r > 0$ is open in the sense of the terminology introduced above and the closed ball is closed. This is not a tautology, since directly from the definition we only see that the center x_0 itself belongs to $B(x_0, r)$ with a ball around it. In order to make sure that for every point $x \in B(x_0, r)$ there exists a suitable radius $r(x) > 0$, we have to take $r(x) < r - d(x, x_0)$ and use the triangle inequality. Note that the open ball $B(x_0, r)$ can be a closed set different from the closed ball $\{x: d(x, x_0) \leqslant r\}$! For example, this is the case in every space of two points x and y with $r = d(x, y)$.

It is clear that in every metric space (X, d) the union of any collection of open sets is open. Hence the intersection of any collection of closed sets is closed. It is readily seen that finite intersections of open sets are open and that finite unions of closed sets are closed.

For every set $A \subset X$, the intersection of all closed sets containing A is called the *closure* of A and is denoted by the symbol $\overline{A}$. It is clear that $\overline{A}$ is the smallest closed set containing A. It coincides with the set of all limit and isolated points of A. Indeed, all limit points belong to the closure of A. If a point $z \notin A$ is not limit, then there exists an open ball $B(z, r)$ without common points with A. Then A belongs to the closed set $X \backslash B(z, r)$, i.e., $\overline{A} \subset X \backslash B(z, r)$ and $z \notin \overline{A}$. A point is called *inner* for a set if it belongs to this set along with some ball of a positive radius centered at the given point. The set of all inner points of the set is called its *interior*. It is clear that the interior can be empty.

Although all the definitions given above coincide with those introduced in elementary calculus for sets on the real line, general metric spaces can substantially differ by their properties from the real line and $\mathbb{R}^n$ and their subsets. We shall have many occasions to see this below, but now we only observe that even subsets of the plane with the usual Euclidean metric can have rather peculiar properties. For example, if we take three points a, b and c on the plane such that $|a - b| = 1$, $|b - c| = 2$ and $|a - c| = 5/2$, then in the obtained metric space the ball of radius 2 centered at b strictly contains the ball of the larger radius $9/4$ centered at a. Let us now define one of the key concepts connected with metric spaces.

1.2.5. Definition. *A metric space X is called complete if every Cauchy sequence in this space converges.*

1.2.6. Example. (i) The semi-closed interval $[0, 1)$ with its usual metric is not a complete space because the sequence $\{1 - 1/n\}$ is Cauchy, but does not converge to a point of $[0, 1)$. However, the same interval becomes a complete space with the metric $d(x, y) = |\tan(\pi x/2) - \tan(\pi y/2)|$.

(ii) The space $B(\Omega)$ with metric (1.2.1) is complete. Indeed, let $\{f_n\}$ be a Cauchy sequence. It is clear that for every fixed x the sequence of real numbers $\{f_n(x)\}$ is Cauchy, hence converges. Its limit will be denoted by $f(x)$. The boundedness of the Cauchy sequence $\{f_n\}$ shows that the function f is bounded. It remains to observe that $\{f_n\}$ converges to f in the metric of $B(\Omega)$, not merely pointwise. To this end, for any fixed $\varepsilon > 0$ we find a number n_ε such that $|f_n(x) - f_k(x)| < \varepsilon$ for all $n, k \geqslant n_\varepsilon$ and $x \in \Omega$. If we now fix $n \geqslant n_\varepsilon$ and let k increase to infinity, we obtain $|f_n(x) - f(x)| \leqslant \varepsilon$ for all $x \in \Omega$, i.e., $d(f_n, f) \leqslant \varepsilon$.

(iii) The space l^2 from Example 1.2.2(iv) is complete: if a sequence of vectors $v_j = (x_{j,n}) \in l^2$ is Cauchy, then, for every n, the sequence of numbers $x_{j,n}$ is Cauchy and has a limit x_n. Since $\sup_j \sum_{n=1}^{\infty} x_{j,n}^2 \leqslant C < \infty$, we have $\sum_{n=1}^{\infty} x_n^2 \leqslant C$, i.e., $v := (x_n) \in l^2$. Finally, $d(v_j, v) \to 0$, since for every $\varepsilon > 0$ there exists j_ε such that $\sum_{n=1}^{\infty} (x_{j,n} - x_{k,n})^2 \leqslant \varepsilon^2$ whenever $j, k \geqslant j_\varepsilon$. Letting $k \to \infty$ we find that $\sum_{n=1}^{\infty} (x_{j,n} - x_n)^2 \leqslant \varepsilon^2$ if $j \geqslant j_\varepsilon$.

The completeness of a number of other important spaces will be established below. It is clear from our definitions that closed subsets of complete spaces are complete as separate spaces.

1.2.7. Definition. (i) *A set A is called everywhere dense in a metric space if the closure of A is the whole space.*

(ii) *A metric space is called separable if it contains a finite or countable everywhere dense set.*

(iii) *A set E is called nowhere dense in a metric space X if every nonempty open set in X contains a ball of a positive radius without points of E.*

Even if A is dense in B, the set $A \cap B$ can be empty. However, if U is open, then $\overline{U \cap A} = U \cap \overline{A}$ for any A, which is easily verified (if $x \in U$ and $x \notin \overline{U \cap A}$, then there exists an open set $V \ni x$ with $V \cap U \cap A = \varnothing$, whence $x \notin \overline{A}$, since $U \cap V$ is open).

The real line with its usual metric is separable: the set of rational numbers is everywhere dense. The spaces $C[a, b]$ and l^2 are also separable. In l^2 there is a dense set of finite sequences with rational components. In $C[a, b]$ we can take a countable everywhere dense set of polynomials with rational coefficients (every continuous function is uniformly approximated by polynomials). We can also take the set of piece-wise linear functions of the following form: for each n we split $[a, b]$ into 2^n equal intervals by points $t_{n,i}$ and split $[-2^n, 2^n]$ into 4^n equal intervals by points $r_{n,j}$, next, we take functions assuming values $r_{n,j}$ at the points $t_{n,i}$ and linearly extended between the points $t_{n,i}$. The space $B(\Omega)$ with an infinite set Ω is nonseparable. Indeed, for every nonempty set $E \subset \Omega$, we take the function f_E on Ω that equals 1 at the points of E and 0 at all other points, i.e., the indicator function of the set E. The family of functions thus obtained is uncountable and the distance between the functions f_E and $f_{E'}$ with distinct sets

E and E' is 1. Hence $B(\Omega)$ contains an uncountable set of pairwise disjoint open balls. Clearly, a countable set cannot intersect all these balls. In particular, the space l^∞ is nonseparable.

The properties indicated in the previous definition can be characterized by means of closures in the following way. A space is separable if it is the closure of a finite or countable set. A set E is nowhere dense if its closure does not contain balls of positive radii. Verification of these assertions is a simple and useful exercise.

1.2.8. Definition. *An isometry of two metric spaces (M_1, d_1) and (M_2, d_2) is a one-to-one mapping J from the space M_1 onto M_2 such that*

$$d_2\big(J(x), J(y)\big) = d_1(x, y) \quad \text{for all } x, y \in M_1.$$

1.2.9. Lemma. *Let (M_1, d_1) and (M_2, d_2) be complete metric spaces, let $E_1 \subset M_1$ and $E_2 \subset M_2$ be everywhere dense sets, and let J be an isometry from E_1 onto E_2. Then J extends uniquely to an isometry between M_1 and M_2.*

PROOF. Each point $x \in M_1 \backslash E_1$ is the limit of a sequence $\{x_n\} \subset E_1$. Since J is an isometry and M_2 is complete, the sequence $\{J(x_n)\}$ converges to some point of M_2, which we take for $J(x)$. Any other sequence of points $y_n \in M_1$ converging to x leads to the same element $J(x)$, since $\{x_1, y_1, \ldots, x_n, y_n, \ldots\}$ also converges to x. It is readily seen that the mapping J preserves distances. Finally, $J(M_1) = M_2$, since every point $y \in M_2$ is the limit of a convergent sequence of points $J(x_n)$ with $x_n \in E_1$, so the sequence $\{x_n\}$ is Cauchy in M_1 and hence converges to some point $x \in M_1$, which gives $J(x) = y$. The uniqueness of an isometric extension is obvious. $\square$

Many properties of metric spaces are invariant with respect to isometries. Among such properties one can mention completeness, separability, and boundedness. A remarkable feature of the spaces of bounded functions $B(\Omega)$, which are very special at the first glance, is that up to isometries they contain all possible metric spaces.

1.2.10. Theorem. *Every metric space (M, d) is isometric to a part of the space $B(M)$.*

PROOF. We assume that M is not empty. Let us fix an element $x_0 \in M$ and, for every $x \in M$, define a function f_x on M by $f_x(y) = d(y, x) - d(y, x_0)$. We observe that by the triangle inequality $|f_x(y)| \leqslant d(x, x_0)$, i.e., the function f_x is bounded. For any fixed $x_1, x_2 \in M$, by the triangle inequality we have

$$|f_{x_1}(y) - f_{x_2}(y)| = |d(y, x_1) - d(y, x_2)| \leqslant d(x_1, x_2),$$

which gives $d(f_{x_1}, f_{x_2}) \leqslant d(x_1, x_2)$. Hence $d(f_{x_1}, f_{x_2}) = d(x_1, x_2)$ by the equality $|f_{x_1}(x_1) - f_{x_2}(x_1)| = d(x_1, x_2)$. Thus, the mapping $J\colon x \mapsto f_x$ from M to $B(M)$ is an isometry between M and $J(M)$. $\square$

As an application of this theorem we establish the existence of a completion of M.

1.2.11. Definition. *A completion of a metric space M is a complete metric space $\widetilde{M}$ such that M is isometric to an everywhere dense set in $\widetilde{M}$.*

1.2.12. Proposition. *Every metric space has a completion and this completion is unique up to an isometry.*

PROOF. By the previous theorem it suffices to consider our set M in the space $B(\Omega)$ with $\Omega = M$. Since $B(M)$ is complete, the desired completion is the closure of the set M in $B(\Omega)$. It remains to observe that if M' is some completion of M, then $\widetilde{M}$ and M' are isometric. Indeed, there is an isometry J' between M and an everywhere dense part $J'(M)$ of M'. By the lemma above this isometry extends to an isometry from $\widetilde{M}$ onto M'. □

Let us establish a useful property of separable metric spaces (called the *Lindelöf property*).

1.2.13. Theorem. *Every collection of open sets in a separable metric space contains an at most countable subcollection with the same union.*

PROOF. Let U_α be open sets in a metric space (X, d). Let us take a countable everywhere dense set $\{x_n\}$ in X. For every point x_n there is a countable collection of open balls $B(x_n, r_k)$ centered at x_n with rational radii $r_k > 0$. If $B(x_n, r_k)$ belongs to some of the sets U_α, we pick precisely one such set $U_{n,k}$. As the result we have picked at most countably many sets (corresponding to different n and k). Let us verify that the union of all picked sets coincides with the union of all sets U_α. Indeed, suppose that a point x is contained in a set U_α. There is $r > 0$ such that $B(x, r) \subset U_\alpha$. Let us take a point x_n for which $d(x, x_n) < r/4$. Clearly, there exists $r_k < r/4$ with $x \in B(x_n, r_k)$. By the triangle inequality $B(x_n, r_k) \subset B(x, r) \subset U_\alpha$. Since $B(x_n, r_k) \subset U_\alpha$, there is a picked set $U_{n,k}$ containing $B(x_n, r_k)$. Thus, $x \in B(x_n, r_k) \subset U_{n,k}$. □

The Lindelöf property is frequently used in functional analysis, measure theory and topology.

1.2.14. Proposition. *Every subspace of a separable metric space is separable with respect to the induced metric.*

PROOF. Let S be an at most countable everywhere dense set in a metric space X and let $Y \subset X$. To every point $s \in S$ we associate a finite or countable subset $Y_s \subset Y$ (possibly, empty) as follows. For every fixed $n \in \mathbb{N}$ we pick an element $y_n \in Y$ with $d(y_n, s) < 1/n$, provided it exists. The family of all elements of all sets Y_s, $s \in S$, is at most countable. It is everywhere dense in Y. Indeed, for every $y \in Y$ and every $n \in \mathbb{N}$ there exists $s(y) \in S$ with $d\big(y, s(y)\big) < 1/n$. Since there are points in Y at the distance less than $1/n$ from $s(y)$, we can choose among them a point y_n from the constructed family, which gives the estimate $d(y, y_n) < 2/n$ showing that $y_n \to y$. □

The reader is warned that this assertion can fail for general topological spaces, even compact (see Exercise 1.9.77).

1.3. Continuous Mappings

Here we consider mappings of metric spaces.

1.3.1. Definition. *A mapping f from a metric space (X, d_X) to a metric space (Y, d_Y) is called continuous at a point $x \in X$ if, for every sequence $\{x_n\}$ converging to x, the sequence $\{f(x_n)\}$ converges to $f(x)$.*
The mapping f is called continuous if it is continuous at every point.

It is clear that the continuity at a point x can be formulated in (ε, δ)-terms: for every $\varepsilon > 0$, there exists $\delta > 0$ such that $d_Y\big(f(z), f(x)\big) < \varepsilon$ for all points $z \in X$ such that $d_X(z, x) < \delta$.

In Exercise 1.9.37 it is suggested to verify that the continuity of the mapping f is equivalent to the property that for every open set $V \subset Y$ the set $f^{-1}(V)$ is open in X (this is also equivalent to the property that for every closed set $Z \subset Y$ the set $f^{-1}(Z)$ is closed in X).

As in the case of the real line, a stronger mode of continuity can be introduced: the uniform continuity.

1.3.2. Definition. *A mapping f from a metric space (X, d_X) to a metric space (Y, d_Y) is called uniformly continuous if, for every $\varepsilon > 0$, there exists $\delta > 0$ such that $d_Y\big(f(x), f(y)\big) \leqslant \varepsilon$ whenever $d_X(x, y) \leqslant \delta$.*

It is clear that uniformly continuous mappings are continuous. Let (X, d_X) and (Y, d_Y) be metric spaces.

1.3.3. Proposition. *Let $f_n \colon (X, d_X) \to (Y, d_Y)$ be continuous mappings uniformly converging to a mapping $f \colon (X, d_X) \to (Y, d_Y)$ in the following sense: for every $\varepsilon > 0$, there exists a number n_ε such that $d_Y\big(f_n(x), f(x)\big) \leqslant \varepsilon$ for all $n \geqslant n_\varepsilon$ and $x \in X$. Then the mapping f is continuous.*

PROOF. Let $x_0 \in X$ and $\varepsilon > 0$. Let us take numbers n_ε and $\delta > 0$ such that $d_Y\big(f_{n_\varepsilon}(x), f_{n_\varepsilon}(x_0)\big) < \varepsilon$ whenever $d_X(x, x_0) < \delta$. Then for such x we obtain

$$d_Y\big(f(x), f(x_0)\big) \leqslant d_Y\big(f(x), f_{n_\varepsilon}(x)\big) + d_Y\big(f_{n_\varepsilon}(x), f_{n_\varepsilon}(x_0)\big)$$
$$+ d_Y\big(f_{n_\varepsilon}(x_0), f(x_0)\big) \leqslant 3\varepsilon,$$

which shows the continuity of f at the point x_0. $\qquad\square$

If uniformly convergent mappings f_n are uniformly continuous, then their limit is also uniformly continuous. This is clear from the proof.

We recall that the set $B(X, Y)$ of all bounded mappings from a metric space X to a metric space Y is a metric space with the metric

$$d(f, g) = \sup_{x \in X} d_Y\big(f(x), g(x)\big).$$

Here a mapping is said to be bounded if it takes values in a ball in Y. In applications one frequently encounters various subsets of spaces of such a form equipped with other natural metrics. Some examples are considered in exercises. The subspace in $B(X, Y)$ consisting of bounded continuous mappings is denoted by the symbol $C_b(X, Y)$ (it is equipped with the same metric as $B(X, Y)$). The set of bounded continuous real functions on a metric space X is denoted by $C_b(X)$.

Taking into account Exercise 1.9.28 we obtain the following fact.

1.3.4. Corollary. *Let X and Y be nonempty metric spaces such that Y is complete. Then the space $C_b(X, Y)$ is complete. In particular, $C_b(X)$ and $C[a, b]$ are complete.*

1.3.5. Definition. *A mapping $f\colon X \to Y$ between metric spaces (X, d_X) and (Y, d_Y) is said to satisfy the Lipschitz condition with constant L (or is Lipschitzian with constant L) if $d_Y(f(x), f(y)) \leqslant L d_X(x, y)$ for all $x, y \in X$.*

Clearly, any Lipschitz mapping is uniformly continuous.

Let us give a simple example of a Lipschitz function: the distance from the point to a given set.

1.3.6. Example. Let A be a nonempty set in a metric space (X, d). The distance from a point x to the set A is defined by the formula

$$\mathrm{dist}\,(x, A) := \inf\{d(x, y)\colon\ y \in A\}.$$

Then the function $x \mapsto \mathrm{dist}\,(x, A)$ is Lipschitz with constant 1 and the set of its zeros coincides with the closure of A. Indeed, for any points x and z we have $d(x, y) \leqslant d(x, z) + d(z, y)$ for all y. Hence $\mathrm{dist}\,(x, A) \leqslant d(x, z) + \mathrm{dist}\,(z, A)$, which yields the inequality $|\mathrm{dist}\,(x, A) - \mathrm{dist}\,(z, A)| \leqslant d(x, z)$. Clearly, we have $\mathrm{dist}\,(x, A) = 0$ precisely when either $x \in A$ or there is a sequence of points $y_n \in A$ converging to x.

The next simple assertion will be used below.

1.3.7. Proposition. *Let (X, d_X) and (Y, d_Y) be metric spaces such that Y is complete and let a mapping f from a subset $X_0 \subset X$ to the space Y satisfy the Lipschitz condition with constant L. Then f uniquely extends to a Lipschitz with constant L mapping from the closure of X_0 to the closure of $f(X_0)$ in the space Y.*

PROOF. Every point x in the closure of X_0 is the limit of a sequence of points $x_n \in X_0$ (not necessarily distinct). It is natural to set $\widehat{f}(x) := \lim_{n \to \infty} f(x_n)$. Indeed, the sequence of points $f(x_n)$ is Cauchy in Y by the Lipschitzness of f and hence converges to some point $y \in Y$. For the same reason y does not depend on our choice of a sequence converging to x. This also shows that $\widehat{f}(x) = f(x)$ if $x \in X_0$. Finally, if x' and x'' are two distinct points in the closure of X_0, then, taking sequences $\{x_n'\}$ and $\{x_n''\}$ in X_0 converging to them, we obtain that $d_Y(f(x_n'), f(x_n'')) \leqslant L d_X(x_n', x_n'')$. Hence $d_Y(f(x'), f(x'')) \leqslant L d_X(x', x'')$, which is the desired estimate. $\square$

Note that f does not always extend to a Lipschitz mapping on all of X. For example, if $X = [0, 1]$, $X_0 = Y = \{0, 1\}$ with the usual metric and $f(x) = x$ on X_0, then there is no even continuous extension to $[0, 1]$ (however, see Exercises 1.9.62, 1.9.63).

Let us mention another useful property of Lipschitz mappings.

1.3.8. Proposition. *Let (X, d_X) and (Y, d_Y) be metric spaces such that Y is complete and let mappings $F_n \colon X \to Y$, where $n \in \mathbb{N}$, satisfy the Lipschitz condition with a common constant L. If for all points x in an everywhere dense set in X there exists a limit $\lim_{n\to\infty} F_n(x)$, then this limit exists for all $x \in X$ and defines a mapping that is Lipschitz with constant L.*

PROOF. Let us fix $x \in X$ and $\varepsilon > 0$. By condition there exists a point y with $d(x, y) < \varepsilon$ for which the sequence of elements $\{F_n(y)\}$ converges. Hence there exists a number n_ε such that $d_Y\big(F_n(y), F_k(y)\big) < \varepsilon$ for all $n, k \geqslant n_\varepsilon$. Then by the triangle inequality we have

$$d_Y\big(F_n(x), F_k(x)\big)$$
$$\leqslant d_Y\big(F_n(x), F_n(y)\big) + d_Y\big(F_n(y), F_k(y)\big) + d_Y\big(F_k(y), F_k(x)\big)$$
$$\leqslant Ld(x, y) + \varepsilon + Ld(x, y) \leqslant \varepsilon(2L + 1),$$

which proves that $\{F_n(x)\}$ is a Cauchy sequence. By the completeness of Y this sequence converges. The Lipschitzness of the limiting mapping with constant L is obvious. $\qquad\square$

1.4. The Contracting Mapping Principle

Lipschitz mappings with constant $L < 1$ are called *contracting mappings* or *contractions*. The next result is frequently used in applications.

1.4.1. Theorem. (THE CONTRACTING MAPPING PRINCIPLE) *Every contraction f of a nonempty complete metric space X has a unique fixed point $\widehat{x}$, i.e., $f(\widehat{x}) = \widehat{x}$. In addition, $d(\widehat{x}, x_n) \leqslant L^n(1 - L)^{-1}d(x_1, x_0)$ for every $x_0 \in X$, where $x_{n+1} := f(x_n)$.*

PROOF. Let $x_0 \in X$. Set $x_n = f(x_{n-1})$, $n \in \mathbb{N}$. We show that the sequence $\{f(x_n)\}$ is Cauchy. To this end, we observe that

$$d(x_{k+1}, x_k) \leqslant Ld(x_k, x_{k-1}) \leqslant \cdots \leqslant L^k d(x_1, x_0).$$

Hence $d(x_{n+m}, x_n)$ is estimated by

$$d(x_{n+m}, x_{n+m-1}) + d(x_{n+m-1}, x_{n+m-2}) + \cdots + d(x_{n+1}, x_n) \leqslant$$
$$\leqslant L^{n+m-1}d(x_1, x_0) + L^{n+m-2}d(x_1, x_0) + \cdots + L^n d(x_1, x_0),$$

which yields $d(x_{n+m}, x_n) \leqslant L^n(1 - L)^{-1}d(x_1, x_0)$. This estimate and the condition $L < 1$ imply that $\{x_n\}$ is a Cauchy sequence and that there exists a limit $\widehat{x} = \lim_{n\to\infty} x_n$. Clearly,

$$f(\widehat{x}) = \lim_{n\to\infty} f(x_n) = \lim_{n\to\infty} x_{n+1} = \widehat{x}$$

by the continuity of f. The uniqueness of a fixed point is seen from the fact that $d(\widehat{x}, y) = d\big(f(\widehat{x}), f(y)\big) \leqslant Ld(\widehat{x}, y)$ for any other fixed point y. The estimate for the rate of convergence has been also obtained. $\qquad\square$

1.4.2. Corollary. *Let X be a complete metric space and let $f\colon X \to X$ be a mapping such that some power f^n is a contracting mapping. Then f has a fixed point.*

PROOF. There is $\widehat{x}$ with $f^n(\widehat{x}) = \widehat{x}$. Then $f^n(f(\widehat{x})) = f(f^n(\widehat{x})) = f(\widehat{x})$, i.e., $f(\widehat{x})$ is also a fixed point of f^n. By the uniqueness of a fixed point of any contracting mapping we have $f(\widehat{x}) = \widehat{x}$. □

In applications one frequently encounters contracting mappings depending on a parameter. Let us prove a simple assertion enabling one to give a quantitative bound for such dependence.

1.4.3. Proposition. *Suppose that X is a complete metric space and two mappings f and g from X to X are Lipschitz with constant $\lambda < 1$. Then for the fixed points x_f and x_g of these mappings we have*

$$d(x_f, x_g) \leqslant (1 - \lambda)^{-1} \sup_{x \in X} d(f(x), g(x)).$$

PROOF. Set $y_n = g^n(x_f)$. It follows from the proof of the fixed point theorem that $x_g = \lim_{n \to \infty} y_n$. Hence $d(x_f, x_g) = \lim_{n \to \infty} d(x_f, y_n)$. As in the proof of the theorem, for every n we have

$$d(x_f, y_n) \leqslant d(x_f, g(x_f))(1 + \lambda + \cdots + \lambda^n)$$
$$\leqslant (1 - \lambda)^{-1} d(x_f, g(x_f)) = (1 - \lambda)^{-1} d(f(x_f), g(x_f)),$$

which yields our assertion. □

The next result gives a continuous dependence of the fixed point on the parameter. In addition, it enables us to obtain fixed points of mappings of open balls. The latter is not always possible in the general case. For example, the contraction $x \mapsto x/2 + 1/2$ of the interval $(-1, 1)$ has no fixed points in this interval.

1.4.4. Theorem. *Let $U = B(x_0, r)$ be the open ball of radius $r > 0$ centered at a point x_0 in a complete metric space (X, d), let T be a metric space, and let $F\colon T \times U \to X$ be a mapping such that for some constant $\lambda \in [0, 1)$ we have*

$$d(F(t, x), F(t, x')) \leqslant \lambda d(x, x') \quad \forall\, t \in T,\, \forall\, x, x' \in U.$$

In addition, let

$$d(F(t, x_0), x_0) < r(1 - \lambda) \quad \forall\, t \in T.$$

Then there exists a mapping $f\colon T \to U$ such that

$$f(t) = F(t, f(t)) \quad \forall\, t \in T.$$

Moreover, such a mapping f is unique.

If F is continuous, then f is also continuous (more precisely, f is continuous at every point t at which all mappings $t \mapsto F(x, t)$ are continuous).

PROOF. We observe that

$$d_X(F(t, x), x_0) \leqslant d_X(F(t, x), F(t, x_0)) + d_X(F(t, x_0), x_0)$$
$$< \lambda d_X(x, x_0) + r(1 - \lambda) \leqslant r$$

for all $x \in U$, i.e., $F(t, x) \in U$ for all $t \in T$ and $x \in U$. By Proposition 1.3.7 we can assume that the mappings $x \mapsto F(t, x)$ are extended to the closure $\overline{U}$ of the set U and that the extensions take values in $\overline{U}$ and are Lipschitz with constant λ. Each of these mappings has a unique fixed point $f(t) \in \overline{U}$. The estimate obtained above is valid for the extensions, hence they take values in U itself. In particular, $f(t) \in U$. The uniqueness of f is clear from the uniqueness of a fixed point of a contracting mapping. In the proof of the contracting mapping theorem we have shown that

$$f(t) = \lim_{n \to \infty} f_n(t), \quad \text{where } f_n(t) = F\big(t, f_{n-1}(t)\big), \ f_0(t) = x_0,$$

and have obtained the estimate

$$d_X\big(f_n(t), f_{n+m}(t)\big) \leqslant \frac{\lambda^n}{1 - \lambda} r(1 - \lambda) = r\lambda^n,$$

giving the uniform convergence of the sequence of mappings f_n. Hence in case of continuous F, the mapping f is also continuous. If all mappings $t \mapsto F(t, x)$ are continuous at a point t_0, then every f_n is also continuous at the point t_0. This is easily verified by induction. If the mapping f_{n-1} is continuous at the point t_0, then f_n is also. This follows from the estimate

$$d_X\big(f_n(t), f_n(t_0)\big) \leqslant d_X\Big(F\big(t, f_{n-1}(t)\big), F\big(t, f_{n-1}(t_0)\big)\Big)$$
$$+ d_X\Big(F\big(t, f_{n-1}(t_0)\big), F\big(t_0, f_{n-1}(t_0)\big)\Big)$$
$$\leqslant \lambda d_X\big(f_{n-1}(t), f_{n-1}(t_0)\big) + d_X\Big(F\big(t, f_{n-1}(t_0)\big), F\big(t_0, f_{n-1}(t_0)\big)\Big)$$

and the equality $F(t_0, x) = \lim_{t \to t_0} F(t, x)$ for all x.   □

As an application of Theorem 1.4.4 we shall show in Chapter 12 (see Theorem 12.3.1) that sufficiently small perturbations of identity mappings of Banach spaces are local homeomorphisms.

1.5. Baire's Category Theorem

The next two simple theorems are the most important general results of the theory of metric spaces.

1.5.1. Theorem. (THE NESTED BALL THEOREM) *Let X be a complete metric space and let $\{B_n\}$ be a sequence of closed balls with radii tending to zero such that $B_{n+1} \subset B_n$ for all n. Then $\bigcap_{n=1}^{\infty} B_n$ is not empty.*

PROOF. Let us take $x_n \in B_n$. Since the balls decrease and their radii tend to zero, the sequence $\{x_n\}$ is Cauchy. By the completeness of X it converges to some point, which belongs to all balls B_n by their closedness.   □

It is clear that in place of balls one can take any decreasing closed sets of diameter $d_n \to 0$. Simple examples show that the completeness of X and the closedness of balls are important (see also Exercise 1.9.33). One cannot omit the condition that the radii tend to zero (Exercise 1.9.34).

1.5.2. Theorem. (BAIRE'S CATEGORY THEOREM) *Let X be a complete metric space such that $X = \bigcup_{n=1}^{\infty} X_n$, where the sets X_n are closed. Then at least one of them contains an open ball of a positive radius.*

If $X = \bigcup_{n=1}^{\infty} A_n$, where A_n are arbitrary sets, then at least one of A_n is everywhere dense in some ball of a nonzero radius, i.e., a complete metric space cannot be the countable union of nowhere dense sets.

PROOF. Suppose the contrary. Then for every n in every open ball U there is an open ball disjoint with X_n, since otherwise U belongs to $\overline{X_n} = X_n$. Hence there exists a closed ball B_1 of radius $r_1 > 0$ disjoint with X_1. The ball B_1 contains a closed ball B_2 of a positive radius $r_2 < r_1/2$ disjoint with X_2. By induction we obtain decreasing closed balls B_n with positive radii tending to zero such that $B_n \cap X_n = \varnothing$. The previous theorem gives a common point for all B_n not belonging to the union of X_n, which is a contradiction. The last assertion of the theorem is obvious from the first one applied to the closures of A_n. □

In spite of the triviality of proofs, both theorems have numerous applications, which we shall see below.

At the first encounter of Baire's theorem one can wonder whether it contradicts the fact that the space of natural numbers is complete with respect to the natural metric of the real line and is the countable union of points. Of course, there is no real contradiction here: every point in this space is a ball of a positive radius.

The word "category" in the name of Baire's theorem is explained by the following terminology: sets that are countable unions of nowhere dense sets are called *first category* sets and all other sets are called *second category* sets. Baire's theorem asserts that a complete metric space is not a first category set. Metric spaces in which all open balls are of second category are called *Baire spaces*. However, there are incomplete Baire metric spaces, for example, the interval $(0,1)$ with its usual metric; this can be easily derived from the fact that the real line is a Baire space. Moreover, there are Baire spaces that have no equivalent complete metrics.

1.5.3. Example. (THE UNIFORM BOUNDEDNESS PRINCIPLE) Suppose that X is a complete metric space and continuous functions $f_n \colon X \to \mathbb{R}^1$ are such that for every $x \in X$ the sequence $\{f_n(x)\}$ is bounded. Then there exists a closed ball B of some positive radius such that $\sup_n \sup_{x \in B} |f_n(x)| < \infty$.

PROOF. Let $X_N := \{x \in X \colon \sup_n |f_n(x)| \leqslant N\}$. These sets are closed by the continuity of f_n. By hypothesis their union is X. According to Baire's theorem some X_N has inner points and hence contains a closed ball of a positive radius. □

1.6. Topological Spaces

A natural and very important generalization of the concept of metric space is a topological space.

1.6.1. Definition. *A set X with a distinguished family τ of its subsets is called a topological space if* 1) $\varnothing, X \in \tau$, 2) *the intersection of every two sets from τ*

belongs to τ, 3) *the union of every collection of sets from* τ *belongs to* τ. *The sets from* τ *are called open and the family* τ *is called a topology.*

A *topology base* is a collection of open set such that their unions give all open sets.

A *neighborhood of a point* is any open set containing it.

The complements of open sets are called *closed sets*. It is clear that any finite unions and arbitrary intersections of closed sets are closed. The empty set and the whole space are simultaneously open and closed.

1.6.2. Example. (i) The family $(\varnothing, X)$ is the minimal topology on a set X. (ii) The family 2^X of all subsets of X is the maximal topology on X. (iii) The collection of open sets in a metric space (X, d) (according to the terminology introduced for metric spaces!) is a topology. This topology is called the *topology generated by the metric* d. A topological space is called *metrizable* if its topology is generated by some metric.

Note that although the metric generates the indicated topology, this topology does not enable us to reconstruct the original metric. For example, the standard metric of the real line generates the same topology as the bounded metric defined by the formula $|\text{arctg}\, x - \text{arctg}\, y|$.

Any part X_0 of a topological space becomes a topological space if we take for the topology in X_0 the intersections of X_0 with the sets open in X. This topology is called the *induced topology* from X. Sometimes sets open in this topology in X_0 are called *relatively open*. Of course, such sets need not be open in X (if X_0 itself is not open). For topological spaces one naturally defines analogs of many concepts introduced earlier for metric spaces. A point x in a topological space X is called a *limit point* for a set $A \subset X$ if each open set containing this point contains also an element of the set A distinct from x. If each open set containing x has uncountably many points from A, then x is called an *accumulation point* or a *condensation point* of the set A. A point $a \in A$ is called an *isolated point* of the set A if it possesses a neighborhood which contains no other points from A. Hence all points in A are split into isolated and limit points, but also limit points of A can exist that do not belong to A. A point $x \in X$ is called an *adherence point* of the set A if every neighborhood of x contains at least one point from A (possibly, the point x itself). The adherence points are either limit points for A or isolated points of A. The *closure* $\overline{A}$ of a set A is the intersection of all closed sets containing A. It is clear that $\overline{A}$ is obtained from A by adding all limit points. If $\overline{A} = X$, then A is called *everywhere dense* in X. A space is *separable* if it contains a finite or countable everywhere dense set.

1.6.3. Definition. *A sequence of points* x_n *in a topological space is called convergent to a point* x *if, for every open set* U *containing* x, *there exists a number* N *such that* $x_n \in U$ *for all* $n \geqslant N$.

It is clear that in case of a metric space we arrive at the same convergence considered earlier. In an arbitrary topological space, a limit of a convergent sequence need not be unique. For example, if the topology consists only of the empty set and the whole space, every sequence in this topology converges to every point. The next concept enables us to avoid such phenomena.

1.6.4. Definition. *A topological space* (X, τ) *is called Hausdorff or separated if, for every two different points* $x, y \in X$, *there exist disjoint open sets* U *and* V *such that* $x \in U$, $y \in V$.

The term "Hausdorff" gives credit to the outstanding German topologist Felix Hausdorff. There are other concepts of separation in topological spaces (see [331] and [182]). If X contains more than one point, its minimal topology is not Hausdorff. The topology generated by a metric is always Hausdorff. Let us give one of the most typical examples of nonmetrizable spaces arising in applications.

1.6.5. Example. Let T be a nonempty set and let $\mathbb{R}^T$ be the space of all real functions on T. For fixed $x \in \mathbb{R}^T$, $t_1, \ldots, t_n \in T$ and $\varepsilon > 0$, let us introduce "basis neighborhoods" of the point x of the following form:

$$U_{x, t_1, \ldots, t_n, \varepsilon} := \{ y \in \mathbb{R}^T : |y(t_i) - x(t_i)| < \varepsilon, \ i = 1, \ldots, n \}. \tag{1.6.1}$$

We declare to be open in $\mathbb{R}^T$ the empty set and all possible unions of basis neighborhoods $U_{x, t_1, \ldots, t_n, \varepsilon}$, where one can vary x, n, t_i and ε. The obtained family of sets forms a Hausdorff topology called the *topology of pointwise convergence*. This topology is metrizable only in the case where T is at most countable. A sequence $\{x_n\} \subset \mathbb{R}^T$ converges to an element x precisely when $\lim\limits_{n \to \infty} x_n(t) = x(t)$ for every $t \in T$.

PROOF. Among the three properties of topologies, the only one that is not completely obvious is the property of the introduced system of sets to be closed with respect to finite intersections. It is clear that it suffices to verify that the intersection of two basis neighborhoods $U_1 = U_{x_1, t_1, \ldots, t_n, \varepsilon_1}$ and $U_2 = U_{x_2, s_1, \ldots, s_k, \varepsilon_2}$ is the union of basis neighborhoods. To this end, it suffices to make sure that every point z in $U_1 \cap U_2$ belongs to $U_1 \cap U_2$ along with some basis neighborhood of the form $U = U_{z, t_1, \ldots, t_n, s_1, \ldots, s_k, \varepsilon}$ for sufficiently small $\varepsilon > 0$. For such ε we take a positive number smaller than all numbers $\varepsilon_1 - |x_1(t_i) - z(t_i)|$, $\varepsilon_2 - |x_2(s_i) - z(s_i)|$. With this choice of $y \in U$ we have

$$|y(t_i) - x_1(t_i)| \leqslant |y(t_i) - z(t_i)| + |z(t_i) - x_1(t_i)| < \varepsilon_1.$$

Similarly $|y(s_i) - x_1(s_i)| < \varepsilon_2$. Thus, we have obtained a topology. It is Hausdorff, since any two distinct functions x and y differ at least at one point t_1, hence they can be separated by neighborhoods of the form $U_{x, t_1, \varepsilon}$ and $U_{y, t_1, \varepsilon}$. The equivalence of the pointwise convergence of sequences of functions to convergence in this topology is obvious: the pointwise convergence is deduced from convergence in the topology by using neighborhoods $U_{x, t, \varepsilon}$, and the pointwise convergence yields convergence on every finite set, which is convergence in the sense of neighborhoods of the form (1.6.1).

If $T = \{t_n\}$ is a countable set, then the space $\mathbb{R}^T$ is metrizable by the metric

$$d(x, y) := \sum_{n=1}^{\infty} 2^{-n} \min \left(|x(t_n) - y(t_n)|, 1 \right). \tag{1.6.2}$$

A proof of this is left as Exercise 1.9.65. If T is uncountable, then the introduced topology cannot be generated by a metric. Indeed, otherwise for every n one can take the open ball $B(0, n^{-1})$ centered at the identically zero function x and having the radius n^{-1} with respect to this metric. This ball contains a neighborhood of the form (1.6.1) defined by points $t_1^n, \ldots, t_{m_n}^n$ and a number $\varepsilon_n > 0$. Let us consider the function y equal to 1 at some point t different from all t_k^n (such a function exists, since T is uncountable) and vanishing at all other points. Then $y \in B(0, n^{-1})$ for all n, whence $y = 0$, which is a contradiction. □

1.6.6. Definition. *A mapping f from a topological space X to a topological space Y is called continuous at a point x if, for every open set V in Y containing $f(x)$, there exists an open set U in X containing x such that $f(U) \subset V$.*

A mapping is called continuous if it is continuous at every point.

The set of all continuous mappings between topological spaces X and Y is denoted by the symbol $C(X, Y)$. The set of all bounded continuous functions on a topological space X is denoted by $C_b(X)$.

Exercise 1.9.37 suggests to show that the continuity of a mapping f is equivalent to the property that the preimage of every open set in Y is open in X (which is also equivalent to the property that the preimage of every closed set is closed). Similarly to Proposition 1.3.3 one proves the following fact.

1.6.7. Proposition. *If X is a topological space and functions f_n continuous on X converge uniformly to f, then the function f is continuous. In particular, the space $C_b(X)$ with the metric $d(f, g) = \sup_{x \in X} |f(x) - g(x)|$ is complete.*

A weaker property than the continuity is the *sequential continuity*. A mapping $f \colon X \to Y$ is called *sequentially continuous* if, for every sequence of points x_n in X converging to x, the sequence $\{f(x_n)\}$ converges to $f(x)$. For metric spaces both notions are equivalent. A justification of the next example is Exercise 1.9.39.

1.6.8. Example. (i) Let $C[0, 1]$ be the space of all continuous functions on the interval $[0, 1]$ with the metric $d(f, g) = \sup_t |f(t) - g(t)|$. Then the function

$$f \colon \ x \mapsto \int_0^1 \sin x(t)\, dt$$

is continuous on $C[0, 1]$.

(ii) Let X be the set of all continuous functions on the interval $[0, 1]$ equipped with the topology of pointwise convergence, i.e., the topology induced from $\mathbb{R}^{[0,1]}$. Then the function f from (i) is sequentially continuous, but is discontinuous at every point. The function

$$g \colon \ x \mapsto \int_0^1 x(t)\, dt$$

is also discontinuous everywhere and continuous with respect to the metric from (i).

The following concept gives a precise meaning to the expression "topological spaces with the same properties".

1.6.9. Definition. *A homeomorphism of topological spaces X and Y is a one-to-one mapping h from the space X onto Y such that h and h^{-1} are continuous. In this case the spaces X and Y are called homeomorphic.*

A topological space is metrizable if and only if it is homeomorphic to a metric space.

Let us discuss a frequently used construction: the product of topological spaces. Let T be a nonempty index set and let X_t, where $t \in T$, be nonempty topological spaces. The product $\prod_{t \in T} X_t$ of the sets X_t is defined to be the family of all collections of the form $(x_t)_{t \in T}$, where $x_t \in X_t$ for each t. If all spaces X_t coincide with one and the same space X, the product is denoted by the symbol X^T and called the power of the space X.

The product of topological spaces is equipped with the following *product topology* (or *Tychonoff's topology*): open sets are the empty set and all possible unions of products of the form

$$\prod_{t \in T_0} U_t \times \prod_{t \notin T_0} X_t,$$

where the sets U_t are open in X_t and $T_0 \subset T$ is finite. The proof of the fact that we have obtained a topology is Exercise 1.9.64. For example, if $X_t = \mathbb{R}^1$ for all t, then we obtain the already considered topology of the space $\mathbb{R}^T$. If T is infinite, then the product of infinitely many open sets $U_t \neq X_t$ will not be open. The product of a countable collection of metric spaces (X_n, d_n) can be equipped with the metric

$$d(x, y) := \sum_{n=1}^{\infty} 2^{-n} \min\big(d_n(x_n, y_n), 1\big), \quad x = (x_n), y = (y_n). \tag{1.6.3}$$

The topology generated by this metric coincides with the product topology (the proof is left as Exercise 1.9.66).

1.7. Compact Sets and Their Properties

A very important role in continuous mathematics is played by the concept of compactness.

1.7.1. Definition. *A set in a Hausdorff space is called compact (or compactum) if in every cover of this set by open sets one can pick a finite subcover.*

It is clear from the definition that a set in a Hausdorff space is compact precisely when it is compact as a separate space with the induced topology. The property to be Hausdorff is not always included in the definition and is required here just for convenience of some subsequent formulations.

This definition is not intuitively motivated and may seem at the first glance to be too technical as compared to the intuitively convincing property of compactness of subsets of the real line formulated as the possibility of finding a convergent subsequence in every sequence. However, already a century long experience shows that the given definition (not equivalent to the definition in terms of sequences in case of general topological spaces, but coinciding with it in metric spaces) turns out to be much more fruitful and leads to a substantially more fruitful theory. A cover of a set by a family of open sets is called an *open cover*.

1.7.2. Proposition. (i) *Any closed subset of a compact set is compact.*

(ii) *Any compact set in a Hausdorff space is closed.*

(iii) *The image of a compact set under a continuous mapping with values in a Hausdorff space is compact.*

(iv) *Any infinite subset of a compact set has a limit point.*

(v) *Every continuous mapping from a compact metric space to a metric space is uniformly continuous.*

PROOF. (i) Let A be a closed subset of a compact set K and let $\{U_\alpha\}$ be an open cover of A. Adding to this family the open complement of A, we obtain an open cover of all of K. Let us pick a finite subcover. It consists of some of the sets $U_{\alpha_1}, \ldots, U_{\alpha_n}$ of the original family and, possibly, the complement of A. It is clear that the first n sets cover A.

(ii) Let K be a compact set in a Hausdorff space X and let $y \in X \backslash K$. Then for every point $x \in K$ there exist a neighborhood U_x of x and a neighborhood V_x of y such that $U_x \cap V_x = \varnothing$. In the cover by neighborhoods U_x we can select a finite subcover $U_{x_1}, \ldots, U_{x_n}$ of the compact set K. Then the neighborhood $V := V_{x_1} \cap \cdots \cap V_{x_n}$ of the point y is disjoint with K, since it is disjoint with $U_{x_1} \cup \cdots \cup U_{x_n}$. Thus, $V \subset X \backslash K$, which shows that $X \backslash K$ is open.

(iii) Let K be compact and let $f(K)$ be its image under a continuous mapping to a Hausdorff space Y. For every cover of $f(K)$ by open sets V_α we obtain a cover of K by open (by the continuity of f) sets $f^{-1}(V_\alpha)$. Selecting among them a finite subcover of K, we obtain a finite cover of $f(K)$ by the corresponding sets V_α.

(iv) Suppose we are given an infinite sequence of points x_n in a compact set K. If no point x in K is a limit point for $\{x_n\}$, then for every point $x \in K$ there exists an open set U_x such that $x \in U_x$ and $U_x \cap \{x_n\}$ is finite. Selecting a finite subcover, we obtain that the sequence $\{x_n\}$ is finite and arrive at a contradiction.

(v) Let (X, d_X) be a compact metric space, let (Y, d_Y) be a metric space, and let $f \colon X \to Y$ be continuous. For any given $\varepsilon > 0$, for every point $x \in X$ there is $\delta_x > 0$ such that $d_Y\big(f(y), f(z)\big) \leqslant \varepsilon$ whenever $d_X(y, x) \leqslant \delta_x$ and $d_X(z, x) \leqslant \delta_x$. The open balls $B(x, \delta_x/2)$ cover X, hence we can select a finite subcover $B(x_1, \delta_{x_1}/2), \ldots, B(x_n, \delta_{x_n}/2)$. Let $\delta := \min(\delta_{x_1}/2, \ldots, \delta_{x_n}/2)$. Let $d_X(x, y) \leqslant \delta$. There is x_i with $d_X(x, x_i) < \delta_{x_i}/2$. Then $d_X(y, x_i) \leqslant \delta_{x_i}$, since $\delta \leqslant \delta_{x_i}/2$. Hence we have $d_Y\big(f(x), f(y)\big) \leqslant \varepsilon$. $\qquad\square$

As explained in §1.9(iii) below, property (iv) does not mean that $\{x_n\}$ contains a convergent subsequence.

1.7.3. Corollary. *Every continuous real function on a compact set attains the smallest and largest values.*

PROOF. Any nonempty compact set on the real line contains the smallest and largest points. For example, the largest point is the precise upper bound (it is either a point of this set or is the limit of a sequence of its points, hence belongs to it in any case). □

Subsets of compact sets are called *relatively compact* sets.

Compact sets in metric spaces are simpler objects. In this case some additional important concepts appear.

1.7.4. Definition. (i) *Let A be a set in a metric space (X, d) and let $\varepsilon > 0$. A set $E \subset X$ is called an ε-net for A if for every point $a \in A$ there exists a point $e \in E$ with $d(a, e) < \varepsilon$.*

(ii) *A set A is called totally bounded if, for every $\varepsilon > 0$, it possesses a finite ε-net.*

In other words, a totally bounded set is a set which for every $\varepsilon > 0$ can be covered by finitely many open balls of radius ε. Note that the centers of such balls need not belong to A, but in A one can always find a finite 2ε-net with the same or smaller number of elements as the number of these balls. To this end in every given ball intersecting A we pick an arbitrary element of A and use the triangle inequality. For every $\varepsilon > 0$, there is the minimal possible number of elements of an ε-net for A, denoted by N_ε. The binary logarithm $\log_2 N_\varepsilon$ is an important metric characteristic of a totally bounded set, called the *metric entropy*. This characteristic is used in several areas, in particular, in information theory. Note that any totally bounded space is separable: taking finitely many points of a $1/n$-net for each n, we obtain an at most countable everywhere dense set. Totally bounded sets are also called *precompact* (but there is no terminological unification here). It is readily seen that the closure of a totally bounded set is totally bounded (an ε-net for A is a 2ε-net for $\overline{A}$). The boundedness of a set does not mean that it is totally bounded.

1.7.5. Example. (i) The space $\mathbb{N}$ in which the distance is 1 between any different points is bounded (and is the closed ball of radius 1 centered at any point), but is not totally bounded: this space cannot be covered by finitely many balls of radius $1/2$. (ii) In the space l^2 balls are not totally bounded. For example, the ball of radius 1 centered at the origin contains the points e_n with the nth coordinate 1 and other coordinates 0. Then $d(e_n, e_k) = \sqrt{2}$ if $n \neq k$. Hence finitely many balls of radius $1/2$ cannot cover all points e_n.

The following property of totally bounded sets is obvious from the definition.

1.7.6. Example. Uniformly continuous mappings take totally bounded sets to totally bounded sets. In particular, this is true for Lipschitz mappings.

Let us prove the following criterion for a set to be totally bounded in terms of Cauchy sequences.

1.7.7. Proposition. *A set A in a metric space X is totally bounded precisely when every infinite sequence of its elements contains a Cauchy subsequence.*

PROOF. Let A be totally bounded and let $\{x_n\} \subset A$ be infinite. Let us cover A by finitely many balls of radius 1. At least one ball U_1 of this cover contains an infinite part of $\{x_n\}$. The set $A \cap U_1$ can be covered by finitely many balls of radius $1/2$. We can find among them a ball U_2 such that $U_1 \cap U_2$ contains an infinite part of $\{x_n\}$. Continuing by induction, for every n we obtain a ball U_n of radius $1/n$ with the property that $V_n := U_1 \cap \cdots \cap U_n$ contains infinitely many points of the original sequence. Now we can find pairwise distinct elements $x_{k_n} \in V_n$. Clearly, we have obtained a Cauchy sequence.

Conversely, suppose that A possesses the indicated property. Suppose that for some $\varepsilon > 0$ there is no finite ε-net in A. By induction we construct a sequence of points $a_n \in A$ with mutual distances at least ε: for a_1 we take an arbitrary element of A; if points $a_1, \ldots, a_n$ are already constructed, there exists a point a_{n+1} with the distances at least ε to all of them, since otherwise the sets $a_1, \ldots, a_n$ would form an ε-net. Such a sequence does not contain a Cauchy subsequence. □

Let us give several equivalent descriptions of compactness in metric spaces. Analogous properties for topological spaces are discussed in §1.9(iii).

1.7.8. Theorem. *Let K be a set in a metric space X. The following conditions are equivalent.*

(i) *The set K is compact.*

(ii) *Every infinite part of K has a limit point in K.*

(iii) *Every infinite sequence in K contains a subsequence converging to a point of K.*

(iv) *The set K is totally bounded and complete.*

In particular, if X is complete, then the compactness of K is equivalent to the total boundedness and closedness.

PROOF. We already know that (i) implies (ii). It is clear that (ii) implies (iii), because if a sequence $\{x_n\}$ has a limit point $x \in K$, then its subsequence converges to x. For this it suffices to take a point $x_{k_n} \in \{x_n\}$ different from x in every ball of radius $1/n$ centered at x. By the previous proposition property (iii) implies that K is totally bounded. In addition, K is complete, because every Cauchy sequence in K contains a subsequence converging in K and hence converges itself.

Let us prove that (iv) implies (i). Suppose we have a cover of K by open sets U_α, $\alpha \in \Lambda$. Since K is separable (as observed above), by Theorem 1.2.13 there exists an at most countable subcover $\{U_{\alpha_n}\}$. Suppose that this subcover contains no finite subcover. Then, for every n, there exists a point $x_n \in K$ not belonging to $\bigcup_{i=1}^{n} U_{\alpha_i}$. By the previous proposition and the completeness of K the sequence $\{x_n\}$ contains a subsequence $\{x_{n_k}\}$ converging to a point $x \in K$. There is a number m such that $x \in U_{\alpha_m}$. Therefore, for some N we obtain $x_{n_k} \in U_{\alpha_m}$ for all $k \geqslant N$. This contradicts the fact that $x_{n_k} \notin \bigcup_{i=1}^{n_k} U_{\alpha_i}$. □

Clearly, the property of compactness is invariant with respect to homeomorphisms. Let us prove a stronger assertion.

1.7.9. Theorem. *Let X be compact and let f be a continuous one-to-one mapping of X onto a Hausdorff space Y. Then f^{-1} is also continuous, i.e., f is a homeomorphism.*

PROOF. Let $g = f^{-1}$. If E is closed in X, then E is compact. Then $g^{-1}(E) = f(E)$ is compact as well. Hence $g^{-1}(E)$ is closed. This gives the continuity of g. $\square$

A good application of compactness is the following useful classical result.

1.7.10. Theorem. (DINI'S THEOREM) *Suppose that a sequence of continuous functions on a compact space decreases pointwise to a continuous function. Then it converges uniformly. The same is true in case of pointwise increasing.*

PROOF. Suppose we are given a compact space K, continuous functions f_n such that $f_n(x) \geqslant f_{n+1}(x)$ and the function $f(x) = \lim_{n\to\infty} f_n(x)$ is also continuous. We have to show that convergence is uniform. Clearly, by subtracting f we reduce the problem to the case $f = 0$. Let $\varepsilon > 0$. For every point x there is a number $n(x)$ such that $f_{n(x)}(x) < \varepsilon$. By the continuity of $f_{n(x)}$ there exists a neighborhood U_x of the point x such that $f_{n(x)}(y) < \varepsilon$ whenever $y \in U_x$. These neighborhoods cover K, hence we can find a finite subcover $U_{x_1}, \ldots, U_{x_m}$. Set $N_\varepsilon := \max(n(x_1), \ldots, n(x_m))$. Then by the estimate $f_{n+1} \leqslant f_n$ for all $n \geqslant N_\varepsilon$ we obtain $f_n(y) < \varepsilon$ for all $y \in K$. Note that the continuity of the limiting function (more precisely, of the differences $f_n - f$) is important: the functions $f_n(x) = x^n$ on $[0,1]$ decrease pointwise, but not uniformly. The same example with $[0,1)$ in place of $[0,1]$ shows that the compactness of K is important. $\square$

In the next section we obtain compactness criteria in some concrete metric spaces.

1.8. Compactness Criteria

In the standard coordinate space $\mathbb{R}^n$ compact sets are precisely closed bounded sets. In calculus this fact is usually deduced from the case $n = 1$, which in turn is established with the aid of basic properties of real numbers. In most of spaces interesting for applications the class of compact sets is strictly contained in the class of closed bounded sets. Hence it is important to have compactness criteria in concrete spaces. Here we consider three typical examples.

1.8.1. Theorem. *A set K in the space l^2 is compact precisely when it is closed and bounded and satisfies the following condition:*

$$\lim_{N\to\infty} \sup_{x\in K} \sum_{n=N}^{\infty} x_n^2 = 0.$$

PROOF. If K is compact, then it is closed and bounded and for every $\varepsilon > 0$ has a finite ε-net $a^1, \ldots, a^m$, where $a^i = (a_1^i, a_2^i, \ldots)$. Let us take N such that $\sum_{n=N}^{\infty} |a_n^i|^2 < \varepsilon^2$ for all $i \leqslant m$. We obtain $\sum_{n=N}^{\infty} x_n^2 < 4\varepsilon^2$ for every $x \in K$, since there exists $i \leqslant m$ with $\sum_{n=1}^{\infty} |x_n - a_n^i|^2 < \varepsilon^2$ and $x_n^2 \leqslant 2|x_n - a_n^i|^2 + 2|a_n^i|^2$.

Conversely, if the indicated condition is fulfilled, then K possesses a finite ε-net for every $\varepsilon > 0$. Indeed, let N be such that $\sup_{x\in K} \sum_{n=N+1}^{\infty} x_n^2 < \varepsilon^2/4$. The set K_N of points of the form $\pi_N x := (x_1, \ldots, x_N, 0, 0, \ldots)$, where $x \in K$, is an $\varepsilon/2$-net for K (since the distance between x and $\pi_N x$ is not larger than $\varepsilon/2$).

The set K_N has a finite $\varepsilon/2$-net (which will be a finite ε-net for K), since the projection of K_N onto $\mathbb{R}^N$ is bounded by the boundedness of K and hence has a finite $\varepsilon/2$-net, which becomes an $\varepsilon/2$-net for K_N after adding zero coordinates starting from the $(N+1)$th position. $\square$

1.8.2. Example. The set $E = \{x \in l^2\colon \sum_{n=1}^{\infty} \alpha_n x_n^2 \leqslant 1\}$, where $\alpha_n > 0$ and $\alpha_n \to +\infty$, is compact in l^2. Indeed, it is easy to verify that it is closed and bounded. In addition,

$$\sup_{x \in E} \sum_{n=N}^{\infty} x_n^2 \leqslant \sup_{x \in E} \sup_{n \geqslant N} \alpha_n^{-1} \sum_{n=N}^{\infty} \alpha_n x_n^2 \leqslant \sup_{n \geqslant N} \alpha_n^{-1} \to 0$$

as $N \to \infty$. Hence the theorem proved above applies.

1.8.3. Theorem. *A set K in the space $B(\Omega)$ is compact if and only if it is closed, bounded and for every $\varepsilon > 0$ the set Ω can be partitioned into finitely many parts $\Omega_1, \ldots, \Omega_n$ with the property that for every $i = 1, \ldots, n$ we have $|f(\omega) - f(\omega')| < \varepsilon$ for all $f \in K$, $\omega, \omega' \in \Omega_i$.*

PROOF. As in the previous proof, if the set K is compact, then it is closed, bounded and for every $\varepsilon > 0$ has a finite ε-net $f_1, \ldots, f_m$. Note that for the finite collection of functions $f_1, \ldots, f_m$ the desired sets Ω_i exist. Indeed, the mapping $F = (f_1, \ldots, f_m)\colon \Omega \to \mathbb{R}^m$ takes values in a bounded set M. Let us partition M into disjoint parts M_i of diameters less than ε and set $\Omega_i := F^{-1}(M_i)$. It remains to observe that for every function $f \in K$ for all $\omega, \omega' \in \Omega_i$ we have $|f(\omega) - f(\omega')| < 3\varepsilon$. Indeed, there exists a function f_j in the picked ε-net such that $d(f, f_j) < \varepsilon$, whence we have

$$|f(\omega) - f(\omega')| \leqslant |f(\omega) - f_j(\omega)| + |f_j(\omega) - f_j(\omega')| + |f_j(\omega') - f(\omega')|.$$

Conversely, suppose that the indicated condition is fulfilled. It suffices to show that K is totally bounded. Let $\varepsilon > 0$. Let us take sets $\Omega_1, \ldots, \Omega_n$ from the hypotheses. There is a number $M > 0$ such that $\sup_{\omega \in \Omega} |f(\omega)| \leqslant M$ for all $f \in K$. The interval $[-M, M]$ can be partitioned by points $a_1 = -M, \ldots, a_m = M$ into intervals of equal length less than ε. Let us consider the finite set F of all functions f on Ω such that on each Ω_i the function f assumes a constant value from $\{a_1, \ldots, a_m\}$. The obtained set is a 3ε-net for K. Indeed, for every function $f \in K$ we can pick a point ω_i in each Ω_i and then take a number $a_{j(i)}$ such that $f(\omega_i) \in [a_{j(i)}, a_{j(i)+1}]$. The function g equal to $a_{j(i)}$ on Ω_i belongs to F. Moreover, $|f(\omega) - g(\omega)| < 2\varepsilon$ for all $\omega \in \Omega$. Indeed, $\omega \in \Omega_i$ for some i, whence we obtain

$$|f(\omega) - g(\omega)| \leqslant |f(\omega) - f(\omega_i)| + |f(\omega_i) - g(\omega_i)| + |g(\omega_i) - g(\omega)| < 2\varepsilon,$$

since $|f(\omega) - f(\omega_i)| < \varepsilon$ and $g(\omega_i) = g(\omega)$. $\square$

1.8.4. Theorem. (THE ASCOLI–ARZELÀ THEOREM) *A set K in the space $C[a, b]$ is compact precisely when it is closed, bounded and uniformly equicontinuous, i.e., for every $\varepsilon > 0$ there exists $\delta > 0$ such that, whenever $|t - s| < \delta$, we have $|f(t) - f(s)| < \varepsilon$ for all $f \in K$. A similar assertion is true in case of the space $C(X)$, where X is a compact metric space.*

PROOF. As in the previous theorem, the necessity of the indicated condition is easily seen from the fact that it holds for finite sets. The sufficiency of this condition follows from the previous theorem. Indeed, the space $C[a, b]$ is isometric to a subspace in the space $B([a, b])$. Hence it suffices to show that K is totally bounded as a subset of $B([a, b])$. For sets Ω_i one can take intervals of length less than δ with union $[a, b]$. The same reasoning applies to every compact metric space X in place of the interval (for Ω_i we take parts of diameter less than δ). $\quad\square$

Exercise 1.9.46 suggests to extend this result to mappings with values in metric spaces. An analog for general compact spaces K is given in Exercise 1.9.78.

1.8.5. Remark. If in the three theorems above we omit the closedness assumption, then we obtain criteria of the total boundedness in the corresponding spaces. This is clear from the proof.

1.8.6. Corollary. *If X is a compact metric space and a sequence of functions $f_n \in C(X)$ converges pointwise and is uniformly equicontinuous, then it converges uniformly.*

PROOF. The conditions yield the uniform boundedness of the sequence $\{f_n\}$ and the uniqueness of a limit point in $C(X)$. $\quad\square$

1.8.7. Example. The set E_C of all functions f differentiable on $[0, 1]$ with $|f(0)| \leqslant C$ and $\sup_t |f'(t)| \leqslant C$ is totally bounded in $C[0, 1]$. Indeed, by the mean value theorem $|f(t) - f(s)| \leqslant C|t - s|$ for all $f \in E_C$, which gives the uniform equicontinuity. In addition, $|f(t)| \leqslant |f(0)| + C|t| \leqslant 2C$, i.e., E_C is bounded.

1.9. Complements and Exercises

(i) Nets in topological spaces (25). (ii) Tychonoff's theorem (28). (iii) Countable and sequential compactness (29). (iv) Functional separation of sets (31). (v) The Stone–Weierstrass theorem (36). (vi) The Cantor set (38). Cardinal characteristics of metric spaces (40). Exercises (40).

1.9(i). Nets in topological spaces

Unlike the case of metric spaces, the most important properties of sets in general topological spaces cannot be described in terms of sequences. For this purpose one has to employ generalized sequences indexed by uncountable sets (or filters, which we do not discuss, see [86]).

A nonempty set T is called *directed* if it is equipped with a partial order (see §1.1 about this concept) satisfying the following condition: for every $t, s \in T$ there exists $u \in T$ with $t \leqslant u$ and $s \leqslant u$.

A directed set can contain incomparable elements. For example, we can partially order $\mathbb{R}^2$ by defining the relation $(x, y) \leqslant (u, v)$ as follows: $x \leqslant u$ and $y \leqslant v$. Not all elements are comparable, but any two are majorized by some third element.

A *net* in X is defined to be a family of elements $\{x_t\}_{t \in T}$ in X indexed by a directed set T. Similarly one defines a net of sets $\{U_t\}_{t \in T}$ in X. A net $\{x_t\}_{t \in T}$ is called a *subnet* of a net $\{y_s\}_{s \in S}$ if there is a mapping $\pi \colon T \to S$ such that

$x_t = y_{\pi(t)}$ and, for every $s_0 \in S$, there exists $t_0 \in T$ with $\pi(t) \geqslant s_0$ for all $t \geqslant t_0$. One should have in mind that even if the index set S is countable, not every countable part gives a subnet. For example, if $S = \mathbb{Z}$ with the usual ordering, then the elements with negative indices do not form a subnet: here the condition that for every index of the original net there is a larger index in the subnet is violated.

1.9.1. Definition. *A net $\{x_t\}_{t \in T}$ in a topological space X converges to an element x if, for every open set U containing x, there exists an index t_0 such that $x_t \in U$ for all $t \in T$ with $t_0 \leqslant t$. Notation:* $\lim_t x_t = x$.

An important distinction of convergent nets from convergent sequences is that the set of indices t such that $t \leqslant t_0$ and x_t does not belong to U can be infinite. Hence the expression "all elements of the net starting from some index belong to U" has a somewhat different meaning and does not mean "all excepting finitely many". Even a countable convergent net is not the same as a convergent sequence. For example, if $T = \mathbb{Z}$ with the usual ordering, $x_t = 1/t$ for all $t > 0$ and $x_t = 1$ for all $t \leqslant 0$, then this net converges to zero. Its part with $t < 0$ is outside the interval $(-1, 1)$ and is not a subnet, as explained above.

1.9.2. Proposition. *A mapping $f \colon X \to Y$ between topological spaces is continuous at a point x precisely when for every net $\{x_\alpha\}$ converging to x the net $\{f(x_\alpha)\}$ converges to $f(x)$.*

PROOF. Suppose that f is continuous at a point x, a net $\{x_\alpha\}$ converges to x and V is a neighborhood of $f(x)$. Let us take a neighborhood U of x such that $f(U) \subset V$. Find an index α_0 with the property that $x_\alpha \in U$ for all $\alpha_0 \leqslant \alpha$. Then $f(x_\alpha) \in f(U) \subset V$ for all such α, i.e., we have $f(x) = \lim_\alpha f(x_\alpha)$.

Conversely, suppose that the indicated condition with nets is fulfilled and V is a neighborhood of the point $f(x)$. Let us take for an index set T the family $\mathcal{U}$ of all neighborhoods of the point x with the following ordering: $U_1 \leqslant U_2$ if $U_2 \subset U_1$. It is clear that we have obtained a directed set. Suppose that for every $U \in \mathcal{U}$ there exists a point $x_U \in U$ with $f(x_U) \notin V$. The net $\{x_U\}$ converges to x. Indeed, let W be a neighborhood of x. Then, whenever $W \leqslant U$, we have $U \subset W$. Hence $x_U \in U \subset W$. Our condition gives a neighborhood U_0 such that $f(x_U) \in V$ if $U_0 \leqslant U$, which contradicts our choice of elements x_U. Therefore, our assumption is false, i.e., there exists a neighborhood U of x such that $f(U) \subset V$. This proves the continuity of f at x. □

Let us give another example of application of nets.

1.9.3. Example. A limit point of a set A in a topological space X is the limit of some net of elements of A distinct from this point. Indeed, let us use the same index set as in the proposition proved above. Then, for every neighborhood U of the given limit point x, there exists an element $x_U \in A \cap U$ distinct from x. The net we obtained converges to x.

Note that a limit point of the set formed by all elements of the net $\{x_t\}$ can fail to be the limit of a subnet in $\{x_t\}$ (although is the limit of a net composed by some elements of $\{x_t\}$). For example, let $T = \mathbb{Z}$ with the usual ordering, $x_n = 1/n$ if $n < 0$, $x_n = n$ if $n \geqslant 0$. Then 0 is a limit point of the set of numbers x_n, but the net $\{x_n\}$ has no subnet converging to zero. Finally, even if a net contains only countably many different points and converges to some limit, there might be no convergent countable sequence of elements in this net.

1.9.4. Example. Let $X = [0,1]^{\mathbb{R}}$ with the topology of pointwise convergence. Set $x_n(t) = \sin nt$. Every neighborhood U of the point $x = 0$ contains some element x_n. Indeed, the neighborhood U contains a neighborhood of the form

$$U_{t_1,\ldots,t_k,\varepsilon} = \{x \colon |x(t_i)| < \varepsilon, i = 1, \ldots, k\}.$$

It is easy to see (this is left to the reader) that for some n we have $|\sin nt_i| < \varepsilon$ whenever $i = 1, \ldots, k$, i.e., $x_n \in U_0$. It follows from the previous example that there exists a net converging to zero and consisting of elements of the given sequence. On the other hand, it is easy to show that there is no sequence of natural numbers n_i such that $\sin n_i t \to 0$ for all t, i.e., no subsequence in $\{x_n\}$ converges to zero.

All these simple examples and assertions also demonstrate very typical methods of dealing with nets. Let us give yet another useful (but a bit more difficult) assertion of a similar nature. We shall say that a net $\{x_t\}$ *frequently visits* a set E if for every t_0 there exists t with $t_0 \leqslant t$ and $x_t \in E$.

1.9.5. Proposition. *Suppose that a net $\{x_t\}_{t\in T}$ of elements of a topological space X and a point $x \in X$ are such that $\{x_t\}$ frequently visits every neighborhood of x. Then $\{x_t\}$ has a subnet converging to x.*

PROOF. The decisive step here is the right choice of an index set for a subnet. For such a set S we take the collection of all pairs of the form $s = (t, U)$, where $t \in T$, U is a neighborhood of x and $x_t \in U$. The set of pairs is partially ordered as follows:

$$(t_1, U_1) \leqslant (t_2, U_2) \quad \text{if } t_1 \leqslant t_2 \text{ and } U_2 \subset U_1.$$

It is clear that this is indeed a partial order. Moreover, the set S is directed. Indeed, since T is directed, for any $s_1 = (t_1, U_1)$ and $s_2 = (t_2, U_2)$ in S there exists $t_0 \in T$ with $t_1 \leqslant t_0$ and $t_2 \leqslant t_0$. Set $U_3 = U_1 \cap U_2$. By assumption there exists $t_3 \in T$ with $t_0 \leqslant t_3$ and $x_{t_3} \in U_3$. Then the element $s_3 := (t_3, U_3)$ belongs to S and majorizes s_1 and s_2.

We now set $x_s := x_t$ if $s = (t, U)$. Then $x = \lim_s x_s$. Indeed, let U be a neighborhood of x. Since $\{x_t\}$ frequently visits U, there exists $t_0 \in T$ with $x_{t_0} \in U$, i.e., $s_0 := (t_0, U) \in S$. If $s = (t, V)$ and $s_0 \leqslant s$, then we obtain $x_s = x_t \in V \subset U$, which proves our assertion.

Finally, we should verify that $\{x_s\}$ is a subnet in the original net. For every $s = (t, U) \in S$ we set $\pi(s) = t$. Let $t_0 \in T$. Since for U we can take the whole space X, the pair $s_0 = (t_0, X)$ belongs to S. If $s_0 \leqslant s = (t, V) \in S$, then $t_0 \leqslant t = \pi(s)$, which completes the proof. $\qquad\qquad\square$

We draw the reader's attention once again that for the validity of the conclusion of this proposition it is not enough to assume that x is a limit point of the set of elements x_t even in the case of the real line.

Nets will be used below in §1.9(iii) for describing compactness in terms of convergence.

1.9(ii). Tychonoff's theorem

Let us give a useful characterization of compactness. A family of sets in a space X is called *centered* if every finite subfamily has a nonempty intersection.

1.9.6. Theorem. *A Hausdorff space X is compact precisely when every centered family of its closed subsets has a nonempty intersection.*

Therefore, X is compact precisely when every centered family of its arbitrary subsets has a common adherence point.

PROOF. Let X be compact and let $\mathcal{F}$ be a centered family of its closed subsets. If the intersection of all sets in $\mathcal{F}$ is empty, then the complements of sets in $\mathcal{F}$ form an open cover of X. By compactness this cover contains a finite subcover by sets $X\backslash F_1, \ldots, X\backslash F_n$, where $F_i \in \mathcal{F}$. Then $\bigcap_{i=1}^{n} F_i$ is empty contrary to our condition.

Conversely, suppose that X satisfies the indicated condition and open sets U_α, where $\alpha \in A$, cover X. Then the closed sets $X\backslash U_\alpha$ have the empty intersection. Therefore, there exists a finite collection of indices $\alpha_1, \ldots, \alpha_n$ for which the set $\bigcap_{i=1}^{n}(X\backslash U_{\alpha_i})$ is empty. This means that $X = \bigcup_{i=1}^{n} U_{\alpha_i}$. In the case of arbitrary subsets we apply the established act to their closures. $\square$

A very important role in topology and functional analysis and applications is played by the following result due to A. N. Tychonoff (Tikhonov).

1.9.7. Theorem. (TYCHONOFF'S THEOREM) *Let T be a nonempty index set such that for every $t \in T$ we are given a nonempty compact space K_t. Then the product $K := \prod_{t \in T} K_t$ is compact in the product topology.*

PROOF. We verify that every centered family S of subsets of the space K has a common adherence point. Then we can apply the previous theorem. We first show that we can assume that the given family is maximal, i.e., there is no strictly larger centered family. To this end we consider the set $\mathfrak{S}$ the elements of which are all possible centered families of subsets of K containing the given family S. It is partially ordered by inclusion, moreover, every linearly ordered part $\mathfrak{S}$ has a majorant: the union of all families in this part. By Zorn's lemma $\mathfrak{S}$ has a maximal element, by which we can replace S. The maximality of S implies that S is closed with respect to finite intersections, since by adding to S all possible finite intersections of sets from S we obtain a centered system.

Now, assuming that the family S is maximal, for every t we take the natural projection $\pi_t \colon K \to K_t$ and consider the family $S_t := \{\pi_t(\Lambda) \colon \Lambda \in S\}$ of subsets of K_t, i.e., the projections on K_t of the original sets. It is clear that this family is also centered. By the compactness of K_t it has a common adherence point. Let us take some of such points x_t. We show that the point $x = (x_t) \in K$ is the desired one. By the definition of the topology in K every neighborhood of this point contains a neighborhood of the form $V := \prod_t U_t$, where for some finite collection $t_1, \ldots, t_n$ the sets U_{t_i} are neighborhoods of x_{t_i} and all remaining U_t coincide with K_t. By our choice of x_t every neighborhood U_t intersects all sets of the family S_t, i.e., $\pi_t^{-1}(U_t)$ intersects all sets of the family S. By the maximality of S we obtain that $\pi_t^{-1}(U_t) \in S$, since by adding $\pi_t^{-1}(U_t)$ to S we obtain a centered system. Since S is closed with respect to finite intersections, we obtain $V \in S$. Hence the set V intersects each set in S, i.e., x is a common adherence point. $\square$

There are also other proofs of this important result.

Tychonoff's theorem is frequently used for the proof of compactness of more special sets by embedding them into products of compact spaces and verifying the closedness of the image. For example, in such a way we shall prove the important Banach–Alaoglu–Bourbaki theorem. It should be also noted that every compact space is homeomorphic to a compact set in some power of the compact interval (see Exercise 1.9.76).

1.9(iii). Countable and sequential compactness

In nonmetrizable spaces compactness is not characterized by sequences.

1.9.8. Definition. *A set in a Hausdorff space is called sequentially compact if every infinite sequence of its elements contains a subsequence converging to a point of this set.*

A set in a Hausdorff space is called countably compact if every countable cover of this set by open sets has a finite subcover.

It is clear that compact sets are countably compact, but the converse is not true, as we shall see below. In addition, a compact space need not be sequentially compact.

1.9.9. Example. Let $X = [-1,1]^{\mathbb{R}} = \{x \colon \mathbb{R} \to [-1,1]\}$ be equipped with the product topology. By Tychonoff's theorem X is compact, but the sequence of elements $x_n \in X$ of the form $x_n(t) = \sin nt$ has no convergent subsequences. Indeed, if $\{x_n\}$ had a convergent subsequence $\{x_{n_k}\}$, then this would mean that the sequence $\sin n_k t$ converges at every $t \in \mathbb{R}$. We leave it to the reader to show that this is impossible.

However, an analog of the characterization of compactness obtained above in metric spaces is possible if in place of sequences we employ nets.

1.9.10. Theorem. *Let X be a Hausdorff space. The following conditions are equivalent:*

(i) *the space X is compact;*

(ii) *every infinite net in X has a convergent subnet.*

PROOF. Let X be compact and let $\{x_t\}_{t \in T}$ be an infinite net in X. We observe that there exists a point $x \in X$ such that $\{x_t\}$ frequently visits every neighborhood of x. Otherwise every point $x \in X$ possesses a neighborhood U_x and an index t_x such that $x_t \notin U_x$ whenever $t_x \leqslant t$. Since the sets U_x cover X, some finite subfamily $U_{x_1}, \ldots, U_{x_n}$ also covers X. There is an index $\tau \in T$ majorizing all t_{x_i} with $i \leqslant n$. Then $x_t \notin U_{x_1} \cup \cdots \cup U_{x_n}$ for every t such that $\tau \leqslant t$, which is a contradiction. Now it remains to use Proposition 1.9.5.

Suppose that (ii) is true and that we are given a cover of X by open sets U_α, $\alpha \in A$. Let us consider the set T consisting of all possible finite subsets of A and partially ordered by inclusion. It is clear that this is a directed set, since the union of two finite sets is finite. Suppose that the given cover has no finite subcover. Then for every $t = \{\alpha_1, \ldots, \alpha_n\} \in T$ there is a point $x_t \in X \backslash \bigcup_{i=1}^n U_{\alpha_i}$. Thus, we obtain a net in X. By assumption it has a subnet $\{x_s\}_{s \in S}$ converging to some point $x \in X$. Let us find $\alpha_0 \in A$ with $x \in U_{\alpha_0}$. By construction there is an element $s_0 \in S$ such that $x_s \in U_{\alpha_0}$ for all $s \in S$ with $s_0 \leqslant s$. The set S contains an element s_1 which exceeds s_0 and the element $\{\alpha_0\} \in T$ consisting of the single point α_0. Then $x_{s_1} \in U_{\alpha_0}$. This leads to a contradiction, since s_1 is a finite collection $\{t_1, \ldots, t_k\}$ such that x_{s_1} does not belong to the union of sets $U_{\alpha_{t_i}}$, $i \leqslant k$, but among these sets there is U_{α_0} due to the fact that $\{\alpha_0\} \leqslant s_1$. Therefore, any open cover of the space X has a finite subcover. □

The reader may observe that we do not mention here an analog of property (ii) from Theorem 1.7.8: every infinite part of X has a limit point. It turns out that this property in general topological spaces is weaker than compactness and is equivalent to the property of countable compactness.

1.9.11. Proposition. *A Hausdorff topological space X is countably compact precisely when every infinite part of X has a limit point.*

PROOF. Suppose that our space X is countably compact and $\{x_n\}$ is an infinite sequence of different points in X. If it has no limit points, for every n, the set F_n of points x_k with $k \geqslant n$ is closed. In addition, $\bigcap_{n=1}^\infty F_n = \varnothing$. The sets $X \backslash F_n$ are open and cover X. Hence finitely many of them cover X, i.e., one of them is all of X, which is impossible.

Suppose that every infinite part of X has a limit point and that we are given open sets U_n, where $n \in \mathbb{N}$, covering X. Suppose that for every n there exists a point x_n not covered by the sets $U_1, \ldots, U_n$. The set of such points is infinite (otherwise there exists a finite subcover) and hence has a limit point x. Let us find n with $x \in U_n$. Next we find $m > n$ with $x_m \in U_n$ (this is possible, since X is Hausdorff and x is a limit point of $\{x_n\}$). We have obtained a contradiction with the fact that x_m is not covered by the sets $U_1, \ldots, U_m$. □

It is clear that in this assertion it suffices to consider only countable subsets.

This proposition also yields the already known fact that every infinite part of a compact space possesses a limit point.

Obviously, every sequentially compact space is countably compact. However, the converse is not true: in Example 1.9.9 the compact space is not sequentially compact. Let us give an example of a noncompact countably compact and sequentially compact space.

1.9.12. Example. Let S be the set of all functions x on the real line with values in $[0, 1]$ for which there exists a countable set E_x outside of which the function x vanishes. Then, with the topology of pointwise convergence (i.e., the topology induced by the compact space $[0, 1]^{\mathbb{R}}$), the space S is not compact, although is countably compact and even sequentially compact. Indeed, if we are given a sequence of functions $x_n \in S$, then there is a countable set $E = \{t_j\}$ outside of which all functions x_n vanish. By using the diagonal method we can pick a subsequence $\{x_{n_k}\}$ converging at every point t_j. It readily seen that S is everywhere dense in $[0, 1]^{\mathbb{R}}$. Hence S is not compact. One can also explicitly construct an open cover without finite subcovers. For this we consider open sets

$$U_{t,k} := \{x \in S \colon |x(t) - t| > 1/k\}, \quad t \in [0, 1], k \in \mathbb{N}.$$

Each point x in S belongs to some set $U_{t,k}$ (it suffices to take $t \neq 0$ with $x(t) = 0$ and k with $1/k < t$), but no finite union of these sets covers S.

Thus, the countable compactness follows from compactness and also from the sequential compactness, but there are no other implications between these three notions in the general case. In metric spaces all the three are equivalent.

1.9(iv). Functional separation of sets

In many applications of general topology an important role is played by continuous functions on compact sets. So far our considerations do not show that on infinite compact spaces there are always nonconstant continuous functions. Of course, on metric compact spaces one can define such a function with the aid of a metric. But what can one do in case of a general topological compact space? The answer is given by the remarkable Urysohn theorem that applies to an even broader class of *normal spaces*, i.e., Hausdorff topological spaces in which for every pair of disjoint closed sets A and B one can find disjoint open sets U and V with $A \subset U$ and $B \subset V$.

1.9.13. Example. Every compact space is normal. In addition, every metric space is normal.

PROOF. Let A and B be disjoint closed sets in a compact space. We first observe that for every point x in the complement of B there exist open sets U_x and V_x such that $x \in U_x$, $B \subset V_x$ and $U_x \cap V_x = \varnothing$, i.e., x and B are separated by open sets. Indeed, for every $y \in B$ we can find disjoint neighborhoods W_y and S_y of the points x and y. The cover of the compact set B by the sets S_y contains a finite subcover $S_{y_1}, \ldots, S_{y_n}$. Let

$$U_x = W_{y_1} \cap \cdots \cap W_{y_n}, \quad V_x = S_{y_1} \cup \cdots \cup S_{y_n}.$$

Since the closed sets A and B in our compact space are disjoint, we conclude that for every point $x \in A$ there is a neighborhood U_x of x and a neighborhood V_x of B with $U_x \cap V_x = \varnothing$. The cover of A by the sets U_x has a finite subcover $U_{x_1}, \ldots, U_{x_n}$. It remains to take

$$U = U_{x_1} \cup \cdots \cup U_{x_n}, \quad V = V_{x_1} \cap \cdots \cap V_{x_n}.$$

In the case of a metric space we can use the sets

$$U := \bigcup_{x \in A} B\big(x, \operatorname{dist}(x, B)/2\big), \quad V := \bigcup_{x \in B} B\big(x, \operatorname{dist}(x, A)/2\big),$$

which is easily verified with the aid of the triangle inequality. It is even simpler to verify the functional separation of A and B with the aid of the formula indicated below. □

Two sets A and B in a topological space X are called *functionally separated* if there exists a continuous function $f \colon X \to [0,1]$ such that $A \subset f^{-1}(0)$ and $B \subset f^{-1}(1)$.

The next important result is due to P. S. Urysohn.

1.9.14. Theorem. (URYSOHN'S THEOREM) *In a normal space any two disjoint closed sets are functionally separated. In particular, this is true for compact spaces and for metrizable spaces.*

PROOF. The difficulty of constructing the desired function is that we have no information about our space X, except for the condition of existence of separating neighborhoods. For this reason the desired function is constructed very implicitly by indicating its level sets. Let A and B be disjoint closed sets. There is an open set U such that $A \subset U \subset V := X \backslash B$. Suppose that for every rational number $r \in [0,1]$ of the form $m2^{-n}$ we have managed to find an open set $U_r \subset X$ such that $U_0 = U$, $U_1 = V$, $\overline{U_t} \subset U_r$ whenever $t < r$. Then we define a function f by the formula

$$f(x) = \inf\{r \colon x \in U_r\} \quad \text{if } x \in V, \ f(x) = 1 \text{ if } x \in X \backslash V.$$

We observe that $f(x) = 0$ for $x \in A \subset U$. By construction $B \subset f^{-1}(1)$. For the proof of the continuity of f it suffices to verify that the sets $f^{-1}\big((a,b)\big)$ are open. The inequality $f(x) < b$ is equivalent to the property that $x \in U_r$ for some $r < b$, i.e., the set $f^{-1}\big([0,b)\big) = \bigcup_{r < b} U_r$ is open. The inequality $f(x) > a$ is equivalent to the property that there exists $r > a$ such that $x \notin U_r$. By the properties of U_r this is fulfilled precisely when for some $q > a$ we have $x \notin \overline{U_q}$. Therefore,

$$f^{-1}\big((a,1]\big) = \bigcup_{q > a} (X \backslash \overline{U_q}),$$

which shows that the set $f^{-1}\big((a,b)\big)$ is open. It remains to find sets U_r for r of the form $r = m2^{-n}$. Let us apply induction on n. Suppose that some sets U_r are already constructed for all $r = m2^{-n}$, where $m = 0, \ldots, 2^n$. At the next step we have to find a set U_r for every new point that is the middle r of the interval $[s, t]$ with $s = m2^{-n}$, $t = (m+1)2^{-n}$. For such a point r the set U_r is constructed as follows: since the closed sets $\overline{U_s}$ and $X \backslash U_t$ are disjoint, there exist open sets U_r and V_r such that $\overline{U_s} \subset U_r$ and $X \backslash U_t \subset V_r$. The latter inclusion implies that $\overline{U_r} \subset U_t$, hence U_r is a suitable set. □

In the case of a metric space the desired function can be defined explicitly: $f = f_1/(f_1 + f_2)$, where $f_1(x) = \operatorname{dist}(x, A)$, $f_2(x) = \operatorname{dist}(x, B)$. In this case an even stronger condition is fulfilled (which cannot be always achieved for general compact spaces): $A = f^{-1}(0)$ and $B = f^{-1}(1)$.

The following Tietze–Urysohn theorem on extension of continuous functions is close in its spirit and range of applications.

1.9.15. Theorem. *Let X be a normal topological space and let Z be a closed subset of X. Then every bounded continuous real function on Z extends to a continuous real function on all of X without changing its infimum and supremum. In particular, this is true if X is compact or metrizable.*

PROOF. Without loss of generality we can assume that $\inf_{z \in Z} f(z) = -1$, $\sup_{z \in Z} f(z) = 1$. The idea of constructing an extension is this: we first verify that for every $c > 0$ and every continuous function $\varphi \colon Z \to [-c, c]$ there exists a continuous function $\psi \colon X \to [-c/3, c/3]$ such that $\sup_{z \in Z} |\varphi(z) - \psi(z)| \leqslant 2c/3$. Indeed, the sets $A := \varphi^{-1}([-c, -c/3])$ and $B := \varphi^{-1}([c/3, c])$ are closed and disjoint. By the Urysohn theorem there exists a continuous function $g \colon X \to [0, 1]$ such that $A \subset g^{-1}(0)$ and $B \subset g^{-1}(1)$. Let $\psi(x) = 2c(g(x) - 1/2)/3$. Then $|\psi(x)| \leqslant c/3$ and $|\varphi(z) - \psi(z)| \leqslant 2c/3$ for all $z \in Z$, since for all $z \in A$ we have $\varphi(z) \in [-c, -c/3]$ and $\psi(z) = -c/3$, for all $z \in B$ we have $\varphi(z) \in [c/3, c]$ and $\psi(z) = c/3$, and for all other $z \in Z$ we have $|\varphi(z)| \leqslant c/3$ and $|\psi(z)| \leqslant c/3$. Now by induction we find continuous functions f_k on X such that
$$\sup_{x \in X} |f_k(x)| \leqslant (2/3)^{k-1}/3, \quad \sup_{z \in Z} \left| f(z) - \sum_{j=1}^{k} f_j(z) \right| \leqslant (2/3)^k.$$
For f_1 we take the function ψ constructed above for the function $\varphi = f$ and $c = 1$. If $f_1, \ldots, f_k$ are already constructed, we apply the assertion proved above to the function $\varphi = f - \sum_{j=1}^{k} f_j$ on Z and $c = (2/3)^k$. The uniformly convergent series $\sum_{k=1}^{\infty} f_k$ defines the desired extension. $\square$

Functional separation is connected with constructing the so-called *partitions of unity*, employed in many problems.

1.9.16. Theorem. *Let X be a normal topological space and let $\{U_t\}_{t \in T}$ be its cover by open sets that is locally finite in the following sense: every point possesses a neighborhood intersecting only finitely many of these sets. Then there exists a family of continuous functions $f_t \colon X \to [0, 1]$ with the following properties: $\sum_{t \in T} f_t(x) = 1$ for all x and $f_t = 0$ outside U_t for each t.*

PROOF. 1. We first show that there exists an open cover $\{V_t\}_{t \in T}$ of X with $\overline{V_t} \subset U_t$ for all t. Let $\mathcal{W}$ be the class of all open sets in X and let $\mathcal{F}$ be the family of all mappings $F \colon T \to \mathcal{W}$ such that $\bigcup_{t \in T} F(t) = X$ and either $F(t) = U_t$ or $\overline{F(t)} \subset U_t$. The original cover gives a mapping from $\mathcal{F}$. The family $\mathcal{F}$ is partially ordered in the following way: $F_1 \leqslant F_2$ provided that $F_2(t) = F_1(t)$ for every t such that $F_1(t) \neq U_t$. It is readily verified that we have obtained a partial order.

We prove that every linearly ordered part $\mathcal{F}_0 \subset \mathcal{F}$ has a majorant F_0, which can be defined by the formula $F_0(t) := \bigcap_{F \in \mathcal{F}_0} F(t)$. To this end we only have to show that $F_0 \in \mathcal{F}$. It is clear that the set $F_0(t)$ either coincides with U_t or has the closure belonging to U_t. Note that $F_0(t)$ is open, since if $F(t) \neq U_t$ for some $F \in \mathcal{F}_0$ and $t \in T$, then for every other $F' \in \mathcal{F}_0$ we have either $F'(t) = U_t$ or $F'(t) = F(t)$. This follows from the fact that either $F \leqslant F'$ or $F' \leqslant F$ due to the condition that $\mathcal{F}_0$ is linearly ordered. In order to verify that

$X = \bigcup_{t \in T} F(t)$, we take $x \in X$ and choose a finite set of indices $t_1, \ldots, t_k \in T$ for which $x \in \bigcup_{i=1}^{k} U_{t_i}$ and $x \notin U_t$ whenever $t \notin \{t_1, \ldots, t_k\}$. This is possible by assumption. If $F_0(t_i) = U_{t_i}$ for some $i \leqslant k$, then our verification is finished. Suppose that $F_0(t_i) \neq U_{t_i}$ for all $i \leqslant k$. Then for every $i \leqslant k$ there exists $F_i \in \mathcal{F}_0$ with $F_i(t_i) \neq U_{t_i}$. By the linear ordering of $\mathcal{F}_0$ there is $j \leqslant k$ such that $F_i \leqslant F_j$ for all $i \leqslant k$, which yields $x \in F_j(t_i)$ for some $i \leqslant k$, since $x \notin F_j(t)$ if $t \notin \{t_1, \ldots, t_k\}$, because $x \notin U_t$, but $x \in \bigcup_{t \in T} F_j(t)$. Since $F_i(t_i) \neq U_{t_i}$, it follows from what has been said above that $F_i(t_i) = F_j(t_i) = F(t_i)$ for all $F \in \mathcal{F}_0$ such that $F(t_i) \neq U_{t_i}$. Therefore, $x \in F_0(t_i) = F_j(t_i)$.

By Zorn's lemma $\mathcal{F}$ has a maximal element F. We show that $\overline{F(t)} \subset U_t$ for all t, which will give the desired cover. Suppose that $\overline{F(t)} \cap (X \backslash U_{t_0}) \neq \varnothing$ for some t_0. Then the definition of $\mathcal{F}$ implies that $F(t_0) = U_{t_0}$. Note that the set $E := X \backslash \bigcup_{t \neq t_0} F(t) \subset F(t_0)$ is closed. By the normality of X there is $V \in W$ such that $E \subset V \subset \overline{V} \subset F(t_0)$. The mapping F' defined by the formula $F'(t_0) = V$, $F'(t) = F(t)$ if $t \neq t_0$, belongs to $\mathcal{F}$, moreover, $F \leqslant F'$ and $F \neq F'$. This contradicts the maximality of F and proves that $\overline{F(t)} \subset U_t$ for all t.

2. Let us take an open set W_t such that $\overline{V_t} \subset W_t$ and $\overline{W_t} \subset U_t$. According to the Urysohn theorem, for every t there exists a continuous function $g_t \colon X \to [0, 1]$ with $g_t(x) = 1$ if $x \in \overline{V_t}$ and $g_t(x) = 0$ if $x \notin W_t$. Hence every point has a neighborhood in which only finitely many functions g_t are different from zero. Then the function $g = \sum_t g_t$ is finite and continuous. In addition, $g(x) \geqslant 1$, since every point x belongs to some V_t and hence $g_t(x) = 1$. It remains to take the function $f_t := g_t/g$. $\qquad \square$

For applications of this theorem it is useful to be able to inscribe in open covers locally finite open covers. This is not always possible. Spaces in which this is possible are called *paracompact*. It is clear that compact spaces are paracompact. The next theorem due to Stone shows that metrizable spaces are also paracompact.

1.9.17. Theorem. *Let (X, d) be a metric space and let $\{U_t\}_{t \in T}$ be its open cover. Then there exists a locally finite cover $\{V_s\}_{s \in S}$ of X inscribed in $\{U_t\}_{t \in T}$, i.e., every element V_s belongs to some element of the original cover (this new cover is called a locally finite refinement of the original one). Moreover, one can take S consisting of countably many parts S_j with the property that, for every j, every point has a neighborhood intersecting at most one of the sets V_s with $s \in S_j$.*

PROOF. We can make T a well-ordered set. We recall that then every nonempty subset T has the smallest element. By induction we define a family $\mathcal{V}_i = \{V_{t,i}\}_{t \in T}$, $i \in \mathbb{N}$, of open subsets of X by the formula $V_{t,i} := \bigcup B(c, 1/2^i)$, where $B(c, r)$ is the open ball of radius r centered at c and the union is taken over all $c \in X$ satisfying the following conditions: (1) t is the smallest element in T for which $c \in U_t$; (2) $c \notin V_{s,j}$ if $j < i$ and $s \in T$; (3) $B(c, 3/2^i) \subset U_t$.

It is clear that $V_{t,i} \subset U_t$. Let $x \in X$. We take the smallest t with $x \in U_t$ and then $i \in \mathbb{N}$ such that $B(x, 3/2^i) \subset U_t$. We observe that either $x \in V_{s,j}$ for some $j < i$ and $s \in T$ or $x \in V_{t,i}$. Thus, $\mathcal{V} := \bigcup_{i=1}^{\infty} \mathcal{V}_i$ is an open cover inscribed in the original one.

We show that for each fixed i we have $d(x, x') > 1/2^i$ if $x \in V_{s,i}$, $x' \in V_{t,i}$ and $s \neq t$. This means that every ball of radius $1/2^{i+1}$ intersects at most one set from $\mathcal{V}_i$. Let $s < t$. By the definition of $V_{s,i}$ and $V_{t,i}$ there exist points c and c' satisfying (1)–(3) such that $x \in B(c, 1/2^i) \subset V_{s,i}$ and $x' \in B(c', 1/2^i) \subset V_{t,i}$. Since $c' \notin U_s$ by (1) and $B(c, 3/2^i) \subset U_s$ by (3), we have $d(c, c') \geqslant 3/2^i$. Hence $d(x, x') \geqslant d(c, c') - d(c, x) - d(c', x') > 1/2^i$, as required.

For completing the proof it suffices to show that for every $t \in T$ and every pair of natural numbers j, k the inclusion $B(x, 1/2^k) \subset V_{t,j}$ implies the equality $B(x, 1/2^{j+k}) \cap V_{s,i} = \varnothing$ for all $i \geqslant j + k$ and $s \in T$. Indeed, if this is established, for every x we can find $k, j \in \mathbb{N}$ and $t \in T$ such that $B(x, 1/2^k) \subset V_{t,j}$, and then the ball $B(x, 1/2^{j+k})$ intersects at most $j + k - 1$ elements of $\mathcal{V}$. For justification of this assertion we observe that property (2) used in the definition of $V_{s,i}$ yields that the points c mentioned there do not belong to $V_{t,j}$ if $i \geqslant j + k$. Since $B(x, 1/2^k) \subset V_{t,j}$, we have $d(x, c) > 1/2^k$ for such c. By the inequalities $j + k \geqslant k + 1$ and $i \geqslant k + 1$ we have $B(x, 1/2^{j+k}) \cap B(c, 1/2^i) = \varnothing$, whence we obtain $B(x, 1/2^{j+k}) \cap V_{s,i} = \varnothing$. $\square$

In terms of functional separation yet another important class of topological spaces is introduced. A Hausdorff space X is called *completely regular* or *Tychonoff* if, for every point $x \in X$ and every neighborhood U of x, there exists a continuous function $f \colon X \to [0, 1]$ with $f(x) = 1$ and $f(y) = 0$ for all $y \notin U$. It is clear from the results above that normal spaces are completely regular. In particular, compact spaces are completely regular, since they are normal. However, there are completely regular spaces that are not normal (Exercise 1.9.71).

For completely regular spaces there is an important procedure of the so-called Stone–Čech compactification. A compact space βX is called the *Stone–Čech compactification* of a completely regular space X if X is homeomorphically embedded as an everywhere dense set in βX such that every bounded continuous function on X extends to a continuous function on βX.

1.9.18. Theorem. *Every completely regular space X has a Stone–Čech compactification βX and this compactification is unique up to a homeomorphism.*

PROOF. Let $\mathcal{F}$ be the class of all continuous functions from X to $[0, 1]$. The space $[0, 1]^{\mathcal{F}}$ of all functions on $\mathcal{F}$ with values in $[0, 1]$ is equipped with the topology of pointwise convergence. The space X is embedded into $[0, 1]^{\mathcal{F}}$ by means of the formula $x \mapsto \Psi_x$, $\Psi_x(\varphi) := \varphi(x)$, $\varphi \in \mathcal{F}$. It is readily verified that X is homeomorphic to its image under this embedding (here the assumption is used). For βX we take the closure $\overline{X}$ of X in $[0, 1]^{\mathcal{F}}$ (more precisely, the closure of its image under this embedding). The space $[0, 1]^{\mathcal{F}}$ is compact by Tychonoff's theorem, hence $\overline{X}$ is also compact. In order to extend a bounded continuous function f from X to $\overline{X}$ it suffices to do this in the case where $f(X) \subset [0, 1]$. Then we set $f(\Psi) := \Psi(f)$, $\Psi \in \overline{X}$, which is possible, since $f \in \mathcal{F}$. The proof of uniqueness of βX up to a homeomorphism is delegated to Exercise 1.9.81. $\square$

1.9.19. Corollary. *If K is a compact subset of a completely regular space X, then every continuous function f on K extends to a continuous function g on X with $\sup_{x \in X} |g(x)| = \sup_{x \in K} |f(x)|$.*

PROOF. The set K is compact also in βX, hence Theorem 1.9.15 applies. $\square$

For a discussion of compactifications, see [**182**]. We only note that the Stone–Čech compactification is a very complicated object even for simple spaces. For example, the Stone–Čech compactification of the open interval $(0, 1)$ is not the compact interval $[0, 1]$. The Stone–Čech compactification $\beta\mathbb{IN}$ of the space of natural numbers $\mathbb{IN}$ does not coincide with the one-point compactification $\mathbb{IN} \cup \{\infty\}$, but is an uncountable nonmetrizable compact space. From the space X to βX one can extend not only bounded continuous functions, but arbitrary continuous mappings to compact spaces (see Exercise 1.9.81). Observe also that we have $\beta X = \beta Y$ if $X \subset Y \subset \beta X$.

1.9(v). The Stone–Weierstrass theorem

It is known from a course of analysis (see [**520**]) that every continuous function on a compact interval can be uniformly approximated by polynomials. The next fundamental Stone–Weierstrass theorem gives a very broad generalization of this fact.

A family $\mathcal{E}$ of real functions on a set X is called an *algebra of functions* if, for all $f, g \in \mathcal{E}$ and all $\lambda \in \mathbb{R}^1$, we have $\lambda f, f + g, fg \in \mathcal{E}$, where the operations over functions are defined pointwise.

We shall say that a family of functions $\mathcal{F}$ *separates points* of the set X if for every pair of different points $x, y \in X$ there exists a function $f \in \mathcal{F}$ such that $f(x) \neq f(y)$.

1.9.20. Theorem. (THE STONE–WEIERSTRASS THEOREM) *Let K be a nonempty compact space and let $\mathcal{E}$ be an algebra of continuous real functions on K containing all constants and separating points. Then the algebra $\mathcal{E}$ is everywhere dense in the space $C(K)$ with its standard sup-metric.*

PROOF. In order not to use the classical Weierstrass theorem, but obtain it as a corollary, we observe that the function $t \mapsto \sqrt{t}$ on $[0, 1]$ can be uniformly approximated by polynomials. To this end, it suffices to be able to approximate by polynomials the functions $t \mapsto \sqrt{t + 1/k}$, but these functions can be approximated by the Taylor series.

The first step of the proof is a verification of the fact that, for all functions $f, g \in \mathcal{E}$, the functions $\min(f, g)$ and $\max(f, g)$ belong to the closure $\overline{\mathcal{E}}$ of the algebra $\mathcal{E}$ in $C(K)$. Since

$$\min(f, g) = \frac{1}{2}(f + g - |f - g|), \quad \max(f, g) = \frac{1}{2}(f + g + |f - g|),$$

by the linearity of $\mathcal{E}$ it suffices to show that $|f| \in \overline{\mathcal{E}}$ for all $f \in \mathcal{E}$. We can assume that $\sup_x |f(x)| \leqslant 1$. Let us take a sequence of polynomials P_j uniformly converging on $[0, 1]$ to the function $t \mapsto \sqrt{t}$, and let $f_j(x) := P_j(f(x)^2)$. It is clear that the functions f_j belong to $\mathcal{E}$ and converge uniformly to $|f| = \sqrt{f^2}$. Note that at this step we have not used that K is compact and that $\mathcal{E}$ separates points. In addition, observe that $\overline{\mathcal{E}}$ is also an algebra of functions.

We now show that for every $f \in C(K)$ and every $\varepsilon > 0$, there exists an element $f_\varepsilon \in \mathcal{E}$ with $\sup_x |f(x) - f_\varepsilon(x)| < \varepsilon$. For every pair of different points $a, b \in K$ by assumption there is a function $h_{a,b} \in \mathcal{E}$ with $h_{a,b}(a) \neq h_{a,b}(b)$. Set

$$f_{a,b}(x) = \frac{f(b) - f(a)}{h_{a,b}(b) - h_{a,b}(a)} \left(h_{a,b}(x) - h_{a,b}(a) \right) + f(a).$$

Then $f_{a,b} \in \mathcal{E}$ and $f_{a,b}(a) = f(a)$, $f_{a,b}(b) = f(b)$. The sets

$$U_{a,b} := \left\{ x \colon f_{a,b}(x) < f(x) + \varepsilon \right\} \quad \text{and} \quad V_{a,b} := \left\{ x \colon f_{a,b}(x) > f(x) - \varepsilon \right\}$$

are neighborhoods of the points a and b, respectively. Let us fix b. The cover of the compact set K by the sets $U_{a,b}$ contains a finite subcover by sets $U_{a_1,b}, \ldots, U_{a_n,b}$. As shown above, the function $f_b := \min(f_{a_1,b}, \ldots, f_{a_n,b})$ belongs to $\overline{\mathcal{E}}$. It is clear that $f_b(x) < f(x) + \varepsilon$ for all $x \in K$ and $f_b(x) > f(x) - \varepsilon$ if $x \in V_b := \bigcap_{i=1}^n V_{a_i,b}$. In addition, the set V_b is an open neighborhood of the point b. The obtained cover of the compact set K contains a finite subcover by sets $V_{b_1}, \ldots, V_{b_m}$. Finally, set $g_\varepsilon = \max(f_{b_1}, \ldots, f_{b_m})$. It is clear that $\sup_x |f(x) - g_\varepsilon(x)| < \varepsilon$. It follows from what has been said at the first step of the proof that $g_\varepsilon \in \overline{\mathcal{E}}$. Hence there exists $f_\varepsilon \in \mathcal{E}$ with $\sup_x |f(x) - f_\varepsilon(x)| < \varepsilon$. $\qquad\square$

This theorem implies the classical Weierstrass theorems on approximation by polynomials (algebraic or trigonometric). For example, the usual polynomials restricted to a compact set K in $\mathbb{R}^n$ form an algebra and separate points. Hence they approximate all continuous real functions on K. Let us consider another example.

1.9.21. Example. (i) If $\mathcal{F}$ is a class of continuous real functions on a compact space K separating points, then every continuous real function on K is uniformly approximated by functions of the form $P(f_1, \ldots, f_n)$, where P is a polynomial in n variables, $f_1, \ldots, f_n \in \mathcal{F}$. In particular, every continuous real function on any power of the interval $[0, 1]^T$ is uniformly approximated by functions of the form $P(x_{t_1}, \ldots, x_{t_n})$, where P is a polynomial in n variables and x_t are coordinate functions.

Indeed, the class of functions of the indicated form is an algebra that separates points.

(ii) Let K_t, where $t \in T$, be a family of nonempty compact sets and let $K = \prod_{t \in T} K_t$ be their product. Then every continuous real function on K is uniformly approximated by functions of the form

$$P\big(f_1(x_{t_1}), \ldots, f_n(x_{t_n})\big),$$

where P is a polynomial in n variables, each f_i is a continuous function on K_{t_i}, $t_i \in T$ and x_t are coordinate functions. For example, continuous real functions on the product of two compact spaces K_1 and K_2 are uniformly approximated by functions of the form $P\big(f_1(x_1), f_2(x_2)\big)$, where P is a polynomial in two variables, f_i is a continuous function on K_i and x_i are coordinate functions for $i = 1, 2$. Indeed, here we also obtain an algebra of functions and the fact that it separates points follows by the Urysohn theorem.

Let us give a complex version of the Stone-Weierstrass theorem. In the formulation given above it does not extend to complex functions. For example, the function $|z|$ on a disc in $\mathbb{C}$ cannot be approximated by polynomials in the complex variable z, because it is not analytic.

1.9.22. Theorem. *Let K be a nonempty compact space and let $\mathcal{E}$ be an algebra of continuous complex-valued functions on K containing all constants and separating points. Suppose also that the algebra $\mathcal{E}$ is closed with respect to the complex conjugation. Then this algebra is everywhere dense in the complex space $C(K)$ with its standard* sup-*metric.*

PROOF. Let $\mathcal{F}_r$ be the class of all real functions in the closure of $\mathcal{E}$ in $C(K)$. Then $\mathcal{F}_r$ is a closed subalgebra in the algebra $C_{(r)}(K)$ of real continuous functions on K, moreover, $\mathcal{F}_r$ contains all real constants. In addition, $\mathcal{F}_r$ separate points. Indeed, let $f \in C(K)$. Then $f_1 = (f + \overline{f})/2$ and $f_2 = (f - \overline{f})/(2i)$ belong to $\overline{\mathcal{E}}$, hence to $\mathcal{F}_r$. Hence if $f(x) \neq f(y)$ for some $x, y \in K$, then at least one of the functions f_1 and f_2 separates the points x and y. Therefore, $\mathcal{F}_r = C_{(r)}(K)$, whence our assertion follows. $\square$

1.9.23. Example. (i) The set of polynomials in z and $\overline{z}$ is dense in the complex space $C(K)$ for every compact set $K \subset \mathbb{C}$.

(ii) In the space $C_{2\pi}$ of continuous complex 2π-periodic functions on the real line the set of finite linear combinations of functions $\exp(ikt)$ with $k \in \mathbb{Z}$ is everywhere dense. Indeed, every continuous function f on $[0, 2\pi]$ with $f(0) = f(2\pi)$ corresponds to a continuous function g on the unit circle in $\mathbb{C}$: $g(\exp(it)) = f(t)$. Approximations of the function g by polynomials in z and $\overline{z}$ on the unit circle generate approximations of f by polynomials in $\exp(it)$ and $\exp(-it)$. In case of a real function f the real parts of approximating polynomials give approximations by linear combinations of the functions $1, \sin(kt), \cos(kt)$.

(iii) Let $\mathcal{F}$ be a family of continuous real functions on a compact space K separating points. Then the set of linear combinations of functions of the form $\exp(ic_1 f_1 + \cdots + ic_n f_n)$, where $c_i \in \mathbb{R}^1$ and $f_1, \ldots, f_n \in \mathcal{F}$, is everywhere dense in the complex space $C(K)$.

The Stone–Weierstrass theorem gives no construction of approximations, it just says that approximations exist. Constructing approximations is a separate important problem.

1.9(vi). The Cantor set

In the theory of metric and topological spaces and many applications an important role is played by the *Cantor set* C (called in honor of the famous German mathematician Georg Cantor). This set can be defined in two different ways. The first construction gives C as a closed set in $[0, 1]$ obtained by deleting from $[0, 1]$ the sequence of open intervals of the following form: at the first step we delete the open middle third of $[0, 1]$, i.e., $(1/3, 2/3)$, at the second step we delete the open middle thirds of the two obtained closed intervals, and then we inductively continue this process. It is readily verified that the set C consists of all points t of the form $t = \sum_{n=1}^{\infty} t_n 3^{-n}$, where the numbers t_n take the values 0 and 2. Note that the indicated expansion on C is unique, since it employs numbers $t_n \neq 1$. For example, $1/3$ is written as $(0, 2, 2, \ldots)$; the non-uniqueness of a representation is only possible due to infinitely repeating 2. The second way involves the compact metric space $\Delta := \{0, 1\}^{\infty}$, i.e., the countable power of a two-point set. This space consists of all possible sequences of 0 and 1; a metric can be defined by the formula

$$d(x, y) = \sum_{n=1}^{\infty} 2^{-n}|x_n - y_n|, \quad x = (x_n), y = (y_n).$$

Although these two sets are very different at the first glance, they are identical as topological spaces: the mapping

$$h \colon \{0, 1\}^{\infty} \to C, \quad (x_1, x_2, \ldots) \mapsto \sum_{n=1}^{\infty} 2x_n 3^{-n},$$

is a homeomorphism. This is clear from the fact that it is continuous and maps Δ one-to-one onto C (see Theorem 1.7.9).

In different situations, one or the other representation of the Cantor set can be useful. For example, it is completely obvious that the space Δ is homeomorphic to its square Δ^2, as well as to its countable power Δ^{∞}. Of course, this extends at once to C, which is not that obvious in the first representation. However, a convenience of the first representation is that it embeds the space Δ homeomorphically into the interval $[0, 1]$.

1.9.24. Proposition. *The homeomorphic sets Δ and C can be continuously mapped onto $[0, 1]$ and $[0, 1]^{\infty}$.*

PROOF. A continuous surjection $\psi \colon \Delta \to [0, 1]$ can be defined by the formula $\psi(x) = \sum_{n=1}^{\infty} x_n 2^{-n}$, $x = (x_n)$. Since the spaces Δ and Δ^{∞} are homeomorphic, it suffices to define a surjection of Δ^{∞} onto $[0, 1]^{\infty}$, which can be done as follows: $(z_n) \mapsto (\psi(z_n))$, $(z_n) \in \Delta^{\infty}$. $\square$

1.9.25. Corollary. *Every nonempty compact metric space is the image of the Cantor set under a continuous mapping.*

PROOF. According to Exercise 1.9.67, any nonempty compact metric space K is homeomorphic to a compact set K_0 in $[0, 1]^{\infty}$. The proposition above and the compactness of the preimage of K_0 under a continuous mapping of C onto $[0, 1]^{\infty}$ show that it suffices to be able to map C continuously onto its nonempty compact parts. Let S be such a part. Let us replace C by the homeomorphic set C_1 consisting of the points in $[0, 1]$ of the form $x = \sum_{k=1}^{\infty} x_k 6^{-k}$, where $x_k = 0$ or $x_k = 5$ (both sets are homeomorphic to $\{0, 1\}^{\infty}$). It is readily seen that if $x, y \in C_1$, then $(x + y)/2 \notin C_1$. Hence for any nonempty compact set $S_1 \subset C_1$ we can consider the mapping h that to every point $x \in C_1$ associates the nearest point $h(x) \in S_1$. This mapping is continuous and obviously surjective. $\square$

The Cantor set has no isolated points (a nonempty closed set without isolated points is called *perfect*).

1.9(vii). Cardinal characteristics of metric spaces

Given an infinite topological space X, one can consider various cardinal characteristics related to X. The simplest one is the cardinality $\mathrm{card}(X)$ of X itself. The weight $w(X)$ is the minimal cardinality of a topology base in X. The density $d(X)$ of X is the minimal cardinality of an everywhere dense set in X. The cellularity $c(X)$ of X is the largest cardinality of a pairwise disjoint family of nonempty open sets in X. A thorough discussion of such characteristics is given in [**277**, Chapter 1], but here we only mention some basic facts in the case of metric spaces, where the situation simplifies. First of all, in this case $w(X) = d(X) = c(X)$. Indeed, obviously, $c(X) \leqslant d(X) \leqslant w(X)$. On the other hand, $c(X) \geqslant w(X)$, since for each $n \in \mathbb{N}$ we can take a maximal family W_n of pairwise disjoin open balls of radius $1/n$ (which exists by Zorn's lemma) and then their union is a topology base and its cardinality does not exceed $c(X)$. For a discrete space, $\mathrm{card}(X)$ can be arbitrary, hence the same is true for $w(X) = d(X) = c(X)$. However, the situation changes for complete spaces in which all nonempty open sets have the same weight. The first assertion in the following theorem is due to A. H. Stone and the second one is due to F. K. Schmidt and A. H. Stone (for a proof, see [**277**, Chapter 1, Theorem 8.3, Theorem 8.2]).

1.9.26. Theorem. *If X is a complete metric space of weight κ, then either* $\mathrm{card}(X) = \kappa$ *or* $\mathrm{card}(X) = \kappa^{\omega}$, $\omega = \mathbb{N}$.

Moreover, if nonempty open subsets of X are infinite and have the same weight κ, then $\mathrm{card}(X) = \kappa^{\omega}$, $\omega = \mathbb{N}$.

In the second case we obtain the equality $\mathrm{card}(X) = \mathrm{card}(X)^{\omega}$, which does not hold for arbitrary uncountable cardinalities (see [**257**, §7] or [**546**, §2.9]).

Exercises

1.9.27. Let X be an infinite set. Prove that it has the same cardinality as the set of all its finite subsets. If X has the cardinality of the continuum, then it has the same cardinality as its countable power.

1.9.28. Prove that the space $B(X, Y)$ of bounded mappings from a nonempty set X to a complete nonempty metric space (Y, d_Y) is complete with respect to the metric $d(f, g) := \sup_{x \in X} d_Y\big(f(x), g(x)\big)$. Show that this is false if Y is incomplete.

1.9.29. Is it true that every metric space consisting of three points can be isometrically embedded into $\mathbb{R}^n$ with some n?

1.9.30. Is it true that the closure of an open ball in a metric space is a closed ball?
HINT: consider the space consisting of two points.

1.9.31. Justify the following construction of a completion of a metric space M. Let $\widetilde{M}$ be the space of equivalence classes of Cauchy sequences $x = (x_n)$ in the space M, where $x = (x_n)$ and $y = (y_n)$ are equivalent if the mixed sequence $x_1, y_1, \ldots, x_n, y_n, \ldots$ is Cauchy; the metric is defined by $d(x, y) := \lim_{n \to \infty} d(x_n, y_n)$. The natural embedding of M into $\widetilde{M}$ is given by the formula $x \mapsto (x, x, \ldots)$.

1.9.32. Let A and B be nonempty compact sets in a metric space (X, d). Prove that there exist points $a \in A$ and $b \in B$ such that $d(a, b) = \inf\{d(x, y): x \in A, b \in B\}$.

1.9.33. Show that in every incomplete metric space there is a sequence of nested closed balls with radii decreasing to zero and the empty intersection.

HINT: consider a Cauchy sequence that has no limit and find a subsequence $\{c_n\}$ in this sequence and numbers $r_n \to 0$ with the follow properties: $\{c_n\}$ has a limit c in the completion, $r_1 = 2d(c_1, c)$, c_2 is an element of the original sequence in the ball of radius $r_1/8$ centered at c, $r_2 = 2d(c_2, c)$, and so on.

1.9.34. Construct an example showing that in the nested ball theorem one cannot omit the assumption that the radii of balls tend to zero.

HINT: define a metric on $\mathbb{N}$ such that the set $\{k: k \neq n\}$ is a closed ball centered at $n + 1$.

1.9.35. Prove that a metric space is separable precisely when it possesses the Lindelöf property.

1.9.36. Suppose that U is an open set in a metric space, F is a closed set, $\overline{U}$ is the closure of U and F^o is the set of inner points of F. Prove that $\overline{U} \backslash U$ and $F \backslash F^o$ are nowhere dense.

1.9.37. Show that the continuity of a mapping f between topological spaces is equivalent to the property that, for every open set $V \subset Y$, the set $f^{-1}(V)$ is open in X (this is also equivalent to the property that, for every closed set $Z \subset Y$, the set $f^{-1}(Z)$ is closed in X).

1.9.38. Prove that a function f on a topological space is continuous precisely when for every $c \in \mathbb{R}^1$ the sets $f^{-1}((-\infty, c))$ and $f^{-1}((c, +\infty))$ are open.

1.9.39. Justify Example 1.6.8.

1.9.40. Give an example of an unbounded uniformly continuous real function on a bounded complete metric space.

HINT: consider $\mathbb{N}$ with the discrete metric.

1.9.41. (i) Let (X, d_X) and (Y, d_Y) be metric spaces and let $A \subset X$ and $B \subset Y$ be nonempty totally bounded sets. Prove that $A \times B$ is totally bounded in the space $X \times Y$ with the metric defined by the formula $d((x_1, y_1), (x_2, y_2)) := d_X(x_1, x_2) + d_Y(y_1, y_2)$.

(ii) Prove directly that the product of two nonempty compact spaces is compact with the product topology.

1.9.42. Does there exist an incomplete metric space in which every contracting mapping has a fixed point?

HINT: consider the subset of $\mathbb{R}^2$ consisting of the origin and the sequence of closed intervals $n^{-1} \times [0, 1]$.

1.9.43. Let (K, d) be a compact metric space and let $f: K \to K$ be a mapping such that $d(f(x), f(y)) < d(x, y)$ if $x \neq y$. Prove that f has a fixed point.

HINT: use that the function $x \mapsto d(f(x), x)$ attains its minimum.

1.9.44. Extend Theorem 1.4.4 to the case where T is a topological space.

1.9.45. (i) Show that a metric space is compact precisely when every continuous real function on this space is bounded. (ii) Show that on the noncompact topological space from Example 1.9.12 every continuous function is bounded.

1.9.46.° Let (X, d_X) be a compact metric space, let (Y, d_Y) be a metric space, and let the space $C(X, Y)$ of continuous mappings from X to Y be equipped with the metric $d(f, g) := \sup_{x \in X} d_Y(f(x), g(x))$.

(i) Prove that $K \subset C(X, Y)$ is totally bounded precisely when the following conditions are fulfilled: (a) for every $\varepsilon > 0$, there exists $\delta > 0$ such that $d_Y(f(x), f(x')) \leqslant \varepsilon$ for all $f \in K$ whenever $d_X(x, x') \leqslant \delta$, (b) there exists a totally bounded set $M \subset Y$ such that $f(x) \in M$ for all $f \in K$ and $x \in X$.

(ii) Give an example showing that (b) cannot be replaced by the uniform boundedness of the mappings in K.

(iii) Show that under condition (a) condition (b) is equivalent to the following one: for every fixed $x \in X$, the set of points $\{f(x): f \in K\}$ is totally bounded. (iv) Show that the compactness of K is equivalent to the closedness of K along with conditions (a) and (b), where the set M in (b) is compact.

1.9.47.° Let (K, d) be a compact metric space. Prove that a set $F \subset C(K)$ is totally bounded precisely when it is bounded and there is a function $\omega \colon [0, +\infty) \to [0, +\infty)$ satisfying the condition $\lim_{t \to 0} \omega(t) = \omega(0) = 0$ such that $|f(x) - f(y)| \leqslant \omega(d(x, y))$ for all $x, y \in K$ and $f \in F$.

1.9.48. Prove that the set of all isometries of a compact metric space K and the set of all mappings Lipschitz with a fixed constant are compact with respect to the sup-metric in the space $C(K, K)$. In particular, if $f \colon K \to K$ is a 1-Lipschitz mapping, then the sequence of iterations f^n contains a uniformly convergent subsequence.

HINT: use Exercise 1.9.46.

1.9.49. Let (K, d) be a compact metric space and $f \colon K \to K$. (i) Prove that if $d(f(x), f(y)) \leqslant d(x, y)$ and $f(K) = K$, then f is an isometry. (ii) Prove that if $d(f(x), f(y)) \geqslant d(x, y)$, then $f(K) = K$. In particular, a nonempty compact metric space cannot be isometric to its proper part. (iii) Prove that if $d(f(x), f(y)) \geqslant d(x, y)$, then f is an isometry.

HINT: (i) assume that $d(f(x), f(y)) \leqslant d(x, y) - \varepsilon$ for some x, y and $\varepsilon > 0$; observe that $d(f^n(x), f^n(y)) \leqslant d(x, y) - \varepsilon$ for all n; use the previous exercise to obtain numbers m and k such that $d(f^{m+k}(u), f^m(v)) < \varepsilon$ for all u, v and obtain a contradiction by picking u and v with $x = f^m(u)$, $y = f^m(v)$. (ii) Show first that K is the closure of $f(K)$, for otherwise one can take $x_1 \in K$ with $\mathrm{dist}(x_1, f(K)) = \alpha > 0$, which yields that the sequence $x_{n+1} = f(x_n)$ has no convergent subsequence. Hence $g = f^{-1} \colon f(K) \to K$ is 1-Lipschitz and extends to a 1-Lipschitz surjection of K, so is an isometry by (i). Then f is an isometry as well, which also gives (iii).

1.9.50. Prove that a metric space is nonseparable precisely when for some $\varepsilon > 0$ it contains an uncountable set of points with mutual distances not less than ε.

HINT: for every $\varepsilon > 0$, by Zorn's lemma there is a maximal family of disjoint balls of radius ε; if all such families are countable, then the space is separable.

1.9.51. Let A_c be the set of condensation points of an uncountable set A in a metric space. Prove that A_c is closed and has no isolated points, and if A is separable, then $A \backslash A_c$ is finite or countable.

1.9.52. Prove that a nonempty closed set without isolated points in a complete metric space has the cardinality not less than that of the continuum.

HINT: see [182, Section 4.5.5].

1.9.53. A set A in a metric space is called a first category set at a point x if x has a neighborhood U for which $A \cap U$ is a first category set. Otherwise A is called a second category set at x. Prove that for every second category set A there is a ball U such that A is a second category set at every point of U.

1.9.54. Let $X = C[0,1]$, let $n \in \mathbb{N}$, and let X_n be the set of functions $f \in X$ such that for some $t_0 \in [0, 1 - 1/n]$ the inequality $|f(t) - f(t_0)| \leqslant n(t - t_0)$ is true for all $t \in [t_0, 1)$. (i) Prove that X_n is a closed nowhere dense set. (ii) Prove that the set of all functions in X having a finite right derivative at least at one point is a first category set. Deduce from this the existence of a nowhere differentiable continuous function.

1.9.55.° Let X be a metric space. The oscillation of a function $f \colon X \to \mathbb{R}^1$ on a set $E \subset X$ is the quantity

$$\mathrm{osc}\,(f, E) := \sup_{x,y \in E} |f(x) - f(y)|.$$

The oscillation of the function f at a point x_0 is the quantity

$$\mathrm{osc}\, f(x_0) := \lim_{\varepsilon \to 0} \mathrm{osc}\,\big(f, B(x_0, \varepsilon)\big),$$

where $B(x_0, \varepsilon)$ is the closed ball of radius ε centered at x_0. Prove that the function f is continuous at the point x_0 precisely when $\mathrm{osc}\, f(x_0) = 0$.

1.9.56.° Let F be a compact set in $C[0,1]$. Prove the continuity of the functions $\psi(x) := \inf_{f \in F} f(x)$ and $\varphi(x) := \sup_{f \in F} f(x)$ on $[0,1]$. Prove the analogous assertion for compact sets in $C_b(T)$, where T is a topological space.
   HINT: observe that this assertion is true for finite sets.

1.9.57. (V. D. Erohin) Prove that every compact set of diameter d in $C[0,1]$ is contained in a closed ball of radius $d/2$.

1.9.58. Let X be a complete metric space and let $f_n \colon X \to \mathbb{R}^1$ be continuous functions such that for every $x \in X$ there exists a finite limit $f(x) = \lim_{n \to \infty} f_n(x)$. Prove that the function f has at least one point of continuity. Deduce from this that f has an everywhere dense set of continuity points.
   HINT: show that f has a point of zero oscillation arguing from the contradiction and assuming that at every point the oscillation is at least $1/n$ for some n; apply Baire's theorem to obtain a ball in which the oscillation is separated from zero. On the other hand, for each $\varepsilon > 0$ again by Baire's theorem one can find a smaller ball U such that for some N one has $|f_n(x) - f_k(x)| \leqslant \varepsilon$ for all $n, k \geqslant N$ and $x \in U$, which leads to a contradiction.

1.9.59. Let X and Y be complete metric spaces and let $f \colon X \times Y \to \mathbb{R}^1$ be a function continuous in every variable separately. Prove that there exists a point of continuity of f.
   HINT: as in the previous exercise, arguing from the contradiction find closed balls $B(x_0, r)$ and $B(y_0, r)$ of positive radius on the product of which the oscillation of f is larger than some $\varepsilon > 0$. For each x there is $\delta_x > 0$ such that $|f(x, y) - f(x, y_0)| \leqslant \varepsilon/4$ if $d(y, y_0) \leqslant \delta_x$. By Baire's theorem and the continuity in x we can find $\delta > 0$ and a smaller ball $B(x_1, r_1) \subset B(x_0, r)$ such that $|f(x, y) - f(x, y_0)| \leqslant \varepsilon/4$ whenever $d(x, x_1) \leqslant r_1$ and $d(y, y_0) \leqslant \delta$. Then $|f(x, y) - f(x_1, y_0)| \leqslant \varepsilon/2$ if $d(x, x_1) \leqslant r_2$, $d(y, y_0) \leqslant \delta$ for $r_2 > 0$ sufficiently small.

1.9.60. (I. Ekeland) Let (X, d) be a complete metric space and let f be a bounded upper semicontinuous function on X. Prove that for every $\varepsilon > 0$ there exists a point $x_\varepsilon \in X$ such that $f(x) \leqslant f(x_\varepsilon) + \varepsilon d(x, x_\varepsilon)$ for all $x \in X$.

1.9.61. Suppose we are given a sequence of functions f_n on a metric space such that if $x_n \to x$, then there exists a finite limit $f(x) = \lim\limits_{n\to\infty} f_n(x_n)$. Prove that the function f is continuous.

HINT: if $x_n \to x$, then there are increasing numbers k_n with $|f(x_n) - f_{k_n}(x_n)| < n^{-1}$.

1.9.62. Let (X, d) be a metric space. (i) Let f be a bounded function on a set $A \subset X$ such that $|f(x) - f(y)| \leqslant d(x, y)$ for all $x, y \in A$. Set

$$g(x) := \max\{\sup\nolimits_{y\in A}(f(y) - d(x, y)), \inf\nolimits_A f\}.$$

Show that $g(x) = f(x)$ if $x \in A$, $\sup_{y\in X}|g(y)| = \sup_{x\in A}|f(x)|$, $|g(x) - g(y)| \leqslant d(x, y)$ for all $x, y \in X$. (ii) Prove that every bounded uniformly continuous function on X can be uniformly approximated by bounded Lipschitz functions.

1.9.63.** Let $A \subset \mathbb{R}^n$ and let $f \colon A \to \mathbb{R}^k$ be Lipschitz. Prove that f has an extension to all of $\mathbb{R}^n$ that is Lipschitz with the same constant. Deduce from this that if $A \subset l^2$ and $f \colon A \to l^2$ is Lipschitz, then f extends to all of l^2 with the same Lipschitz constant.

HINT: see [56].

1.9.64.° Show that the construction described at the end of §1.6 defines indeed a topology on the product of topological spaces.

1.9.65.° Show that formula (1.6.2) defines a metric generating the topology of $\mathbb{R}^T$ for a countable set T.

1.9.66.° Suppose we are given a countable set of metric spaces X_n with metrics d_n. Show that formula (1.6.3) gives a metric on the product of X_n and that the topology generated by this metric coincides with the product topology.

1.9.67.° Prove that every compact metric space (K, d) is homeomorphic to a compact set in the compact space $[0, 1]^\infty$ by defining an embedding of K to $[0, 1]^\infty$ by the formula $h(x) = \big(d(x, x_n)\big)_{n=1}^\infty$, where $\{x_n\}$ is a countable everywhere dense set in K.

HINT: the set $h(K)$ is compact in $[0, 1]^\infty$, hence h is a homeomorphism.

1.9.68. Prove that a compact space K is metrizable precisely when there is a countable family of continuous functions f_n on K separating points in the following sense: if $x \neq y$, then for some n we have $f_n(x) \neq f_n(y)$. In this case for a metric one can take

$$d(x, y) = \sum\nolimits_{n=1}^\infty 2^{-n} \min\big(|f_n(x) - f_n(y)|, 1\big).$$

HINT: if K is metrizable, then one can take a countable dense set $\{a_n\}$ in K; the functions $f_n(x) = d(x, a_n)$ separate points. Conversely, a sequence of continuous functions f_n separating points gives a continuous injection into $\mathbb{R}^\infty$ by setting $h(x) = \big(f_n(x)\big)_{n=1}^\infty$. The set $h(K)$ is compact in the metrizable space $\mathbb{R}^\infty$, hence h is a homeomorphism. Use that (1.6.2) defines a metric.

1.9.69. Prove that a compact space K is metrizable precisely when the space $C(K)$ with the usual sup-metric is separable.

HINT: if $C(K)$ is separable, then the previous exercise applies. If K is metrizable, then find a countable set of continuous functions on K separating points and apply the Stone–Weierstrass theorem to the algebra generated by these functions.

1.9.70. Prove that a compact space K cannot be represented as a countable union of nowhere dense sets.

HINT: see [**182**, Theorem 3.9.3].

1.9.71. Give an example of a completely regular space that is not normal.

HINT: see [**182**, Example 1.5.9] or consider an uncountable power of the real line (see [**32**, Chapter II, Problem 392]).

1.9.72? A function $f\colon M \to \mathbb{R}^1$ on a metric space M is called upper semicontinuous if all sets $\{f \geqslant c\}$ are closed. Prove that this is equivalent to the following property: $f(x) \geqslant \limsup_{n\to\infty} f(x_n)$ whenever $x_n \to x$. Show that if M is compact, then such a function is bounded and achieves its maximum.

1.9.73. A function f on a metric space (X, d) is called lower semicontinuous if all sets $\{f \leqslant c\}$ are closed, i.e., $-f$ is upper semicontinuous. Let also $f \geqslant 0$. Show that

$$f_n(x) = \inf\{f(y) + nd(x,y)\colon y \in Y\}$$

are Lipschitz functions that increase pointwise to f. Show that if X is complete, then f must have continuity points.

1.9.74? Let $\{U_\alpha\}$ be an open cover of a compact metric space K. Prove that there exists $\varepsilon > 0$ with the following property: for each $x \in K$, there is U_α with $B(x, \varepsilon) \subset U_\alpha$.

1.9.75. Let (M, d) be a metric space and let $\mathcal{F}_d$ be the set of all nonempty bounded closed subsets of M. (i) Prove that

$$d_H(A, B) := \max\{\sup_{a\in A} d(a, B), \sup_{b\in B} d(b, A)\}$$

is a metric (the *Hausdorff metric*) on $\mathcal{F}_d$, moreover, M is isometric to the subset of singletons. (ii) Let $x_0 \in M$. Show that $\mathcal{F}_d$ is isometrically embedded into $C_b(M)$ by means of the mapping $A \mapsto f_A$, where $f_A(x) := d(x, A) - d(x, x_0)$. (iii) Show that $(\mathcal{F}_d, d_H)$ is complete precisely when (M, d) is complete. (iv) Show that $(\mathcal{F}_d, d_H)$ is compact precisely when (M, d) is compact.

Hint: use (ii) to prove (iii) and (iv); in addition, in (iv) use that f_A is 1-Lipschitz if M is bounded.

1.9.76. Prove that every compact space is homeomorphic to a compact set in some power of the interval $[0, 1]$.

Hint: take a family of continuous functions with values in $[0, 1]$ separating points and use them to construct a continuous injection into the corresponding power of $[0, 1]$.

1.9.77. Construct an example of a compact separable topological space that possesses a nonseparable subspace.

HINT: consider the set of functions in the product-space $[0, 1]^{[0,1]}$ each of which vanishes outside a finite set. One can also take the complement of $\mathbb{N}$ in its Stone–Čech compactification. There is a compact nonseparable subspace in $[0, 1]^{[0,1]}$ (embed into $[0, 1]^{[0,1]}$ the nonseparable compact space constructed in [**32**, Example 3.1.26]).

1.9.78. (The Ascoli–Arzelà theorem) Let K be a compact topological space. Prove that a set $F \subset C(K)$ has compact closure precisely when it is bounded and uniformly equicontinuous in the following sense: for every $x \in K$ and every $\varepsilon > 0$, there exists a neighborhood $U_{x,\varepsilon}$ of x such that $|f(y) - f(x)| < \varepsilon$ for all $y \in U_{x,\varepsilon}$ and all $f \in F$.

1.9.79. Let X be a completely regular space, let $K \subset X$ be compact, and let U be an open set containing K. Prove that there exists a continuous function $f\colon X \to [0, 1]$ with $f|_K = 1$ and $f|_{X\setminus U} = 0$.

HINT: use that for every $k \in K$ there is a continuous function $f_k\colon X \to [0, 1]$ with $f_k(k) = 1$ and $f_k|_{X\setminus U} = 0$; take a finite subcover of K by the sets $\{f_k > 1/2\}$.

1.9.80. Let X and Y be completely regular spaces, let $K \subset X \times Y$ be compact, and let $f\colon K \to [0, 1]$ be a continuous function. Prove that for every $\varepsilon > 0$ there exists a function g of the form $g(x, y) = \sum_{i=1}^n \varphi_i(x)\psi_i(y)$, where $\varphi_i \in C_b(X)$, $\psi_i \in C_b(Y)$, such that $|g(x, y)| \leqslant 1$ for all $x \in X, y \in Y$ and $|g(x, y) - f(x, y)| \leqslant \varepsilon$ for all $(x, y) \in K$.

HINT: use the Stone–Weierstrass theorem.

1.9.81. Let X be a completely regular space. (i) Prove that every continuous mapping f from X to a compact space K extends to a continuous mapping from the Stone–Čech compactification βX to K. (ii) Prove that the Stone–Čech compactification is unique up to a homeomorphism.

HINT: (i) embed K into a power of $[0,1]$; (ii) if K is another Stone–Čech compactification, then the embeddings $f\colon X \to K$ and $g\colon X \to \beta X$ extend to continuous mappings $\beta X \to K$ and $K \to \beta X$.

1.9.82. (i) Let X be a normal space and let $\{x_n\}$ be a sequence in X such that for every bounded continuous function f on the space X the sequence $\{f(x_n)\}$ converges. Show that the sequence $\{x_n\}$ converges in X. (ii)* Give an example of a completely regular space for which (i) is false.

HINT: (i) $\{x_n\}$ has a limit point. Otherwise one can find disjoint neighborhoods U_n of the points x_n such that U_n contains the closure of a smaller neighborhood W_n of x_n. For every n there is a function f_n with $0 \leqslant f_n \leqslant 1$, equal to 1 at x_{2n+1} and 0 outside W_{2n+1}. The function f equal to f_n in W_{2n+1} and 0 outside the union of the sets W_{2n+1} is bounded and continuous, which gives a contradiction. The same is true for every subsequence in $\{x_n\}$. It is readily seen that there are no different limit points. (ii) Let ω_1 be the minimal uncountable ordinal and let X be the product $[0,\omega_1]\times(\mathbb{N}\cup\{\infty\})$ without the point (ω_1,∞); consider the points $x_n = (\omega_1,n)$ (use [**320**, Chapter 5, Problem C]).

1.9.83. Let (M,d) be a nonempty metric space. A set $E \subset M$ is called an existence set if, for every $x \in M$, in E there is at least one nearest point y, i.e., $d(x,E) = d(x,y)$. If every point $x \in M$ has at most one nearest point, then E is called a uniqueness set. If E is an existence and uniqueness set, then E is called a Chebyshev set. Give examples of (i) a uniqueness set that is not an existence set, (ii) an existence set that is not a uniqueness set, (iii) an infinite-dimensional Chebyshev set.

HINT: in (iii) consider the ball in l^2.

1.9.84. Let (M,d) be a metric space and let $E \subset M$ be a nonempty set. If $x \in M$, $y_n \in E$ and $d(x,y_n) \to d(x,E)$, then $\{y_n\}$ is called minimizing for x. The set E is called approximately compact (the term is due to N. V. Efimov and B. S. Stechkin) if every minimizing sequence in this set contains a subsequence converging to a point of E.

(i) Give an example of a noncompact approximately compact set.

(ii) Prove that every approximately compact set is an existence set.

HINT: in (i) consider the ball in l^2.

1.9.85. Prove that a subset M of a complete metric space is homeomorphic to a complete metric space precisely when M is a G_δ-set, i.e., the intersection of a sequence of open sets.

HINT: see [**182**, Theorems 4.3.23, 4.3.24].

1.9.86. With the aid of a Hamel basis of the real line regarded over the field $\mathbb{Q}$ construct an additive function $f\colon \mathbb{R}^1 \to \mathbb{R}^1$, i.e., $f(t+s) = f(t)+f(s)$ for all $t,s \in \mathbb{R}^1$, that is unbounded on every interval. Show that if an additive function is bounded on an interval, then it is continuous.

HINT: let $\{v_\alpha\}$ be a Hamel basis of $\mathbb{R}$ over $\mathbb{Q}$ and let $\{v_{\alpha_n}\}$ be a countable subset of it; by shifting v_α by rational numbers we can assume that every interval with rational endpoints contains an element of $\{v_{\alpha_n}\}$; define f by $f(v_{\alpha_n}) = n$ and $f(v_\alpha) = 0$ if α does not belong to this countable set and extend by linearity.

CHAPTER 2

Fundamentals of Measure Theory

In this chapter and the next two we present Lebesgue's theory of measure and integral. In addition, we discuss connections between integration and differentiation.

2.1. Introductory Remarks

The problem of measuring length, area and volume goes back to antiquity. Its partial solution given by mathematicians of antiquity and formalized at the end of the XIX century leads to the so-called Jordan measure (which should be called more correctly the Peano–Jordan measure), defined as follows. For simplicity we discuss the one-dimensional case and attempt to define the length λ of subsets of the interval $I = [0,1]$. For intervals J of the form (a,b), $[a,b)$, $[a,b]$ or $(a,b]$ we set $\lambda(J) = |b - a|$. For the finite union of disjoint intervals $J_1,\ldots,J_n$ we set $\lambda(\bigcup_{i=1}^{n} J_i) = \sum_{i=1}^{n} \lambda(J_i)$. Such sets will be called elementary. We now have to extend our measure to non-elementary sets. A natural approach is to approximate non-elementary sets by elementary ones. But how should we approximate? The construction formalized by Jordan and Peano is this: we say that a set $A \subset I$ is Jordan measurable if, for every $\varepsilon > 0$, there exist elementary sets A_ε and B_ε such that $A_\varepsilon \subset A \subset B_\varepsilon$ and $\lambda(B_\varepsilon \backslash A_\varepsilon) < \varepsilon$. It is clear that, as $\varepsilon \to 0$, the lengths of A_ε and B_ε have a common limit, which we take for $\lambda(A)$. The Jordan measure is additive, i.e., $\lambda(A \cup B) = \lambda(A) + \lambda(B)$ for any disjoint sets A and B from the domain of definition. Moreover, it is countably additive in the following sense: if disjoint sets A_n along with their union $A = \bigcup_{n=1}^{\infty} A_n$ are Jordan measurable, then $\lambda(A) = \sum_{n=1}^{\infty} \lambda(A_n)$. However, not all sets gain length after this procedure. For example, the set $\mathbb{Q} \cap I$ of rational numbers in the interval is not Jordan measurable: it contains no elementary sets of positive measure, but all elementary sets containing $\mathbb{Q} \cap I$ have measure 1. This phenomenon looks strange: the set of rational numbers is a countable union of very simple elementary sets. For example, the disc is also composed of countably many elementary sets, moreover, these sets are more massive than points, but the disc is Jordan measurable. What is the problem? The problem is that the class of Jordan measurable sets is not closed with respect to the operation of countable union and the situation with the disc is rather an exception than a rule. Is it possible to extend λ to a broader domain of definition closed with respect to countable operations with preservation of the property of countable additivity? The role of countable additivity is already clear from evaluation of the

© Springer Nature Switzerland AG 2020
V. I. Bogachev, O. G. Smolyanov, *Real and Functional Analysis*,
Moscow Lectures 4, https://doi.org/10.1007/978-3-030-38219-3_2

disc area by means of approximations by unions of rectangles. For this reason the fact that the domain of definition of Jordan's measure is not closed with respect to countable unions is a serious drawback. It was overcome only at the beginning of the XX century by the outstanding French mathematician Henri Lebesgue, who suggested a principally different method of approximation by elementary sets that leads to Lebesgue measure. Namely: first, similarly to the old construction, the outer measure λ^* is introduced for *every* set $A \subset I$ as the infimum of the sums of measures of elementary sets forming countable covers of A. The decisive step is passage to countable covers (this step had already been made by Émil Borel). Next, the set A is called Lebesgue measurable if $\lambda^*(A) + \lambda^*(I \backslash A) = \lambda(I)$. An equivalent description of Lebesgue measurability in terms of approximations by elementary sets is this: for every $\varepsilon > 0$ there exists an elementary set A_ε such that $\lambda^*(A \bigtriangleup A_\varepsilon) < \varepsilon$. It is important that, unlike Jordan's construction, no containment of sets is required, i.e., we admit "crosswise displaced approximations". This nuance (along with countable covers in the definition of the outer measure) leads to substantial enlargement of the class of measurable sets. This enlargement is so great that the answer to the question about the existence of nonmeasurable sets depends on accepting or non accepting some special set-theoretic axioms. The class of Lebesgue measurable sets is closed with respect to countable unions and intersections, not only finite as in Jordan's construction. The outer measure on the class of measurable sets turns to be countably additive, moreover, the measure of any measurable set A equals the limit of measures of elementary sets approximating it in the aforementioned sense. The necessity to restrict the outer measure to the class of measurable sets is due to the fact that on the family of all sets it can fail to be even additive (although this can happen only under certain additional set-theoretic axioms, as noted above). Thus, Lebesgue's approach has two ideological novelties: countable covers in place of finite ones (this idea in more special cases was already employed by predecessors of Lebesgue, in particular, by Borel) and the restriction of the domain of definition of measure by the condition $\lambda^*(A) + \lambda^*(I \backslash A) = \lambda(I)$. It is clear from all that has been said that in the discussion of measures a key role is played by issues connected with domains of definition and extensions. For this reason the next section is concerned with some classes of sets arising as domains of definition of measures. It turns out that special features of length on subsets of the real line play no role, so from the very beginning it is reasonable to consider measures of a general form. Moreover, this point of view becomes necessary for the consideration of measures on spaces of arbitrary nature, for example, on manifolds or functional spaces, which is very important for many areas of mathematics and theoretical physics.

2.2. Algebras and σ-Algebras

One of the central concepts of measure theory is an algebra of sets.

2.2.1. Definition. *An algebra of sets $\mathcal{A}$ is a class of subsets of some fixed set X (called the space) such that*
(i) $\varnothing, X \in \mathcal{A}$; (ii) *if $A, B \in \mathcal{A}$, then $A \cap B \in \mathcal{A}$, $A \cup B \in \mathcal{A}$, $A \backslash B \in \mathcal{A}$.*

In place of the condition $A\backslash B \in \mathcal{A}$ it suffices to have $X\backslash B \in \mathcal{A}$ for all $B \in \mathcal{A}$, since $A\backslash B = A \cap (X\backslash B)$. However, the condition $A\backslash B \in \mathcal{A}$ gives two other conditions in (ii), because $A \cap B = A\backslash(A\backslash B)$.

Note that sometimes in the definition of an algebra in place of the inclusion $X \in \mathcal{A}$ the following broader condition is used: there exists a set $E \in \mathcal{A}$, called the unit of the algebra, such that $A \cap E = A$ for all $A \in \mathcal{A}$. It is clear that replacing X by E, we arrive at our definition on a smaller space. One should bear in mind that not all results presented below extend to the indicated broader concept.

2.2.2. Definition. *An algebra of sets $\mathcal{A}$ is called a σ-algebra if $\bigcup_{n=1}^{\infty} A_n \in \mathcal{A}$ for every sequence of sets A_n in $\mathcal{A}$.*

2.2.3. Definition. *A pair $(X, \mathcal{A})$ consisting of a set X and a σ-algebra $\mathcal{A}$ of its subsets is called a measurable space.*

Sometimes some other classes of sets are useful (semi-algebras, rings, semi-rings, σ-rings, etc., defined below), which slightly differ by admissible operations. Note that in the definition of a σ-algebra in place of the closedness with respect to countable unions one can require the closedness with respect to countable intersections. Indeed, the formula $\bigcup_{n=1}^{\infty} A_n = X\backslash\bigcap_{n=1}^{\infty}(X\backslash A_n)$ and the closedness of algebras with respect to complements show that the two properties are equivalent.

2.2.4. Example. The set of all finite unions of intervals of the form $[a, b]$, $[a, b)$, $(a, b]$, (a, b) in the interval $[0, 1]$ (or only of the form $[a, b) \cap [0, 1)$ in $[0, 1)$) is an algebra, but not a σ-algebra.

Clearly, the set 2^X of all subsets of a fixed set X is an σ-algebra as well as the class consisting only of X and the empty set. Any other σ-algebra of subsets in X is contained between these two trivial examples.

2.2.5. Definition. *Let $\mathcal{F}$ be a family of subsets of a space X. The smallest σ-algebra of subsets of X containing $\mathcal{F}$ is called the σ-algebra generated by $\mathcal{F}$ and is denoted by the symbol $\sigma(\mathcal{F})$. The algebra generated by $\mathcal{F}$ is the smallest algebra containing $\mathcal{F}$.*

The smallest σ-algebra and the smallest algebra mentioned in the definition exist indeed.

2.2.6. Proposition. *Let X be a set. For every family of subsets of X, there exists the unique σ-algebra generated by them. There is also the unique algebra generated by this family.*

PROOF. Set $\sigma(\mathcal{F}) = \bigcap_{\mathcal{F} \subset \mathcal{A}} \mathcal{A}$, where the intersection is taken over all σ-algebras of subsets of the space X containing the system of sets $\mathcal{F}$ (such σ-algebras exist: for example, 2^X). By construction $\mathcal{F} \subset \sigma(\mathcal{F})$. If we are given a sequence of sets $A_n \in \sigma(\mathcal{F})$, then their intersection, union and complements belong to every σ-algebra $\mathcal{A}$ containing $\mathcal{F}$, hence also belong to $\sigma(\mathcal{F})$. Therefore, $\sigma(\mathcal{F})$ is a σ-algebra. The uniqueness is obvious from the fact that the existence of a σ-algebra $\mathcal{B}$ containing $\mathcal{F}$, but not containing $\sigma(\mathcal{F})$, would contradict the definition of $\sigma(\mathcal{F})$, since $\mathcal{B} \cap \sigma(\mathcal{F})$ contains $\mathcal{F}$ and is a σ-algebra. The case of algebras is similar. $\square$

It follows from the definition that the complements of sets in the class $\mathcal{F}$ generate the same σ-algebra as $\mathcal{F}$. A countable class can generate an uncountable σ-algebra. For example, the intervals with rational endpoints generate a σ-algebra containing all singletons of the interval.

The algebra generated by a family of sets $\mathcal{F}$ can be easily described explicitly (see Exercise 2.7.14). However, elements of the σ-algebra generated by a class of sets $\mathcal{F}$ do not usually admit such a simple description. For example, not every set in the σ-algebra generated by all intervals of the real line can be represented in a constructive form with the aid of countable unions or intersections of intervals. The point is that these operations can be repeated in an unlimited number of steps in any order. For example, one can form the class $\mathcal{F}_\sigma$ of countable unions of closed sets in the interval, next the class $\mathcal{F}_{\sigma\delta}$ of countable intersections of sets from $\mathcal{F}_\sigma$ and continue this process inductively. At every step we shall obtain strictly larger classes, but even their union does not exhaust the σ-algebra generated by all closed sets. Let us give two examples when it is possible to describe explicitly the σ-algebra generated by a class of sets.

2.2.7. Example. (i) Let $\mathcal{A}_0$ be a σ-algebra of subsets of a space X and let S be a set in X that does not belong to $\mathcal{A}_0$. Then the σ-algebra $\sigma(\mathcal{A}_0 \cup \{S\})$ generated by $\mathcal{A}_0$ and S is the collection of all sets of the form

$$E = (A \cap S) \cup (B \cap (X \backslash S)), \quad A, B \in \mathcal{A}_0. \qquad (2.2.1)$$

(ii) Let X be an arbitrary set. Then the singleton sets generate the σ-algebra consisting of all at most countable sets and the complements of such sets.

PROOF. (i) All sets of the form (2.2.1) belong to the σ-algebra $\sigma(\mathcal{A}_0 \cup \{S\})$. On the other hand, the sets of the indicated form constitute a σ-algebra. Indeed,

$$X \backslash E = ((X \backslash A) \cap S) \cup ((X \backslash B) \cap (X \backslash S)),$$

since x does not belong to E precisely when either x belongs to S, but not to A, or x belongs neither to S, nor to B. In addition, if sets E_n are of type (2.2.1) with some $A_n, B_n \in \mathcal{A}_0$, then $\bigcap_{n=1}^\infty E_n$ and $\bigcup_{n=1}^\infty E_n$ is also of type (2.2.1). For example, $\bigcap_{n=1}^\infty E_n$ has the form (2.2.1) with $A = \bigcap_{n=1}^\infty A_n$ and $B = \bigcap_{n=1}^\infty B_n$. Finally, any set from $\mathcal{A}_0$ is obtained in the form (2.2.1) with $A = B$, and for obtaining S we take $A = X$, $B = \varnothing$.

(ii) Clearly, the class $\mathcal{A}$ of all at most countable sets and their complements belongs to the σ-algebra $\mathcal{A}_0$ generated by singleton sets. For proving the equality $\mathcal{A} = \mathcal{A}_0$ it suffices to observe that $\mathcal{A}$ is a σ-algebra. This is clear from the fact that if sets A_n are at most countable, then so is their union, and if among them there is at least one set A_{n_1} with an at most countable complement, then we have $X \backslash \bigcup_{n=1}^\infty A_n \subset X \backslash A_{n_1}$. $\qquad \square$

Dealing with σ-algebras it is customary to employ the following readily verified set-theoretic relations. Let $(A_\alpha)_{\alpha \in \Lambda}$ be a family of subsets of a set X and let $f \colon E \to X$ be an arbitrary mapping from some set E to X. Then

$$X \backslash \bigcup_{\alpha \in \Lambda} A_\alpha = \bigcap_{\alpha \in \Lambda} (X \backslash A_\alpha), \; X \backslash \bigcap_{\alpha \in \Lambda} A_\alpha = \bigcup_{\alpha \in \Lambda} (X \backslash A_\alpha),$$
$$f^{-1}\Big(\bigcup_{\alpha \in \Lambda} A_\alpha\Big) = \bigcup_{\alpha \in \Lambda} f^{-1}(A_\alpha), \; f^{-1}\Big(\bigcap_{\alpha \in \Lambda} A_\alpha\Big) = \bigcap_{\alpha \in \Lambda} f^{-1}(A_\alpha).$$

It is seen from these equalities that if $\mathcal{A}$ is a σ-algebra of subsets of a set X and f is an arbitrary mapping from a set E to X, then the class $f^{-1}(\mathcal{A})$ of all sets of the form $f^{-1}(A)$, where $A \in \mathcal{A}$, is a σ-algebra in E.

Next, for a σ-algebra of sets $\mathcal{B}$ in E the class of sets $\{A \subset X : f^{-1}(A) \in \mathcal{B}\}$ is a σ-algebra. Finally, $\sigma\big(f^{-1}(\mathcal{F})\big) = f^{-1}\big(\sigma(\mathcal{F})\big)$ for any class of sets $\mathcal{F}$ in X.

Simple examples show that the class $f(\mathcal{B})$ of all sets of the form $f(B)$, where $B \in \mathcal{B}$, is not always an algebra.

2.2.8. Definition. *The Borel σ-algebra of $\mathbb{R}^n$ is the σ-algebra $\mathcal{B}(\mathbb{R}^n)$ generated by all open sets. The sets in $\mathcal{B}(\mathbb{R}^n)$ are called Borel sets. For an arbitrary set $E \subset \mathbb{R}^n$ let $\mathcal{B}(E)$ denote the class of all sets of the form $E \cap B$, where $B \in \mathcal{B}(\mathbb{R}^n)$.*

The class $\mathcal{B}(E)$ is the σ-algebra generated by the intersections of E with open sets in $\mathbb{R}^n$. This is clear from the following reasoning: if the latter σ-algebra is $\mathcal{E}$, then the class of all sets $B \in \mathcal{B}(\mathbb{R}^n)$ such that $B \cap E \in \mathcal{E}$ is a σ-algebra containing all open sets, i.e., it coincides with $\mathcal{B}(\mathbb{R}^n)$. The sets from $\mathcal{B}(E)$ are called Borel sets of the space E and $\mathcal{B}(E)$ is called the Borel σ-algebra of the space E. One should remember that such sets need not be Borel in $\mathbb{R}^n$ if, of course, E itself is not Borel in $\mathbb{R}^n$. For example, we always have $E \in \mathcal{B}(E)$, since $E \cap \mathbb{R}^n = E$. It is clear that $\mathcal{B}(\mathbb{R}^n)$ is also generated by the class of all closed sets. We shall see below that not all sets are Borel.

2.2.9. Proposition. *The Borel σ-algebra of the real line is generated by each of the following classes of sets:*

(i) *the set of all intervals;*

(ii) *the set all of intervals with rational endpoints;*

(iii) *the set of all rays of the form $(-\infty, c)$, where c is rational;*

(iv) *the set of all rays of the form $(-\infty, c]$, where c is rational;*

(v) *the set of all rays of the form $(c, +\infty)$, where c is rational;*

(vi) *the set of all rays of the form $[c, +\infty)$, where c is rational.*

Finally, the same is true if in place of rational numbers we take points of an everywhere dense set.

PROOF. The listed sets are Borel, because they are either open or closed. Hence the σ-algebras generated by them are contained in $\mathcal{B}(\mathbb{R}^1)$. Since every open set on the real line is the union of at most countably many intervals, it suffices to show that every interval (a, b) belongs to the σ-algebras corresponding to the classes (i)–(vi). This follows from the fact that (a, b) is the union of the intervals of the form (a_n, b_n), where a_n and b_n are rational, and also the union of the intervals of the form $[a_n, b_n)$ with rational endpoints, while such intervals belong to the σ-algebra generated by the rays $(-\infty, c)$, since they are differences of such rays. Similarly, differences of rays of the form (c, ∞) give all intervals $(a_n, b_n]$, from which by means of unions we construct all intervals (a, b). □

It is clear that $\mathcal{B}(\mathbb{R}^1)$ is also generated by compact intervals with rational endpoints. By the way, this demonstrates that disjoint classes of sets can generate the same σ-algebra.

Let us introduce some other classes of sets employed in measure theory.

2.2.10. Definition. (i) *A system $\mathcal{R}$ of subsets of a set X is called a ring if $\varnothing \in \mathcal{R}$ and the sets $A \cap B$, $A \cup B$ and $A\backslash B$ belong to $\mathcal{R}$ for all $A, B \in \mathcal{R}$.*

(ii) *A system $\mathcal{S}$ of subsets of a set X is called a semi-ring if it contains the empty set, $A \cap B \in \mathcal{S}$ for all $A, B \in \mathcal{S}$ and, for every pair of sets $A, B \in \mathcal{S}$ with $A \subset B$, the set $B\backslash A$ is the union of finitely many disjoint sets in $\mathcal{S}$. If $X \in \mathcal{S}$, then $\mathcal{S}$ is called a semi-algebra.*

(iii) *A ring is called a σ-ring if it is closed with respect to countable unions. A ring is called a δ-ring if it is closed with respect to countable intersections.*

All bounded sets on the real line constitute a ring that is not an algebra. The class of all intervals in a given interval gives an example of a semi-ring that is not a ring. Another example of a semi-ring: all intervals of the form $[a, b)$ on the real line (their intersections with $[0, 1]$ or with $[0, 1)$ constitute a semi-algebra in $[0, 1]$ or in $[0, 1)$, respectively). According to the following lemma, the class of all finite unions of elements of a semi-ring is a ring (called the ring generated by the given semi-ring). It is clear that this is the minimal ring containing the given semi-ring.

2.2.11. Lemma. *For every semi-ring $\mathcal{S}$, the collection of all finite unions of sets in $\mathcal{S}$ is a ring $\mathcal{R}$. Every set in $\mathcal{R}$ is a finite union of pairwise disjoint sets in $\mathcal{S}$. If $\mathcal{S}$ is a semi-algebra, then $\mathcal{R}$ is an algebra.*

PROOF. It is clear that the class $\mathcal{R}$ is closed with respect to finite unions. Let $A = A_1 \cup \cdots \cup A_n$, $B = B_1 \cup \cdots \cup B_k$, where $A_i, B_j \in \mathcal{S}$. Then obviously $A \cap B = \bigcup_{i \leqslant n, j \leqslant k} A_i \cap B_j \in \mathcal{R}$. In addition,

$$A\backslash B = \bigcup_{i=1}^{n}\left(A_i\backslash\bigcup_{j=1}^{k} B_j\right) = \bigcup_{i=1}^{n}\bigcap_{j=1}^{k}(A_i\backslash B_j).$$

Since $A_i\backslash B_j = A_i\backslash(A_i \cap B_j)$ is a finite union of sets in $\mathcal{S}$, we have $A\backslash B \in \mathcal{R}$. It is clear that A can be written as a disjoint union of sets in $\mathcal{S}$, because $\mathcal{S}$ is closed with respect to intersections. If $X \in \mathcal{S}$, then $X \in \mathcal{R}$.   □

Note that for every σ-algebra $\mathcal{B}$ in any space X and every set $A \subset X$, the class $\mathcal{B}_A := \{B \cap A : B \in \mathcal{B}\}$ is a σ-algebra in the space A.

We now prove the *monotone class theorem*, a very useful tool for working with σ-algebras.

A family $\mathcal{E}$ of subsets of a set X is called a *monotone class* if for every increasing sequence of sets $E_n \in \mathcal{E}$ (i.e., $E_n \subset E_{n+1}$) we have $\bigcup_{n=1}^{\infty} E_n \in \mathcal{E}$ and also $\bigcap_{n=1}^{\infty} E_n \in \mathcal{E}$ for every decreasing sequence of sets $E_n \in \mathcal{E}$.

For every class $\mathcal{E}$ of subsets of X, there exists the minimal monotone class containing $\mathcal{E}$, which is called the *monotone class generated* by $\mathcal{E}$. This minimal class is the intersection of all monotone classes containing $\mathcal{E}$.

2.2.12. Theorem. *Let $\mathcal{A}$ be an algebra of sets. Then the σ-algebra generated by $\mathcal{A}$ coincides with the monotone class generated by $\mathcal{A}$. Therefore, if the algebra $\mathcal{A}$ is contained in some monotone class $\mathcal{M}$, then $\sigma(\mathcal{A}) \subset \mathcal{M}$.*

PROOF. Let $\mathcal{M}(\mathcal{A})$ be the monotone class generated by $\mathcal{A}$. Since $\sigma(\mathcal{A})$ is a monotone class, we have $\mathcal{M}(\mathcal{A}) \subset \sigma(\mathcal{A})$. Let us prove the inverse inclusion.

For this we show that $\mathcal{M}(\mathcal{A})$ is a σ-algebra. It suffices to show that $\mathcal{M}(\mathcal{A})$ is an algebra. We prove first that the class $\mathcal{M}(\mathcal{A})$ is closed with respect to taking complements. Let

$$\mathcal{M}_0 := \{B\colon B, X\backslash B \in \mathcal{M}(\mathcal{A})\}.$$

The class $\mathcal{M}_0$ is monotone, which is obvious from the monotonicity of the class $\mathcal{M}(\mathcal{A})$ and the equalities

$$X\backslash \bigcap_{n=1}^{\infty} B_n = \bigcup_{n=1}^{\infty}(X\backslash B_n), \quad X\backslash \bigcup_{n=1}^{\infty} B_n = \bigcap_{n=1}^{\infty}(X\backslash B_n).$$

Since $\mathcal{A} \subset \mathcal{M}_0 \subset \mathcal{M}(\mathcal{A})$, we have $\mathcal{M}_0 = \mathcal{M}(\mathcal{A})$.

We now verify that $\mathcal{M}(\mathcal{A})$ is closed with respect to finite intersections. Let $A \in \mathcal{M}(\mathcal{A})$. Set

$$\mathcal{M}_A := \{B \in \mathcal{M}(\mathcal{A})\colon A\cap B \in \mathcal{M}(\mathcal{A})\}.$$

If $B_n \in \mathcal{M}_A$ are increasing sets, then

$$A\cap \left(\bigcup_{n=1}^{\infty} B_n\right) = \bigcup_{n=1}^{\infty}(A\cap B_n) \in \mathcal{M}(\mathcal{A}).$$

Similarly we consider the case when B_n decrease. Hence $\mathcal{M}_A$ is a monotone class. If $A \in \mathcal{A}$, then $\mathcal{A} \subset \mathcal{M}_A \subset \mathcal{M}(\mathcal{A})$, whence we obtain $\mathcal{M}_A = \mathcal{M}(\mathcal{A})$. Let now $A \in \mathcal{A}$ and $B \in \mathcal{M}(\mathcal{A})$. Then $B \in \mathcal{M}_A$. Hence $A\cap B \in \mathcal{M}(\mathcal{A})$, which gives $A \in \mathcal{M}_B$. Thus, $\mathcal{A} \subset \mathcal{M}_B \subset \mathcal{M}(\mathcal{A})$. Therefore, $\mathcal{M}_B = \mathcal{M}(\mathcal{A})$ for all $B \in \mathcal{M}(\mathcal{A})$, which means that $\mathcal{M}(\mathcal{A})$ is closed with respect to taking the intersection of two sets. It follows that $\mathcal{M}(\mathcal{A})$ is an algebra, as required. □

A similar result is contained in Exercise 2.7.45.

2.3. Additivity and Countable Additivity

The term "scalar functions" will be reserved for functions with values in $(-\infty, +\infty)$. In those cases where we consider functions with values in the extended real line $[-\infty, +\infty]$, this will be particularly notified.

2.3.1. Definition. *A scalar set function μ defined on some class of sets $\mathcal{A}$ is called additive if*

$$\mu\left(\bigcup_{i=1}^{n} A_i\right) = \sum_{i=1}^{n}\mu(A_i) \tag{2.3.1}$$

for all finite collections of pairwise disjoint sets $A_i \in \mathcal{A}$ such that $\bigcup_{i=1}^{n} A_i \in \mathcal{A}$.
The function μ is called countably additive if

$$\mu\left(\bigcup_{n=1}^{\infty} A_n\right) = \sum_{n=1}^{\infty}\mu(A_n) \tag{2.3.2}$$

for all countable collections of pairwise disjoint sets A_n in $\mathcal{A}$ with $\bigcup_{n=1}^{\infty} A_n \in \mathcal{A}$.
A countably additive set function defined on an algebra is called a measure.

A countably additive measure μ on a σ-algebra of subsets of a space X is called a probability measure if μ is nonnegative and $\mu(X) = 1$.

A measure defined on the Borel σ-algebra of the whole space $\mathbb{R}^n$ or of its part is called a Borel measure.

It is easily seen from the definition that the series (2.3.2) converges absolutely (since its sum does not depend on rearrangements of the series).

2.3.2. Remark. For set functions with values in $(-\infty, +\infty]$ the following natural rule of addition is established: $+\infty + c = +\infty$ for all $c \in (-\infty, +\infty]$. The value $-\infty$ is excluded on purpose in order to avoid the indeterminant case $+\infty + (-\infty)$. With this agreement one can similarly define the additivity and countable additivity of a set function μ with values in $(-\infty, +\infty]$, however, in this case it is required in addition that $\mu(\varnothing) = 0$ if $\varnothing \in \mathcal{A}$ (for example, if $\mathcal{A}$ is an algebra or a ring). For finite set functions this requirement is fulfilled automatically by the additivity.

A measure μ with values in $[0, +\infty)$ on a σ-algebra $\mathcal{A}$ in X is called σ-finite if $X = \bigcup_{n=1}^{\infty} X_n$, where $\mu(X_n) < \infty$. The simplest example of a measure with values in $[0, +\infty]$ that is not σ-finite is the measure on the set consisting of a single point a defined by $\mu(a) = \infty$ and $\mu(\varnothing) = 0$.

In the case where the class $\mathcal{A}$ is closed with respect to finite unions, the finite additivity is equivalent to the equality

$$\mu(A \cup B) = \mu(A) + \mu(B) \tag{2.3.3}$$

for all disjoint sets $A, B \in \mathcal{A}$. For example, this equivalence holds if $\mathcal{A}$ is an algebra. Similarly, if $\mathcal{A}$ is a σ-algebra, the countable additivity is equality (2.3.2) for all possible disjoint sequences of sets in $\mathcal{A}$. The formulations given above are convenient for the two reasons: first, the validity of the corresponding equalities must be verified only for those collections of sets for which both parts are meaningful, second, as we shall see further, under natural assumptions additive (or countably additive) set functions admit additive (respectively, countably additive) extensions to broader classes of sets closed with respect to taking unions of the corresponding type.

Additive set functions are also called additive measures, but for simplification of terminology we shall call *measures only countably additive measures on algebras* (or rings in those very rare cases when they are considered). In addition, countably additive functions with values in $(-\infty, +\infty]$ defined on algebras will be called measures with values in $(-\infty, +\infty]$ (i.e., the fact that we admit infinite values will be particularly notified). All these conventions are caused by the fact that a substantial part of our main results in case of infinite measures require additional conditions (moreover, different for different theorems). Countably additive measures are also called σ-additive.

It is useful to introduce the property of *subadditivity* (also called the *semiadditivity*):

$$\mu\Big(\bigcup_{i=1}^{n} A_i\Big) \leqslant \sum_{i=1}^{n} \mu(A_i) \tag{2.3.4}$$

for all $A_i \in \mathcal{A}$ with $\bigcup_{i=1}^{n} A_i \in \mathcal{A}$. Any additive nonnegative set function on an algebra is subadditive (see below).

2.3.3. Proposition. *Let μ be an additive scalar set function on an algebra (or a ring) of sets $\mathcal{A}$. Then the following conditions are equivalent:*

(i) *the function μ is countably additive;*

(ii) *the function μ is continuous at zero in the following sense: if $A_n \in \mathcal{A}$, $A_{n+1} \subset A_n$ for all $n \in \mathbb{N}$ and $\bigcap_{n=1}^{\infty} A_n = \varnothing$, then*

$$\lim_{n\to\infty} \mu(A_n) = 0; \tag{2.3.5}$$

(iii) *the function μ is lower continuous in the sense that if $A_n \in \mathcal{A}$ are such that $A_n \subset A_{n+1}$ for all $n \in \mathbb{N}$ and $\bigcup_{n=1}^{\infty} A_n \in \mathcal{A}$, then*

$$\mu\bigl(\textstyle\bigcup_{n=1}^{\infty} A_n\bigr) = \lim_{n\to\infty} \mu(A_n).$$

PROOF. Suppose that μ is countably additive and sets $A_n \in \mathcal{A}$ monotonically decrease to the empty set. Let $B_n = A_n \backslash A_{n+1}$. The sets B_n belong to $\mathcal{A}$ and are disjoint, and their union is A_1. Hence the series $\sum_{n=1}^{\infty} \mu(B_n)$ converges. Then $\sum_{n=N}^{\infty} \mu(B_n)$ tends to zero as $N \to \infty$, but the sum of this series is $\mu(A_N)$, since $\bigcup_{n=N}^{\infty} B_n = A_N$. Thus, we arrive at condition (ii).

Suppose now that condition (ii) is fulfilled and $B_n \in \mathcal{A}$ are pairwise disjoint sets such that $B := \bigcup_{n=1}^{\infty} B_n \in \mathcal{A}$, $A_n = B \backslash \bigcup_{k=1}^{n} B_k$. It is clear that $A_n \in \mathcal{A}$, $A_{n+1} \subset A_n$, $\bigcap_{n=1}^{\infty} A_n = \varnothing$. By assumption, $\mu(A_n) \to 0$. Hence $\lim_{n\to\infty} \sum_{k=1}^{n} \mu(B_k) = \mu(B)$ by the finite additivity. Thus, μ is countably additive. Clearly, (iii) follows from (ii), since if sets $A_n \in \mathcal{A}$ monotonically increase and their union is a set $A \in \mathcal{A}$, then the sets $A \backslash A_n \in \mathcal{A}$ monotonically decrease to the empty set. Finally, by the finite additivity (iii) obviously yields the countable additivity of μ. □

One should bear in mind that the stated equivalence does not hold for semialgebras (Exercise 2.7.29).

2.3.4. Example. Let $\mathcal{A}$ be an algebra of sets $A \subset \mathbb{N}$ such that either A or $\mathbb{N} \backslash A$ is finite. For any finite A, let $\mu(A) = 0$, and for any A with a finite complement let $\mu(A) = 1$. Then μ is an additive, but not countably additive set function.

PROOF. Obviously, $\mathcal{A}$ is an algebra. Relation (2.3.3) is trivial for disjoint sets A and B if A is finite, and disjoint A and B in $\mathcal{A}$ cannot be infinite simultaneously. If μ were countably additive, we would have $\mu(\mathbb{N}) = \sum_{n=1}^{\infty} \mu(\{n\}) = 0$. □

There exist additive not countably additive set functions on σ-algebras (see Example 6.4.10). The simplest countably additive set function is identically zero. Another example: let $X \neq \varnothing$, $a \in X$, and let *Dirac's measure* δ_a be defined on every $A \subset X$ as follows: it equals 1 on A if $a \in A$ and 0 if $a \notin A$. Let us give a bit less trivial example.

2.3.5. Example. Let $\mathcal{A}$ be the σ-algebra of all subsets of $\mathbb{N}$. For $A = \{n_k\}$, let $\mu(A) = \sum_k 2^{-n_k}$. Then μ is a measure on $\mathcal{A}$.

In order to construct deeper examples (say, Lebesgue measure) we need auxiliary technical tools discussed in the next section.

Let us note some simple useful properties of additive and countably additive set functions.

2.3.6. Proposition. *Let μ be a nonnegative additive set function on an algebra or a ring $\mathcal{A}$.*

(i) *If $A, B \in \mathcal{A}$ and $A \subset B$, then $\mu(A) \leqslant \mu(B)$;*

(ii) *if $A_1, \ldots, A_n \in \mathcal{A}$, then $\mu\left(\bigcup_{i=1}^{n} A_i\right) \leqslant \sum_{i=1}^{n} \mu(A_i)$;*

(iii) *the function μ is countably additive precisely when, in addition, it is countably subadditive in the following sense: for every sequence $\{A_n\} \subset \mathcal{A}$ with $\bigcup_{n=1}^{\infty} A_n \in \mathcal{A}$ we have $\mu\left(\bigcup_{n=1}^{\infty} A_n\right) \leqslant \sum_{n=1}^{\infty} \mu(A_n)$.*

PROOF. Assertion (i) follows by the nonnegativity of $\mu(B\backslash A)$. Assertion (ii) is readily verified by induction on account of the nonnegativity of μ and the relation $\mu(A \cup B) = \mu(A\backslash B) + \mu(B\backslash A) + \mu(A \cap B)$.

If μ is countably additive and the union of sets $A_n \in \mathcal{A}$ also belongs to $\mathcal{A}$, then by Proposition 2.3.3 we have convergence $\mu\left(\bigcup_{i=1}^{n} A_i\right) \to \mu\left(\bigcup_{i=1}^{\infty} A_i\right)$, which by (ii) gives the estimate indicated in (iii). Finally, this estimate combined with the additivity gives the countable additivity. Indeed, let B_n be pairwise disjoint sets in $\mathcal{A}$ with the union B also belonging to $\mathcal{A}$. Then for every $n \in \mathbb{N}$ we have

$$\sum_{k=1}^{n} \mu(B_k) = \mu\left(\bigcup_{k=1}^{n} B_k\right) \leqslant \mu(B) \leqslant \sum_{k=1}^{\infty} \mu(B_k),$$

whence it follows that $\sum_{k=1}^{\infty} \mu(B_k) = \mu(B)$. $\qquad\square$

2.3.7. Proposition. *Let $\mathcal{A}_0$ be a semialgebra . Then every additive set function μ on $\mathcal{A}_0$ uniquely extends to an additive set function on the algebra $\mathcal{A}$ consisting of all possible finite unions of sets from $\mathcal{A}_0$ (i.e., on the algebra generated by $\mathcal{A}_0$). This extension is countably additive if μ is countably additive on $\mathcal{A}_0$. The same is true in the case of a semiring $\mathcal{A}$ and the generated ring.*

PROOF. By Lemma 2.2.11 the class of all finite unions of elements from $\mathcal{A}_0$ is an algebra (or a ring when $\mathcal{A}_0$ is a semiring). Every set from $\mathcal{A}$ has the form of a disjoint union of elements from $\mathcal{A}_0$. Let $\mu(A) = \sum_{i=1}^{n} \mu(A_i)$ if sets $A_i \in \mathcal{A}_0$ are pairwise disjoint and their union is A. The indicated extension is obviously additive, but we have to verify that it is well-defined, i.e., is independent of the partition of A into parts from $\mathcal{A}_0$. Indeed, if $B_1, \ldots, B_m$ are pairwise disjoint sets from $\mathcal{A}_0$ whose union is A, then by the additivity of μ on the algebra $\mathcal{A}_0$ we have $\mu(A_i) = \sum_{j=1}^{m} \mu(A_i \cap B_j)$, $\mu(B_j) = \sum_{i=1}^{n} \mu(A_i \cap B_j)$, whence the desired independence follows. Let us verify the countable additivity of the indicated extension in case of the countable additivity on $\mathcal{A}_0$. Let $A, A_n \in \mathcal{A}$, $A = \bigcup_{n=1}^{\infty} A_n$ be such that $A_n \cap A_k = \varnothing$ if $n \neq k$. Then $A = \bigcup_{j=1}^{N} B_j$, $A_n = \bigcup_{i=1}^{N_n} B_{n,i}$, where $B_j, B_{n,i} \in \mathcal{A}_0$. Let $C_{n,i,j} := B_{n,i} \cap B_j$. The sets $C_{n,i,j}$ are disjoint and

$$B_j = \bigcup_{n=1}^{\infty} \bigcup_{i=1}^{N_n} C_{n,i,j}, \quad B_{n,i} = \bigcup_{j=1}^{N} C_{n,i,j}.$$

By the countable additivity of μ on $\mathcal{A}_0$ we have

$$\mu(B_j) = \sum_{n=1}^{\infty} \sum_{i=1}^{N_n} \mu(C_{n,i,j}), \quad \mu(B_{n,i}) = \sum_{j=1}^{N} \mu(C_{n,i,j}).$$

In addition, $\mu(A) = \sum_{j=1}^{N} \mu(B_j)$, $\mu(A_n) = \sum_{i=1}^{N_n} \mu(B_{n,i})$ by the definition of μ on $\mathcal{A}$. These relations yield that $\mu(A) = \sum_{n=1}^{\infty} \mu(A_n)$, since both quantities equal the sum of all $\mu(C_{n,i,j})$. The possibility to interchange summations in n and j is obvious from the fact that the series in n converge and the sums in j and i are finite.   □

Since on algebras there are additive but not countably additive functions, the following efficiently verified sufficient (although not necessary) condition for the countable additivity is very useful in practice.

2.3.8. Definition. *A class of sets $\mathcal{K}$ in X is called compact if, whenever $K_n \in \mathcal{K}$ and $\bigcap_{n=1}^{\infty} K_n = \varnothing$, it follows that $\bigcap_{n=1}^{N} K_n = \varnothing$ for some N.*

The terminology is explained by the fact that any collection of compact sets is a compact class (see Theorem 1.9.6).

2.3.9. Theorem. *Let μ be a nonnegative additive set function on some algebra of sets $\mathcal{A}$ possessing a compact class $\mathcal{K}$ approximating μ in the following sense: for every $A \in \mathcal{A}$ and every $\varepsilon > 0$, there exist sets $K_\varepsilon \in \mathcal{K}$ and $A_\varepsilon \in \mathcal{A}$ such that $A_\varepsilon \subset K_\varepsilon \subset A$ and $\mu(A \backslash A_\varepsilon) < \varepsilon$. Then μ is countably additive.*

In particular, this is true if $\mathcal{A}$ contains a compact class $\mathcal{K}$ such that for every $A \in \mathcal{A}$ we have

$$\mu(A) = \sup_{K \subset A,\ K \in \mathcal{K}} \mu(K).$$

PROOF. Suppose that sets $A_n \in \mathcal{A}$ decrease and their intersection is empty. We show that $\mu(A_n) \to 0$. Let $\varepsilon > 0$. Let us take sets $K_n \in \mathcal{K}$ and $B_n \in \mathcal{A}$ such that $B_n \subset K_n \subset A_n$ and $\mu(A_n \backslash B_n) < \varepsilon 2^{-n}$. It is clear that

$$\bigcap_{n=1}^{\infty} K_n \subset \bigcap_{n=1}^{\infty} A_n = \varnothing.$$

By the compactness of $\mathcal{K}$ there is N such that $\bigcap_{n=1}^{N} K_n = \varnothing$. Then $\bigcap_{n=1}^{N} B_n = \varnothing$ and $A_N = A_N \backslash \bigcap_{n=1}^{N} B_n = \bigcup_{n=1}^{N} (A_N \backslash B_n)$ is contained in $\bigcup_{n=1}^{N} (A_n \backslash B_n)$, hence $\mu(A_N) \leqslant \sum_{n=1}^{N} \mu(A_n \backslash B_n) \leqslant \sum_{n=1}^{N} \varepsilon 2^{-n} \leqslant \varepsilon$. Thus, $\mu(A_n) \to 0$, which implies the countable additivity of μ.   □

2.3.10. Example. (i) Let I be a closed or open interval in $\mathrm{I\!R}^1$ and let $\mathcal{A}$ be the algebra of finite unions of intervals in I (closed, open and semiopen). Then the usual length λ_1 that equals the sum of $b_i - a_i$ on the union of finitely many disjoint intervals with endpoints a_i and b_i is countably additive on the algebra $\mathcal{A}$. The same is true for the semialgebra of all intervals of the form $[a, b)$ in $[0, 1)$.

(ii) Let I be a cube in $\mathrm{I\!R}^n$ of the form $[a, b]^n$ and let $\mathcal{A}$ be the algebra of finite unions of parallelepipeds in I that are products of intervals in $[a, b]$ (open, closed or semiopen). Then the usual volume λ_n is countably additive on $\mathcal{A}$. We shall call λ_n *Lebesgue measure* on the algebra generated by parallelepipeds.

PROOF. Finite unions of compact intervals are compact and approximate from within finite unions of other intervals. The case of a cube is similar.   □

There exist measures not possessing compact approximating classes, but such measure are rather exotic and do not appear in real applications. The next result shows that the presented sufficient condition for the countable additivity has a very universal nature.

2.3.11. Theorem. *Let μ be a nonnegative countably additive measure on the Borel σ-algebra $\mathcal{B}(\mathbb{R}^n)$ of $\mathbb{R}^n$. Then, for every Borel set $B \subset \mathbb{R}^n$ and every $\varepsilon > 0$, there exists an open set U_ε along with a compact set K_ε such that $K_\varepsilon \subset B \subset U_\varepsilon$ and $\mu(U_\varepsilon \backslash K_\varepsilon) < \varepsilon$.*

PROOF. We show that for every $\varepsilon > 0$ there exists a closed set $F_\varepsilon \subset B$ such that $\mu(B \backslash F_\varepsilon) < \varepsilon/2$. Then by the countable additivity of μ the set F_ε can be approximated from within up to $\varepsilon/2$ by the compact set $F_\varepsilon \cap U$, where U is a closed ball of a sufficiently large radius. Let $\mathcal{A}$ denote the class of all sets $A \in \mathcal{B}(\mathbb{R}^n)$ such that, for every $\varepsilon > 0$, there exists a closed set F_ε and an open set U_ε for which $F_\varepsilon \subset A \subset U_\varepsilon$ and $\mu(U_\varepsilon \backslash F_\varepsilon) < \varepsilon$. We observe that every closed set A belongs to $\mathcal{A}$, since for F_ε we can take A itself, and for U_ε we can take some open δ-neighborhood A^δ of A, i.e., the union of all open balls of radius δ centered at points of A (as δ tends to zero, the open sets A^δ decrease to A, hence their measures tend to the measure of A). Let us show that $\mathcal{A}$ is a σ-algebra. Once this is done, the theorem is proved, because closed sets generate the Borel σ-algebra. By construction the class $\mathcal{A}$ is closed with respect to the operation of complementation. Hence it remains to verify that $\mathcal{A}$ is closed with respect to countable unions. Let $A_j \in \mathcal{A}$ and $\varepsilon > 0$. There exist closed sets F_j and open sets U_j such that $F_j \subset A_j \subset U_j$ and $\mu(U_j \backslash F_j) < \varepsilon 2^{-j}$, $j \in \mathbb{N}$. The set $U = \bigcup_{j=1}^\infty U_j$ is open and the set $Z_k = \bigcup_{j=1}^k F_j$ is closed for every $k \in \mathbb{N}$. It remains to observe that $Z_k \subset \bigcup_{j=1}^\infty A_j \subset U$ and for k sufficiently large one has the estimate $\mu(U \backslash Z_k) < \varepsilon$. Indeed, $\mu\left(\bigcup_{j=1}^\infty (U_j \backslash F_j)\right) < \sum_{j=1}^\infty \varepsilon 2^{-j} = \varepsilon$ and $\mu(Z_k) \to \mu\left(\bigcup_{j=1}^\infty F_j\right)$ as $k \to \infty$ by the countable additivity. $\qquad\square$

This theorem shows that the measurability can be defined (as is really done in some textbooks) in the spirit of the Peano–Jordan construction through inner approximations by compact sets and outer approximations by open sets. For this it is necessary first to define measures of open sets (which will also define the measures of compact sets). In the case of an interval this is easy, since any open set is a union of disjoint intervals, which by the countable additivity defines its measure via measures of intervals. However, already in the case of a square there is no such disjoint representation of open sets, and the aforementioned construction is not that efficient here.

Finally, it is worth noting that the Lebesgue measure considered above on the algebra generated by cubes could be defined at once on the Borel σ-algebra by the equality

$$\lambda_n(B) := \inf \sum_{j=1}^\infty \lambda_n(I_j),$$

where inf is taken over all at most countable covers of B by cubes I_j. Actually, exactly this will be done below, but the justification of the fact that the indicated equality gives a countably additive measure is not trivial and will be given by a long detour with the aid of the construction of outer measures, which is the subject of the next section.

2.4. The Outer Measure and the Lebesgue Extension of Measures

Here we show how to extend countably additive measures from algebras to σ-algebras. On extensions from rings, see §2.7(i). We shall consider finite set functions and in the end make a remark about functions with values in $[0, +\infty]$.

For every nonnegative set function μ defined on some class $\mathcal{A}$ of subsets of a space X containing X itself the formula

$$\mu^*(A) = \inf\left\{\sum_{n=1}^{\infty} \mu(A_n) \,\middle|\, A_n \in \mathcal{A}, \ A \subset \bigcup_{n=1}^{\infty} A_n\right\}$$

defines a new set function defined for every subset $A \subset X$. The same construction applies to set functions with values in $[0, +\infty]$. If X does not belong to $\mathcal{A}$, the function μ^* is defined by the indicated formula on all sets A that can be covered by countable sequences of elements of $\mathcal{A}$, and to all other sets one can assign the infinite value (sometimes it is more convenient to assign them the value equal the supremum of values of μ^* on their subsets that can be covered by sequences from $\mathcal{A}$). The function μ^* is called the outer measure generated by μ, although it need not be even additive. In greater detail Caratheodory outer measures, not necessarily generated by additive set functions, are discussed in [73, Chapter 1]; see also §2.7(i) below. Our main example (Lebesgue measure): $A \subset [0,1]$ is covered by intervals A_n and $\mu(A_n)$ is the length of A_n, see §2.5.

2.4.1. Definition. *Let μ be a nonnegative set function on some domain of definition $\mathcal{A} \subset 2^X$. A set A is called μ-measurable (or Lebesgue measurable with respect to μ) if, for every $\varepsilon > 0$, there is a set $A_\varepsilon \in \mathcal{A}$ with $\mu^*(A \bigtriangleup A_\varepsilon) < \varepsilon$.*

The class of μ-measurable sets is denoted by $\mathcal{A}_\mu$.

We shall be interested in the case where μ is a countably additive measure on an algebra $\mathcal{A}$. The definition of measurability given by Lebesgue himself consisted in the equality $\mu^*(A) + \mu^*(X \backslash A) = \mu(X)$ (for a closed interval X). It will be shown below that for additive functions on algebras this definition (possibly, intuitively not that transparent, but simply expressing the additivity for mutually complementing sets) is equivalent to the one given above (see Proposition 2.4.12 and Theorem 2.7.8). In addition, the cited assertions contain a criterion of the Caratheodory measurability, which is also equivalent to our definition in the case of nonnegative additive set functions on algebras, but is much more effective in the general case (in particular, for measures with values in $[0, +\infty]$).

2.4.2. Example. (i) Let $\varnothing \in \mathcal{A}$ and $\mu(\varnothing) = 0$. Then $\mathcal{A} \subset \mathcal{A}_\mu$, since for any $A \in \mathcal{A}$ we take $A_\varepsilon = A$. In addition, every set A with $\mu^*(A) = 0$ is μ-measurable (one can take $A_\varepsilon = \varnothing$).

(ii) Let $\mathcal{A}$ be the algebra of finite unions of intervals from Example 2.3.10 with the usual length λ. Then the λ-measurability of A is equivalent to the property that for every $\varepsilon > 0$ one can find a finite union of intervals E and sets $A'_\varepsilon, A''_\varepsilon$ in I (not necessarily Lebesgue measurable, of course) such that

$$A = (E \cup A'_\varepsilon) \backslash A''_\varepsilon, \quad \lambda^*(A'_\varepsilon) \leqslant \varepsilon, \quad \lambda^*(A''_\varepsilon) \leqslant \varepsilon.$$

(iii) Let $X = [0,1]$, $\mathcal{A} = \{\varnothing, X\}$, $\mu(X) = 1$, $\mu(\varnothing) = 0$. Then μ is a countably additive measure on $\mathcal{A}$ and $\mathcal{A}_\mu = \mathcal{A}$. Indeed, $\mu^*(E) = 1$ for every nonempty set E. Hence among nonempty sets only the whole interval can be approximated by a set from $\mathcal{A}$ up to $\varepsilon < 1$.

Note that even if μ is a countably additive measure on a σ-algebra $\mathcal{A}$, the corresponding outer measure μ^* can fail to be countably additive on the class of all sets.

2.4.3. Example. Let X be the two-point set $\{0,1\}$ and $\mathcal{A} = \{\varnothing, X\}$. Set $\mu(\varnothing) = 0$, $\mu(X) = 1$. Then the class $\mathcal{A}$ is a σ-algebra and μ is countably additive on $\mathcal{A}$, but μ^* is not additive on the σ-algebra of all sets, since $\mu^*(\{0\}) = 1$, $\mu^*(\{1\}) = 1$, $\mu^*(\{0\} \cup \{1\}) = 1$.

The next lemmas contain two important properties of μ^*.

2.4.4. Lemma. *Let μ be a nonnegative set function on a class $\mathcal{A}$. Then the function μ^* is countably subadditive, i.e.,*

$$\mu^*\left(\bigcup_{n=1}^\infty A_n\right) \leqslant \sum_{n=1}^\infty \mu^*(A_n) \tag{2.4.1}$$

for all sets A_n.

PROOF. Let $\varepsilon > 0$ and $\mu^*(A_n) < \infty$. For every n, there exists a collection $\{B_{n,k}\}_{k=1}^\infty \subset \mathcal{A}$ such that $A_n \subset \bigcup_{k=1}^\infty B_{n,k}$ and

$$\sum_{k=1}^\infty \mu(B_{n,k}) \leqslant \mu^*(A_n) + \frac{\varepsilon}{2^n}.$$

Then $\bigcup_{n=1}^\infty A_n \subset \bigcup_{n=1}^\infty \bigcup_{k=1}^\infty B_{n,k}$ and hence

$$\mu^*\left(\bigcup_{n=1}^\infty A_n\right) \leqslant \sum_{n=1}^\infty \sum_{k=1}^\infty \mu(B_{n,k}) \leqslant \sum_{n=1}^\infty \mu^*(A_n) + \varepsilon.$$

Since ε was arbitrary, we arrive at (2.4.1). $\qquad\square$

2.4.5. Lemma. *In the situation of the previous lemma, for any sets A and B with $\mu^*(B) < \infty$, one has*

$$|\mu^*(A) - \mu^*(B)| \leqslant \mu^*(A \triangle B). \tag{2.4.2}$$

PROOF. We observe that $A \subset B \cup (A \triangle B)$, whence by the subadditivity of μ^* we obtain the estimate

$$\mu^*(A) \leqslant \mu^*(B) + \mu^*(A \triangle B),$$

i.e., $\mu^*(A) - \mu^*(B) \leqslant \mu^*(A \triangle B)$. The estimate $\mu^*(B) - \mu^*(A) \leqslant \mu^*(A \triangle B)$ is obtained similarly. Of course, the same is true if $\mu^*(A) < \infty$. $\qquad\square$

2.4.6. Theorem. *Let μ be a finite nonnegative countably additive set function on an algebra $\mathcal{A}$. Then*

(i) *$\mathcal{A} \subset \mathcal{A}_\mu$ and the outer measure μ^* coincides with μ on $\mathcal{A}$;*

(ii) *the collection $\mathcal{A}_\mu$ of all μ-measurable sets is a σ-algebra and the restriction of μ^* to $\mathcal{A}_\mu$ is countably additive;*

(iii) *μ^* is a unique nonnegative countably additive extension of μ to the σ-algebra $\sigma(\mathcal{A})$ generated by $\mathcal{A}$ and also a unique such extension of μ to $\mathcal{A}_\mu$.*

PROOF. (i) As already noted, $\mathcal{A} \subset \mathcal{A}_\mu$. Let $A \in \mathcal{A}$ and $A \subset \bigcup_{n=1}^\infty A_n$, where $A_n \in \mathcal{A}$. Then $A = \bigcup_{n=1}^\infty (A \cap A_n)$. Hence by Proposition 2.3.6(iii) we have

$$\mu(A) \leqslant \sum_{n=1}^\infty \mu(A \cap A_n) \leqslant \sum_{n=1}^\infty \mu(A_n),$$

whence $\mu(A) \leqslant \mu^*(A)$. By definition, $\mu^*(A) \leqslant \mu(A)$. Hence $\mu(A) = \mu^*(A)$.

(ii) We first observe that the complement of any measurable set A is measurable. This is obvious from the formula $(X \backslash A) \bigtriangleup (X \backslash A_\varepsilon) = A \bigtriangleup A_\varepsilon$. Furthermore, the union of measurable sets A and B is measurable. Indeed, let $\varepsilon > 0$ and $A_\varepsilon, B_\varepsilon \in \mathcal{A}$ be such that $\mu^*(A \bigtriangleup A_\varepsilon) < \varepsilon/2$ and $\mu^*(B \bigtriangleup B_\varepsilon) < \varepsilon/2$. Since

$$(A \cup B) \bigtriangleup (A_\varepsilon \cup B_\varepsilon) \subset (A \bigtriangleup A_\varepsilon) \cup (B \bigtriangleup B_\varepsilon),$$

we have

$$\mu^*\Big((A \cup B) \bigtriangleup (A_\varepsilon \cup B_\varepsilon)\Big) \leqslant \mu^*\Big((A \bigtriangleup A_\varepsilon) \cup (B \bigtriangleup B_\varepsilon)\Big) < \varepsilon.$$

Therefore, $A \cup B \in \mathcal{A}_\mu$. In addition, by the already proved facts we have $A \cap B = X \backslash ((X \backslash A) \cup (X \backslash B)) \in \mathcal{A}_\mu$. Thus, $\mathcal{A}_\mu$ is an algebra.

We now establish two less obvious properties of the outer measure. First we verify its additivity on $\mathcal{A}_\mu$. Let $A, B \in \mathcal{A}_\mu$ be such that $A \cap B = \varnothing$. Let us fix $\varepsilon > 0$ and find $A_\varepsilon, B_\varepsilon \in \mathcal{A}$ such that

$$\mu^*(A \bigtriangleup A_\varepsilon) < \varepsilon/2 \quad \text{and} \quad \mu^*(B \bigtriangleup B_\varepsilon) < \varepsilon/2.$$

By Lemma 2.4.5, on account of the coincidence of μ^* on $\mathcal{A}$ with μ we have

$$\mu^*(A \cup B) \geqslant \mu(A_\varepsilon \cup B_\varepsilon) - \mu^*\Big((A \cup B) \bigtriangleup (A_\varepsilon \cup B_\varepsilon)\Big). \qquad (2.4.3)$$

The inclusion $(A \cup B) \bigtriangleup (A_\varepsilon \cup B_\varepsilon) \subset (A \bigtriangleup A_\varepsilon) \cup (B \bigtriangleup B_\varepsilon)$ and the subadditivity of μ^* yield the inequality

$$\mu^*\Big((A \cup B) \bigtriangleup (A_\varepsilon \cup B_\varepsilon)\Big) \leqslant \mu^*(A \bigtriangleup A_\varepsilon) + \mu^*(B \bigtriangleup B_\varepsilon) \leqslant \varepsilon. \qquad (2.4.4)$$

By the inclusion $A_\varepsilon \cap B_\varepsilon \subset (A \bigtriangleup A_\varepsilon) \cup (B \bigtriangleup B_\varepsilon)$ we obtain

$$\mu(A_\varepsilon \cap B_\varepsilon) = \mu^*(A_\varepsilon \cap B_\varepsilon) \leqslant \mu^*(A \bigtriangleup A_\varepsilon) + \mu^*(B \bigtriangleup B_\varepsilon) \leqslant \varepsilon.$$

By the estimates $\mu(A_\varepsilon) \geqslant \mu^*(A) - \varepsilon/2$ and $\mu(B_\varepsilon) \geqslant \mu^*(B) - \varepsilon/2$ we have

$$\mu(A_\varepsilon \cup B_\varepsilon) = \mu(A_\varepsilon) + \mu(B_\varepsilon) - \mu(A_\varepsilon \cap B_\varepsilon) \geqslant \mu^*(A) + \mu^*(B) - 2\varepsilon.$$

On account of relations (2.4.3) and (2.4.4) this gives

$$\mu^*(A \cup B) \geqslant \mu^*(A) + \mu^*(B) - 3\varepsilon.$$

Since ε was arbitrary, we obtain $\mu^*(A \cup B) \geqslant \mu^*(A) + \mu^*(B)$. Furthermore, since $\mu^*(A \cup B) \leqslant \mu^*(A) + \mu^*(B)$, we have

$$\mu^*(A \cup B) = \mu^*(A) + \mu^*(B).$$

The next important step is to verify that the countable unions of measurable sets are measurable. It suffices to prove this for disjoint sets $A_n \in \mathcal{A}_\mu$. Indeed, in the general case we can take $B_n = A_n \backslash \bigcup_{k=1}^{n-1} A_k$. Then the sets B_n are disjoint, are measurable as shown above and have the same union as the sets A_n. Now, dealing with disjoint sets, we observe that by the finite additivity of μ^* on $\mathcal{A}_\mu$ the following relations are valid:

$$\sum_{k=1}^{n} \mu^*(A_k) = \mu^*\left(\bigcup_{k=1}^{n} A_k\right) \leqslant \mu^*\left(\bigcup_{k=1}^{\infty} A_k\right) \leqslant \mu(X).$$

Thus, $\sum_{k=1}^{\infty} \mu^*(A_k) < \infty$. Let $\varepsilon > 0$. Choose n such that

$$\sum_{k=n+1}^{\infty} \mu^*(A_k) < \frac{\varepsilon}{2}.$$

Using the already known measurability of finite unions, we find a set $B \in \mathcal{A}$ such that $\mu^*\left(B \triangle \bigcup_{k=1}^{n} A_k\right) < \varepsilon/2$. Since

$$B \triangle \bigcup_{k=1}^{\infty} A_k \subset \left(B \triangle \bigcup_{k=1}^{n} A_k\right) \cup \left(\bigcup_{k=n+1}^{\infty} A_k\right),$$

we obtain

$$\mu^*\left(B \triangle \bigcup_{k=1}^{\infty} A_k\right) \leqslant \mu^*\left(B \triangle \bigcup_{k=1}^{n} A_k\right) + \mu^*\left(\bigcup_{k=n+1}^{\infty} A_k\right)$$

$$\leqslant \frac{\varepsilon}{2} + \sum_{k=n+1}^{\infty} \mu^*(A_k) < \varepsilon.$$

Thus, $\bigcup_{k=1}^{\infty} A_k$ is measurable. Hence $\mathcal{A}_\mu$ is a σ-algebra. It remains to observe that the additivity and the countable subadditivity of μ^* on $\mathcal{A}_\mu$ yield the countable additivity (see Proposition 2.3.6).

(iii) We observe that $\sigma(\mathcal{A}) \subset \mathcal{A}_\mu$, since $\mathcal{A}_\mu$ is a σ-algebra containing $\mathcal{A}$. Let ν be some nonnegative countably additive extension of μ to $\sigma(\mathcal{A})$. Let $A \in \sigma(\mathcal{A})$ and $\varepsilon > 0$. Since $A \in \mathcal{A}_\mu$, there exists a set $B \in \mathcal{A}$ with $\mu^*(A \triangle B) < \varepsilon$. Hence there exist sets $C_n \in \mathcal{A}$ such that $A \triangle B$ is contained in the union $\bigcup_{n=1}^{\infty} C_n$ and $\sum_{n=1}^{\infty} \mu(C_n) < \varepsilon$. Then we have

$$|\nu(A) - \nu(B)| \leqslant \nu(A \triangle B) \leqslant \sum_{n=1}^{\infty} \nu(C_n) = \sum_{n=1}^{\infty} \mu(C_n) < \varepsilon.$$

Since $\nu(B) = \mu(B) = \mu^*(B)$, we finally obtain

$$|\nu(A) - \mu^*(A)| = |\nu(A) - \nu(B) + \mu^*(B) - \mu^*(A)|$$

$$\leqslant |\nu(A) - \nu(B)| + |\mu^*(B) - \mu^*(A)| \leqslant 2\varepsilon.$$

Since ε was arbitrary we arrive at the equality $\nu(A) = \mu^*(A)$. This reasoning shows the uniqueness of a nonnegative countably additive extension of μ also to $\mathcal{A}_\mu$, since we have used only the inclusion $A \in \mathcal{A}_\mu$ (the nonnegativity is important, see below). □

Important notation: below the measure μ^* on $\mathcal{A}_\mu$ will be denoted by μ.

A control question: where in the proof above is the countable additivity of μ used? It is important indeed.

2.4.7. Example. Let $\mathcal{A}$ be the algebra of all finite subsets of $\mathbb{N}$ and their complements and let the set function μ equal 0 on all finite sets and 1 on their complements. Then μ is additive and the singletons $\{n\}$ cover $\mathbb{N}$, however, $\mu^*(\mathbb{N}) = 0 < \mu(\mathbb{N})$.

Note that in the theorem above μ cannot have signed countably additive extensions from $\mathcal{A}$ to $\sigma(\mathcal{A})$, which follows by (iii) and the Jordan decomposition from Chapter 3, however, it can have signed extensions to $\mathcal{A}_\mu$. For example, let $X = \{0, 1\}$, $\mathcal{A} = \sigma(\mathcal{A}) = \{\varnothing, X\}$, $\mu \equiv 0$, $\nu(\{0\}) = 1$, $\nu(\{1\}) = -1$, $\nu(X) = 0$. Then $\mu = \nu = 0$ on $\mathcal{A}$, the points $0, 1$ belong to $\mathcal{A}_\mu$ and have zero measure with respect to the extension of μ to $\mathcal{A}_\mu$, but the measure ν is a nonzero extension of μ.

2.4.8. Example. An important special case to which one can apply the extension theorem is described in Example 2.3.10. Since the σ-algebra generated by the cubes with edges parallel to the coordinate axes is the Borel σ-algebra, we obtain the countably additive *Lebesgue measure* λ_n on the Borel σ-algebra of the cube (and even on the broader σ-algebra of Lebesgue measurable sets) extending the elementary volume. This measure is discussed in greater detail in §2.5. By Theorem 2.4.6 Lebesgue measure of every Borel (as well as any measurable) set B in the cube is $\lambda_n^*(B)$. The question naturally arises why do not we use this formula at once to define the measure on the σ-algebra of Borel subsets of the cube. The point is that the main trouble is to verify the additivity of the obtained set function. In order to circumvent this difficulty, one has to verify the additivity on the larger class of measurable sets and prove that it is a σ-algebra.

With the aid of the proved theorem we can give a new description of measurable sets.

2.4.9. Corollary. *Let μ be a nonnegative countably additive set function on an algebra $\mathcal{A}$. A set A is μ-measurable precisely when there exist sets $A', A'' \in \sigma(\mathcal{A})$ such that*

$$A' \subset A \subset A'' \quad \text{and} \quad \mu^*(A'' \backslash A') = 0.$$

For A' one can take a set of the form $\bigcup_{n=1}^\infty \bigcap_{k=1}^\infty A_{n,k}$, where $A_{n,k} \in \mathcal{A}$, and for A'' a set of the form $\bigcap_{n=1}^\infty \bigcup_{k=1}^\infty B_{n,k}$, where $B_{n,k} \in \mathcal{A}$.

PROOF. Let $A \in \mathcal{A}_\mu$. For every $\varepsilon > 0$ we can find a set $A_\varepsilon \in \sigma(\mathcal{A})$ such that $A \subset A_\varepsilon$ and $\mu^*(A) \geqslant \mu^*(A_\varepsilon) - \varepsilon$. Indeed, by definition there exist sets $A_n \in \mathcal{A}$ with $A \subset \bigcup_{n=1}^\infty A_n$ and $\mu^*(A) \geqslant \sum_{n=1}^\infty \mu(A_n) - \varepsilon$. Set $A_\varepsilon = \bigcup_{n=1}^\infty A_n$. Clearly, $A \subset A_\varepsilon$, $A_\varepsilon \in \sigma(\mathcal{A}) \subset \mathcal{A}_\mu$. By the countable additivity of μ^* on $\mathcal{A}_\mu$ we have

$\mu^*(A_\varepsilon) \leqslant \sum_{n=1}^\infty \mu(A_n)$. Set $A'' = \bigcap_{n=1}^\infty A_{1/n}$. Then $A \subset A'' \in \sigma(\mathcal{A}) \subset \mathcal{A}_\mu$, moreover, $\mu^*(A) = \mu^*(A'')$, since

$$\mu^*(A) \geqslant \mu^*(A_{1/n}) - 1/n \geqslant \mu^*(A'') - 1/n \quad \text{for all } n.$$

Note that for constructing A'' the measurability of A is not needed. Let us apply this to the complement of A and find a set $B \in \sigma(\mathcal{A}) \subset \mathcal{A}_\mu$ containing $X \backslash A$ and having the same outer measure as $X \backslash A$. Let us set $A' = X \backslash B$. Then obviously $A' \subset A$, moreover, by the additivity of μ^* on the σ-algebra $\mathcal{A}_\mu$ and the inclusions $A, B \in \mathcal{A}_\mu$ we have

$$\mu^*(A') = \mu(X) - \mu^*(B) = \mu(X) - \mu^*(X \backslash A) = \mu^*(A),$$

as required.

Conversely, suppose that such sets A' and A'' exist. Since A is the union of A' and a part of $A'' \backslash A'$, it suffices to verify that every set C in $A'' \backslash A'$ belongs to $\mathcal{A}_\mu$. This is true, since $\mu^*(C) \leqslant \mu^*(A'' \backslash A') = \mu^*(A'') - \mu^*(A') = 0$ by the additivity of μ^* on $\mathcal{A}_\mu$ and the inclusion $A'', A' \in \sigma(\mathcal{A}) \subset \mathcal{A}_\mu$. $\qquad\square$

The uniqueness of extensions yields the following useful result.

2.4.10. Corollary. *For the equality of two nonnegative Borel measures μ and ν on the real line it is necessary and sufficient that they have equal values on all compact intervals (or on all open intervals).*

PROOF. Since a compact interval is the intersection of a decreasing sequence of open intervals and an open interval is the union of an increasing sequence of compact intervals, by the countable additivity the coincidence of values of μ and ν on compact intervals is equivalent to the coincidence of values on open intervals and implies the equality of both measures on the algebra of finite unions of intervals (possibly, unbounded) in $\mathbb{R}^1$. Since this algebra generates $\mathcal{B}(\mathbb{R}^1)$, by the uniqueness of a countably additive extension from an algebra to the generated σ-algebra we obtain the desired assertion. $\qquad\square$

The countably additive extension described in Theorem 2.4.6 is called the *Lebesgue extension* or the *Lebesgue completion* of the measure μ, and the measure space $(X, \mathcal{A}_\mu, \mu)$ is called the Lebesgue completion of $(X, \mathcal{A}, \mu)$. In addition, $\mathcal{A}_\mu$ is called the Lebesgue completion of the algebra (or the σ-algebra) $\mathcal{A}$ with respect to μ. This terminology (usually employed in the case where $\mathcal{A}$ is a σ-algebra) is connected with the fact that the measure μ on $\mathcal{A}_\mu$ is complete in the sense of the following definition.

2.4.11. Definition. *A nonnegative countably additive measure μ on a σ-algebra $\mathcal{A}$ is called complete if $\mathcal{A}$ contains all subsets of every set in $\mathcal{A}$ of μ-measure zero. In this case it is also customary to say that the σ-algebra $\mathcal{A}$ is complete with respect to the measure μ. In the same manner one defines complete measured on algebras or rings.*

It is clear from the definition of the outer measure that if $A \subset B \in \mathcal{A}_\mu$ and $\mu(B) = 0$, then $A \in \mathcal{A}_\mu$ and $\mu(A) = 0$. The completeness of a bounded measure μ on a σ-algebra $\mathcal{A}$ is equivalent to the equality $\mathcal{A} = \mathcal{A}_\mu$ (Exercise 2.7.31). An example of an incomplete countably additive measure on a σ-algebra is the zero measure on the σ-algebra consisting of the empty set and the interval $[0,1]$. A less trivial example: Lebesgue measure on the σ-algebra of all Borel subsets of the interval constructed by means of Example 2.3.10. This measure is considered below in greater detail; we shall see that there exists a compact set of zero Lebesgue measure containing non-Borel subsets. Sets of measure zero are called *negligible*.

The following simple, but useful criterion of measurability of sets in terms of the outer measure was, as already noted, the original Lebesgue definition.

2.4.12. Proposition. *Let μ be a finite nonnegative countably additive measure on an algebra $\mathcal{A}$. A set A belongs to $\mathcal{A}_\mu$ if and only if $\mu^*(A) + \mu^*(X \backslash A) = \mu(X)$. This is also equivalent to the property that $\mu^*(E \cap A) + \mu^*(E \backslash A) = \mu^*(E)$ for every set $E \subset X$.*

PROOF. Let us verify the sufficiency of the first indicated condition (hence the second stronger condition will be sufficient as well). Let us find μ-measurable sets B and C such that $A \subset B$, $X \backslash A \subset C$, $\mu(B) = \mu^*(A)$, $\mu(C) = \mu^*(X \backslash A)$. The existence of such sets is established in the proof of Corollary 2.4.9. It is clear that $D = X \backslash C \subset A$ and

$$\mu(B) - \mu(D) = \mu(B) + \mu(C) - \mu(X) = 0.$$

Therefore, $\mu^*(A \triangle B) = 0$, whence the measurability of A follows.

We now establish the necessity of the second condition mentioned in the theorem. By the subadditivity of the outer measure it suffices to show that we have $\mu^*(E \cap A) + \mu^*(E \backslash A) \leqslant \mu^*(E)$ for all $E \subset X$ and $A \in \mathcal{A}_\mu$. By approximations and estimate (2.4.2) it suffices to prove this inequality for all $A \in \mathcal{A}$. Let $\varepsilon > 0$ and $A_n \in \mathcal{A}$ be sets such that $E \subset \bigcup_{n=1}^\infty A_n$ and $\mu^*(E) \geqslant \sum_{n=1}^\infty \mu(A_n) - \varepsilon$. Then $E \cap A \subset \bigcup_{n=1}^\infty (A_n \cap A)$ and $E \backslash A \subset \bigcup_{n=1}^\infty (A_n \backslash A)$, whence we have

$$\mu^*(E \cap A) + \mu^*(E \backslash A) \leqslant \sum_{n=1}^\infty \mu(A_n \cap A) + \sum_{n=1}^\infty \mu(A_n \backslash A)$$

$$= \sum_{n=1}^\infty \mu(A_n) \leqslant \mu^*(E) + \varepsilon.$$

Since ε was arbitrary, the assertion is proved. $\qquad\square$

The given criterion of measurability of a set A can be formulated as the equality $\mu^*(A) = \mu_*(A)$ provided the *inner measure* is defined by the equality

$$\mu_*(A) := \mu(X) - \mu^*(X \backslash A),$$

as was actually done by Lebesgue. However, one should bear in mind that here one cannot employ the definition of the inner measure in the spirit of the Jordan measure as the supremum of measures of sets from the algebra $\mathcal{A}$ contained in A. Below in §2.7(i) we shall consider abstract outer measures and see that the last property in Proposition 2.4.12 can be taken as the definition of measurability, which leads to very interesting results (the proposition itself will be extended to finitely additive set functions).

2.4.13. Remark. Every set $A \in \mathcal{A}_\mu$ can be turned into a measure space by restricting μ to the class of μ-measurable subsets of A that is a σ-algebra in A. The obtained measure μ_A (or $\mu|_A$) is called the *restriction* of μ to A. See Exercise 2.7.46 concerning restrictions to arbitrary sets.

One should bear in mind that not every countably additive measure on a sub-σ-algebra $\mathcal{A}_0$ of a σ-algebra $\mathcal{A}$ extends to a measure on $\mathcal{A}$ (see Exercise 2.7.43).

2.4.14. Remark. (i) An analog of Theorem 2.4.6 holds for measures with values in $[0, +\infty]$ and also for measures on rings, however, in this case in place of the class $\mathcal{A}_\mu$ one has to take either the class $\mathfrak{M}_{\mu^*}$ of sets measurable in the sense of Caratheodory (see §2.7(i)) or the class of locally measurable sets

$$\mathcal{A}_\mu^{\mathrm{loc}} := \{A \subset X \colon A \cap E \in \mathcal{A}_\mu \text{ for all } E \in \mathcal{A} \text{ with } \mu(E) < \infty\}$$

that coincides with $\mathfrak{M}_{\mu^*}$ in the case of a countably additive measure (see Remark 2.7.11 below). The class $\mathcal{A}_\mu$ from Definition 2.4.1 in the case of a measure with values in $[0, +\infty]$ can fail to be a σ-algebra. Let us consider the following example: let $\mathcal{A}$ be the algebra of finite subsets of $\mathbb{N}$ and their complements and let $\mu(A) = \infty$ if $A \in \mathcal{A}$ is not empty, $\mu(\varnothing) = 0$. Then the measure μ with values in $[0, +\infty]$ is countably additive. Moreover, the inequality $\mu^*(E) < \infty$ yields that $E = \varnothing$. Hence $\mathcal{A}_\mu$ coincides with $\mathcal{A}$ and is not a σ-algebra.

(ii) Suppose that a finite measure $\mu \geqslant 0$ is countably additive on a ring $\mathcal{R}$ and is σ-finite, i.e., $X = \bigcup_{k=1}^\infty R_k$, where $R_k \in \mathcal{R}$ and $\mu(R_k) < \infty$. Then Theorem 2.4.6 enables us to extend μ to a countably additive measure with values in $[0, +\infty]$ on the σ-algebra $\sigma(\mathcal{R})$ generated by $\mathcal{R}$. Indeed, passing to $\bigcup_{j=1}^k R_j$, we can assume that $R_k \subset R_{k+1}$. For every k, the measure μ has a unique countably additive extension to the σ-algebra $\mathcal{A}_k$ of subsets of R_k generated by the intersections $R \cap R_k$, where $R \in \mathcal{R}$, since such intersections form an algebra and μ is finite and countably additive on it. It is clear that such extensions are consistent, i.e., the extension to $\mathcal{A}_{k+1}$ restricted to $\mathcal{A}_k$ is the same as the extension to $\mathcal{A}_k$. It is readily seen that $\sigma(\mathcal{R})$ is the class of all sets E such that $E \cap R_k \in \mathcal{A}_k$ for every k. Now the extension to $\sigma(\mathcal{R})$ is given by the formula

$$\mu(E) = \lim_{k \to \infty} \mu(E \cap R_k),$$

where $\mu(E \cap R_k)$ is the value on $E \cap R_k$ of the extension of μ to $\mathcal{A}_k$. The obtained extension is countably additive. Indeed, for disjoint sets $A_j \in \sigma(\mathcal{R})$, the sets $A_j \cap R_k$ are disjoint for every k, whence $\mu\big(R_k \cap \bigcup_{j=1}^\infty A_j\big) = \sum_{j=1}^\infty \mu(A_j \cap R_k)$. It is readily seen that the equality remains valid in the limit as $k \to \infty$ (including the case of infinite values). By the same formula μ extends to the broader class $\mathcal{R}_\mu^{\mathrm{loc}}$.

2.5. Lebesgue Measure and Lebesgue–Stieltjes Measures

We now return to Lebesgue measure considered in Examples 2.3.10 and 2.4.8. Let I be a cube in $\mathbb{R}^n$ of the form $[a, b]^n$ and let $\mathcal{A}_0$ be the algebra of finite unions of parallelepipeds in I with edges parallel to the coordinate axes. We know that the usual volume λ_n is countably additive on $\mathcal{A}_0$ and extends to a countably additive measure, also denoted by λ_n and called *Lebesgue measure* on the σ-algebra $\mathcal{L}_n(I)$

of all λ_n-measurable sets in I, containing the Borel σ-algebra. Let us write $\mathbb{R}^n$ as the union of increasing cubes $I_k = \{|x_i| \leqslant k, i = 1, \ldots, n\}$ and set

$$\mathcal{L}_n := \{E \subset \mathbb{R}^n \colon E \cap I_k \in \mathcal{L}_n(I_k) \text{ for all } k \in \mathbb{N}\}.$$

It is clear that $\mathcal{L}_n$ is a σ-algebra. The formula

$$\lambda_n(E) = \lim_{k \to \infty} \lambda_n(E \cap I_k)$$

defines on it the σ-finite measure λ_n.

2.5.1. Definition. *The measure λ_n introduced above on $\mathcal{L}_n$ is called Lebesgue measure on $\mathbb{R}^n$. Sets in $\mathcal{L}_n$ are called Lebesgue measurable.*

In those cases where a subset of $\mathbb{R}^n$ is considered with Lebesgue measure, it is customary to use the terms "a set of measure zero", "a measurable set", etc., without explicit mentioning Lebesgue measure. We shall also follow this tradition.

For defining Lebesgue measure of a set $E \in \mathcal{L}_n$ one can also use the formula

$$\lambda_n(E) = \sum_{j=1}^{\infty} \lambda_n(E \cap Q_j),$$

where Q_j are pairwise disjoint cubes that are shifts of $[0, 1)^n$ and give all of $\mathbb{R}^n$ in their union. Since the σ-algebra generated by the parallelepipeds of the form indicated above is the Borel σ-algebra $\mathcal{B}(I)$ of the cube I, all Borel sets in the cube I, hence also in all of $\mathbb{R}^n$, are Lebesgue measurable.

Lebesgue measure can be also considered on the δ-ring $\mathcal{L}_n^0$ of all sets of finite Lebesgue measure.

In the one-dimensional case Lebesgue measure of a set E is the sum of the series of $\lambda_1\big(E \cap (n, n+1]\big)$ over all integer numbers n.

The shift of a set A by a vector h, i.e., the set of all points of the form $a + h$, where $a \in A$, will be denoted by $A + h$.

2.5.2. Lemma. *Let $W \subset I = (-1, 1)^n$ be an open set. There exist at most countably many pairwise disjoint open cubes $Q_j \subset W$ of the form $Q_j = c_j I + h_j$, $c_j > 0$, $h_j \in I$ such that the set $W \backslash \bigcup_{j=1}^{\infty} Q_j$ has Lebesgue measure zero.*

PROOF. We apply Exercise 2.7.16 and represent W as $W = \bigcup_{j=1}^{\infty} W_j$, where W_j are open cubes whose edges are parallel to the coordinate axes and have length $q2^{-p}$ with natural numbers p, q and centers with coordinates of the form $l2^{-m}$ with integer l and natural m. Next we restructure the cubes W_j in the following way. We remove all cubes W_j contained in W_1 and set $Q_1 = W_1$. Let us take the first cube W_{n_2} in the remaining collection and represent the interior of the set $W_{n_2} \backslash Q_1$ as a finite union of open cubes $Q_2, \ldots, Q_{m_2}$ of the same form as the cubes W_j and also some pieces of boundaries of these new cubes. This is possible due to our choice of the original cubes. Next we remove all cubes W_j contained in the union $\bigcup_{i=1}^{m_2} Q_i$, take the first cube in the remaining family and refine as indicated above its part going out of the union of the earlier constructed cubes. Continuing the described process, we obtain pairwise disjoint cubes covering W up to a set of measure zero, namely, the countable union of boundaries of these cubes. $\square$

2.5.3. Theorem. *Let A be a Lebesgue measurable set of finite measure. Then*
(i) $\lambda_n(A + h) = \lambda_n(A)$ *for every vector* $h \in \mathbb{R}^n$;
(ii) $\lambda_n(\alpha A) = |\alpha|^n \lambda_n(A)$ *for every real number* α.
(iii) $\lambda_n(U(A)) = \lambda_n(A)$ *for every orthogonal linear operator* U *on* $\mathbb{R}^n$;

PROOF. It is obvious from the definition of Lebesgue measure that it suffices to prove the stated properties for bounded measurable sets.

(i) Let us take a cube I centered at the origin such that the sets A and $A + h$ are contained in some cube inside I. Let $\mathcal{A}_0$ be the algebra generated by the cubes in I with edges parallel to the coordinate axes. In the definition of the outer measure of A it is enough to consider only sets $B \in \mathcal{A}_0$ such that $B + h \subset I$. Since the volumes of sets from $\mathcal{A}_0$ do not change under shifts, the sets $A + h$ and A have equal outer measures. For every $\varepsilon > 0$ there exists a set $A_\varepsilon \in \mathcal{A}_0$ with $\lambda_n^*(A \bigtriangleup A_\varepsilon) < \varepsilon$. Then

$$\lambda_n^*\big((A + h) \bigtriangleup (A_\varepsilon + h)\big) = \lambda_n^*\big((A \bigtriangleup A_\varepsilon) + h\big) = \lambda_n^*(A \bigtriangleup A_\varepsilon) < \varepsilon,$$

which yields the measurability of $A + h$ and the desired equality. Assertion (ii) is obvious for sets from $\mathcal{A}_0$, hence, as in (i), it extends to all measurable sets.

(iii) As above, it suffices to consider sets from $\mathcal{A}_0$. By (i) any face of a cube has measure zero, since we can take countably many disjoint shifts of this face contained in a cube (which is of finite measure). It remains to show that for every closed cube K with edges parallel to the coordinate axes

$$\lambda_n(U(K)) = \lambda_n(K). \tag{2.5.1}$$

Suppose that this is false for some closed cube K, i.e.,

$$\lambda_n(U(K)) = r\lambda_n(K),$$

where $r \neq 1$. Then for every set Q that is a finite union of cubes (closed or open) with edges parallel to the coordinate axes and disjoint interiors we have

$$\lambda_n(U(Q)) = r\lambda_n(Q). \tag{2.5.2}$$

Indeed, by (ii) this is true for every closed or open cube with edges parallel to the coordinate axes, hence remains true for their finite disjoint unions. Moreover, since all faces are of measure zero, it suffices that only the interiors be disjoint. Suppose that $r < 1$; the case $r > 1$ is similar. Let B be the closed unit ball centered at the origin. Since $U(B) = B$, we have $\lambda_n(U(B)) = \lambda_n(B)$. This contradicts (2.5.2). Indeed, let $0 < \varepsilon < r^{-1} - 1$. We can find a larger ball B_1 centered at the origin such that $\lambda_n(B_1) < (1 + \varepsilon)\lambda_n(B)$. Using the compactness of B, we can find a set Q of the form indicated above such that $B \subset Q \subset B_1$ (first we cover B by finitely many cubes contained in B_1 and then construct a suitable set Q). So $B = U(B) \subset U(Q) \subset U(B_1) = B_1$ and

$$\lambda_n(B) = \lambda_n(U(B)) \leqslant \lambda_n(U(Q)) = r\lambda_n(Q) \leqslant r\lambda_n(B_1) < r(1 + \varepsilon)\lambda_n(B).$$

This yields the bound $r(1 + \varepsilon) > 1$, a contradiction. $\qquad\square$

Note that property (ii) of Lebesgue measure is a corollary of property (i), since by (i) it follows for cubes and $\alpha = 1/m$, where $m \in \mathbb{N}$, and next extends to rational α, which by continuity yields the general case. As is seen from the proof, property (iii) also follows from (i). Moreover, property (i) characterizes Lebesgue measure up to a constant factor (see Exercise 2.7.28). It follows from Theorem 2.5.3 that Lebesgue measure of every rectangular parallelepiped $P \subset I$ (not necessarily with edges parallel to the coordinate axes) equals the product of the edge lengths. This easily yields the following result.

2.5.4. Corollary. *In the previous theorem, for every linear operator T on $\mathbb{R}^n$ we have $\lambda_n(T(A)) = |\det T| \lambda_n(A)$.*

Every countable set has zero Lebesgue measure. As the following example of the Cantor set (described in §1.9(vi)) shows, there exist uncountable sets of Lebesgue measure zero. For the reader's convenience we recall the construction.

2.5.5. Example. Let $I = [0,1]$. Denote by $J_{1,1}$ the interval $(1/3, 2/3)$. Let $J_{2,1}$, $J_{2,2}$ denote the intervals $(1/9, 2/9)$ and $(7/9, 8/9)$ that are the middle thirds of the compact intervals obtained by deleting the interval $J_{1,1}$. Let us continue inductively the process of deleting the middle intervals. At the nth step we obtain 2^n intervals and at the next step we delete their middle thirds $J_{n+1,1}$, $J_{n+1,2}, \ldots, J_{n+1,2^n}$, after which 2^{n+1} intervals remain and the process continues. The set $C = I \backslash \bigcup_{n,j} J_{n,j}$ is called the Cantor set. It is compact, has the cardinality of the continuum, but its Lebesgue measure is zero.

PROOF. The set C is compact, since its complement in the whole interval is open. In order to see that C has the cardinality of the continuum, we write points of the interval $[0,1]$ in the triadic system, i.e., $x = \sum_{j=1}^{\infty} x_j 3^{-j}$, where x_j takes values $0, 1, 2$. As in the case of the decimal expansion, such a representation is not unique, for example, the sequence $(0, 2, 2, \ldots)$ corresponds to the same number as the sequence $(1, 0, 0, \ldots)$. However, such nonuniqueness is only possible for countably many points. It is verified by induction that after the nth step of deletion there remain points x for which $x_j = 0$ or $x_j = 2$ if $j \leqslant n$ (if the points like $1/3$ are written as $(0, 2, 2, \ldots)$). Thus, C consists of points admitting the triadic expansion with 0 and 2 only, whence it follows that C has the cardinality of the set of real numbers. We observe that the expansion using only 0 and 2 is unique (even for points possessing different expansions with the use of 1). In order to show that C has zero measure, it remains to verify that the complement to C has measure 1. By induction it is easily verified that the measure of the set $J_{n,1} \cup \cdots \cup J_{n,2^{n-1}}$ equals $2^{n-1} 3^{-n}$. Moreover, $\sum_{n=1}^{\infty} 2^{n-1} 3^{-n} = 1$. $\qquad \square$

2.5.6. Example. Let $\{r_n\}$ be the set of all rational points in $[0,1]$ and $\varepsilon > 0$. Set

$$K = [0,1] \backslash \bigcup_{n=1}^{\infty} (r_n - \varepsilon 4^{-n}, r_n + \varepsilon 4^{-n}).$$

Then K is a compact set without inner points and its Lebesgue measure is at least $1 - \varepsilon$, since the measure of $[0,1] \backslash K$ does not exceed $\sum_{n=1}^{\infty} 2\varepsilon 4^{-n}$.

Thus, the positivity of the measure of a compact set does not mean that it has inner points.

Note that every subset of the Cantor set has measure zero as well. It follows that the collection of all measurable sets has the cardinality equal to the cardinality of the class of all subsets of the real line. One can show (see [73, Chapter 6]) that the Borel σ-algebra has the cardinality of the continuum. Hence among subsets of the Cantor set there are non-Borel measurable sets.

Naturally the question arises of how large is the class of Lebesgue measurable sets and whether it contains all sets. It turns out that an answer to this question depends on employing additional set-theoretic axioms and cannot be given in the framework of the so-called "naive set theory" without the axiom of choice. In any case, as the following *Vitali example* shows, with the aid of the axiom of choice one easily constructs a Lebesgue non-measurable set.

2.5.7. Example. Let us call points x and y in $[0, 1]$ equivalent if the number $x - y$ is rational. Clearly, we have obtained an equivalence relation, i.e., 1) $x \sim x$, 2) $y \sim x$ if $x \sim y$, 3) $x \sim z$ if $x \sim y$ and $y \sim z$. Thus, equivalence classes arise that are unions of points which differ one from another by rational numbers, and the difference between representatives of different classes is always irrational. Let us now pick in every class precisely one representative and denote the obtained set by E. The possibility of this construction is due exactly to the axiom of choice. The set E cannot be Lebesgue measurable. Indeed, if its measure equals zero, then the measure of $[0, 1]$ also equals zero, since $[0, 1]$ is covered by countably many shifts of E by all possible rational numbers. On the other hand, the measure of E cannot be positive, since for different rational numbers p and q the sets $E + p$ and $E + q$ are disjoint and have equal measure $c > 0$. Since $E + p \subset [0, 2]$ if $p \in [0, 1]$, the interval $[0, 2]$ would have infinite measure.

However, one should bear in mind that in place of the axiom of choice one can add to the standard set-theoretic axioms an assertion which will make all subsets of the real line Lebesgue measurable.

Let us also remark that even if we use the axiom of choice, there still remains the question: does there exist *some* extension of Lebesgue measure to a countably additive measure on the class of all subsets of the interval? The example constructed above only says that such an extension cannot be obtained by means of the Lebesgue completion. An answer to the posed question also depends on using additional set-theoretic axioms (see §1.12(x) in [73]). In any case, the Lebesgue extension is not maximally possible, since by Theorem 2.7.12, for every set $E \subset [0, 1]$ that is Lebesgue nonmeasurable, one can extend Lebesgue measure to a countably additive measure on the σ-algebra generated by all Lebesgue measurable sets in $[0, 1]$ and the set E.

An important property of the Lebesgue measurability in $\mathbb{R}^n$ is its invariance with respect to a broad class of transformations. We recall that a differentiable mapping with a locally bounded derivative is Lipschitz on bounded sets.

2.5.8. Theorem. *Suppose that a mapping $f \colon \mathbb{R}^n \to \mathbb{R}^n$ is Lipschitz on bounded sets. Then f takes Lebesgue measurable sets to measurable sets.*

If f has an inverse that is Lipschitz on bounded sets, then the preimages of Lebesgue measurable sets are measurable.

PROOF. Let $A \subset \mathbb{R}^n$ be Lebesgue measurable. It suffices to consider a bounded set A. Let C be the Lipschitz constant of the mapping f. For every $\varepsilon > 0$ there exists a compact set $K_\varepsilon \subset A$ with $\lambda(A \backslash K_\varepsilon) < \varepsilon$. Then $f(K_\varepsilon)$ is compact in $f(A)$. We show that $\lambda^*\big(f(A) \backslash f(K_\varepsilon)\big) \leqslant 2C^n \varepsilon$; since ε is arbitrary, this will mean the measurability of $f(A)$. There is a sequence of closed balls $B_j := B(x_j, r_j)$ for which $A \backslash K_\varepsilon \subset \bigcup_{j=1}^{\infty} B_j$ and $\sum_{j=1}^{\infty} \lambda(B_j) \leqslant 2\varepsilon$. Then $f(A) \backslash f(K_\varepsilon) \subset \bigcup_{j=1}^{\infty} f(B_j)$, hence $\lambda^*\big(f(A) \backslash f(K_\varepsilon)\big) \leqslant \sum_{j=1}^{\infty} \lambda\big(f(B_j)\big)$. It remains to observe that $f(B_j)$ is contained in the ball $B\big(f(x_j), Cr_j\big)$ of measure $C^n \lambda(B_j)$. The second assertion of the theorem follows from the first one. $\square$

Note that this theorem is true not for every Borel measure. For example, if the measure μ on $[0, 1]$ is given by the formula $\mu(A) = \lambda(A \cap [0, 1/2])$, $A \in \mathcal{B}([0,1])$, then all sets in $[1/2, 1]$ have μ-measure zero. It is clear that under the mapping $x \mapsto x/2$ some of them will be taken to nonmeasurable subsets of $[0, 1/2]$.

Let us say a few words about the Jordan measure (or the Peano–Jordan, which better reflects the history of the question).

2.5.9. Definition. *A bounded set E in $\mathbb{R}^n$ is called Jordan measurable if, for every $\varepsilon > 0$, there exist sets U_ε and V_ε that are finite unions of cubes such that $U_\varepsilon \subset E \subset V_\varepsilon$ and $\lambda_n(V_\varepsilon \backslash U_\varepsilon) < \varepsilon$.*

It is clear that as $\varepsilon \to 0$, there exists a common limit of the measures of U_ε and V_ε, called Jordan measure of the set E. It is seen from the definition that every Jordan measurable set E is Lebesgue measurable and its Lebesgue measure coincides with Jordan measure. However, the converse is false: for example, the set of rational numbers in the interval is not Jordan measurable. The class of Jordan measurable sets is a ring (see Exercise 2.7.30), on which Jordan measure coincides with Lebesgue measure. Of course, Jordan measure is countably additive on its domain of definition and its Lebesgue extension is Lebesgue measure.

We now proceed to consideration of Lebesgue–Stieltjes measures. Let μ be a nonnegative Borel measure on $\mathbb{R}^1$. Then the function

$$t \mapsto F(t) = \mu\big((-\infty, t)\big)$$

is bounded, increasing, left continuous, i.e., $F(t_n) \to F(t)$ as $t_n \uparrow t$, which follows from the countable additivity of μ, in addition, one has $\lim_{t \to -\infty} F(t) = 0$. These conditions turn out to be also sufficient in order that the function F be generated by some measure by the indicated formula. The function F is called the *distribution function* of the measure μ. Note that the distribution function is often defined by the formula $F(t) = \mu\big((-\infty, t]\big)$, which leads to distinctions at the points of positive μ-measure (the jumps of the function F are the points of positive μ-measure).

2.5.10. Theorem. *Let F be a bounded, nondecreasing and left continuous function such that $\lim_{t \to -\infty} F(t) = 0$. Then there exists a unique nonnegative Borel measure on $\mathbb{R}^1$ such that $F(t) = \mu\big((-\infty, t)\big)$ for all $t \in \mathbb{R}^1$.*

PROOF. It is known from calculus that the function F has at most a countable set D of discontinuity points. It is clear that $\mathbb{R}^1 \backslash D$ contains a countable set S that is everywhere dense in $\mathbb{R}^1$. Let us consider the class $\mathcal{A}$ of all sets of the form $A = \bigcup_{i=1}^n J_i$, where J_i is an interval of one of the four types (a, b), $[a, b]$, $(a, b]$ or $[a, b)$, where a and b either belong to S or coincide with $-\infty$ or $+\infty$. It is readily verified that $\mathcal{A}$ is an algebra. Let us define a set function μ on $\mathcal{A}$ as follows: if A is an interval with endpoints a and b and $a \leqslant b$, then $\mu(A) = F(b) - F(a)$, where $F(-\infty) = 0$, $F(+\infty) = \lim_{t \to +\infty} F(t)$, and if A is a finite union of disjoint intervals J_i, then $\mu(A) = \sum_i \mu(J_i)$. Clearly, the function μ is well-defined and additive. For the proof of the countable additivity of μ on $\mathcal{A}$ it suffices to observe that the class of finite unions of compact intervals is compact and approximating. Indeed, if J an open or semiopen interval, for example, $J = (a, b)$, where a and b belong to S (or coincide with the points $+\infty$, $-\infty$), then by the continuity of F at points from S we have $F(b) - F(a) = \lim_{i \to \infty} [F(b_i) - F(a_i)]$, where $a_i \downarrow a$, $b_i \uparrow b$, a_i, $b_i \in S$. If $a = -\infty$, then the same follows from the condition $\lim_{t \to -\infty} F(t) = 0$.

Let us extend μ to a measure on the σ-algebra $\mathcal{B}(\mathbb{R}^1)$ that is generated by the algebra $\mathcal{A}$ due to the density of S. We observe that $F(t) = \mu\big((-\infty, t)\big)$ for all t, not only for $t \in S$, since both functions are left continuous and coincide on S. The uniqueness of μ is clear from the fact that F uniquely defines the values of μ on all intervals. In this proof, by Proposition 2.3.7, in place of $\mathcal{A}$ we could take the semialgebra of intervals of the form $(-\infty, b)$, $[a, b)$, $[a, +\infty)$, where $a, b \in S$. $\square$

The measure μ constructed above by the function F is called the *Lebesgue–Stieltjes measure* with the distribution function F. Sometimes this measure is denoted by dF. Similarly, with the aid of distribution functions of n variables representing the measures of the sets $(-\infty, x_1) \times \cdots \times (-\infty, x_n)$ one defines Lebesgue–Stieltjes measures on $\mathbb{R}^n$.

2.6. Signed Measures

In this section we consider signed measures. The next theorem enables us in many cases to reduce signed measures to nonnegative measures.

2.6.1. Theorem. (THE HAHN DECOMPOSITION) *Let μ be a countably additive real measure on a measurable space $(X, \mathcal{A})$. Then there exists a set $X^- \in \mathcal{A}$ such that, letting $X^+ = X \backslash X^-$, for all sets $A \in \mathcal{A}$, we have*

$$\mu(A \cap X^-) \leqslant 0 \quad and \quad \mu(A \cap X^+) \geqslant 0.$$

PROOF. A set $E \in \mathcal{A}$ will be called negative if $\mu(A \cap E) \leqslant 0$ for all $A \in \mathcal{A}$. Similarly we define positive sets. Let $\alpha = \inf \mu(E)$, where the infimum is taken over all negative sets. Let $\{E_n\}$ be a sequence of negative sets for which $\lim_{n \to \infty} \mu(E_n) = \alpha$. It is clear that $X^- = \bigcup_{n=1}^\infty E_n$ is a negative set and $\mu(X^-) = \alpha$, since $\alpha \leqslant \mu(X^-) \leqslant \mu(E_n)$. We show that $X^+ = X \backslash X^-$ is a positive set. Suppose the contrary. Then there exists a set $A_0 \in \mathcal{A}$ such that $A_0 \subset X^+$ and $\mu(A_0) < 0$. The set A_0 cannot be negative, since in that case we would take the negative set $X^- \cup A_0$ for which $\mu(X^- \cup A_0) < \alpha$, which is impossible. Hence there exists a set $A_1 \subset A_0$ and a natural number k_1 such that $A_1 \in \mathcal{A}$, $\mu(A_1) \geqslant 1/k_1$, and k_1 is the smallest possible natural number k for which A_0 contains a subset of measure at least $1/k$. We observe that $\mu(A_0 \backslash A_1) < 0$. Repeating the reasoning given for A_0 and applying it to $A_0 \backslash A_1$, we obtain a set $A_2 \subset A_0 \backslash A_1$ from $\mathcal{A}$ for which $\mu(A_2) \geqslant 1/k_2$ with the smallest possible natural number k_2. We continue this process inductively. This will give pairwise disjoint sets $A_i \in \mathcal{A}$ with the following property: $A_{n+1} \subset A_0 \backslash \bigcup_{i=1}^n A_i$ and $\mu(A_n) \geqslant 1/k_n$, where k_n is the smallest natural number k for which $A_0 \backslash \bigcup_{i=1}^{n-1} A_i$ contains a subset of measure at least $1/k$. We observe that $k_n \to +\infty$, since otherwise by the disjointness of the sets A_n we would obtain $\mu(A_0) = +\infty$. Let $B = A_0 \backslash \bigcup_{i=1}^\infty A_i$. Then $\mu(B) < 0$, since $\mu(A_0) < 0$, $\mu(\bigcup_{i=1}^\infty A_i) > 0$ and $\bigcup_{i=1}^\infty A_i \subset A_0$. Moreover, B is a negative set. Indeed, if $C \subset B$, $C \in \mathcal{A}$ and $\mu(C) > 0$, then there exists a natural number k with $\mu(C) > 1/k$, which in case $k_n > k$ contradicts our choice of k_n, because $C \subset A_0 \backslash \bigcup_{i=1}^n A_i$. Thus, joining B to X^-, we arrive at a contradiction with the definition of α. Therefore, the set X^+ is positive. $\qquad\square$

The decomposition of the space X into the disjoint union $X = X^+ \cup X^-$ constructed in the previous theorem is called the *Hahn decomposition*. It is clear that the Hahn decomposition is not unique, because adding a set all subsets of which have measure zero does not influence the property of a set to be negative. However, if $X = \widetilde{X}^+ \cup \widetilde{X}^-$ is another Hahn decomposition, then for all $A \in \mathcal{A}$ we have

$$\mu(A \cap X^-) = \mu(A \cap \widetilde{X}^-) \quad \text{and} \quad \mu(A \cap X^+) = \mu(A \cap \widetilde{X}^+). \qquad (2.6.1)$$

Indeed, every set B in $\mathcal{A}$ contained in $X^- \cap \widetilde{X}^+$ or in $X^+ \cap \widetilde{X}^-$, has measure zero, since $\mu(B)$ is simultaneously nonnegative and nonpositive.

2.6.2. Corollary. *Under the hypotheses of Theorem 2.6.1 let*

$$\mu^+(A) = \mu(A \cap X^+), \quad \mu^-(A) = -\mu(A \cap X^-), \quad A \in \mathcal{A}. \qquad (2.6.2)$$

Then μ^+ and μ^- are nonnegative countably additive measures and $\mu = \mu^+ - \mu^-$.

It is clear that $\mu(X^+)$ is the maximal value assumed by the measure μ and $\mu(X^-)$ is its minimal value.

2.6.3. Corollary. *If $\mu \colon \mathcal{A} \to \mathbb{R}^1$ is a countably additive measure on a σ-algebra $\mathcal{A}$, then the set of values of μ is bounded.*

2.6.4. Definition. *The measures μ^+ and μ^- constructed above are called the positive and negative parts of μ, respectively. The measure*

$$|\mu| := \mu^+ + \mu^-$$

is called the total variation of μ. The quantity $\|\mu\| := |\mu|(X)$ is called the variation or the variation norm of the measure μ.

The decomposition $\mu = \mu^+ - \mu^-$ is called the *Jordan decomposition* or the *Jordan–Hahn decomposition*.

The measure $|\mu|$ should not be confused with the set function $A \mapsto |\mu(A)|$, which is usually *not additive* (for example, if $\|\mu\| > \mu(X) = 0$).

Note that the measures μ^+ and μ^- possess the following property which can be taken for their definition:

$$\mu^+(A) = \sup\{\mu(B): \ B \subset A, B \in \mathcal{A}\},$$

$$\mu^-(A) = \sup\{-\mu(B): \ B \subset A, B \in \mathcal{A}\}$$

for all $A \in \mathcal{A}$. In addition,

$$|\mu|(A) = \sup\Big\{\sum_{n=1}^{\infty} |\mu(A_n)|\Big\}, \qquad (2.6.3)$$

where the supremum is taken over all at most countable partitions of A into pairwise disjoint parts from $\mathcal{A}$. One can take finite partitions and replace sup by max, since the supremum is attained at the partition $A_1 = A \cap X^+$, $A_2 = A \cap X^-$. Note that $\|\mu\|$ does not coincide with the quantity $\sup\{|\mu(A)|, A \in \mathcal{A}\}$ if both measures μ^+ and μ^- are nonzero. However, we have

$$\|\mu\| \leqslant 2\sup\{|\mu(A)|: \ A \in \mathcal{A}\} \leqslant 2\|\mu\|. \qquad (2.6.4)$$

All these relations are obvious from the Hahn decomposition.

2.6.5. Remark. It is seen from the proof of Theorem 2.6.1 that it remains in force in the case where μ is a countably additive set function on $\mathcal{A}$ with values in $(-\infty, +\infty]$. In this case the measure μ^- is bounded and the measure μ^+ takes values in $[0, +\infty]$. Thus, in the case under consideration the boundedness of μ is equivalent to that $\mu(X)$ is a finite number.

One can also introduce complex measures. Let $\mathcal{A}$ be a σ-algebra in a space X. A complex measure $\mu\colon \mathcal{A} \to \mathbb{C}$ is a function of the form $\mu = \mu_1 + i\mu_2$, where μ_1 and μ_2 are real measures on $\mathcal{A}$. Its variation $\|\mu\|$ is given by the formula

$$\|\mu\| := \sup\Big\{\sum_{k=1}^{n} |\mu(E_k)|\Big\},$$

where sup is taken over all finite partitions of X into disjoint sets $E_k \in \mathcal{A}$. Since $|\mu(E_k)| \leqslant |\mu_1(E_k)| + |\mu_2(E_k)|$, we have $\|\mu\| \leqslant \|\mu_1\| + \|\mu_2\|$.

2.7. Complements and Exercises

(i) The Caratheodory measurability and extensions of measures (74). Exercises (79).

2.7(i). The Caratheodory measurability and extensions of measures

In this section we discuss a construction of measures by means of the so-called Caratheodory outer measures. The main idea has already been encountered in our discussion of extensions of countably additive measures from algebras to σ-algebras, but now we do not assume that an "outer measure" is generated by an additive measure. In addition, we shall consider set functions with infinite values.

2.7.1. Definition. *A set function $\mathfrak{m}$ defined on the class of all subsets of a given set X and taking values in $[0, +\infty]$ is called an outer measure on X (or a Caratheodory outer measure) if it has the following properties:*

(i) $m(\varnothing) = 0$;

(ii) $m(A) \leqslant m(B)$ whenever $A \subset B$;

(iii) $m\left(\bigcup_{n=1}^{\infty} A_n\right) \leqslant \sum_{n=1}^{\infty} m(A_n)$ for all $A_n \subset X$.

An important example of a Caratheodory outer measure is the function μ^* discussed in §2.4.

2.7.2. Definition. *Let* m *be a set function with values in* $[0, +\infty]$ *defined on the class of all subsets of a space* X *such that* $m(\varnothing) = 0$. *A set* $A \subset X$ *is called Caratheodory measurable with respect to* m *(or Caratheodory* m-*measurable) if, for every set* $E \subset X$, *we have*

$$m(E \cap A) + m(E \backslash A) = m(E). \tag{2.7.1}$$

Let $\mathfrak{M}_m$ *be the class of all Caratheodory* m-*measurable sets.*

Thus, a measurable set splits every set according to the requirement of additivity of m. We note at once that in general the measurability does not follow from the equality

$$m(A) + m(X \backslash A) = m(X) \tag{2.7.2}$$

even in the case of an outer measure with $m(X) < \infty$ (unlike the situation of Proposition 2.4.12). Let us consider the following example.

2.7.3. Example. Let $X = \{1, 2, 3\}$, $m(\varnothing) = 0$, $m(X) = 2$, and let $m(A) = 1$ for all other sets A. It is readily seen that m is an outer measure. Every subset $A \subset X$ satisfies equality (2.7.2), but for the sets $A = \{1\}$ and $E = \{1, 2\}$ equality (2.7.1) is not true (the left-hand side equals 2, the right-hand side equals 1). One can also show that only $\varnothing$ and X are m-measurable.

In this example the class $\mathfrak{M}_m$ of all Caratheodory m-measurable sets is smaller than the class $\mathcal{A}_m$ from Definition 2.4.1, since for the outer measure m on the class of all sets the family $\mathcal{A}_m$ is the class of all sets. However, we shall see that in the case where $m = \mu^*$ is the outer measure generated by a countably additive measure μ on a σ-algebra with values in $[0, +\infty]$, the class $\mathfrak{M}_m$ can be larger than $\mathcal{A}_\mu$ (Exercise 2.7.42). On the other hand, under reasonable assumptions the classes $\mathfrak{M}_{\mu^*}$ and $\mathcal{A}_\mu$ coincide. Below we single out a class of outer measures for which measurability is equivalent to (2.7.2). This class includes, for example, all outer measures generated by countably additive measures on algebras (see Proposition 2.7.7 and Theorem 2.7.8).

2.7.4. Theorem. (THE CARATHEODORY THEOREM) *Let* m *be a set function on the class of all subsets of a space* X *with values in* $[0, +\infty]$ *and let* $m(\varnothing) = 0$. *Then*

(i) $\mathfrak{M}_m$ *is an algebra and the function* m *is additive on* $\mathfrak{M}_m$.

(ii) *If the function* m *is an outer measure on* X, *then the class* $\mathfrak{M}_m$ *is a* σ-*algebra and the function* m *with values in* $[0, +\infty]$ *is countably additive on* $\mathfrak{M}_m$. *In addition, the measure* m *is complete on* $\mathfrak{M}_m$.

For a proof, see [73, §1.11].

Given the countably additive measure $\mu := \mathfrak{m}|_{\mathfrak{M}_{\mathfrak{m}}}$ on $\mathfrak{M}_{\mathfrak{m}}$, where $\mathfrak{m}$ is an outer measure, one can construct the usual outer measure μ^*, as we did earlier. However, this outer measure can differ from the original function $\mathfrak{m}$ (of course, on the sets from $\mathfrak{M}_{\mathfrak{m}}$ both outer measures coincide). Say, in Example 2.7.3 we obtain $\mu^*(A) = 2$ for every nonempty set A different from X.

In applications outer measures are often constructed with the aid of the so-called Method I described in the next example and already used in §2.4, where in Lemma 2.4.4 we established the countable subadditivity.

2.7.5. Example. Let $\mathfrak{X}$ be a family of subsets of a set X and $\varnothing \in \mathfrak{X}$. Suppose that we are given a function $\tau\colon \mathfrak{X} \to [0, +\infty]$ with $\tau(\varnothing) = 0$. Set

$$\mathfrak{m}(A) = \inf\Big\{\sum_{n=1}^{\infty} \tau(X_n)\colon X_n \in \mathfrak{X}, A \subset \bigcup_{n=1}^{\infty} X_n\Big\}, \qquad (2.7.3)$$

where we put $\mathfrak{m}(A) = \infty$ if there are no such sets X_n. Then $\mathfrak{m}$ is an outer measure. It will be denoted by τ^*.

The proof of Lemma 2.4.4 (where we dealt with finite functions) obviously applies here.

Let us emphasize that in the example above it is not asserted that the constructed outer measure extends the function τ. This is false in general. In addition, sets of the original family $\mathfrak{X}$ can be nonmeasurable with respect to $\mathfrak{m}$. Let us consider the corresponding examples. For X we take $\mathbb{N}$ and for $\mathfrak{X}$ we take the family of all singletons and the whole space X. Set $\tau(n) = 2^{-n}$, $\tau(X) = 2$. Then $\mathfrak{m}(X) = 1$ and X is measurable with respect to $\mathfrak{m}$. If for X we take the interval $[0, 1]$ and for τ we take the outer Lebesgue measure defined on the class $\mathfrak{X}$ of all sets, then the constructed function $\mathfrak{m}$ will coincide with the original function τ and the class of $\mathfrak{m}$-measurable sets will coincide with the class of usual Lebesgue measurable sets, i.e., not all sets from $\mathfrak{X}$. One can construct a similar example with an additive function τ on the σ-algebra of all subsets of the interval (see [**73**, Exercise 1.12.108]).

We now single out an important class of outer measures.

2.7.6. Definition. *An outer measure $\mathfrak{m}$ on X is called regular if for every set $A \subset X$ there exists an $\mathfrak{m}$-measurable set B such that $A \subset B$ and $\mathfrak{m}(A) = \mathfrak{m}(B)$.*

For example, the outer measure λ^* constructed by Lebesgue measure on an interval is regular, because for B we can take the set $\bigcap_{n=1}^{\infty} A_n$, where each A_n is measurable, $A \subset A_n$ and $\lambda(A_n) < \lambda^*(A) + 1/n$ (such a set is called a *measurable envelope* of A).

2.7.7. Proposition. *Let $\mathfrak{m}$ be a finite regular outer measure on a set X. Then the $\mathfrak{m}$-measurability of a set $A \subset X$ is equivalent to the equality*

$$\mathfrak{m}(A) + \mathfrak{m}(X\backslash A) = \mathfrak{m}(X). \qquad (2.7.4)$$

PROOF. If $A \in \mathfrak{M}_{\mathfrak{m}}$, then (2.7.4) holds. Conversely, suppose that (2.7.4) is true. Let E be an arbitrary set in X, $C \in \mathfrak{M}_{\mathfrak{m}}$, $E \subset C$, $\mathfrak{m}(C) = \mathfrak{m}(E)$. It suffices

to show that
$$m(E) \geqslant m(E \cap A) + m(E \backslash A), \qquad (2.7.5)$$
because the reverse inequality follows from the subadditivity. We observe that
$$m(A \backslash C) + m((X \backslash A) \backslash C) \geqslant m(X \backslash C). \qquad (2.7.6)$$
By the measurability of C we have
$$m(A) = m(A \cap C) + m(A \backslash C), \quad m(X \backslash A) = m(C \cap (X \backslash A)) + m((X \backslash A) \backslash C).$$
These equalities on account of (2.7.4) give
$$m(A \cap C) + m(A \backslash C) + m(C \cap (X \backslash A)) + m((X \backslash A) \backslash C) = m(X)$$
$$= m(C) + m(X \backslash C).$$
Comparing this equality with (2.7.6) and using that m assumes finite values, we arrive at the estimate
$$m(C \cap A) + m(C \backslash A) \leqslant m(C).$$
From this estimate along with the inclusion $E \subset C$ and monotonicity of m we obtain $m(E \cap A) + m(E \backslash A) \leqslant m(C) = m(E)$. Thus, we have proved (2.7.5). $\square$

Example 2.7.3 shows that Method I from Example 2.7.5 does not always give regular outer measures. However, if $\mathfrak{X} \subset \mathfrak{M}_m$, then this method gives a regular outer measure, since in this case for all $A_n \in \mathfrak{X}$ we have $\bigcup_{n=1}^{\infty} A_n \in \mathfrak{M}_m$. One more result in this direction is contained in the following theorem (for a proof, see [73, §1.11]).

2.7.8. Theorem. *Let* X, $\mathfrak{X}$, τ *and* m *be the same as in Example 2.7.5. Suppose that* $\mathfrak{X}$ *is an algebra or a ring and the function* τ *is additive. Then the outer measure* m *is regular and all sets in the class* $\mathfrak{X}$ *are measurable with respect to* m, *moreover, if* τ *is countably additive, then* m *coincides with* τ *on* $\mathfrak{X}$. *Finally, if* $\tau(X) < \infty$, *then* $\mathfrak{M}_m = \mathfrak{X}_\tau$, *i.e., in this case the definition of the Caratheodory measurability is equivalent to Definition 2.4.1.*

2.7.9. Corollary. *If on a ring (or semiring) a countably additive set function with values in* $[0, +\infty]$ *is defined, then it has a countably additive extension to the* σ-algebra generated by this ring (semiring, respectively).

Unlike the case of an algebra, the aforementioned extension is not always unique. An example: $X = \{0\}$ with the zero measure m on the ring $\mathfrak{X} = \{\varnothing\}$, which can be extended to X by an arbitrary value. It is easy to verify the uniqueness of an extension of a σ-finite measure τ from a ring $\mathfrak{X}$ to the generated σ-ring. For this we represent X as the union of countably many sets $X_n \in \mathfrak{X}$ of finite measure and apply the assertion about uniqueness to the restrictions of τ to the algebra of sets $\{X_n \cap E : E \in \mathfrak{X}\}$ in the spaces X_n. Let us emphasize once again that the outer measure m can differ from τ on $\mathfrak{X}$, but if the function τ on the algebra $\mathfrak{X}$ is countably additive, then the outer measure m constructed by it coincides with τ on $\mathfrak{X}$. For infinite measures it can happen that the class $\mathfrak{X}_\tau$ is strictly contained in $\mathfrak{M}_{\tau^*}$ (see Remark 2.4.14(i)). We now show that the previous corollary can be proved without Caratheodory's construction. For the definition of $\mathcal{A}_\mu^{loc}$ see Remark 2.4.14.

2.7.10. Proposition. *Let μ be a countably additive measure with values in $[0, +\infty]$ defined on a ring $\mathcal{A}$. Then $\mathcal{A}_\mu^{loc}$ is a σ-algebra containing $\mathcal{A}$ and the function μ^* on $\mathcal{A}_\mu^{loc}$ is a countably additive extension of μ. In addition, this extension is complete.*

PROOF. The fact that $\mathcal{A}_\mu^{loc}$ is a σ-algebra follows at once from Theorem 2.4.6 applied to the restrictions of μ to sets from $\mathcal{A}$ of finite measure, since the restriction of μ^* to such a set E coincides with the outer measure on E constructed by the restriction of μ to E (this is clear from the definition of the outer measure). The same theorem gives equality (2.3.2) in the case where the sum of the series is finite. Indeed, it suffices to show that $\mu^*\left(\bigcup_{n=1}^\infty A_n\right) \geqslant \sum_{n=1}^\infty \mu^*(A_n)$ for disjoint sets $A_n \in \mathcal{A}_\mu^{loc}$ (the opposite inequality follows from the definition of μ^* precisely as in the case of finite measures). Hence it suffices to obtain the estimate $\mu^*\left(\bigcup_{n=1}^N A_n\right) \geqslant \sum_{n=1}^N \mu^*(A_n)$ for all $N \in \mathbb{N}$. Let $\varepsilon > 0$. Then there exists a set $E \in \mathcal{A}$ such that $\sum_{n=1}^N \mu^*(E \cap A_n) \geqslant \sum_{n=1}^N \mu^*(A_n) - \varepsilon$. It remains to observe that by Theorem 2.4.6 we have $\mu^*\left(E \cap \bigcup_{n=1}^N A_n\right) = \sum_{n=1}^N \mu^*(E \cap A_n)$. If the sum of the series in (2.3.2) is infinite, it is easy to see that the left-hand side of (2.3.2) is infinite as well. □

2.7.11. Remark. If μ is a countably additive measure with values in $[0, +\infty]$ defined on an algebra $\mathcal{A}$, then the Caratheodory method gives exactly the class $\mathcal{A}_\mu^{loc}$ of locally measurable sets described in Remark 2.4.14. Indeed, it is clear that $\mathfrak{M}_{\mu^*} \subset \mathcal{A}_\mu^{loc}$. On the other hand, if $A \subset \mathcal{A}_\mu^{loc}$, then, for every set E with $\mu^*(E) < \infty$, we have $\mu^*(E) = \mu^*(E \cap A) + \mu^*(E \backslash A)$ by Proposition 2.4.12. If $\mu^*(E) = \infty$, then at least one of the numbers $\mu^*(E \cap A)$ or $\mu^*(E \backslash A)$ is also infinite. Thus, $\mathfrak{M}_{\mu^*} = \mathcal{A}_\mu^{loc}$.

Closing our discussion of extensions we observe that a measure can be always extended to a broader σ-algebra containing an a priori given set if it was defined not on all sets. The proof is delegated to Exercise 2.7.52. The construction makes use of a *measurable envelope* of an arbitrary set E in a measurable space $(X, \mathcal{A})$ with a finite nonnegative measure μ. This is a set $\widetilde{E} \in \mathcal{A}$ for which $E \subset \widetilde{E}$ and $\mu(\widetilde{E}) = \mu^*(E)$. A measurable envelope exists: for example, one can take $\widetilde{E} = \bigcap_{n=1}^\infty E_n$, where $E \subset E_n \in \mathcal{A}$ and $\mu(E_n) \leqslant \mu^*(E) + 1/n$. Similarly one verifies the existence of a *measurable kernel* E, i.e., a set $\underline{E} \in \mathcal{A}$ such that

$$\underline{E} \subset E \quad \text{and} \quad \mu(\underline{E}) = \mu_*(E) := \sup\{\mu(A): A \in \mathcal{A}, A \subset E\}.$$

If M is a measurable envelope of $X \backslash E$, then $X \backslash M$ is a measurable kernel of E.

2.7.12. Theorem. *Let μ be a finite nonnegative measure on a σ-algebra $\mathcal{B}$ in a space X and let S be a set such that $\mu_*(S) = \alpha < \mu^*(S) = \beta$, where $\mu_*(S) = \sup\{\mu(B): B \subset S, B \in \mathcal{B}\}$. Then, for every $\gamma \in [\alpha, \beta]$, there exists a countably additive measure ν on the σ-algebra $\sigma(\mathcal{B} \cup S)$ generated by $\mathcal{B}$ and S such that $\nu = \mu$ on $\mathcal{B}$ and $\nu(S) = \gamma$.*

Exercises

2.7.13.° Prove that the class of all singletons in a space X generates the σ-algebra consisting of all sets that are either at most countable or have at most countable complements. In particular, this σ-algebra is smaller than the Borel one if $X = \mathbb{R}^1$.

2.7.14.° Let $\mathcal{F}$ be some class of sets in a space X. Let us add to $\mathcal{F}$ the empty set and denote by $\mathcal{F}_1$ the class of all sets from this extended collection and their complements. Next, let $\mathcal{F}_2$ denote the class of all intersections of finitely many sets from $\mathcal{F}_1$. Prove that the class $\mathcal{F}_3$ of all finite unions of sets from $\mathcal{F}_2$ coincides with the algebra generated by $\mathcal{F}$.

2.7.15.° Let $\mathcal{F}$ be a collection of sets in a space X. Prove that every set A from the σ-algebra $\sigma(\mathcal{F})$ generated by $\mathcal{F}$ is contained in the σ-algebra generated by some at most countable subfamily $\{F_n\} \subset \mathcal{F}$.

HINT: verify that the union of all σ-algebras $\sigma(\{F_n\})$ generated by at most countable subfamilies $\{F_n\} \subset \mathcal{F}$ is a σ-algebra.

2.7.16.° Let W be a nonempty open set in $\mathbb{R}^n$. Prove that W is the union of an at most countable collection of open cubes with edges parallel to the coordinate axes having length of the form $p2^{-q}$, where $p, q \in \mathbb{N}$, and centers with coordinates of the form $m2^{-k}$, where $m \in \mathbb{Z}$, $k \in \mathbb{N}$.

HINT: the union of all cubes of the indicated form contained in W is the whole set W.

2.7.17. Show that the σ-algebra $\mathcal{B}(\mathbb{R}^1)$ of all Borel subsets of the real line is the smallest class of sets containing all closed sets and admitting countable intersections and countable unions.

HINT: the indicated class is monotone and contains the algebra of finite unions of rays and intervals; alternatively, one can verify that the collection of all sets belonging to the indicated class along with their complements is a σ-algebra and contains all closed sets.

2.7.18. Show that there exists a Borel set B on the real line such that for every nonempty interval J both sets $B \cap J$ and $(\mathbb{R}^1 \backslash B) \cap J$ have positive Lebesgue measure.

2.7.19. Let μ be an arbitrary finite Borel measure on a compact interval I. Show that there exists a first category set E (i.e., a countable union of nowhere dense sets) such that $\mu(I \backslash E) = 0$.

HINT: it suffices to find for every n a compact set K_n without inner points with $\mu(K_n) > \mu(I) - 2^{-n}$. Using that μ has at most countably many points a_j of nonzero measure, one can find a countable everywhere dense set of points s_j of zero μ-measure. Around every point s_j one can take an interval $U_{n,j}$ with $\mu(U_{n,j}) < 2^{-j-n}$. The compact set $K_n = I \backslash \bigcup_{j=1}^{\infty} U_{n,j}$ is a required one.

2.7.20.° Prove that Lebesgue measure of every measurable set $E \subset \mathbb{R}^n$ equals the infimum of the sums $\sum_{k=1}^{\infty} \lambda_n(U_k)$ over all sequences of open balls U_k covering E.

2.7.21.° Show that a set $E \subset \mathbb{R}$ is Lebesgue measurable precisely when for every $\varepsilon > 0$ there exist open sets U and V such that $E \subset U$, $U \backslash E \subset V$ and $\lambda(V) < \varepsilon$.

2.7.22.° Let $A \subset \mathbb{R}^1$ be a set of positive Lebesgue measure. Prove that there exist points $a_1, a_2 \in A$ such that the number $a_1 - a_2$ is irrational.

2.7.23. Let $A \subset \mathbb{R}^1$ be a set of positive Lebesgue measure. Prove that the difference set $A - A := \{a_1 - a_1 : a_1, a_2 \in A\}$ contains an interval. In particular, there exist $a_1, a_2 \in A$ such that the number $a_1 - a_2$ is rational.

2.7.24. Prove without using the continuum hypothesis that every set of positive Lebesgue measure has the cardinality of the continuum.

2.7.25. Show that the additive function on the real line mentioned in Exercise 1.9.86 is unbounded on every set of positive measure.

2.7.26.° Show that every infinite σ-algebra is uncountable.
HINT: show that every infinite σ-algebra contains a sequence of pairwise disjoint sets.

2.7.27.° Let μ be a bounded nonnegative Borel measure on $\mathbb{R}^n$. Prove that for every μ-measurable set E and every $\varepsilon > 0$, there exists a finite union E_ε of open cubes with edges parallel to the coordinate axes such that $\mu(E \bigtriangleup E_\varepsilon) < \varepsilon$.

2.7.28.° Let μ be a probability Borel measure on a cube I of unit volume such that $\mu(A) = \mu(B)$ for all Borel sets $A, B \subset I$ that differ only by a shift. Show that μ coincides with Lebesgue measure λ_n.
HINT: observe that μ vanishes on faces of cubes, hence has the same value as Lebesgue measure on the cubes obtained by any partitioning of I generated by dividing the edges into equal intervals. Then both measures coincide on finite unions of such smaller cubes.

2.7.29. (i) Let $\mathfrak{R}$ be a semiring, let $\mu\colon \mathfrak{R} \to [0, +\infty)$ be a countably additive function, and let $A, A_n \in \mathfrak{R}$ be such that the sets A_n either increase or decrease to A. Show that $\mu(A) = \lim\limits_{n\to\infty} \mu(A_n)$.
(ii) Give an example showing that the properties indicated in (i) do not imply the countable additivity for a nonnegative additive set function on a semiring.
HINT: (ii) consider the semiring of sets of the form $\mathbb{Q} \cap (a, b)$, $\mathbb{Q} \cap (a, b]$, $\mathbb{Q} \cap [a, b)$, $\mathbb{Q} \cap [a, b]$, where $\mathbb{Q}$ is the set of rational numbers; on these sets let μ equal $b - a$.

2.7.30. (i) Show that a bounded set $E \subset \mathbb{R}^n$ is Jordan measurable (see Definition 2.5.9) precisely when the boundary of E (the set of points in every neighborhood of which there is a point of E and a point of the complement of E) has measure zero.
(ii) Show that the collection of all Jordan measurable subsets of an interval (or a cube) is a ring.

2.7.31.° Show that a bounded nonnegative measure μ on a σ-algebra $\mathcal{A}$ is complete precisely when $\mathcal{A} = \mathcal{A}_\mu$; in particular, the Lebesgue extension of any complete measure coincides with the original measure.

2.7.32. Let $E \subset [0, 1]$ be a set of zero Lebesgue measure. Can the difference set $E - E := \{x - y\colon x, y \in E\}$ have a positive measure?

2.7.33. Give an example of a σ-finite measure on a σ-algebra that is not σ-finite on some sub-σ-algebra.
HINT: restrict Lebesgue measure on $\mathbb{R}^1$ to the sub-σ-algebra of sets that are either at most countable or have at most countable complements.

2.7.34. Show that every set of positive Lebesgue measure contains a nonmeasurable subset.

2.7.35. Find Lebesgue measure of the set of points in $[0, 1]$ in the dyadic expansion of which all even positions are occupied by 0.

2.7.36. Prove that there exists no countably additive measure defined on all subsets of the space $X = \{0, 1\}^\infty$ assuming only the values 0 and 1 and vanishing on all singletons.
HINT: let $X_n = \{(x_i) \in X\colon x_n = 0\}$; if such a measure μ exists, then for every n either $\mu(X_n) = 1$ or $\mu(X_n) = 0$; let Y_n denote that one of the two sets X_n and $X\backslash X_n$ which has measure 1; then the set $\bigcap_{n=1}^\infty Y_n$ also has measure 1 and is a point.

2.7.37.° (Young's criterion of measurability) Let $(X, \mathcal{A}, \mu)$ be a space with a finite nonnegative measure. Prove that a set $A \subset X$ belongs to $\mathcal{A}_\mu$ precisely when for every set B disjoint with A the equality $\mu^*(A \cup B) = \mu^*(A) + \mu^*(B)$ holds.

2.7.38. Prove that there exists a sequence of sets $A_n \subset [0, 1]$ such that $A_{n+1} \subset A_n$, $\bigcap_{n=1}^\infty A_n = \varnothing$ and $\lambda^*(A_n) = 1$, where λ is Lebesgue measure.

HINT: let $\{r_n\}$ be some enumeration of rational numbers and let $E \subset [0, 1]$ be the nonmeasurable set from Vitali's example. Show that the sets

$$E_n := \big(E \cup (E + r_1) \cup \cdots \cup (E + r_n)\big) \cap [0, 1]$$

have zero inner measure and take $A_n := [0, 1] \backslash E_n$.

2.7.39. Extend Theorem 2.5.8 to Lipschitz mappings defined on measurable sets.

2.7.40.° Let $(X, \mathcal{A}, \mu)$ be a probability space, $\mathcal{B}$ a sub-σ-algebra in $\mathcal{A}$, and $\mathcal{B}^\mu$ the σ-algebra generated by $\mathcal{B}$ and all sets of measure zero in $\mathcal{A}_\mu$. (i) Show that $E \in \mathcal{B}^\mu$ precisely when there exists a set $B \in \mathcal{B}$ such that $E \triangle B \in \mathcal{A}_\mu$ and $\mu(E \triangle B) = 0$.

(ii) Give an example showing that $\mathcal{B}^\mu$ can be strictly larger than the σ-algebra $\mathcal{B}_\mu$ that is the completion of $\mathcal{B}$ with respect to the measure $\mu|_\mathcal{B}$.

HINT: (i) the sets of the indicated form belong to $\mathcal{B}^\mu$ and form a σ-algebra. (ii) Take Lebesgue measure λ on the σ-algebra of all measurable sets in $[0, 1]$ and $\mathcal{B} = \{\varnothing, [0, 1]\}$. Then $\mathcal{B}_\lambda = \mathcal{B}$.

2.7.41.° Let $\mathcal{F}$ be some class of sets in a space X and let μ and ν be two probability measures on $\sigma(\mathcal{F})$ that agree on $\mathcal{F}$. Is it true that $\mu = \nu$ on $\sigma(\mathcal{F})$?

2.7.42. Let $(X, \mathcal{A}, \mu)$ be a measurable space, where $\mathcal{A}$ is a σ-algebra and μ is a countably additive measure with values in $[0, +\infty]$. Let $\mathcal{L}_\mu$ denote the class of all sets $E \subset X$ for which there exist sets $A_1, A_2 \in \mathcal{A}$ with $A_1 \subset E \subset A_2$ and $\mu(A_2 \backslash A_1) = 0$.

(i) Show that $\mathcal{L}_\mu$ is a σ-algebra, coincides with $\mathcal{A}_\mu$ and is contained in $\mathfrak{M}_{\mu^*}$.

(ii) Show that if the measure μ is σ-finite, then $\mathcal{L}_\mu$ coincides with $\mathfrak{M}_{\mu^*}$.

(iii) Let $X = [0, 1]$, let $\mathcal{A}$ be the σ-algebra generated by all singletons, and let μ with values in $[0, +\infty]$ be defined as follows: $\mu(A)$ is the cardinality of A, $A \in \mathcal{A}$. Show that $\mathfrak{M}_{\mu^*}$ contains all sets, but $[0, 1/2] \notin \mathcal{L}_\mu$.

HINT: (iii) Show that $\mu^*(A)$ is the cardinality of A and that $\mathcal{L}_\mu = \mathcal{A}$, using that nonempty sets have measure at least 1.

2.7.43. Let $\mathcal{A}$ be the class of all first category sets in $[0, 1]$ (i.e., countable unions of nowhere dense sets) and their complements and let $\mu(A) = 0$ if A is a first category set, $\mu(A) = 1$ if $[0, 1] \backslash A$ is a first category set. Prove that $\mathcal{A}$ is a σ-algebra, the measure μ is countably additive, but does not extend to a countably additive measure on $\mathcal{B}([0, 1])$.

HINT: apply Exercise 2.7.19.

2.7.44. Let E be a set on the circumference S having zero measure with respect to the natural linear Lebesgue measure on S such that the complement of E is a first category set. Prove that every countable set in S is contained in the image of E under some rotation.

HINT: the intersection of countably many images of E under rotations is not empty.

2.7.45. A family $\mathcal{S}$ of subsets of a set X is called a σ-*additive class* if (i) $X \in \mathcal{S}$, (ii) $S_2 \backslash S_1 \in \mathcal{S}$ for all S_1, $S_2 \in \mathcal{S}$ with $S_1 \subset S_2$, (iii) $\bigcup_{n=1}^\infty S_n \in \mathcal{S}$ for all pairwise disjoint sets $S_n \in \mathcal{S}$. Prove that if a class of sets $\mathcal{E}$ admits finite intersections, then the σ-additive class generated by $\mathcal{E}$ (i.e., the intersection of all σ-additive classes containing $\mathcal{E}$) coincides with the σ-algebra generated by $\mathcal{E}$.

2.7.46. Let $\mu \geqslant 0$ be a finite measure on $(X, \mathcal{A})$ and let $E \subset X$ be an arbitrary set (not necessarily measurable). On the σ-algebra $\mathcal{A}_E$ of sets of the form $B = E \cap A$, where $A \in \mathcal{A}$, in the space E we define a measure μ_E by the formula $\mu_E(B) := \mu(\widetilde{E} \cap A)$, where $\widetilde{E}$ is a measurable envelop of E. Show that μ_E is well-defined (i.e., $\mu(\widetilde{E} \cap A_1) = \mu(\widetilde{E} \cap A_2)$ whenever $A_1 \cap E = A_2 \cap E$) and countably additive. The measure μ_E is called the *restriction* of μ to the set E.

2.7.47. Prove that every convex set in IR^n (i.e., a set that along with every pair of its points contains the segment connecting them) is Lebesgue measurable and its boundary has measure zero.

HINT: show that the interior and the closure of this set differ by a set of measure zero.

2.7.48. Let $E \subset [0, 1]$ be a set of positive Lebesgue measure. Prove that for every $\alpha < \lambda(E)$ there exists a compact set $K \subset E$ without inner and isolated points for which $\lambda(K) = \alpha$. Prove the same assertion for every atomless Borel measure on $[0, 1]$.

2.7.49. Prove that $[0, 1]$ can be written as the union of a continuum of disjoint sets of outer measure 1.

HINT: the continual set of all compact sets of positive measure can be written in the form $\{K_\omega\}_{\omega \in \Omega_0}$, where Ω_0 is the set of finite and countable ordinals. By the transfinite induction one can construct a continuum of disjoint sets $A_t \subset [0, 1]$ each of which intersects all K_ω. To this end, at the inductive step corresponding to an ordinal $\omega \in \Omega_0$, take in K_ω a continual set C_t of measure zero (about ordinal numbers, see [**331**]).

2.7.50. If a σ-algebra $\mathcal{A}$ in a space X is generated by a finite or countable family of sets, then $\mathcal{A}$ is called *countably generated*. (i) Prove that the σ-algebra in $[0, 1]$ generated by all points is not countably generated, but the larger Borel σ-algebra is.

(ii) Prove that every countably generated σ-algebra $\mathcal{A}$ has the form

$$\mathcal{A} = \{f^{-1}(B) \colon B \in \mathcal{B}([0, 1])\},$$

where $f \colon X \to [0, 1]$ is some function.

(iii) Prove that every countably generated σ-algebra has the cardinality not larger than that of the continuum, using that the family of Borel sets on the real line is continual.

HINT: (i) assuming that this σ-algebra is generated by a sequence of countable sets, consider a point outside their union.

2.7.51. Let μ be a finite nonnegative measure on a measurable space $(X, \mathcal{A})$. A measurable set E is called an *atom* of μ if $\mu(E) > 0$ and there are no measurable subsets $A \subset E$ with $0 < \mu(A) < \mu(E)$. Measures without atoms are called *atomless*. Two atoms E_1 and E_2 are equivalent if $\mu(E_1 \triangle E_2) = 0$. (i) Prove that μ can have only a finite or countable number of pairwise nonequivalent atoms. (ii) Prove that if $\{E_n\}$ are all atoms of μ up to equivalence, then the restriction of μ to $X \setminus \bigcup_{n=1}^{\infty} E_n$ is atomless. (iii) Prove that if the measure μ is atomless, then so is every measure $\nu \ll \mu$. (iv) Prove that the set of values of every atomless measure μ is the whole interval $[0, \mu(X)]$.

2.7.52. Prove Theorem 2.7.12.

HINT: see [**73**, Theorem 1.12.14].

2.7.53. (Nikodym's theorem) Let μ_n be finite measures on a σ-algebra $\mathcal{A}$ such that, for each $A \in \mathcal{A}$, there exists a finite limit $\mu(A) = \lim\limits_{n \to \infty} \mu_n(A)$. Show that $\sup_n \|\mu_n\| < \infty$ and μ is a measure on $\mathcal{A}$. See [**73**, Chapter 4].

CHAPTER 3

The Lebesgue Integral

In this chapter we study the Lebesgue integral and obtain fundamental results about passage to the limit under the integral sign and spaces of integrable functions. In addition, we prove very important theorems due to Radon and Nikodym and Fubini.

3.1. Measurable Functions

In this section we study measurable functions. In spite of the name, the concept of measurability of functions is defined in terms of σ-algebras and *is not connected with measures*. A connection with measures arises when for the basic σ-algebra we take the σ-algebra of all sets measurable with respect to a fixed measure. This important particular case is considered at the end of the section. Measurable functions are needed for constructing the integral.

3.1.1. Definition. *Let $(X, \mathcal{A})$ be a space with a σ-algebra. A function $f\colon X \to \mathbb{R}^1$ is called measurable with respect to $\mathcal{A}$ (or $\mathcal{A}$-measurable) if we have $\{x\colon f(x) < c\} \in \mathcal{A}$ for every $c \in \mathbb{R}^1$.*

The simplest example of an $\mathcal{A}$-measurable function is the *indicator function* I_A of a set $A \in \mathcal{A}$ defined as follows: $I_A(x) = 1$ if $x \in A$, $I_A(x) = 0$ if $x \notin A$. The indicator function of a set A is also called the *characteristic function* of A. The set $\{x\colon I_A(x) < c\}$ is empty if $c \leqslant 0$, equals the complement of A if $c \in (0, 1]$, and coincides with X if $c > 1$. It is clear that the inclusion $A \in \mathcal{A}$ is also necessary for the $\mathcal{A}$-measurability of I_A.

3.1.2. Theorem. *A function f is measurable with respect to a σ-algebra $\mathcal{A}$ precisely when $f^{-1}(B) \in \mathcal{A}$ for all sets $B \in \mathcal{B}(\mathbb{R}^1)$.*

PROOF. Let f be an $\mathcal{A}$-measurable function. Denote by $\mathcal{E}$ the collection of all sets $B \in \mathcal{B}(\mathbb{R}^1)$ such that $f^{-1}(B) \in \mathcal{A}$. We show that $\mathcal{E}$ is a σ-algebra. Indeed, if $B_n \in \mathcal{E}$, then $f^{-1}\left(\bigcup_{n=1}^{\infty} B_n\right) = \bigcup_{n=1}^{\infty} f^{-1}(B_n) \in \mathcal{A}$. In addition, for any $B \in \mathcal{E}$ we have $f^{-1}(\mathbb{R}^1 \backslash B) = X \backslash f^{-1}(B) \in \mathcal{A}$. Since $\mathcal{E}$ contains the rays $(-\infty, c)$, we obtain that $\mathcal{B}(\mathbb{R}^1) \subset \mathcal{E}$, i.e., $\mathcal{B}(\mathbb{R}^1) = \mathcal{E}$. The converse assertion is obvious, because all rays are Borel sets. $\qquad\square$

Let us write f in the form $f = f^+ - f^-$, where
$$f^+(x) := \max\big(f(x), 0\big) \quad \text{and} \quad f^-(x) := \max\big(-f(x), 0\big).$$

© Springer Nature Switzerland AG 2020
V. I. Bogachev, O. G. Smolyanov, *Real and Functional Analysis*,
Moscow Lectures 4, https://doi.org/10.1007/978-3-030-38219-3_3

We observe that the $\mathcal{A}$-measurability of f is equivalent to the $\mathcal{A}$-measurability of both functions f^+ and f^-. For example, for $c > 0$ one has the equality $\{x\colon f(x) < c\} = \{x\colon f^+(x) < c\}$.

Note that the restriction $f|_E$ of any $\mathcal{A}$-measurable function f to a set $E \subset X$ is measurable with respect to the induced σ-algebra $\mathcal{A}_E = \{A \cap E\colon A \in \mathcal{A}\}$.

It is often useful to introduce the following more general definition.

3.1.3. Definition. *Let $(X_1, \mathcal{A}_1)$ and $(X_2, \mathcal{A}_2)$ be two spaces with σ-algebras. A mapping $f\colon X_1 \to X_2$ is called measurable with respect to the pair $(\mathcal{A}_1, \mathcal{A}_2)$, or $(\mathcal{A}_1, \mathcal{A}_2)$-measurable, if $f^{-1}(B) \in \mathcal{A}_1$ for all $B \in \mathcal{A}_2$.*

If $(X_2, \mathcal{A}_2) = \big(\mathbb{R}, \mathcal{B}(\mathbb{R})\big)$, then we arrive at the definition of $\mathcal{A}_1$-measurable functions. In another particular case where X_1 and X_2 are metric (or topological) spaces with Borel σ-algebras $\mathcal{A}_1 = \mathcal{B}(X_1)$ and $\mathcal{A}_2 = \mathcal{B}(X_2)$, we arrive at the concept of a *Borel* (or *Borel measurable*) mapping or a *Borel function* if $X_2 = \mathbb{R}$.

3.1.4. Example. Any continuous function f is Borel, since in this case the sets $\{x\colon f(x) < c\}$ are open, hence are Borel.

An important subclass of $\mathcal{A}$-measurable functions consists of *simple functions*, i.e., $\mathcal{A}$-measurable functions f that assume only finitely many values. Therefore, any simple function f has the form $f = \sum_{i=1}^n c_i I_{A_i}$, where $c_i \in \mathbb{R}^1$, $A_i \in \mathcal{A}$, i.e., is a finite linear combination of the indicator functions of sets from $\mathcal{A}$. Obviously, the converse is also true. A representation of a simple function in the indicated form is not unique, but it is easy to pass to a representation in which all numbers c_i are different and the sets A_i are disjoint. For this it suffices take all different values of f and their full preimages.

The next theorem describes some basic properties of measurable functions.

3.1.5. Theorem. *Suppose that functions f, g, and f_n, where $n \in \mathbb{N}$, are measurable with respect to a σ-algebra $\mathcal{A}$. Then*

(i) the function $\varphi \circ f$ is measurable with respect to $\mathcal{A}$ for every Borel function $\varphi\colon \mathbb{R}^1 \to \mathbb{R}^1$; in particular, this is true if the function φ is continuous;

(ii) the function $\alpha f + \beta g$ is measurable with respect to $\mathcal{A}$ for all $\alpha, \beta \in \mathbb{R}^1$;

(iii) the function fg is measurable with respect to $\mathcal{A}$;

(iv) if $g \neq 0$, then the function f/g is measurable with respect to $\mathcal{A}$;

(v) if there exists a finite limit $f_0(x) = \lim_{n\to\infty} f_n(x)$ for all x, then the function f_0 is measurable with respect to $\mathcal{A}$;

(vi) if the functions $\sup_n f_n(x)$ and $\inf_n f_n(x)$ are finite for all x, then they are measurable with respect to $\mathcal{A}$.

PROOF. Assertion (i) follows from the equality $(\varphi \circ f)^{-1}(B) = f^{-1}\big(\varphi^{-1}(B)\big)$, which is easily verified. Due to (i), for the proof of (ii) it suffices to consider the case $\alpha = \beta = 1$ and observe that the set

$$\{x\colon f(x) + g(x) < c\} = \{x\colon f(x) < c - g(x)\}$$

has the form

$$\bigcup_{r_n} \big(\{x\colon f(x) < r_n\} \bigcap \{x\colon r_n < c - g(x)\}\big),$$

where the union is taken over all rational numbers r_n. The right-hand side belongs to $\mathcal{A}$, since the functions f and g are measurable with respect to $\mathcal{A}$. Assertion (iii) follows from the equality $fg = \left[(f + g)^2 - f^2 - g^2\right]/2$, since the square of a measurable function is measurable by virtue of (i). Observing that the function φ defined by the equality $\varphi(x) = 1/x$ for $x \neq 0$ and $\varphi(0) = 0$ is Borel (a simple verification of this is left as an exercise) we obtain (iv). Assertion (v) is less obvious, however, it follows from the following easily verified relation:

$$\{x:\ f_0(x) < c\} = \bigcup_{k=1}^{\infty} \bigcup_{n=1}^{\infty} \bigcap_{m=n+1}^{\infty} \left\{x:\ f_m(x) < c - \frac{1}{k}\right\}.$$

For the proof of (vi) we observe that

$$\sup_n f_n(x) = \lim_{n\to\infty} \max\big(f_1(x), \ldots, f_n(x)\big).$$

By virtue of (v) it suffices to establish the measurability of $\max(f_1, \ldots, f_n)$. By induction this reduces to $n = 2$. It remains to observe that

$$\{x:\ \max\big(f_1(x), f_2(x)\big) < c\} = \{x:\ f_1(x) < c\} \cap \{x:\ f_2(x) < c\}.$$

The assertion for inf can be verified similarly or deduced from the case of sup by using that $\inf_n f_n = -\sup_n(-f_n)$. □

3.1.6. Remark. For functions f that take values in the extended real line $\overline{\mathbb{R}} = [-\infty, +\infty]$ we define the $\mathcal{A}$-measurability by requiring the inclusions

$$f^{-1}(-\infty), f^{-1}(+\infty) \in \mathcal{A}$$

and the $\mathcal{A}$-measurability of f on the set $f^{-1}(\mathbb{R})$. This is equivalent to the measurability in the sense of Definition 3.1.3 if we equip $\overline{\mathbb{R}}$ with the σ-algebra $\mathcal{B}(\overline{\mathbb{R}})$ consisting of the Borel sets of the usual real line with a possible addition of the points $-\infty, +\infty$. Then for functions with values in $\overline{\mathbb{R}}$ assertions (i), (v), and (vi) of the proved theorem remain valid, and for the validity of assertions (ii), (iii), and (iv) one has to consider functions f and g with values either in $[-\infty, +\infty)$ or in $(-\infty, +\infty]$. For such values the operations are defined in the natural way: $+\infty + c = +\infty$ if $c \in (-\infty, +\infty]$, $+\infty \cdot 0 = 0$, $+\infty \cdot c = +\infty$ if $c > 0$, $+\infty \cdot c = -\infty$ if $c < 0$.

3.1.7. Proposition. *Suppose that functions f and g are measurable with respect to a σ-algebra $\mathcal{A}$ and a function Ψ is continuous on the subset of the plane formed by the values of the mapping (f, g). Then the function $\Psi(f, g)$ is measurable with respect to $\mathcal{A}$.*

PROOF. Let $h := (f, g)\colon X \to \mathbb{R}^2$ and $Y := h(X)$. By the continuity of Ψ on Y for every c there exists an open set $U \subset \mathbb{R}^2$ such that

$$Y \cap \Psi^{-1}\big((-\infty, c)\big) = Y \cap U.$$

Hence $\{x:\ \Psi\big(f(x), g(x)\big) < c\} = h^{-1}(U)$. Let us write U as the union of a countable collection of open rectangles of the form $A_n \times B_n$, where A_n and B_n are open intervals. Hence we obtain $h^{-1}(U) = \bigcup_{n=1}^{\infty} f^{-1}(A_n) \cap g^{-1}(B_n) \in \mathcal{A}$. □

3.1.8. Proposition. *Let $\mathcal{A}$ be a σ-algebra of subsets of a space X. Then for every bounded $\mathcal{A}$-measurable function f there exists a sequence of simple functions f_n converging to f uniformly on X. One can take this sequence increasing.*

PROOF. Dividing by a constant and adding a constant, we can assume that $0 \leqslant f(x) \leqslant 1$. For every $n \in \mathbb{N}$ we split $[0,1)$ into 2^n disjoint intervals

$$I_j = \big[(j-1)2^{-n}, j2^{-n}\big)$$

of length 2^{-n} and include 1 in I_{2^n}. Set $A_j = f^{-1}(I_j)$. It is clear that $A_j \in \mathcal{A}$ and $\bigcup_{j \geqslant 1} A_j = X$. Let us define the function f_n by the equality $f_n(x) = (j-1)2^{-n}$ if $x \in A_j$. Then f_n is a simple function and $\sup_{x \in X} |f(x) - f_n(x)| \leqslant 2^{-n}$, since f takes A_j to I_j and f_n maps A_j to the left end of I_j, which is at the distance not more than 2^{-n} from any point of I_j. It is readily seen that $f_n(x) \leqslant f_{n+1}(x)$. $\square$

3.1.9. Corollary. *Suppose that $\mathcal{A}$ is a σ-algebra of subsets of a space X. Then for every $\mathcal{A}$-measurable function f there exists a sequence of simple functions f_n converging to f at every point. If $f \geqslant 0$, then the sequence $\{f_n\}$ can be chosen increasing.*

PROOF. The function $g := \operatorname{arctg} f$ is measurable and takes values in the interval $(-\pi/2, \pi/2)$. As shown above, there exist simple functions g_n with values in $[-\pi/2, \pi/2)$ pointwise increasing to g. If $f \geqslant 0$, they can be taken with values in $[0, \pi/2)$. In the general case, dropping the monotonicity of $\{g_n\}$, we can choose g_n with values in $(-\pi/2, \pi/2)$; for example, we can redefine every function g_n by the value $-\pi/2 + 1/n$ on the set $\{x \colon g_n(x) \leqslant -\pi/2 + 1/n\}$. The functions $f_n = \tan g_n$ possess the desired properties. $\square$

Let us draw the reader's attention once again to the fact that so far in our discussion of measurable functions no measures show up. Suppose now that on a σ-algebra $\mathcal{A}$ of subsets of a space X a nonnegative countably additive measure μ is defined. Then the completion $\mathcal{A}_\mu$ of $\mathcal{A}$ with respect to μ can be larger than $\mathcal{A}$.

3.1.10. Definition. *A real function f on X is called μ-measurable if it is measurable with respect to the σ-algebra $\mathcal{A}_\mu$ of all μ-measurable sets.*

We shall also call μ-measurable every function f that is not defined on a set Z of μ-measure zero or is infinite on Z and on $X\backslash Z$ is $\mathcal{A}_\mu$-measurable in the previous sense. The set of all μ-measurable functions is denoted by $\mathcal{L}^0(\mu)$.

In the case of a measure with values in $[0, +\infty]$ the definition is similar, but in place of $\mathcal{A}_\mu$ we take $\mathcal{A}_\mu^{\mathrm{loc}}$.

Thus, the μ-measurability of a function f means that f is defined everywhere excepting possibly a set of measure zero and $\{x \colon f(x) < c\}$ belongs to $\mathcal{A}_\mu$ (or to $\mathcal{A}_\mu^{\mathrm{loc}}$ in case of an infinite measure) for all $c \in \mathbb{R}^1$. It is clear that the class of μ-measurable functions can be larger than the class of $\mathcal{A}$-measurable functions. If a μ-measurable function f is not defined on a set Z of measure zero, then for any continuation to Z it will be $\mathcal{A}_\mu$-measurable. From our subsequent discussion it will be clear that it is very reasonable to admit a somewhat broader concept of measurability of a function mentioned in the second part of our definition.

When in concrete situations it is clear whether a considered measurable function is defined everywhere or only almost everywhere, this is not specified. Sometimes such precision is necessary. For example, if we consider a continuum of functions f_α on $[0,1]$, where $\alpha \in [0,1]$ and f_α is not defined at the point α, then from the formal point of view these functions have no common point from domains of definition. When it is clear which measure μ is considered, the μ-measurability is called just measurability. Dealing with Lebesgue measure on $\mathbb{R}^n$, it is common to speak of the Lebesgue measurability or just measurability.

For functions with values in $[-\infty, +\infty]$ (not necessarily finite outside a set of measure zero) the μ-measurability is understood as follows:

$$f^{-1}(-\infty), f^{-1}(+\infty) \in \mathcal{A}_\mu$$

and on the set $\{|f| < \infty\}$ the function f is μ-measurable. Such functions *will not* be included in $\mathcal{L}^0(\mu)$ (and we shall not consider them at all); to avoid confusion it is more convenient to call them not functions, but mappings.

Two μ-measurable functions f and g are called *equivalent* if they are equal outside a set of measure zero. Notation: $f \sim g$. The functions f and g are called *modifications* or *versions* of each other. Since the union of two sets of measure zero is a set of measure zero, this gives an equivalence relation on $\mathcal{L}^0(\mu)$. The space of equivalence classes is denoted by $L^0(\mu)$.

3.1.11. Proposition. *Let $\mu \geqslant 0$ be a countably additive measure on a σ-algebra $\mathcal{A}$. Then for every μ-measurable function f one can find a set $A \in \mathcal{A}$ and a function g measurable with respect to $\mathcal{A}$ such that $f(x) = g(x)$ for all $x \in A$ and $\mu(X \backslash A) = 0$.*

PROOF. We can assume that the function f is defined everywhere and finite. By Corollary 3.1.9 there exists a sequence of $\mathcal{A}_\mu$-measurable simple functions f_n pointwise converging to f. Each function f_n assumes finitely many different values on sets $A_1, \ldots, A_k \in \mathcal{A}_\mu$. Every A_i contains a set B_i from $\mathcal{A}$ such that $\mu(B_i) = \mu(A_i)$. Let us consider the function g_n that coincides with f_n on the union of the sets B_i and equals 0 outside this union. It is clear that g_n is a $\mathcal{A}$-measurable simple function and there exists a set $Z_n \in \mathcal{A}$ of measure zero such that $f_n(x) = g_n(x)$ if $x \notin Z_n$. Set $Y = X \backslash \bigcup_{n=1}^\infty Z_n$. Then $Y \in \mathcal{A}$ and $\mu(X \backslash Y) = 0$. Set $g(x) = f(x)$ if $x \in Y$ and $g(x) = 0$ else. For every $x \in Y$ we have $f(x) = \lim_{n\to\infty} f_n(x) = \lim_{n\to\infty} g_n(x)$. Hence f is $\mathcal{A}$-measurable on Y, which gives the $\mathcal{A}$-measurability of g on X. $\qquad\square$

It follows from what we proved that for a *bounded* μ-measurable function f there exist two $\mathcal{A}$-measurable functions f_1 and f_2 such that

$$f_1(x) \leqslant f(x) \leqslant f_2(x) \text{ for all } x \text{ and } \mu\big(x\colon f_1(x) \neq f_2(x)\big) = 0.$$

Indeed, set $f_1 = f_2 = g$ on Y. Outside Y we set $f_1(x) = \inf f$, $f_2 = \sup f$. However, see Exercise 3.12.47.

The characteristic property of measurable functions to have measurable *preimages* of Borel sets should not be confused with the property of a completely different nature: to have measurable *images* of Borel sets. The following instructive example is worth remembering. It employs the *Cantor function*, which is interesting in many other respects (it will be needed below in our discussion of the connections between the integral and derivative).

3.1.12. Proposition. *There exists a continuous nondecreasing function C_0 on the closed interval $[0,1]$ (the Cantor function or the Cantor staircase) such that $C_0(0) = 0$, $C_0(1) = 1$ and $C_0 = (2k-1)2^{-n}$ on the interval $J_{n,k}$ from the complement of the Cantor set C from Example 2.5.5. In addition, the function*

$$f(x) = [C_0(x) + x]/2$$

is a one-to-one continuous mapping of $[0,1]$ onto itself and there is a set of measure zero E in the Cantor set C such that $f(E)$ is Lebesgue nonmeasurable.

PROOF. The function defined above on the complement of C is nondecreasing. Let $C_0(0) = 0$, $C_0(x) = \sup\{C_0(t)\colon t \notin C, \, t < x\}$ if $x \in C$. The nondecreasing function C_0 assumes all values of the form $p2^{-q}$. Hence it has no jumps and is continuous on $[0,1]$. It is clear that f is a continuous and one-to-one mapping of the interval $[0,1]$ onto itself. On every interval of the complement of C the function f has the form $x/2 + \text{const}$ (where const depends on the interval) and hence takes such interval to an interval of the twice smaller length. Therefore, the complement of C is taken to an open set U of measure $1/2$. The set $[0,1]\backslash U$ of measure $1/2$ has a Lebesgue nonmeasurable subset D. It is obvious that $E = f^{-1}(D) \subset C$ has measure zero and $f(E) = D$. $\qquad\square$

Let g be the function inverse to the function f from the previous example. Then the set $g^{-1}(E)$ is nonmeasurable, although E has measure zero and g is a Borel function. This shows that in the definition of a Lebesgue measurable function the requirement of the measurability of the preimages of Borel sets does not imply the measurability of the preimages of all Lebesgue measurable sets.

3.2. Convergence in Measure and Almost Everywhere

Let $(X, \mathcal{A}, \mu)$ be a space with a nonnegative measure μ. We shall say that a certain property is fulfilled *almost everywhere* (or μ-*almost everywhere*) on X if the set Z of points in X not possessing this property belongs to $\mathcal{A}_\mu$ and has measure zero with respect to the measure μ extended to $\mathcal{A}_\mu$. The complements of sets of measure zero are called *sets of full measure*. We shall use the following abbreviations for "almost everywhere": a.e., μ-a.e. It is clear from the definition of $\mathcal{A}_\mu$ that there exists a set $Z_0 \in \mathcal{A}$ such that $Z \subset Z_0$ and $\mu(Z_0) = 0$, i.e., the corresponding property is fulfilled outside some set from $\mathcal{A}$ of measure zero. This circumstance should be remembered when dealing with incomplete measures.

For example, one can say that a sequence of functions f_n converges a.e. or is Cauchy a.e., a function is nonnegative a.e., etc. It is clear that convergence of $\{f_n\}$ a.e. follows from convergence of $\{f_n(x)\}$ for every x (called the pointwise convergence), and the latter follows from the uniform convergence of $\{f_n\}$. A deeper connection between convergence almost everywhere and the uniform convergence is described by the following theorem due to the celebre Russian mathematician D. F. Egorov.

3.2.1. Theorem. (EGOROV'S THEOREM) *Let $(X, \mathcal{A}, \mu)$ be a space with a finite nonnegative measure μ and let $\{f_n\}$ be a sequence of μ-measurable functions such that μ-almost everywhere there exists a finite limit $f(x) = \lim\limits_{n\to\infty} f_n(x)$. Then, for every $\varepsilon > 0$, there exists a set $X_\varepsilon \in \mathcal{A}$ such that $\mu(X\backslash X_\varepsilon) < \varepsilon$ and the functions f_n converge to f uniformly on X_ε.*

PROOF. Let us extend μ to $\mathcal{A}_\mu$. By hypothesis there is a set $L \in \mathcal{A}_\mu$ such that $\mu(L) = \mu(X)$ and on L all functions f_n are defined, finite and converge to f. Hence our assertion reduces to the case where $\{f_n(x)\}$ converges at every point. Then

$$X_n^m := \bigcap_{i \geqslant n} \Big\{ x\colon |f_i(x) - f(x)| < \frac{1}{m} \Big\} \in \mathcal{A}_\mu.$$

For all m, $n \in \mathbb{N}$ we have $X_n^m \subset X_{n+1}^m$, moreover, $\bigcup_{n=1}^\infty X_n^m = X$, since for every fixed m for every x there exists a number n such that $|f_i(x) - f(x)| < 1/m$ whenever $i \geqslant n$. Let $\varepsilon > 0$. By the countable additivity of μ, for every m there exists a number $k(m)$ with $\mu(X\backslash X_{k(m)}^m) < \varepsilon 2^{-m}$. Let $X_\varepsilon := \bigcap_{m=1}^\infty X_{k(m)}^m$. Then $X_\varepsilon \in \mathcal{A}_\mu$ and

$$\mu(X\backslash X_\varepsilon) = \mu\Big(\bigcup_{m=1}^\infty (X\backslash X_{k(m)}^m)\Big) \leqslant \sum_{m=1}^\infty \mu(X\backslash X_{k(m)}^m) < \varepsilon \sum_{m=1}^\infty 2^{-m}.$$

Finally, for any fixed m we have $|f_i(x) - f(x)| < 1/m$ for all $x \in X_\varepsilon$ and all $i \geqslant k(m)$, which means the uniform convergence of $\{f_n\}$ to f on the set X_ε. It remains to take in X_ε a subset from $\mathcal{A}$ of equal measure, which we denote by the same symbol.   □

Egorov's theorem does not extend to the case $\varepsilon = 0$. For example, the functions $f_n\colon x \mapsto x^n$ on $(0,1)$ converge at every point to zero, but do not converge uniformly on a set $E \subset (0,1)$ of Lebesgue measure 1, since in every neighborhood of the point 1 there are points from E and hence $\sup_{x\in E} f_n(x) = 1$ for every n. The property of convergence established by Egorov is called the *almost uniform convergence*.

Let us proceed to one more important mode of convergence of measurable functions.

3.2.2. Definition. *Suppose we are given a space $(X, \mathcal{A}, \mu)$ with a finite measure μ and a sequence of μ-measurable functions f_n, where μ is extended to $\mathcal{A}_\mu$.*

(i) *A sequence $\{f_n\}$ is called Cauchy in measure if for every $c > 0$ we have*

$$\lim_{N\to\infty} \sup_{n,k \geqslant N} \mu\big(x\colon |f_n(x) - f_k(x)| \geqslant c\big) = 0.$$

(ii) *The sequence $\{f_n\}$ converges in measure to a μ-measurable function f if for every $c > 0$ we have*

$$\lim_{n\to\infty} \mu\big(x\colon |f(x) - f_n(x)| \geqslant c\big) = 0.$$

In the case of an infinite measure μ the definition of convergence in measure is modified as follows: we require convergence in measure on every set of positive measure. This is a weaker and more convenient condition than the straightforward analog of the definition above.

One can verify that convergence in measure is generated by a metric (see Exercise 3.12.29).

Every sequence of measurable functions f_n converging in measure is Cauchy in measure. This follows from the fact that the set $\{x\colon |f_n(x) - f_k(x)| \geqslant c\}$ is contained in the set

$$\{x\colon |f(x) - f_n(x)| \geqslant c/2\} \cup \{x\colon |f(x) - f_k(x)| \geqslant c/2\}.$$

Note also that if a sequence $\{f_n\}$ converges in measure to functions f and g, then $f = g$ almost everywhere , i.e., up to a redefinition of functions on sets of measure zero only one limit can exist in the sense of convergence in measure. Indeed, for every $c > 0$ we have

$$\mu\big(x\colon |f(x) - g(x)| \geqslant c\big) \leqslant \mu\big(x\colon |f(x) - f_n(x)| \geqslant c/2\big)$$
$$+ \mu\big(x\colon |f_n(x) - g(x)| \geqslant c/2\big) \to 0,$$

whence $\mu\big(x\colon |f(x) - g(x)| > 0\big) = 0$, since the set of points where the function $|f - g|$ is positive is the union of the sets of points where it is at least n^{-1}.

We now investigate the connection between convergence in measure and convergence almost everywhere.

3.2.3. Theorem. *Let $(X, \mathcal{A}, \mu)$ be a space with a finite measure. If μ-measurable functions f_n converge almost everywhere to a function f, then they converge to f in measure.*

PROOF. Let $A_n = \bigcap_{i=n}^{\infty}\{x\colon |f(x) - f_i(x)| < c\}$, where $c > 0$. The sets A_n are μ-measurable and $A_n \subset A_{n+1}$. It is clear that the set $\bigcup_{n=1}^{\infty} A_n$ contains all points at which $\{f_n\}$ converges to f. Hence $\mu(X) = \mu(\bigcup_{n=1}^{\infty} A_n)$. By the countable additivity of μ we obtain $\mu(A_n) \to \mu(X)$, i.e., $\mu(X \backslash A_n) \to 0$. It remains to observe that $\{x\colon |f(x) - f_n(x)| \geqslant c\} \subset X \backslash A_n$. $\square$

The converse assertion to this theorem is false: there exists a sequence of measurable functions on $[0, 1]$ which converges to zero in Lebesgue measure, but converges at no point.

3.2.4. Example. For every $n \in \mathbb{N}$, let us partition $[0, 1]$ into 2^n intervals $I_{n,k} = [(k-1)2^{-n}, k2^{-n})$, $k = 1, \ldots, 2^n$, of length 2^{-n}. Set $f_{n,k}(x) = 1$ if $x \in I_{n,k}$ and $f_{n,k}(x) = 0$ if $x \notin I_{n,k}$. We arrange $f_{n,k}$ in a single sequence

$$f_n = (f_{1,1}, f_{1,2}, f_{2,1}, f_{2,2}, \ldots),$$

in which $f_{n+1,k}$ follows the functions $f_{n,j}$. The sequence $\{f_n\}$ converges to zero in Lebesgue measure, since the length of the interval on which the function f_n differs from zero tends to zero as n is growing. However, convergence fails at each point x, since for every fixed x the sequence $\{f_n(x)\}$ contains infinitely many zeros and ones. One can also take $f_n = I_{[r_n, r_{n+1}]}$ with a suitable enumeration of rational numbers r_n in $[0, 1]$ such that $r_{n+1} - r_n \to 0$.

The following result due to F. Riesz gives a partial converse to Theorem 3.2.3.

3.2.5. Theorem. (THE RIESZ THEOREM) *Let $(X, \mathcal{A}, \mu)$ be a space with a finite measure.*

(i) *If a sequence of μ-measurable functions f_n converges to a function f in measure μ, then there exists its subsequence $\{f_{n_k}\}$ which converges to f almost everywhere.*

(ii) *If a sequence of μ-measurable functions f_n is Cauchy in measure μ, then it converges in measure μ to some measurable function f.*

PROOF. Let us extend μ to $\mathcal{A}_\mu$. Let $\{f_n\}$ be Cauchy in measure. We show that there exist natural numbers n_k increasing to infinity for which

$$\mu\big(x\colon\ |f_n(x) - f_j(x)| \geq 2^{-k}\big) \leq 2^{-k} \quad \forall n, j \geq n_k.$$

Indeed, let us find a number n_1 for which

$$\mu\big(x\colon\ |f_n(x) - f_j(x)| \geq 2^{-1}\big) \leq 2^{-1} \quad \forall n, j \geq n_1.$$

Next, we find a number $n_2 > n_1$ for which

$$\mu\big(x\colon\ |f_n(x) - f_j(x)| \geq 2^{-2}\big) \leq 2^{-2} \quad \forall n, j \geq n_2.$$

Continuing this process inductively, we obtain a desired sequence $\{n_k\}$. Let us show that the sequence $\{f_{n_k}\}$ converges a.e. For this it suffices to prove that it is Cauchy a.e. Set

$$E_j = \big\{x\colon\ |f_{n_{j+1}}(x) - f_{n_j}(x)| \geq 2^{-j}\big\}.$$

Since $\mu\big(\bigcup_{j=k}^{\infty} E_j\big) \leq \sum_{j=k}^{\infty} 2^{-j} = 2^{-k+1} \to 0$ as $k \to \infty$, the measurable set $Z = \bigcap_{k=1}^{\infty} \bigcup_{j=k}^{\infty} E_j$ has μ-measure zero. If $x \in X \backslash Z$, then the sequence $\{f_{n_k}(x)\}$ is Cauchy. Indeed, there exists k with $x \notin \bigcup_{j=k}^{\infty} E_j$, i.e., $x \notin E_j$ for all $j \geq k$. This means by definition that $|f_{n_{j+1}}(x) - f_{n_j}(x)| < 2^{-j}$ for all $j \geq k$. Therefore, for every fixed $m \geq k$ for all $i > j > m$ we have

$$|f_{n_i}(x) - f_{n_j}(x)| \leq |f_{n_i}(x) - f_{n_{i-1}}(x)| + |f_{n_{i-1}}(x) - f_{n_{i-2}}(x)| + \cdots$$
$$+ |f_{n_{j+1}}(x) - f_{n_j}(x)| \leq 2^{-j} + 2^{-j-1} + \cdots \leq 2^{-j+1} \leq 2^{-m},$$

which means that $\{f_{n_k}(x)\}$ is Cauchy. Thus, $\{f_{n_k}\}$ converges almost everywhere to some function f. Then $\{f_{n_k}\}$ converges to f in measure, which gives assertion (ii) due to the inclusion of the set $\{x\colon |f(x) - f_n(x)| > c\}$ into the union of the sets $\{x\colon |f(x) - f_{n_k}(x)| > c\}$ and $\{x\colon |f_{n_k}(x) - f_n(x)| > c\}$ and the fact that $\{f_n\}$ is Cauchy in measure. Assertion (i) follows from the property that any sequence converging in measure is Cauchy in measure and the limit of a subsequence converging almost everywhere coincides almost everywhere with the limit of $\{f_n\}$ in measure by the uniqueness of the limit in measure up to equivalence. $\qquad\square$

3.2.6. Corollary. *Let μ be a finite measure. A sequence of μ-measurable functions converges in measure μ precisely when every subsequence of this sequence has a further subsequence converging μ-a.e.*

3.2.7. Corollary. *Let μ be a finite measure and let two sequences of μ-measurable functions f_n and g_n converge in measure μ to functions f and g, respectively. Suppose that Ψ is a continuous function on some set $Y \subset \mathbb{R}^2$ such* *that* $\big(f(x), g(x)\big) \in Y$ *and* $\big(f_n(x), g_n(x)\big) \in Y$ *for all* x *and all* n. *Then the functions* $\Psi(f_n, g_n)$ *converge in measure* μ *to the function* $\Psi(f, g)$. *In particular,* $f_n g_n \to fg$ *and* $\alpha f_n + \beta g_n \to \alpha f + \beta g$ *in measure* μ *for all* $\alpha, \beta \in \mathbb{R}^1$.

PROOF. Note that the functions $\Psi(f, g)$ and $\Psi(f_n, g_n)$ are measurable by Proposition 3.1.7. If our assertion is false, then there exist $c > 0$ and a subsequence j_n such that

$$\mu\Big(x\colon \big|\Psi\big(f(x), g(x)\big) - \Psi\big(f_{j_n}(x), g_{j_n}(x)\big)\big| > c\Big) > c \qquad (3.2.1)$$

for all n. By the Riesz theorem $\{j_n\}$ contains a subsequence $\{i_n\}$ for which $f_{i_n}(x) \to f(x)$ and $g_{i_n}(x) \to g(x)$ a.e. By the continuity of Ψ we obtain $\Psi\big(f_{i_n}(x), g_{i_n}(x)\big) \to \Psi\big(f(x), g(x)\big)$ a.e., whence $\Psi(f_{i_n}, g_{i_n}) \to \Psi(f, g)$ in measure, which contradicts (3.2.1). The remaining assertions follow from the established one applied to the functions $\Psi(x, y) = xy$ and $\Psi(x, y) = \alpha x + \beta y$. $\square$

3.2.8. Lemma. *Let* K *be a compact set on the real line, let* U *be an open set containing* K, *and let* f *be a continuous function on* K. *There is a continuous function* g *on the real line such that* $g = f$ *on* K, $g = 0$ *outside* U, *and*

$$\sup_{x \in \mathbb{R}} |g(x)| = \sup_{x \in K} |f(x)|.$$

PROOF. It suffices to consider the case where U is bounded. The set $U \backslash K$ is a finite or countable union of pairwise disjoint open intervals. Set $g = 0$ outside U, $g = f$ on K, and on every interval (a, b) constituting $U \backslash K$ we define g by the linear interpolation: $g\big(ta + (1 - t)b\big) = tg(a) + (1 - t)g(b)$. $\square$

The structure of Lebesgue measurable functions on an interval is clarified by the following classical result due to N. N. Luzin.

3.2.9. Theorem. (LUZIN'S THEOREM) *A function* f *on an interval* I *is Lebesgue measurable precisely when for every* $\varepsilon > 0$ *there exists a continuous function* f_ε *and a compact set* K_ε *such that* $\lambda(I \backslash K_\varepsilon) < \varepsilon$ *and* $f = f_\varepsilon$ *on* K_ε.

PROOF. If the indicated condition holds, then the set $\{x\colon f(x) < c\}$ up to a set of measure zero coincides with the Borel set $\bigcup_{n=1}^{\infty} \{x \in K_{1/n}\colon f_{1/n}(x) < c\}$. Let us verify the necessity of this condition. We can assume that the function f is bounded, passing to $\arctan f$. Let us take a sequence of simple functions f_n uniformly converging to f. The function f_n assumes k values on disjoint measurable sets $A_{n,1}, \ldots, A_{n,k}$. Each $A_{n,i}$ contains a compact set $K_{n,i}$ such that the sum of the measures of $A_{n,i} \backslash K_{n,i}$ will be less than $\varepsilon 2^{-n}$. The restriction of f_n to the compact set $K_n := K_{n,1} \cup \cdots \cup K_{n,k}$ is continuous by the disjointness of $K_{n,i}$. Hence the restriction of f to the compact set $K_\varepsilon := \bigcap_{n=1}^{\infty} K_n$ is continuous (as a uniform limit of functions continuous on K_ε). By the lemma above f can be extended from K_ε to a continuous function f_ε on I. It remains to observe that the measure of the complement of K_ε is not greater than the sum of the series of the measures of the complements of K_n, which does not exceed $\sum_{n=1}^{\infty} \varepsilon 2^{-n} = \varepsilon$. $\square$

Note that this theorem along with its proof remains valid for an arbitrary bounded Borel measure on an interval.

3.2.10. Corollary. *For each Lebesgue measurable function f on an interval, there exists a sequence of continuous functions f_n converging to f in measure.*

3.3. The Construction of the Lebesgue Integral

There are several equivalent definitions of the Lebesgue integral. We use the one in which the integral is first defined for simple functions and then the integral of a nonnegative function is defined as the supremum of the integrals of simple functions dominated by this function. In §3.6 we establish the equivalence of this definition to several other ones, including the approach of Lebesgue himself.

Let $(X, \mathcal{A}, \mu)$ be a measurable space with a nonnegative measure (possibly, with values in $[0, +\infty]$). As above, let

$$f^+ = \max(f, 0), \quad f^- = \max(-f, 0).$$

Let f be a simple nonnegative function with values $c_i \geq 0$ on pairwise disjoint sets A_i, where $i = 1, \ldots, n$. Set

$$\int_X f(x)\, \mu(dx) := \sum_{i=1}^{n} c_i \mu(A_i),$$

where $0 \cdot \mu\big(x \colon f(x) = 0\big) := 0$ and we admit the value $+\infty$. The additivity of the measure shows that this value is well-defined, since we can pass to the case where all c_i are different.

For the definition of the integral it is convenient to use the extended concept of a measurable function from Definition 3.1.10 and admit functions which are defined almost everywhere (i.e., possibly, on a set of measure zero are not defined or assume an infinite value).

It is clear from the additivity of measure that the integral on nonnegative simple functions is additive and nonnegative. In addition, the integral of $f \geq 0$ is bounded by $\mu(X) \sup_x f(x)$ if μ is finite.

3.3.1. Definition. *Let f be a μ-measurable function and $f(x) \geq 0$ μ-a.e. Set*

$$\int_X f(x)\, \mu(dx) := \sup\Big\{ \int_X \varphi(x)\, \mu(dx) :$$

$$\varphi \geq 0 \text{ is a simple function and } \varphi(x) \leq f(x)\ \mu\text{-a.e.} \Big\}.$$

We shall call f integrable if this quantity is finite.

In the general case of a signed function we shall call f Lebesgue integrable with respect to μ, or μ-integrable, if both functions f^+ and f^- are integrable. Then we set

$$\int_X f(x)\, \mu(dx) := \int_X f^+(x)\, \mu(dx) - \int_X f^-(x)\, \mu(dx).$$

Below we define integrals over subsets, but in the case where integration is taken over the whole space X (as in the definition above), the domain of integration is sometimes not indicated and one uses the notation $\int f\, d\mu$. In the case of Lebesgue measure on $\mathbb{R}^n$ we also denote the integral of the function f by the symbol $\int f(x)\,dx$.

It is readily seen that for simple nonnegative functions this definition gives the previously defined value of the integral.

An important property of the Lebesgue integral is that, by the very definition, every function equivalent to an integrable one is also integrable. In addition, a measurable function f is integrable if and only if the function $|f|$ is integrable, moreover,

$$\left| \int_X f\,d\mu \right| \leqslant \int_X |f|\,d\mu.$$

It is also obvious from the definition that if f and g are μ-measurable functions such that $|g(x)| \leqslant |f(x)|$ μ-a.e. and f is integrable, then g is also integrable and

$$\int_X |g|\,d\mu \leqslant \int_X |f|\,d\mu. \tag{3.3.1}$$

In particular, if the function f is integrable, then for every measurable set A the function fI_A is also integrable. The integral of the function fI_A will be called the *integral of f over the set A* and denoted by one of the following symbols:

$$\int_A f(x)\,\mu(dx) \quad \text{and} \quad \int_A f\,d\mu.$$

If $f(x) \geqslant 0$ for μ-a.e. $x \in A$, then we have $\int_A f\,d\mu \geqslant 0$.

Of course, the integral of fI_A over X is the same as the integral of f over A regarded as a separate measurable space equipped with the restriction of μ.

If $\int_X |f|\,d\mu = 0$, then $f = 0$ a.e., since otherwise for some $n \in \mathbb{N}$ we obtain that $\mu(x\colon |f(x)| \geqslant n^{-1}) = c > 0$, so the integral of $|f|$ is at least cn^{-1}.

A simple corollary of the definition is the following frequently used inequality due to P. L. Chebyshev.

3.3.2. Theorem. (THE CHEBYSHEV INEQUALITY) *For every μ-integrable function f and every $R > 0$ we have*

$$\mu(x\colon |f(x)| \geqslant R) \leqslant \frac{1}{R} \int_X |f(x)|\,\mu(dx). \tag{3.3.2}$$

PROOF. Set $A_R = \{x\colon |f(x)| \geqslant R\}$. It is clear that $R \cdot I_{A_R}(x) \leqslant |f(x)|$ for a.e. x. Therefore, the integral of the function $R \cdot I_{A_R}$ is dominated by the integral of $|f|$, which gives (3.3.2). $\square$

3.3.3. Corollary. *For every μ-integrable function f there exists a sequence of measurable sets A_n of finite measure such that $f(x) = 0$ whenever $x \notin \bigcup_{n=1}^{\infty} A_n$.*

PROOF. The sets $\{x\colon |f(x)| \geqslant 1/n\}$ have finite measure. $\square$

We now show that for evaluation of the integral of a nonnegative measurable function it suffices to use an arbitrary sequence of simple functions increasing to it. This is not completely obvious from the definition.

3.3.4. Lemma. *Let $\{f_n\}$ be a sequence of simple functions $f_n \geqslant 0$ such that $f_n(x) \leqslant f_{n+1}(x)$ μ-a.e. for all n and the integrals of f_n are uniformly bounded. Then the function $f(x) := \lim\limits_{n \to \infty} f_n(x)$ is finite μ-a.e. and integrable, moreover,*

$$\int_X f \, d\mu = \lim_{n \to \infty} \int_X f_n \, d\mu.$$

PROOF. The integrals of f_n increase to a finite limit L. For every $R > 0$ by the Chebyshev inequality we have $\mu(x\colon f_n(x) > R) \leqslant L/R$. As $n \to \infty$ the left-hand side increases to $\mu(x\colon f(x) > R)$, since the sets $\{x\colon f_n(x) > R\}$ increase and their union is $\{x\colon f(x) > R\}$ up to a set of measure zero (on which $\{f_n(x)\}$ is not increasing). Hence $\mu(x\colon f(x) > R) \leqslant L/R$ for all $R > 0$, whence the equality $\mu(x\colon f(x) = +\infty) = 0$ follows.

Let $\varphi \geqslant 0$ be a simple function with $\varphi(x) \leqslant f(x)$ μ-a.e. We show that the integral of φ does not exceed P. The functions $\varphi_n := \min(f_n, \varphi)$ are simple and μ-a.e. increase to φ. Since the integral of φ_n is not greater than the integral of f_n, it suffices to show that the integrals of φ_n increase to the integral of φ. By the additivity of the integral on simple functions it remains to show that the integrals of $\varphi_n I_E$ converge to the integral of φI_E for every set E of the form $E = \varphi^{-1}(c)$. For $c = 0$ this is obvious. Let $c > 0$ and $\varepsilon \in (0, c/2)$. Deleting a set of measure zero we can assume that the sets $E_n := \{x \in E\colon \varphi_n(x) \geqslant c - \varepsilon\}$ increase and cover E, whence $\mu(E_n) \to \mu(E)$. This yields that $\mu(E) < \infty$, because

$$\int_E \varphi_n \, d\mu \geqslant (c - \varepsilon)\mu(E_n) \to (c - \varepsilon)\mu(E),$$

and the left-hand side is not greater than L. Since ε can be taken as small as we wish, the integrals of $\varphi_n I_E$ converge to the integral of φI_E. $\qquad \square$

If $f \geqslant 0$ is a μ-measurable function, then Corollary 3.1.9 gives a sequence of simple functions f_n increasing to f μ-a.e. (these functions are integrable when f is integrable).

Let us find conditions for the integrability of a function with countably many values.

3.3.5. Example. Let f be a μ-measurable function with countably many different values c_n. The integrability of f is equivalent to convergence of the series $\sum_{n=1}^{\infty} |c_n| \mu(x\colon f(x) = c_n)$, moreover, in case of integrability

$$\int_X f(x) \, \mu(dx) = \sum_{n=1}^{\infty} c_n \mu(x\colon f(x) = c_n).$$

Indeed, it suffices to consider the case where $c_n \geqslant 0$. Let us enumerate c_n in the order of increasing. Let $f_n(x) = f(x)$ whenever $f(x) \leqslant c_n$ and $f_n(x) = 0$ whenever $f(x) > c_n$. Then f_n are simple functions increasing to f. Since the integral of f_n equals $\sum_{i=1}^{n} c_i \mu(x\colon f(x) = c_i)$, it remains to apply the proposition above.

The next theorem establishes one of the most important elementary properties of the integral — the linearity.

3.3.6. Theorem. *If functions f and g are integrable with respect to the measure μ, then, for all numbers α and β, the function $\alpha f + \beta g$ is also μ-integrable and*

$$\int_X (\alpha f + \beta g)\, d\mu = \alpha \int_X f\, d\mu + \beta \int_X g\, d\mu.$$

PROOF. For simple functions this equality follows from the additivity of measure. The case $\beta = 0$ is obvious from the definition. Now it suffices to consider the case $\alpha = \beta = 1$. Then $f + g = f^+ + g^+ - (f^- + g^-)$. Hence it suffices to prove our assertion in two cases: when $f \geqslant 0$ and $g \geqslant 0$ and when $f \geqslant 0$ and $g \leqslant 0$. The first case follows from the validity of this theorem for simple functions and Lemma 3.3.4, because taking sequences of simple nonnegative functions f_n and g_n a.e. increasing to f and g, respectively, we obtain a sequence of simple functions $f_n + g_n$ that increases a.e. to $f + g$. In the remaining case $f \geqslant 0$ and $g \leqslant 0$ we observe that $|f + g| \leqslant |f| + |g|$ and hence the function $f + g$ is integrable. In addition, $f + g = (f + g)I_E + (f + g)I_{X \setminus E}$, where $E = \{x \colon f(x) + g(x) \geqslant 0\}$. Since $fI_E = (f + g)I_E - gI_E$ and $-gI_{X \setminus E} = fI_{X \setminus E} + (-f - g)I_{X \setminus E}$, where all terms are nonnegative (all these relations are fulfilled a.e.), by the case proved above we have

$$\int_X fI_E\, d\mu = \int_X (f + g)I_E\, d\mu - \int_X gI_E\, d\mu,$$

$$-\int_X gI_{X \setminus E}\, d\mu = \int_X fI_{X \setminus E}\, d\mu - \int_X (f + g)I_{X \setminus E}\, d\mu.$$

From these equalities on account of the identity $I_E + I_{X \setminus E} = 1$ we obtain the desired equality. $\square$

In particular, if the function f is integrable and A and B are disjoint sets from $\mathcal{A}_\mu$, then

$$\int_{A \cup B} f(x)\, \mu(dx) = \int_A f(x)\, \mu(dx) + \int_B f(x)\, \mu(dx).$$

3.3.7. Proposition. *Every $\mathcal{A}_\mu$-measurable bounded function f is integrable on every set $A \in \mathcal{A}_\mu$ of finite measure and*

$$\left| \int_A f(x)\, \mu(dx) \right| \leqslant \sup_{x \in A} |f(x)|\, \mu(A).$$

PROOF. If φ is a nonnegative simple function such that $\varphi(x) \leqslant |f(x)|$ for a.e. $x \in A$ assuming different values $c_1, \ldots, c_n$; then the integral of φI_A equals

$$\sum_{i=1}^n c_i \mu\big(A \cap \varphi^{-1}(c_i)\big) \leqslant \max_i c_i \sum_{i=1}^n \mu\big(A \cap \varphi^{-1}(c_i)\big) = \max_i c_i \mu(A),$$

which does not exceed $\sup_{x \in A} |f(x)|\, \mu(A)$. $\square$

The next theorem establishes a very important property of the *absolute continuity of the Lebesgue integral*.

3.3.8. Theorem. *Let f be a μ-integrable function. Then, for every $\varepsilon > 0$, there exists $\delta > 0$ such that*

$$\int_A |f(x)|\, \mu(dx) < \varepsilon \quad \text{if } \mu(A) < \delta.$$

PROOF. There is a simple function $\varphi \geqslant 0$ such that $\varphi \leqslant |f|$ a.e. and the difference of the integrals of the functions $|f|$ and φ is less than $\varepsilon/2$. Let $\delta = \varepsilon(2\max_x \varphi(x) + 2)^{-1}$. Whenever $\mu(A) < \delta$, we obtain

$$\int_A |f|\, d\mu = \int_A \varphi\, d\mu + \int_A [|f| - \varphi]\, d\mu$$

$$\leqslant \sup_x \varphi(x)\mu(A) + \int_X [|f| - \varphi]\, d\mu \leqslant \frac{\varepsilon}{2} + \frac{\varepsilon}{2} = \varepsilon,$$

as required. $\square$

3.3.9. Remark. We do not assume the completeness of the measure μ, but it is clear that for $\mathcal{A}$ one can take $\mathcal{A}_\mu$. Replacing $\mathcal{A}$ by $\mathcal{A}_\mu$ we obtain the same class of integrable functions, which is obvious from the definition. However, if in place of the σ-algebra $\mathcal{A}$ we take its subalgebra $\mathcal{A}_0$ (even generating $\mathcal{A}$) and in the definition of the integral consider only simple functions corresponding to sets from $\mathcal{A}_0$, then the supply of integrable functions can enlarge so greatly that will be meaningless. For example, let $\mathcal{A}_0$ be the algebra of finite unions of intervals in $[0, 1]$. Let us take a set $E \subset [0, 1]$ such that both E and its complement intersect every nonempty interval by a set of positive measure (see Exercise 2.7.18). Then every measurable nonnegative function f vanishing on E will obtain the zero integral with this class of "simple" functions (in this class only a.e. zero functions φ satisfy the condition $\varphi \leqslant f$ a.e.), although in the sense of our usual definition the integral of f is infinite once it is infinite over the complement of E.

3.3.10. Remark. The integrability of a function f with respect to a signed measure $\mu = \mu^+ - \mu^-$ is defined as its integrability with respect to μ^+ and μ^- (and its integral is the difference of these integrals). The measurability and integrability of a complex function f will be understood, respectively, as the measurability and integrability of the real part $\mathrm{Re}\, f$ and the imaginary part $\mathrm{Im}\, f$ of f. Set

$$\int f\, d\mu := \int \mathrm{Re}\, f\, d\mu + i \int \mathrm{Im}\, f\, d\mu.$$

For mappings with values in $\mathrm{I\!R}^n$ the measurability and integrability will be understood similarly, i.e., component-wise. Thus, the integral of the vector mapping $f = (f_1, \ldots, f_n)$ with integrable components f_i is the vector the coordinates of which are the integrals of f_i.

3.3.11. Remark. For the integral with respect to the classical Lebesgue measure on $\mathrm{I\!R}^n$ the change of variables formula

$$\int_{\mathrm{I\!R}^n} f(Tx + a)\, dx = |\det T|^{-1} \int_{\mathrm{I\!R}^n} f(y)\, dy$$

holds for every invertible linear operator T and every $a \in \mathrm{I\!R}^n$. In view of Corollary 2.5.4, for justification it suffices to verify this formula for simple functions with bounded support, which reduces everything to indicator functions of bounded sets and further to indicator functions of cubes. Below in §3.12(ii) we discuss more general change of variables formulas for Lebesgue measure and for abstract measures.

Let us say a few words about the Lebesgue–Stieltjes integral. In Chapter 2 we defined Lebesgue–Stieltjes measures on the real line: to every left continuous nondecreasing function F with limit 0 on $-\infty$ and limit 1 on $+\infty$ a probability Borel measure μ was associated such that $F(t) = \mu\big((-\infty, t)\big)$. Let g be a μ-integrable function. We shall define the *Lebesgue–Stieltjes integral* of the function g with respect to the function F as the number

$$\int_{-\infty}^{\infty} g(t)\, dF(t) := \int_{\mathbb{R}} g(t)\, \mu(dt). \tag{3.3.3}$$

This definition can be easily extended to include all functions F of the form $F = c_1 F_1 + c_2 F_2$, where F_1, F_2 are the distribution functions of probability measures μ_1 and μ_2 and c_1, c_2 are constants. Then for μ we take the measure $c_1\mu_1 + c_2\mu_2$. Similarly one defines the Lebesgue–Stieltjes integral over a compact or closed interval. In some applications the distribution function F is known, but not the measure μ itself, so the notation for the integral by means of the left-hand side of (3.3.3) becomes useful and visual in calculations. If g assumes finitely many values c_i on intervals $[a_i, b_i)$ and vanishes outside these intervals, then

$$\int g(t)\, dF(t) = \sum_{i=1}^{n} c_i [F(b_i) - F(a_i)].$$

For a continuous function g on a compact interval $[a, b]$, the Lebesgue–Stieltjes integral can be obtained as the limit of such Riemann type sums with $c_i = g(a_i)$, in which also $b_n > b$ (or $F(b_n) = F(b+)$) is taken if b is an atom of μ. Usually in courses of analysis this is the definition of the Stieltjes integral of a continuous function. Indeed, g is uniformly continuous on $[a, b]$, so, for every $\varepsilon > 0$, there is $\delta > 0$ such that for any points $a = t_0 < t_1 < \cdots < t_n = b$ with $|t_i - t_{i-1}| \leqslant \delta$ the step function h equal to $g(t_{i-1})$ on $[t_{i-1}, t_i)$ differs from g by at most ε. Hence the difference of the integrals of g and h against the measure μ does not exceed $\varepsilon \|\mu\|$ in absolute value. Note that with the aid of Riemann type sums one can define the Stieltjes integral of a continuous function g with respect to an increasing function F that is not necessarily left continuous. For this it suffices to verify that when we refine partitions, the corresponding sums will converge to the same value that is obtained if F is redefined on the countable set of discontinuity points to make it left continuous. The Lebesgue–Stieltjes integral is also considered in the next chapter in connection with functions of bounded variation.

3.4. Passage to the Limit under the Integral Sign

In this section we present three main theorems on convergence of integrable functions bearing the names of Lebesgue, Beppo Levi and Fatou. As usual, we suppose that μ is a nonnegative measure (possibly, with values in $[0, +\infty]$) on a measurable space $(X, \mathcal{A})$.

Our next important result is the Beppo Levi monotone convergence theorem.

3.4.1. Theorem. (THE BEPPO LEVI THEOREM) *Let* $\{f_n\}$ *be a sequence of* μ*-integrable functions such that* $f_n(x) \leqslant f_{n+1}(x)$ *a.e. for every* n. *Suppose that*

$$\sup_n \int_X f_n(x)\,\mu(dx) < \infty.$$

Then the function $f(x) = \lim\limits_{n\to\infty} f_n(x)$ *is almost everywhere finite, integrable and*

$$\int_X f(x)\,\mu(dx) = \lim_{n\to\infty} \int_X f_n(x)\,\mu(dx). \tag{3.4.1}$$

PROOF. Subtracting f_1 from all functions we arrive at the case $f_n \geqslant 0$. The integrals of f_n increase to some number J. For every n, there is a sequence of simple functions $\varphi_{n,k} \geqslant 0$ increasing to f_n a.e. as $k \to \infty$. The functions $g_n := \max(\varphi_{1,n}, \ldots, \varphi_{n,n})$ are simple, $g_n \leqslant g_{n+1}$, since $\varphi_{j,n} \leqslant \varphi_{j,n+1}$, and $g_n \leqslant f_n$, since $\varphi_{j,n} \leqslant f_j \leqslant f_n$ if $j \leqslant n$. Hence $g = \lim\limits_{n\to\infty} g_n \leqslant f$ a.e. On the other hand, $g \geqslant \varphi_{n,k}$ if $n \leqslant k$, whence $g \geqslant f_n$ a.e., so $g = f$ a.e. By Lemma 3.3.4 the function g is integrable and its integral equals the limit of the integrals of g_n, which does not exceed J. Hence the integral of f equals J. $\square$

Of course, in place of increasing of $f_n(x)$ to $f(x)$ one can require decreasing. The Beppo Levi theorem can be formulated as follows:

if the series of the integrals of nonnegative integrable functions f_n *converges, then the series* $\sum_{n=1}^{\infty} f_n(x)$ *converges almost everywhere and the integral of its sum equals the sum of the series of the integrals of* f_n.

The next frequently used result of the theory of integration is Fatou's theorem (sometimes it is called Fatou's lemma). It shows that in case of nonnegative functions f_n (or at least having a common integrable minorant) for the integrability of the limit function f it suffices to have the uniform boundedness of the integrals of f_n. However, in this case in place of equality (3.4.1) only some inequality is guaranteed.

3.4.2. Theorem. (FATOU'S THEOREM) *Let* $\{f_n\}$ *be a sequence of nonnegative* μ*-integrable functions which converge to the function* f *almost everywhere and let*

$$\sup_n \int_X f_n(x)\,\mu(dx) \leqslant K < \infty.$$

Then the function f *is* μ*-integrable and*

$$\int_X f(x)\,\mu(dx) \leqslant K.$$

Moreover,

$$\int_X f(x)\,\mu(dx) \leqslant \liminf_{n\to\infty} \int_X f_n(x)\,\mu(dx).$$

PROOF. Set $g_n(x) = \inf_{k\geqslant n} f_k(x)$. Then

$$0 \leqslant g_n \leqslant f_n, \quad g_n \leqslant g_{n+1}.$$

Hence the functions g_n are integrable and form a monotone sequence and their integrals are bounded from above by K. By the monotone convergence theorem, almost everywhere there exists a finite limit $g(x) = \lim\limits_{n\to\infty} g_n(x)$ and the function g is integrable, moreover, its integral equals the limit of the integrals of the functions g_n and does not exceed K. It remains to observe that $f(x) = g(x)$ a.e. due to convergence of $\{f_n(x)\}$ a.e. The second assertion of the theorem follows from the proved one by passing to a suitable subsequence. □

3.4.3. Corollary. *Let $\{f_n\}$ be a sequence of μ-integrable functions and let g be a μ-integrable function.*
  (i) *Let $f_n \geqslant g$ a.e. for every n and*

$$\liminf_n \int_X f_n(x)\,\mu(dx) < \infty.$$

Then the function $\liminf\limits_{n\to\infty} f_n$ is μ-integrable and

$$\int_X \liminf_{n\to\infty} f_n(x)\,\mu(dx) \leqslant \liminf_{n\to\infty} \int_X f_n(x)\,\mu(dx).$$

  (ii) *Let $f_n \leqslant g$ a.e. for every n and*

$$\limsup_n \int_X f_n(x)\,\mu(dx) > -\infty.$$

Then the function $\limsup\limits_{n\to\infty} f_n$ is μ-integrable and

$$\int_X \limsup_{n\to\infty} f_n(x)\,\mu(dx) \geqslant \limsup_{n\to\infty} \int_X f_n(x)\,\mu(dx).$$

PROOF. (i) Passing to $f_n - g$, we reduce the assertion to the case $f_n \geqslant 0$. We observe that $\liminf\limits_{n\to\infty} f_n(x) = \lim\limits_{k\to\infty} \inf_{n\geqslant k} f_n(x)$ and apply Fatou's theorem to the functions $\inf_{n\geqslant k} f_n(x)$. Assertion (ii) follows from (i) by passing to $-f_n$. □

The most important in the theory of integral is the following result due to Lebesgue.

3.4.4. Theorem. (THE LEBESGUE DOMINATED CONVERGENCE THEOREM) *Suppose that μ-integrable functions f_n converge almost everywhere to a function f. If there exists a μ-integrable function Φ such that*

$$|f_n(x)| \leqslant \Phi(x) \quad \text{a.e. for every } n,$$

then the function f is integrable and

$$\int_X f(x)\,\mu(dx) = \lim_{n\to\infty} \int_X f_n(x)\,\mu(dx).$$

In addition,

$$\lim_{n\to\infty} \int_X |f(x) - f_n(x)|\,\mu(dx) = 0.$$

PROOF. The first equality follows from the second one, and the latter follows from assertion (ii) of the previous corollary. Indeed, for $g_n := |f_n - f|$ a.e. we have $g_n \to 0$ and $0 \leqslant g_n \leqslant 2\Phi$. Another proof follows from Egorov's theorem combined with Theorem 3.3.8 applied to Φ (the details are left to the reader). $\square$

3.4.5. Theorem. *The Lebesgue and Fatou theorems remain valid if in place of convergence almost everywhere in their hypotheses we require convergence of $\{f_n\}$ to f in measure μ (in the case of an unbounded measure we require convergence in measure on every set of finite measure).*

PROOF. Let $\mu(X) < \infty$. Since $\{f_n\}$ has a subsequence converging to f almost everywhere, we obtain the analog of the Fatou theorem for convergence in measure and also the conclusion of the Lebesgue theorem for the selected subsequence. Then the assertion is true for the whole sequence $\{f_n\}$, since otherwise we could find a subsequence $\{f_{n_k}\}$ such that $\int_X |f_{n_k} - f|\, d\mu \geqslant c > 0$ for all k, but this is impossible by the already proved assertion, since $\{f_{n_k}\}$ contains a subsequence converging a.e. In the case of an unbounded measure, each function f_n vanishes outside some countable union of sets of finite measure. Hence there exists a countable collection of disjoint sets E_j of finite measure outside the union E of which all functions f_n are zero. Then $f = 0$ a.e. outside E. Now we can extract a subsequence in $\{f_n\}$ converging almost everywhere on E. $\square$

Consideration of functions $f_n(x) = nI_{(0,1/n]}(x)$, pointwise converging to zero on $(0, 1]$, shows that in the Lebesgue theorem one cannot omit the hypothesis of the existence of a common integrable majorant, and in the Fatou theorem it is not always possible to interchange the limit and integral. However, it can happen that the integrals of $|f_n - f|$ converge to zero without a common integrable majorant (Exercise 3.12.12(ii)).

With the aid of the Lebesgue dominated convergence theorem one proves the following assertion about continuity and differentiability of the integral with respect to a parameter.

3.4.6. Corollary. *Let μ be a nonnegative measure (possibly, with values in $[0, +\infty]$) on a space X and let $f \colon X \times (a, b) \to \mathbb{R}^1$ be a function such that for every $\alpha \in (a, b)$ the function $x \mapsto f(x, \alpha)$ is integrable.*

(i) Suppose that for a.e. x the function $\alpha \mapsto f(x, \alpha)$ is continuous and there exists a μ-integrable function Φ such that for every fixed α almost everywhere we have $|f(x, \alpha)| \leqslant \Phi(x)$. Then the function

$$J \colon \alpha \mapsto \int_X f(x, \alpha)\, \mu(dx)$$

is continuous.

(ii) Suppose that for a.e. x the function $\alpha \mapsto f(x, \alpha)$ is differentiable and there is a μ-integrable function Φ such that $|\partial f(x, \alpha)/\partial \alpha| \leqslant \Phi(x)$ for a.e. x for all α simultaneously. Then the function J is differentiable and

$$J'(\alpha) = \int_X \frac{\partial f(x, \alpha)}{\partial \alpha}\, \mu(dx).$$

PROOF. Assertion (i) is obvious from the Lebesgue theorem. (ii) Let α be fixed and $t_n \to 0$. Then by the mean value theorem for a.e. x there exists $\xi = \xi(x, \alpha, n)$ such that

$$\left| t_n^{-1}\big(f(x, \alpha + t_n) - f(x, \alpha)\big) \right| = |\partial f(x, \xi)/\partial \alpha| \leqslant \Phi(x).$$

The indicated difference quotient converges to $\partial f(x, \alpha)/\partial \alpha$. By the Lebesgue theorem $\lim\limits_{n \to \infty} t_n^{-1}\big(J(\alpha + t_n) - J(\alpha)\big)$ equals the integral of $\partial f(x, \alpha)/\partial \alpha$. $\square$

In Exercise 3.12.21 one can find a modification of assertion (ii) that ensures the differentiability at a given point.

Note that we deduced the Lebesgue theorem from the monotone convergence theorem, but if we first derive the Lebesgue theorem from Egorov's theorem as mentioned above, the monotone convergence theorem becomes a corollary (by applying the Lebesgue theorem to the functions $\min(f_n, N)$ for fixed N).

3.5. The Space L^1

Let $(X, \mathcal{A}, \mu)$ be a measurable space with a nonnegative measure (possibly, with values in $[0, +\infty]$).

The collection of all μ-integrable functions will be denoted by $\mathcal{L}^1(\mu)$.

It is often useful to identify functions that are equal almost everywhere (we recall that such functions are called equivalent). For this in place of the space $\mathcal{L}^1(\mu)$ the space $L^1(\mu)$ is considered (an alternative notation: $L^1(X, \mu)$) the elements of which are equivalence classes in $\mathcal{L}^1(\mu)$ consisting of almost everywhere equal functions. Thus, $L^1(\mu)$ is the part of $L^0(\mu)$ corresponding to the classes with integrable representatives.

Here we prove that the space of Lebesgue integrable functions possesses the property of completeness, i.e., all sequences fundamental in mean converge in mean (this important property fails for the Riemann integral, see Exercise 3.12.25).

3.5.1. Definition. (i) *A sequence of functions f_n integrable with respect to a measure μ (possibly, with values in $[0, +\infty]$) is called Cauchy, or fundamental, in mean if, for every $\varepsilon > 0$, there exists a number N such that*

$$\int_X |f_n(x) - f_k(x)|\, \mu(dx) < \varepsilon \quad \forall\, n, k \geqslant N.$$

(ii) *We shall say that a sequence of μ-integrable functions f_n converges to a μ-integrable function f in mean if*

$$\lim_{n \to \infty} \int_X |f(x) - f_n(x)|\, \mu(dx) = 0.$$

Sequences that are fundamental or convergent in mean will be also called fundamental or convergent in $L^1(\mu)$, respectively.

These properties are just the respective properties in the natural metric of the space $L^1(\mu)$ given by the formula

$$d(f, g) = \int_X |f(x) - g(x)|\, \mu(dx),$$

where in the right-hand side arbitrary representatives of the equivalence classes of f and g are meant (it is obvious that the value of the integral is independent of their choice). The triangle inequality is obvious from the pointwise estimate $|f(x) - h(x)| \leqslant |f(x) - g(x)| + |g(x) - h(x)|$. The same formula defines a semimetric on $\mathcal{L}^1(\mu)$. The passage to the space of equivalence classes is needed to guarantee that the condition $d(f, g) = 0$ implies the equality $f = g$. Let us also observe that the indicated metric corresponds to the norm

$$\|f\|_{L^1(\mu)} := \int_X |f(x)|\, \mu(dx)$$

on $L^1(\mu)$, about which more is said in Chapter 5.

It is readily seen that if a sequence of integrable functions f_n is Cauchy in mean, then $\sup_n \|f_n\|_{L^1(\mu)} < \infty$. In addition, if a sequence of integrable functions f_n converges in the metric of $L^1(\mu)$ to functions f and g, then $f(x) = g(x)$ μ-a.e.

3.5.2. Theorem. *If a sequence of μ-integrable functions f_n is Cauchy in mean, then it converges in mean to some μ-integrable function f. In other words, the space $L^1(\mu)$ with the indicated metric is complete.*

PROOF. Let us find a sequence $\{n_k\}$ such that $\|f_{n_{k+1}} - f_{n_k}\|_{L^1(\mu)} \leqslant 2^{-k}$. The Lebesgue monotone convergence theorem gives convergence μ-a.e. of the series $\sum_{k=1}^{\infty} |f_{n_{k+1}}(x) - f_{n_k}(x)|$. Hence the sequence $\{f_{n_k}(x)\}$ has a finite limit $f(x)$ for μ-a.e. x. Then μ-a.e. $|f(x)| = \lim_{n \to \infty} |f_{n_k}(x)|$. By Fatou's theorem the function f is integrable. Finally, for every $\varepsilon > 0$ there exists a number N_ε such that

$$\int_X |f_n - f_m|\, d\mu \leqslant \varepsilon \quad \text{for all } n, m \geqslant N_\varepsilon.$$

Substituting $n = n_k$ and applying Fatou's theorem once again, we arrive at the estimate $\|f - f_m\|_{L^1(\mu)} \leqslant \varepsilon$ for all $m \geqslant N_\varepsilon$, which completes the proof. $\square$

3.5.3. Corollary. *If a sequence of μ-integrable functions f_n is Cauchy in mean and converges almost everywhere to a function f, then the function f is integrable and the sequence $\{f_n\}$ converges to it in mean.*

Employing the idea of completion, one could define integrable functions as follows (see, for example, [73, Chapter 2]).

3.5.4. Proposition. *A function f is integrable with respect to a measure μ precisely when there exists a sequence of simple functions f_n which converges to f almost everywhere and is Cauchy in mean. In this case $\|f - f_n\|_{L^1(\mu)} \to 0$.*

PROOF. If such a sequence exists, then $f \in \mathcal{L}^1(\mu)$ and $\|f - f_n\|_{L^1(\mu)} \to 0$, which follows from the proof of the previous theorem. For the proof of the converse we take a sequence of simple functions f_n such that $f_n(x) \to f(x)$ a.e. and $|f_n(x)| \leqslant |f(x)|$ a.e. (this is easily done, see the proof of Corollary 3.1.9). By the Lebesgue theorem $\|f - f_n\|_{L^1(\mu)} \to 0$. $\square$

If μ is a signed measure, by definition $L^1(\mu) := L^1(|\mu|)$, $\mathcal{L}^1(\mu) := \mathcal{L}^1(|\mu|)$. For $f \in \mathcal{L}^1(|\mu|)$ we set

$$\int_X f \, d\mu := \int_X f(x) \, \mu(dx) := \int_X f(x) \, \mu^+(dx) - \int_X f(x) \, \mu^-(dx).$$

Introducing the function ξ equal to 1 on X^+ and -1 on X^-, we obtain

$$\int_X f(x) \, \mu(dx) = \int_X f(x) \xi(x) \, |\mu|(dx).$$

3.6. Criteria of Integrability

The definition of the integral is almost never used for establishing the integrability of concrete functions. Very efficient and frequently used sufficient conditions for integrability are provided by the Beppo Levi and Fatou theorems. In real problems one often applies the most obvious criterion for the integrability of a measurable function: majorization in absolute value by an a priori integrable function. In this section from this trivial criterion we derive a number of less obvious corollaries and obtain some criteria of integrability in terms of convergence of series or Riemann integrals over the real line.

3.6.1. Theorem. *Let $(X, \mathcal{A}, \mu)$ be a space with a finite nonnegative measure and let f be a μ-measurable function. Then the integrability of the function f with respect to μ is equivalent to convergence of the series*

$$\sum_{n=1}^{\infty} n\mu(x\colon n \leqslant |f(x)| < n+1), \qquad (3.6.1)$$

and is also equivalent to convergence of the series

$$\sum_{n=1}^{\infty} \mu(x\colon |f(x)| \geqslant n). \qquad (3.6.2)$$

PROOF. Let $A_0 = \{x\colon |f(x)| < 1\}$. Set $A_n := \{x\colon n \leqslant |f(x)| < n+1\}$ for $n \in \mathbb{N}$. The sets A_n are μ-measurable, pairwise disjoint and give in union the whole space (up to a set of measure zero on which f can be undefined). The function g given by the equalities $g|_{A_n} = n$, where $n = 0, 1, \ldots$, is obviously μ-measurable and $g(x) \leqslant |f(x)| \leqslant g(x) + 1$. Therefore, the function g is integrable precisely when the function f is integrable. According to Example 3.3.5 the integrability of g is equivalent to convergence of the series (3.6.1). It remains to observe that the series (3.6.1) and (3.6.2) converge or diverge simultaneously. Indeed, $\{x\colon |f(x)| \geqslant n\} = \bigcup_{k=n}^{\infty} A_k$, whence $\mu(x\colon |f(x)| \geqslant n) = \sum_{k=n}^{\infty} \mu(A_k)$. Thus, in the sum in n the number $\mu(A_n)$ enters the right-hand side n times. $\qquad \square$

3.6.2. Example. (i) A function f measurable with respect to a bounded nonnegative measure μ is integrable to every power $p \in (0, \infty)$ precisely when $\mu(x\colon |f(x)| > t)$ decreases faster than every negative power of t as $t \to +\infty$. This is seen from the equality $\mu(x\colon |f(x)|^p > t) = \mu(x\colon |f(x)| > t^{1/p})$.

(ii) The function $|\ln x|^p$ on $(0, 1)$ is Lebesgue integrable for all $p > -1$, and the function x^α is integrable if $\alpha > -1$.

(iii) The function $\|x\|^{-p}$ on $\mathbb{R}^d$ is integrable in a neighborhood of zero if and only if $p < d$.

For infinite measures the stated criteria do not work, because they do not take into account sets of small values of $|f|$. They can be modified also for infinite measures, but we give instead a universal criterion, an advantage of which is the reduction of the problem to some Riemann integral.

3.6.3. Theorem. *Let μ be a countably additive measure with values in $[0, +\infty]$ and let f be a μ-measurable function. Then the μ-integrability of f is equivalent to the integrability of the function $t \mapsto \mu(x\colon |f(x)| > t)$ on $(0, +\infty)$ against Lebesgue measure (which is also its Riemann integrability). Moreover,*

$$\int_X |f(x)|\, \mu(dx) = \int_0^\infty \mu(x\colon |f(x)| > t)\, dt. \tag{3.6.3}$$

PROOF. Here we prove only the equivalence of the integrability of f to convergence of the indicated integral and a short derivation of equality (3.6.3) will be given below with the aid of Fubini's theorem. If f is integrable, then the set $\{f > 1\}$ has a finite measure. The integral of $g(t) := \mu(x\colon |f(x)| > t)$ over $[1, +\infty)$ does not exceed the sum of the series of $g(n)$ and hence is finite. Conversely, convergence of this integral yields the integrability of $fI_{\{f \geqslant 1\}}$. Hence we can assume that $0 \leqslant f \leqslant 1$. Similarly to the previous theorem one verifies that the integrability of g over $(0, 1]$ is equivalent to convergence of the series of $2^{-n}\mu(x\colon |f(x)| > 2^{-n})$, which, as is easy to see, reduces to convergence of the series of $2^{-n}\mu(x\colon 2^{-n} < |f(x)| \leqslant 2^{-n+1})$, which is equivalent to the integrability of $fI_{\{0 \leqslant f \leqslant 1\}}$. $\square$

One of the original constructions of Lebesgue himself can be also used as a convenient and visual criterion for integrability.

Let $(X, \mathcal{A}, \mu)$ be a space with a finite nonnegative measure and let f be a μ-measurable function. Let us consider the Lebesgue sums

$$\Lambda(\varepsilon) := \sum_{k=-\infty}^{+\infty} k\varepsilon\, \mu(x\colon k\varepsilon \leqslant f(x) < (k+1)\varepsilon), \quad \varepsilon > 0. \tag{3.6.4}$$

3.6.4. Proposition. *A measurable function f is Lebesgue integrable precisely when for some $\varepsilon > 0$ both parts of the series (3.6.4) in $k > 0$ and $k < 0$ converge. In this case the indicated series converges for all $\varepsilon > 0$ and the quantities $\Lambda(\varepsilon)$ as $\varepsilon \to 0$ tend to the integral of f.*

PROOF. Let us consider the function f_ε that equals the number $k\varepsilon$ on the set $\{x\colon k\varepsilon \leqslant f(x) < (k+1)\varepsilon\}$ for all integer k. It is clear that this function is measurable and the absolute convergence of the series (3.6.4) is the condition for its integrability. In the case of convergence the sum of the series equals the integral of f_ε. It remains to observe that $\sup_x |f(x) - f_\varepsilon(x)| \leqslant \varepsilon$. Since $\mu(X) < \infty$, either both functions f and f_ε are integrable or both are not integrable. In the case of integrability the absolute value of the difference of the integrals does not exceed $\varepsilon\mu(X)$, whence the last two assertions follow. $\square$

3.7. Connections with the Riemann Integral

We assume to be known the definition of the Riemann integral (see, e.g., [520] or [659]). For piecewise constant functions on the interval the Riemann integral coincides with the Lebesgue one. The Dirichlet function (the indicator function of the set of rational numbers) is not Riemann integrable, but is simple and has zero Lebesgue integral. See also Exercises 3.12.22 and 3.12.23.

3.7.1. Theorem. *If a function f is Riemann integrable in the proper sense on $I = [a, b]$, then it is Lebesgue integrable on I and its Riemann and Lebesgue integrals are equal.*

PROOF. We can assume that $b - a = 1$. For each $n \in \mathbb{N}$ we partition the interval $I = [a, b]$ into disjoint intervals $[a, a+2^{-n}), \ldots, [b-2^{-n}, b]$ of length 2^{-n}, denoted by $I_1, \ldots, I_{2^n}$. Let $m_k = \inf_{x \in I_k} f(x)$, $M_k = \sup_{x \in I_k} f(x)$. Let us consider step functions f_n and g_n defined in the following way: $f_n = m_k$ on I_k, $g_n = M_k$ on I_k, $k = 1, \ldots, 2^n$. It is clear that $f_n(x) \leqslant f(x) \leqslant g_n(x)$. In addition, $f_n(x) \leqslant f_{n+1}(x)$, $g_{n+1}(x) \leqslant g_n(x)$. Hence there exist limits $\varphi(x) = \lim_{n \to \infty} f_n(x)$, $\psi(x) = \lim_{n \to \infty} g_n(x)$, and $\varphi(x) \leqslant f(x) \leqslant \psi(x)$. It is known from the course of calculus that the Riemann integrability of f implies the equality

$$\lim_{n \to \infty} \int_a^b f_n(x)\, dx = \lim_{n \to \infty} \int_a^b g_n(x)\, dx = R(f), \qquad (3.7.1)$$

where $R(f)$ denotes the Riemann integral of f (here we use the aforementioned coincidence of the Riemann and Lebesgue integrals for piecewise constant functions). The functions φ and ψ are bounded and Lebesgue measurable (being pointwise limits of step functions), so they are Lebesgue integrable. It is clear that

$$\int_a^b f_n(x)\, dx \leqslant \int_a^b \varphi(x)\, dx \leqslant \int_a^b \psi(x)\, dx \leqslant \int_a^b g_n(x)\, dx$$

for all n. It follows from (3.7.1) that the integrals of the functions φ and ψ are equal to $R(f)$, whence $\varphi(x) = \psi(x)$ a.e., since $\varphi(x) \leqslant \psi(x)$. Hence $\varphi = f = \psi$ a.e., which yields the desired assertion. $\qquad \square$

There exist functions Riemann integrable in the improper sense which are not Lebesgue integrable (see Exercise 3.12.24). However, the existence of the absolute improper Riemann integral implies the Lebesgue integrability.

3.7.2. Theorem. *Suppose that a function f is integrable on an interval I (bounded or unbounded) in the improper Riemann sense along with the function $|f|$. Then f is Lebesgue integrable on I and its improper Riemann integral equals the Lebesgue integral.*

PROOF. We shall consider the case where the interval $I = (a, b]$ is bounded and f is Riemann integrable in the proper sense on all intervals $[a + \varepsilon, b]$, $\varepsilon > 0$. The case of $a = -\infty$ is similar, and the general case is a combination of finitely many considered cases. Let $f_n = f I_{[a+n^{-1}, b]}$. Then $f_n(x) \to f(x)$ if $x \in (a, b]$. By the previous theorem the functions f_n are Lebesgue measurable. Hence f is also measurable. In addition, the Lebesgue integrals of the functions $|f_n|$ equal their Riemann integrals, which are uniformly bounded by our assumption. By Fatou's theorem the function $|f|$ is Lebesgue integrable. By the Lebesgue theorem the integrals of the functions f_n over $(a, b]$ tend to the Lebesgue integral of f, which yields its coincidence with the improper Riemann integral. □

Both theorems above remain valid in the multidimensional case with similar proofs. The Lebesgue integral in $\mathbb{R}^n$ can be defined with the aid of some generalized Riemann-type sums (see §5.7 in [73], where the integrals of Henstock–Kurzweil and McShane are discussed). Even the absolute improper Riemann integral has no property of completeness from §3.5, i.e., it can happen that a sequence of Riemann integrable functions (even step functions) is Cauchy in mean, but has no limit among Riemann integrable functions (see Exercise 3.12.25).

3.8. The Hölder, Jensen and Minkowski Inequalities

Let $(X, \mathcal{A}, \mu)$ be a space with a nonnegative measure μ (finite or with values in $[0, +\infty]$) and $p \in (0, +\infty)$. Denote by $\mathcal{L}^p(\mu)$ the set of all μ-measurable functions f for which $|f|^p$ is a μ-integrable function. In particular, $\mathcal{L}^1(\mu)$ is the set of all μ-integrable functions. As in §3.1, let $\mathcal{L}^0(\mu)$ denote the class of all μ-a.e. finite μ-measurable functions. Let $L^p(\mu)$ denote the quotient space of $\mathcal{L}^p(\mu)$ with respect to the equivalence relation introduced in $\mathcal{L}^0(\mu)$. Thus, $L^p(\mu)$ is the set of equivalence classes of μ-measurable functions f for which $|f|^p$ is integrable. The same notation is used for complex functions. In case of Lebesgue measure on $\mathbb{R}^n$ or on a set $E \subset \mathbb{R}^n$ the symbols $\mathcal{L}^p(\mathbb{R}^n)$, $L^p(\mathbb{R}^n)$, $\mathcal{L}^p(E)$, and $L^p(E)$ are used without specifying the measure. In place of $L^p([a, b])$ and $L^p([a, +\infty))$ it is customary to write $L^p[a, b]$ and $L^p[a, +\infty)$. Sometimes it is necessary to indicate the space X explicitly in this notation, then the symbols $\mathcal{L}^p(X, \mu)$ and $L^p(X, \mu)$ are used. It is customary to deliberately admit some abuse of terminology in expressions like "the function f from L^p", in which one should formally speak of "the function f from $\mathcal{L}^p$" or "the equivalence class of the function f in L^p". Usually such usage does not lead to confusion, moreover, it contributes to shortening formulations, implicitly indicating that the assertion in question is true not only for an individual function, but for all representatives of its equivalence class. Whenever $1 \leqslant p < \infty$, we set

$$\|f\|_p := \|f\|_{L^p(\mu)} := \left(\int_X |f|^p \, d\mu \right)^{1/p}, \quad f \in \mathcal{L}^p(\mu).$$

The same notation is used for elements of $L^p(\mu)$. Finally, let $\mathcal{L}^\infty(\mu)$ be the set of all bounded *everywhere defined* μ-measurable functions. For $f \in \mathcal{L}^\infty(\mu)$ set $\|f\|_{L^\infty(\mu)} := \|f\|_\infty := \inf_{\tilde{f} \sim f} \sup_{x \in X} |\tilde{f}(x)|$. A function f is called *essentially bounded* if it equals a bounded function μ-a.e. The quantity $\|f\|_\infty$ is defined as above (it does not depend on the representative of the equivalence class of f). An alternative notation: $\operatorname{ess\,sup} |f|$ or $\operatorname{vrai\,sup} |f|$. The space of equivalence classes of essentially bounded functions is denoted by $L^\infty(\mu)$. Note that $\|f\|_\infty$ can be strictly smaller than $\sup_{x \in X} |f(x)|$. The space $L^\infty(\mu)$ becomes a metric space with the metric $\|f - g\|_{L^{\infty}(\mu)}$. It is readily seen that it is complete. Below we introduced a natural metric on $L^p(\mu)$.

For this we need the following Hölder inequality, which is of great importance: this is one of the most frequently used inequalities of the theory of integral.

3.8.1. Theorem. (HÖLDER'S INEQUALITY) *Let* $1 < p < \infty$, $q = p(p-1)^{-1}$ *and* $f \in \mathcal{L}^p(\mu)$, $g \in \mathcal{L}^q(\mu)$. *Then* $fg \in \mathcal{L}^1(\mu)$ *and* $\|fg\|_1 \leqslant \|f\|_p\|g\|_q$, *i.e.,*

$$\int_X |fg|\,d\mu \leqslant \left(\int_X |f|^p\,d\mu\right)^{1/p} \left(\int_X |g|^q\,d\mu\right)^{1/q}. \qquad (3.8.1)$$

PROOF. The function fg is defined a.e. and measurable. It is easy to show (Exercise 3.12.32) that $ab \leqslant a^p/p + b^q/q$ for all nonnegative a and b. Then

$$\frac{|f(x)|}{\|f\|_p} \frac{|g(x)|}{\|g\|_q} \leqslant \frac{1}{p}\frac{|f(x)|^p}{\|f\|_p^p} + \frac{1}{q}\frac{|g(x)|^q}{\|g\|_q^q}.$$

The right-hand side is integrable and its integral is 1, so the left-hand side is also integrable and its integral does not exceed 1, which is equivalent to (3.8.1). □

3.8.2. Corollary. *Under the hypotheses of the previous theorem*

$$\int_X fg\,d\mu \leqslant \left(\int_X |f|^p\,d\mu\right)^{1/p} \left(\int_X |g|^q\,d\mu\right)^{1/q}. \qquad (3.8.2)$$

Exercise 3.12.33 gives conditions for the equality in inequality (3.8.2). An immediate corollary of Hölder's inequality is the following Cauchy–Bunyakovskii inequality, which, however, can be easily proved directly (see Chapter 5).

3.8.3. Corollary. *If* $f, g \in \mathcal{L}^2(\mu)$, *then* $fg \in \mathcal{L}^1(\mu)$ *and*

$$\int_X fg\,d\mu \leqslant \left(\int_X |f|^2\,d\mu\right)^{1/2} \left(\int_X |g|^2\,d\mu\right)^{1/2}. \qquad (3.8.3)$$

For probability measures the following more general inequality holds.

3.8.4. Theorem. (JENSEN'S INEQUALITY) *Let* μ *be a probability measure on* $(X, \mathcal{A})$ *and let* f *be a* μ-*integrable function with values in an interval* U *on which a convex function* V *is defined (i.e.,* $V(tu + (1-t)w) \leqslant tV(u) + (1-t)V(w)$ *for all* $t \in [0,1]$ *and* $u, w \in U$) *such that the function* $V(f)$ *is integrable. Then*

$$V\left(\int_X f(x)\,\mu(dx)\right) \leqslant \int_X V(f(x))\,\mu(dx). \qquad (3.8.4)$$

In particular, if $\exp f$ *is integrable, then*

$$\exp\left(\int_X f(x)\,\mu(dx)\right) \leqslant \int_X \exp f(x)\,\mu(dx).$$

PROOF. It is easy to see that the integral u_0 of f belongs to U. One can derive from the convexity of V that there is a number α with $V(u) \geqslant V(u_0) + \alpha(u - u_0)$ for all $u \in U$. Hence $V(f(x)) \geqslant V(u_0) + \alpha[f(x) - u_0]$. We integrate this inequality and observe that the integral of the second summand on the right is zero. □

3.8.5. Theorem. (THE MINKOWSKI INEQUALITY) *Let $p \in [1, +\infty)$ and let $f, g \in \mathcal{L}^p(\mu)$. Then $f + g \in \mathcal{L}^p(\mu)$ and*

$$\left(\int_X |f + g|^p \, d\mu \right)^{1/p} \leqslant \left(\int_X |f|^p \, d\mu \right)^{1/p} + \left(\int_X |g|^p \, d\mu \right)^{1/p}. \qquad (3.8.5)$$

PROOF. The function $f + g$ is defined a.e. and measurable. For $p = 1$ inequality (3.8.5) is obvious. For $p > 1$ we have $|f + g|^p \leqslant 2^p(|f|^p + |g|^p)$, hence $|f + g|^p \in \mathcal{L}^1(\mu)$. We observe that

$$|f + g|^p \leqslant |f + g|^{p-1} |f| + |f + g|^{p-1} |g|. \qquad (3.8.6)$$

Since $|f + g|^{p-1} \in \mathcal{L}^{p/(p-1)}(\mu) = \mathcal{L}^q(\mu)$, by Hölder's inequality

$$\int_X |f + g|^{p-1} |f| \, d\mu \leqslant \left(\int_X |f + g|^p \, d\mu \right)^{1/q} \left(\int_X |f|^p \, d\mu \right)^{1/p}.$$

Estimating similarly the integral of the second term in the right-hand side of (3.8.6), we arrive at the estimate

$$\int_X |f + g|^p \, d\mu \leqslant \left(\int_X |f + g|^p \, d\mu \right)^{1/q} \left[\left(\int_X |f|^p \, d\mu \right)^{1/p} + \left(\int_X |g|^p \, d\mu \right)^{1/p} \right].$$

Observing that $1 - 1/q = 1/p$, we obtain $\|f + g\|_p \leqslant \|f\|_p + \|g\|_p$. $\square$

The Minkowski inequality means that, whenever $1 \leqslant p < \infty$, the formula $d(f, g) := \|f - g\|_p$ defines a metric on $L^p(\mu)$, since

$$\|g - h\|_p = \|g - f + f - h\|_p \leqslant \|g - f\|_p + \|f - h\|_p.$$

3.8.6. Theorem. *The space $L^p(\mu)$, where $1 \leqslant p < \infty$, is complete with respect to the metric $d(f, g) = \|f - g\|_p$.*

PROOF. As in Theorem 3.5.2 we take n_k such that $\|f_{n_{k+1}} - f_{n_k}\|_p^p \leqslant 2^{-pk}$ and observe that the function $|f_{n_{k+1}}(x) - f_{n_k}(x)|$ is obviously estimated by the sum $2^{-k/(p-1)} + 2^k |f_{n_{k+1}}(x) - f_{n_k}(x)|^p$ if $p > 1$. $\square$

Similarly to Proposition 3.5.4 we obtain the following result.

3.8.7. Proposition. *A function f belongs to $\mathcal{L}^p(\mu)$ precisely when there exists a sequence of simple functions f_n that converges to f almost everywhere and is fundamental in $L^p(\mu)$. In this case $\|f - f_n\|_{L^p(\mu)} \to 0$.*

3.8.8. Example. The spaces $L^p[a, b]$ and $L^p(\mathbb{R}^n)$, $p \in [1, +\infty)$, are separable. The sets of infinitely differentiable functions with compact support are dense in these spaces. These assertions are true for every bounded Borel measure μ on $\mathbb{R}^n$. The spaces $L^\infty[a, b]$ and $L^\infty(\mathbb{R}^n)$ are nonseparable.

PROOF. Simple functions are dense in these spaces. For every set E of finite measure and $\varepsilon > 0$ one can find a finite union E_ε of open cubes with rational centers and edges of rational length parallel to the coordinate axes for which $\lambda_n(E \bigtriangleup E_\varepsilon) < \varepsilon$. Hence linear combinations with rational coefficients of the indicator functions of such cubes are dense in $L^p(\mathbb{R}^n)$ if $p < \infty$. The same is true for every bounded Borel measure on $\mathbb{R}^n$. Thus, we have a countable everywhere dense set. For approximating by functions of class $C_0^\infty(\mathbb{R}^n)$ we observe that the indicator function of a cube can be approximated in L^p by such functions (we take functions f_j in this class with $0 \leqslant f_j \leqslant 1$ pointwise converging to the indicator function of the cube). The nonseparability of $L^\infty[a, b]$ is clear from the fact that if $t \neq s$, then the distance between the indicators of $[a, t]$ and $[a, s]$ in L^∞ is 1. □

The presented reasoning shows that for the separability of $L^p(\mu)$ with $p < \infty$ it suffices that the σ-algebra $\mathcal{A}$ on which the measure μ is defined be generated by a countable family of sets. One should bear in mind that there are probability measures μ for which all spaces $L^p(\mu)$ are nonseparable (for example, the product of the continuum of copies of Lebesgue measure on $[0, 1]$; see §3.5 in [73] about infinite products of measures). The previous example yields the following fact.

3.8.9. Lemma. *If the integrals with respect to two bounded Borel measures μ and ν on $\mathbb{R}^n$ coincide on $C_0^\infty(\mathbb{R}^n)$, then $\mu = \nu$.*

Hölder's inequality is useful to verify membership in L^p.

3.8.10. Example. Let μ be a finite nonnegative measure. Suppose that a μ-measurable function f satisfies the following condition: there exist $p \in (1, \infty)$ and $M \geqslant 0$ such that for every function $\varphi \in L^\infty(\mu)$ we have $f\varphi \in L^1(\mu)$ and

$$\int_X f\varphi \, d\mu \leqslant M\|\varphi\|_{L^p(\mu)}.$$

Then $f \in L^q(\mu)$, where $q = p(p-1)^{-1}$, and $\|f\|_{L^q(\mu)} \leqslant M$.
Indeed, substituting $\varphi_n := \operatorname{sgn} f|f|^{q-1}I_{\{|f|\leqslant n\}}$ for φ, we obtain

$$\int_{\{|f|\leqslant n\}} |f|^q \, d\mu \leqslant M\left(\int_{\{|f|\leqslant n\}} |f|^q \, d\mu\right)^{1/p},$$

so $\|fI_{\{|f|\leqslant n\}}\|_{L^q(\mu)} \leqslant M$, since $1 - 1/p = 1/q$. Fatou's theorem yields our claim. It is also true for infinite measures if the bound holds for $\varphi \in L^\infty(\mu) \cap L^p(\mu)$.

Although functions from the space $L^p(\mu)$ can be summed and multiplied by numbers (on sets of full measure), these spaces *are not linear*, since the mentioned operations are not associative: for example, if the function f is not defined at a point x, then so is $f + (-f)$, but this function must be everywhere zero, since in a linear space there is a unique zero element. Of course, one can take in $L^p(\mu)$ the subset of everywhere defined finite functions, which is a linear space, but it is preferable to pass to the quotient space $L^p(\mu)$.

3.9. The Radon–Nikodym Theorem

Let f be a function integrable with respect to a measure μ (possibly, signed or with values in $[0, +\infty]$) on a space $(X, \mathcal{A})$. Then we obtain a set function by letting

$$\nu(A) = \int_A f \, d\mu. \tag{3.9.1}$$

The Lebesgue theorem yields that ν is countably additive on $\mathcal{A}$. Indeed, if sets $A_n \in \mathcal{A}$ are pairwise disjoint, then the series $\sum_{n=1}^\infty I_{A_n}(x)f(x)$ converges for every x to $I_A(x)f(x)$, since only one term can be nonzero due to the disjointness of A_n. Moreover,

$$\left| \sum_{n=1}^{N} I_{A_n}(x)f(x) \right| \leqslant I_A(x)|f(x)|.$$

Hence this series admits a term-by-term integration.

We denote ν by $f \cdot \mu$. The function f is denoted by the symbol $d\nu/d\mu$ and called the *density* of the measure ν with respect to μ (the *Radon–Nikodym density* or *Radon–Nikodym derivative*). The measure ν is absolutely continuous with respect to μ in the sense of the following definition.

3.9.1. Definition. *Let μ and ν be two countably additive measures on a measurable space $(X, \mathcal{A})$.*

(i) *The measure ν is called absolutely continuous with respect to the measure μ if $|\nu|(A) = 0$ for every set A with $|\mu|(A) = 0$. Notation: $\nu \ll \mu$.*

(ii) *The measure ν is called singular with respect to μ if there exists a set $\Omega \in \mathcal{A}$ such that $|\mu|(\Omega) = 0$ and $|\nu|(X\backslash\Omega) = 0$. Notation: $\nu \perp \mu$.*

This definition is meaningful also for measures with values in $[0, +\infty]$.

Note that if the measure ν is singular with respect to μ, then μ is singular with respect to ν, i.e., $\mu \perp \nu$. So the measures μ and ν are called mutually singular. If $\nu \ll \mu$ and $\mu \ll \nu$, then the measures μ and ν are called equivalent. The corresponding notation: $\mu \sim \nu$.

The next result, called the Radon–Nikodym theorem, is one of the fundamental facts in measure theory.

3.9.2. Theorem. (THE RADON–NIKODYM THEOREM) *Let μ and ν be finite measures on a space $(X, \mathcal{A})$. The measure ν is absolutely continuous with respect to the measure μ precisely when there exists a μ-integrable function f such that ν is given by formula (3.9.1).*

PROOF. Since $\mu = f_1 \cdot |\mu|$ and $\nu = f_2 \cdot |\nu|$, where $|f_1(x)| = |f_2(x)| = 1$, it suffices to prove the theorem for nonnegative measures μ and ν. Let $\nu \ll \mu$ and

$$\mathcal{F} := \left\{ f \in \mathcal{L}^1(\mu) \colon f \geqslant 0, \int_A f \, d\mu \leqslant \nu(A) \quad \text{for all } A \in \mathcal{A} \right\}.$$

Set

$$M := \sup\left\{ \int_X f \, d\mu \colon f \in \mathcal{F} \right\}.$$

We show that $\mathcal{F}$ contains a function f at which this supremum is attained. We find a sequence of functions $f_n \in \mathcal{F}$ the integrals of which tend to M. Let $g_n(x) = \max(f_1(x), \ldots, f_n(x))$. We observe that $g_n \in \mathcal{F}$. Indeed, every set $A \in \mathcal{A}$ can be written in the form $A = \bigcup_{k=1}^{n} A_k$, where $A_k \in \mathcal{A}$ are pairwise disjoint and $g_n(x) = f_k(x)$ for all $x \in A_k$. Then

$$\int_A g_n \, d\mu = \sum_{k=1}^{n} \int_{A_k} g_n \, d\mu \leqslant \sum_{k=1}^{n} \nu(A_k) = \nu(A).$$

The sequence $\{g_n\}$ is increasing and the integrals of the functions g_n are bounded by $\nu(X)$. By the monotone convergence theorem the function $f := \lim_{n \to \infty} g_n$ is integrable. It is also clear that $f \in \mathcal{F}$ and that the integral of f against the measure μ is M. We show that f satisfies (3.9.1). By the choice of f the set function

$$\eta(A) := \nu(A) - \int_A f \, d\mu$$

is a nonnegative measure and $\nu \ll \mu$. We have to show that $\eta = 0$. Suppose that this is false. Let us consider the signed measures $\eta - n^{-1}\mu$ and take their Hahn decompositions $X = X_n^+ \cup X_n^-$. Let $X_0^- := \bigcap_{n=1}^{\infty} X_n^-$. Then from the definition of X_n^- we have $\eta(X_0^-) \leqslant n^{-1}\mu(X_0^-)$ for all n, whence $\eta(X_0^-) = 0$. Hence there exists n such that $\eta(X_n^+) > 0$, since otherwise $\eta(X) = \eta(X_n^-)$ for all n and then $\eta(X) = \eta(X_0^-) = 0$. For every measurable set $E \subset X_n^+$ we have $n^{-1}\mu(E) \leqslant \eta(E)$. Hence, setting $h(x) := f(x) + n^{-1}I_{X_n^+}(x)$, for every $A \in \mathcal{A}$ we obtain

$$\int_A h \, d\mu = \int_A f \, d\mu + n^{-1}\mu(A \cap X_n^+) \leqslant \int_A f \, d\mu + \eta(A \cap X_n^+)$$

$$= \int_{A \setminus X_n^+} f \, d\mu + \nu(A \cap X_n^+) \leqslant \nu(A \setminus X_n^+) + \nu(A \cap X_n^+) = \nu(A).$$

Thus, $h \in \mathcal{F}$ contrary to the fact that the integral of h against the measure μ is greater than M, since $\mu(X_n^+) > 0$. Therefore, $\eta = 0$. $\qquad\square$

It is clear that the function $d\nu/d\mu$ is uniquely determined up to a set of measure zero, since the function with zero integral over every set equals zero a.e. In Chapter 6 (Example 6.5.4) we shall give another proof of the Radon–Nikodym theorem.

Note that if the measures μ and ν are finite and nonnegative and $\nu \ll \mu$, then $\nu \sim \mu$ precisely when $d\nu/d\mu > 0$ a.e. with respect to μ. It is readily seen that if we are given three measures μ_1, μ_2 and μ_3 for which $\mu_1 \ll \mu_2$ and $\mu_2 \ll \mu_3$, then $\mu_1 \ll \mu_3$ and

$$d\mu_1/d\mu_3 = (d\mu_1/d\mu_2)(d\mu_2/d\mu_3).$$

This follows from the next assertion.

3.9.3. Proposition. *Suppose that on a measurable space $(X, \mathcal{A})$ we are given two bounded measures μ and ν and $\nu \ll \mu$. Let $\varrho := d\nu/d\mu$. Then for every ν-integrable function f the function $f\varrho$ is integrable with respect to μ and*

$$\int_X f \, d\nu = \int_X f\varrho \, d\mu. \tag{3.9.2}$$

Conversely, if a μ-measurable function f is such that $f\varrho \in \mathcal{L}^1(\mu)$, then $f \in \mathcal{L}^1(\nu)$ and equality (3.9.2) is fulfilled.

PROOF. Using the Hahn decomposition and the definition of the integral for signed measures one can pass to the case of nonnegative measures. In addition, it suffices to consider a nonnegative function f. Equality (3.9.2) is true for simple functions by the definition of the Radon–Nikodym density. By means of uniform approximations it extends at once to bounded functions. Finally, the validity of this equality for $\min(f, n)$ yields its validity for f by the monotone convergence theorem. This reasoning gives also the converse assertion. $\qquad \square$

Obviously, this proposition remains in force for σ-finite nonnegative measures.

3.9.4. Example. (i) Let μ be a bounded nonnegative measure and let ν be the measure given by a density $f \in \mathcal{L}^1(\mu)$ with respect to μ. Then the measure $|\nu|$ is given by the density $|f|$ with respect to μ. This is clear by considering ν on the sets $\{f < 0\}$ and $\{f \geqslant 0\}$.

(ii) Let μ be a signed bounded measure and let ν be the measure given by a density $f \in \mathcal{L}^1(\mu)$ with respect to μ. Then the measure $|\nu|$ is given by the density $|f|$ with respect to $|\mu|$. This follows from (i), since the measure μ is given by the density ξ with $|\xi| = 1$ with respect to $|\mu|$.

From the Radon–Nikodym theorem one can obtain the following *Lebesgue decomposition*.

3.9.5. Theorem. *Let μ and ν be finite measures on a σ-algebra $\mathcal{A}$. Then there exist a measure μ_0 on $\mathcal{A}$ and a μ-integrable function f such that*

$$\nu = f \cdot \mu + \mu_0, \quad \mu_0 \perp \mu.$$

PROOF. Let us consider the measure $\lambda := |\mu| + |\nu|$. By the Radon–Nikodym theorem we obtain $\mu = f_\mu \cdot \lambda$, $\nu = f_\nu \cdot \lambda$, where $f_\mu, f_\nu \in L^1(\lambda)$. Let us set $Y = \{x \colon f_\mu(x) \neq 0\}$. For $x \in Y$ set $f(x) = f_\nu(x)/f_\mu(x)$. Finally, we set $\mu_0(A) := \nu(A \cap (X \backslash Y))$. For the restrictions μ_Y and ν_Y of the measures μ and ν to the set Y we have $\nu_Y = f \cdot \mu_Y$. This is the desired decomposition. $\qquad \square$

If μ is a finite or σ-finite nonnegative measure on a σ-algebra $\mathcal{A}$ in a space X, then every finite measurable nonnegative function f (not necessarily integrable) defines a σ-finite measure $\nu := f \cdot \mu$ by formula (3.9.1), since X is the union of sets of finite measure $\{x \colon f(x) \leqslant n\} \cap X_n$, where $\mu(X_n) < \infty$. It is clear that in such a form the Radon–Nikodym theorem is true also for σ-finite measures. For the measures $\mu(\{0\}) = 1$, $\nu(\{0\}) = \infty$ (or vice-versa) it fails (with finite f).

3.10. Products of Measure Spaces

Let $(X_1, \mathcal{A}_1, \mu_1)$ and $(X_2, \mathcal{A}_2, \mu_2)$ be spaces with finite nonnegative measures. On the set $X_1 \times X_2$ we consider sets $A_1 \times A_2$, where $A_i \in \mathcal{A}_i$, called *measurable rectangles*. Let $\mu_1 \times \mu_2(A_1 \times A_2) := \mu_1(A_1)\mu_2(A_2)$. Extending the function $\mu_1 \times \mu_2$ by additivity to finite unions of pairwise disjoint measurable rectangles, we obtain a finite additive function on the algebra $\mathcal{R}$ generated by such rectangles. Note that $\mu_1 \times \mu_2$ is well-defined on $\mathcal{R}$ (independent of the partition of the set into pairwise disjoint measurable rectangles) by the additivity of μ_1 and μ_2. Finally, let $\mathcal{A}_1 \otimes \mathcal{A}_2$ denote the σ-algebra generated by the rectangles indicated above and called the *product* of the σ-algebras $\mathcal{A}_1$ and $\mathcal{A}_2$.

The next result is a typical application of the monotone class theorem.

3.10.1. Proposition. (i) *Let* $(X_1, \mathcal{A}_1)$ *and* $(X_2, \mathcal{A}_2)$ *be measurable spaces and let* $\mathcal{A}_1 \otimes \mathcal{A}_2$ *be the σ-algebra generated by all sets* $A_1 \times A_2$, *where* $A_1 \in \mathcal{A}_1$, $A_2 \in \mathcal{A}_2$. *Then, for every set* $A \in \mathcal{A}_1 \otimes \mathcal{A}_2$ *and every* $x_1 \in X_1$, *the set*

$$A_{x_1} := \{x_2 \in X_2 : (x_1, x_2) \in A\}$$

is contained in $\mathcal{A}_2$. *In addition, for every* $\mathcal{A}_1 \otimes \mathcal{A}_2$-*measurable function* f, *for every* $x_1 \in X_1$ *the function* $x_2 \mapsto f(x_1, x_2)$ *is* $\mathcal{A}_2$-*measurable.*

(ii) *If* $\mathcal{A}_2$ *is equipped with a finite measure* ν, *then the function* $x_1 \mapsto \nu(A_{x_1})$ *is* $\mathcal{A}_1$-*measurable.*

PROOF. (i) If A is the product of two sets from $\mathcal{A}_1$ and $\mathcal{A}_2$, then our assertion is true. Let $\mathcal{E}$ be the class of all sets $A \in \mathcal{A}_1 \otimes \mathcal{A}_2$ for which it is true. Since to intersections, unions and complements of sets there correspond the same operations on their sections at the point x_1, the class $\mathcal{E}$ is obviously a σ-algebra. Hence $\mathcal{E} = \mathcal{A}_1 \otimes \mathcal{A}_2$. The measurability of the function $x_2 \mapsto f(x_1, x_2)$ follows from the already proved fact applied to the sets $\{x_2 : f(x_1, x_2) < c\}$.

(ii) The function $f_A(x_1) = \nu(A_{x_1})$ is well-defined according to assertion (i). Let $\mathcal{E}$ denote the class of sets $A \in \mathcal{A}_1 \otimes \mathcal{A}_2$ for which this function is $\mathcal{A}_1$-measurable. This class contains all sets $A_1 \times A_2$ with $A_i \in \mathcal{A}_i$. Hence $\mathcal{E}$ contains all finite disjoint unions of such rectangles, i.e., the algebra generated by them. Next, $\mathcal{E}$ is a monotone class, which follows from the Lebesgue dominated convergence theorem and the fact that if sets A^j increase to A, then $A_{x_1}^j$ increase to A_{x_1} (and similarly with decreasing sets). By Theorem 2.2.12 the class $\mathcal{E}$ contains $\mathcal{A}_1 \otimes \mathcal{A}_2$ and hence coincides with $\mathcal{A}_1 \otimes \mathcal{A}_2$. $\qquad \square$

3.10.2. Corollary. *In the situation of assertion* (ii) *of the proposition above, for every bounded* $\mathcal{A}_1 \otimes \mathcal{A}_2$-*measurable function* f *on* $X_1 \times X_2$, *the function*

$$x_1 \mapsto \int_{X_2} f(x_1, x_2)\, \nu(dx_2)$$

is defined and $\mathcal{A}_1$-*measurable.*

PROOF. It suffices to consider the case where f is the indicator function of the set $A \in \mathcal{A}_1 \otimes \mathcal{A}_2$, since by linear combinations of such functions one can uniformly approximate every bounded $\mathcal{A}_1 \otimes \mathcal{A}_2$-measurable function and the corresponding integrals in one variable will converge uniformly. Thus, this assertion is a corollary of the proposition. $\qquad \square$

Let us apply this result for constructing product measures.

3.10.3. Theorem. *The set function* $\mu_1 \times \mu_2$ *is countably additive on the algebra generated by all measurable rectangles and uniquely extends to a countably additive measure* $\mu_1 \otimes \mu_2$ *on the Lebesgue completion of the indicated algebra denoted by the symbol* $\mathcal{A}_1 \overline{\otimes} \mathcal{A}_2$.

PROOF. By the proposition we have a well-defined set function

$$A \mapsto \int_{X_1} \mu_2(A_{x_1})\, \mu_1(dx_1), \quad A \in \mathcal{A}_1 \otimes \mathcal{A}_2.$$

It is countably additive since if $A = \bigcup_{n=1}^{\infty} A_n$, where $A_n \in \mathcal{A}_1 \otimes \mathcal{A}_2$ are disjoint, then $A_{x_1} = \bigcup_{n=1}^{\infty} (A_n)_{x_1}$, where $(A_n)_{x_1} \in \mathcal{A}_2$ are disjoint, whence we obtain $\mu_2(A_{x_1}) = \sum_{n=1}^{\infty} \mu_2((A_n)_{x_1})$. This equality can be integrated termwise with respect to the measure μ_1, since the partial sums of the series equal $\mu_2(\bigcup_{n=1}^{N}(A_n)_{x_1})$ and do not exceed $\mu_2(X_2)$. If $A = A_1 \times A_2$, we obtain $\mu_1(A_1)\mu_2(A_2)$, because $A_{x_1} = A_2$ if $x_1 \in A_1$ and $A_{x_1} = \varnothing$ if $x_1 \notin A_1$. $\qquad\square$

The measure $\mu_1 \otimes \mu_2$ obtained above is called the *product of the measures* μ_1 and μ_2. By construction the measure $\mu_1 \otimes \mu_2$ is complete. The products of measures are sometimes called *product measures*.

Note that the Lebesgue completion of the σ-algebra $\mathcal{A}_1 \otimes \mathcal{A}_2$ generated by all rectangles $A_1 \times A_2$, where $A_1 \in \mathcal{A}_1$, $A_2 \in \mathcal{A}_2$, is, as a rule, broader than this σ-algebra. For example, if for $\mathcal{A}_1 = \mathcal{A}_2$ we take the Borel σ-algebra of $[0, 1]$ and for $\mu_1 = \mu_2$ we take Lebesgue measure, then $\mathcal{A}_1 \otimes \mathcal{A}_2$ coincides with the Borel σ-algebra of the square (every open set in the square is a countable union of open squares). Of course, there exist measurable non-Borel sets in the square. It will not help if we replace the Borel σ-algebra of the interval by the σ-algebra of Lebesgue measurable sets. In this case, as one can see from Proposition 3.10.1, the σ-algebra $\mathcal{A}_1 \otimes \mathcal{A}_2$ will not contain nonmeasurable subsets of the interval considered as subsets of the square (they obviously have measure zero in the square and belong to the completion of $\mathcal{A}_1 \otimes \mathcal{A}_2$). Of course, the measure $\mu_1 \otimes \mu_2$ can be considered on the σ-algebra $\mathcal{A}_1 \otimes \mathcal{A}_2$ that is not necessarily complete.

With the aid of the Jordan–Hahn decomposition one can naturally define the product of signed measures (though, this can be done directly). Let $\mu = \mu^+ - \mu^-$, $\nu = \nu^+ - \nu^-$, $X = X^+ \cup X^-$, and let $Y = Y^+ \cup Y^-$ be the Jordan–Hahn decomposition of the measures μ and ν on the spaces X and Y, respectively. Set $\mu \otimes \nu := \mu^+ \otimes \nu^+ + \mu^- \otimes \nu^- - \mu^+ \otimes \nu^- - \mu^- \otimes \nu^+$. Clearly, the measures $\mu^+ \otimes \nu^+ + \mu^- \otimes \nu^-$ and $\mu^+ \otimes \nu^- + \mu^- \otimes \nu^+$ are mutually singular, since the first one is concentrated on the set $(X^+ \times Y^+) \cup (X^- \times Y^-)$ and the second one is concentrated on the set $(X^+ \times Y^-) \cup (X^- \times Y^+)$.

By induction we define the product of finitely many measures μ_n on spaces $(X_n, \mathcal{A}_n)$, $n = 1, \ldots, N$. This product is associative, i.e., one has the equality

$$\mu_1 \otimes (\mu_2 \otimes \mu_3) = (\mu_1 \otimes \mu_2) \otimes \mu_3.$$

One can also define the product of σ-finite nonnegative measures μ and ν on σ-algebras $\mathcal{A}$ and $\mathcal{B}$. Let X be the union of an increasing sequence of sets X_n of finite μ-measure and let Y be the union of an increasing sequence of sets Y_n of finite ν-measure. The formula

$$\mu \otimes \nu(E) = \lim_{n \to \infty} \big(\mu|_{X_n} \otimes \nu|_{Y_n}\big)\big(E \cap (X_n \times Y_n)\big)$$

defines a σ-finite measure on $\mathcal{A} \otimes \mathcal{B}$.

One could reduce this at once to finite measures by taking finite measures μ_0 and ν_0 such that $\mu = \varrho_\mu \cdot \mu_0$, $\nu = \varrho_\nu \cdot \nu_0$, where ϱ_μ and ϱ_ν are positive measurable functions. Then one can set $\mu \otimes \nu := \big(\varrho_\mu(x)\varrho_\nu(y)\big) \cdot (\mu_0 \otimes \nu_0)$. It is easy to verify that this gives the same measure as above.

Note that one can define the product of an arbitrary family of probability measures (see [73, Chapter 3, §3.5]).

3.10.4. Proposition. *Let $(X, \mathcal{A})$ and $(Y, \mathcal{B})$ be two measurable spaces and let $f\colon X \to \mathbb{R}^1$ and $g\colon Y \to \mathbb{R}^1$ be measurable functions. Then the mapping $(f, g)\colon X \times Y \to \mathbb{R}^2$ is measurable with respect to the pair $\mathcal{A} \otimes \mathcal{B}$ and $\mathcal{B}(\mathbb{R}^2)$. In particular, the graph of the function f and the sets $\{(x, y)\colon y \leqslant f(x)\}$ and $\{(x, y)\colon y \geqslant f(x)\}$ belong to $\mathcal{A} \otimes \mathcal{B}(\mathbb{R}^1)$.*

PROOF. For any open rectangle $\Pi = I \times J$ in $\mathbb{R}^2$ the set

$$\{(x, y)\colon \big(f(x), g(y)\big) \in \Pi\}$$

is the product of elements of $\mathcal{A}$ and $\mathcal{B}$ and belongs to $\mathcal{A} \otimes \mathcal{B}$. The class of sets $E \in \mathcal{B}(\mathbb{R}^2)$ the preimages of which with respect to the mapping (f, g) belong to $\mathcal{A} \otimes \mathcal{B}$ is a σ-algebra. Since this class contains all rectangles of the indicated form, it contains the σ-algebra generated by them, which is $\mathcal{B}(\mathbb{R}^2)$. In the case where $(Y, \mathcal{B}) = \big(\mathbb{R}^1, \mathcal{B}(\mathbb{R}^1)\big)$ and $g(y) = y$, we obtain the measurability of the mapping $(x, y) \mapsto \big(f(x), y\big)$ from $X \times \mathbb{R}^1$ to $\mathbb{R}^2$, which shows that the preimages of Borel sets belong to $\mathcal{A} \otimes \mathcal{B}(\mathbb{R}^1)$. For example, the graph of f is the preimage of the straight line $y = x$, and the two other sets mentioned in the formulation are preimages of half-planes. $\qquad\square$

To this theme belong also Exercises 3.12.37 and 3.12.38.

3.11. Fubini's Theorem

Suppose that μ and ν are finite nonnegative measures on spaces $(X, \mathcal{A})$ and $(Y, \mathcal{B})$, respectively. For every set $A \subset X \times Y$ we define the sections

$$A_x = \{y\colon (x, y) \in A\}, \quad A^y = \{x\colon (x, y) \in A\}.$$

3.11.1. Theorem. *Suppose that the set $A \subset X \times Y$ is measurable with respect to the measure $\mu \otimes \nu$, i.e., belongs to $(\mathcal{A} \otimes \mathcal{B})_{\mu \otimes \nu}$. Then, for μ-a.e. x, the set A_x is ν-measurable and the function $x \mapsto \nu(A_x)$ is μ-measurable, for ν-a.e. y, the set A^y is μ-measurable and the function $y \mapsto \mu(A^y)$ is ν-measurable, and*

$$\mu \otimes \nu(A) = \int_X \nu(A_x)\, \mu(dx) = \int_Y \mu(A^y)\, \nu(dy). \qquad (3.11.1)$$

PROOF. It is seen from the proof of Theorem 3.10.3 that the measure $\mu \otimes \nu$ on $\mathcal{A} \otimes \mathcal{B}$ coincides with the measure

$$\zeta_1(A) := \int_X \nu(A_x)\, \mu(dx), \quad \zeta_2(A) := \int_Y \mu(A^y)\, \nu(dy).$$

Hence our assertion is true for $A \in \mathcal{A} \otimes \mathcal{B}$. It remains true for every set E of $\mu \otimes \nu$-measure zero. Indeed, there exists a set $\widehat{E} \in \mathcal{A} \otimes \mathcal{B}$ which contains E and has measure zero with respect to $\mu \otimes \nu$. Then $E_x \subset \widehat{E}_x$ and $\nu(\widehat{E}_x) = 0$ for μ-a.e. x by the already established equality

$$\int_X \nu(\widehat{E}_x)\, \mu(dx) = 0.$$

Similarly, $\mu(E^y) = \mu(\widehat{E}^y) = 0$ for ν-a.e. y. Finally, every $\mu\otimes\nu$-measurable A has the form $A = B \cup E$, where $B \in \mathcal{A}\otimes\mathcal{B}$, $B \cap E = \varnothing$ and $\mu\otimes\nu(E) = 0$. Hence the theorem is true for A as well. □

3.11.2. Corollary. *The previous theorem is true in the case where μ and ν are σ-finite measures if the set A has a finite measure.*

PROOF. We write X and Y in the form $X = \bigcup_{n=1}^{\infty} X_n$, $Y = \bigcup_{n=1}^{\infty} Y_n$, where X_n and Y_n are increasing sets of finite measure, and then use the established assertion for $X_n \times Y_n$ and the monotone convergence theorem. □

3.11.3. Corollary. *Let $Y = \mathbb{R}^1$, let λ be Lebesgue measure on $\mathbb{R}^1$, and let $f \geqslant 0$ be an integrable function on a space $(X, \mathcal{A}, \mu)$ with a σ-finite measure μ. Then*

$$\int_X f \, d\mu = \mu\otimes\lambda\big(\{(x,y)\colon \ 0 \leqslant y \leqslant f(x)\}\big). \qquad (3.11.2)$$

PROOF. The set $A = \{(x,y)\colon \ 0 \leqslant y \leqslant f(x)\}$ is measurable with respect to $\mu\otimes\lambda$ by Proposition 3.10.4. It remains to observe that $\lambda(A_x) = f(x)$. □

Note that with the aid of equality (3.11.2) one could define the integral of a nonnegative function (Lebesgue called this the "geometric definition"). The facts established above imply the following general result.

3.11.4. Theorem. (FUBINI'S THEOREM) *Let μ and ν be σ-finite nonnegative measures and let f be a function integrable with respect to $\mu\otimes\nu$. Then, for μ-a.e. x, the function $y \mapsto f(x,y)$ is integrable with respect to ν, for ν-a.e. y, the function $x \mapsto f(x,y)$ is integrable with respect to μ, the functions $x \mapsto \int_Y f(x,y)\,\nu(dy)$ and $y \mapsto \int_X f(x,y)\,\mu(dx)$ are integrable on the corresponding spaces and*

$$\int_{X\times Y} f \, d(\mu\otimes\nu) = \int_Y \int_X f(x,y)\,\mu(dx)\,\nu(dy)$$

$$= \int_X \int_Y f(x,y)\,\nu(dy)\,\mu(dx). \qquad (3.11.3)$$

PROOF. It is clear that it suffices to prove the theorem for nonnegative functions f. Let us consider the space $X \times Y \times \mathbb{R}^1$ and the measure $\mu\otimes\nu\otimes\lambda$, where λ is Lebesgue measure. Set

$$A = \{(x,y,z)\colon \ 0 \leqslant z \leqslant f(x,y)\}.$$

Then by Corollary 3.11.3 we obtain

$$\int_{X\times Y} f \, d(\mu\otimes\nu) = \mu\otimes\nu\otimes\lambda(A).$$

Applying Theorem 3.11.1 and using Corollary 3.11.3 once again, we arrive at the equality

$$\mu\otimes\nu\otimes\lambda(A) = \int_X \nu\otimes\lambda(A_x)\,\mu(dx) = \int_X \left(\int_Y f(x,y)\,\nu(dy)\right)\mu(dx).$$

We also obtain the integrability of all functions entering these relations (their measurability is clear from Theorem 3.11.1 and the equality $f(x, y) = \lambda(A_{(x,y)})$). The second equality in (3.11.3) is proved similarly. $\square$

Exercise 3.12.39 suggests to construct examples showing that the existence and equality of the repeated integrals in (3.11.3) do not guarantee the $\mu \otimes \nu$-integrability of the measurable function f. In addition, it can happen that both repeated integrals exist, but are not equal. Finally, there exist measurable functions f such that one of the repeated integrals exists, but the other does not. There is, however, an important special case where the existence of the repeated integral implies the integrability of the function on the product.

3.11.5. Theorem. (TONELLI'S THEOREM) *Let f be a nonnegative $\mu \otimes \nu$-measurable function on $X \times Y$, where μ and let ν be σ-finite measures. Then $f \in \mathcal{L}^1(\mu \otimes \nu)$ if*

$$\int_Y \int_X f(x, y)\, \mu(dx)\, \nu(dy) < \infty.$$

PROOF. It suffices to prove our assertion for finite measures. Let us set $f_n = \min(f, n)$. Then the functions f_n are bounded and measurable with respect to $\mu \otimes \nu$, hence are integrable. It is clear that $f_n \to f$ at every point. It remains to observe that by Fubini's theorem applied to f_n we have the estimate

$$\int_{X \times Y} f_n\, d(\mu \otimes \nu) = \int_Y \left(\int_X f_n\, d\mu \right) d\nu \leqslant \int_Y \left(\int_X f\, d\mu \right) d\nu,$$

since $f_n(x, y) \leqslant f(x, y)$. By Fatou's theorem f is integrable. $\square$

Note that the existence of repeated integrals of a function f on $X \times Y$ does not imply its measurability (Exercise 3.12.41).

It is readily seen that Fubini 's theorem extends to signed measures, but Tonelli's theorem does not.

Let us give another useful corollary of Fubini's theorem.

3.11.6. Corollary. *Suppose that a function f on $X \times Y$ is measurable with respect to $\mu \otimes \nu$, where both measures are σ-finite. Suppose that, for μ-a.e. x, the function $y \mapsto f(x, y)$ is integrable with respect to ν. Then the function*

$$\Psi\colon\ x \mapsto \int_Y f(x, y)\, \nu(dy)$$

is measurable with respect to μ.

PROOF. Assume first that the measures μ and ν are bounded. Let us set $f_n(x, y) = f(x, y)$ if $|f(x, y)| \leqslant n$, $f_n(x, y) = n$ if $f(x, y) \geqslant n$, $f_n(x, y) = -n$ if $f(x, y) \leqslant -n$. Then the functions f_n are measurable with respect to $\mu \otimes \nu$ and bounded, hence are integrable. By Fubini's theorem the functions

$$\Psi_n(x) = \int_Y f_n(x, y)\, \nu(dy)$$

are μ-measurable. Since $f_n \to f$ pointwise and $|f_n| \leqslant |f|$, by the Lebesgue theorem we obtain that $\Psi_n(x) \to \Psi(x)$ for all points x for which the function

$y \mapsto |f(x,y)|$ is integrable with respect to ν, i.e., for μ-a.e. x. Therefore, Ψ is a μ-measurable function. In the general case we find an increasing sequence of measurable sets $X_n \times Y_n \subset X \times Y$ of finite $\mu \otimes \nu$-measure such that the measure $\mu \otimes \nu$ is concentrated on their union. Next we use the proved facts for the functions $\Phi_n(x) = \displaystyle\int_{Y_n} f(x,y)\,\nu(dy)$ and observe that $\Phi_n(x) \to \Psi(x)$ for μ-a.e. x by the Lebesgue dominated convergence theorem. $\qquad\square$

As an application of Fubini's theorem we derive a useful identity expressing the Lebesgue integral oven an abstract space via the Riemann integral over the halfline, which in case $p = 1$ was already announced in Theorem 3.6.3.

3.11.7. Theorem. *Let f be a measurable function on a measurable space $(X, \mathcal{A})$ with a measure μ with values in $[0, +\infty]$. Let $1 \leqslant p < \infty$. The function $|f|^p$ is integrable with respect to the measure μ if and only if the function $t \mapsto t^{p-1}\mu(x\colon |f(x)| > t)$ is integrable on $[0, +\infty)$ with respect to Lebesgue measure (equivalently, is Riemann integrable in the improper sense). Moreover,*

$$\int_X |f|^p \, d\mu = p \int_0^\infty t^{p-1}\mu(x\colon |f(x)| > t)\, dt. \qquad (3.11.4)$$

PROOF. Let $p = 1$. Suppose that the function f is integrable. Then the assertion reduces to the case of a σ-finite measure, since so is the measure μ on the set $\{f \neq 0\}$. Next, with the help of the monotone convergence theorem we can pass to the case of a finite measure. Let λ be Lebesgue measure on $[0, +\infty)$ and let

$$S = \{(x,y) \in X \times [0, +\infty)\colon\ y \leqslant |f(x)|\}.$$

The integral of $|f|$ coincides with the measure of the set S with respect to $\mu \otimes \lambda$, as we have already shown (Corollary 3.11.3). Let us calculate this measure by Fubini's theorem. For any fixed t we have

$$S_t = \{x\colon\ (x,t) \in S\} = \{x\colon\ t \leqslant |f(x)|\}.$$

Since the integral of $\mu(S_t)$ in the variable t over $[0, +\infty)$ gives the integral of $|f|$, we arrive at (3.11.4) with $(x\colon |f(x)| \geqslant t)$ in place of $(x\colon |f(x)| > t)$. However, for almost all t these two sets have equal μ-measures, since the set of points t such that $\mu(x\colon |f(x)| = t) > 0$ is at most countable. Indeed, if this set were uncountable, then for some number $k \in \mathbb{N}$ we could find an infinite set of points t with $\mu(x\colon |f(x)| = t) \geqslant k^{-1}$, which contradicts the integrability of f.

Conversely, if the integral in the right-hand side of (3.11.4) is finite, then the sets $(x\colon |f(x)| > t)$ have finite measures for all $t > 0$. Hence, for every $n \in \mathbb{N}$, the function f_n equal to $|f|$ whenever $n^{-1} \leqslant |f| \leqslant n$ and 0 otherwise is integrable. Since $\mu(x\colon |f_n(x)| > t) \leqslant \mu(x\colon |f(x)| > t)$, by the already verified assertion the functions f_n have uniformly bounded integrals. By Fatou's theorem the function f is integrable. The case $p > 1$ reduces to the case $p = 1$ by the change of variable $t = s^p$ and the equality $(x\colon |f(x)|^p > t) = (x\colon |f(x)| > t^{1/p})$. Here it suffices to apply the change of variables formula for the Riemann integral, it is also possible to apply the change of variables formula (4.4.3) for the Lebesgue integral. $\qquad\square$

Fubini's theorem is applied for constructing the convolution.

3.11.8. Lemma. *Let f be a Lebesgue measurable function on $\mathbb{R}^n$. Then the function $(x, y) \mapsto f(x - y)$ is Lebesgue measurable on the space $\mathbb{R}^{2n}$.*

PROOF. It suffices to observe that $f(x-y) = f_0\big(F(x, y)\big)$, where the function $f_0(x, y) = f(x)$ is Lebesgue measurable on $\mathbb{R}^{2n}$ and F is the invertible linear transformation $(x, y) \mapsto (x - y, y)$. The preimages of measurable sets under F are measurable (Theorem 2.5.8). $\qquad\square$

3.11.9. Theorem. (i) *Let $f, g \in \mathcal{L}^1(\mathbb{R}^n)$. Then the function*

$$f * g(x) = \int_{\mathbb{R}^n} f(x - y)g(y)\, dy, \qquad (3.11.5)$$

called the convolution of f and g, is defined for almost all x and integrable. In addition, $f * g = g * f$ almost everywhere. Moreover,

$$\|f * g\|_{L^1(\mathbb{R}^n)} \leqslant \|f\|_{L^1(\mathbb{R}^n)} \|g\|_{L^1(\mathbb{R}^n)}. \qquad (3.11.6)$$

(ii) *Let $f \in \mathcal{L}^\infty(\mathbb{R}^n)$, $g \in \mathcal{L}^1(\mathbb{R}^n)$. Then the function*

$$f * g(x) = \int_{\mathbb{R}^n} f(x - y)g(y)\, dy$$

is defined for all x and uniformly continuous, moreover,

$$\|f * g\|_{L^\infty(\mathbb{R}^n)} \leqslant \|f\|_{L^\infty(\mathbb{R}^n)} \|g\|_{L^1(\mathbb{R}^n)}. \qquad (3.11.7)$$

In addition, $f * g(x) = g * f(x)$.

PROOF. (i) We know that the function

$$\psi \colon (x, y) \mapsto |f(x - y)g(y)|$$

is measurable on $\mathbb{R}^{2n}$. Since

$$\int_{\mathbb{R}^n} \int_{\mathbb{R}^n} |f(x - y)|\, |g(y)|\, dx\, dy = \int_{\mathbb{R}^n} \left(\int_{\mathbb{R}^n} |f(z)|\, dz \right) |g(y)|\, dy < \infty,$$

by Theorem 3.11.5 the function ψ is integrable on $\mathbb{R}^{2n}$ and

$$\|\psi\|_{L^1(\mathbb{R}^{2n})} \leqslant \|f\|_{L^1(\mathbb{R}^n)} \|g\|_{L^1(\mathbb{R}^n)}.$$

By Fubini's theorem the function $\varphi \colon x \mapsto \int \psi(x, y)\, dy$ is defined for almost all x and integrable. Then $f * g$ is integrable as well, because $|f * g(x)| \leqslant \varphi(x)$ and the measurability of $f * g$ follows from Lemma 3.11.8 and the assertion about measurability in Fubini's theorem. For all points x for which the function $f(x - y)g(y)$ is integrable in y we use the change of variable $z = x - y$ (see Remark 3.3.11) to obtain the equality $f * g(x) = g * f(x)$.

(ii) Estimate (3.11.7) is obvious. The equality $f * g(x) = g * f(x)$ is proved as in (i). If the function g is continuous and vanishes outside some cube, then $g * f$ is obviously uniformly continuous. In the general case we take continuous functions g_j with bounded supports such that $\|g - g_j\|_{L^1(\mathbb{R}^n)} \to 0$ as $j \to \infty$ (Example 3.8.8). Estimate (3.11.7) gives the uniform convergence of $g_j * f$ to the function $g * f$, whence our claim follows. $\qquad\square$

3.11.10. Corollary. *Let* $g \in \mathcal{L}^1(\mathbb{R}^n)$. *If a function* $f \in \mathcal{L}^\infty(\mathbb{R}^n)$ *has bounded partial derivatives up to order* k, *then the function* $f * g$ *also has this property and*

$$\partial_{x_{i_1}} \ldots \partial_{x_{i_m}} (f * g) = (\partial_{x_{i_1}} \ldots \partial_{x_{i_m}} f) * g \quad \text{for all } m \leqslant k$$

and these derivatives are uniformly continuous.

PROOF. The equality $\partial_{x_i}(f * g) = \partial_{x_i} f * g$ follows from the theorem on differentiation of the Lebesgue integral with respect to a parameter (see Corollary 3.4.6). By induction our assertion extends to higher order derivatives, and their uniform continuity follows from the theorem above. $\qquad\square$

The convolution can be also defined for functions from $L^p(\mathbb{R}^n)$. We consider two important special cases and the general result is delegated to Exercise 3.12.42.

3.11.11. Proposition. (i) *Let* $f \in \mathcal{L}^p(\mathbb{R}^n)$, $g \in \mathcal{L}^q(\mathbb{R}^n)$, *where* $p^{-1} + q^{-1} = 1$. *Then the function* $f * g$ *defined by formula* (3.11.5) *is uniformly continuous and bounded and*

$$\|f * g\|_{L^\infty(\mathbb{R}^n)} \leqslant \|f\|_{L^p(\mathbb{R}^n)} \|g\|_{L^q(\mathbb{R}^n)}. \tag{3.11.8}$$

(ii) *Let* $f \in \mathcal{L}^p(\mathbb{R}^n)$, $g \in \mathcal{L}^1(\mathbb{R}^n)$. *Then the convolution* $f * g$ *is defined almost everywhere and*

$$\|f * g\|_{L^p(\mathbb{R}^n)} \leqslant \|f\|_{L^p(\mathbb{R}^n)} \|g\|_{L^1(\mathbb{R}^n)}. \tag{3.11.9}$$

PROOF. (i) For every fixed x the function $y \mapsto f(x - y)$ belongs to $\mathcal{L}^p(\mathbb{R}^n)$. By Hölder's inequality

$$\left| \int_{\mathbb{R}^n} f(x - y) g(y) \, dy \right| \leqslant \|f(x - \cdot)\|_{L^p(\mathbb{R}^n)} \|g\|_{L^q(\mathbb{R}^n)}$$
$$= \|f\|_{L^p(\mathbb{R}^n)} \|g\|_{L^q(\mathbb{R}^n)},$$

the integral in (3.11.5) exists for all x and defines a bounded function. For any $f \in C_0^\infty(\mathbb{R}^n)$ the uniform continuity of $f * g$ is obvious. In the general case, if $p < \infty$, we take continuous functions f_j with bounded supports converging to f in $L^p(\mathbb{R}^n)$ (Example 3.8.8). By the obtained estimate the functions $f_j * g$ converge uniformly to $f * g$. The case $p = \infty$, $q = 1$ has already been considered.

(ii) Hölder's inequality applied to the measure $|g| \, dy$ gives the estimate

$$\left| \int_{\mathbb{R}^n} f(x - y) g(y) \, dy \right|^p \leqslant \int_{\mathbb{R}^n} |f(x - y)|^p |g(y)| \, dy \, \|g\|_{L^1(\mathbb{R}^n)}^{p/q}.$$

Integrating in x and using (3.11.6) with $|f|^p$ and $|g|$ in place of f and g, we get

$$\|f * g\|_{L^p}^p \leqslant \|g\|_{L^1}^{p/q} \big\| |f|^p * |g| \big\|_{L^1} \leqslant \|g\|_{L^1}^{p/q} \|f\|_{L^p}^p \|g\|_{L^1(\mathbb{R}^n)},$$

which by the equality $1 + p/q = p$ gives the desired bound. $\qquad\square$

The convolution can be used for constructing smooth approximations of functions.

3.11.12. Corollary. *Let* $f \in \mathcal{L}^p(\mathbb{R}^n)$. *Suppose that a function* ϱ *of class* $C_0^\infty(\mathbb{R}^n)$ *has the integral* 1 *over the whole space. Set* $\varrho_\varepsilon(x) := \varepsilon^{-n} \varrho(x/\varepsilon)$ *if* $\varepsilon > 0$. *Then* $f * \varrho_\varepsilon \in C_b^\infty(\mathbb{R}^n)$ *and* $f * \varrho_\varepsilon \to f$ *in* $L^p(\mathbb{R}^n)$ *as* $\varepsilon \to 0$ *if* $p < \infty$.

PROOF. It remains to verify convergence in L^p. For $f \in C_0(\mathbb{R}^n)$ the functions $f * \varrho_\varepsilon$ have uniformly bounded supports and converge to f uniformly, which gives convergence in L^p. In the general case for any fixed $\delta > 0$ we choose $\varphi \in C_0(\mathbb{R}^n)$ such that $\|f - \varphi\|_{L^p} \leqslant \delta$. Next we find $\varepsilon_0 > 0$ such that $\|\varphi * \varrho_\varepsilon - \varphi\|_{L^p} \leqslant \delta$ whenever $\varepsilon < \varepsilon_0$. For such ε we have

$$\|f * \varrho_\varepsilon - f\|_{L^p} \leqslant \|f * \varrho_\varepsilon - \varphi * \varrho_\varepsilon\|_{L^p} + \|\varphi * \varrho_\varepsilon - \varphi\|_{L^p} + \|\varphi - f\|_{L^p}$$
$$\leqslant \|\varphi - f\|_{L^p}\|\varrho_\varepsilon\|_{L^1} + 2\delta \leqslant 3\delta,$$

since $\|\varrho_\varepsilon\|_{L^1} = 1$. $\qquad\square$

To consideration of convolutions we return in Chapter 9, where convolutions of generalized functions are discussed.

3.12. Complements and Exercises

(i) The Riemann integrability criterion (122). (ii) The image of a measure under a mapping (123). (iii) Uniform integrability (124). (iv) Liftings (128). Exercises (128).

3.12(i). The Riemann integrability criterion

For the reader's convenience we prove Lebesgue's criterion for Riemann's integrability (often included as a supplementary material in courses of calculus).

3.12.1. Theorem. *A function f on $[a, b]$ is Riemann integrable precisely when it is bounded and the set of its discontinuity points has measure zero.*

PROOF. The continuity of the function f at a point x is equivalent to the property that at this point the oscillation of f is zero, where the oscillation is defined by the formula

$$\omega_f(x) := \lim_{\delta \to 0} \sup\{|f(x') - f(x'')|:\ |x - x'| < \delta, |x - x''| < \delta\}.$$

For any function f and any $\varepsilon > 0$ the set $W_\varepsilon = \{x: \omega_f(x) \geqslant \varepsilon\}$ is closed. Hence the set D_f of all discontinuity points of f, which coincides with the set $\{\omega_f > 0\}$, is always Borel. Let f be Riemann integrable. Then it is bounded. Suppose that the set D_f has a positive measure. Then for some $\delta > 0$ the closed set $A := \{x: \omega_f(x) \geqslant \delta\}$ has a positive measure α. If the interval $[a, b]$ is partitioned into finitely many intervals I_j of equal length, then the total length of those intervals inside of which there are points from A is not less than α. Let $I_{j_1}, \ldots, I_{j_k}$ be all such intervals. Then inside of every I_{j_k} there are points x'_k and x''_k such that $|f(x'_k) - f(x''_k)| \geqslant \delta/2$, i.e., either $f(x') - f(x'') \geqslant \delta/2$ or $f(x''_k) - f(x'_k) \geqslant \delta/2$. We can assume that at least for the half of the intervals I_{j_k} the first inequality is fulfilled. In all other intervals I_j we take the middles x_j. Then the Riemann sums using x'_k differ from the sums using x''_k by at least $\alpha\delta/4$, which contradicts the Riemann integrability.

Conversely, let $\sup_{x \in [a,b]} |f(x)| \leqslant M$ and let D_f have measure zero. Let $\varepsilon > 0$. Then the compact set $K_\varepsilon := \{x: \omega_f(x) \geqslant \varepsilon\}$ of zero measure can be covered by finitely many intervals of the total length less than ε. Suppose we are given points $a = t_0 < t_1 < \cdots < t_n = b$ including the endpoints of all intervals of our covering. Refining the partition, we can assume that $|f(x') - f(x'')| \leqslant \varepsilon$ for all $x', x'' \in [t_i, t_{i+1}]$ for every partition interval not intersecting any other intervals of the covering. Hence the absolute value of the difference between the two Riemann sums for the given partition does not exceed $\varepsilon(b - a) + 2C\varepsilon$. Since ε was arbitrary, the function f is Riemann integrable. $\square$

3.12(ii). The image of a measure under a mapping

Suppose we are given two spaces X and Y with σ-algebras $\mathcal{A}$ and $\mathcal{B}$ and some $(\mathcal{A}, \mathcal{B})$-measurable mapping $f \colon X \to Y$. Then for every bounded (or bounded from below) measure μ on $\mathcal{A}$, the formula

$$\mu \circ f^{-1}(B) := \mu\big(f^{-1}(B)\big), \quad \text{where } B \in \mathcal{B},$$

defines the measure $\mu \circ f^{-1}$ on $\mathcal{B}$ called the *image of the measure* μ under the mapping f. The countable additivity of $\mu \circ f^{-1}$ follows from the countable additivity of μ. The following general *change of variables formula* holds.

3.12.2. Theorem. *Let μ be a bounded nonnegative measure. A function g on Y measurable with respect to $\mathcal{B}$ is integrable with respect to the measure $\mu \circ f^{-1}$ precisely when the function $g \circ f$ is integrable with respect to μ. In addition,*

$$\int_Y g(y)\, \mu \circ f^{-1}(dy) = \int_X g\big(f(x)\big)\, \mu(dx). \tag{3.12.1}$$

PROOF. For the indicator functions of sets from $\mathcal{B}$ formula (3.12.1) is the definition of the image measure, hence by linearity it extends to simple functions. From simple functions this formula extends to bounded $\mathcal{B}$-measurable functions, since such functions are uniform limits of simple functions. If g is a nonnegative $\mathcal{B}$-measurable function integrable with respect to $\mu \circ f^{-1}$, then for the functions $g_n = \min(g, n)$ equality (3.12.1) is already established. The monotone convergence theorem yields its validity for g, since the integrals of the functions $g_n \circ f$ with respect to the measure μ are uniformly bounded. Our reasoning shows also the necessity of the μ-integrability of $g \circ f$ for the integrability of $g \geqslant 0$ with respect to $\mu \circ f^{-1}$. By the linearity of (3.12.1) in g we obtain the general case. $\square$

It is clear that equality (3.12.1) remains in force for every function g that is measurable with respect to the Lebesgue completion of the measure $\mu \circ f^{-1}$ and $\mu \circ f^{-1}$-integrable. This follows from the fact that every such function is equivalent to a $\mathcal{B}$-measurable function. The condition of the $\mathcal{B}$-measurability of the function g can be replaced by the measurability with respect to the σ-algebra

$$\mathcal{A}^f = \{E \subset Y \colon f^{-1}(E) \in \mathcal{A}\},$$

on which we define the measure $\mu \circ f^{-1}$ by the same formula as on $\mathcal{B}$. However, one should bear in mind that the σ-algebra $\mathcal{A}^f$ can be strictly larger than the Lebesgue completion of $\mathcal{B}$ with respect to $\mu \circ f^{-1}$ (see [73, §7.5]).

In the case of a signed measure μ equality (3.12.1) remains in force if the function $g \circ f$ is integrable with respect to μ (this is clear from the Jordan decomposition for μ). However, the integrability of g with respect to the measure $\mu \circ f^{-1}$ now does not imply the integrability of $g \circ f$ (for example, the measure $\mu \circ f^{-1}$ can be zero for a nonzero measure μ).

If ψ is a $\mathcal{B}$-measurable function on X, then by formula (3.12.1) the integral of $\psi \circ f$ can be represented as the integral of ψ with respect to the measure $\mu \circ f^{-1}$ on the real line. For example,

$$\int_X |f(x)|^p \, \mu(dx) = \int_{\mathbb{R}} |t|^p \, \mu \circ f^{-1}(dt).$$

Let us introduce the distribution function of the function f:

$$\Phi_f(t) := \mu\big(x \colon f(x) < t\big), \quad t \in \mathbb{R}^1. \tag{3.12.2}$$

It is clear that $\Phi_f(t) = \mu \circ f^{-1}\big((-\infty, t)\big)$, i.e., Φ_f coincides with the distribution function $F_{\mu \circ f^{-1}}$ of the measure $\mu \circ f^{-1}$. In the case where μ is a probability measure, the function Φ_f is nondecreasing, left continuous, has right limits at every point and

$$\lim_{t \to -\infty} \Phi_f(t) = 0, \quad \lim_{t \to \infty} \Phi_f(t) = 1.$$

Recalling the definition of the Lebesgue–Stieltjes integral with respect to an increasing function (see formula (3.3.3)), we can write

$$\int_X \psi\big(f(x)\big) \, \mu(dx) = \int_{\mathbb{R}} \psi(t) \, d\Phi_f(t). \tag{3.12.3}$$

Let us explain the connection of the established general change of variables formula in the integral with the one known from the course of calculus. Let Ω be an open set in $\mathbb{R}^n$ and let $f \colon U \to \mathbb{R}^n$ be a mapping continuously differentiable and injective. Then the set $f(U)$ is measurable (it is a countable union of compact sets) and for every Lebesgue integrable function φ on $f(U)$ we have

$$\int_U \varphi\big(f(x)\big) |\det f'(x)| \, dx = \int_{f(U)} \varphi(y) \, dy. \tag{3.12.4}$$

In particular, the function $\varphi\big(f(x)\big) |\det f'(x)|$ is integrable on U; for $\varphi = 1$ we obtain that the volume of $f(U)$ is the integral of $|\det f'|$ over U. For justification it suffices to consider the case where the sets U and $f(U)$ are bounded (then the general case follows from this one with with the aid of decomposing φ into the difference of nonnegative functions and representing U in the form of a union of increasing domains with compact closures in U). From the point of view of the general change of variables formula equality (3.12.4) means that the mapping f transforms the measure μ given by density $|\det f'|$ with respect to Lebesgue measure on U to Lebesgue measure λ_n on $f(U)$. For the coincidence of the measures $\mu \circ f^{-1}$ and λ_n on $f(U)$ it suffices to have the equality of the integrals with respect to these measures for every smooth function with compact support in $f(U)$ (Exercise 3.12.43). But this is the fact known from the course of calculus.

3.12(iii). Uniform integrability

Here we discuss the property of uniform integrability, which is closely connected with the property of absolute continuity and passage to the limit under the integral sign.

Let $(X, \mathcal{A}, \mu)$ be a measurable space with a *finite* nonnegative measure μ.

3.12.3. Definition. *A set of functions $\mathcal{F} \subset \mathcal{L}^1(\mu)$ (or a set $\mathcal{F} \subset L^1(\mu)$) is called uniformly integrable if*

$$\lim_{C \to +\infty} \sup_{f \in \mathcal{F}} \int_{\{|f|>C\}} |f|\, d\mu = 0. \tag{3.12.5}$$

The set consisting of a single integrable function is uniformly integrable by the absolute continuity of the Lebesgue integral. Hence for every integrable function f_0 the set of all measurable functions f with $|f| \leqslant |f_0|$ is uniformly integrable.

3.12.4. Definition. *A set of functions $\mathcal{F} \subset \mathcal{L}^1(\mu)$ (or a set $\mathcal{F} \subset L^1(\mu)$) has uniformly absolutely continuous integrals if for every $\varepsilon > 0$ there exists $\delta > 0$ such that*

$$\int_A |f|\, d\mu < \varepsilon \quad \text{for all } f \in \mathcal{F} \text{ whenever } \mu(A) < \delta.$$

3.12.5. Proposition. *A set $\mathcal{F} \subset \mathcal{L}^1(\mu)$ is uniformly integrable precisely when it is bounded in $L^1(\mu)$ and has uniformly absolutely continuous integrals. In case of Lebesgue measure on a set $X \subset \mathbb{R}^n$ of finite measure uniform integrability is equivalent to the uniform absolute continuity of integrals.*

PROOF. Let $\mathcal{F}$ be uniformly integrable and $\varepsilon > 0$. Find $C > 0$ such that

$$\int_{\{|f|>C\}} |f|\, d\mu < \frac{\varepsilon}{2} \quad \forall f \in \mathcal{F}.$$

Set $\delta = \varepsilon(2C)^{-1}$. Let $\mu(A) < \delta$. Then for all $f \in \mathcal{F}$ we have

$$\int_A |f|\, d\mu = \int_{A \cap \{|f| \leqslant C\}} |f|\, d\mu + \int_{A \cap \{|f| > C\}} |f|\, d\mu \leqslant \frac{C\varepsilon}{2C} + \frac{\varepsilon}{2} = \varepsilon.$$

In addition,

$$\int_X |f|\, d\mu \leqslant C\mu(X) + \int_{\{|f|>C\}} |f|\, d\mu < C\mu(X) + \frac{\varepsilon}{2}.$$

Suppose now that the set $\mathcal{F}$ is bounded in $L^1(\mu)$ and has uniformly absolutely continuous integrals. Let $\varepsilon > 0$. Let us take δ from the definition of the uniform absolute continuity of integrals and observe that by Chebyshev's inequality there exists $C_1 > 0$ such that

$$\mu(\{|f| > C\}) \leqslant C^{-1} \|f\|_{L^1(\mu)} < \delta \quad \text{for all } f \in \mathcal{F} \text{ and } C > C_1.$$

Finally, in case of Lebesgue measure on a set $X \subset \mathbb{R}^n$ of finite measure the uniform absolute continuity of integrals implies the boundedness in $L^1(\mu)$, since for $\varepsilon = 1$ the set X can be partitioned into finitely many sets (say, $N(\delta)$) with measures less than the corresponding δ, so $\|f\|_{L^1(\mu)} \leqslant N(\delta)$ for all $f \in \mathcal{F}$. $\square$

The proof shows that the last assertion is true for every measure without atoms (see Exercise 2.7.51). If μ is a probability measure at the point 0, then the uniform absolute continuity of integrals does not imply the boundedness in $L^1(\mu)$, since the values of functions from $\mathcal{F}$ at the point 0 can be arbitrary large.

The next important result — the *Lebesgue–Vitali theorem* — is often applied in the theory of integration and its applications.

3.12.6. Theorem. *Let f be a μ-measurable function and let $\{f_n\}$ be a sequence of μ-integrable functions. Then the following assertions are equivalent:*

(i) *the sequence $\{f_n\}$ converges to f in measure and is uniformly integrable;*

(ii) *the function f is integrable and the sequence $\{f_n\}$ converges to f in the space $L^1(\mu)$.*

PROOF. Let condition (i) be fulfilled. Then the set $\{f_n\}$ is bounded in $L^1(\mu)$. By Fatou's theorem applied to the functions $|f_n|$, the function f is integrable. For the proof of convergence of $\{f_n\}$ to f in $L^1(\mu)$ it suffices to show that every subsequence $\{g_n\}$ in $\{f_n\}$ contains a subsequence $\{g_{n_k}\}$ converging to f in $L^1(\mu)$. For $\{g_{n_k}\}$ we take a subsequence in $\{g_n\}$ converging to f almost everywhere, which is possible by the Riesz theorem. Let $\varepsilon > 0$. By Proposition 3.12.5 there exists $\delta > 0$ such that

$$\int_A |f_n|\, d\mu \leqslant \varepsilon$$

for every n and every set A with $\mu(A) < \delta$. By Fatou's theorem we obtain

$$\int_A |f|\, d\mu \leqslant \varepsilon \quad \text{if } \mu(A) < \delta.$$

By Egorov's theorem there exists a set A with $\mu(A) < \delta$ such that convergence of $\{g_{n_k}\}$ to f on $X \backslash A$ is uniform. Let N be such that for all $k \geqslant N$ one has the inequality $\sup_{X \backslash A} |g_{n_k} - f| \leqslant \varepsilon$. Then

$$\int_X |g_{n_k} - f|\, d\mu \leqslant \varepsilon\mu(X) + \int_A |g_{n_k}|\, d\mu + \int_A |f|\, d\mu \leqslant \varepsilon\big(2 + \mu(X)\big),$$

whence we obtain convergence of $\{g_{n_k}\}$ to f in the space $L^1(\mu)$.

If condition (ii) is fulfilled, then the sequence $\{f_n\}$ is bounded in $L^1(\mu)$ and converges in measure to f. By Proposition 3.12.5 it remains to observe that the sequence $\{f_n\}$ has uniformly absolutely continuous integrals. This follows from the estimate

$$\int_A |f_n|\, d\mu \leqslant \int_A |f_n - f|\, d\mu + \int_A |f|\, d\mu$$

and the absolute continuity of the Lebesgue integral. $\qquad\square$

We now prove the following useful criterion of uniform integrability due to de la Vallée-Poussin. Even applied to a single function it gives a useful conclusion about "improvement of integrability".

3.12.7. Theorem. *Suppose that μ is a finite nonnegative measure. A family $\mathcal{F}$ of μ-integrable functions is uniformly integrable if and only if there exists a nonnegative increasing function G on $[0, +\infty)$ such that*

$$\lim_{t \to +\infty} \frac{G(t)}{t} = \infty \quad \text{and} \quad \sup_{f \in \mathcal{F}} \int G\big(|f(x)|\big)\, \mu(dx) < \infty. \tag{3.12.6}$$

In this case the function G can be taken convex increasing.

PROOF. Let condition (3.12.6) be fulfilled and let a number M dominate the integrals of the functions $G \circ |f|$ for all $f \in \mathcal{F}$. For a given number $\varepsilon > 0$ we find C such that $G(t)/t \geqslant M/\varepsilon$ for all $t \geqslant C$. Then for all $f \in \mathcal{F}$ we have $|f(x)| \leqslant \varepsilon G(|f(x)|)/M$ whenever $|f(x)| \geqslant C$. Therefore,

$$\int_{\{|f| \geqslant C\}} |f| \, d\mu \leqslant \frac{\varepsilon}{M} \int_{\{|f| \geqslant C\}} G \circ |f| \, d\mu \leqslant \frac{\varepsilon}{M} M = \varepsilon.$$

Thus, $\mathcal{F}$ is uniformly integrable. Let us prove the inverse implication. The function G will be found in the form

$$G(t) = \int_0^t g(s) \, ds,$$

where $g \geqslant 0$ is a nondecreasing step function, $g(t) \to +\infty$ as $t \to +\infty$, $g(t) = \alpha_n$ if $t \in (n, n+1]$, where $n = 0, 1, \ldots$. In order to find suitable numbers α_n, for each $f \in \mathcal{F}$ we set $\mu_n(f) = \mu(x \colon |f(x)| > n)$. By the uniform integrability of $\mathcal{F}$ there exists a sequence of natural numbers C_n increasing to infinity for which

$$\sup_{f \in \mathcal{F}} \int_{\{|f| \geqslant C_n\}} |f| \, d\mu \leqslant 2^{-n}. \tag{3.12.7}$$

We observe that

$$\int_{\{|f| \geqslant C_n\}} |f| \, d\mu \geqslant \sum_{j=C_n}^{\infty} j\mu\big(x \colon \; j < |f(x)| \leqslant j+1\big) \geqslant \sum_{k=C_n}^{\infty} \mu_k(f).$$

It follows from (3.12.7) that $\sum_{n=1}^{\infty} \sum_{k=C_n}^{\infty} \mu_k(f) \leqslant 1$ for all $f \in \mathcal{F}$. For $n < C_1$ we set $\alpha_n = 0$. For $n \geqslant C_1$ we set $\alpha_n = \max\{k \in \mathbb{N} \colon C_k \leqslant n\}$. It is clear that $\alpha_n \to +\infty$. For each $f \in \mathcal{F}$ the integral of $G(|f(x)|)$ does not exceed

$$\alpha_1 \mu\big(x \colon 1 < |f(x)| \leqslant 2\big) + (\alpha_1 + \alpha_2)\mu\big(x \colon 2 < |f(x)| \leqslant 3\big) + \ldots,$$

where the right-hand side equals $\sum_{n=1}^{\infty} \alpha_n \mu_n(f) = \sum_{n=1}^{\infty} \sum_{k=C_n}^{\infty} \mu_k(f) \leqslant 1$. It remains to observe that the function G is nonnegative, increasing, convex, and satisfies the condition $G(t)/t \to +\infty$ as $t \to +\infty$. □

3.12.8. Example. Let μ be a finite nonnegative measure. (i) For every integrable function f there exists a convex increasing positive function G on $[0, +\infty)$ such that $G(t)/t \to +\infty$ as $t \to +\infty$ and the function $G \circ |f|$ is integrable.

(ii) A family $\mathcal{F} \subset \mathcal{L}^1(\mu)$ is uniformly integrable if

$$\sup_{f \in \mathcal{F}} \int |f| \ln |f| \, d\mu < \infty,$$

where we set $0 \ln 0 := 0$. In order to apply de la Vallée–Poussin's criterion we take the function $G(t) = t \ln t$ for $t \geqslant 1$, $G(t) = 0$ for $t < 1$, observing that $G(|f|) \leqslant |f| \ln |f| + 1$. Another sufficient condition:

$$\sup_{f \in \mathcal{F}} \int |f|^p \, d\mu < \infty \quad \text{for some } p > 1.$$

3.12.9. Example. Let $A^1(U)$ be the Bergmann space of all complex functions on the open unit disc U in $\mathbb{C}$ that are Lebesgue integrable on U and analytic inside of U, hence are representable by the series $f(z) = \sum_{n=0}^{\infty} c_n z^n$ converging in U. Then every function from $A^1(U)$ is approximated in the metric of $L^1(U)$ by a sequence of polynomials in z.

PROOF. If the function f is analytic in a disc of a larger radius, then its Taylor series converges to it uniformly on U, hence in $L^1(U)$. In the general case uniform convergence of the Taylor series can fail. Let us show that the functions $f_\theta(z) = f(\theta z)$ with $\theta \in (0,1)$ converge to f in $L^1(U)$ as $\theta \to 1$. Since the function f_θ is analytic in the disc of radius $1/\theta$, it can be approximated by partial sums of its Taylor series on U. For every z we have $f_\theta(z) \to f(z)$ as $\theta \to 1$. It remains to take $\theta_n = 1 - 2^{-n}$ and observe that the functions f_{θ_n} are uniformly integrable. This is seen from de la Vallée-Poussin's criterion: taking the corresponding function G for f, we obtain the uniform boundedness of the integrals of $G(f_{\theta_n})$ with the aid of a change of variables. The same change of variables can be used at once to verify that the functions f_{θ_n} possess uniformly absolutely continuous integrals. $\square$

3.12(iv). Liftings

Let μ be a bounded nonnegative measure on a measurable space $(X, \mathcal{A})$. The set of all μ-measurable everywhere defined functions with values in $\mathbb{R}^1$ is a linear space and also possesses the multiplication operation. It is natural to ask whether in every μ-equivalence class one can choose a representative in such a way that the algebraic operations on these representatives would correspond to the operations on the classes. Such a choice is called a *lifting*. It turns out that a lifting exists on the set of bounded functions, but in nontrivial cases there is no lifting on L^p. Let us formulate the Maharam theorem (see [**73**, Chapter 10] for the proof).

3.12.10. Theorem. *There exists a mapping L, called a lifting, from $L^\infty(\mu)$ to $\mathcal{L}^\infty(\mu)$ with the following properties*:

(i) *for every element $f \in L^\infty(\mu)$, the function Lf is a representative of the equivalence class of f;*

(ii) *for all x we have $L(1)(x) = 1$;*

(iii) *for all $f, g \in L^\infty(\mu)$, $\alpha, \beta \in \mathbb{R}$ and all x we have*

$$L(\alpha f + \beta g)(x) = \alpha L(f)(x) + \beta L(g)(x), \quad L(fg) = L(f)(x)L(g)(x).$$

Exercises

3.12.11. Let $\{f_n\}$ be a sequence of functions measurable with respect to a σ-algebra $\mathcal{A}$ in a space X. Prove that the set L of all points $x \in X$ at which there exists a finite limit $\lim\limits_{n\to\infty} f_n(x)$ is an element of $\mathcal{A}$. The same is true for the sets L^- and L^+ of points where the limit equals $-\infty$ and $+\infty$.

3.12.12. (i) Construct a sequence of nonnegative functions f_n on $[0,1]$ the integrals of which tend to zero and which tend to zero at every point, but the function $\Phi(x) = \sup\limits_n f_n(x)$ is not integrable. In particular, the functions f_n have no common integrable majorant.

(ii) Construct a sequence of functions $f_n \geqslant 0$ on $[0,1]$ the integrals of which tend to zero, but $\sup_n f_n(x) = +\infty$ for every x.

3.12.13. For which α and β is the function $x^\alpha |\sin x|^\beta$ integrable on $[0,1]$? On $[0, 2\pi]$?

3.12.14. A sequence of nonnegative functions f_n on $[0,1]$ converges to a function f in $L^1[0,1]$. Is it true that $\sqrt{f_n} \to \sqrt{f}$ in $L^2[0,1]$?

3.12.15. A sequence of functions f_n on $[0, 1]$ converges to a function f in $L^1[0, 1]$. Prove that the functions $\sin f_n$ converge to $\sin f$ in $L^2[0, 1]$.

3.12.16. Let $f_n \geqslant 0$ be Lebesgue integrable functions on the interval $[0, 1]$ such that $f_n(x) \to f(x)$ a.e. Prove that

$$\int f(x)^2 e^{-f(x)} \, dx = \lim_{n \to \infty} \int f_n(x)^2 e^{-f_n(x)} \, dx.$$

3.12.17. Let $f_n \to f$ in $L^1(\mu)$. Prove that there exists a subsequence $\{n_k\}$ and a function $\Phi \in \mathcal{L}^1(\mu)$ such that $|f_{n_k}(x)| \leqslant \Phi(x)$ for all k for μ-a.e. x.

3.12.18. Let $\mu \geqslant 0$ be a finite measure and $\{f_n\} \subset L^1(\mu)$ a uniformly integrable sequence such that $\|f_n\|_{L^1(\mu)} \geqslant \delta > 0$ for all n. Prove that if numbers $\alpha_n \geqslant 0$ are such that $\sum_{n=1}^{\infty} \alpha_n |f_n(x)| < \infty$ a.e., then $\sum_{n=1}^{\infty} \alpha_n < \infty$. Deduce from this that if E_n are sets measurable with respect to μ, $\mu(E_n) \geqslant \delta > 0$ and $\sum_{n=1}^{\infty} \alpha_n I_{E_n}(x) < \infty$ a.e., then we have $\sum_{n=1}^{\infty} \alpha_n < \infty$.

3.12.19. Let $f \in \mathcal{L}^1(\mathbb{R}^n)$. Prove that

$$\lim_{h \to 0} \int_{\mathbb{R}^n} |f(x + h) - f(x)| \, dx = 0.$$

3.12.20. (i) Prove the following Young theorem. Suppose that three sequences of μ-integrable functions φ_n, f_n and ψ_n converge a.e. to μ-integrable functions φ, f and ψ, respectively, and $\varphi_n \leqslant f_n \leqslant \psi_n$ a.e. If the integrals of φ_n and ψ_n converge to the integrals of φ and ψ, respectively, then the integrals of f_n converge to the integral of f.

(ii) Deduce from (i) the following Vitali result. If integrable functions $f_n \geqslant 0$ converge a.e. to an integrable function f and the integrals of f_n converge to the integral of f, then $\{f_n\}$ converges to f in the metric of $L^1(\mu)$.

3.12.21. Suppose that $(X, \mathcal{A}, \mu)$ is a space with a nonnegative measure and a function $f \colon X \times (a, b) \to \mathbb{R}^1$ is integrable in x for all t and, for every x, is differentiable in t at a fixed point $t_0 \in (a, b)$ (independent of x). Suppose in addition that there exists a μ-integrable function Φ such that for every t there exists a set Z_t for which $\mu(Z_t) = 0$ and $|f(x, t) - f(x, t_0)| \leqslant \Phi(x)|t - t_0|$ whenever $x \notin Z_t$. Show that the integral of $f(x, t)$ against the measure μ is differentiable in t at the point t_0 and

$$\frac{d}{dt} \int_X f(x, t) \, \mu(x) \bigg|_{t = t_0} = \int_X \frac{\partial f(x, t_0)}{\partial t} \, \mu(x).$$

3.12.22. Construct a measurable set in $[0, 1]$ such that every function on $[0, 1]$ almost everywhere equal to its indicator function is discontinuous almost everywhere (and is not Riemann integrable, see Exercise 3.12.23).

HINT: take a set such that the set itself and its complement intersect every interval by a set of positive measure.

3.12.23. Prove that a function f on $[a, b]$ is Riemann integrable if and only if for every $\varepsilon > 0$ there exist step functions g and h such that $|f(x) - g(x)| \leqslant h(x)$ and $\int_a^b h(x) \, dx \leqslant \varepsilon$. Prove a similar assertion with continuous g and h.

HINT: apply Theorem 3.7.1 and Chebyshev's inequality.

3.12.24. Let $\{J_n\}$ be a sequence of disjoint intervals in $[0, 1]$ converging to the origin with $|J_n| = 4^{-n}$ and let $f = n^{-1}/|J_{2n}|$ on J_{2n}, $f = -n^{-1}/|J_{2n+1}|$ on J_{2n+1}, and at all other points let $f = 0$. Show that the function f is Riemann integrable in the improper sense, but is not Lebesgue integrable.

3.12.25.° Construct a sequence of nonnegative step functions that is Cauchy in L^1 and converges in L^1 to a bounded function that has no Riemann integrable modifications.

3.12.26.° Let $K = [0, 1)^n$, let $f \in \mathcal{L}^1(K)$ have the integral I_f, and let $v_1, \ldots, v_k \in \mathbb{R}^n$. Prove that there is x for which $I_f \leqslant \sum_{i=1}^k f(x + v_i)/k$, where f is extended to $\mathbb{R}^n$ periodically (show that I_f is the integral of $f(x + v_i)$ over K).

3.12.27.° Let $f \in \mathcal{L}^2(\mathbb{R}^1)$ and let the function $xf(x)$ also belong to $\mathcal{L}^2(\mathbb{R}^1)$. Prove that $f \in \mathcal{L}^1(\mathbb{R}^1)$. Obtain an analog for $f \in \mathcal{L}^p(\mathbb{R}^1)$.

3.12.28.° Let $f \in \mathcal{L}^p(\mathbb{R}^n)$, $1 \leqslant p < \infty$. Prove that $\lim_{h \to 0} \|f(\cdot + h) - f\|_{L^p} = 0$.

3.12.29. Let μ be a finite nonnegative measure on a space X. For $f, g \in L^0(\mu)$ set

$$d_0(f, g) := \int_X \frac{|f - g|}{1 + |f - g|}\, d\mu, \quad d_1(f, g) := \int_X \min(|f - g|, 1)\, d\mu.$$

Prove that d_0 and d_1 are metrics with respect to which $L^0(\mu)$ is complete. Prove that a sequence converges in one of these metrics precisely when it converges in measure.

3.12.30. Show that convergence almost everywhere on an interval I with Lebesgue measure cannot be generated by a topology, i.e., there is no topology on the set of all measurable functions on I (or on the set of all continuous functions on I) such that a sequence of functions converges in this topology precisely when it converges almost everywhere.

HINT: use that convergence generated by a topology possesses the following property: if an element f is such that every subsequence of the sequence $\{f_n\}$ contains a further subsequence converging to f, then $f_n \to f$; find a sequence of continuous functions which converges un measure, but does not converge at any point.

3.12.31. Let $f \in \mathcal{L}^1[a, b]$ and let φ be a bounded increasing function on (a, b). Prove that there exists a point $\xi \in [a, b]$ for which

$$\int_a^b \varphi(x) f(x)\, dx = \varphi(a + 0) \int_a^\xi f(x)\, dx + \varphi(b - 0) \int_\xi^b f(x)\, dx,$$

where $\varphi(a + 0)$ and $\varphi(b - 0)$ denote the right and left limits, respectively. If, in addition, φ is nonnegative, then there exists a point $\eta \in [a, b]$ such that

$$\int_a^b \varphi(x) f(x)\, dx = \varphi(b - 0) \int_\eta^b f(x)\, dx.$$

3.12.32.° Let $1 < p < \infty$, $p^{-1} + q^{-1} = 1$. Show that for all nonnegative numbers a and b one has $ab \leqslant \frac{a^p}{p} + \frac{b^q}{q}$.

3.12.33.° Let $1 < p < \infty$, $p^{-1} + q^{-1} = 1$, $f \in \mathcal{L}^p(\mu)$, $g \in \mathcal{L}^q(\mu)$, and

$$\int fg\, d\mu = \|f\|_p \|g\|_q > 0.$$

Prove that $g = \operatorname{sgn} f \cdot |f|^{p-1}$ a.e.

HINT: the proof of Hölder's inequality and Exercise 3.12.32 show that $|g| = |f|^{p-1}$.

3.12.34. (Generalized Hölder's inequality) Suppose that $1 \leqslant r, p_1, \ldots, p_n < \infty$, $1/p_1 + \cdots + 1/p_n = 1/r$, $f_1 \in \mathcal{L}^{p_1}(\mu), \ldots, f_n \in \mathcal{L}^{p_n}(\mu)$. Then $f_1 \cdots f_n \in \mathcal{L}^r(\mu)$ and

$$\left(\int_X |f_1 \cdots f_n|^r\, d\mu \right)^{1/r} \leqslant \left(\int_X |f_1|^{p_1}\, d\mu \right)^{1/p_1} \cdots \left(\int_X |f_n|^{p_n}\, d\mu \right)^{1/p_n}.$$

HINT: pass to the case $r = 1$, replacing p_i by p_i/r; next use induction on n and Hölder's inequality setting $1/p_2 + \cdots + 1/p_n = 1/q$.

3.12.35. Let $f \in \mathcal{L}^1(\mathbb{R}^1)$ and $f_h(x) := (2h)^{-1} \int_{x-h}^{x+h} f(t)\, dt$, where $h > 0$. Prove that $\|f_h - f\|_{L^1} \to 0$ as $h \to 0$.

3.12.36. Suppose that for every $k \in \mathbb{N}$ we are given a partition of $\mathbb{R}^n$ into countably many measurable parts $E_{k,i}$ of diameter less than δ_k and $\delta_k \to 0$. For every $f \in \mathcal{L}^1(\mathbb{R}^n)$, set $J_k(f)(x) := \lambda_n(E_{k,i})^{-1} \int_{E_{k,i}} f(y)\, dy$ if $x \in E_{k,i}$. Prove that $\|J_k(f) - f\|_{L^1} \to 0$.

3.12.37.° Prove that the graph of a measurable real function on a measurable space $(X, \mathcal{A}, \mu)$ with a finite measure is a measurable set of measure zero with respect to $\mu \otimes \lambda$, where λ is Lebesgue measure.

3.12.38.° Let $(X, \mathcal{A}_X)$ and $(Y, \mathcal{A}_Y)$ be measurable spaces and let f be a mapping from X to Y. Construct examples in which
 (i) f is $(\mathcal{A}_X, \mathcal{A}_Y)$-measurable, but the graph of f does not belong to $\mathcal{A}_X \otimes \mathcal{A}_Y$;
 (ii) the graph of f belongs to $\mathcal{A}_X \otimes \mathcal{A}_Y$, but f is not measurable.
Prove that if the set $\{(y, y), y \in Y\}$ belongs to $\mathcal{A}_Y \otimes \mathcal{A}_Y$, then the graph of every $(\mathcal{A}_X, \mathcal{A}_Y)$-measurable mapping belongs to $\mathcal{A}_X \otimes \mathcal{A}_Y$.
 HINT: (i) consider $f(x) = x$ from $[0,1]$ with the σ-algebra generated by singletons to the same space; (ii) consider $f(x) = x$ from $[0,1]$ with the standard Borel σ-algebra to $[0,1]$ with the σ-algebra of Lebesgue measurable sets. The last assertion follows from the measurability of $(x, y) \mapsto (f(x), y)$ with respect to the pair $(\mathcal{A}_X \otimes \mathcal{A}_Y, \mathcal{A}_Y \otimes \mathcal{A}_Y)$.

3.12.39.° Construct examples showing that (a) the existence and equality of the repeated integrals in (3.11.3) do not guarantee the $\mu \otimes \nu$-integrability of a measurable function f; (b) it can happen that both repeated integrals exist for a measurable function f, but are not equal; (c) there exists a measurable function f such that one of the repeated integrals exists, but the other one does not.

3.12.40.* (W. Sierpiński) (i) Show that the square in the plane contains a Lebesgue nonmeasurable set all intersections of which with the straight lines parallel to the coordinate axes consist of at most one point. (ii) Show that the square in the plane contains also a nonmeasurable set such that every straight line intersects it by at most two points.
 HINT: see [**73**, p. 232].

3.12.41. Show that there exists a Lebesgue nonmeasurable bounded nonnegative function f on $[0,1] \times [0,1]$ such that there exist finite and equal to zero repeated integrals
$$\int_0^1 \int_0^1 f(x, y)\, dxdy, \quad \int_0^1 \int_0^1 f(x, y)\, dydx.$$
 HINT: use the previous exercise.

3.12.42. (Young's inequality) Suppose that $1 \leqslant p \leqslant q \leqslant \infty$, $q^{-1} = r^{-1} + p^{-1} - 1$. Prove that for any $f \in \mathcal{L}^p(\mathbb{R}^n)$ and $g \in \mathcal{L}^r(\mathbb{R}^n)$, the function $f * g$ is defined almost everywhere (everywhere if $q = \infty$), belongs to $\mathcal{L}^q(\mathbb{R}^n)$, and $f * g = g * f$ almost everywhere and $\|f * g\|_{L^q(\mathbb{R}^n)} \leqslant \|f\|_{L^p(\mathbb{R}^n)} \|g\|_{L^r(\mathbb{R}^n)}$.
 HINT: see [**73**, Theorem 3.9.4].

3.12.43.° Prove that two bounded Borel measures on $\mathbb{R}^n$ are equal if they assign equal integrals to every smooth function with compact support.

3.12.44. Let μ be a probability measure on the Borel σ-algebra $\mathcal{B}(X)$ of a completely regular topological space X such that μ is Radon, which means that for all $B \in \mathcal{B}(X)$ one has $\mu(B) = \sup\{\mu(K) \colon K \subset B \text{ is compact}\}$. Prove that the set $C_b(X)$ of bounded continuous functions is dense in $L^2(\mu)$.

HINT: for approximating I_B, where B is a Borel set, for $\varepsilon > 0$ take compact sets $K_1 \subset B$ and $K_2 \subset X \backslash B$ with $\mu(B \backslash K_1) + \mu((X \backslash B) \backslash K_2) < \varepsilon$; apply Corollary 1.9.19 to the function f equal 1 on K_1 and 0 on K_2.

3.12.45. Let $\Omega \subset \mathbb{R}^n$ be a bounded domain with smooth boundary. Suppose that $f, g \colon \Omega \to \mathbb{R}^n$ are two mappings continuously differentiable in a neighborhood of the closure of Ω such that $f|_{\partial\Omega} = g|_{\partial\Omega}$. Prove that

$$\int_\Omega \det Df(x)\, dx = \int_\Omega \det Dg(x)\, dx.$$

HINT: the mappings $f_t(x) = tf(x) + x$ and $g_t(x) = tg(x) + x$ for small $t > 0$ are injective on Ω and $f_t(\Omega) = g_t(\Omega)$, since $f_t|_{\partial\Omega} = g_t|_{\partial\Omega}$; hence the volumes of $f_t(\Omega)$ and $g_t(\Omega)$ coincide, which gives the equality of the integrals of $\det Df_t$ and $\det Dg_t$, which are polynomials in t; compare their coefficients at t^n.

3.12.46. (i) Deduce from the previous exercise that there is no mapping f that is continuously differentiable in a neighborhood of the closed ball B such that $f(B) \subset \partial B$ and $f(x) = x$ whenever $x \in \partial B$. (ii) Deduce from (i) the *Bohl–Brouwer theorem*: every continuous mapping $F \colon B \to B$ has a fixed point, i.e., a point x_0 such that $F(x_0) = x_0$.

HINT: (i) take $g(x) = x$, then $\det Df(x) = 0$, $\det Dg(x) = 1$. (ii) Let B be the unit ball. If F has no fixed points and is continuously differentiable in a neighborhood of B, there exists $\varepsilon > 0$ such that, whenever $1 \leqslant q \leqslant 1 + \varepsilon$, the equation $F(x) = qx$ has no solutions in B; let φ be a smooth function on B, $\varphi|_{\partial B} = 0$, $\varphi(x) = (1 + \varepsilon)^{-1}$ whenever $\|x\| \leqslant 1 - \varepsilon/4$, $0 \leqslant \varphi \leqslant (1 + \varepsilon)^{-1}$; then $g(x) = x - \varphi(x)F(x)$ has no zeros on B, $f(x) = g(x)/\|g(x)\|$ takes values in ∂B and is the identity on ∂B. The case of a continuous map F is deduced from the considered one by means of smooth approximations.

3.12.47. Prove that there is a Lebesgue measurable function $f \colon [0,1] \to \mathbb{R}^1$ such that for no Borel function $g \colon [0,1] \to \mathbb{R}^1$ the inequality $f(x) \leqslant g(x)$ holds for all $x \in [0,1]$.

HINT: use Exercise 2.7.25 and the fact that the Cantor set is the image of the real line under some one-to-one Borel function.

3.12.48. (P. Sjögren) Let f and K be nonnegative integrable functions on $\mathbb{R}^n$ and let C be a constant such that $f(x) \leqslant f * K(x) + C$ a.e. Then $f \in L^\infty(\mathbb{R}^n)$.

HINT: let $\|f\| \leqslant 1$, $C \leqslant 1$, take M such that the integral of K over $A = \{K > M\}$ is less than $1/2$. Set $K_1 = I_A K$. Then $f \leqslant K_1 * f + M + 1 \leqslant K_1 * K_1 * f + C/2 + C$. By induction $f \leqslant K_1^{*n} * f + (M + 1)\sum_{i=0}^{n-1} 2^{-i}$, where K_1^{*n} is the n-fold convolution of K_1. Since $\|K_1^{*n}\|_1 \leqslant 2^{-n}$, we have $\|K_1^{*n} * f\|_1 \to 0$, so $f \leqslant (M + 1)\sum_{i=0}^{n-1} 2^{-i}$ a.e.

3.12.49. Let μ be an atomless probability measure on a space $(X, \mathcal{A})$. Prove that there is a $\mathcal{A}$-measurable function $f \colon X \to [0,1]$ such that $\mu \circ f^{-1}$ is Lebesgue measure.

HINT: see [73, Proposition 9.1.11].

3.12.50. (Lyapunov's theorem) Let $\mu_1, \ldots, \mu_n$ be atomless probability measures on a measurable space $(X, \mathcal{A})$. Prove that there exists a $\mathcal{A}$-measurable function $f \colon X \to [0,1]$ such that $\mu_i \circ f^{-1}$ is Lebesgue measure for every $i = 1, \ldots, n$.

HINT: see [73, p. 326].

3.12.51. Let $\mu_1, \ldots, \mu_n$ be atomless probability measures on a measurable space $(X, \mathcal{A})$ and let numbers $\alpha_1, \ldots, \alpha_k \geqslant 0$ be such that $\sum_{j=1}^k \alpha_j = 1$. Prove that there exist sets $A_1, \ldots, A_k \in \mathcal{A}$ such that $\mu_i(A_j) = \alpha_j$ for all $i = 1, \ldots, n$ and $j = 1, \ldots, k$.

HINT: take f from the previous exercise, divide $[0,1]$ into disjoint intervals I_j of length α_j and set $A_j = f^{-1}(A_j)$.

Connections between the Integral and Derivative

In this chapter we briefly discuss integration by parts for the Lebesgue integral and some related differentiation problems.

4.1. Differentiable Functions

We recall that a function f defined in a neighborhood of a point $x \in \mathbb{R}^1$ is called differentiable at this point if there exists a finite limit

$$\lim_{h \to 0} \frac{f(x+h) - f(x)}{h},$$

which is called the derivative of f at the point x and denoted by $f'(x)$. In analysis and diverse applications an important role is played by the problem of recovering a function from its derivative. The main formula of calculus — the famous Newton–Leibniz formula — gives the following expression for the function f on $[a, b]$ through its derivative f':

$$f(x) = f(a) + \int_a^x f'(y)\, dy. \tag{4.1.1}$$

For continuously differentiable functions f the integral in formula (4.1.1) exists in Riemann's sense, so there is no problem with interpretation of this equality. Problems do arise when we attempt to extend the Newton–Leibniz formula to broader classes of functions. If the derivative exists everywhere or almost everywhere, the questions arise about its integrability in a certain sense and about validity of equality (4.1.1). In order to explain the nature of potential difficulties here, we consider some examples. First we construct a function f that is differentiable at every point of the real line and such that f' is not integrable on $[0, 1]$ in Lebesgue's sense and in the improper Riemann sense.

4.1.1. Example. Let $f(x) = x^2 \sin \frac{1}{x^2}$ if $x \neq 0$ and $f(0) = 0$. Then the function f is everywhere differentiable, but the function f' is not Lebesgue integrable on the interval $[0, 1]$.

PROOF. The equality $f'(0) = 0$ follows from the boundedness of sin. For $x \neq 0$ we have $f'(x) = 2x \sin \frac{1}{x^2} - 2\frac{1}{x} \cos \frac{1}{x^2}$. It suffices to show that the function $\psi(x) = \frac{1}{x} \cos \frac{1}{x^2}$ is not Lebesgue integrable on $[0, 1]$. Suppose the opposite. Then the function $\frac{1}{x} \cos \frac{1}{2x^2}$ is integrable as well, which is verified with the aid of

© Springer Nature Switzerland AG 2020
V. I. Bogachev, O. G. Smolyanov, *Real and Functional Analysis*,
Moscow Lectures 4, https://doi.org/10.1007/978-3-030-38219-3_4

change of variables $y = \sqrt{2}x$. Therefore, the function $\varphi(x) = \frac{1}{x}\cos^2\frac{1}{2x^2}$ must be integrable. Since $\psi(x) = 2\varphi(x) - \frac{1}{x}$, we obtain the integrability of $\frac{1}{x}$, which is a contradiction. $\qquad\square$

In the considered example the function f' is integrable in the improper Riemann sense. However, now it is easy to spoil this property too. Let us take a compact set $K \subset [0,1]$ of positive Lebesgue measure without inner points (see Example 2.5.6). The set $[0,1]\backslash K$ has the form $\bigcup_{n=1}^{\infty}(a_n, b_n)$, where the intervals (a_n, b_n) are pairwise disjoint. Take a smooth function θ such that $\theta(x) = 1$ if $x \leqslant 1/2$ and $\theta(x) = 0$ if $x \geqslant 1$. Set $g(x) = \theta(x)f(x)$ if $x \geqslant 0$ and $g(x) = 0$ if $x < 0$. Then $g'(0) = g'(1) = 0$ and $|g(x)| \leqslant C\min\{x^2, (1-x)^2\}$ for some C.

4.1.2. Example. Set $F(x) = \sum_{n=1}^{\infty}(b_n - a_n)^2 g\left(\frac{x-a_n}{b_n-a_n}\right)$. The function F is everywhere differentiable, but F' is not Lebesgue integrable on $[0,1]$ and is discontinuous at every point of the set K (hence is not integrable in the improper Riemann sense).

PROOF. It is clear that the series defining F converges uniformly, because the function g is bounded. It suffices to show that $F'(x) = 0$ at every point $x \in K$, since in (a_n, b_n) the function F equals $(b_n - a_n)^2 g\big((x - a_n)/(b_n - a_n)\big)$. By our construction, $F(x) = 0$ if $x \in K$. Let $h > 0$. If $x + h \in K$, then $F(x + h) - F(x) = 0$. If $x + h \notin K$, then we can find an interval (a_n, b_n) containing $x + h$. Then $x + h - a_n < h$ and hence

$$\left|\frac{F(x+h) - F(x)}{h}\right| = \left|\frac{F(x+h)}{h}\right| = (b_n - a_n)^2 \frac{1}{h}\left|g\left(\frac{x+h-a_n}{b_n-a_n}\right)\right|$$
$$\leqslant \frac{(b_n - a_n)^2}{h}C\frac{h^2}{(b_n - a_n)^2} = Ch,$$

which tends to zero as $h \to 0$. The case $h < 0$ is similar. It is obvious that the function F' is not bounded in the right neighborhood of the point a_n, since F on (a_n, b_n) is an affine transformation of g on $(0,1)$. Therefore, F' is discontinuous at every point of the closure of $\{a_n\}$. This closure coincides with K due to the absence of inner points of K. $\qquad\square$

Thus, neither the Lebesgue integral nor the improper Riemann integral solve the problem of recovery of an everywhere differentiable function from its derivative. In Chapter 5 of the book [73] one can read about more general (non absolute) integrals solving this problem (but only in a certain sense). Note, however, that in applications of the theory of integral it is much more typical to encounter the problem of recovery of a function that has a derivative only almost everywhere. Of course, without additional assumptions such a recovery is impossible. For example, the Cantor function considered above (Proposition 3.1.12) has the zero derivative almost everywhere, but is not constant. Lebesgue described the class of all functions which are almost everywhere differentiable and can be recovered from the derivative by means of the Newton–Leibniz formula for the Lebesgue integral. It turned out that these are absolutely continuous functions. Before we start their discussion, we consider a broader class of functions which are also almost everywhere differentiable, although are not necessarily integrals.

In the study of derivative it is useful to consider the so-called derivates of a function f which take values in the extended real line and are defined by

$$D^+ f(x) = \limsup_{h \to +0} \frac{f(x+h) - f(x)}{h},$$

$$D_+ f(x) = \liminf_{h \to +0} \frac{f(x+h) - f(x)}{h},$$

$$D^- f(x) = \limsup_{h \to -0} \frac{f(x+h) - f(x)}{h},$$

$$D_- f(x) = \liminf_{h \to -0} \frac{f(x+h) - f(x)}{h}.$$

If $D^+ f(x) = D_+ f(x)$, then $f'_+(x) := D^+ f(x) = D_+ f(x)$ is called the right derivative of f at x; if $D^- f(x) = D_- f(x)$, then $f'_-(x) := D^- f(x) = D_- f(x)$ is called the left derivative of f at x. It is clear that the existence of a finite derivative of f at the point x is equivalent to the property that at this point the right and left derivatives coincide and are finite. Finally, if $f'_-(x) = f'_+(x) = +\infty$, then one writes $f'(x) = +\infty$ (and similarly for $-\infty$).

Let us also define the upper and lower derivatives $\overline{D}f(x)$ and $\underline{D}f(x)$ as, respectively, the upper and lower limits of the ratio $[f(x+h) - f(x)]/h$ as $h \to 0$, where $h \neq 0$.

4.1.3. Lemma. *For every function f on the interval $[a,b]$, the set of all points at which the right and left derivatives of f exist and are different is at most countable.*

PROOF. Let $D := \{x \colon f'_-(x) < f'_+(x)\}$ and let $\{r_n\}$ be the set of all rational numbers. For every point $x \in D$ there exists the smallest number k such that $f'_-(x) < r_k < f'_+(x)$. Next, there exists the smallest number m such that $r_m < x$ and for all $t \in (r_m, x)$ one has the inequality $(f(t) - f(x))/(t - x) < r_k$. Finally, there exists the smallest number n such that $r_n > x$ and for all $t \in (x, r_n)$ one has the inequality $(f(t) - f(x))/(t - x) > r_k$. By our choice of m and n we obtain

$$f(t) - f(x) > r_k(t - x) \quad \text{if } t \neq x \text{ and } t \in (r_m, r_n). \tag{4.1.2}$$

Thus, every point $x \in D$ is associated with a triple of natural numbers (k, m, n). Different points are associated with different triples. Indeed, suppose that two points x and y are associated with the same triple (k, m, n). Letting $t = y$ in (4.1.2) we obtain $f(y) - f(x) > r_k(y - x)$. If in (4.1.2) in place of x we take y and set $t = x$, then we obtain the opposite inequality. Thus, D is at most countable. Similarly one proves that the set $\{f'_+ < f'_-\}$ is at most countable. $\square$

4.2. Functions of Bounded Variation

4.2.1. Definition. *We shall say that a function* f *on a set* $T \subset \mathbb{R}^1$ *has bounded variation if*

$$V(f, T) := \sup \sum_{i=1}^{n} |f(t_{i+1}) - f(t_i)| < \infty,$$

where sup *is taken over all collections* $t_1 \leqslant t_2 \leqslant \cdots \leqslant t_{n+1}$ *in* T. *If* $T = [a, b]$, *then we set* $V_a^b(f) := V(f, [a, b])$.

If a function f has bounded variation, then it is bounded and for every $t_0 \in T$ one has

$$\sup_{t \in T} |f(t)| \leqslant |f(t_0)| + V(f, T).$$

We shall be mainly interested in the case where T is an interval $[a, b]$ or (a, b) (possibly, unbounded).

The simplest example of a function of bounded variation is a nondecreasing function f on $[a, b]$ (in case of an open or unbounded interval we also require that the limits at the endpoints be finite). Here $V_a^b(f) = V(f, [a, b]) = f(b) - f(a)$. It is clear that any nonincreasing function is of bounded variation. The same is true for the difference of two nondecreasing functions, since the space $BV[a, b]$ of all functions of bounded variation is obviously linear. Moreover,

$$V_a^b(\alpha f + \beta g) \leqslant |\alpha| V_a^b(f) + |\beta| V_a^b(g)$$

for every two functions f and g of bounded variation and arbitrary scalars α and β. This is obvious from the estimate

$$|\alpha f(t_{i+1}) + \beta g(t_{i+1}) - \alpha f(t_i) - \beta g(t_i)|$$
$$\leqslant |\alpha| |f(t_{i+1}) - f(t_i)| + |\beta| |g(t_{i+1}) - g(t_i)|.$$

4.2.2. Proposition. *Let* f *be a function of bounded variation on* $[a, b]$. *Then*

(i) *the functions* $V: x \mapsto V(f, [a, x])$ *and* $U: x \mapsto V(x) - f(x)$ *are nondecreasing on* $[a, b]$;

(ii) *the function* V *is continuous at a point* $x_0 \in [a, b]$ *precisely when the function* f *is continuous at this point*;

(iii) *for every* $c \in (a, b)$ *one has*

$$V(f, [a, b]) = V(f, [a, c]) + V(f, [c, b]). \tag{4.2.1}$$

PROOF. Since adding a new point to a partition of $[a, b]$ we do not decrease the corresponding sum of absolute values of increments of the function, for evaluation of $V_a^b(f)$ we can consider partitions containing the point c. Then

$$V(f, [a, b]) = \sup \left[\sum_{i=1}^{k} |f(t_{i+1}) - f(t_i)| + \sum_{i=k+1}^{n} |f(t_{i+1}) - f(t_i)| \right],$$

where sup is taken over all partitions with $t_{k+1} = c$. From this equality we obtain (4.2.1), whence it follows that V is a nondecreasing function. The function $U = V - f$ is also nondecreasing, since whenever $x \geqslant y$ we have

$$V(x) - V(y) = V_y^x(f) \geqslant |f(x) - f(y)| \geqslant f(x) - f(y).$$

Hence $|V(x) - V(y)| \geqslant |f(x) - f(y)|$, whence we obtain at once the continuity of f at any point x at which the function V is continuous. It remains to verify the continuity of V at the point x where f is continuous. Let $\varepsilon > 0$. Let us choose $\delta_0 > 0$ such that $|f(x + h) - f(x)| \leqslant \varepsilon/2$ if $|h| \leqslant \delta_0$. By definition there exist collections of points $a = t_1 < \cdots < t_{n+1} = x$ and $x = s_1 < \cdots < s_{n+1} = b$ such that

$$\left| V(f, [a, x]) - \sum_{i=1}^{n} |f(t_{i+1}) - f(t_i)| \right| + \left| V(f, [x, b]) - \sum_{i=1}^{n} |f(s_{i+1}) - f(s_i)| \right| \leqslant \frac{\varepsilon}{2}.$$

Let $|h| < \delta := \min(\delta_0, x - t_n, s_2 - x)$ and $h > 0$. Then

$$V(x + h) - V(x) = V_x^b(f) - V_{x+h}^b(f) \leqslant \sum_{i=1}^{n} |f(s_{i+1}) - f(s_i)|$$

$$+ \frac{\varepsilon}{2} - V_{x+h}^b(f) \leqslant |f(x) - f(x + h)| + |f(x + h) - f(s_2)|$$

$$+ \sum_{i=2}^{n} |f(s_{i+1}) - f(s_i)| + \frac{\varepsilon}{2} - V_{x+h}^b(f) \leqslant |f(x) - f(x + h)| + \frac{\varepsilon}{2} \leqslant \varepsilon,$$

since $|f(x + h) - f(s_2)| + \sum_{i=2}^{n} |f(s_{i+1}) - f(s_i)| \leqslant V_{x+h}^b(f)$. A similar estimate holds for $h < 0$. □

The variation of a function f is not always an additive set function. For example, $V(f, [0, 1]) = 1 > V(f, [0, 1)) = 0$ if $f(x) = 0$ on $[0, 1)$ and $f(1) = 1$.

4.2.3. Corollary. *Any continuous function of bounded variation is the difference of two continuous nondecreasing functions.*

4.2.4. Corollary. *Any function of bounded variation has at most countably many points of discontinuity.*

PROOF. By virtue of the results proved above it suffices to consider a nondecreasing function f. In this case its points of discontinuity are points x such that $\lim_{h \to 0+} f(x - h) < \lim_{h \to 0+} f(x + h)$. Clearly, the number of such points is at most countable. □

In the proof of the next important theorem we need a technical lemma that is of independent interest.

4.2.5. Lemma. *Let E be a set in $(0, 1)$. Suppose that we are given some family $\mathcal{I}$ of intervals which for every $x \in E$ and every $\delta > 0$ contains an interval $(x, x + h)$ with $0 < h < \delta$. Then for every $\varepsilon > 0$ one can choose a finite subfamily of disjoint intervals $I_1, \ldots, I_k$ in this family such that*

$$\lambda\left(\bigcup_{j=1}^{k} I_j\right) < \lambda^*(E) + \varepsilon \quad \text{and} \quad \lambda^*\left(E \bigcap \bigcup_{j=1}^{k} I_j\right) > \lambda^*(E) - \varepsilon.$$

Moreover, for any given open set U containing E these intervals can be taken in U.

PROOF. Let us find an open set $G \supset E$ for which $\lambda(G) < \lambda^*(E) + \varepsilon$. If we are given an open set $U \supset E$, then G is taken in U. Deleting from $\mathcal{I}$ all intervals not contained in G, we can assume from the very beginning that all intervals in $\mathcal{I}$ belong to G. Then the measure of their union is not greater than $\lambda^*(E) + \varepsilon$. Let E_n be the set of all points $x \in E$ for which $\mathcal{I}$ contains an interval $(x, x+h)$ with $h > 1/n$. Since E is the union of the increasing sets E_n, there exists n with $\lambda^*(E_n) > \lambda^*(E) - \varepsilon/2$. Let $\delta = \varepsilon/(2n+2)$. Let a_1 be the infimum of E_n. Let us take a point $x_1 \in E_n$ in $[a_1, a_1 + \delta]$. Let $I_1 = (x_1, x_1 + h_1) \in \mathcal{I}$ be an interval with $h_1 > 1/n$. If the set $E_n \cap (x_1 + h_1, 1)$ is not empty, then let a_2 be its infimum. Let us take a point $x_2 \in E_n$ in $[a_2, a_2 + \delta]$ and find $I_2 = (x_2, x_2 + h_2) \in \mathcal{I}$ with $h_2 > 1/n$. Continuing this process, we obtain $k \leqslant n$ intervals $I_j = (x_j, x_j + h_j)$ with $h_j > 1/n$ such that there are no points of E_n to the right of $x_k + h_k$, $x_j \in [a_j, a_j + \delta]$, where a_j is the infimum of $E_n \cap (x_{j-1} + h_{j-1}, 1)$. It is clear that all points of E_n not covered by $\bigcup_{j=1}^k I_j$ are contained in the union of the intervals $[a_j, a_j + \delta]$, $j = 1, \ldots, k$. Hence the outer measure of the set of such points does not exceed $n\delta < \varepsilon/2$. By the subadditivity of the outer measure

$$\lambda^*\left(E \cap \bigcup_{j=1}^k I_j\right) \geqslant \lambda^*(E_n) - \lambda^*\left(E_n \setminus \bigcup_{j=1}^k I_j\right) > \lambda^*(E) - \varepsilon.$$

In addition, $\lambda\left(\bigcup_{j=1}^k I_j\right) \leqslant \lambda(G) < \lambda^*(E) + \varepsilon$. $\square$

4.2.6. Theorem. *Every function of bounded variation on an interval has a finite derivative almost everywhere.*

PROOF. It suffices to prove this assertion for a nondecreasing function f. Let $S = \{x \colon D_+ f(x) < D^+ f(x)\}$. Let us show that $\lambda(S) = 0$. For this it suffices to show that for every pair of rational numbers $u < v$ the set

$$S(u, v) = \{x \colon D_+ f(x) < u < v < D^+ f(x)\}$$

has measure zero. Suppose that $\lambda^*\big(S(u,v)\big) = c > 0$. Every point x in $S(u,v)$ is the left end of arbitrarily small intervals $(x, x+h)$ with $f(x+h) - f(x) < hu$. By Lemma 4.2.5 for any fixed $\varepsilon > 0$ there exists a finite collection of pairwise disjoint intervals $(x_i, x_i + h_i)$ such that for their union U we have the estimates

$$\lambda^*(U \cap S(u,v)) > c - \varepsilon, \quad \lambda(U) = \sum_i h_i < c + \varepsilon.$$

It is clear that $\sum_i [f(x_i + h_i) - f(x_i)] < \sum_i h_i u < u(c + \varepsilon)$. On the other hand, every point $y \in U \cap S(u,v)$ is the left end of arbitrarily small intervals $(y, y+r)$ with $f(y+r) - f(y) > rv$. Hence by the same Lemma 4.2.5 we can find a finite collection of pairwise disjoint intervals $(y_j, y_j + r_j)$ in U whose union W satisfies the estimate

$$\lambda^*\big(W \cap S(u,v)\big) > \lambda^*\big(U \cap S(u,v)\big) - \varepsilon > c - 2\varepsilon.$$

Then $\sum_j [f(y_j + r_j) - f(y_j)] > v \sum_j r_j > v(c - 2\varepsilon)$. Since f is nondecreasing and every interval $(y_j, y_j + r_j)$ belongs to some interval $(x_i, x_i + h_i)$, we have

$$\sum_j [f(y_j + r_j) - f(y_j)] \leqslant \sum_i [f(x_i + h_i) - f(x_i)].$$

Thus, $v(c - 2\varepsilon) < u(c + \varepsilon)$. Using that $\varepsilon > 0$ was arbitrary we obtain that $v \leqslant u$, a contradiction. Therefore, $c = 0$ and the right derivative of f exists almost everywhere. Similarly we prove that the left derivative of f exists almost everywhere. Moreover, the set E of all points x where $f'_+(x) = +\infty$ has measure zero. Indeed, for every $\varepsilon > 0$ and $N \in \mathbb{N}$ there exists $h(x) > 0$ such that $f(x + h) - f(x) > Nh$ whenever $0 < h < h(x)$. By Lemma 4.2.5 there is a finite collection of disjoint intervals $(x_i, x_i + h_i)$, where $h_i = h(x_i)$, of total measure $L \geqslant \lambda^*(E) - \varepsilon$. Then the intervals $\big(f(x_i), f(x_i + h_i)\big)$ are disjoint and the measure of their union is not less than NL, whence $\lambda^*(E) \leqslant \varepsilon + L \leqslant \varepsilon + V(f, [a, b])/N$. Now our assertion follows from Lemma 4.1.3. $\qquad\Box$

4.2.7. Corollary. *Every nondecreasing function f on the interval $[a, b]$ almost everywhere on $[a, b]$ has a finite derivative f', moreover, the function f' is integrable on $[a, b]$ and*

$$\int_a^b f'(x)\, dx \leqslant f(b) - f(a). \qquad (4.2.2)$$

PROOF. Set $f(x) = f(b)$ if $x \geqslant b$, $h_n = n^{-1}$, $f_n(x) = \dfrac{f(x + h_n) - f(x)}{h_n}$. Then $f_n \geqslant 0$ and a.e. we have $f_n(x) \to f'(x)$. In addition,

$$\int_a^b f_n(x)\, dx = \frac{1}{h_n} \int_{a+h_n}^{b+h_n} f(y)\, dy - \frac{1}{h_n} \int_a^b f(x)\, dx$$

$$= \frac{1}{h_n} \int_b^{b+h_n} f(x)\, dx - \frac{1}{h_n} \int_a^{a+h_n} f(x)\, dx \leqslant f(b) - f(a),$$

since $f = f(b)$ on $[b, b + h_n]$ and $f \geqslant f(a)$ on $[a, a + h_n]$. It remains to apply Fatou's theorem. $\qquad\Box$

4.2.8. Corollary. *For every function f of bounded variation on $[a, b]$ the derivative f' exists almost everywhere and is integrable.*

The Cantor function C_0 (see Example 3.1.12) shows that in (4.2.2) there might be no equality even for continuous functions. Indeed, $C'_0(x) = 0$ almost everywhere, but $C_0(x) \neq$ const. In the next section we consider a subclass of the class of functions of bounded variation leading to the equality in (4.2.2).

In connection to Lebesgue–Stieltjes measures discussed in §2.5 it is the right place here to prove the following assertion.

4.2.9. Proposition. *Let μ be a bounded Borel measure on $\mathbb{R}$ (possibly, signed) with the distribution function $F_\mu(t) := \mu\big((-\infty, t)\big)$. Then $\|\mu\| = V(F_\mu, \mathbb{R})$. If μ is defined on $[a, b)$, then $\|\mu\| = V_a^b(F_\mu)$.*

PROOF. If semi-open intervals $[a_1, b_1), \ldots, [a_n, b_n)$ are pairwise disjoint, then the sum of $|F_\mu(b_i) - F_\mu(a_i)| = \big|\mu\big([a_i, b_i)\big)\big|$ does not exceed $\|\mu\|$. On the other hand, let X^+ and X^- be Borel sets from the Hahn decomposition for μ. For any given $\varepsilon > 0$ we can find compact sets $K_1 \subset X^+$ and $K_2 \subset X^-$ with

$|\mu|(X^+ \backslash K_1) < \varepsilon$ and $|\mu|(X^- \backslash K_2) < \varepsilon$. Next we can find disjoint semi-open intervals $[a_1, b_1), \ldots, [a_k, b_k)$ and $[a_{k+1}, b_{k+1}), \ldots, [a_m, b_m)$ such that

$$\mu(K_1) \leqslant \sum_{i=1}^{k} \mu\big([a_i, b_i)\big) + \varepsilon, \quad \mu(K_2) \geqslant \sum_{i=k+1}^{m} \mu\big([a_i, b_i)\big) - \varepsilon.$$

This is clear from Theorem 2.3.11. Then we obtain

$$\|\mu\| = \mu(X^+) - \mu(X^-) \leqslant \mu(K_1) - \mu(K_2) + 2\varepsilon$$

$$\leqslant \sum_{i=1}^{m} |F_\mu(b_i) - F_\mu(a_i)| + 4\varepsilon \leqslant V(F_\mu, \mathbb{R}) + 4\varepsilon.$$

Since $\varepsilon > 0$ was arbitrary, we have $\|\mu\| \leqslant V(F_\mu, \mathbb{R})$. □

4.3. Absolutely Continuous Functions

In this section we consider functions on bounded intervals.

4.3.1. Definition. *A function f on an interval $[a, b]$ is called absolutely continuous if, for every $\varepsilon > 0$, there exists $\delta > 0$ such that for every collection of pairwise disjoint intervals (a_i, b_i) in $[a, b]$ with $\sum_{i=1}^{n} |b_i - a_i| < \delta$ one has $\sum_{i=1}^{n} |f(b_i) - f(a_i)| < \varepsilon$.*

It is seen from the definition that an absolutely continuous function is uniformly continuous. The converse is not true: for example, the function f on the interval $[0, 1]$, equal to n^{-1} at $(2n)^{-1}$, vanishing at $(2n+1)^{-1}$ and defined by continuity linearly between these points is not absolutely continuous. This is clear from divergence of the series $\sum_{n=1}^{\infty} |f((2n)^{-1})|$ and convergence to zero of the quantities $\sum_{n=m}^{\infty} [(2n)^{-1} - (2n+1)^{-1}]$.

Denote by $AC[a, b]$ the class of all absolutely continuous functions on the closed interval $[a, b]$.

4.3.2. Lemma. *Suppose that functions $f_1, \ldots, f_n$ are absolutely continuous on the interval $[a, b]$, a function φ is defined and satisfies the Lipschitz condition on a set $U \subset \mathbb{R}^n$ such that for all x in the interval $[a, b]$ we have $\big(f_1(x), \ldots, f_n(x)\big) \in U$. Then the function $\varphi(f_1, \ldots, f_n)$ is absolutely continuous on the interval $[a, b]$.*

PROOF. By assumption for some $C > 0$ we have $|\varphi(x) - \varphi(y)| \leqslant C\|x - y\|$ for all $x, y \in U$. In addition, for any given $\varepsilon > 0$ there exists $\delta > 0$ such that for every $j = 1, \ldots, n$ we obtain $\sum_{i=1}^{k} |f_j(b_i) - f_j(a_i)| < \varepsilon n^{-1}(C+1)^{-1}$ for every collection of pairwise disjoint intervals $(a_1, b_1), \ldots, (a_k, b_k)$ in $[a, b]$ such that $\sum_{i=1}^{k} |b_i - a_i| < \delta$. Now the estimate

$$\sum_{i=1}^{k} |\varphi\big(f_1(b_i), \ldots, f_n(b_i)\big) - \varphi\big(f_1(a_i), \ldots, f_n(a_i)\big)|$$

$$\leqslant \sum_{i=1}^{k} C\Big(\sum_{j=1}^{n} |f_j(b_i) - f_j(a_i)|^2\Big)^{1/2} \leqslant C \sum_{i=1}^{k} \sum_{j=1}^{n} |f_j(b_i) - f_j(a_i)| < \varepsilon$$

gives our assertion. □

4.3.3. Corollary. *If functions f and g are absolutely continuous, then so are fg and $f + g$ and if $g \geqslant c > 0$, then f/g is also absolutely continuous.*

4.3.4. Proposition. *Every function f that is absolutely continuous on $[a, b]$ has bounded variation on this interval.*

PROOF. Let us take δ corresponding to $\varepsilon = 1$ in the definition of an absolutely continuous function. Next we take a natural number $M > |b - a|\delta^{-1}$. Suppose we have a partition $a = t_1 \leqslant \cdots \leqslant t_n = b$. Let us add to the points t_i all points of the form $s_j = a + (b - a)jM^{-1}$, where $j = 0, \ldots, M$. The elements of the obtained partition will be denoted by z_i, $i = 1, \ldots, k$. Then

$$\sum_{i=1}^{n-1} |f(t_{i+1}) - f(t_i)| \leqslant \sum_{i=1}^{k-1} |f(z_{i+1}) - f(z_i)|$$

$$= \sum_{j=1}^{M} \sum_{i:\, z_{i+1} \in (s_{j-1}, s_j]} |f(z_{i+1}) - f(z_i)| \leqslant M,$$

since the sum of lengths of the intervals (z_i, z_{i+1}) with $z_{i+1} \in (s_{j-1}, s_j]$ does not exceed $s_j - s_{j-1} = |b - a|M^{-1} < \delta$. Thus, $V(f, [a, b]) \leqslant M$. $\square$

4.3.5. Corollary. *For every absolutely continuous function f on $[a, b]$, its derivative f' exists almost everywhere and is integrable. In particular, this is true if f is Lipschitz.*

PROOF. We can apply Corollary 4.2.8. $\square$

4.3.6. Proposition. *Let a function f be absolutely continuous on $[a, b]$. Then the function $V: x \mapsto V(f, [a, x])$ is also absolutely continuous and hence f is the difference of two nondecreasing absolutely continuous functions V and $V - f$.*

PROOF. Let $\varepsilon > 0$. Find $\delta > 0$ such that the sum of absolute values of the increments of f on every finite collection of disjoint intervals (a_i, b_i) of the total length less than δ is estimated by $\varepsilon/2$. It remains to observe that the sum of absolute values of the increments of V on the intervals (a_i, b_i) is estimated by ε. Indeed, suppose we are given such collection of k intervals (a_i, b_i). For every i we can find a partition of $[a_i, b_i]$ by points $a_i = t_1^i \leqslant \cdots \leqslant t_{N_i}^i = b_i$ such that

$$V(f, [a_i, b_i]) < \sum_{j=1}^{N_i-1} |f(t_{j+1}^i) - f(t_j^i)| + \varepsilon 4^{-i}.$$

Then

$$\sum_{i=1}^{k} |V(b_i) - V(a_i)| = \sum_{i=1}^{k} V(f, [a_i, b_i]) < \sum_{i=1}^{k} \sum_{j=1}^{N_i-1} |f(t_{j+1}^i) - f(t_j^i)| + \frac{\varepsilon}{2} < \varepsilon,$$

since the intervals (t_j^i, t_{j+1}^i) are pairwise disjoint and the sum of their lengths does not exceed δ. $\square$

For a Lebesgue integrable function f on $[a, b]$ and a number C the function

$$F(x) = C + \int_a^x f(t)\, dt$$

is called the *indefinite integral* of f. It turns out that functions of this form are precisely absolutely continuous functions.

4.3.7. Theorem. (LEBESGUE'S THEOREM) *A function f is absolutely continuous on $[a, b]$ if and only if there exists a function g integrable on $[a, b]$ such that*

$$f(x) = f(a) + \int_a^x g(y)\, dy \quad \forall\, x \in [a, b]. \tag{4.3.1}$$

PROOF. If f has the form (4.3.1), then by the absolute continuity of the Lebesgue integral for every $\varepsilon > 0$ there exists $\delta > 0$ such that

$$\int_D |g(x)|\, dx < \varepsilon$$

for every set D of measure less than δ. It remains to observe that

$$\sum_{i=1}^n |f(b_i) - f(a_i)| = \sum_{i=1}^n \left| \int_{a_i}^{b_i} g(x)\, dx \right| \leqslant \int_U |g(x)|\, dx < \varepsilon$$

for every union $U = \bigcup_{i=1}^n [a_i, b_i]$ of pairwise disjoint intervals of the total length less than δ.

Let us prove the converse assertion. It suffices to prove it for a nondecreasing function f, since by Proposition 4.3.6 the function f is the difference of two nondecreasing absolutely continuous functions. Let $f(a) = 0$. According to Theorem 2.5.10 there exists a nonnegative Borel measure μ on $[a, b]$ such that $f(x) = \mu([a, x))$ for all $x \in [a, b]$. Now it suffices to show that the measure μ is given by some integrable density g with respect to Lebesgue measure λ, which by the Radon–Nikodym theorem is equivalent to the absolute continuity of μ with respect to λ. Let E be a Borel set of Lebesgue measure zero in $[a, b]$. We have to verify that $\mu(E) = 0$. Fix $\varepsilon > 0$. By assumption there exists $\delta > 0$ such that the sum of absolute values of the increments of f on any disjoint intervals of the total length less than δ is estimated by ε. We can find an open set U containing E such that $\lambda(U) < \delta$. The set U is the finite or countable union of pairwise disjoint intervals (a_i, b_i). Due to our choice of δ, for every finite union of (a_i, b_i) we have

$$\mu\left(\bigcup_{i=1}^n (a_i, b_i) \right) = \sum_{i=1}^n |f(b_i) - f(a_i)| < \varepsilon,$$

whence by the countable additivity of μ we obtain that $\mu(U) \leqslant \varepsilon$. Therefore, $\mu(E) \leqslant \mu(U) \leqslant \varepsilon$, since $\mu \geqslant 0$. Thus, $\mu(E) = 0$. $\qquad\square$

4.3.8. Corollary. *If* (4.3.1) *holds, then*

$$V(f, [a, b]) = \int_a^b |g(x)|\, dx = \|g\|_{L^1[a, b]}. \tag{4.3.2}$$

PROOF. Since for every interval $[s, t]$ in $[a, b]$ we have

$$|f(t) - f(s)| = \left| \int_s^t g(x)\, dx \right| \leqslant \int_s^t |g(x)|\, dx,$$

it follows that $V(f, [a, b]) \leqslant \|g\|_{L^1[a,b]}$. Let us prove the inverse inequality. We can assume that $f(a) = 0$. Fix $\varepsilon > 0$. Using the absolute continuity of the Lebesgue integral, we can find $\delta > 0$ such that

$$\int_D |g(x)|\, dx < \frac{1}{8}\varepsilon$$

for every set D of measure less than δ. Set

$$\Omega_+ := \{x\colon g(x) \geqslant 0\}, \quad \Omega_- := \{x\colon g(x) < 0\}.$$

Let us find in $[a, b]$ a finite collection of pairwise disjoint intervals (a_1, b_1), ...,(a_n, b_n) for which

$$\lambda\Big(\Omega_+ \,\triangle\, \bigcup_{i=1}^{n}(a_i, b_i)\Big) < \delta. \tag{4.3.3}$$

Next we pick in $[a, b]\backslash \bigcup_{i=1}^{n}(a_i, b_i)$ a finite collection of pairwise disjoint intervals $(c_1, d_1), \ldots, (c_k, d_k)$ for which

$$\lambda\Big(\Omega_- \,\triangle\, \bigcup_{i=1}^{k}(c_i, d_i)\Big) < \delta.$$

Set $\Delta_i := (a_i, b_i)\backslash\{g > 0\}$. Then

$$f(b_i) - f(a_i) = \int_{a_i}^{b_i} g(x)\, dx = \int_{a_i}^{b_i} |g(x)|\, dx + \int_{a_i}^{b_i}\Big[g(x) - |g(x)|\Big]\, dx$$

$$= \int_{a_i}^{b_i} |g(x)|\, dx - 2\int_{\Delta_i} |g(x)|\, dx.$$

From (4.3.3) we have $\sum_{i=1}^{n}\lambda(\Delta_i) < \delta$. Hence

$$\sum_{i=1}^{n} |f(b_i) - f(a_i)| \geqslant \sum_{i=1}^{n}\int_{a_i}^{b_i} |g(x)|\, dx - \frac{1}{4}\varepsilon \geqslant \int_{\Omega_+} |g(x)|\, dx - \frac{1}{2}\varepsilon.$$

Similarly, $\sum_{i=1}^{k} |f(d_i) - f(c_i)| \geqslant \int_{\Omega_-} |g(x)|\, dx - \frac{1}{2}\varepsilon$. Thus,

$$V(f, [a, b]) \geqslant \sum_{i=1}^{n} |f(b_i) - f(a_i)| + \sum_{i=1}^{k} |f(d_i) - f(c_i)| \geqslant \|g\|_{L^1[a,b]} - \varepsilon,$$

which completes the proof. $\qquad\square$

4.4. The Newton–Leibniz Formula

As we shall see in this section, the Lebesgue integral gives the Newton–Leibniz formula (4.1.1) in its natural generality.

4.4.1. Lemma. *Suppose that a function f is integrable on $[a, b]$ and for all* $x \in [a, b]$ *we have* $\displaystyle\int_a^x f(t)\, dt = 0$. *Then $f = 0$ almost everywhere.*

PROOF. The equality to zero of the integral with a variable upper limit yields that the integral of f over any interval in $[a, b]$ is zero, which gives the equality to zero of the integrals of f over finite unions of intervals. This yields that the integral of f over the set $\Omega = \{x \colon f(x) > 0\}$ vanishes. Indeed, let $\varepsilon > 0$. By the absolute continuity of the Lebesgue integral there exists $\delta > 0$ such that $\int_D |f|\, dx < \varepsilon$ for every set D of measure less than δ. Let us find a set A equal to a finite union of intervals for which $\lambda(\Omega \bigtriangleup A) < \delta$. Then

$$\int_\Omega f(x)\, dx \leqslant \int_A f(x)\, dx + \int_{\Omega \bigtriangleup A} |f(x)|\, dx \leqslant \int_A f(x)\, dx + \varepsilon = \varepsilon.$$

Since $\varepsilon > 0$ was arbitrary, the left-hand side of this inequality vanishes, hence Ω has measure zero. Similarly, the set $\{f < 0\}$ has measure zero. An alternative justification: the Borel measure $\mu := f \cdot \lambda$ vanishes on all intervals, hence on the generated σ-algebra, which is the Borel σ-algebra. In other words, the integrals of f over all Borel sets are zero, whence $f = 0$ a.e. $\square$

4.4.2. Theorem. *Let $f \in \mathcal{L}^1[a, b]$. Then*

$$\frac{d}{dx} \int_a^x f(t)\, dt = f(x) \quad \text{almost everywhere on } [a, b].$$

PROOF. Set $f(x) = 0$ if $x \notin [a, b]$. Let

$$F(x) = \int_a^x f(t)\, dt.$$

Suppose first that $|f(x)| \leqslant M < \infty$. Let $h_n \to 0$. The function F is Lipschitz. By Corollary 4.3.5 it is almost everywhere differentiable on $[a, b]$. Then $\lim_{n \to \infty} h_n^{-1}[F(x + h_n) - F(x)] = F'(x)$ for a.e. $x \in [a, b]$. Since

$$\left| \frac{F(x + h_n) - F(x)}{h_n} \right| = \left| \frac{1}{h_n} \int_x^{x+h_n} f(t)\, dt \right| \leqslant M,$$

by the dominated convergence theorem we obtain

$$\lim_{n \to \infty} \int_a^x \frac{F(y + h_n) - F(y)}{h_n}\, dy = \int_a^x F'(y)\, dy$$

for every $x \in [a, b]$. We observe that

$$\int_a^x \frac{F(y + h_n) - F(y)}{h_n}\, dy = \frac{1}{h_n} \int_{a+h_n}^{x+h_n} F(y)\, dy - \frac{1}{h_n} \int_a^x F(y)\, dy$$

$$= \frac{1}{h_n} \int_x^{x+h_n} F(y)\, dy - \frac{1}{h_n} \int_a^{a+h_n} F(y)\, dy,$$

which tends to zero $F(x) - F(a)$ as $n \to \infty$ by the continuity of F. Thus,

$$F(x) = F(x) - F(a) = \int_a^x F'(y)\, dy.$$

Therefore, we arrive at the identity

$$\int_a^x \left[F'(y) - f(y) \right] dy = 0 \quad \forall x \in [a, b].$$

According to Lemma 4.4.1 this means that $F'(x) - f(x) = 0$ a.e. on $[a, b]$.

Let us proceed to the general case. We can assume that $f \geqslant 0$, since f is the difference of two nonnegative integrable functions. Let $f_n = \min(f, n)$. Since $f - f_n \geqslant 0$, the function

$$\int_a^x \left(f(t) - f_n(t) \right) dt$$

is nondecreasing, hence its derivative exists almost everywhere and is nonnegative. Thus,

$$\frac{d}{dx} \int_a^x f(y)\, dy \geqslant \frac{d}{dx} \int_a^x f_n(y)\, dy \quad \text{a.e.}$$

By the boundedness of f_n and the already established facts we obtain the estimate $F'(x) \geqslant f_n(x)$ a.e. Hence $F'(x) \geqslant f(x)$ a.e., whence

$$\int_a^b F'(x)\, dx \geqslant \int_a^b f(x)\, dx.$$

On the other hand, by Corollary 4.2.7 we have

$$\int_a^b F'(x)\, dx \leqslant F(b) - F(a) = \int_a^b f(x)\, dx,$$

whence

$$\int_a^b \left[F'(x) - f(x) \right] dx = 0,$$

which is only possible if $F'(x) - f(x) = 0$ a.e., because $F'(x) - f(x) \geqslant 0$ a.e., as shown above.   □

4.4.3. Corollary. *Let f be a function integrable on $[a, b]$. Then for almost every point $x \in [a, b]$ we have*

$$\lim_{h \to 0} \frac{1}{2h} \int_{x-h}^{x+h} |f(y) - f(x)|\, dy = 0, \quad f(x) = \lim_{h \to 0} \frac{1}{2h} \int_{x-h}^{x+h} f(y)\, dy.$$

Points x with such a property are called Lebesgue points of the function f.

PROOF. The second of these equalities is fulfilled almost everywhere by the above theorem. Applying it to the function $|f(x) - c|$, we obtain that for every rational number c almost everywhere one has the equality

$$|f(x) - c| = \lim_{h \to 0} \frac{1}{2h} \int_{x-h}^{x+h} |f(y) - c|\, dy.$$

Then for almost all x this equality is true simultaneously for all rational numbers c. For such x it remains in force for every $c \in \mathrm{IR}^1$: take rational numbers $c_k \to c$ and observe that $\big| |f(y) - c_k| - |f(y) - c_m| \big| \leqslant |c_k - c_m|$. Finally, we take $c = f(x)$.   □

If the function f is absolutely continuous, then, combining the previous theorem with Theorem 4.3.7, we see that the integrable function g whose indefinite integral is $f - f(a)$ equals f' almost everywhere. Hence we obtain the following *Newton–Leibniz formula* for absolutely continuous functions.

4.4.4. Corollary. *Suppose that a function f is absolutely continuous on $[a, b]$. Then its derivative exists almost everywhere, is integrable and satisfies the equality*

$$f(x) = f(a) + \int_a^x f'(t)\, dt, \quad x \in [a, b]. \tag{4.4.1}$$

The Newton–Leibniz formula yields the following *integration by parts formula*.

4.4.5. Corollary. *Let f and g be absolutely continuous functions on the interval $[a, b]$. Then*

$$\int_a^b f'(x)g(x)\, dx = f(b)g(b) - f(a)g(a) - \int_a^b f(x)g'(x)\, dx. \tag{4.4.2}$$

PROOF. Since the function fg is absolutely continuous (Corollary 4.3.3), we can apply the Newton–Leibniz formula and the fact that $(fg)' = f'g + fg'$ almost everywhere (i.e., at all point where f and g are differentiable). □

Yet another useful corollary of the Newton–Leibniz formula is the *change of variable formula*.

4.4.6. Corollary. *Let φ be a monotone absolutely continuous function on the interval $[c, d]$ and let $\varphi([c, d]) \subset [a, b]$. Then, for every function f Lebesgue integrable on the interval $[a, b]$, the function $f(\varphi)\varphi'$ is integrable on $[c, d]$ and*

$$\int_{\varphi(c)}^{\varphi(d)} f(x)\, dx = \int_c^d f(\varphi(y))\varphi'(y)\, dy. \tag{4.4.3}$$

This assertion remains in force also for intervals of the form $(-\infty, d]$, $[c, +\infty)$, and $(-\infty, +\infty)$.

PROOF. We can assume that φ is nondecreasing and $a = \varphi(c)$, $b = \varphi(d)$. By the integrability of φ' on $[c, d]$ the nonnegative Borel measure $\mu = \varphi' \cdot \lambda$, where λ is Lebesgue measure, is finite on $[c, d]$. Denote by ν the Borel measure $\mu \circ \varphi^{-1}$ on $[a, b]$, i.e., $\nu(B) = \mu(\varphi^{-1}(B))$. By the general change of variables formula (Theorem 3.12.2) equality (4.4.3) for all Borel integrable functions f is equivalent to the property that the measure ν coincides with Lebesgue measure λ_1 on $[a, b]$. Hence for the proof in the case of Borel f it suffices to establish the equality of the values of ν and λ_1 on every interval $[\alpha, \beta]$ in $[a, b]$ (see Corollary 3.5.3). There is an interval $[\gamma, \delta] \subset [c, d]$ such that $\varphi(\gamma) = \alpha$, $\varphi(\delta) = \beta$, $[\gamma, \delta] = \varphi^{-1}([\alpha, \beta])$. It remains to observe that

$$\nu([\alpha, \beta]) = \mu([\gamma, \delta]) = \int_\gamma^\delta \varphi'(y)\, dy = \varphi(\delta) - \varphi(\gamma) = \beta - \alpha.$$

In order to extend the proved equality from Borel functions to all Lebesgue measurable ones we verify that the measure ν is absolutely continuous. This is equivalent to the property that $\varphi^{-1}(E) \cap \{y \colon \varphi'(y) > 0\}$ has Lebesgue measure zero for every set E of Lebesgue measure zero. Since E can be covered by a Borel set of Lebesgue measure zero, we can assume that E is Borel. Then it remains to apply equality (4.4.3) to the function $f = I_E$. The case of an unbounded interval follows from the considered one. $\square$

Formula (4.4.3) is true for a not necessarily monotone function φ if it is known in addition that the function $f(\varphi)\varphi'$ is integrable, but unlike the considered case this is not fulfilled automatically (see [**73**, Exercises 5.8.63–5.8.65]).

As a corollary of the established facts we obtain the following *Lebesgue decomposition* of a monotone function.

4.4.7. Proposition. *Let F be a nondecreasing left continuous function on the interval $[a, b]$. Then $F = F_{\mathrm{ac}} + F_{\mathrm{sing}}$, where F_{ac} is an absolutely continuous nondecreasing function and F_{sing} is a nondecreasing left continuous function with $F'_{\mathrm{sing}}(t) = 0$ a.e. Moreover, $F_{\mathrm{sing}} = F_{\mathrm{a}} + F_{\mathrm{c}}$, where F_{c} is a continuous nondecreasing function and F_{a} is a left continuous nondecreasing jump function, i.e., $F_{\mathrm{a}}(t) = \sum_{n \colon t_n < t} h_n$, where $\{t_n\} \subset (a, b)$, $h_n > 0$, $\sum_{n=1}^{\infty} h_n < \infty$.*

PROOF. We know that F' exists a.e., is integrable and, whenever $a \leqslant x \leqslant y \leqslant b$, one has $F(y) - F(x) \geqslant \displaystyle\int_x^y F'(t)\, dt$. Hence the function

$$F_{\mathrm{sing}}(x) := F(x) - \int_a^x F'(t)\, dt$$

is increasing. It is clear that $F'_{\mathrm{sing}}(x) = 0$ a.e. Set $F_{\mathrm{ac}}(x) := \displaystyle\int_a^x F'(t)\, dt$. The function F_{sing} has at most countably many discontinuity points t_n. The jump of F_{sing} at the point t_n will be denoted by h_n. Set $F_{\mathrm{a}}(t) := \sum_{n \colon t_n < t} h_n$. The function F_{a} (*a jump function*) has jumps at the points t_n of size h_n, i.e., $\lim_{t \to t_n+} F_{\mathrm{a}}(t) = F_{\mathrm{a}}(t_n) + h_n$; at all other points it is continuous. It is straightforward to verify that we have obtained an increasing left continuous function and the function $F_{\mathrm{c}} := F_{\mathrm{sing}} - F_{\mathrm{a}}$ is increasing and continuous. $\square$

The Lebesgue decomposition can be also regarded from another side (which gives another justification). If F is the distribution function of a bounded nonnegative Borel measure μ on the interval $[a, b]$ (every left continuous increasing function has such a form), then the Lebesgue decomposition for measures gives the equality $\mu = \mu_{\mathrm{ac}} + \mu_{\mathrm{sing}}$, where μ_{ac} is an absolutely continuous measure and μ_{sing} is a measure singular with respect to Lebesgue measure. Then F_{ac} and F_{sing} are the corresponding distribution functions. Next, in the singular measure μ_{sing} we can extract a purely atomic (concentrated on a countable set) and an atomless components, which gives the decomposition of F_{sing}. A continuous monotone function $f \neq \mathrm{const}$ for which $f'(t) = 0$ almost everywhere, is called *singular*.

For an application of the Newton–Leibniz formula to differentiation with respect to a parameter, see Exercise 4.5.12.

4.5. Complements and Exercises

(i) Integration by parts in the Stieltjes integral (148). (ii) Convergence of Fourier series (148). Exercises (157).

4.5(i). Integration by parts in the Stieltjes integral

Let μ and ν be bounded Borel measures on an interval (a,b) (possibly, unbounded) with the distribution functions F_μ and F_ν, i.e., $F_\mu(t) := \mu\big((a,t)\big)$ and similarly for ν. These functions are bounded and Borel. Hence they are integrable with respect to all Borel measures.

4.5.1. Theorem. *For all $a, b \in [-\infty, +\infty]$ we have*

$$\int_{(a,b)} F_\mu(t)\,\nu(dt) = F_\mu(b)F_\nu(b) - \int_{(a,b)} F_\nu(t+)\,\mu(dt). \qquad (4.5.1)$$

If at least one of the functions F_μ and F_ν is continuous (or they have no common points of jumps), then in place of $F_\nu(t+)$ in this equality one can write $F_\nu(t)$.

PROOF. By Fubini' theorem we have

$$\int_{(a,b)} F_\mu(t)\,\nu(dt) = \int_{(a,b)} \int_{(a,b)} I_{(a,t)}(s)\,\mu(ds)\,\nu(dt)$$

$$= \int_{(a,b)} \nu\big((s,b)\big)\,\mu(ds) = \int_{(a,b)} [F_\nu(b) - F_\nu(s+)]\,\mu(ds),$$

as required. The last assertion is obvious. $\qquad\square$

If F is an absolutely continuous function on $[a,b]$, then

$$\int_{(a,b)} F(t)\,\nu(dt) = F(b)F_\nu(b) - \int_{(a,b)} F_\nu(t)F'(t)\,dt,$$

which follows from (4.5.1) applied to the measure μ with density F'.

4.5(ii). Convergence of Fourier series

Here we discuss the pointwise convergence of Fourier series. In the next chapter we shall consider convergence in mean square for more general orthogonal expansions. However, the specific features of trigonometric series are such that it is more natural to discuss them in this chapter connected with real analysis.

For a function $f \in \mathcal{L}^1[0, 2\pi]$ and $n = 0, 1, \ldots$, let

$$a_n := \frac{1}{\pi} \int_0^{2\pi} f(x)\cos nx\,dx, \quad b_n := \frac{1}{\pi} \int_0^{2\pi} f(x)\sin nx\,dx. \qquad (4.5.2)$$

These numbers are called the *Fourier coefficients* of the function f, and the formal series

$$\frac{a_0}{2} + \sum_{n=1}^\infty [a_n \cos nx + b_n \sin nx]$$

is called the *Fourier series* of the function f with respect to the trigonometric system. This series can be considered for arbitrary given coefficients a_n and b_n, not necessarily generated by an integrable function f. Then this series is called a *trigonometric series*. One can also introduce the coefficients

$$c_n := \frac{1}{\sqrt{2\pi}} \int_0^{2\pi} f(x)e^{-inx}\, dx, \quad n \in \mathbb{Z},$$

leading to the formal series $\sum_{n \in \mathbb{Z}} c_n e^{inx}$.

Such series are useful in applications, because they provide a convenient tool for representations or approximations of functions. Hence we arrive at the questions about convergence of Fourier series, the properties of the Fourier coefficients and about a possibility to recover a function or determine its properties by its Fourier coefficients. For functions on the whole real line there is a close object: the Fourier transform, studied in Chapter 9. Fourier series and transforms are the classics of analysis and remain an important object of research. All the listed questions turned out to be very subtle; substantial contributions in their investigation are due to outstanding mathematicians, such as B. Riemann, G. Cantor, A. Lebesgue, N. N. Luzin, A. N. Kolmogorov, and L. Carleson. Below we give a brief account of their main achievements.

Convergence of a Fourier series can be understood in different ways. First, one can consider convergence of the Fourier series at some given point and (where there is convergence) pose the question about the relation of the sum of the series to the value of the original function. In this case it is reasonable to assume that there is some natural modification of the function (otherwise by redefining f at the regarded point we obtain another value of the function, but the same Fourier series). For example, such a question sounds quite natural for continuous 2π-periodic functions. Second, one can be interested in convergence of the Fourier series almost everywhere (and in the relation of the sum to the function f almost everywhere), but not take care of single points. Finally, it can be useful to study convergence of the series in the metric of L^1 (or L^p for $f \in L^p$). This does not exhaust all possibilities; for example, even convergence at a point can be understood in a generalized way, by using diverse modes of summation (this will be briefly mentioned below). In addition, other questions arise. Say, if we are given a sequence of numbers $\{c_n\}$ (or two sequences $\{a_n\}$ and $\{b_n\}$), then it is interesting to know when there is a function with such Fourier coefficients. Before presenting more special results, we inform the reader about some fundamental general facts.

First of all, even in case of a 2π-periodic continuous function f the Fourier series can fail to converge at some points. This will be established in Chapter 6 (see Example 6.1.11). In order to guarantee convergence at a given point, one has to impose certain additional conditions in a neighborhood (for example, the Dini condition). However, the Cesàro summation gives convergence for every 2π-periodic continuous function and enables us to uniformly approximate such functions by trigonometric polynomials.

Secondly, as shown by A. N. Kolmogorov, there exists a function $f \in \mathcal{L}^1[0, 2\pi]$ such that its Fourier series diverges at *every* point. However, as we shall see below, if we sum such a series not in the usual sense, but in the Cesàro or Abel sense, then almost everywhere its sum coincides with f. In particular, two integrable functions with the same Fourier coefficients are equal almost everywhere.

Thirdly, if a function f belongs to $\mathcal{L}^p[0, 2\pi]$ with some $p > 1$, then its Fourier series converges to it almost everywhere. This fact was proved in 1966 by L. Carleson for $p = 2$ and in 1968 by R. A. Hunt for all $p > 1$. A detailed proof can be read in Arias de Reyna [31], Fremlin [197], Jørboe, Mejlbro [295], Grafakos [235], Lacey [693], and Mozzochi [440]. In particular, the Fourier series of a continuous function f converges to it almost everywhere (in spite of the fact that points can exist at which there is no convergence). Before appearance of a more general result of Carleson even this remained an open problem for half a century. At the same time, for every function $f \in \mathcal{L}^2[0, 2\pi]$ its Fourier series converges to it in the metric of $L^2[0, 2\pi]$, which is proved very simply for every orthonormal basis (see Chapter 5).

One of the simplest fundamental facts is contained in the following *Riemann–Lebesgue theorem*.

4.5.2. Theorem. *Let* $f \in \mathcal{L}^1[a, b]$ *and* $\alpha_n \to \infty$. *Then*

$$\lim_{n \to \infty} \int_a^b f(t) \sin \alpha_n t \, dt = \lim_{n \to \infty} \int_a^b f(t) \cos \alpha_n t \, dt = 0.$$

In particular, the Fourier coefficients of any integrable function tend to zero.

PROOF. If $f = I_{[c,d)}$, then the equality is verified directly. It remains valid for finite linear combinations of indicator functions of intervals. Finally, for every $\varepsilon > 0$ there exists a linear combination g of indicator functions of intervals such that the integral of the function $|f - g|$ over $[a, b]$ is less than ε. Then the integral of $f(t) \sin \alpha_n t$ differs from the integral of $g(t) \sin \alpha_n t$ less than by ε, and the latter is less than ε in absolute value for all sufficiently large n. For $\cos \alpha_n t$ the reasoning is similar. $\square$

Further in this subsection integrable functions on $[0, 2\pi]$ will be *redefined at* 2π *by* $f(2\pi) = f(0)$ *and extended periodically to the whole real line*. In the study of convergence of trigonometric Fourier series one often employs the following useful representation of partial sums obtained with the aid of the identity

$$\frac{1}{2} + \cos z + \cos 2z + \cdots + \cos kz = \frac{\sin \frac{2k+1}{2} z}{2 \sin \frac{z}{2}}$$

and elementary transformations:

$$S_n(x) := \frac{a_0}{2} + \sum_{k=1}^n [a_k \cos kx + b_k \sin kx]$$

$$= \frac{1}{\pi} \int_0^{2\pi} f(x+t) \frac{\sin \frac{2n+1}{2} t}{2 \sin \frac{t}{2}} \, dt. \qquad (4.5.3)$$

Indeed, from the equality $\cos kx \cos kt + \sin kx \sin kt = \cos k(t - x)$ we find

$$S_n(x) = \frac{1}{\pi} \int_0^{2\pi} f(t) \left[\frac{1}{2} + \sum_{k=1}^n \cos k(t - x) \right] dt,$$

which by the identity indicated above gives the representation

$$S_n(x) = \frac{1}{\pi} \int_0^{2\pi} f(t) \frac{\sin \frac{2n+1}{2}(t-x)}{2\sin\frac{t-x}{2}}\, dt = \frac{1}{\pi}\int_{-x}^{2\pi-x} f(x+u) \frac{\sin\frac{2n+1}{2}u}{2\sin\frac{u}{2}}\, du.$$

By the 2π-periodicity of the function under the integral sign we obtain the desired equality. If we substitute $f = 1$, then $a_n = b_n = 0$ for all $n \geqslant 1$ and $a_0 = 2$, hence we arrive at the useful identity

$$\frac{1}{\pi}\int_0^{2\pi} \frac{\sin\frac{2n+1}{2}t}{2\sin\frac{t}{2}}\, dt = 1.$$

The function

$$D_n(z) = \frac{1}{\pi} \frac{\sin\frac{2n+1}{2}z}{2\sin\frac{z}{2}}$$

is called the Dirichlet kernel. Formula (4.5.3) is a starting point for various sufficient conditions for the pointwise convergence of Fourier series. By using the 2π-periodicity of f and D_n and the property that D_n is even we obtain

$$\int_\pi^{2\pi} f(x+t) \frac{\sin\frac{2n+1}{2}t}{\sin\frac{t}{2}}\, dt = \int_{-\pi}^0 f(x+t)\frac{\sin\frac{2n+1}{2}t}{\sin\frac{t}{2}}\, dt = \int_0^\pi f(x-t)\frac{\sin\frac{2n+1}{2}t}{\sin\frac{t}{2}}\, dt.$$

This gives the representation

$$S_n(x) = \frac{1}{2\pi}\int_0^\pi [f(x+t)+f(x-t)]\frac{\sin\frac{2n+1}{2}t}{\sin\frac{t}{2}}\, dt. \qquad (4.5.4)$$

4.5.3. Example. (Dini's condition) Suppose that f is 2π-periodic and is integrable on $[0, 2\pi]$ and that the function $\psi\colon t \mapsto t^{-1}[f(x+t)-2f(x)+f(x-t)]$ is integrable on $[0, \pi]$ for some x. Then $\lim_{n\to\infty} S_n(x) = f(x)$. For example, this is true if the function $t \mapsto t^{-1}[f(x+t)-f(x)]$ is integrable on $[-\delta, \delta]$ for some $\delta > 0$. In particular, this is true if f is differentiable at the point x. Indeed, from (4.5.4) we have

$$S_n(x) - f(x) = \frac{1}{2\pi}\int_0^\pi [f(x+t)+f(x-t)-2f(x)]\frac{\sin\frac{2n+1}{2}t}{\sin\frac{t}{2}}\, dt$$

$$= \frac{1}{2\pi}\int_0^\pi \psi(t)\frac{t}{\sin\frac{t}{2}}\sin\frac{2n+1}{2}t\, dt.$$

By the Riemann–Lebesgue theorem the right-hand side tends to zero as $n \to \infty$ by the integrability of ψ and boundedness of $t/\sin(t/2)$ on $[0, \pi]$.

The next *de la Vallée-Poussin condition* also employs (4.5.4).

4.5.4. Theorem. *Let x be fixed. Suppose that $\varphi(t) = f(x+t) + f(x-t)$ satisfies the following condition: the function*

$$\psi(u) = \frac{1}{u}\int_0^u \varphi(t)\, dt$$

on $(0, \pi]$ has bounded variation. Let $\lim_{u\to 0} \psi(u) = 2S$. Then $\lim_{n\to\infty} S_n(x) = S$.

PROOF. We observe that $\varphi(t) = \psi(t) + t\psi'(t)$ for almost all $t \in [0, 2\pi]$. We can assume that $S = 0$. By (4.5.4) the equality $\lim_{n\to\infty} S_n(x) = 0$ is equivalent to

$$\lim_{n\to\infty} \int_0^\pi \varphi(t) \frac{\sin \frac{2n+1}{2}t}{\sin \frac{t}{2}} \, dt = 0,$$

which by the Riemann–Lebesgue theorem reduces to the equality

$$\lim_{n\to\infty} \int_0^\pi \psi(t) \frac{\sin \frac{2n+1}{2}t}{t} \, dt = 0,$$

since the function $\varphi(t)/\sin(t/2) - 2\varphi(t)/t$ is integrable on $[0, \pi]$ by the boundedness of the functions $1/\sin(t/2) - 2/t$ and $\psi' \in L^1[0, \pi]$. We have $\psi = \psi_1 - \psi_2$, where ψ_1 and ψ_2 are increasing. We can assume that $\lim_{t\to 0} \psi_1(t) = \lim_{t\to 0} \psi_2(t) = 0$. For a given $\varepsilon > 0$ we find $\eta > 0$ such that $\psi_1(\eta) < \varepsilon$. Then by the second mean value theorem (see Exercise 3.12.31) we obtain

$$\int_0^\eta \psi_1(t) \frac{\sin \frac{2n+1}{2}t}{t} \, dt = \psi_1(\eta) \int_\xi^\eta \frac{\sin \frac{2n+1}{2}t}{t} \, dt = \psi_1(\eta) \int_{(n+1/2)\xi}^{(n+1/2)\eta} \frac{\sin v}{v} \, dv,$$

where $\xi \in [0, \eta]$. Since the function $\sin v/v$ is improper Riemann integrable on $(0, +\infty)$, there exists a constant C such that the expression in the right-hand side of the obtained equality is estimated in absolute value by $C\varepsilon$. It remains to observe that for any fixed η by the Riemann–Lebesgue theorem we have

$$\lim_{n\to\infty} \int_\eta^\pi \psi_1(t) \frac{\sin \frac{2n+1}{2}t}{t} \, dt = 0$$

The same reasoning applies to ψ_2, which completes the proof. □

Note that the function ψ is absolutely continuous and has bounded variation on every interval $[\delta, 2\pi]$ with $\delta > 0$. Hence the de La Vallée-Poussin condition is the boundedness of variation in a neighborhood of zero.

4.5.5. Example. (i) (Jordan's condition) If a function f has bounded variation in a neighborhood of the point x, then the Fourier series of f converges at the point x to $[f(x+0) + f(x-0)]/2$. If the function f has bounded variation on $[0, 2\pi]$ and is continuous, then its Fourier series converges to it at every point.

Indeed, if f in a neighborhood of x has the form $f_1 - f_2$, where the functions f_1 and f_2 are positive and increasing, then ψ has bounded variation too. This follows from the fact that for a monotone function g (we actually consider here $g(t) = f_1(x+t) - f_2(x-t)$ and $g(t) = f_1(x-t) - f_2(x+t)$) the function

$$G(x) = \frac{1}{x} \int_0^x g(u) \, du$$

is also monotone (Exercise 4.5.25).

(ii) (Dini's condition in terms of φ). For convergence of the Fourier series of f to S it suffices that the function $h(t) = \varphi(t)/t$ be integrable in a neighborhood of zero. Indeed, we have

$$\psi(u) = \chi(u) - \frac{1}{u} \int_0^u \chi(t) \, dt, \quad \chi(t) = \int_0^t h(s) \, ds.$$

This is clear from our discussion above, since the function χ has bounded variation.

Note that the Dini and Jordan conditions do not follow from each other. For example, if $f(x) = (\ln x)^{-1}$ for $x \in (0, 1/2)$ and $f(x) = 0$ for $x \in [1/2, 2\pi)$, then at the point 0 Jordan's condition is fulfilled, but Dini's condition is not. If $f(x) = \sqrt{x} \sin(1/x)$ for $x \in (0, \pi)$ and $f(x) = 0$ for $x \in [\pi, 2\pi)$, then at the point 0 Dini's condition is fulfilled, but Jordan's condition is not.

Let us make some remarks about the rate of decreasing of Fourier coefficients.

(1) If the function f satisfies Hölder's condition of order $\alpha \in (0, 1]$, i.e., $|f(x + h) - f(x)| \leqslant C|h|^\alpha$, then

$$a_n = O(n^{-\alpha}), \quad b_n = O(n^{-\alpha}), \quad c_n = O(n^{-\alpha}).$$

Indeed, the change of variable $t = s + \pi/n$ gives

$$\int_0^{2\pi} f(t) \exp(-int)\, dt = -\int_{-\pi/n}^{2\pi - \pi/n} f(s + \pi/n) \exp(-isn)\, ds$$

$$= -\int_0^{2\pi} f(s + \pi/n) \exp(-isn)\, ds.$$

Therefore,

$$c_n = \frac{1}{2\sqrt{2\pi}} \int_0^{2\pi} [f(t) - f(t + \pi/n)] \exp(-int)\, dt,$$

whence we obtain our claim.

(2) If the function f has bounded variation, then

$$a_n = O(n^{-1}), \quad b_n = O(n^{-1}), \quad c_n = O(n^{-1}).$$

Indeed, we can assume that the function f is left continuous, redefining it on a countable set of discontinuity points. By the integration by parts formula for the Stieltjes integral we have

$$\int_0^{2\pi} f(t) \exp(-int)\, dt = -in^{-1} \int_0^{2\pi} \exp(-int)\, df(t),$$

which in absolute value does not exceed $n^{-1} V_0^{2\pi}(f)$.

(3) If the function f is absolutely continuous, then

$$a_n = o(n^{-1}), \quad b_n = o(n^{-1}), \quad c_n = o(n^{-1}).$$

This is seen from the previous case, because after integration by parts we obtain the Fourier coefficients of the integrable function f'.

Let us give a simple sufficient condition for the uniform convergence of a Fourier series (its proof is delegated to Exercise 4.5.23).

4.5.6. Theorem. *Let $f \in \mathcal{L}^1[0, 2\pi]$ be extended 2π-periodically. Assume that for every $\varepsilon > 0$ there exists $\delta > 0$ such that*

$$\int_0^\delta \frac{|f(x + t) - 2f(x) + f(x - t)|}{t}\, dt \leqslant \varepsilon \quad \forall x \in [0, 2\pi],$$

i.e., the functions $t \mapsto t^{-1}[f(x+t) - 2f(x) + f(x-t)]$, $x \in [0, 2\pi]$, are uniformly integrable. Then the Fourier series converges to f uniformly.

4.5.7. Example. If f satisfies Hölder's condition $|f(x + t) - f(x)| \leqslant C|t|^\alpha$, where $\alpha > 0$, $f(0) = f(2\pi)$, then the Fourier series converges to f uniformly. In particular, it suffices (Exercise 4.5.22) that the function f be absolutely continuous and $f' \in L^p[0, 2\pi]$ for some $p > 1$. Actually, the Fourier series converges to f uniformly also in the case $p = 1$, i.e., for an absolutely continuous function (see [**578**, Exercise 6(d), p. 519]).

For improving convergence of series one often uses the Cesàro summation, i.e., the passage from the series with the general term α_n and the partial sums $s_n = \sum_{k=1}^n \alpha_k$ to the sequence $\sigma_n := (s_1 + \cdots + s_n)/n$. If the series $\sum_{n=1}^\infty \alpha_n$ converges to a number s, then the sequence σ_n converges to s as well, but this transformation can produce a convergent sequence from a divergent series (for example, $\alpha_n = (-1)^n$). Yet another method of summation of series is called Abels' summation. It involves the power series $S(r) = \sum_{n=1}^\infty \alpha_n r^n$, $r \in (0,1)$. If the sums $S(r)$ are defined and have a finite limit s as $r \to 1$, then s is called the sum of the series $\sum \alpha_n$ in Abel's sense. If a series is Cesàro summable to a number s, then it is Abel summable to s (Exercise 4.5.14). Applying to Fourier series the Cesàro summation by means of representation (4.5.3) and the equality

$$\sum_{k=0}^{n-1} \sin(2k + 1)z = \sin^2 nz / \sin z$$

(it follows from the identity $2 \sin z \sin(2k+1)z = \cos 2kz - \cos 2(k+1)z$) we arrive at the following *Fejér sums*:

$$\sigma_n(x) := \frac{S_0(x) + \cdots + S_{n-1}(x)}{n} = \int_0^{2\pi} f(x + z)\Phi_n(z)\, dz, \qquad (4.5.5)$$

where the function

$$\Phi_n(z) := \frac{1}{2\pi n} \sum_{k=0}^{n-1} \frac{\sin \frac{2k+1}{2} z}{\sin \frac{z}{2}} = \frac{1}{2\pi n}\left(\sin \frac{nz}{2} / \sin \frac{z}{2}\right)^2$$

is called the *Fejér kernel*. The Fejér kernels are nonnegative, which radically improves the convergence properties.

With the aid of (4.5.4) the Fejér sums can be written in the form

$$\sigma_n(x) = \int_0^\pi [f(x + z) + f(x - z)]\Phi_n(z)\, dz. \qquad (4.5.6)$$

For the sequel we note the equality

$$\int_0^{2\pi} \Phi_n(z)\, dz = 1, \qquad \int_0^\pi \Phi_n(z)\, dz = \frac{1}{2}. \qquad (4.5.7)$$

For the proof it suffices to set $f = 1$ in (4.5.5) and (4.5.6).

4.5.8. Theorem. *If f is a 2π-periodic function integrable on $[0, 2\pi]$, then for every Lebesgue point x (see Corollary 4.4.3) we have*

$$f(x) = \lim_{n \to \infty} \sigma_n(x) = \frac{a_0}{2} + \lim_{r \to 1-} \sum_{n=1}^\infty [a_n \cos nx + b_n \sin nx] r^n.$$

PROOF. According to Exercise 4.5.14 it suffices to prove the first equality. We can assume that $f(x) = 0$. In addition, the general case obviously reduces to the case $x = 0$. We use representation (4.5.6). For all points $z \in [0, \pi]$ we have $\Phi_n(z) \leqslant n/2$, $\Phi_n(z) \leqslant \pi/(2nz^2)$. The first estimate is obvious from the fact that Φ_n is the sum of n terms not exceeding in absolute value $1/2$, because

$$|\sin(z/2)| \geqslant |z|/\pi, \quad |\sin(k+1/2)z| \leqslant n|z|.$$

The second estimate follows from the same inequality for $\sin(z/2)$ and the inequality $|\sin(nz/2)|^2 \leqslant 1$. Using the first estimate for $z \in [0, 1/n]$ and the second one for $z > 1/n$, we obtain

$$\int_0^\pi \big[|f(z)| + |f(-z)|\big] \Phi_n(z)\, dz$$
$$\leqslant \frac{n}{2} \int_0^{1/n} \big[|f(z)| + |f(-z)|\big]\, dz + \frac{\pi}{2n} \int_{1/n}^\pi \frac{[|f(z)| + |f(-z)|]}{z^2}\, dz.$$

The first term in the right-hand side tends to zero by our assumption that $x = 0$ is a Lebesgue point and $f(0) = 0$. Introducing the function

$$F(z) = \int_0^z \big[|f(t)| + |f(-t)|\big]\, dt$$

and integrating by parts, the second term is transformed to

$$\frac{\pi}{2n}\big[F(\pi)\pi^{-2} - F(n^{-1})n^2\big] + \frac{\pi}{n} \int_{1/n}^\pi F(z) z^{-3}\, dz.$$

This expression tends to zero as $n \to \infty$. Indeed, for the reason indicated above $F(z) \leqslant \psi(z)z$, where ψ is a bounded positive measurable function with $\lim_{z \to 0} \psi(z) = 0$. Hence $nF(n^{-1}) \to 0$, $F(z)z^{-3} \leqslant \psi(z)z^{-2}$. For a given $\varepsilon > 0$ we find $\delta > 0$ such that $\psi(z) \leqslant \varepsilon$ if $z \leqslant \delta$. Next we pick $N \geqslant \delta^{-1}$ such that $n^{-1} \int_\delta^\pi \psi(z)z^{-2}\, dz \leqslant \varepsilon$ for all $n \geqslant N$. For such n we have

$$n^{-1} \int_{1/n}^\pi \psi(z)z^{-2}\, dz \leqslant n^{-1} \int_{1/n}^\delta \psi(z)z^{-2}\, dz + \varepsilon \leqslant 2\varepsilon,$$

which completes the proof. $\square$

We now show that for a continuous function the Fejér sums converge to it uniformly on $[0, 2\pi]$, and for an integrable function convergence holds in the metric of $L^1[0, 2\pi]$. The Fourier series has no such properties (see Exercise 6.10.87).

4.5.9. Theorem. *Let f be a 2π-periodic function integrable on $[0, 2\pi]$. Then*

$$\lim_{n \to \infty} \int_0^{2\pi} |f(x) - \sigma_n(x)|\, dx = 0.$$

If the function f is continuous, then $\lim_{n \to \infty} \sup_{x \in [0, 2\pi]} |f(x) - \sigma_n(x)| = 0.$

PROOF. Suppose first that the function f is continuous. Let us estimate the difference $|f(x) - \sigma_n(x)|$. From (4.5.6) and (4.5.7) we find

$$f(x) - \sigma_n(x) = \int_0^\pi [2f(x) - f(x+z) - f(x-z)]\Phi_n(z)\,dz.$$

It suffices to estimate the integral of $|2f(x) - f(x+z) - f(x-z)|\Phi_n(z)$. Set $C := \sup_{y \in [0,2\pi]} |f(y)|$. Let $\varepsilon > 0$. By the uniform continuity of f there exists $\delta \in (0,1)$ such that $|f(x) - f(y)| \leqslant \varepsilon$ whenever $|x - y| \leqslant \delta$. Hence the integral we consider does not exceed

$$2\varepsilon + \int_\delta^\pi |2f(x) - f(x+z) - f(x-z)|\Phi_n(z)\,dz \leqslant 2\varepsilon + 4C \int_\delta^\pi \Phi_n(z)\,dz$$

$$\leqslant 2\varepsilon + \frac{2\pi C}{n} \int_\delta^\pi z^{-2}\,dz \leqslant 2\varepsilon + \frac{2\pi C}{n\delta},$$

where we have used the estimate $\Phi_n(z) \leqslant 2^{-1}\pi n^{-1} z^{-2}$ from the proof of the previous theorem. For n sufficiently large the right-hand side does not exceed 3ε, whence the uniform convergence of σ_n to f follows. For any integrable function f, given $\varepsilon > 0$, we find a continuous function g such that $g(0) = g(2\pi)$ and

$$\int_0^{2\pi} |f(x) - g(x)|\,dx \leqslant \varepsilon.$$

As we have proved, the Fejér sums σ_n^g for g converge to g uniform, hence in the metric of L^1, i.e., there exists N such that

$$\int_0^{2\pi} |g(x) - \sigma_n^g(x)|\,dx \leqslant \varepsilon \quad \text{as } n \geqslant N.$$

It remains to compare the Fejér sums σ_n^f for f with σ_n^g. We have

$$\int_0^{2\pi} |\sigma_n^f(x) - \sigma_n^g(x)|\,dx \leqslant \int_0^{2\pi}\int_0^{2\pi} |f(x+z) - g(x+z)|\Phi_n(z)\,dz\,dx$$

$$= \int_0^{2\pi}\int_0^{2\pi} |f(x+z) - g(x+z)|\Phi_n(z)\,dx\,dz \leqslant \varepsilon \int_0^{2\pi} \Phi_n(z)\,dz = \varepsilon.$$

Using the equality $f - \sigma_n^f = f - g + g - \sigma_n^g + \sigma_n^g - \sigma_n^f$ and the triangle inequality we estimate the integral of $|f - \sigma_n^f|$ by 3ε for all $n \geqslant N$. □

It is known that if a trigonometric series converges everywhere except for countably many points to a finite Lebesgue integrable function, then this series is the Fourier series of its sum (see [**255**, Theorem 100]). We recall that the Fourier series of an integrable function can diverge at every point. Nevertheless the Fourier series possess the following surprising integrability property.

4.5.10. Theorem. *Let $f \in \mathcal{L}^1[0, 2\pi]$. Then, for every interval $[a, b] \subset [0, 2\pi]$,*

$$\int_a^b f(x)\,dx = \lim_{n \to \infty} \int_a^b S_n(x)\,dx.$$

Thus, the Fourier series of every integrable function can be integrated term-by-term independently of the property whether it converges or not.

PROOF. We can assume that $a = 0$. Set
$$F(x) = \int_0^x f(y)\, dy.$$
The function F is absolutely continuous, hence it can be expanded in the convergent Fourier series $F(x) = A_0/2 + \sum_{n=1}^{\infty}(A_n \cos nx + B_n \sin nx)$. Then
$$A_n = \frac{1}{\pi}\int_0^{2\pi} F(x)\cos nx\, dx = -\frac{1}{\pi n}\int_0^{2\pi} f(x)\sin nx\, dx = -\frac{b_n}{n}$$
by the integration by parts formula. Similarly $B_n = a_n/n$. Thus,
$$F(x) = A_0/2 + \sum_{n=1}^{\infty} n^{-1}(a_n \sin nx - b_n \cos nx).$$
For $x = 0$ we find $A_0/2 = \sum_{n=1}^{\infty} b_n/n$. We finally obtain
$$F(x) = \sum_{n=1}^{\infty} n^{-1}[a_n \sin nx - b_n(1 - \cos nx)],$$
which coincides with the limit of the integrals of S_n over $[0, x]$. Since this is true for every point x, the theorem is proved. $\square$

On the way we have established the following non-obvious fact.

4.5.11. Corollary. *For every function f integrable on $[0, 2\pi]$ the series $\sum_{n=1}^{\infty} b_n/n$ converges.*

About trigonometric and orthogonal series, see Akhiezer [4], Bary [49], Edwards [172], Hardy, Rogosinski [255], Garsia [204], Grafakos [235], Kashin, Saakyan [316], Kufner, Kadlec [361], Olevskiĭ [463], Suetin [579], and Zygmund [661], where one can find additional references.

Exercises

4.5.12. Let μ be a measure on a space X and let f be a function on $X \times [a, b]$ such that the functions $x \mapsto f(x, t)$ are integrable, the functions $t \mapsto f(x, t)$ are absolutely continuous and the function $\partial f/\partial t$ is integrable with respect to $\mu \otimes \lambda$, where λ is Lebesgue measure. Prove that the function $t \mapsto \int_X f(t, x)\, \mu(dx)$ is absolutely continuous and its derivative a.e. equals $\int_X \dfrac{\partial f(x, t)}{\partial t}\, \mu(dx)$.
HINT: apply the Newton–Leibniz formula and Fubini's theorem.

4.5.13. Construct an example of a continuous increasing function on $[0, 1]$ with a dense set of points of nondifferentiability.

4.5.14. Prove that if a series is Cesàro summable to a number s, then it is Abel summable to s (see §4.5(ii)).

4.5.15. Let $E \subset \mathbb{R}$ be a set of Lebesgue measure zero. Construct a Lipschitz function on $\mathbb{R}$ that is not differentiable at the points of E.
HINT: see [185, Exercise 8.23, p. 266].

4.5.16. Let f_n be increasing absolutely continuous functions on $[0, 1]$, $n \in \mathbb{N}$, such that $f(t) = \sum_{n=1}^{\infty} f_n(t) < \infty$. Prove that the function f is absolutely continuous.
HINT: observe that the series of $f_n' \geqslant 0$ converges in L^1.

4.5.17° Let f be a function on $[a, b]$ such that for every $\varepsilon > 0$ there exists $\delta > 0$ such that $\left| \sum_{i=1}^{n} [f(b_i) - f(a_i)] \right| < \varepsilon$ for every finite collection of pairwise disjoint intervals $[a_i, b_i]$ in $[a, b]$ with $\sum_{i=1}^{n} |b_i - a_i| < \delta$. Prove that the function f is absolutely continuous.

4.5.18. Let f be a function on $[a, b]$ such that for every $\varepsilon > 0$ there exists $\delta > 0$ such that for every finite collection of intervals $[a_i, b_i]$ in $[a, b]$ with $\sum_{i=1}^{n} |b_i - a_i| < \delta$ we have $\sum_{i=1}^{n} |f(b_i) - f(a_i)| < \varepsilon$ (these intervals can overlap). Prove that f is Lipschitz.

4.5.19. Let f be a function on $[a, b]$ such that for every countable collection of intervals $[a_i, b_i]$ in $[a, b]$ with $\sum_{i=1}^{\infty} |b_i - a_i| < \infty$ we have $\sum_{i=1}^{\infty} |f(b_i) - f(a_i)| < \infty$ (these intervals can overlap). Prove that the function f is Lipschitz.

4.5.20° Prove the following Helly theorem: every uniformly bounded sequence of increasing functions on an interval contains a pointwise converging subsequence.

HINT: using Cantor's diagonal method pick a subsequence converging at all rational points, show that it converges at all continuity points of the increasing function defined as its limsup, pick a subsequence converging at the remaining at most countable set.

4.5.21° Prove that a function f on $[0, 1]$ has bounded variation precisely when there is an increasing function g such that $|f(t) - f(s)| \leqslant g(t) - g(s)$ whenever $t > s$.

4.5.22. Let f be an absolutely continuous function on $[0, 1]$ and let $f' \in L^p[0, 1]$, where $p > 1$. Prove that $|f(t) - f(s)| \leqslant C|t - s|^{\alpha}$, where $C = \|f'\|_{L^p}$, $\alpha = 1 - 1/p$.

HINT: apply the Newton–Leibniz formula and Hölder's inequality.

4.5.23. Prove Theorem 4.5.6.

4.5.24. Prove the following Chebyshev inequality for monotone functions: if φ and ψ are nondecreasing finite functions on $[0, 1]$ and ϱ is a probability density on $[0, 1]$, then

$$\int_0^1 \varphi(x)\psi(x)\varrho(x)\,dx \geqslant \int_0^1 \varphi(x)\varrho(x)\,dx \int_0^1 \psi(x)\varrho(x)\,dx.$$

If φ is increasing and ψ is decreasing, then the opposite inequality is true.

HINT: observe that in the first case $\big(\varphi(x) - \varphi(y)\big)\big(\psi(x) - \psi(y)\big) \geqslant 0$ for all x, y, hence the integral of this expression with respect to $\varrho(x)\varrho(y)\,dx\,dy$ is nonnegative.

4.5.25. Let g be a monotone function on $[0, 1]$. Prove that the function

$$G(x) = \frac{1}{x} \int_0^x g(u)\,du$$

is also monotone.

4.5.26. Let F be a left continuous function of bounded variation on $[0, 1]$. Proposition 4.4.7 gives the decomposition $F = F_{\mathrm{ac}} + F_{\mathrm{sing}}$, where F_{ac} is absolutely continuous and $F'_{\mathrm{sing}}(t) = 0$ a.e. Prove that $V(F, [0, 1]) = V(F_{\mathrm{ac}}, [0, 1]) + V(F_{\mathrm{sing}}, [0, 1])$.

4.5.27. Let $\xi = \sum_{n=1}^{\infty} 2^{-n}\xi_n$, where $\{\xi_n\}$ is a sequence of independent random variables assuming the values 0 and 1 with probability $p \in (0, 1/2)$ and $1 - p$, respectively. Show that the distribution function $F(t) = \mathrm{P}(\xi < t)$ is strictly increasing on $[0, 1]$ and Hölder continuous, but $F'(t) = 0$ almost everywhere.

HINT: consider the ratio $2^n[F(t + 2^{-n}) - F(t)]$ at a point of differentiability t written as $\sum_{k=1}^{\infty} 2^{-k}t_k$, where $t_k \in \{0, 1\}$; estimate $2^n\mathrm{P}(\xi \in [t, t + 2^{-n}))$ by using that $2^n p^{m_n}(1 - p)^{n - m_n}$, where m_n is the number of zeros among $t_1, \ldots, t_n$, tends to zero as $n \to \infty$ for a.e. t by the law of large numbers for $\{t_n\}$: $(t_1 + \cdots + t_n)/n \to 1/2$ a.e.

4.5.28° Let a function f be continuous and strictly increasing on $[0, 1]$. Prove that f^{-1} is absolutely continuous precisely when the set $\{x \colon f'(x) = 0\}$ has measure zero.

HINT: see [73, Exercise 5.8.54].

CHAPTER 5

Normed and Euclidean Spaces

In this chapter we discuss some basic geometric and topological properties of normed and Euclidean spaces, which are the most important types of spaces of functional analysis.

5.1. Normed Spaces

We assume that the concept of a linear (or vector) space is known from linear algebra.

5.1.1. Definition. *A linear space X over the field of real or complex numbers is called normed if it is equipped with a function $\| \cdot \| \to [0, +\infty)$ called a norm and satisfying the following conditions:*

(i) *$\|x\| = 0$ if and only if $x = 0$,*

(ii) *$\|\alpha x\| = |\alpha| \|x\|$ for all $x \in X$ and all scalars α,*

(iii) *(the triangle inequality) $\|x + y\| \leq \|x\| + \|y\|$ for all vectors $x, y \in X$.*

It is clear from conditions (i) and (iii) that every normed space becomes a metric space if for the distance between x and y we take $\|x - y\|$. It is readily verified that the mappings $\{x, y\} \mapsto x + y$ and $\{\alpha, x\} \mapsto \alpha x$ are continuous.

The class of normed spaces contains an important subclass of Euclidean spaces (called in honor of the outstanding mathematician of antiquity Euclid).

5.1.2. Definition. *A linear space X over the field of real or complex numbers is called Euclidean if $X \times X$ is equipped with a function $(\cdot, \cdot)$ with values in the respective field such that:*

(i) *$(x, x) \geq 0$ and, in addition, $(x, x) = 0$ only for $x = 0$,*

(ii) *$(x, y) = \overline{(y, x)}$ for all $x, y \in X$ (in the real case: $(x, y) = (y, x)$),*

(iii) *$(\alpha x + \beta y, z) = \alpha(x, z) + \beta(y, z)$ for all $x, y, z \in X$ and all scalars α, β.*

A function with the stated properties is called an inner product. If $(a, b) = 0$, then we write $a \perp b$ and call the vectors a and b mutually orthogonal.

Note that in the complex case $(x, \alpha y) = \overline{\alpha}(x, y)$. In this case we say that the inner product is *conjugate-linear* in the second argument.

Letting $\|x\| := \sqrt{(x, x)}$, we obtain a norm on the Euclidean space X (called a *Euclidean norm*). The triangle inequality after obvious transformations reduces to the *Cauchy inequality*

$$|(x, y)|^2 \leq (x, x)(y, y),$$

© Springer Nature Switzerland AG 2020
V. I. Bogachev, O. G. Smolyanov, *Real and Functional Analysis*,
Moscow Lectures 4, https://doi.org/10.1007/978-3-030-38219-3_5

following from the fact that $(x + ty, x + ty) \geqslant 0$ for all $t \in \mathbb{R}^1$. The complex case is similar.

Since normed and Euclidean spaces are metric spaces, we can speak of their completeness (with respect to the metric generated by the norm).

5.1.3. Definition. *A complete normed space is called a Banach space. A complete Euclidean space is called a Hilbert space.*

The terminology is connected with the names of the two outstanding mathematicians Stefan Banach and David Hilbert.

Let us consider the most important examples of Euclidean spaces.

5.1.4. Example. (i) The set l^2 of all infinite real sequences $x = (x_n)$ such that $\sum_{n=1}^{\infty} |x_n|^2 < \infty$ is a Euclidean space with the inner product

$$(x, y) = \sum_{n=1}^{\infty} x_n y_n.$$

In the complex case we set

$$(x, y) = \sum_{n=1}^{\infty} x_n \overline{y_n}.$$

The space l^2 is complete, i.e., is Hilbert (hence Banach).

(ii) Let us consider an example assuming some acquaintance with the material of Chapters 2 and 3. Let μ be a nonnegative countably additive measure on a space Ω and let $L^2(\mu)$ be the equivalence class of square integrable functions introduced in Chapter 3. The inner product in $L^2(\mu)$ is defined by the formula

$$(f, g) = \int_{\Omega} f(\omega)\overline{g(\omega)}\, \mu(d\omega),$$

where under the integral sign any representatives of equivalence classes can be used. The space $L^2(\mu)$ is complete. However, if we take only Riemann integrable functions, then we obtain an incomplete space. This is one of many advantages of the Lebesgue integral.

The sets $\{x \colon \|x\| \leqslant 1\}$ and $\{x \colon \|x\| = 1\}$ in a normed space are called the *unit ball* and the *unit sphere*, respectively.

Not every norm can be obtained from an inner product. The simplest example is the following norm on the plane: $\|x\| = \max(|x_1|, |x_2|)$. The unit ball in this norm is the square $[-1, 1]^2$. There is the following criterion (the classical *Jordan-von Neumann theorem*) that a given norm $\| \cdot \|$ on a linear space X is generated by an inner product: the *parallelogram equality*

$$\|x + y\|^2 + \|x - y\|^2 = 2\|x\|^2 + 2\|y\|^2$$

must be fulfilled for all $x, y \in X$. The necessity of this condition is trivial, but its sufficiency is not obvious at all (see Exercise 5.6.38).

The existence of a linear structure enables us to consider in normed spaces not only convergence of sequence, but also convergence of series. We shall say that the series $\sum_{n=1}^{\infty} x_n$ of vectors of a normed space X converges to a vector $x \in X$ if $\lim_{m \to \infty} \|x - \sum_{n=1}^{m} x_n\| = 0$.

5.1.5. Proposition. *The completeness of a normed space X is equivalent to the property that convergence of the series of norms $\sum_{n=1}^{\infty} \|x_n\|$ implies convergence of the series of vectors x_n.*

PROOF. If X is complete, then the series of vectors converges, since by the triangle inequality the sequence of partial sums is Cauchy. On the other hand, suppose that convergence of the series of norms implies convergence of the series of vectors. Let $\{x_n\}$ be a Cauchy sequence in X. Let us take a subsequence $\{x_{n_k}\}$ with $\|x_{n_k} - x_{n_{k+1}}\| \leqslant 2^{-k}$. Then the series of $x_{n_k} - x_{n_{k+1}}$ converges, which means convergence of the sequence $x_{n_1} - x_{n_k}$. Hence $\{x_n\}$ converges as well. See also Exercise 5.6.15. $\qquad\square$

5.1.6. Definition. *A completion of a normed space X is a Banach space $\widetilde{X}$ such that X is embedded linearly and isometrically as an everywhere dense linear subspace in $\widetilde{X}$.*

A completion of a Euclidean space X is a Hilbert space $\widetilde{X}$ such that X is embedded linearly with preservation of the inner product as an everywhere dense linear subspace in $\widetilde{X}$.

5.1.7. Remark. In the next chapter (Proposition 6.4.6 and Remark 6.5.2) we describe a simple and natural construction of the completion of an arbitrary normed or Euclidean space. Here we only mention that similarly to the known procedure of completing the set of rational numbers (see Exercise 1.9.31) we can take for a completion of a normed space X the space $\widetilde{X}$ of equivalence classes of all Cauchy sequences $x = (x_n)$ in the space X, where $x = (x_n)$ and $y = (y_n)$ are equivalent if the sequence $x_1, y_1, \ldots, x_n, y_n, \ldots$ is Cauchy, moreover, the linear operations are defined componentwise and the norm is introduced as follows: $\|x\| := \lim_{n\to\infty} \|x_n\|$. In case of a Euclidean space we set $(x, y) := \lim_{n\to\infty} (x_n, y_n)$. A simple justification is offered as Exercise 5.6.16. Note that the completion in the category of general metric spaces does not solve our problem, since now the consistency with the linear structure is required.

The smallest linear subspace containing a set A in a linear space X is called the *linear span* of A or the linear subspace generated by A. If X is normed, then the *closed linear span* of A is the smallest closed linear subspace containing A (it coincides with the closure of the linear span of A).

Given normed spaces $X_1, \ldots, X_n$, their product $X_1 \times X_2 \times \cdots \times X_n$, consisting of all collections of the form $x = (x_1, \ldots, x_n)$ with $x_i \in X_i$, is equipped with the natural linear structure (with the componentwise operations) and the norm

$$\|x\| = \|x_1\|_{X_1} + \cdots + \|x_n\|_{X_n}.$$

This space is also called the *direct sum* of the spaces $X_1, \ldots, X_n$ and denoted by $X_1 \oplus \cdots \oplus X_n$. In case of Euclidean spaces $X_1, \ldots, X_n$, the space X is equipped with the inner product $(x, y) := (x_1, y_1)_{X_1} + \cdots + (x_n, y_n)_{X_n}$ and the corresponding norm

$$\|x\| := \left(\|x_1\|_{X_1}^2 + \cdots + \|x_n\|_{X_n}^2 \right)^{1/2}.$$

It is readily seen that if all spaces X_i are complete, then their sum X is also complete.

Given a normed space X and its closed linear subspace Y, we can consider the *quotient space* X/Y (called also the *factor-space* of X by the subspace Y) consisting of equivalence classes $[x] = x + Y$. It becomes a normed space with respect to the norm

$$\|[x]\| := \inf\{\|z\|: z \in [x]\}.$$

Indeed, if $\|[x]\| = 0$, then there exist vectors $z_n \in [x]$ with $\|z_n\| \to 0$. We have $z_n = x + y_n$, where $y_n \in Y$. Hence $y_n \to -x$ and $x \in Y$, since Y is closed, i.e., $[x] = 0$. If $z \in [x]$, $z' \in [x']$, then $z + z' \in [x + x']$. Hence $\|[x] + [x']\| \leqslant \|z + z'\| \leqslant \|z\| + \|z'\|$. Therefore, $\|[x] + [x']\| \leqslant \|[x]\| + \|[x']\|$. Similarly we verify that $\|\lambda[x]\| = |\lambda| \, \|[x]\|$.

If X is complete, then X/Y is also complete with respect to the introduced norm. Indeed, suppose that $[x_n]$ are elements such that the series of their norms converges. Let us take $x'_n \in [x_n]$ such that $\|x'_n\| \leqslant \|[x_n]\| + 2^{-n}$ for all n. Then the series of x'_n converges to some element $x \in X$. It is readily seen that the series of $[x_n]$ converges in X/Y to $[x]$.

One should bear in mind that the Banach space X need not be linearly homeomorphic to the product $Y \times (X/Y)$, where Y is a closed linear subspace in X. Moreover, the space X can fail to contain a subspace isomorphic to X/Y (see Exercise 6.10.105).

Let us also introduce the *complexification* of a normed space X and a Euclidean space E as follows. The complex space $X_{\mathbb{C}}$ is the space $X \times X$, the vectors in which are denoted by $x + iy$, where $x, y \in X$, and the addition operation is defined in the natural way, but the multiplication by complex numbers $\alpha + i\beta$, where $\alpha, \beta \in \mathbb{R}^1$, is defined as follows:

$$(\alpha + i\beta)(x + iy) := \alpha x - \beta y + i(\alpha y + \beta x).$$

The norm on $X_{\mathbb{C}}$ is given by the formula $\|x + iy\| := \|x\| + \|y\|$.

In the case of a Euclidean space E the construction is similar, but in place of the indicated norm we defined the inner product by the formula

$$(x + iy, u + iv) := (x, u) + (y, v) + i(y, u) - i(x, v).$$

It is readily seen that $X_{\mathbb{C}}$ and $E_{\mathbb{C}}$ are a normed space and a Euclidean space, respectively, and that they are complete in case of complete X and E. Note that $\|x + iy\| = \sqrt{\|x\|^2 + \|y\|^2}$ in the Euclidean case.

Finally, we note that if in the definition of the norm we omit the condition $\|x\| > 0$ for $x \neq 0$, then we obtain the definition of a *seminorm*. Seminorms play an important role in functional analysis and are discussed on various occasions below (see also Chapter 8). A linear space X with a seminorm p is sometimes called a pre-normed space. The set $X_0 := p^{-1}(0)$ is a linear subspace. The factor-space X/X_0 also becomes a normed space if we set $\|[x]\| := p(x)$ for every equivalence class $[x]$ with a representative x. This norm is well-defined, since by the triangle inequality we have $p(x + z) = p(x)$ whenever $p(z) = 0$.

5.2. Examples

Let us give a list of the most frequently used Banach spaces of functional analysis. These spaces play an important role in the theory. In addition, they arise in applications for constructing other spaces (either as "building blocks" for constructing more complicated objects or as models of methods of constructing).

5.2.1. Example. (i) The set $B(\Omega)$ of all bounded real (or complex in the complex case) functions on a nonempty set Ω with the pointwise defined linear structure, i.e., $(f + g)(\omega) = f(\omega) + g(\omega)$, $(\lambda f)(\omega) = \lambda f(\omega)$, is a Banach space with the so-called sup-norm

$$\|f\| = \sup_{x \in \Omega} |f(x)|.$$

(ii) For $\Omega = \mathbb{N}$ we obtain in (i) the space l^∞ of bounded real sequences with the norm

$$\|x\| = \sup_n |x_n|.$$

(iii) The space c_0 consists of all real sequences converging to zero and is equipped with the same norm as l^∞ in (ii); it is also Banach, since is closed in l^∞.

(iv) The space $C[a, b]$ of all continuous (real or complex depending on the field of scalars) functions on the interval $[a, b]$ is Banach with the same sup-norm as on the space of bounded functions: $\|f\| = \sup_{t \in [a,b]} |f(t)|$.

(v) Let $1 \leqslant p < \infty$. The space l^p of infinite sequences $x = (x_1, x_2, \ldots)$ (real or complex) for which $\sum_{n=1}^\infty |x_n|^p < \infty$ is equipped with the linear structure

$$(x_1, x_2, \ldots) + (y_1, y_2, \ldots) = (x_1 + y_1, x_2 + y_2, \ldots),$$

$\lambda(x_1, x_2, \ldots) = (\lambda x_1, \lambda x_2, \ldots)$ and is Banach with the norm

$$\|x\| = \left(\sum_{n=1}^\infty |x_n|^p\right)^{1/p}.$$

(vi) Let $(\Omega, \mathcal{A}, \mu)$ be a space with a nonnegative measure, $1 \leqslant p < \infty$ and let $L^p(\mu)$ be the space of equivalence classes of μ-measurable functions (real or complex depending on the field of scalars) that are integrable to power p. The addition of elements and multiplication by scalars are defined by means of representatives of equivalence classes (the sum of two equivalence classes with representatives f and g is given by a representative $f + g$). Then $L^p(\mu)$ is Banach (and $L^2(\mu)$ is Hilbert, see Example 5.1.4) with the norm

$$\|f\| = \left(\int_\Omega |f(\omega)|^p \, \mu(d\omega)\right)^{1/p},$$

where in the integral we take an arbitrary representative of the equivalence class of f, denoted by the same symbol.

Justifications are delegated to Exercise 5.6.13.

Let us give a less elementary example showing how norms in standard spaces are used for constructing more complicated spaces (its particular case has already been discussed in Example 3.12.9). Yet another important example of this are Sobolev spaces discussed in Chapter 9.

5.2.2. Example. Let $\Omega \subset \mathbb{C}$ be open, let $1 \leqslant p < \infty$, and let $A^p(\Omega)$ be the Bergman space consisting of all holomorphic complex functions f on Ω such that $|f| \in L^p(\Omega)$, where Ω is considered with the usual flat Lebesgue measure. Then the space $A^p(\Omega)$ with the norm from $L^p(\Omega)$ is Banach (unlike the space of real-analytic functions on $(0,1)$ with the norm from $L^p(0,1)$), moreover, $A^2(\Omega)$ is Hilbert. In addition, the space $A^p(\Omega)$ is separable, and if Ω is a disc, then the set of polynomials is dense in $A^p(\Omega)$.

PROOF. Let $\{f_n\}$ be a Cauchy sequence in $A^p(\Omega)$. Then it converges in $L^p(\Omega)$ to some function $f \in L^p(\Omega)$. However, we have to show that this function f has a version holomorphic in Ω. We show that on every closed disc $K \subset \Omega$ the sequence $\{f_n\}$ converges uniformly. It is known from complex analysis that this yields that the limit is holomorphic. It is also clear that the limit coincides with f almost everywhere (since some subsequence in $\{f_n\}$ converges to f almost everywhere). Let us take a closed disc $S \subset \Omega$ with the same center as K and a radius larger by $\delta > 0$ than the radius of K. The desired convergence follows from the following estimate for every function $\varphi \in A^p(\Omega)$ and $z_0 \in K$:

$$|\varphi(z_0)| \leqslant \frac{1}{\pi \delta^2} \int_S |\varphi(x + iy)| \, dx \, dy \leqslant \pi^{-1/p} \delta^{-2/p} \|\varphi\|_{L^p(\Omega)}. \qquad (5.2.1)$$

This estimate is deduced from the Cauchy formula

$$\varphi(z_0) = \frac{1}{2\pi i} \int_{\gamma_r} \frac{\varphi(w)}{w - z_0} \, dw,$$

where γ_r is the circumference of radius $r \leqslant \delta$ centered at z_0. Writing this formula in polar coordinates and multiplying both parts by $r > 0$, we arrive at the estimate

$$r|\varphi(z_0)| \leqslant \frac{r}{2\pi} \int_0^{2\pi} |\varphi(z_0 + re^{i\theta})| \, d\theta.$$

After integration in r from 0 to δ we obtain the inequality

$$\delta^2 |\varphi(z_0)| \leqslant \frac{1}{\pi} \int_{K(z_0, \delta)} |\varphi(x + iy)| \, dx \, dy,$$

where $K(z_0, \delta)$ is the disc of radius δ centered at z_0. This yields estimate (5.2.1), since the L^1-norm of φ on $K(z_0, \delta)$ is estimated by $(\pi \delta^2)^{1-1/p} \|\varphi\|_{L^p(\Omega)}$ by Hölder's inequality. The separability of $A^p(\Omega)$ is clear from the fact that it is a subspace in the separable space $L^p(\Omega)$ (see Example 3.8.8). The fact that the set of polynomials is dense in $A^p(\Omega)$ in the case where Ω is a disc is proved in the same way as in Example 3.12.9. One can show that in case $1 < p < \infty$ any function from $A^p(U)$, where U is the unit disc, is approximated with respect to the norm by its Taylor expansions, but for $p = 1$ this is not true (see Duren, Schuster [168]). Since $L^2(\mu)$ is Hilbert, the space $A^2(\Omega)$ is also Hilbert. $\qquad \square$

On a linear space there exist many different norms. Two norms p and q are called *equivalent* if there exist numbers c_1 and c_2 such that

$$c_1 p(x) \leqslant q(x) \leqslant c_2 p(x) \quad \text{for all } x.$$

Not all norms are comparable. For example, let us consider the following two norms on the space of polynomials on $[0,1]$:

$$p(x) = \int_0^1 |x(t)|\, dt + |x'(0)|, \quad q(x) = \max_{t \in [0,1]} |x(t)|.$$

Neither of the two norms can be estimated by the other.

Note that a norm can be introduced on every linear space X. For this it suffices to take an algebraic basis (a Hamel basis) $\{v_\alpha\}$ in X. It is shown in §1.1 with the aid of the axiom of choice that every linear space possesses an algebraic basis. Set

$$p(x) = |x_{\alpha_1}| + \cdots + |x_{\alpha_n}| \quad \text{if} \quad x = x_{\alpha_1} v_{\alpha_1} + \cdots + x_{\alpha_n} v_{\alpha_n}.$$

We obtain a norm on X. However, it is worth noting that such norms on infinite-dimensional spaces are used only for constructing various counter-examples and are not encountered in real applications.

5.3. Balls in Normed Spaces

The following theorem gives a characteristic property of finite-dimensional linear spaces.

5.3.1. Theorem. *On every finite-dimensional linear space all norms are equivalent.*

PROOF. Let us take a basis $e_1, \ldots, e_n$ in the given space X and with the aid of the expansion $x = x_1 e_1 + \cdots + x_n e_n$ introduce the following norm:

$$p(x) = |x_1| + \cdots + |x_n|.$$

Let q be some other norm on X. Set $c = \max_i q(e_i)$. Then

$$q(x) \leqslant \sum_{i=1}^{n} |x_i| q(e_i) \leqslant cp(x).$$

In particular, $|q(x) - q(y)| \leqslant q(x-y) \leqslant cp(x-y)$, i.e., the function q is Lipschitz, hence is continuous on X with respect to the norm p. It is readily seen that the unit sphere $S = \{x \colon p(x) = 1\}$ in X with the norm p is compact. Hence the function q attains its minimum m on this sphere. Clearly, $m > 0$, because q does not vanish outside the origin. The estimate $q(x) \geqslant m$ on S yields the estimate $q(x) \geqslant mp(x)$ on X, since $q(tx) = |t|q(x)$ and $p(tx) = |t|p(x)$. Thus, $mp(x) \leqslant q(x) \leqslant cp(x)$. $\square$

Using the obvious fact that any finite-dimensional space with the norm indicated in the proof p is complete and its closed balls are compact, we obtain a number of useful corollaries.

5.3.2. Corollary. *In any finite-dimensional linear space all closed balls and closed spheres are compact.*

5.3.3. Corollary. *Every finite-dimensional normed space is complete. Every finite-dimensional linear subspace of a normed space is closed.*

The situation is different in infinite-dimensional spaces: here balls of positive radii cannot be compact. This follows from the next result.

5.3.4. Theorem. *Let X_0 be a closed linear subspace in a normed space X different from X. Then, for every $\varepsilon > 0$, there exists a vector $x_\varepsilon \in X$ such that $\|x_\varepsilon\| = 1$ and $\|x_\varepsilon - y\| \geqslant 1 - \varepsilon$ for all $y \in X_0$.*

PROOF. By assumption, there exists an element $z \in X \backslash X_0$. Set

$$\delta = \inf\{\|z - y\| \colon y \in X_0\}.$$

By the closedness of X_0 we have $\delta > 0$. We pick $\varepsilon_0 > 0$ with $\delta/(\delta + \varepsilon_0) > 1 - \varepsilon$. Let us take $y_0 \in X_0$ with $\|z - y_0\| < \delta + \varepsilon_0$. Let $x_\varepsilon = (z - y_0)/\|z - y_0\|$. Then for every $y \in X_0$ we obtain

$$\|x_\varepsilon - y\| = \frac{1}{\|z - y_0\|} \Big\| z - y_0 - \|z - y_0\| \cdot y \Big\| \geqslant \frac{\delta}{\delta + \varepsilon_0} > 1 - \varepsilon,$$

since $v := y_0 + \|z - y_0\| \cdot y \in X_0$ and hence $\|z - v\| \geqslant \delta$. $\qquad\square$

5.3.5. Remark. If the subspace X_0 is finite-dimensional, there exists an element x such that $\|x\| = 1$ and $\|x - y\| \geqslant 1$ for all $y \in X_0$. Indeed, in the proof by the compactness of balls in X_0 we can choose y_0 with $\|z - y_0\| = \delta$.

5.3.6. Corollary. *In every infinite-dimensional normed space X there exists an infinite sequence of vectors x_n with $\|x_n\| = 1$ and $\|x_n - x_k\| \geqslant 1$ whenever $n \neq k$. Hence all balls in X of positive radii are noncompact.*

PROOF. By the previous remark one can construct by induction a sequence of vectors x_n of unit length with $\|x_n - x_k\| \geqslant 1$ whenever $n \neq k$. Of course, to prove the last assertion we can take vectors with $\|x_n - x_k\| \geqslant 1 - \varepsilon$. $\qquad\square$

Let us note one more useful fact in the circle of ideas of this section. The *algebraic sum* of two sets X_1 and X_2 in a linear space is defined as

$$X_1 + X_2 := \{x_1 + x_2 \colon x_1 \in X_1, x_2 \in X_2\}.$$

5.3.7. Proposition. *Let X_0 be a closed linear subspace in a normed space X. Then, for every finite-dimensional linear subspace X_1 in X, the algebraic sum $X_0 + X_1$ is closed.*

PROOF. The assertion reduces to the one-dimensional space X_1 generated by a vector v (the n-dimensional case is obtained by the n-fold application of the one-dimensional case). We can assume that v does not belong to X_0. Then by the closedness of X_0 we have $\mathrm{dist}(v, X_0) = \inf\{\|v - x\| \colon x \in X_0\} > 0$. Suppose that elements of the form $y_n = x_n + \lambda_n v$, where $x_n \in X_0$, converge to $y \in X$. We observe that

$$\mathrm{dist}(\lambda v, X_0) = |\lambda| \mathrm{dist}(v, X_0)$$

for all scalars λ. Since $y_n - y_k = (\lambda_n - \lambda_k)v - (x_k - x_n)$, where $x_k - x_n \in X_0$, we have $\|y_n - y_k\| \geqslant |\lambda_n - \lambda_k| \mathrm{dist}(v, X_0)$. Thus, the sequence $\{\lambda_n\}$ is Cauchy and converges to some number λ. Then there exists $x = \lim\limits_{n\to\infty} x_n \in X_0$. Hence we have $y = x + \lambda v$. $\qquad\square$

5.4. Orthonormal Systems, Bases, and Projections

The most important concepts connected with Euclidean spaces are orthogonal projections and bases.

5.4.1. Definition. *A system of mutually orthogonal vectors of unit length in a Euclidean space X is called orthonormal.*

Such a system $\{e_\alpha\}$ is called an orthonormal basis in X if, for every $x \in X$, there exists an at most countable subsystem $\{e_{\alpha_n}\} \subset \{e_\alpha\}$ and a finite or countable collection of scalars $\{c_n\}$ for which $x = \sum_{n=1}^{\infty} c_n e_{\alpha_n}$, where the series converges in X.

A system of vectors is called *complete* if its linear span is dense. An orthonormal basis is a complete system.

The numbers c_n, called the Fourier coefficients of x, are uniquely determined by the equalities

$$c_n = (x, e_{\alpha_n}).$$

This follows from the continuity of the inner product: if $u_n \to u$ and $v_n \to v$ with respect to the norm in X, then $(u_n, v_n) \to (u, v)$ by the estimate

$$|(u_n, v_n) - (u, v)| = |(u_n, v_n) - (u_n, v) + (u_n, v) - (u, v)|$$
$$\leqslant \|u_n\| \, \|v_n - v\| + \|v\| \, \|u_n - u\|.$$

This also implies the *Parseval equality*:

$$\|x\|^2 = \sum_{n=1}^{\infty} |c_n|^2. \tag{5.4.1}$$

Therefore, if a series $\sum_{n=1}^{\infty} c_n e_n$ with respect to an orthonormal system $\{e_n\}$ converges in norm, then $\sum_{n=1}^{\infty} |c_n|^2 < \infty$. Conversely, convergence of the latter series implies that the sequence of sums $\sum_{n=1}^{N} c_n e_n$ is Cauchy, which in case of complete X yields their convergence.

5.4.2. Theorem. (BESSEL'S INEQUALITY) *Let $\{e_\alpha\}$ be an orthonormal system in a Euclidean space X. Then for every $x \in X$ the set of all indices α for which $(x, e_\alpha) \neq 0$ is at most countable and*

$$\sum_\alpha |(x, e_\alpha)|^2 \leqslant (x, x). \tag{5.4.2}$$

PROOF. For every finite orthonormal collection $e_1, \ldots, e_n$ in the space X we have $x - \sum_{i=1}^{n} (x, e_i) e_i \perp e_j$ for every $j = 1, \ldots, n$. Hence

$$\left(x, \sum_{i=1}^{n} (x, e_i) e_i \right) = \left(\sum_{i=1}^{n} (x, e_i) e_i, \sum_{i=1}^{n} (x, e_i) e_i \right) = \sum_{i=1}^{n} |(x, e_i)|^2.$$

The left-hand side is estimated by $\|x\| \left(\sum_{i=1}^{n} |(x, e_i)|^2 \right)^{1/2}$ by the Cauchy–Bunyakovskii inequality, which gives the estimate

$$\sum_{i=1}^{n} |(x, e_i)|^2 \leqslant \|x\|^2.$$

Therefore, Bessel's inequality is also true for every countable orthonormal system. In particular, for every natural number k there are at most $k\|x\|^2$ indices α with $|(x, e_\alpha)|^2 \geqslant 1/k$. This shows that the numbers (x, e_α) can be nonzero only for a finite or countable family of indices α. $\qquad\square$

5.4.3. Lemma. *Let E_0 be a linear subspace of a Euclidean space E, $a \in E$, $a \notin E_0$. Then the following conditions are equivalent for a vector $b \in E_0$:*
(i) $a - b \perp E_0$, *i.e.,* $(a - b, x) = 0$ *for all* $x \in E_0$;
(ii) $\|a - b\| = \inf\{\|a - x\|\colon \ x \in E_0\}$.

PROOF. Let $a - b \perp E_0$. For any $x \in E_0$ we have
$$\|a - x\|^2 = \|a - b + b - x\|^2$$
$$= (a - b, a - b) + 2\mathrm{Re}\,(a - b, b - x) + (b - x, b - x)$$
$$= \|a - b\|^2 + \|b - x\|^2 \geqslant \|a - b\|^2.$$

Let (ii) be fulfilled. Suppose that there exists $x_0 \in E_0$ for which $(a - b, x_0) \neq 0$. We can assume that $\|x_0\| = 1$. Then
$$a - b = \widetilde{b} + (a - b - \widetilde{b}), \quad \widetilde{b} = (a - b, x_0)x_0.$$
We observe that $\widetilde{b} \perp a - b - \widetilde{b}$, since $\widetilde{b}$ is proportional to x_0 and
$$(x_0, a - b - \widetilde{b}) = (x_0, a - b) - (a - b, x_0)(x_0, x_0) = 0.$$
Thus, $\|a - b\|^2 = \|\widetilde{b}\|^2 + \|a - b - \widetilde{b}\|^2$, whence $\|a - b - \widetilde{b}\|^2 < \|a - b\|^2$ and $b + \widetilde{b} \in E_0$, i.e., we have found a vector in E_0 with a smaller distance to a than b. The obtained contradiction completes the proof. $\qquad\square$

The next simple fact follows directly from the previous lemma.

5.4.4. Corollary. *If we are given a finite orthonormal family $e_1, \ldots, e_n$, then the vector $\sum_{i=1}^{n}(x, e_i)e_i$ is the nearest element to x in the linear subspace generated by the vectors $e_1, \ldots, e_n$.*

In case of a closed subspace of a Hilbert space the previous lemma gives the existence of the orthogonal projection and orthogonal decomposition.

For every set M in a Euclidean space E its *orthogonal complement* is defined by the equality
$$M^\perp := \big\{x \in X\colon (x, m) = 0 \quad \text{for all } m \in M\big\}.$$
Then $M^\perp$ is a closed linear subspace in E. Indeed, if vectors $x_n \in M^\perp$ converge to x, then
$$(x, m) = \lim_{n \to \infty}(x_n, m) = 0 \quad \text{for all } m \in M,$$
i.e., $x \in M^\perp$. Hence $M^\perp$ is closed. Similarly we verify that $M^\perp$ is a linear subspace. It is also clear that $M \cap M^\perp = 0$, since $m \perp x$ for all $m \in M$, $x \in M^\perp$. Finally, if L is the linear span of M, then $L^\perp = M^\perp$, whence it follows that also for the closed linear span $\overline{L}$ of the set M we have $\overline{L}^\perp = M^\perp$.

The next theorem on the orthogonal decomposition is one of the most important in the geometry of Hilbert spaces.

5.4.5. Theorem. *Let X_0 be a closed linear subspace in a Hilbert space X. Then $X_0^\perp$ is a closed linear subspace in X and $X = X_0 \oplus X_0^\perp$, where the summands are mutually orthogonal.*

In particular, for every vector $x \in X$ there exists a unique vector $x_0 \in X_0$ with the property that $x - x_0 \perp X_0$, i.e., $x - x_0 \perp y$ for all $y \in X_0$. In addition, x_0 is the nearest to x element of X_0. The vector x_0 is called the orthogonal projection of the vector x on the subspace X_0 and denoted by $P_{X_0} x$.

PROOF. As explained above, $X_0^\perp$ is a closed subspace orthogonal to X_0. We show that every element $x \in X$ belongs to the sum $X_0 \oplus X_0^\perp$. Let

$$\delta := \inf\{\|x - y\| : \ y \in X_0\}.$$

There is a sequence $\{y_n\} \subset X_0$ with $\|x - y_n\| \to \delta$. Let us show that this sequence is Cauchy. Let $\varepsilon > 0$. There is a number N such that $\|x - y_n\|^2 < \delta^2 + \varepsilon^2/4$ for all $n \geqslant N$. Let $n, k \geqslant N$. Then by the parallelogram equality

$$\|y_n - y_k\|^2 = 2\|x - y_n\|^2 + 2\|x - y_k\|^2 - 4\|x - (y_n + y_k)/2\|^2$$
$$\leqslant 4\delta^2 + \varepsilon^2 - 4\delta^2 = \varepsilon^2.$$

By the completeness of X and closedness of X_0 the sequence $\{y_n\}$ converges to some vector $x_0 \in X_0$. It is readily seen that

$$\|x - x_0\| = \lim_{n \to \infty} \|x - y_n\| = \delta.$$

According to the lemma above we have $x - x_0 \perp X_0$. Completeness in this theorem is important: see Exercise 5.6.22; it is clear that it suffices that X_0 be complete.   □

5.4.6. Corollary. *Let X_0 be a closed linear subspace in a Hilbert space X. Then the mapping $P_{X_0} : x \mapsto P_{X_0} x$, called the orthogonal projection onto X_0, is linear and continuous, $\|P_{X_0} x\| \leqslant \|x\|$, moreover, $P_{X_0}^2 = P_{X_0}$.*

PROOF. Let $x \in X$. For every scalar λ the vector $\lambda P_{X_0} x$ belongs to X_0 and the difference $\lambda x - \lambda P_{X_0} x$ is orthogonal to X_0. By the uniqueness of the projection we have $P_{X_0}(\lambda x) = \lambda P_{X_0} x$. Similarly $P_{X_0}(x + y) = P_{X_0} x + P_{X_0} y$, because $P_{X_0} x + P_{X_0} y \in X_0$ and $x + y - P_{X_0} x - P_{X_0} y \perp X_0$. Since $x - P_{X_0} x \perp P_{X_0} x$, we have $\|x\|^2 = \|P_{X_0} x\|^2 + \|x - P_{X_0} x\|^2$, whence $\|P_{X_0} x\| \leqslant \|x\|$. By the linearity of P_{X_0} we obtain

$$\|P_{X_0} x - P_{X_0} y\| \leqslant \|x - y\|,$$

which shows the continuity of P_{X_0}. Finally, $P_{X_0} x = x$ for all $x \in X_0$.   □

Orthogonal projections are also called *operators of orthogonal projection*.

5.4.7. Corollary. *Let $\{e_\alpha\}$ be an orthonormal family in a Hilbert space X. Then, for every $x \in X$, the series $\sum_\alpha (x, e_\alpha) e_\alpha$ converges and its sum is the closest to x element of the closed linear subspace generated by the vectors e_α.*

PROOF. Bessel's inequality yields convergence of the series $\sum_\alpha (x, e_\alpha) e_\alpha$, in which at most countably many elements are nonzero. Its sum is the projection of x onto the aforementioned subspace.   □

5.4.8. Corollary. *Let X be a Hilbert space and let $\{e_\alpha\}$ be an orthonormal family. Bessel's inequality for this system becomes the equality for an element x precisely when x belongs to the closed linear span of $\{e_\alpha\}$.*

PROOF. The sum of the series of $|(x, e_\alpha)|^2$ equals the square of the norm of the projection of x onto the closed linear span of $\{e_\alpha\}$. Hence the equality in Bessel's inequality is only possible when the element x coincides with the indicated projection. $\square$

5.4.9. Remark. Thus, an orthonormal system $\{e_\alpha\}$ in a Hilbert space is complete if and only if the Parseval equality holds for this system for every x. The latter property is called the *closedness* of the given system. Therefore, the closedness is equivalent to the completeness. It is clear that this is also equivalent to the *totality* of the system, which is the equality $\{e_\alpha\}^\perp = 0$. In the general case an orthonormal system is a basis in the closure of its linear span. In an incomplete Euclidean space a total orthonormal sequence can fail to be a basis (take the space E from Exercise 5.6.22 and a basis in the subspace E_0).

From any finite or countable sequence of vectors x_n one can obtain an orthonormal sequence with the same linear span with the aid of the standard *Gram–Schmidt orthogonalization procedure*. To this end, assuming that $x_1 \neq 0$, we set $e_1 = x_1/\|x_1\|$, then take in $\{x_n\}$ the first vector x_{k_2} linearly independent with x_1 and in the two-dimensional space generated by e_1 and x_{k_2} we take a unit vector e_2 orthogonal to e_1. This construction continues by induction: if vectors $e_1, \ldots, e_n$ are already constructed and their linear span E_n does not coincide with the linear span of $\{x_n\}$, then we take the first vector $x_{k_{n+1}}$ not belonging to E_n and in the linear space generated by $x_{k_{n+1}}$ and E_n we find a unit vector e_{n+1} orthogonal to E_n. The result will be the required orthonormal system. In a separable space this procedure enables us to obtain an orthonormal basis. We shall see below that an incomplete nonseparable Euclidean space can fail to possess an orthonormal basis.

5.4.10. Theorem. *Every separable Euclidean space $E \neq 0$ possesses a finite or countable orthonormal basis. Moreover, a basis can be chosen in the linear span of an arbitrary everywhere dense countable set.*

PROOF. Let us take an everywhere dense countable set $\{x_n\}$ and apply to it the orthogonalization procedure. The obtained orthonormal sequence $\{e_n\}$ is a basis.

Indeed, for every vector x and every $\varepsilon > 0$, there exists a vector x_n such that $\|x - x_n\| < \varepsilon$. By construction x_n belongs to the linear span of the vectors $e_1, \ldots, e_N$ for some $N \leqslant n$. Therefore, $\left\| x - \sum_{i=1}^N (x, e_i)e_i \right\| \leqslant \|x - x_n\| < \varepsilon$. Then for all $k \geqslant N$ we have $\left\| x - \sum_{i=1}^k (x, e_i)e_i \right\| \leqslant \left\| x - \sum_{i=1}^N (x, e_i)e_i \right\| < \varepsilon$, which means that the series of $(x, e_i)e_i$ converges in norm to x. $\square$

The existence of a basis implies the following classical result about the isomorphism of infinite-dimensional separable Hilbert spaces.

5.4.11. Theorem. (THE RIESZ–FISCHER THEOREM) *Every infinite-dimensional separable Hilbert space is linearly isometric to the space l^2 (over the corresponding field). In particular, all infinite-dimensional separable Hilbert spaces (over the same field) are linearly isometric between themselves.*

PROOF. Let $\{e_n\}$ be an orthonormal basis in a separable Hilbert space H. Set
$$Jx = (x_n)_{n=1}^\infty, \quad x_n = (x, e_n).$$
We obtain a linear mapping to l^2, moreover, $\|Jx\| = \|x\|$. In addition, $J(H) = l^2$. Indeed, for every element $(x_n)_{n=1}^\infty$ in the space l^2 the series $\sum_{n=1}^\infty x_n e_n$ converges in norm in H, since
$$\left\| \sum_{n=m}^{m+k} x_n e_n \right\|^2 = \sum_{n=m}^{m+k} |x_n|^2,$$
which shows that the sequence of partial sums is Cauchy. For the sum x of this series we have $Jx = (x_n)_{n=1}^\infty$. $\qquad\square$

Let us give examples of orthonormal bases.

5.4.12. Example. (i) Let H be the real space $L^2[0, 2\pi]$. Then the trigonometric functions
$$(2\pi)^{-1/2}, \pi^{-1/2} \cos(nx), \pi^{-1/2} \sin(nx), \text{ where } n \in \mathbb{N},$$
form an orthonormal basis. In the complex case there is a basis formed by the functions $(2\pi)^{-1/2} \exp(inx)$, $n \in \mathbb{Z}$. It is easy to check that these systems are orthonormal. For verification of the completeness of the trigonometric system it suffices to use the fact known from calculus (proved in §1.9(v)): every 2π-periodic continuous function is uniformly approximated by finite linear combinations of these functions, hence their linear span is everywhere dense in $L^2[0, 2\pi]$.

(ii) Let μ be a nonnegative finite Borel measure on the interval $[a, b]$. Then the set of polynomials is everywhere dense in $L^2(\mu)$. Therefore, the process of orthogonalization of the functions $1, x, x^2, \ldots$ leads to a sequence of polynomials p_n of degree n that form a basis in $L^2(\mu)$. For measures with finite support this sequence is also finite. Choosing various intervals and measures, we obtain diverse orthonormal systems consisting of polynomials. This method goes back to P. L. Chebyshev. For example, in case of the interval $[-1, 1]$ with Lebesgue measure we obtain *Legendre polynomials* $L_n(x) = c_n \frac{d^n}{dx^n}(x^2 - 1)^n$, $L_0 = 1$, where c_n are normalizing constants.

(iii) In the space $L^2(\gamma)$, where γ is the standard Gaussian measure on the real line with density $\exp(-x^2/2)/\sqrt{2\pi}$, there is an orthonormal basis consisting of the *Chebyshev–Hermite polynomials*
$$H_n(x) = \frac{(-1)^n}{\sqrt{n!}} e^{x^2/2} \frac{d^n}{dx^n} e^{-x^2/2}, \quad H_0(x) = 1.$$
The linear spans of $\{H_n\}$ and $\{x^n\}$ coincide. One can verify by induction that $\{H_n\}$ is orthonormal, the completeness of this system will be proved in §9.1.

(iv) The functions $(2\pi)^{-1/4} H_n(x) \exp(-x^2/4)$ form an orthonormal basis in $L^2(\mathbb{R}^1)$. This is equivalent to (iii).

Nonseparable Hilbert spaces also have orthonormal bases. Below we construct an example of a nonseparable Hilbert space.

5.4.13. Theorem. *Every nonzero Hilbert space possesses an orthonormal basis.*

PROOF. Let $\mathcal{B}$ be the set of all orthonormal systems in a Hilbert space X partially ordered by inclusion. Every chain $\mathcal{B}_0 \subset \mathcal{B}$ has a majorant, for which we can take the union $\mathcal{V}$ of all vectors belonging to the families in $\mathcal{B}_0$. Any two different vectors x and y in $\mathcal{V}$ are orthogonal, since $x \in \mathcal{V}_1 \in \mathcal{B}_0$, $y \in \mathcal{V}_2 \in \mathcal{B}_0$, and either $\mathcal{V}_1 \subset \mathcal{V}_2$ or $\mathcal{V}_2 \subset \mathcal{V}_1$ by the linear ordering of $\mathcal{B}_0$. By Zorn's lemma there is a maximal element in $\mathcal{B}$, i.e., an orthonormal family $\{e_\alpha\}$ that is not a part of a larger orthonormal system. This means that there is no nonzero vector orthogonal to all e_α. By the completeness of X and Theorem 5.4.5 we conclude that the linear span of $\{e_\alpha\}$ is dense in X. Hence every vector x is the limit of a sequence of linear combinations of some countable subfamily $\{e_{\alpha_n}\}$. It is clear from the proof of Theorem 5.4.10 that $x = \sum_{n=1}^{\infty} (x, e_{\alpha_n}) e_{\alpha_n}$. $\qquad \square$

5.4.14. Example. Let Γ be an uncountable set and let $l^2(\Gamma)$ be the linear space of all real functions x on Γ such that $\sum_{\gamma \in \Gamma} |x(\gamma)|^2 < \infty$, i.e., the set of points $\gamma \in \Gamma$ with $x(\gamma) \neq 0$ is at most countable and the sum of the indicated series over such points is finite. Set $(x, y) := \sum_{\gamma \in \Gamma} x(\gamma) y(\gamma)$. It is clear that this makes $l^2(\Gamma)$ a Euclidean space, moreover, it is complete due to the completeness of l^2 (for every countable sequence of elements $x_n \in l^2(\Gamma)$ there is a common countable set outside of which all elements x_n vanish). Functions e_γ of the form $e_\gamma(\nu) = 0$ for $\nu \neq \gamma$ and $e_\gamma(\gamma) = 1$ form an uncountable orthonormal basis in $l^2(\Gamma)$. Hence $l^2(\Gamma)$ is nonseparable. Note that there are probability measures μ for which $L^2(\mu)$ is nonseparable (for example, any uncountable power of Lebesgue measure on $[0, 1]$).

A nonseparable incomplete Euclidean space can fail to have an orthonormal basis.

5.4.15. Example. Let $T = [0, 1]$, $H_1 = l^2$, $H_2 = l^2(T)$, and $X := H_1 \oplus H_2$. The space l^2 has a Hamel basis $\{v_t\}$ of the cardinality of the continuum. We shall consider H_1 and H_2 as the closed subspaces $(H_1, 0)$ and $(0, H_2)$ in X. Let us take in X a linear subspace L that is the linear span of the vectors $v_t + e_t$, where $\{e_t\}$ is an orthonormal basis in $l^2(T)$ from the previous example. Then L has no orthonormal basis.

PROOF. Let us take an orthonormal basis $\{e_n\}$ in the separable space H_1. Suppose that L has an orthonormal basis $\{u_\alpha\}$. By Bessel's inequality for every n we have at most countably many vectors u_α with $(e_n, u_\alpha) \neq 0$. Hence there exists a countable set of indices $\{\alpha_k\}$ such that $(e_n, u_\alpha) = 0$ for all n and $\alpha \notin \{\alpha_k\}$. Hence $u_\alpha \in H_2$ for all such α, but H_2 contains no nonzero vectors from L by the linear independence of v_t and the orthogonality of H_1 and H_2. Hence the basis $\{u_\alpha\}$ is countable, i.e., L is separable. Then the projection of L onto H_2 is also separable, which is impossible, since this projection contains a basis from H_2 and hence is dense in H_2, but H_2 is nonseparable. $\qquad \square$

Note that it actually follows from our reasoning that the Euclidean space L is nonseparable, but contains no uncountable orthonormal systems.

5.5. Convex Sets and the Schauder Theorem

In this section we prove one of the most important results in nonlinear analysis: Schauder's fixed point theorem. We start by introducing the concept of a convex set, which is of independent great importance.

5.5.1. Definition. *A set V in a linear space is called convex if, for every two elements $x, y \in V$ and every number $\lambda \in [0, 1]$, we have $\lambda x + (1 - \lambda)y \in V$.*

A set V is called balanced or circled if $\lambda v \in V$ for all $v \in V$ and all scalars λ with $|\lambda| \leqslant 1$.

The set of points of the form $\lambda x + (1 - \lambda)y$, where λ is running through $[0, 1]$, is called the line segment (or interval) with the endpoints x and y. Thus, a convex set contains all line segments with endpoints in this set.

For every set V in a linear space X there is the minimal convex set containing V (the intersection of all convex sets containing V). This set is called the *convex hull* or the *convex envelope* of V and denoted by $\operatorname{conv} V$. It is easy to verify (Exercise 5.6.32) that the convex envelope of V consists of all sums of the form $\sum_{i=1}^{n} t_i v_i$, where $n \in \mathbb{N}$, $v_i \in V$, $t_i \in [0, 1]$ and $\sum_{i=1}^{n} t_i = 1$.

Note that for constructing the convex envelope it is not sufficient to take all line segments with endpoints in the given set (for example, in case of three vertices of a triangle this gives not the whole triangle, but only its boundary).

The intersection of all convex balanced sets containing V is called the *balanced convex envelope* of V (also the terms *balanced convex hull, circled convex hull* are used) and denoted by $\operatorname{absconv} V$. It consists (Exercise 5.6.32) of all sums of the form $\sum_{i=1}^{n} \lambda_i v_i$, where $n \in \mathbb{N}$, $v_i \in V$ and $\sum_{i=1}^{n} |\lambda_i| \leqslant 1$.

Finally, for a set V in a normed space, the intersection of all closed convex sets containing V is called the *closed convex envelope* of V (or the *closed convex hull*) and the intersection of all closed balanced convex sets containing V is called the *closed balanced convex envelope* of V. These sets are denoted by $\overline{\operatorname{conv}} V$, and $\overline{\operatorname{absconv}} V$, respectively.

5.5.2. Proposition. *The closed convex envelope of V equals the closure of the convex envelope of V, and the closed balanced convex envelope of V equals the closure of the balanced convex envelope of V.*

PROOF. It suffices to verify the convexity of the closure of the convex envelope of V. If x and y belong to the closure of $\operatorname{conv} V$, then $x = \lim\limits_{n \to \infty} x_n$, $y = \lim\limits_{n \to \infty} y_n$, where $x_n, y_n \in \operatorname{conv} V$. Then for every $t \in [0, 1]$ we have $tx + (1 - t)y = \lim\limits_{n \to \infty} tx_n + (1 - t)y_n$, where $tx_n + (1 - t)y_n \in \operatorname{conv} V$. The case of the closed absolutely convex envelope is similar. □

5.5.3. Theorem. (SCHAUDER'S THEOREM) *Let K be a convex compact set in a normed space X and let $f \colon K \to K$ be a continuous mapping. Then there exists $x \in K$ with $f(x) = x$.*

PROOF. We shall assume to be known the finite-dimensional case, which is the Bohl–Brouwer theorem. Diverse elementary proofs can be found in Exercise 3.12.46 and also in [164], [312]. Let us show that for every $\varepsilon > 0$ there exists $x_\varepsilon \in K$ with $\|f(x_\varepsilon) - x_\varepsilon\| \leqslant \varepsilon$. This gives the existence of a fixed point. Indeed, extract from $\{x_{1/n}\}$ a sequence $\{z_n\}$ converging to some point $x \in K$. Then $f(z_n) \to f(x)$, whence we obtain $f(x) = x$, since $\|f(z_n) - z_n\| \to 0$.

Let us fix $\varepsilon > 0$ and pick in K an $\varepsilon/2$-net $x_1, \ldots, x_N$. Let S be the convex envelope of these points. By the convexity of K we have $S \subset K$. For every $i = 1, \ldots, N$ we set

$$\beta_i(x) = \varepsilon - \|x - x_i\| \text{ if } \|x - x_i\| \leqslant \varepsilon, \quad \beta_i(x) = 0 \text{ if } \|x - x_i\| > \varepsilon.$$

It is readily seen that the functions β_i are continuous. We observe that

$$\sum_{i=1}^{N} \beta_i(x) > 0 \quad \text{for all } x \in K,$$

since for every $x \in K$ there exists a number i such that $\|x - x_i\| \leqslant \varepsilon/2$. Hence the functions

$$\alpha_i(x) = \frac{\beta_i(x)}{\sum_{j=1}^{N} \beta_j(x)}$$

are continuous on K. We also observe that

$$0 \leqslant \alpha_i(x) \leqslant 1 \text{ and } \sum_{i=1}^{N} \alpha_i(x) = 1 \quad \text{for all } x \in K.$$

We now construct a continuous mapping g on K by the formula

$$g(x) = \sum_{i=1}^{N} \alpha_i(x) x_i.$$

The mapping $\psi = g \circ f \colon K \to K$ is also continuous, moreover, $\psi(S) \subset S$ by the convexity of S and the condition $f(K) \subset K$. By the aforementioned Bohl–Brouwer theorem there exists a point $z \in S$ with $\psi(z) = z$. Let us estimate the distance $\|f(z) - z\|$. We have

$$\|f(z) - z\| = \left\|f(z) - g\big(f(z)\big)\right\|$$

$$= \left\|\sum_{i=1}^{N} \alpha_i\big(f(z)\big) f(z) - \sum_{i=1}^{N} \alpha_i\big(f(z)\big) x_i\right\|$$

$$\leqslant \sum_{i=1}^{N} \alpha_i\big(f(z)\big) \|f(z) - x_i\| \leqslant \varepsilon,$$

because whenever $\|f(z) - x_i\| > \varepsilon$ we have $\alpha_i\big(f(z)\big) = 0$. □

5.5.4. Proposition. *The convex envelope and the convex balanced envelope of a totally bounded set in a normed space are totally bounded. The closed convex envelope and the closed balanced convex envelope of a compact set in a Banach space are compact.*

PROOF. Let $x_1, \ldots, x_n$ be an ε-net in the set V. Then the convex envelope S of the finite set $\{x_1, \ldots, x_n\}$ is an ε-net in the convex envelope $\widetilde{V}$ of the set V. Indeed, for every $x \in \widetilde{V}$ there exist vectors $v_1, \ldots, v_k \in V$ and points $t_1, \ldots, t_k \in [0, 1]$ such that $\sum_{j=1}^{k} t_j = 1$ and $x = \sum_{j=1}^{k} t_j v_j$. There are points x_{i_j} for which $\|x_{i_j} - v_j\| < \varepsilon$. Then $z = \sum_{j=1}^{k} t_j x_{i_j} \in S$ and $\|x - z\| < \varepsilon$. It remains to observe that the set S in a finite-dimensional space is compact. The case of the convex balanced envelope is similar. The second assertion follows from the compactness of any closed totally bounded set in a Banach space. $\square$

5.5.5. Corollary. *Let V be a closed convex set in a Banach space X and let $f \colon V \to V$ be a continuous mapping such that $f(V)$ is contained in a compact set. Then f has a fixed point.*

PROOF. Let K be the closure of the convex envelope of $f(V)$. By assumption and the proposition above K is convex compact and $f(K) \subset K$. $\square$

Schauder's theorem is a powerful tool for proving the solvability of integral and differential equations.

5.5.6. Example. Let $\varphi \colon \mathbb{R}^1 \to \mathbb{R}^1$ and $\psi \colon \mathbb{R}^2 \to \mathbb{R}^1$ be bounded continuous functions. Then there exists a continuous function u on the interval $[0, 1]$ satisfying the integral equation

$$u(t) = \varphi(t) + \int_0^t \psi\big(u(s), s\big)\, ds.$$

Indeed, let us define a mapping f on the space $C[0, 1]$ by the formula

$$f(u)(t) = \varphi(t) + \int_0^t \psi\big(u(s), s\big)\, ds.$$

The set of values of f is uniformly bounded. In addition, this set is uniformly equicontinuous due to the uniform continuity of φ on $[0, 1]$ and boundedness of ψ:

$$|f(u)(t) - f(u)(t')| \leqslant |\varphi(t) - \varphi(t')| + |t - t'| \sup_{x, y} |\psi(x, y)|.$$

Thus, the set of values of f is contained in a compact set. By the obvious continuity of f there exists a fixed point.

In connection with Schauder's theorem it is useful to have in mind the following result describing the structure of compact sets in Banach spaces.

5.5.7. Proposition. *Let K be a compact set in a Banach space X. Then K is contained in the closed convex envelope of some sequence converging to zero.*

PROOF. For every n we find a finite 4^{-n-2}-net $K_n \subset K$ for K. The set $\bigcup_{n=1}^{\infty} K_n$ is dense in K. Set $S_1 := 2K_1$. Next, for each $n > 1$ we choose a finite set S_n in X as follows: for every $v \in K_n$ we find an element $u \in K_{n-1}$ with $\|v - u\| \leqslant 4^{-n}$ and take the element $x := 2^n(v - u) \in X$. The set S_n consisting of such elements has the cardinality not exceeding the cardinality of K_n. Since $\|2^n(v - u)\| \leqslant 2^{-n}$, the sequence $\{x_n\}$ obtained by the consecutive enumeration of the points from $S_1, S_2, \ldots$ converges to zero with respect to the norm. We observe that every element $v \in K_n$ has the form $v = 2^{-1}x_{i_1} + \cdots + 2^{-n}x_{i_n}$ and hence is contained in the balanced convex envelope of $\{x_i\} = \bigcup_{n=1}^{\infty} S_n$. Hence $\bigcup_{n=1}^{\infty} K_n$ belongs to the closed convex envelope of $\{x_n\} \cup \{-x_n\}$. It remains to recall that K is the closure of $\bigcup_{n=1}^{\infty} K_n$. $\square$

Thus, compact sets are precisely closed subsets of closed convex envelopes of sequences converging to zero. Exercise 5.6.33 gives a simple explicit description of such closed convex envelopes.

In many problems connected with convex sets an important role is played by *extreme points*. A point x of a set M in a linear space is called extreme if it is not an inner point of any line segment with endpoints in M, i.e., the equality $x = tm_1 + (1 - t)m_2$, where $t \in (0,1)$ and $m_1, m_2 \in M$, is only possible for $m_1 = m_2 = x$. We shall see in §8.6(vi) that convex compact sets are closed convex envelopes of their extreme points.

5.6. Complements and Exercises

(i) Balls and ellipsoids (176). (ii) The Kadec and Milyutin theorems (177). (iii) Ordered vector spaces and vector lattices (178). Exercises (180).

5.6(i). Balls and ellipsoids

In the geometry of Banach spaces there is an interesting direction dealing with the structure of finite-dimensional subspaces (the so-called local theory). We mention here two results about comparisons of balls with ellipsoids. A Banach space X is called *finitely representable* in a Banach space Y if for every $\varepsilon > 0$ and every finite-dimensional subspace $L \subset X$ there exists an injective linear mapping $T_L \colon L \to Y$ such that

$$(1 - \varepsilon)\|x\| \leqslant \|T_L x\| \leqslant (1 + \varepsilon)\|x\|, \quad x \in L.$$

The following remarkable result was obtained by A. Dvoretzky.

5.6.1. Theorem. *The Hilbert space l^2 is finitely representable in every infinite-dimensional Banach space.*

The geometric meaning of the Dvoretzky theorem is that for every $\varepsilon > 0$ for every n one can find an n-dimensional subspace L_n in X with a Euclidean ball U_n (in some linear coordinates) such that U_n and the unit ball in L_n with respect to the norm from X are in the ε-neighborhoods of each other. Hence the ball in X has "almost Euclidean" sections of every finite dimension. Of course, there might be no precisely Euclidean sections. A more precise quantitative estimate is this. There is an absolute constant $C > 0$ with the following property. Suppose that $\mathbb{R}^n$ is equipped with a norm $\| \cdot \|_X$ such that $\|x\|_X \leqslant \|x\|$, where $\| \cdot \|$ is the usual Euclidean norm. Let θ_X denote the integral of $\|x\|_X$ over the Euclidean unit sphere with the standard surface measure. Let $0 < \varepsilon < 1/3$ and $k \leqslant n$ be such that $k \leqslant C\theta_X^2 n^2 \varepsilon^2 |\ln \varepsilon|^{-1}$. Then there exists a k-dimensional subspace $L \subset \mathbb{R}^n$ such that

$$(1 - \varepsilon)\theta_X \|x\| \leqslant \|x\|_X \leqslant (1 + \varepsilon)\theta_X \|x\|, \quad x \in L.$$

If the dimension of X is finite, the balls with respect to Euclidean norms on the whole space can differ substantially from the balls with respect to the given norm of X. However, also here there is a useful positive result: John's theorem. Before formulating this theorem, we observe that X can be identified with $\mathbb{R}^n$ equipped with some norm $\| \cdot \|$. The unit ball U with respect to this norm contains ellipsoids. It is not difficult to show that among them there is an ellipsoid U_J of the maximal volume (with respect to the standard Lebesgue measure on $\mathbb{R}^n$), which is called *John's ellipsoid*. John's ellipsoid generates a Euclidean norm $\| \cdot \|_J$ on X.

5.6.2. Theorem. *We have* $U \subset \sqrt{n} U_J$.

For example, if X is $\mathbb{R}^2$ with the norm $\max(|x_1|, |x_2|)$, then the unit ball is the square and John's ellipsoid is the usual unit disc. Here $\sqrt{2}$ is the radius of the smallest disc containing this square. Proofs of the stated results can be found in [**9**, Chapter 12].

5.6(ii). The Kadec and Milyutin theorems

Here we mention two deep results connected with classification of Banach spaces. The first result is the famous Kadec theorem asserting that in the category of topological spaces all infinite-dimensional separable Banach spaces are homeomorphic.

5.6.3. Theorem. *Every two infinite-dimensional separable Banach space are homeomorphic. In particular, they are homeomorphic to* l^2.

According to a more general result of Anderson–Kadec, the same is true for infinite-dimensional separable Fréchet spaces, and Toruńczyk extended this theorem to spaces of equal topological weight. Proofs and references can be found in Bessaga, Pełczyński [**64**].

The situation is completely different when we consider only linear homeomorphisms (or only linear isometries). In the next chapter we present results that enable one to distinguish many Banach spaces from the point of view of the linear topological classification. For example, by using the adjoint spaces one shows at once that the spaces $C[0, 1]$ and $L^2[0, 1]$ are not linearly homeomorphic. On the other hand, we shall see that every separable Banach space is linearly isometric to a closed linear subspace in $C[0, 1]$. The following remarkable theorem is due to A. A. Milyutin (for a proof, see [**473**]).

5.6.4. Theorem. *Let K be an uncountable compact metric space. Then the spaces $C(K)$ and $C[0, 1]$ are linearly homeomorphic.*

However, isometries between these spaces exist quite seldom, as the following Banach–Stone theorem shows (its proof is Exercise 6.10.176 in Chapter 6).

5.6.5. Theorem. *Let K_1 and K_2 be compact spaces. The spaces $C(K_1)$ and $C(K_2)$ are linearly isometric precisely when K_1 and K_2 are homeomorphic.*

For example, the spaces $C[0, 1]$ and $C([0, 1]^2)$ are are linearly homeomorphic, but not isometric.

5.6(iii). Ordered vector spaces and vector lattices

Many functional spaces encountered in applications possess natural orderings consistent with their linear structure. Here we briefly discuss this question. Let E be a real linear space equipped with some partial order $\leqslant$ (about partially ordered sets see §1.1) that agrees with the linear structure in the following sense: if $x \leqslant y$, then $x + z \leqslant y + z$ for all $z \in E$ and $\lambda x \leqslant \lambda y$ for all numbers $\lambda \geqslant 0$. Then E is called an *ordered vector space*. We emphasize that E is only partially ordered. One often includes in the definition the *anti-symmetry* of the order, i.e., the equality $x = y$ if $x \leqslant y$ and $y \leqslant x$, but we do not do this. It is clear that $x \leqslant y$ precisely when $x - y \leqslant 0$. Set $y \geqslant x$ if $x \leqslant y$. Thus, the order is uniquely determined by the set

$$K := \{z \in E \colon\ z \geqslant 0\}.$$

This set is called the *positive cone*. It possesses the following two properties: $z_1 + z_2 \in K$ if $z_1, z_2 \in K$ and $\lambda z \in K$ if $z \in K$ and $\lambda \geqslant 0$. Conversely, every set with the indicated properties defines an order that agrees with the linear structure if we set $x \leqslant y$ whenever $y - x \in K$. In case $K \cap (-K) = 0$ the order is anti-symmetric.

On the space $B(\Omega)$ of all real functions on a nonempty set Ω there is a natural partial order given by the pointwise inequality $x(\omega) \leqslant y(\omega)$. The same order is defined on the space $C(T)$ of continuous functions on a topological space T.

On the space $L^p(\mu)$ a partial order $f \leqslant g$ on equivalence classes is defined by the inequality $f(x) \leqslant g(x)$ μ-a.e. for representatives of equivalence classes. On the space $\mathcal{M}(\mathcal{A})$ of all bounded real measures on a σ-algebra $\mathcal{A}$ the relation $\mu \leqslant \nu$ is naturally defined by the inequality $\mu(A) \leqslant \nu(A)$ for all $A \in \mathcal{A}$. Of course, every real vector space can be equipped with the partial order by setting $x \leqslant y$ only for $x = y$, i.e., by taking the trivial cone $K = \{0\}$.

A linear subspace of an ordered vector space is also an ordered vector space with the same partial order.

A majorant of a set F in a partially ordered set $(E, \leqslant)$ is an element $m \in E$ such that $f \leqslant m$ for all $f \in F$. A majorant m is called a precise upper bound of F if $m \leqslant \widetilde{m}$ for every other majorant $\widetilde{m}$ of the set F. A precise upper bound is unique if the order is anti-symmetric.

Similarly one defines the terms a *minorant* and a *precise lower bound*. A partially ordered set $(E, \leqslant)$ is called a *lattice* (or a *structure*) if for every $x, y \in E$ there is a precise upper bound $x \vee y$ and a precise lower bound $x \wedge y$ (not necessarily unique). A lattice E is called *complete* if every subset of E having a majorant has also a precise upper bound. If this condition is fulfilled for countable sets, then E is called a σ-complete lattice. A precise upper bound of a set F in a lattice E is denoted by $\bigvee F$. An ordered vector space with an anti-symmetric order that is a lattice is called a *vector lattice* or a *Riesz space*.

5.6.6. Example. (i) Let $\mathcal{F}$ be a linear space of real functions on a nonempty set Ω such that $\max(f, g) \in \mathcal{F}$ for all $f, g \in \mathcal{F}$, where $\max(f, g)$ and $\min(f, g)$ are defined pointwise. Then $\min(f, g) = -\max(-f, -g) \in \mathcal{F}$, i.e., $\mathcal{F}$ is a vector lattice with respect to the partial order given by the pointwise comparison, where

$f \vee g := \max(f, g)$, $f \wedge g := \min(f, g)$. Observe also that $|f| \in \mathcal{F}$ for all $f \in \mathcal{F}$. Since $\max(f, g) = (|f - g| + f + g)/2$, it suffices to require the closedness of $\mathcal{F}$ with respect to taking absolute values.

(ii) The set $\widetilde{\mathcal{L}}^0(\mu)$ of real μ-measurable everywhere defined functions is a lattice with its natural ordering: $f \leqslant g$ if $f(x) \leqslant g(x)$ μ-a.e. For $f \vee g$ and $f \wedge g$ we take pointwise defined $\max(f, g)$ and $\min(f, g)$, respectively. Recall that the whole set $\mathcal{L}^0(\mu)$ of μ-measurable functions (which may be undefined or infinite on a measure zero set) is not a linear space. But we obtain lattices with the same order relation considering the subsets of real everywhere defined functions in $\mathcal{L}^p(\mu)$, $p \in (0, \infty]$, and the spaces $L^p(\mu)$ of equivalence classes.

A sequence $\{x_n\}$ in a vector lattice E is called *order convergent* to x if there is a sequence $\{y_n\} \subset E$ such that $|x_n - x| \leqslant y_n$ and $y_n \downarrow 0$, i.e., $y_{n+1} \leqslant y_n$ and $\inf\{y_n\} = 0$.

Let us mention a number of useful results about lattice properties of the space L^p; for their proofs, see [73, §4.7(i)].

5.6.7. Theorem. *Let $(X, \mathcal{A}, \mu)$ be a measurable space with a σ-finite measure μ. Then the sets $\widetilde{\mathcal{L}}^0(\mu)$ and $L^0(\mu)$ are complete lattices with the aforementioned order relations. In addition, if a set $\mathcal{F} \subset \widetilde{\mathcal{L}}^0(\mu)$ has a majorant h, then there exists an at most countable set $\{f_n\} \subset \mathcal{F}$ such that $f \leqslant \sup_n f_n \leqslant h$ for all functions $f \in \mathcal{F}$.*

5.6.8. Corollary. *For any σ-finite measure μ, the sets $\widetilde{\mathcal{L}}^p(\mu)$ and $L^p(\mu)$ for all $p \in [0, +\infty]$ are complete lattices with the aforementioned order relations. In addition, if a set $\mathcal{F}$ in $\widetilde{\mathcal{L}}^p(\mu)$ has a majorant h, then its precise upper bound in $\widetilde{\mathcal{L}}^p(\mu)$ coincides with its precise upper bound in $\widetilde{\mathcal{L}}^0(\mu)$, moreover, there exists an at most countable set $\{f_n\} \subset \mathcal{F}$ such that $f \leqslant \sup_n f_n \leqslant h$ for all $f \in \mathcal{F}$.*

5.6.9. Corollary. *Let μ be a σ-finite nonnegative measure on a space $(X, \mathcal{A})$ and let A_t, $t \in T$, be a family of measurable sets. Then it contains an at most countable subfamily $\{A_{t_n}\}$ such that $\mu\big(A_t \backslash \bigcup_{n=1}^{\infty} A_{t_n}\big) = 0$ for every t.*

Note that a precise upper bound $\bigvee \mathcal{F}$ of a set $\mathcal{F}$ in $\widetilde{\mathcal{L}}^p(\mu)$ need not coincide with the pointwise defined function $\sup_{f \in \mathcal{F}} f(x)$. For example, let F be a set in $[0, 1]$. For each $t \in F$ set $f_t(s) = 1$ if $s = t$, $f_t(s) = 0$ if $s \neq t$, where $s \in [0, 1]$. Then $\sup_{t \in F} f_t(s) = I_F(s)$, although the precise upper bound of the family $\{f_t\}$ in $\widetilde{\mathcal{L}}^p[0, 1]$ is the zero function. If F is a nonmeasurable set, then the function $\sup_{t \in F} f_t(s) = I_F(s)$ is also nonmeasurable. As an example of an incomplete lattice consider the space $C[0, 1]$ of continuous functions with its natural order. In this lattice the set of all continuous functions vanishing on $[0, 1/2)$ and not exceeding 1 on $[1/2, 1]$ has a majorant 1, but has no precise upper bound. If on the set of measurable functions on $[0, 1]$ in place of the comparison almost everywhere (as was done above) we introduce the order generated by the inequality $f(x) \leqslant g(x)$ for every x, then we also obtain an incomplete lattice.

It is worth noting that the results above do not extend to arbitrary infinite measures, although there exist non σ-finite measures for which they hold.

5.6.10. Corollary. *Let μ_t, $t \in T$, be a family of probability measures on a σ-algebra $\mathcal{A}$ absolutely continuous with respect to some fixed probability measure μ on $\mathcal{A}$. Then there exists an at most countable set of indices t_n such that all measures μ_t are absolutely continuous with respect to the measure $\sum_{n=1}^{\infty} 2^{-n} \mu_{t_n}$.*

The space $\mathcal{M}(X, \mathcal{A})$ of all bounded signed measures on $\mathcal{A}$ is also a complete lattice. We have the following natural order on $\mathcal{M}(X, \mathcal{A})$: $\mu \leqslant \nu$ if and only if $\mu(A) \leqslant \nu(A)$ for all $A \in \mathcal{A}$. For any $\mu, \nu \in \mathcal{M}(X, \mathcal{A})$ we set

$$\mu \vee \nu := \mu + (\nu - \mu)^+, \quad \mu \wedge \nu := \mu - (\nu - \mu)^-.$$

If μ and ν are given by densities f and g with respect to a nonnegative measure λ (for example, $\lambda = |\mu| + |\nu|$), then $\mu \vee \nu = \max(f, g) \cdot \lambda$, $\mu \wedge \nu = \min(f, g) \cdot \lambda$. It is readily seen that $\mu \vee \nu$ is the minimal measure majorizing μ and ν (i.e., is a precise upper bound of μ and ν). Indeed, if a measure η is such that $\mu \leqslant \eta$ and $\nu \leqslant \eta$, then we take a nonnegative measure λ such that $\mu = f \cdot \lambda$, $\nu = g \cdot \lambda$, $\eta = h \cdot \lambda$. Then $h \geqslant f$ and $h \geqslant g$ λ-a.e., whence $h \geqslant \max(f, g)$ λ-a.e. Thus, $\mathcal{M}(X, \mathcal{A})$ is a lattice.

5.6.11. Theorem. *The lattice $\mathcal{M}(X, \mathcal{A})$ is complete.*

For a vector lattice E we set $|x| := x \vee (-x)$, $x^+ := x \vee 0$, $x^- := (-x) \vee 0$. The proofs of the following facts are delegated to Exercise 5.6.67.

5.6.12. Proposition. *Let E be a vector lattice. Then $x + y = x \vee y + x \wedge y$, $x = x^+ - x^-$, $|x| = x^+ + x^-$, $|\lambda x| = |\lambda||x|$, $|x + y| \leqslant |x| + |y|$ for all $x, y \in X$ and $\lambda \in \mathbb{R}^1$.*

A normed lattice is a vector lattice E equipped with a norm $\| \cdot \|$ such that $\|y\| \leqslant \|x\|$ whenever $|y| \leqslant |x|$. If E is complete with respect to this norm, then E is called a *Banach lattice*. We observe that $\|x\| = \||x|\|$, because $|x| = ||x||$. The spaces $C(K)$, $\mathcal{M}(X, \mathcal{A})$, $L^p(\mu)$ with $1 \leqslant p \leqslant \infty$ are Banach lattices. We observe that a Banach lattice need not be a complete lattice (a typical example is $C[0, 1]$).

Exercises

5.6.13. Prove that the spaces described in Example 5.2.1 are linear and complete with respect to the indicated norms.

5.6.14. Prove that the closure of a linear subspace of a normed space is a linear subspace.

5.6.15. Prove that a normed space is complete precisely when every series $\sum_{n=1}^{\infty} x_n$ such that $\|x_n\| \leqslant 2^{-n}$ converges in this space.

HINT: use that the completeness of a metric space follows from convergence of all sequences $\{x_n\}$ such that the distance between x_{n+1} and x_n is estimated by 2^{-n}.

5.6.16. Justify the construction of the completions of normed and Euclidean spaces presented in Remark 5.1.7.

5.6.17. Show that a normed space is separable if and only if its unit sphere is separable.

HINT: observe that if the unit sphere is separable, then all spheres are separable, so one can take countable dense subsets in spheres of rational radii.

5.6.18. Show that the spaces $C_b(\mathbb{R})$ and $C_b((0,1))$ are nonseparable.
HINT: consider the family of function $f_\alpha(x) = \sin(\alpha/x)$ on $(0,1)$.

5.6.19. Is it true that any two incomplete infinite-dimensional separable Euclidean spaces are isometric?
HINT: consider the subspace of finite sequences $l_0^2 \subset l^2$ and its direct sum with l^2.

5.6.20. Let Γ be an infinite set, $1 \leqslant p < \infty$, and let $l^p(\Gamma)$ be the space of all functions $x\colon \Gamma \to \mathbb{R}^1$ for which

$$\|x\|_p := \left(\sum_{\gamma \in \Gamma} |x(\gamma)|^p\right)^{1/p} < \infty,$$

where convergence of the series means that at most countably many numbers $x(\gamma)$ are nonzero and the sum over them is finite. Show that $\| \cdot \|_p$ is a norm with respect to which $l^p(\Gamma)$ is complete. Show that $l^p(\Gamma)$ is nonseparable if Γ is uncountable.

5.6.21. In the space l^2 consider the following two subspaces of codimension 1 and 2: $Y_1 := \{x\colon x_1 = 0\}$, $Y_2 := \{x\colon x_1 = x_2 = 0\}$. Show that Y_1 and Y_2 are linearly isometric, but the factor-spaces l^2/Y_1 and l^2/Y_2 are not even linearly isomorphic.

5.6.22. Let E be the space of all sequences of the form $(x_1, \ldots, x_n, 0, 0, \ldots)$ with the inner product from l^2 and let $E_0 := \{x \in E\colon \sum_{n=1}^{\infty} n^{-1}x_n = 0\}$. Prove that E_0 is a closed subspace in E of codimension 1, but $E_0^\perp = 0$.
HINT: show that the orthogonal complement of this subspace in l^2 is generated by the vector with coordinates $1/n$.

5.6.23. Prove that on every infinite-dimensional Banach space X there exists a norm not equivalent to the original norm, but also making X a complete space.
HINT: let X be separable; take in X a Hamel basis $\{v_\alpha\}$ of the cardinality of the continuum and use it to establish linear isomorphisms between X and the spaces l^2 and l^1 (which are not linearly homeomorphic, as will be clear from the next chapter, since the dual to l^1 is nonseparable). This gives on X two nonequivalent norms with respect to which X is complete. In the general case use the spaces of the form $l^2(\Gamma)$ and $l^1(\Gamma)$.

5.6.24. Prove that on every infinite-dimensional Banach space there is a larger norm with respect to which it is not complete.
HINT: use a Hamel basis of vectors of unit norm and the Banach theorem about the equivalence of comparable complete norms.

5.6.25. Suppose that an infinite-dimensional linear space X is equipped with two norms with respect to each of which X is complete. Is it true that X is complete with respect to the sum of these norms?
HINT: consider the case where these norms are not equivalent.

5.6.26. Let Y be a closed subspace of a Banach spaces X. Show that X is separable if and only if Y and X/Y are separable.

5.6.27. Let $\{B_n\}$ be a sequence of closed balls in a Banach space and $B_{n+1} \subset B_n$. Prove that their intersection is not empty.
HINT: consider the sequence of their centers and observe that the radii are decreasing.

5.6.28. Show that the subspace of $L^1[0,1]$ corresponding to proper Riemann integrable functions is a first category set.

5.6.29. Give an example of a sequence of decreasing bounded closed convex sets in the space $C[0,1]$ with the empty intersection.

HINT: consider the subsets of the unit ball in $C[0,1]$ that consist of nonnegative functions f with $f(0) = 0$ and $f(t) = 1$ if $t \geqslant 1/n$.

5.6.30.° Let A and B be closed sets in a normed space such that B is compact. Prove that $A + B$ is closed. Can one omit the compactness of B?

5.6.31.° Let A and B be totally bounded sets in a normed space X. Prove that their algebraic sum $A + B$ is totally bounded.

5.6.32. (i) Prove that the convex hull of a set V in a linear space consists of all finite sums of the form $\sum_{i=1}^{n} t_i v_i$, where $n \in \mathbb{N}$, $v_i \in V$, $t_i \in [0,1]$ and $\sum_{i=1}^{n} t_i = 1$, and the balanced convex hull consists of the sums of the form $\sum_{i=1}^{n} \lambda_i v_i$, where $n \in \mathbb{N}$, $v_i \in V$ and $\sum_{i=1}^{n} |\lambda_i| \leqslant 1$ (the numbers λ_i are real or complex depending on the field of scalars).

(ii) Show that the convex hull of a compact set in $\mathbb{R}^n$ is compact.

(iii) Show that the convex hull of an open set is open. Is it true that the convex hull of a closed set in the plane is closed?

5.6.33. Let X be a Banach space and let $\{x_n\} \subset X$ be such that $\|x_n\| \to 0$. Prove that $\overline{\operatorname{conv}}\{x_n\} = \left\{\sum_{n=1}^{\infty} t_n x_n \colon t_n \geqslant 0, \sum_{n=1}^{\infty} t_n \leqslant 1\right\}$.

HINT: show that the set on the right is convex and closed and that for each n the convex hull of x_n and the origin is contained in the set on the left.

5.6.34. Let K be a nonempty convex compact set in a Banach space. Prove that for every $\varepsilon > 0$ one can find a convex compact set $C \subset K$ such that set $K \backslash C$ is not empty and has diameter less than ε.

HINT: see [**51**, p. 232].

5.6.35. Give an example of a closed linear subspace L and a point f in $C[0,1]$ such that in L there is no nearest point to f.

HINT: consider the hyperplane that consists of functions with equal integrals over $[0, 1/2]$ and $[1/2, 1]$.

5.6.36. Let A be a closed convex set in a Hilbert space H and $x \in H$.

(i) Prove that a point $p \in A$ is the nearest point to x in the set A precisely when $(a - p, x - p) \leqslant 0$ for all $a \in A$.

(ii) Prove that A has a unique nearest point to x.

5.6.37. Let X be a real normed space of finite dimension n with the unit sphere S. (i) Prove that there exist points $v_1, \ldots, v_{2n} \in S$ with $\|v_i - v_j\| \geqslant 1$ for all $i \neq j$. (ii) Prove that there exist points $u_1, \ldots, u_n \in S$ with $\|u_i - u_j\| > 1$ for all $i \neq j$. (iii)* Prove that there exist points $v_1, \ldots, v_{2n} \in S$ with $\|v_i - v_j\| > 1$ for all $i \neq j$.

HINT: (i) take $2n - 2$ such points in the hyperplane P and a point $v_{2n-1} \in S$ outermost from P, then set $v_{2n} = -v_{2n-1}$; (ii) pick $x_1, \ldots, x_{n-1}$, take a linear function f with $x_i \in f^{-1}(0)$ having the maximum 1 on S; let g be a linear function on X equal 1 on $x_1, \ldots, x_{n-1}$; take x_n such that $g(x_n) = \min\{g(x) \colon \|x\| = 1, f(x) = 1\}$; (iii)* show that P above can be taken such that only two points of S will be outermost from P.

5.6.38. Prove that the parallelogram identity implies that the given norm is generated by some inner product.

HINT: see [**331**, Chapter III, §4].

5.6.39.° Prove that the set $\{x \in l^2 \colon |x_n| \leqslant c_n\}$, where $c_n > 0$, is compact in l^2 precisely when $(c_n) \in l^2$.

5.6.40. Let H be a separable Hilbert space with an orthonormal basis $\{e_n\}$ and let $K \subset H$ be a compact set. Prove that K is contained in a compact ellipsoid of the form $Q := \{x \in H \colon \sum_{n=1}^{\infty} \alpha_n |(x, e_n)|^2 \leqslant 1\}$, where $\alpha_n > 0$ and $\alpha_n \to +\infty$.

HINT: find increasing numbers m_k such that $\sum_{n=m_k}^{\infty} |(x, e_n)|^2 \leqslant 4^{-k}$ for all $x \in K$.

5.6.41.° Prove that the space

$$\mathrm{Lip}_0[0,1] := \{f \in C[0,1] \colon f(0) = 0, \ \|f\|_{\mathrm{Lip}} := \sup_{t \neq s} |f(t) - f(s)|/|t - s| < \infty\}$$

is linearly isometric to the space $L^{\infty}[0,1]$ and that for an isometry one can take the mapping

$$V \colon L^{\infty}[0,1] \to \mathrm{Lip}_0[0,1], \quad Vf(t) = \int_0^t f(s)\, ds.$$

5.6.42. Let X be a Hilbert space. Prove that there exists a probability measure μ such that X is linearly isometric to $L^2(\mu)$.

HINT: let $\{e_\alpha\}_{\alpha \in A}$ be an orthonormal basis in X. Take for μ the product of A copies of Lebesgue measure on $[0,1]$ (see [73, §3.5]).

5.6.43.° Let $0 < \alpha \leqslant 1$ and let H^α be the Hölder space of functions φ on $[0,1]$ such that $\|\varphi\|_\alpha := \sup_{s \neq t} |\varphi(t) - \varphi(s)|/|t - s|^\alpha + \sup_s |\varphi(s)| < \infty$. (i) Prove that the space $(H^\alpha, \|\cdot\|_\alpha)$ is Banach and nonseparable. (ii) Prove that the closed balls of the space $(H^\alpha, \|\cdot\|_\alpha)$ are compact in the space $(H^\beta, \|\cdot\|_\beta)$ whenever $\alpha > \beta$.

HINT: (i) consider the functions $f_t(x) = (x - t)^\alpha$, $x > t$, $f_t(x) = 0$, $x \leqslant t$. (ii) Observe that if a sequence of functions f_n is bounded in H^β and converges uniformly to f, then it also converges in H^α.

5.6.44.° Prove that the set $C_0^\infty(\mathbb{R}^n)$ of infinitely differentiable functions with bounded support is dense in every space $L^p(\mathbb{R}^n)$ with $1 \leqslant p < \infty$.

5.6.45. The Chebyshev–Laguerre polynomials are defined by the formula

$$L_k(t) := e^t \frac{d^k}{dt^k}(t^k e^{-t}), \quad k = 0, 1, 2, \ldots$$

Prove that the Chebyshev–Laguerre functions $\psi_k(t) := e^{-t/2} L_k(t)/k!$ form an orthonormal basis in the space $L^2[0, +\infty)$.

5.6.46. Let μ be a measure on $[0, +\infty)$ with density $\varrho(x) = x^{-\ln x}$ with respect to Lebesgue measure. Show that for any $n = 0, 1, \ldots$ the function x^n belongs to $L^2(\mu)$ and is orthogonal to the function $\sin(2\pi \ln x)$, i.e., polynomials are not dense in $L^2(\mu)$.

5.6.47. Let s_1 be the space of all real sequences $x = (x_n)$ for which the series $\sum_{n=1}^{\infty} x_n$ converges (not necessarily absolutely). Set

$$\|x\|_{s_1} := \sup_n \left| \sum_{i=1}^{n} x_i \right|.$$

Prove that $\|\cdot\|_{s_1}$ is a norm with respect to which s_1 is complete.

5.6.48.° Prove that the set of nonnegative functions is closed and nowhere dense in the space $L^1[0,1]$.

5.6.49.° Suppose that a real function f on a normed space X is uniformly continuous. Prove that there exists $C \geqslant 0$ such that $|f(x)| \leqslant |f(0)| + C\|x\|$ for all $x \in X$.

5.6.50.° Prove that a set $K \subset c_0$ is totally bounded precisely when there is $y \in c_0$ such that $|x_n| \leqslant |y_n|$ for all $n \in \mathbb{N}$ and $x \in K$.

5.6.51.° Let A be an everywhere dense set in a normed space X. Prove that for every $x \in X$ there exist points $a_i \in A$ such that $x = \sum_{i=1}^{\infty} a_i$ and $\sum_{i=1}^{\infty} \|a_i\| < \infty$, moreover, the latter sum can be made as close to $\|x\|$ as we wish.

HINT: given $\varepsilon > 0$, we can take $a_1 \in A$ with $\|x - a_1\| < \varepsilon/2$ and pick $a_i \in A$ such that $\|x - a_1 - \cdots - a_i\| < \varepsilon 2^{-i}$.

5.6.52. (Müntz's theorem) Suppose that we are given a sequence of real numbers $p_i > -1/2$ with $\lim_{i \to \infty} p_i = +\infty$. Then $\sum_{i:\, p_i \neq 0} 1/p_i = \infty$ precisely when the linear span of the functions x^{p_i} is everywhere dense in $L^2[0,1]$. If $p_i > 0$, then this condition is necessary and sufficient in order that the sequence $\{1, x^{p_i}\}$ have a dense linear span in $C[0,1]$.

HINT: see Achieser [**4**, Chapter 1, Section 27].

5.6.53. Prove that any two orthonormal bases in a Hilbert space have the same cardinality and their cardinality equals the least possible cardinality of an everywhere dense set in this space.

HINT: observe that the set of finite rational linear combinations of vectors from a basis has the same cardinality as the basis and is dense; on the other hand, the cardinality of a dense set cannot be smaller than that of an orthonormal basis, since the balls of radius $1/2$ centered at the vectors of this basis are pairwise disjoint.

5.6.54. Show that in every infinite-dimensional separable Banach space there is an everywhere dense countable set of linearly independent vectors.

HINT: observe that if suffices to construct such a set in l^2 and then take a continuous linear injection of l^2 with a dense range; in case of l^2, we partition the standard basis into countably many countable parts $\{e_j^n\}_{j \geqslant 1}$, $n \in \mathbb{N}$, in each $\mathbb{R}^n$ take a dense set $\{a_j^n\}$ and take vectors $a_{n,j} + e_j^n/j$.

5.6.55. (i) Prove that any Hamel basis of an infinite-dimensional Banach space has the same cardinality as the space itself.

(ii) Prove that an infinite cardinal κ equals the cardinality of a Hamel basis of a Banach (or Hilbert) space if and only if it equals its countable power: $\kappa = \kappa^{\infty}$. Hence not every uncountable set has the cardinality of a Hamel basis of a Banach space.

HINT: use Theorem 1.9.26 or define explicitly an injective mapping from the countable power of the unit sphere (which has the same cardinality as the whole space) to the space itself by mapping each sequence of unit vectors u_n to $\sum_{n=1}^{\infty} 3^{-n} u_n$; in case of a Hilbert space with an orthonormal basis $\{e_\alpha\}$ of cardinality κ observe that the cardinality of the space is κ^ω, $\omega = \mathbb{N}$, since to every sequence of elements e_{α_n} of the basis (not necessarily different) one can associate the element $\sum_n 2^{-n} e_{\alpha_n}$. For the existence of uncountable cardinals that differ from their countable powers, see [**257**, §7] or [**546**, §2.9].

5.6.56. Show that an infinite-dimensional Banach space cannot be covered by a sequence of balls with radii tending to zero.

HINT: observe that the ball of radius 4 contains a ball of radius 1 that does not intersect the balls from the given sequence with radii between 1 and $1/4$.

5.6.57. In the space l^1 consider the sets

$$E_1 := \{(x_n) \colon x_{2n} = 0 \,\forall n \in \mathbb{N}\}, \quad E_2 := \{(x_n) \colon x_{2n-1} = n x_{2n} \,\forall n \in \mathbb{N}\}.$$

Prove that E_1 and E_2 are closed linear subspaces whose direct algebraic sum is not closed.

HINT: to see that $E_1 + E_2$ is not closed, observe that this sum is dense in l^1, but does not coincide with l^1, for example, does not contain the element (z_n) with $z_{2n} = n^{-2}$.

5.6.58. Prove that l^2 contains a nonclosed linear subspace that is not a first category set. This gives an example of an incomplete Euclidean space that is a Baire space (i.e., is not representable as a countable union of nowhere dense sets).

HINT: let $\{a_n\} \subset l^2$ be an everywhere dense countable set of linearly independent vectors and let Γ be some Hamel basis containing $\{a_n\}$; take a countable part $\{b_n\}$ in $\Gamma\backslash\{a_n\}$ and consider the linear span L_k of the vectors from $\Gamma\backslash\{b_{k+1}, b_{k+2}, \ldots\}$; observe that for some k the intersection of L_k with the closed unit ball is not a first category set; verify that the corresponding L_k is incomplete.

5.6.59. Let μ and ν be nonnegative measures on spaces X and Y, $F \in \mathcal{L}^2(\mu \otimes \nu)$, and let $\{\varphi_n\}$ and $\{\psi_k\}$ be orthonormal systems in $L^2(\mu)$ and $L^2(\nu)$, respectively, such that F belongs to the closed linear span of the functions $\varphi_n(x)\psi_k(y)$ in $L^2(\mu \otimes \nu)$. Prove that for ν-a.e. y the function $x \mapsto F(x, y)$ belongs to the closed linear span of the functions φ_n in $L^2(\mu)$.

HINT: observe that if a sequence of functions $f_n(x, y)$ converges to zero in $L^2(\mu \otimes \nu)$, then there is a subsequence $\{n_k\}$ for which the functions $x \mapsto f_{n_k}(x, y)$ converge to zero in $L^2(\mu)$ for ν-a.e. y.

5.6.60. Let $1 \leqslant p < \infty$ and let $\mathcal{K}$ be a bounded set in $L^p(\mathbb{R}^n)$. (i) (A. N. Kolmogorov; for $p = 1$ A. N. Tulaikov) Prove that $\mathcal{K}$ has compact closure in $L^p(\mathbb{R}^n)$ precisely when

(a) $\displaystyle \lim_{C \to \infty} \sup_{f \in \mathcal{K}} \int_{|x|>C} |f(x)|^p \, dx = 0,$

(b) for every $\varepsilon > 0$ there exists $r > 0$ such that $\sup_{f \in \mathcal{K}} \|f - S_r f\|_p \leqslant \varepsilon$, where $S_r f$ is Steklov's function defined by the equality

$$S_r f(x) := \lambda_n\big(B(x, r)\big)^{-1} \int_{B(x,r)} f(y) \, dy,$$

$B(x, r)$ is the ball of radius r in $\mathbb{R}^n$ centered at x.

(ii) (M. Riesz) Show that the compactness of the closure of $\mathcal{K}$ is also equivalent to the property that (a) is fulfilled along with

(b$'$) $\displaystyle \lim_{h \to 0} \sup_{f \in \mathcal{K}} \int_{\mathbb{R}^n} |f(x + h) - f(x)|^p \, dx = 0.$

(iii) (V. N. Sudakov) Show that conditions (a) and (b) (or (a) and (b$'$)) automatically yield the boundedness of $\mathcal{K}$ in $L^p(\mathbb{R}^n)$, hence it need not be required in advance.

HINT: (i) if $\mathcal{K}$ has compact closure, then $\mathcal{K}$ is bounded and for every $\varepsilon > 0$ has a finite ε-net; hence the necessity of (a) and (b) follows from the fact that both conditions are fulfilled for every single function f. For the proof of sufficiency observe that $S_r(\mathcal{K})$ has compact closure. Indeed, S_r is the operator of convolution with the bounded function $g = I_{B(0,r)}/\lambda_n\big(B(0, r)\big)$. For every $\delta > 0$ there is a function $g_\delta \in C_0^\infty(\mathbb{R}^n)$ with $\|g_\delta - g\|_1 \leqslant \delta$, which by the Young inequality reduces everything to the operator of convolution with g_δ. Then the functions $g_\delta * f$, where $f \in \mathcal{K}$, are equicontinuous on balls, whence one can easily see that every sequence in this set has a subsequence converging in L^p. Next, (b$'$) implies (b), i.e., (ii) follows from (i). Finally, (iii) is verified in [**709**].

5.6.61. Let Φ be a bounded and continuous function on $[0, 1] \times [0, 1] \times \mathbb{R}^1$. Prove that for every continuous function ψ on $[0, 1]$, there exists a continuous function x on $[0, 1]$ satisfying the equation

$$x(t) = \psi(t) + \int_0^1 \Phi\big(t, s, x(s)\big) \, ds, \quad t \in [0, 1].$$

HINT: use Schauder's fixed point theorem.

5.6.62. Construct examples of continuous mappings in l^2 and c_0 taking the closed unit ball into itself, but having no fixed points.

HINT: consider the mapping $x \mapsto ((1 - \|x\|^2)^{1/2}, x_1, x_2, \ldots)$ on the unit balls in the spaces l^2 and c_0.

5.6.63. The Kuratowski measure of noncompactness $\alpha(A)$ of a set A in a metric space is the infimum of numbers $\varepsilon > 0$ for which A can be partitioned into finitely many parts of diameter less than ε. The Hausdorff measure of noncompactness $\chi(A)$ is the infimum of numbers $\varepsilon > 0$ for which A has a finite ε-net. Note that $\alpha(A) = \chi(A) = 0$ precisely when A is totally bounded. (i) Show that $\chi(A) \leqslant \alpha(A) \leqslant 2\chi(A)$, moreover, for the unit ball U in an infinite-dimensional normed space we have $\chi(U) = 1$, $\alpha(U) = 2$. (ii) Let $A_n \neq \varnothing$ be closed sets in a Banach space, $A_{n+1} \subset A_n$ and $\chi(A_n) \to 0$. Show that $\bigcap_{n=1}^{\infty} A_n \neq \varnothing$.

HINT: take any points $a_n \in A_n$; assuming that they are different (otherwise the assertion is obvious) pick a Cauchy subsequence arguing as follows: A_1 can be covered by finitely many closed balls of radius $2\chi(A_1)$, at least one of these balls contains infinitely many points in $\{a_n\}$; dealing further with these points continue by induction.

5.6.64. Let $V \neq \varnothing$ be a closed convex set in a Banach space and let $f\colon V \to V$ be a continuous mapping taking V to a bounded set.

(i) (Darbo's theorem) Suppose that for some $\kappa < 1$ for all bounded sets $A \subset V$ we have $\alpha(f(A)) \leqslant \kappa\alpha(A)$. Prove that f has a fixed point.

(ii) (Sadovskii's theorem) Suppose that $\chi(f(A)) < \chi(A)$ for every A with $\chi(A) > 0$. Prove that f has a fixed point.

HINT: (i) let V_n be the closure of the convex hull of $f(X_{n-1})$, $V_0 = V$. Consider $W := \bigcap_{n=1}^{\infty} V_n$ and apply Schauder's fixed point theorem.

5.6.65. (The Mazur–Ulam theorem) (i) Let X be a real normed space, $x_1, x_2 \in X$ and $x_0 = (x_1 + x_2)/2$. Set

$$E_1 := \{x\colon \|x - x_1\| = \|x - x_2\| = \|x_1 - x_2\|/2\},$$

$$E_{n+1} := E_n \cap \{x\colon \operatorname{dist}(x, E_n) \leqslant d_n/2\},$$

where d_n is the diameter of E_n. Prove that $x_0 = \bigcap_{n=1}^{\infty} E_n$.

(ii) Let T be an isometry from the space X onto a real normed space Y and $T(0) = 0$. Prove that the mapping T is linear.

HINT: see [**46**, p. 160] or [**138**, Chapter VII, §1].

5.6.66. A normed space X is called a Efimov–Stechkin space if every nonempty closed convex set in this space is approximately compact (see Exercise 1.9.84). Prove that the indicated condition is equivalent to the following one: if $\{x_n\} \subset X$, $\|x_n\| = 1$ and there are $f \in X^*$ with $f(x_n) \to 1$, then $\{x_n\}$ contains a convergent subsequence. See also Exercise 6.10.162.

5.6.67. Prove Proposition 5.6.12.

HINT: see [**534**, Proposition 1.4].

5.6.68. (The Kakutani theorem) Let K be a nonempty convex compact set in a normed space. Prove that all isometries $h\colon K \to K$ have a common fixed point.

HINT: see [**304**, p. 471].

5.6.69. Let X be a Banach space and let L be its linear subspace that is a G_δ-set, i.e., a countable intersection of open sets. Prove that L is closed.

HINT: see [**185**, p. 34].

CHAPTER 6

Linear Operators and Functionals

In this chapter we discuss one of the central concepts of functional analysis — linear operators. We first establish the three most important results about general linear operators: the Banach–Steinhaus theorem, Banach's inverse mapping theorem, and the closed graph theorem. Next we proceed to considering linear functionals, i.e., operators with scalar values. The main results about linear functionals are connected with the Hahn–Banach theorem and its corollaries. A discussion of compact operators completes the main part of this chapter.

6.1. The Operator Norm and Continuity

Let $(X, \| \cdot \|_X)$ and $(Y, \| \cdot \|_Y)$ be normed spaces and let $A \colon X \to Y$ be a linear mapping, i.e., $A(\alpha x + \beta y) = \alpha A x + \beta A y$. Linear mappings are also called *linear operators*. If $X = Y$, then the operator $I \colon x \mapsto x$ is called the *identity* or the *unit operator*. Linear mappings with values in $Y = \mathbb{R}$ or $Y = \mathbb{C}$ are called *linear functionals*. Set

$$\|A\| := \sup_{\|x\|_X \leqslant 1} \|Ax\|_Y$$

if this quantity is finite. We shall call $\|A\|$ the *operator norm* of A. For example, the norm of a linear functional l is defined by the equality

$$\|l\| := \sup_{\|x\|_X \leqslant 1} |l(x)|.$$

A linear operator with finite norm is called *bounded*. It should be noted that this terminology is not consistent with the terminology for the case of general mappings, when a mapping is called bounded provided that its image is bounded. For a bounded linear operator A only the image of the ball is bounded, but the image $\operatorname{Ran} A := A(X)$ of the whole space , called the *range of the operator A*, can be bounded only when $A(X) = 0$.

We observe that $\|A\|$ is the smallest number M such that $\|Ax\|_Y \leqslant M\|x\|_X$ for all $x \in X$. It is also clear that in the definition of $\|A\|$ we can take sup over the unit sphere in place of the unit ball (if $X \neq 0$).

If X, Y, Z are normed spaces, $A \colon Y \to Z$ and $B \colon X \to Y$ are bounded linear operators, then the linear operator $AB \colon X \to Z$ is obviously bounded and

$$\|AB\| \leqslant \|A\| \, \|B\|.$$

© Springer Nature Switzerland AG 2020
V. I. Bogachev, O. G. Smolyanov, *Real and Functional Analysis*,
Moscow Lectures 4, https://doi.org/10.1007/978-3-030-38219-3_6

This inequality can be strict; it is easy to construct an example in $\mathbb{R}^2$ or $\mathbb{C}^2$ (say, the composition of nonzero operators can be zero).

The set $\mathcal{L}(X, Y)$ of all bounded linear operators acting from a normed space X to a normed space Y is a normed space with the operator norm $A \mapsto \|A\|$. This space is obviously linear, since the algebraic sum of two bounded sets is bounded and the product of a bounded set by a scalar is bounded. It is proved below that a linear operator is bounded if and only if it is continuous (which is obviously false in both directions for nonlinear mappings).

The fact that the space of operators is normed follows at once from the relations $\|(A + B)x\|_Y \leqslant \|Ax\|_Y + \|Bx\|_Y$ and $\|\lambda Ax\|_Y = |\lambda| \, \|Ax\|_Y$. The main results of this chapter are connected with the operator norm. A particular role is played by the case where Y is the scalar field.

6.1.1. Definition. *Let X be a normed space. The space $X^* := \mathcal{L}(X, \mathbb{R})$ (or $X^* := \mathcal{L}(X, \mathbb{C})$ in the complex case) of all continuous linear functionals on the space X is called the dual (or topological dual) to the space X. The space X' of all linear functions on X is called the algebraic dual.*

The value of a linear functional f on a vector x is frequently denoted by $\langle f, x \rangle$. On concrete spaces it is easy to construct explicit examples of nonzero continuous functionals. It turns out that such functionals exist on every nonzero normed space. This highly non-obvious, but very important for the whole theory fact will be established below in §6.4 with the aid of the Hahn–Banach theorem. Algebraic duals are used relatively seldom.

6.1.2. Theorem. *For a linear operator $A\colon X \to Y$, the following conditions are equivalent:*

(i) *the operator A is bounded;*
(ii) *the operator A is continuous;*
(iii) *the operator A is continuous at some point.*

PROOF. If the operator A is bounded, then $\|Ax - Ay\| \leqslant \|A\| \|x - y\|$, i.e., the mapping A satisfies the Lipschitz condition with constant $\|A\|$ and hence is continuous. Suppose that the operator A is continuous at some point x_0. The equality $Ax = A(x - x_0) + Ax_0$ yields the continuity of A at the origin. Hence there exists $r > 0$ such that $\|Ax\| \leqslant 1$ whenever $\|x\| \leqslant r$. This gives the estimate $\|Ax\| \leqslant r^{-1}$ whenever $\|x\| \leqslant 1$. Thus, $\|A\| \leqslant r^{-1}$. $\square$

6.1.3. Corollary. *A linear mapping between normed spaces is continuous precisely when it takes sequences converging to zero to bounded sequences.*

PROOF. The necessity of this condition is obvious, its sufficiency follows from the fact that if $\|x_n\| \leqslant 1$ and $\|Ax_n\| \to \infty$, then $y_n := \|Ax_n\|^{-1/2} x_n \to 0$ and $\|Ay_n\| = \|Ax_n\|^{1/2} \to \infty$. $\square$

According to the established properties, a linear mapping discontinuous at one point is discontinuous everywhere. On a finite-dimensional normed space all linear operators are continuous, hence are bounded. In the infinite-dimensional case the situation is different.

6.1.4. Example. On every infinite-dimensional normed space there exists a discontinuous linear functional. Indeed, let $\{v_\alpha\}$ be a Hamel basis consisting of vectors of unit length. Let us pick in this basis a countable part $\{v_n\}$, set $l(v_n) = n$ for every n, on the remaining elements of the basis make l zero and extend by linearity to the whole space. It is clear that we have obtained an unbounded linear functional. It is discontinuous at every point by the previous theorem.

On some incomplete normed spaces one can construct explicitly (without using Hamel bases) discontinuous linear functionals. For example, on the space of continuous functions on $[0, 1]$ one can take the norm from $L^2[0, 1]$ and define l by the formula $l(x) = x(0)$. However, there are no explicit examples of unbounded linear functionals on Banach spaces.

Let us consider some examples of evaluation of norms of functionals and operators.

6.1.5. Example. (i) Let $X = C[0, 1]$ be equipped with the usual sup-norm and let $l(f) = f(0)$. Then $\|l\| = 1$, since $|l(f)| \leqslant 1$ if $\|f\| \leqslant 1$ and $l(1) = 1$.

(ii) Let $X = C[0, 1]$ be equipped with the usual sup-norm and let us set $l(f) = f(0) - f(1)$. Then $\|l\| = 2$, since $|l(f)| \leqslant 2$ if $\|f\| \leqslant 1$ and $l(f) = 2$ for the function $f \colon t \mapsto 1 - 2t$.

(iii) Let $X = C[0, 1]$ be equipped with the usual norm and let

$$l(f) = \int_0^1 f(t)g(t)\, dt, \tag{6.1.1}$$

where $g(t) = -1$ if $t \leqslant 1/2$, $g(t) = 1$ if $t > 1/2$. Then $\|l\| = 1$, since $|l(f)| \leqslant 1$ if $\|f\| \leqslant 1$, and for every $\varepsilon > 0$ there exists a continuous function f with $|f(t)| \leqslant 1$ and $l(f) > 1 - \varepsilon$. We observe that in this example there is no element f with $\|f\| \leqslant 1$ such that $l(f) = \|l\|$, i.e., a continuous linear functional can fail to attain its maximum on a closed ball.

More generally, for every integrable function g on $[0, 1]$ for the functional defined in (6.1.1) we have

$$\|l\| = \|g\|_{L^1} = \int_0^1 |g(t)|\, dt.$$

The bound $\|l\| \leqslant \|g\|_{L^1}$ is obvious, and for the proof of the equality it suffices to observe that for every $\varepsilon > 0$ there exists a step function g_ε on $[0, 1]$ such that $\|g - g_\varepsilon\|_{L^1} \leqslant \varepsilon$. Similarly to the previous case we observe that there exists a function $f_\varepsilon \in C[0, 1]$ with $\|f_\varepsilon\| \leqslant 1$ and

$$\int_0^1 f_\varepsilon(t)g_\varepsilon(t)\, dt \geqslant \|g_\varepsilon\|_{L^1} - \varepsilon.$$

This gives

$$l(f_\varepsilon) \geqslant \int_0^1 f_\varepsilon(t)g_\varepsilon(t)\, dt - \varepsilon \geqslant \|g\|_{L^1} - 3\varepsilon,$$

because we have $\|g_\varepsilon\|_{L^1} \geqslant \|g\|_{L^1} - \varepsilon$. Since ε was arbitrary, we obtain the opposite estimate $\|l\| \geqslant \|g\|_{L^1}$.

(iv) Let X be a Euclidean space and let $l(x) = (x,a)$, where $a \in X$. Then we have $\|l\| = \|a\|$, since $|l(x)| \leqslant \|x\|\|a\|$, which gives $\|l\| \leqslant \|a\|$. On the other hand, if $a \neq 0$, then $l(a/\|a\|) = \|a\|$.

(v) Let $X = L^2[0,1]$ be equipped with the usual norm and let

$$l(f) = \int_0^1 f(t)g(t)\,dt,$$

where $g \in L^2[0,1]$. Then $\|l\| = \|g\|_{L^2}$.

(vi) Let $X = C[0,1]$ and $Y = L^2[0,1]$ be equipped with their usual norms and let

$$Ax(t) = \left(\int_0^1 x(s)y(s)\,ds\right)\psi(t),$$

where $y \in L^1[0,1]$, $\psi \in L^2[0,1]$. Then $\|A\| = \|y\|_{L^1}\|\psi\|_{L^2}$. This follows from (iii) and the equality

$$\|Ax\| = \left|\int_0^1 x(s)y(s)\,ds\right|\|\psi\|.$$

(vii) A *diagonal operator* on a separable Hilbert space H is an operator of the form

$$Ax = \sum_{n=1}^{\infty} \alpha_n(x, e_n)e_n,$$

where $\{e_n\}$ is an orthonormal basis in H and $\{\alpha_n\}$ is a bounded sequence in $\mathbb{C}$. Then $\|A\| = \sup_n |\alpha_n|$, since $\|Ax\| \leqslant \sup_n |\alpha_n|$ and $\|Ae_n\| = |\alpha_n|$, whence one has $\|A\| \geqslant \sup_n |\alpha_n|$.

One should bear in mind that according to (iii) even in the case of a linear functional on an infinite-dimensional space the norm is not always attained on the unit ball, so sup cannot be always replaced by max.

We now prove an important result according to which any pointwise bounded family of continuous operators is uniformly bounded on the unit ball, i.e., is norm bounded.

6.1.6. Theorem. (THE BANACH–STEINHAUS THEOREM OR THE UNIFORM BOUNDEDNESS PRINCIPLE) *Suppose we are given a family $\{A_\alpha\}$ of bounded linear operators on a Banach space X with values in a normed space Y. Suppose that*

$$\sup_\alpha \|A_\alpha x\| < \infty \quad \text{for every } x \in X.$$

Then $\sup_\alpha \|A_\alpha\| < \infty$.

PROOF. As in Example 1.5.3 (we could refer to it at once), we consider the sets

$$M_n = \{x \in X \colon \|A_\alpha x\| \leqslant n \quad \text{for all } \alpha\}.$$

By the continuity of the operators A_α these sets are closed. By our hypothesis they cover X. According to Baire's theorem there exists n such that the set M_n contains some ball $U(a,r)$ centered at a with radius $r > 0$. Since $A_\alpha x = A_\alpha(x+a) - A_\alpha a$

and $\sup_\alpha \|A_\alpha a\| < \infty$, we obtain the uniform boundedness of the operators A_α on the ball $U(0,r)$, which gives their uniform boundedness on the unit ball. $\quad\square$

The completeness of X is essential in this theorem (although the requirement of completeness can be relaxed to the Baire property of X, which is obvious from the proof). For example, on the linear subspace in $C[0,1]$ consisting of functions vanishing in a neighborhood of zero (its own neighborhood for every function), the bounded functionals $l_n(x) = nx(1/n)$ are pointwise bounded (at every fixed element x they are zero starting from some number), but their norms are not uniformly bounded: $\|l_n\| = n$. The same is true for the functionals $l_n(x) = nx_n$ on the linear subspace in l^2 consisting of all vectors with finitely many nonzero coordinates.

6.1.7. Corollary. *Let X and Y be Banach spaces and let $A_n\colon X \to Y$ be continuous linear operators such that for every vector x there exists a limit $Ax = \lim_{n\to\infty} A_n x$ in Y. Then A is a continuous operator.*

PROOF. It is clear that A is a linear mapping. By the Banach–Steinhaus theorem we have $\sup_n \|A_n\| \leqslant C < \infty$. Then $\|Ax\| = \lim_{n\to\infty} \|A_n x\| \leqslant C\|x\|$, hence $\|A\| \leqslant C$. $\quad\square$

6.1.8. Corollary. *In the situation of the previous corollary for every compact set $K \subset X$ we have*

$$\lim_{n\to\infty} \sup_{x\in K} \|Ax - A_n x\| = 0.$$

PROOF. We already know that there exists $C > 0$ such that $\|A_n\| \leqslant C$ for all n and that $\|A\| \leqslant C$. Let $\varepsilon > 0$. Let us take in K a finite $\varepsilon(4C)^{-1}$-net $x_1, \dots, x_k$ and take N such that $\|Ax_i - A_n x_i\| \leqslant \varepsilon/2$ for $i = 1, \dots, k$ and for all $n \geqslant N$. Then, for such n, for every $x \in K$ we have $\|Ax - A_n x\| \leqslant \varepsilon$, since there exists x_i with $\|x - x_i\| \leqslant \varepsilon(4C)^{-1}$, which by the triangle inequality gives

$$\|Ax - A_n x\| \leqslant \|Ax - Ax_i\| + \|Ax_i - A_n x_i\| + \|A_n x_i - A_n x\|$$
$$\leqslant 2C\|x - x_i\| + \varepsilon/2 \leqslant \varepsilon,$$

as required. $\quad\square$

6.1.9. Theorem. *Let Y be a Banach space. Then for every normed space X the space of operators $\mathcal{L}(X,Y)$ is complete with respect to the operator norm.*

In particular, the space X^* is complete for every normed space X (not necessarily complete).

PROOF. Let $\{A_n\} \subset \mathcal{L}(X,Y)$ be a Cauchy sequence. For every $x \in X$ the sequence $\{A_n x\}$ is Cauchy in Y, since $\|A_n x - A_k x\| \leqslant \|A_n - A_k\|\|x\|$. Hence there exists $Ax = \lim_{n\to\infty} A_n x$. It is clear that $A \in \mathcal{L}(X,Y)$, $\|A\| \leqslant \sup_n \|A_n\|$. However, we have to show that $\|A - A_n\| \to 0$. Let $\varepsilon > 0$. Let us find N such that $\|A_n - A_k\| \leqslant \varepsilon$ for all $n, k \geqslant N$. For all $n \geqslant N$ and each vector x of unit norm we have $\|Ax - A_n x\| = \lim_{k\to\infty} \|A_k x - A_n x\| \leqslant \varepsilon$, so $\|A - A_n\| \leqslant \varepsilon$. $\quad\square$

Note that in this theorem and in the Banach–Steinhaus theorem we require the completeness of different spaces. For a beginner it is much easier to remember both theorems requiring the completeness of both spaces. However, then a moment reflexion shows that in the Banach–Steinhaus theorem the completeness of Y is not needed because one can pass to the completion of Y (which preserves the pointwise boundedness), while in the previous theorem we do not need the completeness of X, since all operators A_n can be extended to the completion of X giving a Cauchy sequence of operators on the completion.

The Banach–Steinhaus theorem can be also applied for obtaining negative results (which is sometimes called the "principle of condensation of singularities").

6.1.10. Example. Suppose that a sequence of continuous linear operators A_n from a Banach space X to a normed space Y is not norm bounded. Then there exists an element $x \in X$ such that $\sup_n \|A_n x\| = \infty$.

6.1.11. Example. For every $a \in [0, 2\pi]$ there exists a continuous 2π-periodic function for which the partial sums of the Fourier series at the point a are not uniformly bounded, in particular, have no finite limit.

PROOF. It suffices to consider $a = 0$. If for every function f in the space $C_{2\pi}$ of continuous functions f on $[0, 2\pi]$ with $f(0) = f(2\pi)$ the partial sums of the Fourier series at zero are bounded, then by virtue of representation (4.5.3) for the partial sums we have the pointwise boundedness of the sequence of functionals

$$l_n(f) := \int_0^{2\pi} f(t) \frac{\sin \frac{2n+1}{2}t}{2 \sin \frac{t}{2}} \, dt.$$

According to Example 6.1.5 the norm of this functional on $C_{2\pi}$ equals

$$\int_0^{2\pi} \left| \frac{\sin \frac{2n+1}{2}t}{2 \sin \frac{t}{2}} \right| dt.$$

Hence

$$\|l_n\| \geqslant \int_0^{2\pi} |\sin(n+1/2)t| \frac{1}{t} \, dt = \int_0^{(2n+1)\pi} |\sin s| \frac{1}{s} \, ds,$$

which tends to infinity as $n \to \infty$.   □

In applications it is often useful to approximate in a suitable sense or replace infinite-dimensional operators by finite-dimensional ones. In particular, this is necessary for numeric methods. However, here there are many subtleties connected with the character of approximation. For example, the identity operator on an infinite-dimensional space cannot be approximated by finite-dimensional ones with respect to the operator norm. In §6.9 below we discuss compact operators, which on nice spaces (such as a Hilbert space) are approximated by finite-dimensional ones in the operator norm. But it is often sufficient to have some weaker approximations, for example, pointwise. This becomes possible for every bounded operator on a space with a Schauder basis; this question is discussed in §6.10(iv). In a Hilbert space with an orthonormal basis $\{e_n\}$, for every bounded operator A we have $P_n A x \to A x$ for all x, where P_n is the orthogonal projection onto the linear span of $e_1, \ldots, e_n$, i.e., $P_n x = (x, e_1)e_1 + \cdots + (x, e_n)e_n$. Let us give a less trivial example.

6.1.12. Example. Let $C_{2\pi}$ be the space of all continuous 2π-periodic functions on the real line with the sup-norm and let σ_n be the operator taking a function f to its Fejér sum

$$\sigma_n(f)(x) := \int_0^{2\pi} f(x+z)\Phi_n(z)\,dz,$$

where Φ_n is the nth Fejér kernel (see (4.5.5)). Then by Theorem 4.5.9 we have $\|f - \sigma_n(f)\| \to 0$ for all $f \in C_{2\pi}$. Therefore, for every bounded operator A with values in $C_{2\pi}$, we obtain the pointwise convergence $\|Ax - \sigma_n \circ Ax\| \to 0$. A similar fact is true for the space $C[0,1]$. For this it suffices to observe that the space $C[0,\pi]$ can be embedded into $C_{2\pi}$ by the mapping $f \mapsto \widetilde{f}$, where $\widetilde{f}(t) = f(-t)$ for all $t \in [-\pi, 0]$, next $\widetilde{f}$ extends periodically.

We recall that according to Corollary 6.1.8 the pointwise convergence of operators on a Banach space yields the uniform convergence on compact sets, which increases the effect of such approximations.

6.2. The Closed Graph Theorem

The next result due to Banach and Schauder is fundamental for many other important results connected with operator ranges.

6.2.1. Lemma. *Let X and Y be Banach spaces with open unit balls U_X and U_Y and let $A: X \to Y$ be a continuous linear operator such that U_Y is contained in the closure of $A(U_X)$. Then $U_Y \subset A(U_X)$. In particular, $A(X) = Y$.*

PROOF. It follows from our assumption that

$$A(sU_X) \cap sU_Y \text{ is dense in } sU_Y \text{ for every } s > 0. \qquad (6.2.1)$$

Let $y \in U_Y$ and $0 < \varepsilon < 1 - \|y\|$. Then $\|(1-\varepsilon)^{-1}y\| < 1$. Hence there is a vector $x_1 \in U_X$ for which $\|(1-\varepsilon)^{-1}y - Ax_1\| < \varepsilon$, i.e., $(1-\varepsilon)^{-1}y - Ax_1 \in \varepsilon U_Y$. By condition (6.2.1) there is a vector $x_2 \in \varepsilon U_X$ with $\|(1-\varepsilon)^{-1}y - Ax_1 - Ax_2\| < \varepsilon^2$. By induction with the aid of (6.2.1) we find $x_n \in \varepsilon^{n-1}U_X$ with

$$\|(1-\varepsilon)^{-1}y - Ax_1 - \ldots - Ax_n\| < \varepsilon^n.$$

Then $y = (1-\varepsilon)\sum_{n=1}^{\infty} Ax_n$. By the estimate $\|x_n\| < \varepsilon^{n-1}$ and the completeness of X the series $(1-\varepsilon)\sum_{n=1}^{\infty} x_n$ converges to some element $x \in X$. We have $Ax = y$ and $\|x\| < (1-\varepsilon)\sum_{n=1}^{\infty} \varepsilon^{n-1} = 1$, that is, $x \in U_X$. Thus, we have proved the inclusion $U_Y \subset A(U_X)$. $\qquad \square$

6.2.2. Remark. We have used only the completeness of X. On the way we have obtained the following fact: if a set S is dense in U_Y, then every vector $y \in U_Y$ has the form $y = \sum_{n=1}^{\infty} c_n s_n$, where $s_n \in S$, $\sum_{n=1}^{\infty} |c_n| < 1$.

The next important theorem was obtained by Banach and Schauder.

6.2.3. Theorem. (THE OPEN MAPPING THEOREM) *Let X and Y be Banach spaces, $A \in \mathcal{L}(X, Y)$, and $A(X) = Y$. Then for every set V open in X the set $A(V)$ is open in Y.*

PROOF. Let U_X and U_Y be open unit balls in X and Y, respectively. Since $Y = \bigcup_{n=1}^{\infty} A(nU_X)$, by Baire's theorem there exists k such that the set $A(kU_X)$ is dense in some open ball $a + rU_Y$ of radius $r > 0$ in Y. Since we have $A(kU_X) = -A(kU_X)$, the set $A(kU_X)$ is dense in the ball $-a + rU_Y$. Hence $A(kU_X)$ is dense in the ball rU_Y. Indeed, if $\|y\|_Y \leqslant r$ and $u_n, v_n \in U_X$ are such that $A(ku_n) \to a + y$ and $A(kv_n) \to -a + y$, then $w_n := (u_n + v_n)/2 \in U_X$ and $A(kw_n) \to y$. Replacing A by $r^{-1}kA$, we can assume that $A(U_X)$ is dense in the ball U_Y. By the lemma proved above $U_Y \subset A(U_X)$. Therefore, we have $Ax + rU_Y \subset A(x + rU_X)$, $x \in X$, $r > 0$.

Suppose now that V is a nonempty open set in X. Let $y \in A(V)$, i.e., $y = Ax$, $x \in V$. Find $\varepsilon > 0$ such that $x + \varepsilon U_X \subset V$. Then

$$y + \varepsilon U_Y \subset A(x + \varepsilon U_X) \subset A(V).$$

Thus, $A(V)$ is open. For a generalization, see Exercise 12.5.31. □

6.2.4. Remark. (i) It is seen from the proof that in place of the surjectivity of A it suffices that $A(X)$ be a second category set in Y (then $A(U_X)$ will be dense in some ball in Y). In place of completeness of Y it suffices to have the Baire property for Y, but for arbitrary normed spaces Y the theorem is not valid: take the diagonal operator A on $X = l^2$ with eigenvalues n^{-1} and $Y = A(X)$ with the norm from l^2.

(ii) Since the image of the unit ball from X contains some ball $U_Y(0, \varepsilon)$ centered at the origin, we obtain that for every $y \in Y$ there exists a vector $x \in X$ such that

$$Ax = y \quad \text{and} \quad \|x\| \leqslant \varepsilon^{-1} \|y\|.$$

Of course, such a vector is not always unique.

An important corollary is the following result of Banach.

6.2.5. Theorem. (THE INVERSE MAPPING THEOREM) *Let A be a one-to-one continuous linear mapping of a Banach space X onto a Banach space Y. Then the inverse mapping A^{-1} is continuous.*

PROOF. The preimage under A^{-1} of an open set V in X coincides with $A(V)$ (because A is one-to-one) and is open in Y by the previous theorem. Hence A^{-1} is continuous. □

For nonlinear mappings A this theorem is false (Exercise 6.10.109).

6.2.6. Corollary. *Let X be a linear space that is complete with respect to two norms p_1 and p_2. Suppose that there exists a number c such that $p_1(x) \leqslant cp_2(x)$ for all $x \in X$. Then exists a number M such that $p_2(x) \leqslant Mp_1(x)$ for all $x \in X$.*

PROOF. By assumption the identity mapping of X with norm p_2 to X with norm p_1 is continuous. Hence the inverse mapping is continuous as well, i.e., it has finite norm, which means the existence of a desired number M. □

For formulating yet another important corollary of the open mapping theorem we introduce a new object.

The *graph of a mapping* $A\colon X \to Y$ is the set

$$\Gamma(A) := \{(x, Ax)\colon x \in X\} \subset X \times Y.$$

If X and Y are Banach spaces, then the product $X \times Y$ is equipped with the natural structure of a linear space and the natural norm $\|(x, y)\| := \|x\| + \|y\|$. It is clear that $X \times Y$ is complete with respect to this norm.

6.2.7. Theorem. (THE CLOSED GRAPH THEOREM) *A linear mapping between Banach spaces is continuous precisely when its graph is closed.*

PROOF. It is obvious that the graph of every continuous mapping is closed. The converse is false for nonlinear mappings. For a linear mapping $A\colon X \to Y$ with a closed graph we observe that this graph is a linear subspace in $X \times Y$ and hence is a Banach space. The operator $T\colon \Gamma(A) \to X$, $(x, Ax) \mapsto x$ is linear, continuous and maps $\Gamma(A)$ one-to-one onto X. By the inverse mapping theorem the operator $x \mapsto (x, Ax)$ is continuous. This yields the continuity of A. $\qquad\square$

6.2.8. Corollary. *Let X, Y, Z be Banach spaces and let $j\colon Y \to Z$ be an injective continuous linear operator. Suppose that $A\colon X \to Y$ is a linear mapping such that the composition $j \circ A\colon X \to Z$ is continuous, i.e.,*

$$X \xrightarrow{A} Y \xrightarrow{j} Z$$

is a continuous mapping. Then A is continuous.

PROOF. We verify that the graph of A is closed. Let $x_n \to x$ in X and $Ax_n \to y$ in Y. It follows from our condition that

$$j(Ax_n) \to j(y), \quad j(Ax_n) = j \circ A(x_n) \to j \circ A(x).$$

Hence $j(y) = j \circ A(x)$, whence $y = Ax$. $\qquad\square$

Let us give some typical examples of using the results obtained above.

6.2.9. Example. Let $A\colon L^2[a, b] \to L^2[a, b]$ be a continuous linear operator such that $A(L^2[a, b]) \subset C[a, b]$. Then the operator A is continuous as a mapping from $L^2[a, b]$ to $C[a, b]$.

6.2.10. Example. Suppose that a Banach space X is represented as a direct algebraic sum of its closed subspaces X_1 and X_2. Then the operators of algebraic projections $P_1\colon X \to X_1$ and $P_2\colon X \to X_2$ are continuous.

PROOF. Let us consider X as the Banach direct sum $X_1 \oplus X_2$ with the norm $(x_1, x_2) \mapsto \|x_1\| + \|x_2\|$. Since we always have $\|x_1 + x_2\| \leqslant \|x_1\| + \|x_2\|$, Corollary 6.2.6 shows that the new norm is equivalent to the original one. The projections are obviously continuous with respect to the new norm, hence they are continuous also with respect to the original one. $\qquad\square$

One should bear in mind that the algebraic direct sum of two closed linear subspaces of a Banach space need not be closed (see Exercise 6.10.98). The previous example has the following interesting generalization.

6.2.11. Proposition. *Let X_1 and X_2 be closed subspaces of a Banach space X such that $X = X_1 + X_2$ (the sum is not supposed to be direct unlike the previous example). Then there exists a number $c > 0$ such that every element $x \in X$ admits a representation*

$$x = x_1 + x_2, \quad \text{where } \|x_1\| + \|x_2\| \leqslant c\|x\|.$$

PROOF. Denote by Y the direct sum of the Banach spaces X_1 and X_2. The mapping $T: Y \to X$ defined by the formula $T(x_1, x_2) = x_1 + x_2$ is linear, continuous and surjective. The open mapping theorem yields that the image of the unit ball in Y contains a ball in X of some radius $r > 0$. Now it suffices to take the number $c := r^{-1}$. $\square$

Let us prove one more useful result following from the already established facts. We observe that if X and Y are normed spaces, then any continuous linear operator $A: X \to Y$ with the *kernel* $\operatorname{Ker} A := A^{-1}(0)$ generates an injective linear operator

$$\widetilde{A}: X/\operatorname{Ker} A \to Y,$$

called the *factorization of A by its kernel* and defined by the formula $\widetilde{A}[x] := Ax$, where $[x]$ is the equivalence class in $X/\operatorname{Ker} A$ with a representative x. We have $\widetilde{A}(X/\operatorname{Ker} A) = A(X)$ and $\|\widetilde{A}\| = \|A\|$. The first equality is obvious, the second one us verified as follows: since $\|[x]\| \leqslant \|x\|$, we have $\|\widetilde{A}\| \geqslant \|A\|$. On the other hand, if $\|[x]\| = 1$ and $\varepsilon > 0$, then there exists a representative y of the equivalence class $[x]$ with $\|y\| \leqslant 1 + \varepsilon$, whence $\|\widetilde{A}[x]\| = \|Ay\| \leqslant (1 + \varepsilon)\|A\|$. Hence $\|\widetilde{A}\| \leqslant (1 + \varepsilon)\|A\|$ for all $\varepsilon > 0$ and so $\|\widetilde{A}\| \leqslant \|A\|$.

6.2.12. Proposition. *Let X and Y be Banach spaces and $A \in \mathcal{L}(X, Y)$. If the range of A has a finite codimension in Y, then it is closed.*

PROOF. Since $A(X) = \widetilde{A}(\widetilde{X})$, where $\widetilde{X} = X/\operatorname{Ker} A$ and $\widetilde{A}: \widetilde{X} \to Y$ is the operator generated by A, we can assume that the operator A is injective. By assumption there exists a finite-dimensional linear subspace Y_0 in Y such that Y is the algebraic direct sum of Y_0 and $A(X)$. Let $X \oplus Y_0$ denote the Banach direct sum of X and Y_0 (we recall that finite-dimensional normed spaces are complete). The operator $B: X \oplus Y_0 \to Y$, $(x, y) \mapsto Ax + y$ is continuous. This operator is injective, since we deal with the injective operator A. In addition, the operator B is surjective, since Y is the algebraic sum of $A(X)$ and Y_0. By the Banach theorem the operator B has a continuous inverse. Hence B takes closed sets to closed sets. In particular, the closed subspace X of the space $X \oplus Y_0$ (i.e., the set of pairs $(x, 0)$, $x \in X$) is taken to the closed set $A(X)$, which completes the proof. $\square$

6.3. The Hahn–Banach Theorem

In this section we prove the most important result in linear analysis: the Hahn–Banach theorem. This result has numerous applications in mathematics as well as in applications, in particular, in economics. Unlike most of other assertions in this book, the Hahn–Banach theorem is nontrivial also in the finite-dimensional case.

6.3.1. Definition. *Let X be a linear space. A function $p\colon X \to [0,+\infty)$ is called a seminorm on X if for all scalars α and all vectors x, y we have*

$$p(\alpha x) = |\alpha| p(x) \quad \text{and} \quad p(x + y) \leqslant p(x) + p(y).$$

6.3.2. Definition. *A function $p\colon X \to (-\infty, +\infty)$, where X is a real linear space, is called convex if*

$$p(tx + (1 - t)y) \leqslant tp(x) + (1 - t)p(y) \quad \forall\, x, y \in X,\ \forall\, t \in [0, 1].$$

A function $p\colon X \to (-\infty, +\infty)$ is called positively homogeneous convex if

$$p(\alpha x) = \alpha p(x) \quad \text{and} \quad p(x + y) \leqslant p(x) + p(y) \quad \forall\, \alpha \geqslant 0, \forall\, x, y \in X.$$

It is clear that all seminorms and all linear functions are positively homogeneous convex. It follows from the definition that a positively homogeneous convex function p is convex:

$$p(tx + (1 - t)y) \leqslant p(tx) + p((1 - t)y) = tp(x) + (1 - t)p(y)$$

whenever $t \in [0, 1]$. If also $p(-x) = p(x)$, then p is a seminorm.

6.3.3. Theorem. (THE HAHN–BANACH THEOREM) *Let X be a real linear space, p a convex function on X, X_0 a linear subspace in X, and l_0 a linear function on X_0 satisfying the condition*

$$l_0(x) \leqslant p(x) \quad \text{for all } x \in X_0.$$

Then l_0 can be extended to a linear function l on all of X satisfying the condition $l(x) \leqslant p(x)$ for all $x \in X$.

PROOF. Note that this theorem is not trivial even in the two-dimensional case. As we shall now see, the main problem consists in extending the functional to a larger space in which X_0 is a hyperplane. Suppose first that X is the linear span of X_0 and a vector z not belonging to X_0. Every vector x in X has the form $x = x_0 + tz$. Every linear extension is uniquely determined by our choice of the number $c = l(z)$. Then $l(x) = l_0(x_0) + tc$. We have to pick c in such a way that $l \leqslant p$. Thus, we have to ensure the inequality

$$l_0(x_0) + tc \leqslant p(x_0 + tz). \tag{6.3.1}$$

For $t > 0$ this inequality is equivalent to the inequality

$$c \leqslant t^{-1} p(x_0 + tz) - l_0(t^{-1} x_0).$$

Similarly, for $t = -s < 0$ inequality (6.3.1) is equivalent to the estimate

$$c \geqslant -s^{-1} p(x_0 - sz) + l_0(s^{-1} x_0).$$

We show that there exists a number c satisfying both inequalities for all x_0 and t. To this end we prove the inequality

$$c' := \sup_{y \in X_0,\, s > 0} \left[-s^{-1} p(sy - sz) + l_0(y) \right]$$

$$\leqslant c'' := \inf_{y \in X_0,\, t > 0} \left[t^{-1} p(ty + tz) - l_0(y) \right]. \tag{6.3.2}$$

For c we then take any number between c' and c''. Inequality (6.3.2) is equivalent to the estimate

$$-s^{-1}p(sy' - sz) + l_0(y') \leqslant t^{-1}p(ty'' + tz) - l_0(y'') \quad \forall\, y', y'' \in X_0,\ s, t > 0,$$

which can be written as

$$l_0(y'') + l_0(y') \leqslant t^{-1}p(ty'' + tz) + s^{-1}p(sy' - sz).$$

Multiplying by the number $\lambda = ts/(t + s)$, for which $\lambda t^{-1} + \lambda s^{-1} = 1$, we write the latter as

$$l_0(\lambda y'' + \lambda y') \leqslant \lambda t^{-1}p(ty'' + tz) + \lambda s^{-1}p(sy' - sz).$$

This estimate is true, since

$$l_0(\lambda y'' + \lambda y') = l_0\big((\lambda t^{-1}(ty'' + tz) + s^{-1}(sy' - sz)\big)$$
$$\leqslant p\big((\lambda t^{-1}(ty'' + tz) + s^{-1}(sy' - sz)\big) \leqslant \lambda t^{-1}p(ty'' + tz) + \lambda s^{-1}p(sy' - sz).$$

Thus, in the considered case the theorem is proved. In the general case an extension is constructed step-by-step by adding an independent vector, which is done with the aid of Zorn's lemma. Let $\mathfrak{M}$ denote the collection of all possible extensions of l_0 to larger subspaces satisfying the condition of domination by p. Every such extension l' has a linear domain of definition L', on which $l' \leqslant p$, and $l'|_{X_0} = l_0$. We declare an extension l' subordinated to an extension l'' if for the corresponding domains of definition we have $L' \subset L''$ and $l''|_{L'} = l'$. It is clear that we obtain a partial order. The chain condition is fulfilled: if we are given a chain of extensions l_α with domains L_α, then a majorant $l \in \mathfrak{M}$ for it can be constructed as follows. The union L of all L_α is a linear space, since for every $x, y \in L$ there exist L_α and L_β with $x \in L_\alpha$ and $y \in L_\beta$, but by the definition of a chain either $L_\alpha \subset L_\beta$ or $L_\beta \subset L_\alpha$, i.e., in any case $x + y \in L$. It is clear that $tx \in L$ for all scalars t. By the same reasoning the function $l(x) = l_\alpha(x_\alpha)$ for $x = x_\alpha \in L_\alpha$ is well-defined on L, i.e., $l_\alpha(x_\alpha) = l_\beta(x_\beta)$ if $x_\alpha = x_\beta$. Moreover, $l \leqslant p$ on L. Thus, $l \in \mathfrak{M}$ is a majorant for all l_α. By Zorn's lemma $\mathfrak{M}$ contains a maximal element l. According to the first step, the domain of definition of l coincides with the whole space X: otherwise l could be linearly extended to a larger subspace with subordination to p contrary to the maximality of l. $\square$

With the aid of a Hamel basis (see Proposition 1.1.1) it is easy to find a *linear* extension of l, but it is not always subordinated to p.

Usually various versions of the Hahn–Banach theorem are established for seminorms or positively homogeneous functions p (which is sufficient for most of applications), but the case of convex functions was also considered in the literature, see Altman [665], Bittner [672], it can be found in books, see, e.g., Barbu, Precupanu [47]. Mazur and Orlicz [697] initiated a study of more general extension problems for linear functionals and linear operators with several restrictions, there is an extensive literature on this direction.

6.3.4. Corollary. *Let X be a real or complex linear space, p a seminorm on the space X, X_0 a linear subspace in X, and let l_0 be a linear function on X_0*

satisfying the following condition:
$$|l_0(x)| \leqslant p(x) \quad \forall\, x \in X_0.$$
Then l_0 can be extended to a linear function l on all of X satisfying the condition $|l(x)| \leqslant p(x)$ for all $x \in X$.

PROOF. In the real case this assertion follows directly from the Hahn–Banach theorem. In the complex case, let $X_{\mathbb{R}}$ denote the realification of X, i.e., X over the field $\mathbb{R}$. Let us apply the Hahn–Banach theorem to the function $\operatorname{Re} l_0$ on the realification $X_{0,\mathbb{R}}$ of the space X_0. It is clear that $|\operatorname{Re} l_0| \leqslant p$ on $X_{0,\mathbb{R}}$. We obtain a real linear function l_1 on $X_{\mathbb{R}}$ with $l_1|_{X_{0,\mathbb{R}}} = \operatorname{Re} l_0$ and $l_1 \leqslant p$. We observe that $|l_1| \leqslant p$. We now set
$$l(x) = l_1(x) - i l_1(ix), \quad x \in X,$$
which is possible, since $ix \in X$ and X coincides with $X_{\mathbb{R}}$ as a set. For all $x \in X_0$ we have
$$l_1(x) - i l_1(ix) = \operatorname{Re} l_0(x) - i\operatorname{Re} l_0(ix) = \operatorname{Re} l_0(x) + i\operatorname{Im} l_0(x) = l_0(x).$$
Finally, for every $x \in X$ there exists a real number θ such that $l(x) = e^{i\theta}|l(x)|$. Set $y = e^{-i\theta}x$. Then $l(y) = |l(x)|$, i.e., $l(y) = l_1(y)$ and $l_1(iy) = 0$, since $l_1(iy), l(y) \in \mathbb{R}$. Hence $|l(x)| = l(y) = l_1(y) \leqslant p(y) = p(x)$. $\qquad\square$

6.3.5. Corollary. *Let X_0 be a linear subspace of a normed space X (not necessarily closed) and let l_0 be a continuous linear function on X_0. Then l_0 can be extended to a continuous linear function on all of X with the same norm as the functional l_0 on X_0.*

PROOF. By assumption $|l_0(x)| \leqslant \|l_0\|\|x\|$ if $x \in X_0$. Set $p(x) = \|l_0\|\|x\|$. Applying the previous corollary, we extend l_0 to a linear function l on X with the bound $|l| \leqslant p$. This gives $\|l\| \leqslant \|l_0\|$. Since $\|l_0\| \leqslant \|l\|$, one has $\|l\| = \|l_0\|$. $\qquad\square$

Let us give a geometric form of the Hahn–Banach theorem connected with separation of convex sets.

Let X be a real linear space. We shall say that a linear function l *separates* two sets $A, B \subset X$ if
$$\inf_{x \in A} l(x) \geqslant \sup_{x \in B} l(x).$$
In other words, there exists a number c such that
$$B \subset \{x\colon l(x) \leqslant c\} \quad \text{and} \quad A \subset \{x\colon l(x) \geqslant c\}.$$
Geometrically this means that A and B are on different sides from the affine subspace $l^{-1}(c)$.

If $l \neq 0$, then the set $\{x\colon l(x) \leqslant c\}$ is called a *halfspace* and the set $\{x\colon l(x) = c\}$ is called a *hyperplane*.

The *algebraic kernel* of a set A in a linear space X is defined as the set of all points $x \in A$ such that for every $v \in X$ there exists a number $\varepsilon = \varepsilon(v) > 0$ for which $x + tv \in A$ whenever $|t| < \varepsilon$. If X is a normed space, then every inner point of A belongs to the algebraic kernel, but the algebraic kernel can be larger than the interior. For example, if we take for X the space of polynomials on $[0,1]$ with the norm from $C[0,1]$, then the set of polynomials x with $\max_t |x'(t)| < 1$ has no interior, but coincides with its algebraic kernel. For an arbitrary infinite-dimensional normed space X, the set $l^{-1}\{(-1,1)\}$, where l is a discontinuous linear function, also has no inner points, but coincides with its algebraic kernel.

Let V be a convex set in a linear space X. Suppose that the algebraic kernel of V contains the point 0.

The *Minkowski functional* of the set V is the function

$$p_V(x) := \inf\{t > 0\colon t^{-1}x \in V\}.$$

The condition that 0 belongs to the algebraic kernel of V is needed to guarantee that the functional p_V be with finite values.

6.3.6. Theorem. *Under the stated assumptions, the functional p_V is positively homogeneous convex and nonnegative. If the set V is balanced (i.e., $\theta V \subset V$ whenever $|\theta| \leqslant 1$), then p_V is a seminorm.*

Conversely, for every positively homogeneous convex nonnegative function p the set $U := \{x\colon p(x) \leqslant 1\}$ is convex, its algebraic kernel is the set $\{x\colon p(x) < 1\}$, and $p = p_U$.

PROOF. As we have already noted, $0 \leqslant p_V(x) < \infty$. For every $\alpha > 0$ and every $x \in X$ we have

$$p_V(\alpha x) = \inf\{t > 0\colon t^{-1}\alpha x \in V\} = \alpha \inf\{s > 0\colon s^{-1}x \in V\} = \alpha p_V(x).$$

Let $x, y \in X$. Let us fix $\varepsilon > 0$ and choose $s, t > 0$ such that

$$p_V(x) < s < p_V(x) + \varepsilon, \quad p_V(y) < t < p_V(y) + \varepsilon.$$

Then $x/s \in V$, $y/t \in V$. Set $r = s + t$. The point $(x+y)/r = \frac{s}{r}x/s + \frac{t}{r}y/t$ belongs to the interval with the endpoints x/s and y/t, hence by the convexity of the set V belongs to V. Thus, $p_V\big((x+y)/r\big) \leqslant 1$, whence

$$p(x+y) \leqslant r < p_V(x) + p_V(y) + 2\varepsilon.$$

Since ε was arbitrary, we obtain $p_V(x+y) \leqslant p_V(x) + p_V(y)$. Finally, if V is balanced, then $p_V(\theta x) = p_V(x)$ whenever $|\theta| = 1$.

Let $p \geqslant 0$ be a positively homogeneous convex function. By the convexity of p the set $U = \{x\colon p(x) \leqslant 1\}$ is convex. Every element $x \in X$ with $p(x) < 1$ belongs to the algebraic kernel U, because

$$p(x+ty) \leqslant 1 \quad \text{whenever } |t| < (1 - p(x))/\max(p(y), 1).$$

If $p(x) \geqslant 1$, then x does not belong to the algebraic kernel of U, since

$$p(x + \varepsilon x) = (1 + \varepsilon)p(x) \geqslant 1 + \varepsilon > 1$$

for all $\varepsilon > 0$. It is clear from the definition that $p_U = p$. $\qquad\square$

6.3.7. Theorem. *Let U and V be convex sets in a real linear space X such that the algebraic kernel of U is not empty and does not intersect V. Then there exists a nonzero linear function separating U and V.*

PROOF. We can assume that 0 belongs to the algebraic kernel of U (otherwise we can pass to the sets $U - a$ and $V - a$ for some point a in the algebraic kernel of U). Let us take a point $v_0 \in V$ and set $W = U - V + v_0$. It is straightforward to verify that W is a convex set and its algebraic kernel W^0 contains 0. It is easy to derive from the definition that $\lambda u \in U^0$ and $\lambda w \in W^0$ for all $u \in U^0$, $w \in W^0$, $|\lambda| < 1$. Finally, we observe that $v_0 \notin W^0$. Otherwise the algebraic kernel of $U - V$ contains 0. Then there exists a nonzero element $x \in U \cap V$ and $tx \in U - V$ for some $t \in (0,1)$, i.e., $tx = u - v$, $u \in U$, $v \in V$ and $(t+1)^{-1}u = t(t+1)^{-1}x + (t+1)^{-1}v \in U^0 \cap V$, which is impossible. Let p be the Minkowski functional of the set W^0. Then $p(v_0) \geqslant 1$. On the one-dimensional subspace of vectors of the form tv_0 we define a linear function $l_0(tv_0) = tp(v_0)$. By the Hahn–Banach theorem l_0 extends to a linear function l on X such that $l \leqslant p$. Then $l(w) \leqslant 1$ if $w \in W^0$, whence $l(w) \leqslant 1$ for all $w \in W$, because $\lambda w \in W^0$ if $\lambda \in (0,1)$. Since $l(v_0) = l_0(v_0) = p(v_0) \geqslant 1$, the functional l separates the set W and v_0. Hence l separates $U - V$ and $\{0\}$, but then l separates the sets U and V. $\square$

Note that in the previous theorem no topology was used: the algebraic kernel is defined in purely algebraic terms. Applying to normed spaces, we obtain the following assertion (see also Example 8.3.11).

6.3.8. Corollary. *Suppose that two convex sets U and V in a real normed space X are disjoint and U is open. Then there exists a nonzero continuous linear function separating U and V.*

PROOF. We observe that the algebraic kernel of U coincides with U by the assumption that U is open. Hence there exists a nonzero linear function l separating U and V. This function is automatically continuous in case of an open set U. This follows from an obvious observation: if a linear function l is bounded from above or below on a nonempty open set, then it is continuous. Indeed, the function l is bounded from above or below on some ball of radius $r > 0$. Hence it is bounded from above or below on the ball of radius r centered at the origin, which gives the boundedness in absolute value on this ball. $\square$

6.3.9. Corollary. *If V is a closed convex set in a real normed space X and $x \notin V$, then there exists $l \in X^*$ with $l(x) > \sup_{v \in V} l(v)$. If V is a linear subspace, then one can take l such that $l(x) = 1$ and $l|_V = 0$.*

PROOF. We can assume that $x = 0$. There is an open ball U centered at the origin such that $U \cap V = \varnothing$. By the previous theorem there exists a nonzero functional $l \in X^*$ with $\inf_{u \in U} l(u) \geqslant \sup_{v \in V} l(v)$. Then $\inf_{u \in U} l(u) < l(0) = 0$, since otherwise $l = 0$. If V is linear, then $l|_V = 0$ (otherwise the supremum is infinite). $\square$

6.4. Applications of the Hahn–Banach Theorem

Some interesting applications of the Hahn–Banach theorem have already been discussed above: the separation theorem. Another important application is the proof of the fact (not a priori obvious) that the topological dual to the infinite-dimensional normed space is nonzero.

6.4.1. Theorem. *For every nonzero element x of a normed space X there exists a functional l such that $\|l\| = 1$ and $l(x) = \|x\|$.*

PROOF. On the one-dimensional space generated by x we set $l_0(tx) = t\|x\|$. Then $l_0(x) = \|x\|$ and $\|l_0\| = 1$. It remains to extend l_0 to X with the preservation of its norm. □

With the aid of a similar reasoning it is easy to establish that in the case of an infinite-dimensional space X for every n there exist vectors $x_1, \ldots, x_n \in X$ and functionals $l_1, \ldots, l_n \in X^*$ such that $l_i(x_j) = \delta_{ij}$. In particular, the dual space is also infinite-dimensional.

6.4.2. Corollary. *Let X_0 be a finite-dimensional subspace of a normed space X. Then X_0 is topologically complemented in X, i.e., there exists a closed linear subspace X_1 such that X is the direct algebraic sum of X_0 and X_1 and the natural algebraic projections P_0 and P_1 to X_0 and X_1 are continuous.*

PROOF. As noted above, one can find a basis $x_1, \ldots, x_n$ in the space X_0 and elements $l_i \in X^*$ with $l_i(x_j) = \delta_{ij}$. Set

$$X_1 := \bigcap_{i=1}^{n} \operatorname{Ker} l_i, \quad P_0 x := \sum_{i=1}^{n} l_i(x)x_i, \quad P_1 x := x - P_0 x.$$

For every j we have $P_0 x_j = l_j(x_j)x_j = x_j$. Note that $P_0|_{X_1} = 0$, $X_0 \cap X_1 = \{0\}$. In addition, we have $X = X_0 \oplus X_1$, because $x - P_0 x \in X_1$ by the equalities $l_j(x - P_0 x) = l_j(x) - l_j(x)l_j(x_j) = 0$. The continuity of P_0 and P_1 is obvious from their definitions. It is also clear that P_0 and P_1 coincide with the algebraic projections to X_0 and X_1. □

We now construct an isometric embedding of any normed space X into its second dual X^{**}. For every $x \in X$, we consider the functional $J_x \colon f \mapsto f(x)$ on the space X^*.

6.4.3. Proposition. *The mapping $J \colon x \mapsto J_x$ is a linear isometric embedding of X into X^{**}.*

PROOF. The linearity of J is obvious. Since

$$|J_x(f)| = |f(x)| \leqslant \|x\| \quad \text{whenever } \|f\| \leqslant 1,$$

we have $\|J_x\| \leqslant \|x\|$. On the other hand, as shown above, if $x \neq 0$, there exists $f \in X^*$ with $\|f\| = 1$ and $f(x) = \|x\|$, i.e., $J_x(f) = \|x\|$, whence we obtain that $\|J_x\| \geqslant \|x\|$. □

If $J(X) = X^{**}$, then the space X is called *reflexive*. Below we give examples of reflexive and nonreflexive spaces. One should bear in mind that the reflexivity of a space X is not equivalent to the existence of an isometry between X and X^{**} (Exercise 6.10.180 contains a counter-example: the famous James space); it is required that the canonical mapping J be an isometry onto all of X^{**}.

Combining this proposition with the Banach–Steinhaus theorem, we obtain an important assertion about the boundedness of weakly bounded sets.

6.4.4. Definition. *A set A in a normed space is called weakly bounded if*

$$\sup_{x \in A} |l(x)| < \infty$$

for every continuous linear functional l.

In the real case it suffices to have such an estimate without absolute value, since $-l$ is a continuous functional as well.

6.4.5. Theorem. *A set in a normed space is weakly bounded precisely when it is bounded in norm.*

PROOF. The weak boundedness of the set A means that the family of functionals J_x, where $x \in A$, is bounded on every element of X^*. Since X^* is Banach, this family is bounded in the norm of X^{**} by the Banach–Steinhaus theorem. According to the proposition above the set A is norm bounded. The converse assertion is obvious. $\square$

Yet another useful corollary of the existence of an isometric embedding of X into X^{**} is the following result.

6.4.6. Proposition. *Every normed space X possesses a unique (up to a linear isometry) completion that is a Banach space.*

PROOF. For such a completion we can take the closure of the image of X under the embedding into X^{**}. It should be noted that using the completion constructed earlier in the category of general metric spaces, one can obtain a space that is not linear. The uniqueness of a completion up to a linear isometry is easily verified. $\square$

In the case of a separable normed space, by using the Hahn–Banach theorem it is easy to obtain a countable set of functionals separating points.

6.4.7. Proposition. *Let X be a separable normed space. Then there exists a countable set of functionals $l_n \in X^*$ such that the equality $l_n(x) = 0$ for all n implies the equality $x = 0$.*

PROOF. Let $\{x_n\}$ be a countable everywhere dense set in X. Assuming that $X \neq 0$, for every n we find $l_n \in X^*$ with $l_n(x_n) = \|x_n\|$ and $\|l_n\| = 1$. Let $l_n(x) = 0$ for all n. Let us fix $\varepsilon > 0$ and find x_m with $\|x - x_m\| \leqslant \varepsilon$. Then $\|x_m\| = l_m(x_m) = l_m(x_m - x) \leqslant \|x_m - x\| \leqslant \varepsilon$, whence $\|x\| \leqslant 2\varepsilon$. Hence we have $x = 0$. $\square$

The existence of functionals separating points will be used in the theorem on universality of the space $C(K)$ proved in §6.7.

By using the Hahn–Banach theorem we constructed functionals having maxima on the unit ball. On the other hand, we have encountered examples of functionals that do not attain maxima. In this connection we mention the following result due to Bishop and Phelps (its proof can be found, for example, in Diestel [**147**, Chapter 1]; a stronger assertion can be found in Bollobás [**77**, Chapter 8]).

6.4.8. Theorem. *Let C be a nonempty closed bounded convex set in a real Banach space X. Then the set of functionals in X^* attaining their maximum on C is everywhere dense in X^*.*

In particular, this assertion is true for closed balls. Not all functionals attain their maxima on balls: this is true only in reflexive spaces (see Theorem 6.10.10).

Note that for every normed space X with the completion $\overline{X}$, the duals X^* and $\overline{X}^*$ coincide in the sense that every functional from X^* extends uniquely by continuity to a functional from $\overline{X}^*$, moreover, every element of $\overline{X}^*$ is obtained in this way from its restriction to X. If Banach spaces X and Y are linearly homeomorphic by means of an operator J, then the mapping $l \mapsto l \circ J$ is a homeomorphism from Y^* onto X^*. However, one should bear in mind that there exist non-isomorphic (linearly topologically) Banach spaces X and Y with linearly homeomorphic duals. As an example one can take l^1 and $L^1[0,1]$. The absence of linear isomorphisms between them is the subject of Exercise 6.10.104 and the fact that their duals l^∞ and $L^\infty[0,1]$ are isomorphic is Pełczyński's theorem, a proof of which can be read in Albiac, Kalton [**9**, Theorem 4.3.10].

Let us give less obvious examples of positively homogeneous convex functions that are useful for constructing some interesting linear functions.

6.4.9. Example. The following functions p are positively homogeneous convex:

(i) let X be the space of all bounded real sequences $x = (x_n)$ and let

$$p(x) = \inf S(x, a_1, \ldots, a_n), \quad S(x, a_1, \ldots, a_n) := \sup_{k \geq 1} \frac{1}{n} \sum_{i=1}^{n} x_{k+a_i},$$

where inf is taken over all natural numbers n and all finite collections of numbers $a_1, \ldots, a_n \in \mathbb{N}$;

(ii) let X be the space of all bounded real functions on the real line and let

$$p(f) = \inf S(f, a_1, \ldots, a_n), \quad S(f, a_1, \ldots, a_n) := \sup_{t \in \mathbb{R}} \frac{1}{n} \sum_{i=1}^{n} f(t + a_i),$$

where inf is taken over all natural numbers n and all finite collections of numbers $a_1, \ldots, a_n \in \mathbb{R}$;

(iii) let X be the space of all bounded real sequences $x = (x_n)$ and let

$$p(x) = \inf S(x, a_1, \ldots, a_n), \quad S(x, a_1, \ldots, a_n) := \limsup_{k \to \infty} \frac{1}{n} \sum_{i=1}^{n} x_{k+a_i},$$

where inf is taken over all natural numbers n and all finite collections of numbers $a_1, \ldots, a_n \in \mathbb{N}$.

PROOF. Assertion (i) follows from (ii), so we prove the latter. It is clear that $|p(f)| < \infty$ and $p(\alpha f) = \alpha p(f)$ if $\alpha \geq 0$. Let $f, g \in X$ and $\varepsilon > 0$. Find $a_1, \ldots, a_n, b_1, \ldots, b_m$ such that

$$\sup_{t \in \mathbb{R}} \frac{1}{n} \sum_{i=1}^{n} f(t + a_i) < p(f) + \varepsilon, \quad \sup_{t \in \mathbb{R}} \frac{1}{m} \sum_{i=1}^{m} g(t + b_i) < p(g) + \varepsilon.$$

We observe that the quantity $\sup_{t\in\mathbb{R}}(mn)^{-1}\sum_{j=1}^{m}\sum_{i=1}^{n}(f+g)(t+a_i+b_j)$ does not exceed the sum

$$\sup_{t\in\mathbb{R}}\frac{1}{m}\sum_{j=1}^{m}\frac{1}{n}\sum_{i=1}^{n}f(t+a_i+b_j)+\sup_{t\in\mathbb{R}}\frac{1}{n}\sum_{i=1}^{n}\frac{1}{m}\sum_{j=1}^{m}g(t+a_i+b_j).$$

We have $n^{-1}\sum_{i=1}^{n}f(t+a_i+b_j)\leqslant S(f,a_1,\dots,a_n)$ for any fixed t and b_j, whence

$$\sup_{t\in\mathbb{R}}\frac{1}{m}\sum_{j=1}^{m}\frac{1}{n}\sum_{i=1}^{n}f(t+a_i+b_j)\leqslant S(f,a_1,\dots,a_n).$$

From a similar estimate for g we obtain

$$p(f+g)\leqslant S(f,a_1,\dots,a_n)+S(g,b_1,\dots,b_m)<p(f)+p(g)+2\varepsilon,$$

which gives $p(f+g)\leqslant p(f)+p(g)$ because ε was arbitrary. The proof of assertion (iii) is completely similar. $\qquad\square$

We now apply the established facts for constructing some curious set functions.

6.4.10. Example. On the σ-algebra of all subsets of $\mathbb{N}$ there exists a nonnegative additive function ν that equals zero on all finite sets and equals 1 on $\mathbb{N}$; in particular, the function ν is not countably additive.

PROOF. In the space X of bounded sequences with the function p from assertion (iii) in the previous example we take the subspace X_0 of sequences having a limit. Set $l(x)=\lim\limits_{n\to\infty}x_n$ if $x\in X_0$. Then $l(x)=p(x)$, since $\limsup\limits_{k\to\infty}n^{-1}\sum_{i=1}^{n}x_{k+a_i}=\lim\limits_{k\to\infty}x_k$ for all fixed n and $a_1,\dots,a_n$. Let us extend l to a linear function $\widehat{l}$ on X with $\widehat{l}\leqslant p$. If $x\in X$ and $x_n\leqslant 0$ for all n, then $p(x)\leqslant 0$ and hence $\widehat{l}(x)\leqslant 0$. Hence $\widehat{l}(x)\geqslant 0$ if $x_n\geqslant 0$. If $x=(x_1,\dots,x_n,0,0,\dots)$, then $\widehat{l}(x)=l(x)=0$. Finally, $\widehat{l}(1,1,\dots)=1$. For every $E\subset\mathbb{N}$ set $\nu(E):=\widehat{l}(I_E)$, where I_E is the indicator function of E, i.e., the sequence on the nth position of which one has 1 or 0 depending on whether n belongs to E or not. To finite sets there correspond finite sequences, so ν vanishes on them. On all of $\mathbb{N}$ the value of ν is 1, and the additivity of ν follows from the additivity of $\widehat{l}$ and the equality $I_{E_1\cup E_2}=I_{E_1}+I_{E_2}$ for disjoint E_1 and E_2. The failure of the countable additivity is obvious. $\qquad\square$

6.4.11. Example. On the space X of all bounded real functions with bounded support on the real line there exists a linear function L that coincides with the Lebesgue integral on all Lebesgue integrable functions and has the following properties: $L(f)\geqslant 0$ if $f\geqslant 0$, $|L(f)|\leqslant\sup_t|f(t)|$, $L(f(\cdot+h))=L(f)$ for all $f\in X$ and $h\in\mathbb{R}^1$, where $f(\cdot+h)(t)=f(t+h)$.

PROOF. First we construct such functional on the space X_1 of bounded functions with period 1. Let us consider on X_1 the function p from Example 6.4.9(ii). On the subspace X_0 of functions integrable over the period we define L as the Lebesgue integral and observe that $L(f) \leqslant p(f)$ if $f \in X_0$ by Exercise 3.12.26. Let us extend L to X by the Hahn–Banach theorem. We have $L(-f) = -L(f) \leqslant p(-f)$, whence

$$-p(-f) \leqslant L(f) \leqslant p(f) \quad \forall f \in X_1.$$

If $f \geqslant 0$, then $p(-f) \leqslant 0$ by the definition of p and hence $L(f) \geqslant 0$. It is clear that $|L(f)| \leqslant \sup_t |f(t)|$, since $p(f) \leqslant \sup_t |f(t)|$. We show that $L(f) = L(f(\cdot + h))$ for all $f \in X_1$, $h \in \mathbb{R}^1$. Let us set $g(t) = f(t+h) - f(t)$ and verify that $L(g) = 0$. Let $a_k = (k-1)h$ if $k = 1, \ldots, n$. Then $\sum_{i=1}^{n} g(t + a_i) = f(t + nh) - f(t)$, whence

$$p(g) \leqslant S(g, a_1, \ldots, a_n) = \sup_t n^{-1}[f(t + nh) - f(t)] \to 0 \text{ as } n \to \infty.$$

Thus, $p(g) \leqslant 0$. Similarly $p(-g) \leqslant 0$, whence $L(g) = 0$. Now for a bounded function f with support in $[-n, n)$ we set $L(f) := \sum_{j=-n}^{n-1} L(f_j)$, where f_j is the 1-periodic extension of the restriction of the function f to $[j, j+1)$. It is not difficult to verify that we have obtained the desired functional. $\qquad\square$

Both examples are due to Banach.

Letting $\zeta(A) := L(I_A)$ for a bounded set A, we obtain an additive nonnegative set function extending Lebesgue measure to bounded sets and invariant with respect to translations. By the way, if we take for X_0 the one-dimensional space of constants, then the same reasoning gives yet another linear functional on X_1 equal to 1 at 1, nonnegative on nonnegative functions and invariant with respect to shifts of the argument of functions. By a similar reasoning, on every set T on which there is an action of a commutative group of bijections $\mathcal{G}$, one can construct a nonnegative additive function ζ with $\zeta(T) = 1$ invariant with respect to $\mathcal{G}$. In this way Banach constructed an additive area on the class of all bounded sets that coincides with Lebesgue measure on measurable sets and is invariant with respect to isometries. However, for $\mathbb{R}^3$ such an extension does not exist, which was proved by Hausdorff. Hence the commutativity of the group $\mathcal{G}$ is important.

Here is one more example of application of the Hahn–Banach theorem.

6.4.12. Example. Let X, Y be normed spaces, let $D \subset X$ be a linear subspace, $T \colon D \to Y$ a linear mapping (not necessarily continuous), and let l be a linear function on D such that $|l(x)| \leqslant \|Tx\|_Y$ for all $x \in D$. Then there exists a functional $f \in Y^*$ for which $l(x) = f(Tx)$ for all $x \in D$.

PROOF. If $\operatorname{Ker} T = 0$, then $l_0(y) = l(T^{-1}y)$ is a linear functional on $T(D)$ and $|l_0(y)| \leqslant \|y\|_Y$. Let us extend l_0 to a functional $f \in Y^*$. Then $l(x) = f(Tx)$ for all $x \in D$. If $\operatorname{Ker} T \neq 0$, then we take a linear subspace D_1 in D such that $D = \operatorname{Ker} T \oplus D_1$. As we have proved, there exists $f \in Y^*$ for which $l(x) = f(Tx)$ for all $x \in D_1$. This is also true for $x \in \operatorname{Ker} T$, since $|l(x)| \leqslant \|Tx\|_Y$. Hence this is true for all $x \in D$. $\qquad\square$

This result will be used below in Exercise 6.5.3 for representing linear functionals by means of inner products.

6.5. Duals to Concrete Spaces

Here we describe the dual spaces to the most important Banach spaces. We start with the Riesz theorem on a general form of a continuous linear functional on a Hilbert space.

6.5.1. Theorem. (THE RIESZ THEOREM) *Let H be a real or complex Hilbert space. Then for every $v \in H$ the formula*

$$l_v(x) = (x, v)$$

defines a continuous linear functional on H and $\|l_v\| = \|v\|$.
Conversely, every functional $l \in H^*$ is represented in this way, and the mapping $v \mapsto l_v$ is an isometry that is linear in the real case and conjugate-linear in the complex case.

PROOF. The equality $\|l_v\| = \|v\|$ is obvious from the Cauchy inequality $|(x, v)| \leqslant \|x\|\|v\|$ and the equality $l_v(v) = \|v\|^2$. Let $l \in H^*$. If $l = 0$, then $v = 0$. We shall assume that $H_0 := l^{-1}(0) \neq H$. Let us take a unit vector $e \perp H_0$ and set $v := \overline{l(e)}e$. Then for all $x \in H$ we have $l(x) = (x, v)$, since this true for all $x \in H_0$ and for $x = v$ by the equalities $l(v) = |l(e)|^2$, $(v, v) = |l(e)|^2$. The last assertion of the theorem is obvious. □

In the complex case the Riesz identification of H with H^* is conjugate-linear, but not linear. However, there is also a linear isometry between H and H^*. Indeed, by using an orthonormal basis $\{e_\alpha\}$ in H we can define the conjugation $v \mapsto \overline{v}$ by sending $\sum_\alpha c_\alpha e_\alpha$ to $\sum_\alpha \overline{c_\alpha} e_\alpha$. Then $v \mapsto l_{\overline{v}}$ is the desired linear isometry.

It follows from the Riesz theorem that every continuous linear functional F on H^* has the form $F(l) = l(a)$, where $a \in H$. Indeed, the function $v \mapsto \overline{F(l_v)}$ is linear and continuous. Hence $\overline{F(l_v)} = (v, a)$ for some $a \in H$, i.e., we have $F(l_v) = (a, v) = l_v(a)$. Therefore, H is reflexive.

6.5.2. Remark. For every real Euclidean space E (not necessarily complete) the linear mapping $J \colon E \to E^*$, $v \mapsto l_v$, where $l_v(x) = (x, v)$, preserves the distances. Since E^* is always complete, this mapping gives a completion of E in the category of Euclidean spaces. For this it suffices to observe that on the closure of $J(E)$ in E^* we have not only the norm, but also the inner product: if $f = \lim_{n \to \infty} J(v_n)$ and $g = \lim_{n \to \infty} J(w_n)$, then the limit $(f, g) := \lim_{n \to \infty} (w_n, v_n)$ exists, does not depend on the choice of convergent sequences, is linear in each argument (since we have $J(\lambda v_n) = \lambda J(v_n)$ and $(\lambda w_n, v_n) = \lambda(w_n, v_n)$). We can also set $(f, g) := [\|f + g\|^2 - \|f\|^2 - \|g\|^2]/2$ and use v_n and w_n to make sure that we have obtained an inner product. Thus, the closure $\overline{J(E)}$ of $J(E)$ is a Hilbert space. Indeed, $\overline{J(E)} = E^*$, since after identification of E with $J(E)$ and $\overline{J(E)}^*$ with $\overline{J(E)}$ we obtain $E^* = \overline{J(E)}^* = \overline{J(E)}$. In a similar way one can construct completions of complex Euclidean spaces.

6.5.3. Example. Let D be a linear subspace in a Hilbert space H, $T \colon D \to H$ a linear mapping, l a linear function on D such that $|l(x)| \leqslant \|Tx\|$ for all $x \in D$. Then there exists $v \in H$ for which $l(x) = (Tx, v)$ for all $x \in D$. This follows from Example 6.4.12.

Let us apply the Riesz theorem to prove the Radon–Nikodym theorem.

6.5.4. Example. Let $\mu \geqslant 0$ and $\nu \geqslant 0$ be finite measures on a space $(X, \mathcal{A})$ and $\nu \ll \mu$. Let us consider the measure $\lambda = \mu + \nu$. Every function φ from $\mathcal{L}^1(\lambda)$ belongs to $\mathcal{L}^1(\mu)$ and its integral against the measure μ does not change if we redefine φ on a set of λ-measure zero. Therefore, the linear function

$$L(\varphi) = \int_X \varphi \, d\mu$$

is well-defined on the real space $L^2(\lambda)$ (does not depend on the choice of a representative of φ). By the Cauchy–Bunyakovskii inequality

$$|L(\varphi)| \leqslant \int_X |\varphi| \, d\lambda \leqslant \|1\|_{L^2(\lambda)} \|\varphi\|_{L^2(\lambda)}.$$

The Riesz theorem gives a function $\psi \in L^2(\lambda)$ such that

$$\int_X \varphi \, d\mu = \int_X \varphi\psi \, d\lambda \tag{6.5.1}$$

for all $\varphi \in L^2(\lambda)$. We shall deal with its $\mathcal{A}$-measurable version. Substituting $\varphi = I_A$, where $A \in \mathcal{A}$, we find that $\mu = \psi \cdot \lambda$, $\nu = (1 - \psi) \cdot \lambda$. Let us prove that the function $(1 - \psi)/\psi$ can be taken for $d\nu/d\mu$. Let $\Omega = \{x \colon \psi(x) \leqslant 0\}$, $\Omega_1 = \{x \colon \psi(x) > 1\}$. Then $\Omega, \Omega_1 \in \mathcal{A}$. Substituting in (6.5.1) the functions $\varphi = I_\Omega$ and $\varphi = I_{\Omega_1}$, we obtain

$$\mu(\Omega) = \int_\Omega \psi \, d\lambda \leqslant 0, \quad \mu(\Omega_1) = \int_{\Omega_1} \psi \, d\lambda > \lambda(\Omega_1)$$

if $\mu(\Omega_1) > 0$, whence $\mu(\Omega) = 0$ and $\mu(\Omega_1) = 0$, since $\mu(\Omega_1) \leqslant \lambda(\Omega_1)$. Then the function f defined by the equality

$$f(x) = \frac{1 - \psi(x)}{\psi(x)} \quad \text{if } x \notin \Omega \cup \Omega_1, \quad f(x) = 0 \quad \text{if } x \in \Omega \cup \Omega_1$$

is nonnegative and $\mathcal{A}$-measurable. The function f is integrable with respect to μ. Indeed, the functions $f_n = f I_{\{\psi \geqslant 1/n\}}$ are bounded and increase pointwise to f, moreover,

$$\int_X f_n \, d\mu = \int_X I_{\{\psi \geqslant 1/n\}} (1 - \psi) \, d\lambda = \int_X I_{\{\psi \geqslant 1/n\}} \, d\nu \leqslant \nu(X).$$

The Beppo Levi theorem yields convergence of $\{f_n\}$ to f in $L^1(\mu)$. Finally, for every $A \in \mathcal{A}$ we have $I_A I_{\{\psi \geqslant 1/n\}} \to I_A$ μ-a.e., hence also ν-a.e. (only here we use the absolute continuity of ν with respect to μ). By convergence of $\{f_n\}$ to f in $L^1(\mu)$ and the equality $I_{\{\psi \geqslant 1/n\}} \cdot \nu = I_{\{\psi \geqslant 1/n\}}(1 - \psi) \cdot \lambda = I_{\{\psi \geqslant 1/n\}} f \cdot \mu$ we find that

$$\nu(A) = \lim_{n \to \infty} \int_X I_A I_{\{\psi \geqslant 1/n\}} \, d\nu = \lim_{n \to \infty} \int_X I_A I_{\{\psi \geqslant 1/n\}} f \, d\mu = \int_A f \, d\mu.$$

Let us turn to another theorem due to F. Riesz about a general form of a continuous linear functional on the space $C[a, b]$.

6.5.5. Theorem. (THE RIESZ THEOREM FOR C) *The general form of a continuous linear function on the real or complex space $C[a, b]$ with its sup-norm* *is given by the following integral formula:*

$$l(x) = \int_{[a,b]} x(t)\,\mu(dt),$$

where μ is a bounded Borel measure on $[a,b]$ (complex in the complex case), moreover, $\|l\| = \|\mu\|$.

PROOF. For simplification of formulas we consider the real case. By the Hahn–Banach theorem every continuous linear functional l on $C[a,b]$ extends with the same norm to the space $B[a,b]$ of all bounded functions on $[a,b]$. The extension is also denoted by l. Set

$$F(s) := l(I_{[a,s)}) \text{ if } s \in (a,b], \ F(s) = 0 \text{ if } s \leqslant a, \ F(s) = l(I_{[a,b]}) \text{ if } s > b.$$

We show that the function F is of bounded variation. Let $a = t_0 < \cdots < t_n$, where $t_{n-1} \leqslant b$ and $t_n \geqslant b$. Set $\varepsilon_i := \operatorname{sign}\big(F(t_i) - F(t_{i-1})\big)$ and $[t_{n-1}, t_n) = [t_{n-1}, b]$ if $t_n > b$. Then

$$\sum_{i=1}^{n} |F(t_i) - F(t_{i-1})| = \sum_{i=1}^{n} \varepsilon_i\big(F(t_i) - F(t_{i-1})\big)$$

$$= l\Big(\sum_{i=1}^{n} \varepsilon_i I_{[t_{i-1},t_i)}\Big) \leqslant \|l\| \Big\|\sum_{i=1}^{n} \varepsilon_i I_{[t_{i-1},t_i)}\Big\| \leqslant \|l\|,$$

since the function $\sum_{i=1}^{n} \varepsilon_i I_{[t_{i-1},t_i)}$ can assume only the values $-1, 1, 0$. Thus, we have $V(F, \mathbb{R}) \leqslant \|l\|$. The function F has at most a countable set T of discontinuity points (see §4.2), and these points are jumps. We redefine F at these points in order to obtain a left continuous function F_0. It is readily seen that $V(F_0, \mathbb{R}) \leqslant V(F, \mathbb{R})$.

Let now x be a function continuous on $[a,b]$. We fix $\varepsilon > 0$ and take $\delta > 0$ with the following properties: $|x(t) - x(s)| \leqslant \varepsilon$ whenever $|t - s| \leqslant \delta$ and

$$\left| \int_a^b x(t)\,dF_0(t) - \sum_{i=1}^{n} x(t_i)\big(F_0(t_i) - F_0(t_{i-1})\big) \right| \leqslant \varepsilon$$

if points $a = t_0 < t_1 < \cdots < t_n$ are such that $|t_i - t_{i-1}| \leqslant \delta$, $t_{n-1} \leqslant b$, $t_n > b$. We take such points outside T (at these points F_0 and F are equal) and take a step function ψ equal to $x(t_{i-1})$ on $[t_{i-1}, t_i)$, $i = 1, \ldots, n$. Clearly, we have $\|x - \psi\| \leqslant \varepsilon$. Hence we obtain $|l(x) - l(\psi)| \leqslant \varepsilon \|l\|$. Finally, we have $l(\psi) = \sum_{i=1}^{n} x(t_{i-1})\big(F_0(t_i) - F_0(t_{i-1})\big)$, whence

$$\left| \int_a^b x(t)\,dF_0(t) - l(x) \right| \leqslant \varepsilon(1 + \|l\|).$$

Since ε was arbitrary, $l(x)$ is the Stieltjes integral of x against F_0. This integral equals the Lebesgue integral with respect to the Borel measure μ generated by F_0 (see §4.2). It is readily seen that the Stieltjes integral of x against F_0 is not greater than $V(F_0, \mathbb{R}) \sup_t |x(t)|$, whence $\|l\| \leqslant V(F_0, \mathbb{R})$, so $\|l\| = V(F_0, \mathbb{R})$. The equality $\|\mu\| = V(F_0, \mathbb{R})$ is proved in Proposition 4.2.9. $\square$

Actually, as shown in §6.10(vii), the Riesz theorem remains in force for every compact space, so it is possible to use more advanced techniques as compared to the crude construction presented above. The reader may ask the following question: is not it possible to obtain the required measure μ at once as the action of the extension of l on the indicator functions? It turns out that the answer is negative: the obtained set function can fail to be countably additive (see Exercise 6.10.100).

6.5.6. Theorem. (i) *The general form of a continuous linear function on c_0 is given by the series*

$$l(x) = \sum_{n=1}^{\infty} y_n x_n, \quad \text{where } y = (y_n) \in l^1 \text{ and } \|l\| = \|y\|_{l^1}.$$

(ii) *Let $1 < p < \infty$. The general form of a continuous linear function on l^p is given by the series*

$$l(x) = \sum_{n=1}^{\infty} y_n x_n, \quad \text{where } y = (y_n) \in l^q, \ 1/p + 1/q = 1 \text{ and } \|l\| = \|y\|_{l^q}.$$

(iii) *The general form of a continuous linear function on l^1 is given by the series*

$$l(x) = \sum_{n=1}^{\infty} y_n x_n, \quad \text{where } y = (y_n) \in l^{\infty} \text{ and } \|l\| = \|y\|_{l^{\infty}}.$$

PROOF. It is clear that every element $y \in l^1$ defines a functional l on c_0 by this formula and $\|l\| \leqslant \|y\|_{l^1}$. Since for the vector

$$x = (\operatorname{sgn} y_1, \ldots, \operatorname{sgn} y_n, 0, 0, \ldots)$$

we have $\|x\| = 1$ and $l(x) = \sum_{i=1}^{n} |y_i|$, we actually have $\|l\| = \|y\|_{l^1}$.

Conversely, let l be a continuous linear functional on c_0. Set $y_n := l(e_n)$, where e_n is the sequence with 1 at the nth position and 0 at all other positions. Then $\sum_{i=1}^{n} |y_i| = l(x) \leqslant \|l\|$ for $x = (\operatorname{sgn} y_1, \ldots, \operatorname{sgn} y_n, 0, 0, \ldots)$, whence $y \in l^1$. The vector y defines the functional l_0 that coincides with l on finite linear combinations of the vectors e_n. Since such combinations are dense in c_0 and both functionals are continuous, we obtain the equality $l = l_0$. This proves assertion (i). Assertions (ii) and (iii) are proved completely similarly, one only has to use the inequality $\sum_{n=1}^{\infty} |x_n y_n| \leqslant \|x\|_{l^p} \|y\|_{l^q}$. $\qquad \square$

Thus, $c_0^{**} = l^{\infty}$. This equality gives one of a rather few known examples in which the second dual space is explicitly calculated for a non-reflexive space. Of course, forming direct sums of c_0 with reflexive spaces one can increase the number of examples, but without using c_0 one can hardly proceed. Note that the dual $(l^{\infty})^*$ is not isomorphic to l^1, since is not separable (see Exercise 6.10.86).

We now consider the space L^p.

6.5.7. Theorem. *Let μ be a nonnegative σ-finite measure on a measurable space $(\Omega, \mathcal{A})$.*

(i) *Let $1 < p < \infty$. The general form of a continuous linear function on the real or complex space $L^p(\mu)$ is given by the formula*

$$l(x) = \int_{\Omega} x(t) y(t) \, \mu(dt), \quad \text{where } y \in L^q(\mu), \ 1/p + 1/q = 1.$$

Moreover, $\|l\| = \|y\|_{L^q(\mu)}$.

(ii) *The general form of a continuous linear function on the real or complex space $L^1(\mu)$ is given by the formula*

$$l(x) = \int_\Omega x(t)y(t)\,\mu(dt), \quad \text{where } y \in \mathcal{L}^\infty(\mu).$$

Moreover, $\|l\| = \|y\|_{L^\infty(\mu)}$.

PROOF. To simplify calculations we consider the real case. Hölder's inequality shows that every function $y \in L^q(\mu)$ defines on $L^p(\mu)$ a linear functional the norm of which does not exceed $\|y\|_{L^q(\mu)}$. If $p > 1$, we take for x the function $x(\omega) = \operatorname{sgn} y(\omega)|y(\omega)|^{q/p}$ and on account of the equality $q/p = q - 1$ obtain

$$\|x\|_{L^p(\mu)}^p = \|y\|_{L^q(\mu)}^q \quad \text{and} \quad \int_\Omega x(\omega)y(\omega)\,\mu(d\omega) = \|y\|_{L^q(\mu)}^q,$$

which shows that $\|l\| = \|y\|_{L^q(\mu)}$. In the case $p = 1$ we set $c := \|y\|_{L^\infty(\mu)}$. If $c = 0$, then $l = 0$. Let $c > 0$. The sets $E_n := \{\omega\colon c - 1/n \leqslant |y(\omega)| \leqslant c\}$ are of positive measure. Take the functions $x_n := I_{E_n} \operatorname{sgn} y/\mu(E_n)$ with norm 1. Then

$$l(x_n) = \int_\Omega x_n(\omega)y(\omega)\,\mu(d\omega) \geqslant c - 1/n$$

and hence $\|l\| \geqslant c - 1/n$, i.e., $\|l\| \geqslant c$. The opposite inequality is obvious.

Let l be a continuous linear functional on $L^p(\mu)$. Partitioning the space Ω into parts of finite measure, it is easy to reduce the general case to the case of a finite measure. Hence we further assume that $\mu(\Omega) < \infty$. The function

$$\nu(A) := l(I_A), \quad A \in \mathcal{A},$$

is a countably additive measure on $\mathcal{A}$. If $\mu(A) = 0$ for some set $A \in \mathcal{A}$, then $\nu(A) = 0$, i.e., the measure ν is absolutely continuous with respect to the measure μ. By the Radon–Nikodym theorem there exists a μ-integrable function y such that

$$l(I_A) = \nu(A) = \int_\Omega I_A(\omega)y(\omega)\,\mu(d\omega), \quad A \in \mathcal{A}.$$

Hence for every simple function x the quantity $l(x)$ equals the integral of xy. Therefore,

$$\left| \int_\Omega x(\omega)y(\omega)\,\mu(d\omega) \right| \leqslant \|l\|\,\|x\|_{L^p(\mu)}. \tag{6.5.2}$$

By a limiting procedure we extend this estimate to all bounded μ-measurable functions x. Let $p > 1$. We take for x the function $x(\omega) = \operatorname{sgn} y(\omega)|y(\omega)|^{q/p}I_{\{|y|\leqslant n\}}$, where $n \in \mathbb{N}$. This gives the estimate

$$\int_\Omega |y(\omega)|^q I_{\{|y|\leqslant n\}}(\omega)\,\mu(d\omega) \leqslant \|l\|\,\|yI_{\{|y|\leqslant n\}}\|_{L^q(\mu)}^{q/p}.$$

Thus, $\|yI_{\{|y|\leqslant n\}}\|_{L^q(\mu)}^q \leqslant \|l\|\,\|yI_{\{|y|\leqslant n\}}\|_{L^q(\mu)}^{q/p}$, hence by the equality $q = 1 + q/p$ we obtain the estimate $\|yI_{\{|y|\leqslant n\}}\|_{L^q(\mu)} \leqslant \|l\|$. By Fatou's theorem $\|y\|_{L^q(\mu)} \leqslant \|l\|$. The function y defines on $L^p(\mu)$ a continuous linear functional l_0 which coincides with l on all simple functions. By the continuity of both functionals and density of the set of simple functions in $L^p(\mu)$ with $p < \infty$ we obtain the equality $l = l_0$. Finally, in the case $p = 1$ from (6.5.2) we obtain the inequality

$$\int_A |y(\omega)|\, \mu(d\omega) \leqslant \|l\|\mu(A), \quad A \in \mathcal{A}.$$

Then $\mu\big(\omega\colon |y(\omega)| > \|l\|\big) = 0$, hence $\|y\|_{L^\infty(\mu)} \leqslant \|l\|$.   □

Again, the dual $L^\infty[0,1]^*$ is not isomorphic to $L^1[0,1]$, since is not separable (see Exercise 6.10.86).

Note that there are cases when it is not convenient to identify the dual to a Hilbert space with the space itself, in spite of the existence of the natural isomorphism. For example, this is the case when one deals with duals to Sobolev spaces (see Chapter 9). Here is yet another typical example. Let Ω be the open unit disc in the complex plane and let $A^2(\Omega)$ be the Bergman space of functions holomorphic in Ω and belonging to $L^2(\Omega)$ (see Example 5.2.2). For every $\lambda \in \Omega$, the linear functional $\varphi \mapsto \varphi(\lambda)$ is continuous on $A^2(\Omega)$ (see estimate (5.2.1) in the aforementioned example). This functional is not written in the form (φ, ψ), although, of course, in accordance with the general theorem it can be represented in this form with some element $\psi \in A^2(\Omega)$. A similar situation arises for more general functionals of the form

$$\varphi \mapsto \int_K \varphi(z)\, \mu(dz),$$

where μ is a measure with compact support K in Ω.

6.6. The Weak and Weak-∗ Topologies

Let E be a linear space and let F be some linear space of linear functions on E separating points in E in the following sense: for every $x \neq 0$, there exists $f \in F$ with $f(x) \neq 0$. In other words, for every pair of different points $x, y \in X$, there exists an element $f \in F$ with $f(x) \neq f(y)$. In this situation, when there are no norms or topologies, in many applications it is useful to introduce convergence on E in the following way: $x_n \to x$ if $f(x_n) \to f(x)$ for every $f \in F$.

Similarly, on F it is useful to consider the pointwise convergence, i.e., convergence $f_n(x) \to f(x)$ for every x. We now introduce a natural topology in which convergence of sequences has the indicated form. We warn the reader at once that a topology with such a property is not unique. However, the topology $\sigma(E, F)$ introduced below is natural in many respects. The objects defined and studied in this section belong actually to the theory of locally convex spaces, fundamentals of which are discussed in Chapter 8. However, the special features of weak topologies are so important that the violation of the deductive order of presentation undertaken here seems fully justified, moreover, it is even useful for the subsequent acquaintance with more general concepts.

We first define the following basis of neighborhoods of zero:

$$U_{f_1,\ldots,f_n,\varepsilon} := \big\{x \in E\colon |f_1(x)| < \varepsilon, \ldots, |f_n(x)| < \varepsilon\big\},$$

where $n \in \mathbb{N}$, $f_i \in F$, $\varepsilon > 0$. Next we introduce a basis of neighborhoods of an arbitrary point $a \in E$:

$$U_{f_1,\ldots,f_n,\varepsilon}(a) := U_{f_1,\ldots,f_n,\varepsilon} + a$$
$$= \left\{ x \in E : |f_1(x-a)| < \varepsilon, \ldots, |f_n(x-a)| < \varepsilon \right\}.$$

Finally, we declare to be open the empty set and all possible unions of neighborhoods $U_{f_1,\ldots,f_n,\varepsilon}(a)$, i.e., in these neighborhoods one can vary points a as well as functionals f_i along with numbers n and ε. This topology is the restriction of the familiar topology of the pointwise convergence on $\mathbb{R}^F$ if points of E are regarded as functions on F.

6.6.1. Proposition. *The obtained class of sets is a Hausdorff topology, denoted by $\sigma(E,F)$ and called the weak topology generated by F. A sequence $\{x_n\}$ converges in this topology to x precisely when $f(x_n) \to f(x)$ for every $f \in F$.*

PROOF. We could refer to Example 1.6.5, but we repeat the reasoning. The indicated class contains E and the empty set and admits arbitrary unions. Let us verify that it admits finite intersections. For this it suffices to show that the intersection $V = U_{f_1,\ldots,f_n,\varepsilon}(a) \cap U_{g_1,\ldots,g_m,\delta}(b)$ belongs to $\sigma(E,F)$, i.e., every point $v \in V$ belongs to V along with a neighborhood of the form $U_{f_1,\ldots,f_n,g_1,\ldots,g_m,r}(v)$. To this end we pick $r > 0$ as follows:

$$r = \frac{1}{2} \min_{i,j} \left\{ \varepsilon - |f_i(v-a)|, \delta - |g_j(v-b)| \right\}.$$

Then we obtain $U_{f_1,\ldots,f_n,g_1,\ldots,g_m,r}(v) \subset V$. Indeed, let $x \in U_{f_1,\ldots,f_n,g_1,\ldots,g_m,r}(v)$. We have

$$|f_i(x-a)| = |f_i(x-v) + f_i(v-a)| < r + |f_i(v-a)| < \varepsilon,$$

i.e., $x \in U_{f_1,\ldots,f_n,\varepsilon}(a)$. Similarly, $x \in U_{g_1,\ldots,g_m,\delta}(b)$. The Hausdorff property of this topology follows from the fact that F separates points of E: two points $a \neq b$ can be separated by a simple neighborhood of the form $U_{f,\varepsilon}(a)$ and $U_{f,\varepsilon}(b)$.

Suppose that a sequence $\{x_n\}$ converges in the topology $\sigma(E,F)$ to x. For every $f \in F$ and every $\varepsilon > 0$, there exists a number N such that $x_n \in U_{f,\varepsilon}(x)$ whenever $n \geqslant N$. This means that $f(x_n) \to f(x)$. Conversely, if, for every functional $f \in F$, this is fulfilled, then every neighborhood $U_{f_1,\ldots,f_n,\varepsilon}(x)$ contains all elements x_n starting from some number. □

A similar assertion is true for nets.

The topology $\sigma(E,F)$ possesses the following remarkable property.

6.6.2. Theorem. *The set of all linear functions on E continuous in the topology $\sigma(E,F)$ coincides with F, i.e., one has the equality $\left(E, \sigma(E,F) \right)^* = F$.*

PROOF. All functions from F are continuous on E by our construction of the topology. Let us show that every linear function l on E continuous in the topology $\sigma(E,F)$ is an element of F. By our assumption the set $\{x : |l(x)| < 1\}$ is open and contains the origin. Hence it contains some neighborhood of zero $U_{f_1,\ldots,f_n,\varepsilon}$. This means that l vanishes on the intersection of the kernels of the functionals f_i. We show that l is a linear combination of f_i. We use induction on n. Let $l = 0$ on the set $L_1 = f_1^{-1}(0)$. If $L_1 = E$, then $l = f_1 = 0$. If there is a vector $v \notin L_1$, then $l = l(v)f_1(v)^{-1}f_1$ by the linearity of the functionals l and f_1 and the fact that E is the sum of L_1 and the linear span of v. Suppose that our assertion is true for some $n - 1 \geqslant 1$. Let us consider l on the kernel L_n of the functional f_n. By the inductive assumption there exist numbers $c_1, \ldots, c_{n-1}$ such that $l(x) = c_1 f_1(x) + \cdots + c_{n-1}f_{n-1}(x)$ for all $x \in L_n$. If $L_n = E$, then everything is proved. If there exists $v \notin L_n$, then $l = c_1 f_1 + \cdots + c_n f_n$, where we set $c_n := f_n(v)^{-1}\big(l(v) - c_1 f_1(v) - \cdots - c_{n-1}f_{n-1}(v)\big)$. $\square$

Similarly we define the topology $\sigma(F, E)$ on F. To this end the elements of E must be considered as linear functionals on F, i.e., every $x \in E$ generates the functional $f \mapsto f(x)$. According to what we have proved above, convergence of sequences in the topology $\sigma(F, E)$ is the pointwise convergence. The space of continuous linear functions on F with the topology $\sigma(F, E)$ coincides with the original space E in the following sense: every linear function l continuous in the topology $\sigma(F, E)$ has the form $f \mapsto f(x)$ for some $x \in E$.

Let us consider two important examples.

6.6.3. Example. Let X be a normed space.

(i) Letting $E = X$ and $F = X^*$, we obtain the *weak topology* $\sigma(X, X^*)$ on X and the *weak convergence*. The dual to X with the topology $\sigma(X, X^*)$ remains X^*. However, in an infinite-dimensional space X the weak topology is *always strictly weaker* than the norm topology. This is seen from the fact that every basis neighborhood of zero $\{x \in X \colon |f_1(x)| < \varepsilon, \ldots, |f_n(x)| < \varepsilon\}$, $f_i \in X^*$, contains a linear subspace $\bigcap_{i=1}^{n} f_i^{-1}(0)$, which is infinite-dimensional in the case of any infinite-dimensional X. In particular, an open ball in X cannot contain weakly open sets. It is interesting that in spite of the difference of the weak topology and the norm topology, they can possess equal supplies of convergent sequences. This is the case for $X = l^1$ (Exercise 6.10.104).

(ii) Letting $E = X^*$ and taking X for F, we obtain the so-called *weak-∗ topology* $\sigma(X^*, X)$ on X^* and the *weak-∗ convergence*. The set of linear functions continuous in the topology $\sigma(X^*, X)$ is naturally identified with X, i.e., this set in the general case is smaller than the space X^{**}, as, for example, is the case for the space $X = c_0$, where we have $X^{**} = l^\infty$.

From the results of §6.4 we obtain the following assertion.

6.6.4. Proposition. *Every weakly convergent sequence in a normed space X is norm bounded.*

If X is complete, any sequence in X^* that converges in the weak-∗ topology is norm bounded.

Of particular importance for applications is weak convergence in Hilbert spaces. Since here the dual can be identified with the space itself, the weak topology can be identified with the weak-∗ topology.

6.6.5. Example. Let H be a separable Hilbert space with an orthonormal basis $\{e_n\}$. A sequence of vectors $h_k \in H$ converges weakly to a vector h

precisely when it is norm bounded and for every n the sequence of numbers (h_k, e_n) converges to (h, e_n).

PROOF. Let $\sup_n \|h_n\| < \infty$ and $(h_k, e_n) \to (h, e_n)$ for every n. It is clear that $(h_k, x) \to (h, x)$ for every x that is a finite linear combination of vectors e_n. This gives convergence on every element $x \in H$, since for every $\varepsilon > 0$ there exists a finite linear combination z of basis vectors with $\|x - z\| \leqslant \varepsilon$, whence we have $|(h_k, x) - (h_k, z)| \leqslant \varepsilon \sup_n \|h_n\|$. □

6.6.6. Example. A sequence of functions $f_n \in C[a, b]$ converges weakly to a function $f \in C[a, b]$ in and only if $\sup_n \|f_n\| < \infty$ and $f_n(t) \to f(t)$ for every point $t \in [a, b]$. The sufficiency of this condition is obvious from the Lebesgue dominated convergence theorem and the fact that $C[a, b]^*$ is the space of bounded Borel measures on $[a, b]$. The necessity is obvious from consideration of functionals $\varphi \mapsto \varphi(t)$ and norm boundedness of weakly convergent sequences.

The weak topology in the infinite-dimensional case is not metrizable. This has many different appearances.

6.6.7. Example. Let us consider the vectors $u_n = \ln(n + 1) e_n$ in l^2, where $\{e_n\}$ is the standard basis in l^2. The point 0 belongs to the closure of $\{u_n\}$ in the weak topology, but no subsequence in $\{u_n\}$ can converge weakly, since it is not norm bounded. In order to see that every weak neighborhood of zero contains points from $\{u_n\}$, we observe that for every fixed finite collection of vectors $v_1 = (v_{1,1}, \ldots, v_{1,j}, \ldots), \ldots, v_m = (v_{m,1}, \ldots, v_{m,j}, \ldots)$ in l^2 and every $\varepsilon > 0$, there exists a number n satisfying the condition $\sum_{i=1}^m |v_{i,n}|^2 < \varepsilon |\ln(n + 1)|^{-2}$, which is obvious from divergence of the series of $|\ln(n + 1)|^{-2}$. Then we obtain the bound $|(v_1, u_n)| < \varepsilon, \ldots, |(v_m, u_n)| < \varepsilon$.

Let us note the following fact that is easily verified, but is unexpected at the first glance.

6.6.8. Theorem. *Let A be a linear mapping between normed spaces X and Y. The following conditions are equivalent:*

(i) *the mapping A is continuous;*

(ii) *the mapping A is continuous with respect to the weak topologies, i.e., as a mapping $A\colon \big(X, \sigma(X, X^*)\big) \to \big(Y, \sigma(Y, Y^*)\big)$;*

(iii) *if $x_n \to 0$ weakly, then $\{Ax_n\}$ is weakly convergent.*

PROOF. Let $\|A\| < \infty$. For a weak neighborhood of zero in Y of the form $V = \{y \in Y\colon |g_i(y)| < \varepsilon, g_1, \ldots, g_n \in Y^*\}$ we take the following weak neighborhood of zero: $U = \{x \in X\colon |f_i(x)| < \varepsilon, i = 1, \ldots, n\}$ in X, where $f_i := g_i \circ A \in X^*$. This gives the inclusion $A(U) \subset V$ and shows the weak continuity at zero, which obviously yields the continuity at all other points. It is clear that (ii) implies (iii). Suppose now that (iii) is fulfilled. Then $\|A\| < \infty$ by Corollary 6.1.3, since if $\|x_n\| \to 0$, then $x_n \to 0$ weakly, hence $\{Ax_n\}$ converges weakly by condition (iii), whence $\sup_n \|Ax_n\| < \infty$. □

It has already been noted that the weak topology is not metrizable if X is infinite-dimensional. Hence for nonlinear mappings (even with values in the real line) the weak sequential continuity can be strictly weaker than the continuity in the weak topology (see Exercise 6.10.167). However, on balls in l^2 the weak topology is metrizable (see §6.10(ii)).

We warn the reader that the established equivalence of continuity in the norm topology and the weak topology does not extend to intermediate topologies (Exercise 8.6.47).

Yet another remarkable property of the weak topology of a normed space is the fact that the supply of convex weakly closed sets is the same as in the original topology, although the supply of norm closed sets is larger than that of weakly closed sets (in the infinite-dimensional case).

6.6.9. Theorem. *A convex set V in a normed space X is closed in the weak topology precisely when it is norm closed. In addition, V is the intersection of all closed halfspaces of the form $\{x\colon l(x) \leqslant c\}$, $l \in X^*$ containing V.*

PROOF. Let V be norm closed and $u \notin V$. By Corollary 6.3.9 there exists $l \in X^*$ with $l(u) > c := \sup_{v \in V} l(v)$, i.e., $V \subset \Pi := \{x\colon l(x) \leqslant c\}$ and $u \notin \Pi$, which proves the last assertion that implies the first one at once. □

A convex set open in the norm topology can fail to have inner points in the weak topology (as, for example, an open ball). One should bear in mind that the established theorem does not extend to the weak-$*$ topology in the dual space (see Exercise 6.10.161).

6.6.10. Corollary. *Suppose that a sequence of vectors x_n in a normed space X converges weakly to $x \in X$. Then there exists a sequence of vectors v_n in the convex envelope of $\{x_n\}$ converging to x in norm.*

PROOF. The point x belongs to the closure V of the convex envelope of the sequence $\{x_n\}$ in the weak topology. It remains to observe that V is convex. Indeed, let $u, v \in V$, $t \in [0, 1]$. For every basis neighborhood of zero U in the weak topology there exist points $u_1, v_1 \in \text{conv}\{x_n\}$ with $u - u_1, v - v_1 \in U$. Then $tu_1 + (1-t)v_1 \in \text{conv}\{x_n\}$ and $tu + (1-t)v - [tu_1 + (1-t)v_1] \in U$. Since U was arbitrary, we obtain $tu + (1 - t)v \in V$, which proves the weak closedness of the set V. □

By analogy with the topology of the pointwise convergence on the space of functionals one can introduce a topology on the space of operators. Let X and Y be normed spaces. The space of operators $\mathcal{L}(X, Y)$ has, of course, the weak topology of a normed space. However, the *weak operator topology* is an even weaker topology in which the basis of neighborhoods of zero has the form

$$\{A \in \mathcal{L}(X, Y)\colon |l_i(Ax_i)| < \varepsilon, \ i = 1, \ldots, n\}, \quad x_i \in X, \ l_i \in Y^*.$$

If $X = Y$ are Hilbert spaces, then such neighborhoods are determined by "matrix elements" (Au_i, v_i). In the infinite-dimensional case, the weak operator topology is weaker than the topology $\sigma(\mathcal{L}(X), \mathcal{L}(X)^*)$, because not every continuous functional on $\mathcal{L}(X)$ is a finite linear combination of matrix elements. One more useful topology on $\mathcal{L}(X, Y)$ corresponds to the pointwise convergence of operators. It is called the *strong operator topology*. The corresponding neighborhoods of zero have the form

$$\{A \in \mathcal{L}(X, Y) \colon \|Ax_i\| < \varepsilon, \, i = 1, \ldots, n\}, \quad x_i \in X.$$

The strong and weak operator topologies are also considered in Exercises 6.10.187 and 7.10.118.

6.7. Compactness in the Weak-∗ Topology

This section contains two important results connected with compactness in the weak and weak-∗ topologies.

6.7.1. Theorem. *Let X be a separable normed space. Then every bounded sequence of linear functionals on X contains a weak-∗ convergent subsequence.*

PROOF. Let $f_n \in X^*$ and $\|f_n\| \leqslant C$. Let us take an everywhere dense countable set $\{x_k\}$ in X. Extract in $\{f_n\}$ a subsequence $\{f_{1,n}\}$ for which the sequence $\{f_{1,n}(x_1)\}$ converges. Next we extract a subsequence $\{f_{2,n}\}$ in $\{f_{1,n}\}$ for which the sequence $\{f_{2,n}(x_2)\}$ converges. Continuing by induction, we construct embedded sequences $\{f_{k,n}\}$ with $k \in \mathbb{N}$ for which the sequences $\{f_{k,n}(x_k)\}$ converge. It is clear that the sequence $\{f_{n,n}\}$ converges at every element x_k. This sequence also converges at every element $x \in X$, since for every $\varepsilon > 0$ there exists a vector x_k such that $\|x - x_k\| \leqslant \varepsilon$, which gives $|f_{n,n}(x) - f_{n,n}(x_k)| \leqslant C\varepsilon$ for all n. It is clear that the equality $f(x) = \lim\limits_{n\to\infty} f_{n,n}(x)$ defines an element of the dual space X^* with $\|f\| \leqslant C$. $\qquad \square$

The established property means the *sequential* compactness of the ball in the dual space to any separable normed space with respect to the weak-∗ topology (in §6.10(ii) we verify the metrizability of this topology on balls). This property does not follows from the usual compactness, and without the separability condition it cannot be guaranteed. For example, the sequence of functionals $f_n(x) = x_n$ on l^∞ does not contain a pointwise convergent subsequence (for every subsequence $\{f_{n_k}\}$ there is an element $x \in l^\infty$ such that $\{x_{n_k}\}$ has no limit). In the general case one has compactness in the weak-∗ topology. The proof given below employs some facts from the complementary material of Chapter 1. However, for most of applications it is enough to have the previous elementary result.

6.7.2. Theorem. (THE BANACH–ALAOGLU–BOURBAKI THEOREM) *In the space dual to a normed space, closed balls are compact in the weak-∗ topology.*

PROOF. It suffices to consider the unit ball S of the space X^* dual to a normed space X with the closed unit ball U. According to Tychonoff's theorem, the product of U copies of the closed interval $[-1, 1]$ is compact in the product topology, i.e., the space of all functions on U with values in $[-1, 1]$ is compact in the topology of pointwise convergence. Let us embed S into $K := [-1, 1]^U$ with the aid of the mapping $J(l)(x) = l(x)$, $x \in U$. It is easy to see that this mapping is a homeomorphism between the set S and its image in K. Hence it suffices to verify that $J(S)$ is closed in K. Let y be an element of K that is a limit point for $J(S)$. This means (see 1.9(i)) that there exists a net of elements $y_\alpha \in S$ such that the net $\{y_\alpha(x)\}$ converges to $y(x)$ for every $x \in U$. By the linearity of all y_α convergence holds for every $x \in X$. The function z defined by the formula $z(x) = \lim\limits_\alpha y_\alpha(x)$ is linear on X and coincides with y on U. Hence $z \in X^*$. Thus, $y = J(z)$. Of course, a similar reasoning can be given in terms of neighborhoods in the weak-$*$ topology without nets.   □

In the case of a Hilbert space H we obtain analogous assertions for the weak topology, since the canonical isomorphism between H and H^* identifies the weak topology on H with the weak-$*$ topology on H^*.

6.7.3. Theorem. *Let H be a Hilbert space. Then every bounded sequence in H contains a weakly convergent subsequence.*

In particular, the closed unit ball in H is sequentially compact in the weak topology.

PROOF. Let $\|h_n\| \leqslant C$. The closure of the linear span of $\{h_n\}$ is a separable Hilbert space. Denote it by H_0. In H_0, according to what has been proved, we can extract a weakly convergent subsequence in $\{h_n\}$. It will converge weakly also in the whole space H, since every functional $l \in H^*$ is represented by a vector $v \in H$, which can be decomposed into the sum $v = v_0 + v'$, where $v' \perp H_0$. Hence the action of l on H_0 coincides with the action of the functional generated by the vector v_0.   □

A similar assertion is true for any reflexive Banach space (Exercise 6.10.118). Here we consider the following particular case.

6.7.4. Theorem. *If $1 < p < \infty$, then every bounded sequence in $L^p(\mathbb{R}^n)$ contains a weakly convergent subsequence.*

PROOF. We know that $L^p(\mathbb{R}^n)$ can be identified with the dual to $L^q(\mathbb{R}^n)$, where $q^{-1} + p^{-1} = 1$. Under this identification the weak topology of L^p corresponds to the weak-$*$ topology of the dual space. It remains to apply Theorem 6.7.1.   □

Unlike Hilbert spaces (and some other spaces), in the general case the weak topology does not possess the property of the weak-$*$ topology established in Theorem 6.7.1. For example, the sequence of functions $x_n(t) = t^n$ in $C[0,1]$ does not contain a weakly convergent subsequence (although the sequence itself is fundamental in the weak topology). The sequence of functions $x_n(t) = \sin(\pi n t)$ does not even contain a subsequence that would be fundamental in the weak topology (i.e., the values of every continuous linear functional on it would form a Cauchy sequence of numbers). In will be shown in §6.10(iii) that balls in nonreflexive spaces are never weakly compact.

Now we can show that every Banach space is a closed linear subspace in some space $C(K)$, where K is a compact space. Any separable space can be embedded into $C[0,1]$: this fact is proved below in Theorem 6.10.24.

6.7.5. Theorem. *Every Banach space is linearly isometric to a closed linear subspace of the space $C(K)$, where K is a compact space.*

PROOF. For K it is natural to take the closed unit ball of the space X^* with the weak-$*$ topology. Now to every $x \in X$ we associate the function $\psi_x \in C(K)$ by the formula $\psi_x(f) = f(x)$, $f \in K$. Then $\sup_{f \in K} |f(x)| = \|x\|$. Thus, $x \mapsto \psi_x$ is a linear isometry. □

The next result (*Goldstine's theorem*) is yet another example of application of the Hahn–Banach theorem.

6.7.6. Theorem. *Let X be a normed space, U_X and $U_{X^{**}}$ the closed unit balls in X and X^{**}, and let $J \colon X \to X^{**}$ be the canonical embedding. Then the set $J(U_X)$ is everywhere dense in $U_{X^{**}}$ in the topology $\sigma(X^{**}, X^*)$. Hence $J(X)$ is everywhere dense in X^{**} in the topology $\sigma(X^{**}, X^*)$.*

PROOF. By Theorem 6.7.2 the ball $U_{X^{**}}$ is compact in $\sigma(X^{**}, X^*)$. Let V be the closure of $J(U_X)$ in this topology. It is also compact. If $V \neq U_{X^{**}}$, then there is $x^{**} \in U_{X^{**}} \backslash V$. By Corollary 6.3.9 applied to the weak-$*$ topology, there exists an element $l \in X^*$ such that $x^{**}(l) > \sup_{u \in U_X} Ju(l)$. Such a form of the corollary of the Hahn–Banach theorem will be proved in Corollary 8.3.8, so here we give a direct proof of the existence of l. To this end, using the $\sigma(X^{**}, X^*)$-compactness of V, we find a $\sigma(X^{**}, X^*)$-neighborhood of zero W in X^{**} of the form
$$W = \{z^{**} \in X^{**} \colon |z^{**}(l_i)| < 1, i = 1, \ldots, n\}, \quad l_i \in X^*,$$
for which $(x^{**} + W) \cap V = \varnothing$. Set $P \colon X^{**} \to \mathbb{R}^n$, $Pz^{**} = \big(z^{**}(l_1), \ldots, z^{**}(l_n)\big)$. The convex compact set $P(V)$ does not contain the point Px^{**}. Applying the Hahn–Banach theorem to the finite-dimensional space $P(X^{**})$, we obtain a linear functional f on $P(X^{**})$ with $f(Px^{**}) > \sup_{h \in P(U_X)} f(h)$. The functional $f \circ P$ is generated by an element of X^*, since it is a linear combination of $l_1, \ldots, l_n$, because it vanishes on the intersection of the kernels of l_i. Thus, the desired functional $l \in X^*$ is found. It remains to observe that $\|x^{**}\| \leqslant 1$ and $\sup_{u \in U_X} |Ju(l)| = \sup_{u \in U_X} |l(u)| = \|l\|$, so that the inequality $x^{**}(l) > \|l\|$ is impossible. □

With the aid of this theorem it is easy to prove that the reflexivity of a Banach space is equivalent to the weak compactness of its closed unit ball (see Theorem 6.10.10).

6.8. Adjoint and Selfadjoint Operators

Let X, Y be normed spaces. For every operator $T \in \mathcal{L}(X, Y)$ and every functional $y^* \in Y^*$, the function $T^* y^*$ on X given by the formula
$$\langle T^* y^*, x \rangle := \langle y^*, Tx \rangle, \quad \text{i.e.,} \quad T^* y^*(x) := y^*(Tx),$$
is linear and continuous on X. The obtained linear mapping
$$T^* \colon Y^* \to X^*$$

is called the *adjoint operator*. It is continuous and satisfies the equality

$$\|T^*\| = \|T\|.$$

Indeed,

$$\|T^*y^*\| = \sup_{\|x\| \leqslant 1} |T^*y^*(x)| = \sup_{\|x\| \leqslant 1} |y^*(Tx)| \leqslant \|T\|\|y^*\|,$$

since $\|Tx\| \leqslant \|T\|$. On the other hand, for every $\varepsilon > 0$ there exists $x \in X$ with $\|x\| = 1$ and $\|Tx\| > \|T\| - \varepsilon$. By the Hahn–Banach theorem there exists a functional $y^* \in Y^*$ with $\|y^*\| = 1$ and

$$|T^*y^*(x)| = |y^*(Tx)| = \|Tx\| > \|T\| - \varepsilon,$$

which gives $\|T^*\| \geqslant \|T\|$.

For all $A, B \in \mathcal{L}(X)$ we have

$$(A + B)^* = A^* + B^*, \quad (AB)^* = B^*A^*, \tag{6.8.1}$$

which is easy to verify.

There is the following connection between the range of the operator and the kernel of its adjoint.

6.8.1. Lemma. *Let $A \in \mathcal{L}(X, Y)$. Then*

$$\overline{A(X)} = \{y \in Y\colon f(y) = 0 \quad \forall f \in \operatorname{Ker} A^*\} = \bigcap_{f \in \operatorname{Ker} A^*} \operatorname{Ker} f.$$

PROOF. Let $y = Ax$ and $f \in \operatorname{Ker} A^*$. Then

$$f(y) = f(Ax) = (A^*f)(x) = 0.$$

Thus, $A(X)$ belongs to the right-hand side of the relation in question. Since the latter is closed, it contains the closure of $A(X)$. Conversely, suppose that a vector $y \in Y$ belongs to the right-hand side, but does not belong to $Y_1 := \overline{A(X)}$. By Corollary 6.3.9 there exists $f \in Y^*$ with $f(y) = 1$ and $f|_{Y_1} = 0$. For every $x \in X$ we have $(A^*f)(x) = f(Ax) = 0$. Hence $f \in \operatorname{Ker} A^*$. Then $f(y) = 0$ by our condition on y, which gives a contradiction. $\square$

In the case of a Hilbert space H (real or complex) for an operator $A \in \mathcal{L}(H)$ we define the adjoint operator A^* by the equality

$$(Ax, y) = (x, A^*y).$$

Since the left-hand side is continuous in x, by the Riesz theorem there is a uniquely defined vector A^*y satisfying the indicated equality. It is clear that the operator A^* is linear. The difference from the case of a Banach space is that the adjoint operator acts on the same space as the original one. This definition is consistent with the general case of a Banach space: identifying the functional $l\colon x \mapsto (x, v)$ with the vector v, we obtain $(A^*l)(x) = l(Ax) = (Ax, v) = (x, A^*v)$. Note, however, the following nuance arising in the complex case: for a Hilbert space we have $(\lambda A)^* = \overline{\lambda}A^*$, but for a Banach space $(\lambda A)^* = \lambda A^*$. Thus, in case of a complex Hilbert space the adjoint operator in the category of Hilbert spaces does not coincide with the adjoint operator in the category of Banach spaces. This is explained by the fact that the natural isometry between H^* and H is conjugate-linear, but not linear.

In the case of a Hilbert space X we have the obvious equality

$$(A^*)^* = A.$$

Hence here every bounded operator is adjoint to a bounded operator. For general Banach spaces the situation is different, which is discussed in Exercises 6.10.151 and 6.10.152.

The next definition introduces a very important class of operators on complex or real Hilbert spaces.

6.8.2. Definition. *A bounded linear operator A on a Hilbert space H is called selfadjoint if $A^* = A$, i.e., $(Ax, y) = (x, Ay)$ for all $x, y \in H$.*

Sometimes bounded selfadjoint operators are called *Hermitian operators* or symmetric operators, but in case of unbounded not everywhere defined operators one has to distinguish selfadjoint and symmetric operators (see Chapter 10), so in this chapter we do not use the term "symmetric operator".

6.8.3. Example. (i) Let P be the operator of orthogonal projection onto a closed linear subspace H_0 in a Hilbert space H (see §5.4). Then P is selfadjoint. Indeed, $(Px, y) = (Px, Py) = (x, Py)$, because $Px, Py \in H_0$, $x - Px \perp H_0$ and $y - Py \perp H_0$.

(ii) The diagonal operator from Example 6.1.5(vii) is selfadjoint precisely when all α_n are real.

Given a bounded operator A on a real Hilbert space H, one can take the complexification $H_{\mathbb{C}}$ of the space and define the complexification $A_{\mathbb{C}}$ of the operator A by the formula $A_{\mathbb{C}}(x + iy) := Ax + iAy$, $x, y \in H$. We observe that the operator A in a real space is selfadjoint precisely when the operator $A_{\mathbb{C}}$ is selfadjoint. Indeed, if $A = A^*$, then

$$\big(A_{\mathbb{C}}(x + iy), u + iv\big) = (Ax + iAy, u + iv)$$
$$= (Ax, u) + (Ay, v) + i(Ay, u) - i(Ax, v)$$
$$= (x, Au) + (y, Av) + i(y, Au) - i(x, Av) = \big(x + iy, A_{\mathbb{C}}(u + iv)\big).$$

The complexification of a selfadjoint operator on a real space is actually the direct sum of two copies of this operator. Most of the results of the spectral theory are valid for complex spaces, but in case of selfadjoint operators many facts remain in force in the real case.

In the case of a Hilbert space H, for the adjoint operator in the sense of Hilbert spaces Lemma 6.8.1 can be restated in the following way.

6.8.4. Lemma. *Let $A \in \mathcal{L}(H)$. Then*

$$\overline{A(H)} = (\operatorname{Ker} A^*)^\perp, \quad \overline{A^*(H)} = (\operatorname{Ker} A)^\perp,$$

moreover, one has the orthogonal decomposition

$$H = \overline{A(H)} \oplus \operatorname{Ker} A^* = \overline{A^*(H)} \oplus \operatorname{Ker} A.$$

If the operator A is selfadjoint, then $\overline{A(H)} \perp \operatorname{Ker} A$ and

$$H = \overline{A(H)} \oplus \operatorname{Ker} A.$$

PROOF. It is clear from Lemma 6.8.1 that the subspaces $\overline{A(H)}$ and $\operatorname{Ker} A^*$ are mutually orthogonal and $\overline{A(H)} = (\operatorname{Ker} A^*)^\perp$, which also gives the orthogonal decomposition of H. Since $A^{**} = A$, we obtain the remaining equalities. $\square$

It is seen from these equalities that the operator A maps $\overline{A^*(H)}$ one-to-one onto $A(H)$, and if the sets $A^*(H)$ and $A(H)$ are closed, then the first one is mapped one-to-one onto the second one.

With the aid of the adjoint operator one can give the following condition for the surjectivity of an operator.

6.8.5. Proposition. *Let X, Y be Banach spaces and $A \in \mathcal{L}(X, Y)$. The equality $A(X) = Y$ is equivalent to the property that for some $c > 0$ we have*

$$\|A^* y^*\|_{X^*} \geqslant c \|y^*\|_{Y^*} \quad \forall y^* \in Y^*. \tag{6.8.2}$$

PROOF. Let $A(X) = Y$ and $y^* \in Y^*$. According to Remark 6.2.4 there exists $\varepsilon > 0$ such that for every $y \in Y$ there is $x \in X$ with $Ax = y$ and $\|x\|_X \leqslant \varepsilon^{-1} \|y\|_Y$. Taking y such that $\|y\|_Y = 1$ and $|y^*(y)| \geqslant \|y^*\|_{Y^*}/2$, we obtain

$$\|A^* y^*\|_{X^*} \|x\|_X \geqslant |A^* y^*(x)| = |y^*(Ax)| = |y^*(y)| \geqslant \|y^*\|_{Y^*}/2,$$

whence $\|A^* y^*\|_{X^*} \geqslant \varepsilon \|y^*\|_{Y^*}/2$.

Conversely, suppose we have (6.8.2). In view of Lemma 6.2.1 it suffices to verify that the closure of $A\big(U_X(0,1)\big)$ contains the ball $U_Y(0,c)$. If this is false, then there exists a vector $y \in Y$ with $\|y\|_Y \leqslant c$ not belonging to the indicated closure. According to a corollary of the Hahn–Banach theorem, there is a functional $y^* \in Y^*$ such that $|y^*(y)| > 1$ and $|y^*(Ax)| \leqslant 1$ whenever $\|x\|_X \leqslant 1$. Then $|A^* y^*(x)| = |y^*(Ax)| \leqslant 1$ whenever $\|x\|_X \leqslant 1$, i.e., $\|A^* y^*\|_{X^*} \leqslant 1$. Now from (6.8.2) we have $\|y^*\|_{Y^*} \leqslant c^{-1}$. Since $\|y\|_Y \leqslant c$, we obtain $|y^*(y)| \leqslant 1$, which is a contradiction. $\square$

6.8.6. Corollary. *Let X, Y be Banach spaces and $A \in \mathcal{L}(X, Y)$. (i) The set $A(X)$ is closed if and only if the set $A^*(Y^*)$ is closed.*

(ii) The operator A maps X one-to-one onto Y if and only if A^* maps the space Y^* one-to-one onto X^*.

PROOF. (i) Let the subspace $Z := A(X)$ be closed. Denote by A_0 the operator A considered with values in Z. According to the proposition above, for some $c > 0$ we have $\|A_0^* z^*\|_{X^*} \geqslant c \|z^*\|_{Z^*}$. Suppose that a sequence $\{y_n^*\} \subset Y^*$ is such that $A^* y_n^* \to x^*$ in X^*. The restriction of y_n^* to Z will be denoted by z_n^*. We observe that $A^* y_n^* = A_0^* z_n^*$, since both functionals give $y_n(Ax)$ on any vector $x \in X$. Therefore, the functionals z_n^* converge in Z^* to some $z^* \in Z^*$. We extend z^* to Y and obtain an element $y^* \in Y^*$. Then $A^* y^* = A_0^* z^* = x^*$, which proves the closedness of $A^*(Y^*)$.

If the closedness of $A^*(Y^*)$ is given, then the subspace $A_0^*(Z^*)$ is also closed. Indeed, if $z_n^* \in Z^*$ and $A_0^* z_n^* \to x^*$, then we can extend z_n^* to functionals $y_n^* \in Y^*$.

As above, we have $A^*y_n^* = A_0^*z_n^*$. By the closedness of $A^*(Y^*)$ there exists $y^* \in Y^*$ with $A^*y_n^* \to A^*y^*$. Let $z^* := y^*|_Z$. Then $A_0^*z^* = A^*y^* = \lim_{n\to\infty} A_0^*z_n^*$. The operator A_0^* is injective, because the set $A(X)$ is dense in Z. Since the operator A_0^* has a closed range, its inverse is continuous. Hence (6.8.2) is true, which gives the equality $A(X) = Z$ and the closedness of $A(X)$.

(ii) If the operator A is an isomorphism, then obviously A^* is also an isomorphism, since for every $x^* \in X^*$ the functional $y^*(y) = x^*(A^{-1}y)$ is continuous and $A^*y^* = x^*$. Moreover, A^* has the zero kernel by Lemma 6.8.1. If A^* is an isomorphism, then by the proposition above $A(X) = Y$. In addition, A has the zero kernel. Indeed, if $Ax = 0$ and $x \neq 0$, then there is $x^* \in X^*$ with $x^*(x) = 1$, which for $y^* = (A^*)^{-1}x^*$ gives the contradictory equality $1 = x^*(x) = A^*y^*(x) = y^*(Ax) = 0$. $\qquad\square$

6.8.7. Example. (THE LAX–MILGRAM LEMMA) Let H be a real Hilbert space and let $A \in \mathcal{L}(H)$ be such that $(Ax, x) \geqslant c(x, x)$ for some $c > 0$. Then $A(H) = H$, since $\|A^*y\| \geqslant c\|y\|$ by the estimate $(A^*y, y) = (y, Ay) \geqslant c(y, y)$.

The adjoint operator can be defined for not necessarily bounded linear mappings (see Chapter 10), but such adjoint will not be defined on the whole space.

6.8.8. Example. Let A be a linear mapping from a Banach space X to a Banach space Y such that there exists a linear mapping $A^*: Y^* \to X^*$ for which $l(Ax) = (A^*l)(x)$ for all $x \in X$ and $l \in Y^*$. Then A is continuous.

In particular, if X is a Hilbert space and an everywhere defined linear mapping $A: X \to X$ is such that for all $x, y \in X$ we have $(Ax, y) = (x, Ay)$, then A is continuous.

PROOF. The graph of A is closed, because if $x_n \to x$ and $Ax_n \to y$, then $l(Ax_n) = (A^*l)(x_n) \to (A^*l)(x) = l(Ax)$ for all $l \in Y^*$, whence $l(y) = l(Ax)$, which means that $Ax = y$. $\qquad\square$

6.9. Compact Operators

In this section we begin our study of one special class of operators that is very important for applications.

6.9.1. Definition. *Let X and Y be Banach spaces. A linear operator $K: X \to Y$ is called compact if it takes the unit ball to a set with compact closure. The class of compact operators from X to Y is denoted by the symbol $\mathcal{K}(X, Y)$.*

In terms of sequences compactness of the operator K means that for every bounded sequence $\{x_n\}$ in X the sequence $\{Kx_n\}$ must contain a convergent subsequence.

It is clear from the definition that any compact operator is bounded. A similar definition can be introduced in case of not necessarily complete normed spaces, but here there is another, somewhat more general definition (equivalent to the given one in case of complete Y): the image of the unit ball is totally bounded. Such operators are called *completely bounded*.

The simplest example of a compact operator is the zero operator. Another obvious example is a bounded operator with a finite-dimensional range. It is important here that in any finite-dimensional normed space every bounded set is totally bounded. The reader should be warned: not every linear operator with a finite-dimensional range is compact, since there are unbounded finite-dimensional operators (for example, discontinuous linear functionals). The simplest example of an operator that is not compact is the identity mapping of any infinite-dimensional Banach spaces, i.e., the unit operator.

For the sequel we note some elementary properties of totally bounded sets in normed spaces.

6.9.2. Lemma. (i) *A bounded linear operator takes any totally bounded set to a totally bounded set.*

(ii) *If sets A and B in normed spaces X and Y are totally bounded, then $A \times B$ is totally bounded in $X \times Y$.*

(iii) *If sets A and B in a normed space are totally bounded, then the set $\alpha A + \beta B$ is totally bounded for all numbers α and β. If A and B are compact, then this set is compact as well.*

PROOF. The Lipschitzness of a bounded linear operator gives (i) (see Example 1.7.6). Assertion (ii) follows from Exercise 1.9.41. The first assertion in (iii) follows from (i) and (ii), since αA and βB are obviously totally bounded and the operator $(x, y) \mapsto x + y$ from $X \times X$ to X is continuous. The same reasoning gives compactness of $\alpha A + \beta B$ in case of compact A and B. $\qquad \square$

The main properties of compact operators are collected in the following theorem.

6.9.3. Theorem. *Let X, Y and Z be Banach spaces.*

(i) *The class $\mathcal{K}(X, Y)$ is a closed linear subspace in the space $\mathcal{L}(X, Y)$.*

(ii) *If $A \in \mathcal{K}(X, Y)$ and $B \in \mathcal{L}(Y, Z)$ or if $A \in \mathcal{L}(X, Y)$ and $B \in \mathcal{K}(Y, Z)$, then $BA \in \mathcal{K}(X, Z)$.*

(iii) *An operator $K \in \mathcal{L}(X, Y)$ is compact precisely when the adjoint operator $K^*\colon Y^* \to X^*$ is compact.*

PROOF. (i) Let $A, B \in \mathcal{K}(X, Y)$ and let U be the unit ball in X. Then $(A + B)(U) \subset A(U) + B(U)$, $(\lambda A)(U) = \lambda A(U)$. It remains to recall that the algebraic sum of two totally bounded sets and a homothetic image of a totally bounded set are totally bounded (Lemma 6.9.2).

Let $K_n \in \mathcal{K}(X, Y)$, $K \in \mathcal{L}(X, Y)$ and $\|K_n - K\| \to 0$. For every $\varepsilon > 0$ there exists a number N such that $\|K_n - K\| \leqslant \varepsilon$ for all $n \geqslant N$. This means that the set $K_n(U)$ is an ε-net for $K(U)$. Then a finite ε-net existing in $K_n(U)$ serves as a 2ε-net for $K(U)$.

(ii) The set $A(U)$ is totally bounded in Y, so its image under B is totally bounded in Z.

(iii) Let $K \in \mathcal{K}(X, Y)$. Let V be the unit ball in Y^*. Let us verify that the set $K^*(V)$ is totally bounded in X^*. Suppose that we are given a sequence of functionals $f_n \in V$. We have to show that the sequence of functionals $K^* f_n$ contains a subsequence uniformly converging on the unit ball U of the space X. For this we apply the Ascoli–Arzelà theorem (see Theorem 1.8.4). Since $K^* f_n(x) = f_n(Kx)$ and the set $K(U)$ has compact closure, denoted by S, we only need to observe that the functions f_n are uniformly bounded on S and uniformly Lipschitz, which follows from the bound $\|f_n\| \leqslant 1$. Thus, every sequence in $K^*(V)$ contains a convergent subsequence, which means the compactness of $K^*(V)$.

Suppose now that $K^* \in \mathcal{K}(Y^*, X^*)$. It follows from what we have proved that $K^{**} \colon X^{**} \to Y^{**}$ is a compact operator. Moreover,

$$K^{**} J_1 x = J_2 K x \quad \text{for all } x \in X,$$

where $J_1 \colon X \to X^{**}$ and $J_2 \colon Y \to Y^{**}$ are the canonical isometric embeddings. Indeed, for every $f \in Y^*$ we have

$$(K^{**} J_1 x)(f) = (J_1 x)(K^* f) = (K^* f)(x) = f(Kx) = (J_2 K x)(f).$$

By the isometry of the embedding we obtain the compactness of the closure of the set $K(U)$ in Y. $\qquad \square$

Yet another simple property of a compact operator K on any space X is the separability of $K(X)$, which follows obviously from the separability of the images of all balls of radius n (since these images are also totally bounded).

Let us give some examples of compact operators.

6.9.4. Example. (i) Let $\{\alpha_n\}$ be a bounded sequence of numbers. The diagonal operator

$$A \colon l^2 \to l^2, \quad (x_n) \mapsto (\alpha_n x_n),$$

is compact precisely when $\lim_{n \to \infty} \alpha_n = 0$.

(ii) Let $\mathcal{K} \in C([0,1]^2)$. Then the *integral operator*

$$K x(t) = \int_0^1 \mathcal{K}(t, s) x(s) \, ds$$

on the space $C[0,1]$ is compact.

(iii) Let $\mathcal{K} \in L^2([0,1]^2)$. Then the *integral operator*

$$K x(t) = \int_0^1 \mathcal{K}(t, s) x(s) \, ds$$

on the space $L^2[0,1]$ is compact. The function $\mathcal{K}$ defining the integral operator K is called the *integral kernel*.

(iv) The *Volterra operator*

$$V x(t) = \int_0^t x(s) \, ds$$

is compact as an operator from $L^1[0,1]$ to $L^p[0,1]$ with $1 \leqslant p < \infty$ and also as an operator from $L^p[0,1]$ to $C[0,1]$ with $p > 1$. However, the operator $V \colon L^1[0,1] \to C[0,1]$ is not compact.

PROOF. (i) Let $\lim_{n\to\infty} \alpha_n = 0$. Let us consider the finite-dimensional operators $K_n\colon (x_n) \mapsto (\alpha_1 x_1, \ldots, \alpha_n x_n, 0, 0, \ldots)$. We have $\|K - K_n\| \leqslant \sup_{i>n} |\alpha_i| \to 0$ as $n \to \infty$. Thus, K is a compact operator.

If $\{\alpha_n\}$ contains a subsequence $\{\alpha_{n_i}\}$ such that $|\alpha_{n_i}| \geqslant c > 0$, then the sequence of vectors $K e_{n_i} = \alpha_{n_i} e_{n_i}$, where e_n is the vector with 1 at the nth place and 0 at all other places, contains no Cauchy subsequence. Hence the operator K is not compact.

(ii) The set M of functions Kx, where $\|x\| \leqslant 1$, is totally bounded by the Ascoli–Arzelà theorem. Indeed, this set is bounded by the boundedness of $\mathcal{K}$. In addition, M is equicontinuous: by the uniform continuity of $\mathcal{K}$, for every $\varepsilon > 0$ there is $\delta > 0$ such that $|\mathcal{K}(t, s) - \mathcal{K}(t', s)| \leqslant \varepsilon$ if $|t - t'| \leqslant \delta$, so

$$|Kx(t) - Kx(t')| \leqslant \int_0^1 |\mathcal{K}(t, s) - \mathcal{K}(t', s)|\, |x(s)|\, ds \leqslant \varepsilon \quad \text{if } \|x\| \leqslant 1.$$

(iii) For every function $x \in L^2[0, 1]$ the function $\mathcal{K}(t, s)x(s)$ is integrable in s for almost all t, since by Fubini's theorem $\mathcal{K}(t, \cdot) \in \mathcal{L}^2[0, 1]$ for almost all t. Hence the function Kx is defined almost everywhere. By the Cauchy inequality

$$\left| \int_0^1 \mathcal{K}(t, s)x(s)\, ds \right|^2 \leqslant \int_0^1 |\mathcal{K}(t, s)|^2\, ds \int_0^1 |x(s)|^2\, ds,$$

which after integration in t over $[0, 1]$ gives the estimate

$$\int_0^1 |Kx(t)|^2\, dt \leqslant \|x\|^2 \int_0^1 \int_0^1 |\mathcal{K}(t, s)|^2\, ds\, dt.$$

Thus, $\|K\| \leqslant \|\mathcal{K}\|_{L^2([0,1]^2)}$. Now we take a sequence of functions $\mathcal{K}_n$ on $[0,1]^2$ of the form $\mathcal{K}_n(t, s) = \sum_{i,j \leqslant n} c_{ij} \varphi_i(t) \psi_j(s)$, where $\varphi_i, \psi_j \in L^2[0, 1]$, such that $\|\mathcal{K}_n - \mathcal{K}\|_{L^2([0,1]^2)}$. The operators K_n defined by the functions $\mathcal{K}_n$ converge in the operator norm to K due to the estimate obtained above. It remains to observe that these operators are finite-dimensional: the range of K_n is contained in the linear span of the functions $\varphi_1, \ldots, \varphi_n$.

(iv) The image of the unit ball U from $L^1[0, 1]$ is bounded in $C[0, 1]$. If $\{f_n\} \subset U$, then $V f_n = V(f_n^+) - V(f_n^-)$. The functions $V(f_n^+)$ are monotone and uniformly bounded. Hence one can extract a subsequence pointwise converging on $[0, 1]$ (see Exercise 4.5.20). Selecting yet another subsequence $\{f_{n_k}\}$ for which the monotone functions $V(f_{n_k}^-)$ converge everywhere, we obtain a uniformly bounded pointwise convergent sequence $\{V f_{n_k}\}$. By the Lebesgue dominated convergence theorem it converges in all $L^p[0, 1]$. This gives compactness of V as an operator with values in $L^p[0, 1]$. But there is no compactness of V with values in $C[0, 1]$: the sequence $V f_n$, where $f_n = n I_{[0,1/n]}$, is not equicontinuous, since it converges pointwise to the indicator function of $(0, 1]$. Finally, for $p > 1$ the operator $V\colon L^p[0, 1] \to C[0, 1]$ is compact, since in addition to the uniform boundedness of the image of the unit ball from $L^p[0, 1]$ one has the equicontinuity of this image, which follows from the estimate

$$|Vx(t) - Vx(t')| \leqslant \left| \int_{t'}^t |x(s)|\, ds \right| \leqslant |t - t'|^{1-1/p},$$

fulfilled by Hölder's inequality.   □

Quite often (but not always) for proving the compactness of an operator its approximations by finite-dimensional operators are constructed. In many concrete spaces the class of compact operators coincides with the closure of the set of finite-dimensional operators.

6.9.5. Proposition. *Let H be a Hilbert space. Then the set $\mathcal{K}(H)$ of compact operators coincides with the closure of the set of bounded finite-dimensional operators with respect to the operator norm. If H is separable and $\{e_n\}$ is an orthonormal basis, then for every $K \in \mathcal{K}(H)$ we have $\|K - P_n K\| \to 0$, where P_n is the orthogonal projection onto the linear span of the vectors $e_1, \ldots, e_n$.*

PROOF. We have to show that every operator $K \in \mathcal{K}(H)$ can be approximated by finite-dimensional ones. As noted above, the image of any compact operator is separable, so we can deal with a separable space H and prove the last assertion of the proposition. This assertion follows from the criterion of compactness in H, since the image of the ball under K is contained in a compact set S, whence $\|K - P_n K\|^2 \leqslant \sup_{y \in S} \sum_{i=n+1}^{\infty} |(y, e_i)|^2 \to 0$. □

A similar assertion is true for $C[0, 1]$ and more generally for spaces with Schauder bases (see §6.10(iv)). For a long time it was unknown whether this is true for all Banach spaces; only in 1973 did P. Enflo publish a disproving counterexample.

Let us give a simple sufficient condition for the noncompactness of an operator (in case of a Hilbert space this condition is also necessary).

6.9.6. Example. Let X and Y be Banach spaces and $A \in \mathcal{L}(X, Y)$. If $A(X)$ contains an infinite-dimensional closed subspace, then A is not compact. Indeed, if E is a closed subspace in $A(X)$ and U is the unit ball in X, then by Baire's theorem there is $n \in \mathbb{N}$ such that the closure of $A(nU) \cap E$ contains a ball from E. Since this closure is totally bounded, the space E is finite-dimensional.

Note that the image of the closed unit ball under a compact operator can fail to be closed (hence it need not be compact).

6.9.7. Example. (i) Let us take the continuous linear functional l on $C[0, 1]$ from Example 6.1.5(iii), which does not attain its maximum on the ball. The image of the closed unit ball under l is the interval $(-1, 1)$.

(ii) The image of the closed unit ball in $C[-1, 1]$ under the Volterra operator

$$Vx(t) = \int_{-1}^{t} x(s) \, ds$$

is not closed in $C[-1, 1]$, since the function $x(t) = |t|$ belongs to the closure of this image, but does not belong to the image itself.

On the other hand, there is a positive result about compactness of the image of the ball.

6.9.8. Proposition. *Let X be a reflexive Banach space (for example, a Hilbert space), let Y be a normed space, and let $K: X \to Y$ be a completely bounded operator. Then the image of every closed ball in X is compact in the space Y.*

PROOF. Let B be a closed ball and $y_n = Kx_n$, $x_n \in B$. Passing to a subsequence, we can assume that the sequence $\{y_n\}$ is Cauchy. Using the reflexivity of X, we can pass to a subsequence in $\{x_n\}$ that converges weakly to some element $x \in B$ (see Exercise 6.10.118). We assume again that this is the whole original sequence. Then the vectors $y_n = Kx_n$ converge weakly to Kx. It is readily seen (Exercise 6.10.96) that then $\|y_n - Kx\| \to 0$. $\square$

In diverse concrete spaces compact operators may have additional interesting properties. For example, the following *Daugavet theorem* holds.

6.9.9. Example. For every compact operator K on the space $C[0,1]$ one has

$$\|K + \lambda I\| = \|K\| + |\lambda| \quad \forall \lambda \in \mathbb{C}.$$

PROOF. It suffices to consider the case $\lambda = 1$. We first consider K of the form $Kx = \sum_{i=1}^{n} l_i(x)x_i$, where $l_i \in C[0,1]^*$, $x_i \in C[0,1]$. Due to the bound $\|K + \lambda I\| \leqslant \|K\| + |\lambda|$, it suffices to establish the opposite inequality. We prove that for every $\varepsilon > 0$ the inequality $\|K + \lambda I\| \geqslant \|K\| + |\lambda| - \varepsilon$ holds. Let us find a function x for which $\|x\| = 1$ and $\|Kx\| \geqslant \|K\| - \varepsilon/3$. Next we find a point $t \in [0,1]$ such that $\left|\sum_{i=1}^{n} l_i(x)x_i(t)\right| = \|Kx\|$. By the Riesz theorem the functionals l_i have the form

$$l_i(x) = \int_{[0,1]} x(s)\,\mu_i(ds),$$

where μ_i are bounded Borel measures on $[0,1]$. We replace the taken point by a point t such that $\mu_i(t) = 0$ for all i (if this was not fulfilled at once) and also $r := \left|\sum_{i=1}^{n} l_i(x)x_i(t)\right| \geqslant \|Kx\| - \varepsilon/3$. Let $\sum_{i=1}^{n} l_i(x)x_i(t) = re^{i\theta}$, $\theta \in \mathbb{R}^1$. We redefine the function x in a small neighborhood of t in order to obtain a continuous function y with $\|y\| = 1$, $y(t) = e^{i\theta}$ and $|l_i(y) - l_i(x)| \leqslant \varepsilon(3n \max_i \|x_i\|)^{-1}$, $i = 1, \ldots, n$ (we assume that not all functions x_i are zero). Then

$$\|Ky + y\| \geqslant \left|\sum_{i=1}^{n} l_i(y)x_i(t) + y(t)\right|$$

$$\geqslant \left|\sum_{i=1}^{n} l_i(x)x_i(t) + y(t)\right| - \sum_{i=1}^{n} |l_i(y) - l_i(x)|\,\|x_i\|$$

$$\geqslant |re^{i\theta} + e^{i\theta}| - \varepsilon/3 = r + 1 - \varepsilon/3 \geqslant \|Kx\| + 1 - 2\varepsilon/3$$

$$\geqslant \|K\| + 1 - \varepsilon.$$

In the general case we find a sequence of finite-dimensional operators K_n with $\|K - K_n\| \to 0$, which is possible by the compactness of K and Example 6.1.12. Then $\|K_n\| \to \|K\|$ and $\|K_n + I\| \to \|K + I\|$. $\square$

Operators on Hilbert spaces do not possess such a property (it suffices to take a diagonal operator on $\mathbb{C}^2$ with eigenvalues 0 and -1). The equality fails for the noncompact operator $-I$. Werner [711] gives a survey on the Daugavet property. Compact operators are also discussed in Chapter 7.

6.10. Complements and Exercises

(i) Operator ranges and factorization (229). (ii) Weak compactness in Banach spaces (231). (iii) The Banach–Saks property and uniform convexity (240). (iv) Bases, approximations and complements (241). (v) Operators on ordered vector spaces (247). (vi) Vector integration (253). (vii) The Daniell integral (257). (viii) Interpolation theorems (263). Exercises (264).

6.10(i). Operator ranges and factorization

In this subsection we present a number of useful results connected with the properties of the images of continuous linear mappings and with a close question about the possibility of representing one of the two given operators in the form of the composition of the second operator with some third operator. First we discuss conditions under which a given linear subspace L in a Banach space X coincides with the image of some operator $T \in \mathcal{L}(Z, X)$ on a Banach space Z. If in this situation we are allowed to take arbitrary Banach spaces Z, then it suffices to consider only injective operators T, since $T(Z) = \widetilde{T}(Z/\mathrm{Ker}\, T)$, where $\widetilde{T}$ is the factorization of T by its kernel, i.e., the operator on the quotient space taking the equivalence class $[z]$ of the element $z \in Z$ to Tz. The image $T(Z)$ of an injective operator T can be equipped with the norm

$$\|x\|_T := \|T^{-1}x\|_Z, \quad x \in T(Z).$$

Then $T(Z)$ with this norm turns out to be a Banach space the identity embedding of which into X is continuous. Indeed, we have $\|x\| \leqslant \|T\|\,\|T^{-1}x\|_Z$. If a sequence $\{x_n\} \subset T(Z)$ is Cauchy with respect to the norm above, then the sequence $\{T^{-1}x_n\}$ is Cauchy in Z and hence converges to some element $z \in Z$. Hence we obtain $x_n \to Tz$ and $\|Tz - x_n\|_T \to 0$.

6.10.1. Definition. *A linear subspace E in a Banach space X is called a continuously embedded Banach space if E is equipped with a norm $\| \cdot \|_E$ with respect to which E is complete and the identity mapping $(E, \| \cdot \|_E) \to (E, \| \cdot \|_X)$ is continuous.*

If the balls with respect to the norm $\| \cdot \|_E$ are totally bounded in X, then E is called compactly embedded.

Our discussion above leads to the following conclusion.

6.10.2. Proposition. *A linear subspace L of a Banach space X is the image of some continuous linear operator from a Banach space precisely when L can be equipped with a norm with respect to which it will be a Banach space continuously embedded into X.*

The situation will change if we impose restrictions on Z. In this case it can happen that there are no continuous operators on Z the image of which contains L. For example, there is no continuous operator from l^2 onto l^1: otherwise l^1 would be linearly homeomorphic to a quotient space of l^2, hence to a Hilbert space (and then the space $l^\infty = (l^1)^*$ would be separable). For the same reason $L^2[0, 1]$ cannot be mapped onto $C[0, 1]$ by means of a bounded operator. However, $C[0, 1]$ can be mapped onto $L^2[0, 1]$ by means of a bounded operator (Exercise 6.10.174).

Let us note the following property of operator ranges, or, which is the same, continuously embedded Banach spaces.

6.10.3. Proposition. *Let X and Y be Banach spaces and let an operator $A \in \mathcal{L}(X,Y)$ have a nonclosed range. Then the algebraic dimension of the algebraic complement of $A(X)$ in Y is uncountable.*

PROOF. Otherwise there is a finite or countable set of vectors y_n the linear span of which algebraically complements $A(X)$. Denote by E_n the linear span of $y_1, \ldots, y_n$. It is clear that $A(X) + E_n$ is the image of the continuous operator A_n acting from the Banach space $X_n := X \oplus E_n$ to Y by the formula $(x, y) \mapsto Ax + y$. By Baire's theorem for some n the image of the ball of radius n in X_n is dense in some ball in Y. According to Lemma 6.2.1 this gives the equality $A_n(X_n) = Y$. Hence $A(X)$ has a finite codimension. According to Proposition 6.2.12, the image of A is closed, which contradicts the assumption of the proposition. $\square$

Let us mention the following nontrivial result due to V. V. Shevchik, the proof of which can be found in [**704**] (see also Exercise 6.10.165 for the case of Hilbert spaces).

6.10.4. Theorem. *Let X be a separable Banach space and let $E_1 \neq X$ be a Banach space continuously embedded into X and everywhere dense in X. Then there exists a separable Banach space E_2 continuously embedded into X such that E_2 is also everywhere dense and $E_1 \cap E_2 = 0$.*

We now prove a useful result about factorization.

6.10.5. Theorem. *Let X, Y, Z be Banach spaces and let $A \colon X \to Z$ and $B \colon Y \to Z$ be continuous linear operators such that $A(X) \subset B(Y)$. If $\operatorname{Ker} B = 0$, then there exists a continuous linear operator $C \colon X \to Y$ such that $A = BC$.*

If the operator B is not injective, then there exists a continuous linear operator $C \colon X \to Y/\operatorname{Ker} B$ for which $A = \widetilde{B}C$, where $\widetilde{B} \colon Y/\operatorname{Ker} B \to Z$ is the factorization of B by the kernel.

PROOF. If $\operatorname{Ker} B = 0$, we have a well-defined linear mapping $C \colon X \to Y$, $Cx := B^{-1}Ax$. This mapping has a closed graph: if $x_n \to x$ and $Cx_n \to y$, then $Ax_n \to Ax$ and $Ax_n = BCx_n \to By$, whence $Ax = By$, i.e., $y = Cx$. Therefore, the operator C is continuous. If the kernel of B is nontrivial, then we pass to the injective operator $\widetilde{B}$. $\square$

6.10.6. Corollary. *If in the situation of the previous theorem the operator B is compact, then A is also compact.*

6.10.7. Example. Since the natural embedding of $C[0, 1]$ into $L^2[0, 1]$ is not a compact operator (it suffices to consider functions $x_n(t) = \sin(2\pi n t)$), then by the previous corollary there is no compact operator on $L^2[0, 1]$ with the range containing $C[0, 1]$.

The next result due to Banach and Mazur shows that every separable Banach space is isomorphic to some factor-space of l^1.

6.10.8. Theorem. *For every separable Banach space X, there exists an operator $A \in \mathcal{L}(l^1, X)$ with $A(l^1) = X$.*

PROOF. Let $\{x_n\}$ be a dense sequence in the unit ball of X. The operator

$$A: l^1 \to X, \quad A\xi = \sum_{n=1}^{\infty} \xi_n x_n, \; \xi = (\xi_n)$$

is well-defined, since the series converges in norm. We have $\|A\xi\| \leqslant \|\xi\|$. The surjectivity of A follows from Lemma 6.2.1, since the image of the unit ball from l^1 is dense in the unit ball from X by construction. □

Note that if X is a Banach space and $A: l^2 \to X$ is a continuous linear surjection, then X has an equivalent Hilbert norm.

Let us give a typical example of using weak convergence simultaneously in $C[a, b]$ and $L^2[a, b]$ for establishing the compactness of an operator by means of information about its range.

6.10.9. Example. Let $A: L^2[a, b] \to L^2[a, b]$ be a bounded linear operator such that $A(L^2[a, b]) \subset C[a, b]$. Then the operator A is compact. The same is true if in place of an interval with Lebesgue measure we take an arbitrary topological space T with a bounded Borel measure and replace $C[a, b]$ by $C_b(T)$.

PROOF. We show that for every sequence $\{x_n\}$ bounded in L^2 the sequence $\{Ax_n\}$ contains a subsequence converging in L^2. We know that passing to a subsequence we can assume that $\{x_n\}$ converges weakly to some element x from L^2. By Corollary 6.2.8 the operator A is continuous as an operator with values in the Banach space $C[a, b]$. Hence by weak convergence $x_n \to x$ the sequence $\{Ax_n\}$ is norm bounded and converges weakly in $C[a, b]$ to Ax. Hence for every point $t \in [a, b]$ we have $Ax_n(t) \to Ax(t)$. By the Lebesgue dominated convergence theorem we obtain convergence of Ax_n to Ax with respect to the norm of L^2, as required. It is clear that in this reasoning no special features of the interval are used, so it works for every bounded Borel measure on a topological space. □

Below in Theorem 7.10.27 we establish an even stronger property of operators on L^2 with images in C. We recall that the identity embedding $C[0, 1] \to L^2[0, 1]$ is not a compact operator.

6.10(ii). Weak compactness in Banach spaces

In a Hilbert space the weak topology coincides with the weak-* topology after identification of the space with its dual. In general Banach spaces (even in dual spaces) there is no this phenomenon. The properties of the weak topology of a Banach space can differ substantially from the properties of the weak-* topology of its dual (but in a reflexive space the weak topology can be identified with the weak-* topology of the dual to its dual). For example, the ball in the weak topology need not be compact. Since weak topologies are often used in applications, we make a short excursion in their theory, which is not included in basic courses. In particular, we prove the most important theorems about weak topologies: the Eberlein–Shmulian and Krein–Shmulian theorems.

First we recall that the closed unit ball of a Banach space is not always compact in the weak topology. We have actually encountered such example: in §6.1 we constructed a continuous linear functional on $C[0, 1]$ not attaining its maximum on the closed ball. It turns out that this example exhibits the general picture.

6.10.10. Theorem. *Let X be a Banach space. The following assertions are equivalent*:

(i) *closed balls in X are weakly compact*;

(ii) *every continuous linear functional on X attains its maximum on the closed unit ball*;

(iii) *the space X is reflexive.*

PROOF. Assertion (i) implies (ii), and (iii) implies (i) by the Banach–Alaoglu–Bourbaki theorem, since the weak topology of a reflexive space X can be identified with the weak-$*$ topology of the space X^{**}. The implication (ii)$\Rightarrow$(iii) is a very deep result due to James (see also the next theorem); it is proved, for example, in the book Diestel [**147**, Chapter 1]. We confine ourselves to the proof of the elementary implication (i)$\Rightarrow$(iii). This assertion follows from Theorem 6.7.6, since the weak compactness of the closed unit ball U_X in X gives its $\sigma(X^{**}, X^*)$-compactness and hence $\sigma(X^{**}, X^*)$-closedness in X^{**}. □

6.10.11. Corollary. (i) *A Banach space X is reflexive precisely when X^* is reflexive.*

(ii) *Closed subspaces of a reflexive Banach space are reflexive.*

PROOF. (i) If X is reflexive, then on X^* the weak topology coincides with the weak-$*$ topology, which by the Banach–Alaoglu–Bourbaki theorem gives the weak compactness of closed balls in X^*. If X^* is reflexive, then by the already proven assertion X^{**} is reflexive. Hence the ball $U_{X^{**}}$ is weakly compact. The ball U_X is closed in $U_{X^{**}}$ in the norm topology and hence is weakly closed. Hence it is weakly compact in X^{**}. Then it is weakly compact in X, since the weak topology X^{**} is stronger than the weak topology of X.

(ii) Let E be a closed subspace of a reflexive Banach space X. Then the unit ball of E is the set $E \cap U_X$, which is weakly compact in X by the weak closedness of E and the weak compactness of U_X. It remains to observe that the weak topology of E is the restriction of the weak topology of X to E. This follows from the fact that every element $l \in E^*$ is the restriction to E of some functional $\widetilde{l} \in X^*$. □

Actually James proved the following even more general fact (see [**147**, p. 19]).

6.10.12. Theorem. *If B is a weakly closed bounded set in a Banach space X, then the weak compactness of B is equivalent to the property that every continuous linear functional attains its maximum on B.*

For working with weak topologies the following Eberlein–Shmulian theorem is extremely important.

6.10.13. Theorem. *Let A be a set in a Banach space X. Then the following conditions are equivalent*:

(i) *the set A has compact closure in the weak topology;*

(ii) *every sequence in A has a subsequence weakly converging in X;*

(iii) *every infinite sequence in A has a limit point in X in the weak topology (i.e., a point every neighborhood of which contains infinitely many elements of this sequence).*

In particular, for sets in a Banach space with the weak topology compactness is equivalent to the sequential compactness and is also equivalent to the countable compactness.

PROOF. 1. First we show that any infinite sequence $\{a_n\}$ in a weakly compact set A in a Banach space X contains a weakly convergent subsequence. For this we note the following simple fact (delegated to Exercise 6.10.119): if X^* contains a countable set $\{f_n\}$ separating points in X, then $(A, \sigma(X, X^*))$ is metrizable. Let A_0 be the weak closure of $\{a_n\}$ and let E be the norm closure of the linear span of $\{a_n\}$. Then the set E is weakly closed (being convex) and hence A_0 is weakly compact in E. This enables us to pass to the separable space E. Then E^* contains a countable family of functionals separating points. Hence $(A_0, \sigma(E, E^*))$ is a metrizable compact space. Therefore, there is a subsequence in $\{a_n\}$ that converges weakly in E, hence, as is easily seen, also in X.

2. Suppose now that every infinite sequence in A has a limit point in the weak topology. Let us prove that A is contained in a weakly compact set. We need the following fact: if $Z \subset X^{**}$ is a finite-dimensional subspace, then the unit sphere S_{X^*} in X^* contains a finite set Λ such that

$$\|z^{**}\| \leqslant 2\max\{|z^{**}(l)| : \ l \in \Lambda\} \quad \forall z^{**} \in Z.$$

To this end, using the norm compactness of the unit sphere S_Z in Z, we choose a finite $1/4$-net $z_1^{**}, \ldots, z_n^{**}$ for S_Z. Next we take elements $l_i \in S_{X^*}$ with $|z_i^{**}(l_i)| > 3/4$. Then for every $z^{**} \in S_Z$ there is an element z_k^{**} with the property $\|z^{**} - z_k^{**}\| < 1/4$, which gives the relation

$$z^{**}(l_k) = z_k^{**}(l_k) + z^{**}(l_k) - z_k^{**}(l_k) > 3/4 - 1/4 = 1/2,$$

whence the desired estimate for all $z^{**} \in Z$ follows.

We observe that A is norm bounded. Otherwise we could find a functional $l \in X^*$ such that $\sup_{a \in A} |l(a)| = \infty$. This would give a sequence $\{a_n\}$ with $|l(a_n)| > n$, hence this sequence cannot have limit points in the weak topology. We shall regard A as a set in X^{**} and denote by B its closure in the topology $\sigma(X^{**}, X^*)$, i.e., the weak-$*$ topology of the space $X^{**} = (X^*)^*$. By the Banach–Alaoglu–Bourbaki theorem B is compact in the indicated topology. Our goal is to show that actually $B \subset X$. Then it will turn out that A is contained in a weakly compact set. Let $x^{**} \in B$ and $l_1 \in S_{X^*}$. The neighborhood $\{y^{**} \in X^{**} : |(x^{**} - y^{**})(l_1)| < 1\}$ contains an element $a_1 \in A$. Then

$$|(x^{**} - a_1)(l_1)| < 1.$$

Let us consider the linear span Z_1 of the vectors x^{**} and $x^{**} - a_1$. The observation made above gives elements $l_2, \ldots, l_{k_2} \in S_{X^*}$ with the property

$$\|z^{**}\| \leqslant 2\max\{|z^{**}(l_i)| : \ i = 1, \ldots, k_2\} \quad \forall z^{**} \in Z_1.$$

Now we take the weak-$*$ neighborhood of x^{**} generated by $l_1, l_2, \ldots, l_{k_2}$ and the number $1/2$ and find an element $a_2 \in A$ in this neighborhood, which gives the bounds

$$|(x^{**} - a_2)(l_i)| < 1/2, \quad i = 1, \ldots, k_2.$$

Next we take the linear span Z_2 of the vectors $x^{**}, x^{**} - a_1, x^{**} - a_2$ and with the aid of our observation find functionals $l_{k_2+1}, \ldots, l_{k_3} \in S_{X^*}$ with the property

$$\|z^{**}\| \leqslant 2 \max\{|z^{**}(l_i)| : \ i = 1, \ldots, k_3\} \quad \forall z^{**} \in Z_2.$$

Continuing this process by induction we obtain a sequence of points $a_n \in A$ and functionals $l_i \in S_{X^*}$, $k_{n-1} < i \leqslant k_n$ for which

$$|(x^{**} - a_n)(l_i)| < 1/2^n, \quad i = 1, \ldots, k_n,$$
$$\|z^{**}\| \leqslant 2 \max\{|z^{**}(l_i)| : \ i = 1, \ldots, k_{n+1}\} \quad \forall z^{**} \in Z_n,$$

where Z_n is the linear span of the elements $x^{**}, x^{**} - a_1, \ldots, x^{**} - a_n$. According to our assumption, the sequence $\{a_n\}$ has a limit point $x \in X$ in the weak topology. Since the norm closed linear span E of the sequence $\{a_n\}$ is weakly closed, we have $x \in E$. In the space X^{**} the element $x^{**} - x$ is a limit point of the sequence $x^{**}, x^{**} - a_1, x^{**} - a_2, \ldots$ in the weak-$*$ topology and hence belongs to the closure F of the linear span of this sequence in the weak-$*$ topology. By our construction

$$\|z^{**}\| \leqslant 2 \sup_i |z^{**}(l_i)| \tag{6.10.1}$$

for all z^{**} from the linear span of $x^{**}, x^{**} - a_1, x^{**} - a_2, \ldots$, which obviously extends (6.10.1) to all $z^{**} \in F$. In particular, this inequality is fulfilled for $z^{**} = x^{**} - x$. However, the construction of $\{a_n\}$ and $\{l_i\}$ and the fact that x is a weak limit point of $\{a_n\}$ imply that $|(x^{**} - x)(l_m)| = 0$ for all m. Thus, $\|x^{**} - x\| = 0$, as required. $\square$

6.10.14. Corollary. *A Banach space is reflexive if and only if every separable closed subspace in this space is reflexive.*

PROOF. We have already seen that any closed subspace of a reflexive space is reflexive. Suppose that separable subspaces in a space X are reflexive. Then every sequence $\{x_n\}$ from the unit ball is contained in a separable closed subspace Y, which is reflexive by assumption, hence its closed ball is weakly compact. By the Eberlein–Shmulian theorem $\{x_n\}$ contains a subsequence weakly converging in Y. Then this subsequence also converges weakly in X, so the ball in X is sequentially compact. Applying the Eberlein–Shmulian theorem again, we conclude that the ball in X is weakly compact. An alternative justification follows from the James theorem. $\square$

The next theorem due to Krein and Shmulian is a deep analog of the already known fact (see Proposition 5.5.4) about compactness of the closed convex envelope of a compact set in a Banach space.

6.10.15. Theorem. *Suppose that a set A in a Banach space X is compact in the weak topology. Then the closed convex envelope of A (the intersection of all closed convex sets containing A) is also compact in the weak topology (we recall* *that the closed convex envelope in the norm topology coincides with the closed convex envelope in the weak topology, see Theorem 6.6.9).*

PROOF. We apply Theorem 6.10.12, although there are other proofs. Let V be the closed convex envelope of A with respect to the norm and let $f \in X^*$. By the compactness of A there exists a point $a \in A$ with $f(a) = \sup_{x \in A} f(x)$. We observe that $a \in V$ and $\sup_{x \in V} f(x) = \sup_{x \in A} f(x)$, so the cited theorem applies.   □

The Eberlein–Shmulian theorem is very useful for establishing conditions for the weak compactness and weak convergence in concrete spaces. Let us mention a number of typical results, the proofs of which can be found in §4.7(iv) and §4.7(v) in Chapter 4 of [73].

6.10.16. Theorem. *Let μ be a finite measure on a measurable space $(\Omega, \mathcal{A})$ and let $\mathcal{F}$ be some set of μ-integrable functions. Then the set $\mathcal{F}$ is uniformly integrable precisely when it has compact closure in the weak topology of $L^1(\mu)$.*

6.10.17. Corollary. *Suppose that $\{f_n\}$ is a uniformly integrable sequence on a space with a finite measure μ. Then there exists a subsequence $\{f_{n_k}\}$ that converges in the weak topology of $L^1(\mu)$ to some function $f \in L^1(\mu)$.*

6.10.18. Corollary. *Let μ be a bounded nonnegative measure and let M be a norm bounded set in $L^1(\mu)$. The closure of M in the weak topology is compact precisely when for every sequence of μ-measurable sets A_n with $A_{n+1} \subset A_n$ and $\bigcap_{n=1}^{\infty} A_n = \varnothing$ one has*

$$\lim_{n \to \infty} \sup_{f \in M} \int_{A_n} |f| \, d\mu = 0.$$

Let now $(\Omega, \mathcal{A})$ be a measurable space and let $\mathcal{M}(\Omega, \mathcal{A})$ be the Banach space of all real countably additive measures of bounded variation with its natural variation norm $\mu \mapsto \|\mu\|$.

6.10.19. Theorem. *For every set $M \subset \mathcal{M}(\Omega, \mathcal{A})$ the following conditions are equivalent.*

(i) *The closure of M in the topology $\sigma(\mathcal{M}, \mathcal{M}^*)$ is compact.*

(ii) *The set M is bounded in variation and there exists a nonnegative measure $\nu \in \mathcal{M}(\Omega, \mathcal{A})$ (a probability measure if $M \neq \{0\}$) such that the family M is uniformly ν-continuous, i.e., for every $\varepsilon > 0$ there exists $\delta > 0$ with the property that*

$$|\mu(A)| \leqslant \varepsilon \quad \text{for all } \mu \in M \text{ whenever } A \in \mathcal{A} \text{ and } \nu(A) \leqslant \delta.$$

In this case the measures from M are absolutely continuous with respect to ν, the closure of the set $\{d\mu/d\nu: \mu \in M\}$ is compact in the weak topology of $L^1(\nu)$, and for ν one can choose the measure $\sum_{n=1}^{\infty} c_n |\mu_n|$ for some finite or countable collection $\{\mu_n\} \subset M$ and numbers $c_n > 0$.

(iii) *Every sequence in M contains a subsequence converging on every set from $\mathcal{A}$.*

6.10.20. Corollary. *A sequence of measures $\mu_n \in \mathcal{M}(\Omega, \mathcal{A})$ converges in the topology $\sigma(\mathcal{M}, \mathcal{M}^*)$ if and only if it converges on every set from $\mathcal{A}$. An equivalent condition*:

$$\lim_{n \to \infty} \int_\Omega f(\omega)\, \mu_n(d\omega) = \int_X f(\omega)\, \mu(d\omega) \qquad (6.10.2)$$

for every bounded $\mathcal{A}$-measurable function f.

Let us consider an application to passage to the limit in the integral.

6.10.21. Corollary. *Suppose that a sequence of measures $\mu_n \in \mathcal{M}(\Omega, \mathcal{A})$ converges to a measure μ on every set from $\mathcal{A}$ and a sequence of $\mathcal{A}$-measurable functions f_n is uniformly bounded and $\lim_{n \to \infty} f_n(\omega) = f(\omega)$ for every ω. Then*

$$\lim_{n \to \infty} \int_\Omega f_n(\omega)\, \mu_n(d\omega) = \int_X f(\omega)\, \mu(d\omega).$$

PROOF. By Theorem 6.10.19 there exists a probability measure ν on $\mathcal{A}$ such that μ_n and μ are uniformly ν-continuous. Let $|f_n(\omega)| \leqslant C$, $\|\mu_n\| \leqslant C$ and $\varepsilon > 0$. Let us find $\delta > 0$ such that the bound $\nu(A) < \delta$ yields that $|\mu|(A) < \varepsilon$ and $|\mu_n|(A) < \varepsilon$ for all n. By the Egorov theorem there exists a set A with $\nu(A) > 1 - \delta$ on which convergence $f_n \to f$ is uniform. Hence there is $N \in \mathbb{N}$ such that $|f_n(\omega) - f(\omega)| \leqslant \varepsilon$ if $\omega \in A$ and $n \geqslant N$. It remains to observe that

$$\left| \int_\Omega (f_n - f)\, d\mu_n \right| \leqslant \int_A |f_n - f|\, d|\mu_n| + \int_{\Omega \setminus A} |f_n - f|\, d|\mu_n| \leqslant C\varepsilon + 2C\varepsilon$$

and that the integrals of f with respect to the measures μ_n converge to the integral of f with respect to the measure μ. □

In the case where, for example, $\Omega = [0, 1]$ and $\mathcal{A}$ is the Borel σ-algebra, the space $\mathcal{M}(\Omega, \mathcal{A})$ coincides with the dual to $C[0, 1]$ by the Riesz theorem, so on $\mathcal{M}(\Omega, \mathcal{A})$ there is also the weak-$*$ topology. It should not be confused with the weak topology $\sigma(\mathcal{M}, \mathcal{M}^*)$. Convergence of measures μ_n to μ in the weak-$*$ topology means the equality (6.10.2) for every continuous function f, while convergence in the topology $\sigma(\mathcal{M}, \mathcal{M}^*)$ is (6.10.2) for all bounded Borel functions f. For example, the Dirac measures $\delta_{1/n}$ at the points $1/n$ tend to the Dirac measure δ_0 at zero in the weak-$*$ topology of the space of measures, but not in the weak topology of the Banach space $C[0, 1]^*$. This circumstance becomes especially important, because in many applications (in particular, in probability theory and in the theory of random processes) the weak-$*$ topology of the space of measures is traditionally called merely the weak topology (see, for example, [73, Chapter 8]). Corollary 6.10.21 is false for the weak-$*$ convergence of measure. It suffices to take the measures $\delta_{1/n}$ and the functions f_n defined as follows: $0 \leqslant f_n \leqslant 1$, $f_n(1/n) = 1$, $f_n(t) = 0$ if $|t - 1/n| > 1/(2n)$.

Let us give several additional results and remarks in connection with the weak and weak-$*$ topologies.

6.10.22. Proposition. *Let E be a normed space.*

(i) *The space E with the weak topology is metrizable precisely when E is finite-dimensional.*

(ii) *The unit ball U_E of the space E with the weak topology is metrizable precisely when E^* is separable.*

PROOF. (i) If the topology $\sigma(E, E^*)$ is metrizable by a metric d and E is infinite-dimensional, then the ball of radius n^{-1} centered at zero contains a weakly open set, which is unbounded and hence contains vectors x_n with $\|x_n\| \geqslant n$. Then $d(x_n, 0) \to 0$, but $\{x_n\}$ cannot converge weakly to zero being unbounded, which is a contradiction.

(ii) If E^* is separable, then one can take a countable everywhere dense set $\{f_n\}$ in the unit ball of E^*. It is readily seen that the metric

$$d(x, y) := \sum_{n=1}^{\infty} 2^{-n} |f_n(x - y)|$$

generates the weak topology on the ball U_E. Conversely, suppose that the ball U_E is metrizable in the weak topology. Let us take a countable collection of neighborhoods of zero $U_{f_{n,1}, \ldots, f_{n,k_n}}$ the intersections of which with the ball U_E give a basis of neighborhoods of zero in U_E. Let us show that the linear span of $f_{n,i}$ is dense in E^*. Let Y be the closure of this linear span. If $f \in E^* \backslash Y$, then by the Hahn–Banach theorem there exists an element $x^{**} \in E^{**}$ with $\|x^{**}\| = 1$, $x^{**}|_Y = 0$ and $x^{**}(f) = c > 0$. The weak neighborhood of zero $V := \{x \in U_E \colon |f(x)| < c/2\}$ in U_E contains some set $U_{f_{n,1}, \ldots, f_{n,k_n}} \cap U_E$. By Theorem 6.7.6 there is a vector $x \in U_E$ for which $|x^{**}(f) - f(x)| < c/2$ and $|x^{**}(f_{n,i}) - f_{n,i}(x)| < 1$ for each $i = 1, \ldots, k_n$. Since $x^{**}(f_{n,i}) = 0$, we have $|f_{n,i}(x)| < 1$, i.e., $x \in U_{f_{n,1}, \ldots, f_{n,k_n}} \cap U_E$. However, $|f(x)| > c/2$, because $x^{**}(f) = c$. Therefore, $x \notin V$, which is a contradiction. □

The dual to an infinite-dimensional Banach space cannot be metrizable in the weak-∗ topology, but the dual to an incomplete infinite-dimensional normed space can be metrizable in the weak-∗ topology (see Exercise 6.10.148).

6.10.23. Theorem. *Let X be a separable normed space. Then the closed unit ball U^* of the space X^* with the weak-∗ topology is metrizable and compact. For a metric generating the weak-∗ topology on the ball one can take*

$$d(f, g) := \sum_{n=1}^{\infty} 2^{-n} |f(x_n) - g(x_n)|,$$

where $\{x_n\}$ is a sequence dense in the unit ball from X. In particular, U^* contains a countable set separating points of X.

PROOF. This assertion is a special case of the assertion from Exercise 1.9.68, but it can be derived directly from the facts proved above. To this end we first observe that from Theorem 6.7.1 one can easily deduce the compactness of the ball in X^* with the metric d (the fact that d is a metric is obvious). The identity mapping from the ball with the metric d to the ball with the weak-∗ topology is continuous, since convergence $f_n \to f$ in the metric d on the ball yields the weak-∗ convergence. Hence by Theorem 1.7.9 this mapping is a homeomorphism. Of course, it is not difficult to verify directly the continuity of the identity mapping from the weak-∗ topology to the metric. □

The next remarkable theorem due to Banach and Mazur establishes some universality of the space $C[0, 1]$ in the category of separable Banach spaces.

6.10.24. Theorem. *Every separable Banach space is linearly isometric to some closed linear subspace in* $C[0, 1]$.

PROOF. In case of separable X the space K from the proof of Theorem 6.7.5 (the unit ball in X^* with the weak-* topology) is a compact metric space, as shown above. According to Exercise 1.9.67, the compact space K is homeomorphic to a compact set in $[0, 1]^\infty$. Hence we can assume that $K \subset [0, 1]^\infty$. Proposition 1.9.24 gives a continuous surjection π of the Cantor set in $[0, 1]$, denoted here by K_1, onto $[0, 1]^\infty$. Let $K_0 := \pi^{-1}(K)$. The mapping $f \mapsto f \circ \pi$ is a linear isometry of the space $C(K)$ to $C(K_0)$, since

$$\sup_{t \in K_0} \left| f(\pi(t)) \right| = \sup_{x \in [0,1]^\infty} |f(x)|.$$

It remains to embed $C(K_0)$ into $C[0, 1]$ by a linear isometry. For this we extend every function from $C(K_0)$ to a continuous function on $[0, 1]$ by a linear interpolation on every interval constituting the complement of K_0 in $[0, 1]$. □

Thus, separable normed spaces can be thought as linear subspaces in $C[0, 1]$.

6.10.25. Remark. From the Banach–Steinhaus theorem we know that if a sequence of continuous linear functionals f_n on a Banach space X is such that for every $x \in X$ the sequence $\{f_n(x)\}$ converges, then there is an element $f \in X^*$ to which $\{f_n\}$ converges in the weak-* topology. Thus, the space X^* is sequentially complete in the weak-* topology. One can ask an analogous question about the weak topology in X. Suppose that a sequence of vectors $x_n \in X$ is such that for every $f \in X^*$ the sequence $\{f_n(x)\}$ converges. Is it true that $\{x_n\}$ converges weakly to some vector $x \in X$? Generally speaking, this is false. For example, in the space c_0 the sequence of vectors $x_n = (1, \ldots, 1, 0, 0, \ldots)$, where 1 is at the first n positions, has no weak limit, but for every element $f \in c_0^* = l^1$ the sequence $\{f(x_n)\}$ converges. A space X in which convergence of $\{f(x_n)\}$ for every $f \in X^*$ yields weak convergence of $\{x_n\}$ is called *weakly sequentially complete*. It follows from what has been said at the beginning of this remark that every reflexive Banach space is weakly sequentially complete, since its weak topology can be identified with the weak-* topology of this space regarded as the dual to its dual. For example, a Hilbert space is weakly sequentially complete. There exist non-reflexive spaces with this property: for example, l^1 and $L^1(\mu)$.

Let us give without proof (which can be found in [**75**, Theorem 3.8.15] or [**533**, Chapter IV, §6.4]) the following deep result due to Krein and Shmulian.

6.10.26. Theorem. *Let X be a Banach space and let $V \subset X^*$ be a convex set. If the intersection of V with every closed ball of radius n centered at zero is closed in the topology $\sigma(X^*, X)$, then V is also closed in the topology $\sigma(X^*, X)$.*

If X is separable, then for the closedness of V in the topology $\sigma(X^*, X)$ is suffices that V contain the limits of all its weak- convergent sequences.*

See Exercise 8.6.74 in Chapter 8 about some topology on X^* connected with this theorem.

6.10.27. Corollary. *Let X be a Banach space and let F be a linear function on X^*. The following conditions are equivalent:*

(i) *the function F is continuous in the topology $\sigma(X^*, X)$;*

(ii) *there exists $x \in X$ with $F(l) = l(x)$ for all $l \in X^*$;*

(iii) *the restriction of F to the unit ball U_{X^*} in X^* is continuous in the topology $\sigma(X^*, X)$;*

(iv) *the set $F^{-1}(0) \cap U_{X^*}$ is closed in the topology $\sigma(X^*, X)$.*

Finally, if X is separable, then this is also equivalent to the property that $\lim_{n \to \infty} F(l_n) = 0$ for every sequence $\{l_n\} \subset X^*$ that is weak-$*$ convergent to zero.

A Banach space X is called a space with the *Dunford–Pettis property* if convergence $x_n \to 0$ in the weak topology of X and convergence $l_n \to 0$ in the topology $\sigma(X^*, X^{**})$ in the space X^* implies convergence $l_n(x_n) \to 0$. For example, an infinite-dimensional Hilbert space does not possess this property: it suffices to take for $x_n = l_n$ an orthonormal sequence. The space c_0 has the Dunford–Pettis property, since weak convergence in $l^1 = (c_0)^*$ yields convergence in norm (Exercise 6.10.104). Let us give a less obvious example.

6.10.28. Example. The space $C[0, 1]$ possesses the Dunford–Pettis property.

PROOF. Since $C[0, 1]^*$ is the space of bounded Borel measures on $[0, 1]$, we can apply Corollary 6.10.21. □

One more example of a space with the Dunford–Pettis property is $L^1[0, 1]$ (Exercise 6.10.178).

One should bear in mind that although the weak topology of an infinite-dimensional Banach space is always strictly weaker than its norm topology (Example 6.6.3), it can happen that the supplies of convergent sequences in the weak topology and the norm topology coincide. This is the case in l^1 (Exercise 6.10.104). The situation is different with the weak-$*$ convergence, as the following Josefson–Nissenzweig theorem shows (its proof can be found in Diestel [**148**, Chapter XII]).

6.10.29. Theorem. *If X is an infinite-dimensional Banach space, then there exists a sequence of functionals $f_n \in X^*$ with $\|f_n\| = 1$ that is weak-$*$ convergent to zero. Moreover, every functional $f \in X^*$ with $\|f\| \leqslant 1$ is the limit of some sequence of elements $f_n \in X^*$ of unit norm converging in the weak-$*$ topology.*

Let us mention a couple of interesting results connected with l^1. Proofs and references can be found in Albiac, Kalton [**9**]. The first result is due to H. Rosental.

6.10.30. Theorem. *Let $\{x_n\}$ be a bounded sequence in an infinite-dimensional Banach space X. Then it contains either a weakly fundamental subsequence or a subsequence $\{x_{n_k}\}$ such that the mapping $T\colon l^1 \to X$, $(\xi_k) \mapsto \sum_{k=1}^{\infty} \xi_k x_{n_k}$, is a homeomorphism from l^1 onto the closed subspace generated by $\{x_{n_k}\}$.*

The next result is due to E. Odell and H. Rosental.

6.10.31. Theorem. *A separable Banach space X has no closed subspaces isomorphic to l^1 precisely when every element $x^{**} \in X^{**}$ is the limit of some sequence $\{x_n\} \subset X$ in the topology $\sigma(X^{**}, X^*)$.*

All concrete infinite-dimensional Banach spaces we have encountered so far possess the property that they contain subspaces isomorphic to some of the simplest spaces l^p with $1 \leqslant p < \infty$ or c_0. For a long time it was an open question whether there exist infinite-dimensional spaces not containing l^p and c_0. Finally, in 1974 such example was constructed by a Soviet mathematician B. S. Tsirelson.

6.10(iii). The Banach–Saks property and uniform convexity

We shall say that a Banach space X possesses the *Banach–Saks property* if every norm bounded sequence $\{x_n\}$ in X contains a subsequence $\{x_{n_k}\}$ such that the sequence of arithmetic means

$$\frac{x_{n_1} + \cdots + x_{n_k}}{k}$$

converges in norm.

A normed space E with a norm $\| \cdot \|$ is called *uniformly convex* if for every $\varepsilon > 0$ there exists $\delta > 0$ such that

$$\text{if} \quad \|x\| = 1, \ \|y\| = 1 \quad \text{and} \quad \left\|\frac{x+y}{2}\right\| \geqslant 1 - \delta, \quad \text{then} \quad \|x - y\| \leqslant \varepsilon.$$

The spaces $L^p(\mu)$ with $1 < p < \infty$ are uniformly convex (for a proof, see §4.7(iii) in [**73**]).

6.10.32. Theorem. *All uniformly convex Banach spaces possess the Banach–Saks property.*

For a proof, see Diestel [**147**, Chapter 3, §7]. For example, the space $L^p(\mu)$ with $1 < p < \infty$ possesses the Banach–Saks property. The validity of this property for a Hilbert space can be easily verified directly.

6.10.33. Example. Hilbert spaces possess the Banach–Saks property.

PROOF. Passing to a subsequence we can assume that $\{x_n\}$ converges weakly to some vector x. In addition, we can assume that $x = 0$. Set $n_1 = 1$. Since $(x_{n_1}, x_n) \to 0$, there exists a number $n_2 > n_1$ with $|(x_{n_1}, x_{n_2})| \leqslant 1$. If numbers $n_1 < n_2 < \cdots < n_k$ are already chosen, we find a number $n_{k+1} > n_k$ such that

$$|(x_{n_j}, x_{n_{k+1}})| \leqslant k^{-1}, \quad j = 1, \ldots, k.$$

This is obviously possible by weak convergence of $\{x_n\}$ to zero. We observe that $\sup_n \|x_n\| = M < \infty$. Hence

$$\frac{\|x_{n_1} + \cdots + x_{n_k}\|^2}{k^2} \leqslant \frac{kM^2 + 2 \cdot 1 + \cdots + 2(k-1)(k-1)^{-1}}{k^2} \leqslant \frac{M^2 + 2}{k},$$

which shows norm convergence of the arithmetic means. $\qquad \square$

The space $L^1[0, 1]$ does not have the Banach–Saks property, which is obvious from consideration of the functions $nI_{[0,1/n]}$.

6.10.34. Theorem. *Any space with the Banach–Saks property is reflexive.*

PROOF. We show that every continuous linear functional f on a space E with the Banach–Saks property attains its maximum on the closed unit ball U. Let us find $u_n \in U$ such that $f(u_n) \to \|f\|$. Passing to a subsequence, we can assume that the elements $s_n := (u_1 + \cdots + u_n)/n$ converge in norm to some $u \in U$. It is clear that $f(s_n) \to \|f\|$. Hence $f(u) = \|f\|$. $\square$

Any uniformly convex space X possesses the following property: if $x_n \to x$ weakly and $\|x_n\| \to \|x\|$, then $\|x - x_n\| \to 0$. Indeed, we can assume that $\|x_n\| = \|x\| = 1$. Then $\|x_n + x\|/2 \to 1$, since if $\|x + x_n\|/2 \leqslant q < 1$, then $|l(x + x_n)/2| \leqslant q$ whenever $\|l\| \leqslant 1$. Hence $|l(x)| \leqslant q$, which yields that $\|x\| \leqslant q$, a contradiction. However, the indicated property is weaker than the uniform convexity. In this connection we mention the following theorem due to Kadec and Klee (see Diestel [147, Chapter IV, §4]).

6.10.35. Theorem. *Let X be a Banach space with the separable dual X^*. Then X possesses an equivalent norm that is Fréchet differentiable outside the origin and generates the norm $\| \cdot \|_*$ on X^* with the following property: if $f_n \to f$ in the weak-$*$ topology and $\|f_n\|_* \to \|f\|_*$, then $\|f_n - f\|_* \to 0$.*

The proof of the following interesting result can be read in Fabian, Habala, Hájek, Montesinos Santalucía, Pelant, Zizler [185, p. 259]; see also Exercise 6.10.172 for the Hilbert case.

6.10.36. Theorem. *Suppose that the norm of a Banach space X is Fréchet differentiable outside the origin. Then every bounded closed convex set in X is some intersection of closed balls. In particular, this is true for Hilbert spaces.*

6.10(iv). Bases, approximations and complements

As we have seen above, the most important attribute of Hilbert spaces are orthogonal bases. Many Banach space possess topological bases.

6.10.37. Definition. *Let X be a separable Banach space. A sequence $\{h_n\}$ in X is called a Schauder basis or a topological basis if for every $x \in X$ there is a unique sequence of numbers $\{c_n(x)\}$ such that $x = \sum_{n=1}^{\infty} c_n(x) h_n$, where the series converges in norm.*

It is clear that a Schauder basis is a linearly independent set. Note that in an infinite-dimensional Banach space a topological basis cannot be an algebraic basis (a Hamel basis), since the latter is always uncountable.

By the uniqueness of expansions the functionals $l_n \colon x \mapsto c_n(x)$ are linear. It turns out that they are automatically continuous! Note that $l_i(h_j) = \delta_{ij}$ also by the uniqueness of expansions.

6.10.38. Proposition. *All functionals l_i are continuous. Hence the finite-dimensional mappings $P_n \colon x \mapsto \sum_{i=1}^{n} l_i(x) h_i$ are continuous. In addition, the norm $\|x\|_\infty := \sup_n \|P_n x\|$ is equivalent to the original norm.*

PROOF. Since $\|P_n x\| \to \|x\|$, we have $\|x\| \leqslant \|x\|_\infty$. Let us show that X is complete with the norm $\| \cdot \|_\infty$. Then by the Banach inverse mapping theorem we obtain the equivalence of both norms. This will give the boundedness of all mappings P_n with respect to the original norm and the estimate $\sup_n \|P_n\| < \infty$. This yields the continuity of all l_n, since $P_n x - P_{n-1} x = l_n(x) h_n$. Suppose that a sequence $\{x_j\}$ is Cauchy with respect to the new norm. Then it is Cauchy with respect to the original norm and hence converges in it to some $x \in X$. We have to show that $\|x - x_j\|_\infty \to 0$. We observe that for every fixed n the sequence of vectors $P_n x_j$ is also Cauchy with respect to the new norm and hence converges in the original norm to some vector $y_n \in X$. The whole sequence $\{P_n x_j\}$ is contained in the finite-dimensional space X_n generated by $h_1, \ldots, h_n$. On finite-dimensional subspaces the functionals l_i are continuous, so for every $i = 1, \ldots, n$ there exists a limit

$$l_i(y_n) = \lim_{j \to \infty} l_i(P_n x_j) = \lim_{j \to \infty} l_i(x_j) =: c_i,$$

independent of n by the second equality. Let us verify the equality $x = \sum_{i=1}^{\infty} c_i h_i$ with respect to the original norm. For a given number $\varepsilon > 0$ we find n such that $\|x_n - x_m\|_\infty \leqslant \varepsilon$ for all $m \geqslant n$. Let us take k_0 such that $\|x_n - P_k x_n\| \leqslant \varepsilon$ for all $k \geqslant k_0$. For such k we have

$$\|y_k - x\| = \lim_{m \to \infty} \|P_k x_m - x_m\|$$

$$\leqslant \limsup_{m \to \infty} \Big[\|P_k x_m - P_k x_n\| + \|P_k x_n - x_n\| + \|x_n - x_m\| \Big]$$

$$\leqslant \limsup_{m \to \infty} \Big[\|x_m - x_n\|_\infty + \varepsilon + \|x_n - x_m\|_\infty \Big] \leqslant 3\varepsilon.$$

Hence $\|y_k - x\| \to 0$. By the uniqueness of expansions we have the equality $y_k = P_k x$. Therefore,

$$\|x_n - x\|_\infty = \sup_{k \geqslant 1} \|P_k x_n - P_k x\| \leqslant \limsup_{m \to \infty} \sup_{k \geqslant 1} \|P_k x_n - P_k x_m\|$$

$$= \limsup_{m \to \infty} \|x_n - x_m\|_\infty \to 0$$

as $n \to \infty$. Thus, X is complete with the new norm. $\qquad \square$

With respect to the new norm $\| \cdot \|_\infty$ the projections P_n have unit norms (which is not always true for the original norm).

Schauder bases were constructed in many concrete Banach spaces. For example, the Haar functions χ_n defined by the formulas $\chi_1(t) = 1$,

$$\chi_{2^k+l}(t) = \begin{cases} 1 & \text{if } t \in [(2l-2)2^{-k-1}, (2l-1)2^{-k-1}], \\ -1 & \text{if } t \in ((2l-1)2^{-k-1}, (2l)2^{-k-1}], \\ 0 & \text{else}, \end{cases}$$

where $k = 0, 1, 2, \ldots$ and $l = 1, 2, \ldots, 2^k$, form a Schauder basis in $L^p[0,1]$ if $1 \leqslant p < \infty$. The Faber–Schauder functions

$$\varphi_0(t) = 1, \quad \varphi_n(t) = \int_0^t \chi_n(s) \, ds, \quad n \geqslant 1,$$

form a Schauder basis in $C[0,1]$ (these functions were introduced in 1910 by H. Faber, who discovered that they form a basis). The function $\varphi_0 = 1$ is constant, the function $\varphi_1(t) = t$ is linear, and the subsequent functions φ_n have graphs that are equilateral triangles of height 1 and bases of the form $[2^{-k}, 2^{1-k}]$, vanishing outside these bases. The partial sum with number n of the expansion of f with respect to the functions φ_i is the result of the linear interpolation between the values of f at the points $0, 2^{1-n}, 2^{2-n}, \ldots, 1$ for $n \geqslant 1$. For example, we have $S_0 f(t) = c_0 \varphi_0(t)$, where $c_0 = f(0)$, $S_1 f(t) = c_0 \varphi_0(t) + c_1 \varphi_1(t)$, where $c_1 = f(1) - f(0)$, next, $S_2 f(t) = c_0 \varphi_0(t) + c_1 \varphi_1(t) + c_2 \varphi_2$, where $c_2 = f(1/2) - f(0)/2 - f(1)/2$ and so on. From this one can readily derive the uniform convergence of $S_n f$ to f.

In some spaces the attempts to construct a basis failed for a long time. For example, in 1974 S.V. Bochkarev solved the problem posed by Banach and constructed a Schauder basis in the space of functions analytic in the unit disc and continuous on the closed disc (equipped with the sup-norm). The problem of existence of a Schauder basis in every separable Banach space remained open for several decades. This was one of the most famous problems in the theory of Banach spaces. Finally, in 1973 a Swedish mathematician P. Enflo published his celebrated counterexample. Simultaneously he solved negatively another very difficult old problem about the existence of a separable Banach spaces without the approximation property.

A Banach space X is said to possesses the *approximation property* if for every compact set $K \subset X$ and every $\varepsilon > 0$ there exists a continuous finite-dimensional operator T such that $\|x - Tx\| < \varepsilon$ for all $x \in K$. It is known that this is equivalent to the property that, for every Banach space Z, the set of finite-dimensional operators is dense with respect to the operator norm in the space $\mathcal{K}(Z, X)$ of compact operators. If X has a Schauder basis $\{h_n\}$, then the projections $x \mapsto \sum_{i=1}^n x_i h_i$ converge to the identity operator uniformly on compact sets, hence X has the approximation property. However, there exist spaces with the approximation property, but without Schauder bases. It is now known that spaces without the approximation property (hence without Schauder bases) exist even among closed subspaces of c_0 and l^p with $p > 2$ (they certainly also exist among closed subspaces of the universal space $C[0,1]$).

If a linearly independent sequence $\{h_n\}$ is a Schauder basis of the closure of its linear span, then it is called a *basic sequence*. This is equivalent to the property that for some $C > 0$ for all $n < m$ and all scalars $\alpha_1, \ldots, \alpha_m$ we have $\left\|\sum_{i=1}^n \alpha_i h_i\right\| \leqslant C \left\|\sum_{i=1}^m \alpha_i h_i\right\|$. Indeed, if $\{h_n\}$ is a Schauder basis of the closure Y of its linear span, then the indicated estimate with $C = 1$ is true for the equivalent norm $\|\cdot\|_\infty$. Conversely, suppose that this estimate holds. Then, for any convergent series $\sum_{i=1}^\infty \alpha_i h_i$ with the zero sum it follows that all α_i are zero. If this series converges to some x, then we set $x_i := \alpha_i$. It remains to observe that the set of all vectors x representable in this form is a closed subspace. This is easily seen from the estimate $\left\|\sum_{i=1}^n \alpha_i h_i\right\| \leqslant C\|x\|$ following from our condition as $m \to \infty$. Hence this closed subspace is Y.

Basic sequences exist in all spaces. The next fact was already known in Banach's time.

6.10.39. Theorem. *In every infinite-dimensional Banach space there is an infinite basic sequence.*

PROOF. We employ the following fact: let Y be a finite-dimensional subspace of an infinite-dimensional Banach space X. Then for every $\varepsilon > 0$ there exists a vector $h \in X$ with $\|h\| = 1$ such that $\|y\| \leqslant (1 + \varepsilon)\|y + \lambda h\|$ for all $y \in Y$ and $\lambda \in \mathbb{R}^1$. Since we have $y/\lambda \in Y$, this inequality can be restated as follows: $\|y\| \leqslant (1 + \varepsilon)\|y + h\|$ for all $y \in Y$, or, equivalently, $(1 + \varepsilon)\|y + h\| \geqslant 1$ for all vectors $y \in Y$ of unit norm. For the proof we assume that $\varepsilon \in (0, 1)$ and $y_1, \ldots, y_m$ is an $(\varepsilon/2)$-net in the unit sphere of Y. Pick functionals $f_i \in X^*$ with $\|f_i\| = 1$ and $f_i(y_i) = 1$. Since X is infinite-dimensional, there is a vector h of unit norm such that $f_i(h) = 0$ for each i. Let $y \in Y$ and $\|y\| = 1$. There is y_i with $\|y - y_i\| < \varepsilon/2$. Then for every number λ we have

$$\|y + \lambda h\| \geqslant \|y_i + \lambda h\| - \frac{\varepsilon}{2} \geqslant f_i(y_i + \lambda h) - \frac{\varepsilon}{2} = 1 - \frac{\varepsilon}{2} \geqslant (1 + \varepsilon)^{-1},$$

as required.

We now take $\varepsilon > 0$ and numbers $\varepsilon_n > 0$ such that $\prod_{n=1}^{\infty}(1 + \varepsilon_n) < 1 + \varepsilon$. Let $\|h_1\| = 1$. By induction we obtain vectors h_n with

$$\|y\| \leqslant (1 + \varepsilon_n)\|y + \lambda h_{n+1}\|$$

for all y from the linear span of $h_1, \ldots, h_n$ and all $\lambda \in \mathbb{R}^1$. It is readily seen that $\{h_n\}$ is a basic sequence and $\|P_n\| < 1 + \varepsilon$ for all n. $\square$

Not every linearly independent sequence with a dense linear span in a Banach space is a Schauder basis.

6.10.40. Example. The functions $1, \sin nt, \cos nt$ do not form a Schauder basis in the space $C_{2\pi}$ of 2π-periodic functions, which is clear from the existence of a function in $C_{2\pi}$ with the Fourier series divergent at zero. The functions $x_n(t) = t^n$ do not form a Schauder basis in $C[0, 1]$ and $L^2[0, 1]$. For the proof we observe that the expansion $x = \sum_{n=1}^{\infty} c_n(x)x_n$ in $C[0, 1]$ or in $L^2[0, 1]$ ensures the real analyticity of the function x on $[0, 1)$, since we have convergence of the series of the integrals, i.e., the series $\sum_{n=1}^{\infty} c_n(x)(n + 1)^{-1}$, which yields convergence of the power series $\sum_{n=1}^{\infty} c_n(x)t^n$ if $|t| < 1$.

Nevertheless, one can construct a Schauder basis in $C[0, 1]$ consisting of polynomials. As was shown by H. Faber back in 1914, in $C[0, 1]$ there is no Schauder basis consisting of polynomials h_n of degree n. In 1990 A. A. Privalov proved that if polynomials h_n form a Schauder basis in $C[0, 1]$, then for some $\varepsilon > 0$ and for all sufficiently large n one has $\deg h_n \geqslant (1 + \varepsilon)n$. In addition, for every $\varepsilon > 0$ there is a Schauder basis $\{h_n\}$ with $\deg h_n \leqslant (1 + \varepsilon)n$. It was shown by M. A. Skopina in 2001 that such polynomials can be chosen even orthogonal (for references, see Odinets, Yakubson [461]).

The next result due to Krein–Milman–Rutman shows some stability of bases.

6.10.41. Proposition. *Let $\{u_n\}$ be a Schauder basis in a Banach space X, $\|u_n\| = 1$ and $\sup_n \|P_n\| = K$, where P_n is the projection on the linear span of $u_1, \ldots, u_n$. If a sequence $\{v_n\} \subset X$ is such that $\sum_{n=1}^{\infty} \|u_n - v_n\| < (2K)^{-1}$, then $\{v_n\}$ is also a Schauder basis.*

PROOF. For any $x = \sum_{n=1}^{\infty} x_n u_n$, let us set $Tx = \sum_{n=1}^{\infty} x_n v_n$. Since $|x_n| = \|P_n x - P_{n-1} x\| \leqslant 2K\|x\|$, the series converges by convergence of the series of $x_n u_n$ and $x_n(v_n - u_n)$, moreover,

$$\|x - Tx\| \leqslant \sum_{n=1}^{\infty} |x_n| \, \|u_n - v_n\| \leqslant q\|x\|, \quad q = 2K \sum_{n=1}^{\infty} \|u_n - v_n\| < 1.$$

Hence $\|I - T\| < 1$. So T is invertible (see Theorem 7.1.3), whence our assertion follows. $\qquad \square$

6.10.42. Corollary. *If a Banach space has a Schauder basis, then such a basis can be picked in every everywhere dense set.*

In some problems the lack of a Schauder basis is compensated by biorthogonal systems and Markushevich bases.

6.10.43. Definition. *Let X be a nonzero Banach space. A pair of sequences $\{x_n\} \subset X$ and $\{l_n\} \subset X^*$ is called biorthogonal if $l_i(x_j) = \delta_{ij}$.*

If, in addition, the linear span of $\{x_n\}$ is dense in X and the functionals l_n separate points in X, then $\{x_n\}$ is called a Markushevich basis in X.

Note that if the pair of sequences $\{x_n\} \subset X$ and $\{l_n\} \subset X^*$ is biorthogonal, then the sequence $\{x_n\}$ is *minimal* in the following sense: no element x_n belongs to the closure of the linear span of the remaining elements x_k, $k \neq n$ (otherwise we obtain $l_n(x_n) = 0$). For $X = 0$ zero is a basis.

The next result was obtained by A. I. Markushevich.

6.10.44. Theorem. *Every separable Banach space has a Markushevich basis. Moreover, such a basis can be found in every dense linear subspace.*

PROOF. Let us embed the given space X as a closed subspace in $C[0,1]$ and take a linearly independent sequence $\{y_n\}$ whose linear span is dense in X. Let $\{x_n\}$ be the result of othogonalization of $\{y_n\}$ in $L^2[0,1]$. Set $l_n(x) = (x, x_n)_{L^2}$, $x \in X$. If $l_n(x) = 0$ for some $x \in X$ for all n, then the element x is orthogonal to the closed linear span of $\{x_n\}$ in $L^2[0,1]$ and hence is orthogonal to X in $L^2[0,1]$. Therefore, $(x, x)_{L^2} = 0$. Thus, $x = 0$. $\qquad \square$

A simple example of a Markushevich basis that is not a Schauder basis is the system of functions $\exp(int)$, $n \in \mathbb{Z}$, in the complex space $C[0, 2\pi]$ of continuous functions with $x(0) = x(2\pi)$ equipped with the sup-norm. In the general case it can happen that $\sup_n \|x_n\| \, \|l_n\| = \infty$. However, it is known that for every $\varepsilon > 0$ one can find a Markushevich basis with $\sup_n \|x_n\| \, \|l_n\| \leqslant 1 + \varepsilon$. The question whether this is true for $\varepsilon = 0$ remains open.

In finite-dimensional spaces there are biorthogonal systems (*Auerbach systems*) with the following property.

6.10.45. Proposition. *In a Banach space X of finite dimension n one can find vectors $x_1, \ldots, x_n$ and linear functionals $l_1, \ldots, l_n$ such that $\|x_i\| = \|l_i\| = 1$ and $l_i(x_j) = \delta_{ij}$.*

PROOF. We can assume that $X = \mathbb{R}^n$ with some norm. Let B be the closed unit ball in this norm. For every collection of vectors $y_1, \ldots, y_n \in B$ we denote by $V(y_1, \ldots, y_n)$ the determinant of the matrix $(y_i, y_j)_{i,j \leqslant n}$, where $y_i = (y_{i,1}, \ldots, y_{i,n})$. The function V attains its maximum on B at some collection of vectors $x_1, \ldots, x_n \in B$. It is clear that $\|x_1\| = \cdots = \|x_n\| = 1$. Set

$$l_i(x) := V(x_1, x_2, \ldots, x_{i-1}, x, x_{i+1}, \ldots, x_n)/V(x_1, \ldots, x_n).$$

Then $l_i(x_i) = 1$, $|l_i(x)| \leqslant 1$ for all $x \in B$, i.e., $\|l_i\| = 1$. Finally, $l_i(x_j) = 0$ if $i \neq j$, since $V(y_1, \ldots, y_n) = 0$ if there are two equal vectors among y_i. $\qquad \square$

Yet another important geometric property in Banach spaces is connected with the existence of bounded projections. We shall say that a closed subspace E of a Banach space X is *complemented* in X if there exists a closed subspace $D \subset X$ such that $X = E \oplus D$. This gives a bounded operator $P \colon X \to E$ with $P(X) = E$ and $Px = x$ for all $x \in E$, i.e., a bounded projection onto E. Conversely, the existence of such projection gives a complement to E in the form $D = P^{-1}(0)$. As we know, in a Hilbert space every closed subspace has the orthogonal complement and hence is complemented. It turns out that there are no other spaces with such a property: J. Lindenstrauss and L. Zafriri proved that if in a Banach space X every closed subspace is complemented, then X is linearly homeomorphic to a Hilbert space. Let us give an interesting concrete example of a subspace that is not complemented.

6.10.46. Example. The space $C[0,1]$ has a closed subspace E linearly isometric to $L^2[0,1]$. This subspace is not complemented.

PROOF. We know that $C[0,1]$ possesses the Dunford–Pettis property (Example 6.10.28), but E does not. Let us show that if E were complemented, then it would also have the Dunford–Pettis property. Let P be a bounded projection onto E. If $x_n \to 0$ in the weak topology of E, then the same is true also for the weak topology of $C[0,1]$. Let $l_n \in E^*$ be such that $l_n \to 0$ in the topology $\sigma(E^*, E^{**})$. Set $f_n = l_n \circ P$. Then $f_n \in C[0,1]^*$. Let $F \in C[0,1]^{**}$. Set $G(l) := F(l \circ P)$, $l \in E^{**}$. Clearly, $G \in E^{**}$. Then $F(f_n) = G(l_n) \to 0$ by our assumption. Therefore, $f_n(x_n) \to 0$, i.e., $l_n(x_n) \to 0$, which gives the Dunford–Pettis property in E. Thus, we have obtained a contradiction. $\qquad \square$

The space c_0 also has no complement in l^∞ (Exercise 6.10.117).

Let us prove an interesting result due to A. Sobszyk on extensions of pointwise converging sequences of functionals.

6.10.47. Proposition. *Let X be a separable normed space, let $Y \subset X$ be a linear subspace, and let $\{f_n\} \subset Y^*$ be such that $\lim_{n \to \infty} f_n(y) = 0$ for all $y \in Y$*

and $\|f_n\| \leqslant 1$. *Then there exist functionals* $\widetilde{f}_n \in X^*$ *with* $\widetilde{f}_n|_Y = f_n$, $\|\widetilde{f}_n\| \leqslant 2$ *and* $\lim\limits_{n\to\infty} \widetilde{f}_n(x) = 0$ *for all* $x \in X$.

PROOF. Let $\{x_i\}$ be a countable everywhere dense set in the unit ball of X with $\{x_i\} \cap Y = \varnothing$. It suffices to construct extensions with norms at most 2 pointwise converging to zero on every vector x_i.

This can be easily done if we prove the following fact: for every fixed k and every $\varepsilon > 0$ there exists $N(k, \varepsilon)$ such that for all $n \geqslant N(k, \varepsilon)$ the functionals f_n have extensions $g_n \in X^*$ such that $\|g_n\| \leqslant 2$ and $|g_n(x_i)| \leqslant \varepsilon$, $i = 1, \ldots, k$. Let L denote the linear span of Y and $x_1, \ldots, x_k$. If the aforementioned fact is false, then $\{f_n\}$ contains a subsequence $\{f_{n_j}\}$ for which all f_{n_j} have no extensions with the indicated properties. We can assume that this is the whole original sequence. We know that f_n has an extension $\varphi_n \in X^*$ with $\|\varphi_n\| \leqslant 1$. The sequence of vectors $\big(g_n(x_1), \ldots, g_n(x_k)\big) \in \mathbb{R}^k$ is bounded and hence has a convergent subsequence. Again we can assume that the whole sequence converges. Then the limit $l(x) := \lim\limits_{n\to\infty} g_n(x)$ exists for all $x \in L$. It is clear that l is a linear functional on L and $\|l\| \leqslant 1$. In addition, $l|_Y = 0$. We extend l to a functional $l_0 \in X^*$ with $\|l_0\| \leqslant 1$. Set $\psi_n := f_n - l_0$. Then $\psi_n|_Y = f_n$, $\|\psi_n\| \leqslant 2$ and for all sufficiently large n we have $|\psi_n(x_i)| \leqslant \varepsilon$, $i = 1, \ldots, k$, which contradicts our supposition and completes the proof. $\qquad\square$

This result can be restated as follows.

6.10.48. Corollary. *Let X be a separable normed space, let $Y \subset X$ be a linear subspace, and let $T \colon Y \to c_0$ be a continuous operator. Then T extends to an operator $S \colon X \to c_0$ such that $\|S\| \leqslant 2\|T\|$.*

Therefore, c_0 is complemented in every separable Banach space isometrically containing it.

The separability is important here: as we have already noted, c_0 is not complemented in l^∞ (Exercise 6.10.117). Up to an isomorphism c_0 is the unique separable Banach space that is complemented in every separable Banach space into which it is embedded as a closed subspace (this result is due to M. Zippin, see [293]).

6.10(v). Operators on ordered vector spaces

In the theory of ordered vector spaces, briefly touched upon in §5.6(iii), an important role is played by positive functionals and operators. Let E be an ordered vector space. We say that a linear functional f on E is *positive* (and write $f \geqslant 0$) if $f(x) \geqslant 0$ whenever $x \geqslant 0$. This is equivalent to the property that $f(x) \leqslant f(y)$ whenever $x \leqslant y$. Similarly one defines positive operators between ordered spaces.

For positive functionals there are some analogs of the Hahn–Banach theorem.

6.10.49. Theorem. *Let p be a positively homogeneous convex functional on an ordered vector space E. If on a linear subspace $E_0 \subset E$ we are given a linear functional f satisfying the condition*

$$f(x) \leqslant p(x + z) \quad \text{for all } x \in E_0 \text{ and all } z \geqslant 0, \tag{6.10.3}$$

then f extends to a linear functional on all of E satisfying the indicated condition for all $x \in E$.

PROOF. Set $q(x) = \inf\{p(x + z): z \geqslant 0\}$, $x \in E$. Since

$$0 = f(0) \leqslant p(0 + z) = p(z) \quad \text{if } z \geqslant 0,$$

for any $z \geqslant 0$ we have $p(x+z)+p(-x) \geqslant p(z) \geqslant 0$, i.e., $q(x) \leqslant -p(-x) > -\infty$. It is readily seen that $q(\lambda x) = \lambda q(x)$ for all $x \in E$, $\lambda \geqslant 0$. The inequality $q(x + y) \leqslant q(x) + q(y)$ follows from the fact that for every $\varepsilon > 0$ there exist $z_1, z_2 \geqslant 0$ such that $q(x) > p(x + z_1) - \varepsilon$ and $q(y) > p(y + z_2) - \varepsilon$. Indeed, the latter gives

$$q(x)+q(y) > p(x+z_1)+p(y+z_2)-2\varepsilon \geqslant p(x+y+z_1+z_2)-2\varepsilon \geqslant q(x+y)-2\varepsilon,$$

since $z_1 + z_2 \geqslant 0$. By assumption $f \leqslant q$ on E_0. It remains to apply the usual Hahn–Banach theorem to f and q.   □

6.10.50. Corollary. *Condition (6.10.3) is necessary and sufficient for the existence of a linear extension $\widetilde{f}$ of the function f to E such that $\widetilde{f} \geqslant 0$ and $\widetilde{f} \leqslant p$ on E. In particular, a linear functional f on a linear subspace E_0 has a positive linear extension to E precisely when there exists a positively homogeneous convex function p satisfying condition (6.10.3).*

PROOF. If this condition is fulfilled, then the extension constructed in the theorem is nonnegative, since for all $z \geqslant 0$ we have

$$-\widetilde{f}(z) = \widetilde{f}(-z) \leqslant p(-z) = 0.$$

Conversely, if such an extension $\widetilde{f}$ exists, then we have

$$f(x) = \widetilde{f}(x) \leqslant \widetilde{f}(x) + \widetilde{f}(z) = \widetilde{f}(x + z) \leqslant p(x + z) \quad \text{for all } x \in E_0 \text{ and } z \geqslant 0,$$

i.e., (6.10.3) is fulfilled. Finally, if f extends to a positive linear functional $\widetilde{f}$ on E, then (6.10.3) is fulfilled with $p = \widetilde{f}$.   □

Let us consider some simple examples of positive linear functionals without positive extensions.

6.10.51. Example. (i) Let E be the space of all bounded real functions on the real line with its natural partial order and let E_0 be the linear span of indicator functions of bounded intervals. The functional f on E_0 defined as the Riemann integral is linear and positive. If it had an extension to a positive linear functional $\widetilde{f}$ on E, then we would have $\alpha := \widetilde{f}(1) \in \mathbb{R}^1$. Since $I_J \leqslant 1$ for every interval J, we obtain $\widetilde{f}(1 - I_J) \geqslant 0$, whence $\widetilde{f}(I_J) \leqslant \alpha$, which is impossible if the length of J is greater than α.

(ii) Let $E = \mathbb{R}^\infty$ be the space of all real sequences with the partial order defined by the coordinate-wise comparison and let $E = c$ be the subspace of all sequences with a finite limit. Then the functional $f(x) := \lim_{n\to\infty} x_n$ has no positive linear extensions to E. Indeed, if such an extension $\widetilde{f}$ exists, then the element x with $x_n = n$ is mapped to some number $\alpha \in [0, +\infty)$. Let us take a natural number $k > \alpha$ and an element y with $y_n = k$ for all $n \geqslant k$ and $y_j = 0$ if $j < k$. Then $k = \widetilde{f}(y) \leqslant \widetilde{f}(x) = \alpha$, which is a contradiction.

Let us give two positive results.

6.10.52. Corollary. *Let f be a positive linear functional on a linear subspace $E_0 \subset E$. Suppose that for every $x \in E$ there exists $y \in E_0$ such that $x \leqslant y$. Then f extends to a positive linear functional on E. In particular, this is true if there is a point in the algebraic kernel of the positive cone of E contained in E_0.*

PROOF. Set $U := \{x \in E_0 \colon f(x) < 1\}$ and

$$W := \{w \in E \colon w \leqslant u \text{ for some } u \in U\}.$$

We observe that the set W is convex and $0 \in W$. In addition, for every $x \in E$ there exists $\varepsilon > 0$ such that $\varepsilon x \in W$. Indeed, by condition there exists $y \in E_0$ with $x \leqslant y$. One can take $\varepsilon > 0$ such that $f(\varepsilon y) < 1$. Then $\varepsilon x \in W$. Let us take for p the Minkowski functional of the set W. It follows from what has been said above that p is a positively homogeneous convex function. Let us verify condition (6.10.3). If it fails, then there exist elements $x \in E_0$ and $z \geqslant 0$ such that $f(x) > 1$ and $p(x + z) < 1$. Then for some $\lambda \in (0, 1)$ we have $(x + z)/\lambda \in W$, i.e., $(x + z)/\lambda \leqslant u$, where $u \in U$. Hence $x + z \leqslant \lambda u$ for some $x \leqslant \lambda u$. Hence $f(x) \leqslant \lambda f(u) < 1$, which is a contradiction. Applying Theorem 6.10.49 we complete the proof of the first assertion. If the positive cone has a nonempty algebraic kernel intersecting E_0, then we take a point $z_0 \in E_0$ in this kernel. Then for every $x \in E$ there exists $\alpha > 0$ with $z_0 - \alpha x \geqslant 0$, i.e., $x \leqslant z_0/\alpha \in E_0$. □

6.10.53. Corollary. *Let z_0 belong to the algebraic kernel of the positive cone K. Then the existence of a positive linear functional f with $f(z_0) = 1$ is equivalent to the condition $-z_0 \notin K$.*

PROOF. Let $-z_0 \notin K$. On the one-dimensional space generated by z_0 we define f by the formula $f(\lambda z_0) = \lambda$. If $\lambda z_0 \in K$, then by condition we have $\lambda \geqslant 0$ and hence $f(\lambda z_0) \geqslant 0$. Now we can apply the previous corollary. If f exists, then in case $-z_0 \in K$ we would obtain $f(-z_0) \geqslant 0$, i.e., $f(z_0) \leqslant 0$ contrary to the equality $f(z_0) = 1$. □

Note that if an ordered vector space E is equipped with a norm (or a topology making it a topological vector space) and the positive cone has inner points in the corresponding topology, then every positive linear functional is automatically continuous, being bounded from below on an nonempty open set.

6.10.54. Example. On the spaces $L^\infty(\mu)$ and $C_b(T)$, where T is a topological space, all positive linear functionals are continuous.

A simple example of a discontinuous positive linear functional on an ordered normed space is the function $f(\varphi) = \varphi(0)$ on the space of all polynomials on $[0, 1]$ (or on the space of all continuous functions) equipped with the norm from $L^2[0, 1]$.

However, the existence of inner points in the positive cone is not necessary for the automatic continuity of all positive linear functionals on a given ordered normed space. For the definition of a Banach lattice, see §5.6(iii).

6.10.55. Proposition. *Every positive linear functional f on a Banach lattice E is automatically continuous. In addition,*

$$\|f\| = \sup\{|f(x)|\colon\ x \geqslant 0,\ \|x\| = 1\}.$$

In particular, this is true for $L^p(\mu)$, $1 \leqslant p \leqslant \infty$. Here, in the case, for example, of Lebesgue measure μ the positive cone in $L^p(\mu)$ with $p < \infty$ has no inner points.

A similar assertion is true for positive operators on E with values in a normed lattice.

PROOF. Let f be a discontinuous positive functional. Then there exists a sequence of elements $x_n \in E$ such that $\|x_n\| \leqslant 2^{-n}$ and $f(x_n) \geqslant n$. We can assume that $x_n \geqslant 0$, since we have $|f(x)| \leqslant |f(|x|)|$. Let $x = \sum_{n=1}^{\infty} x_n$. Then $f(x) \geqslant f(x_n) \geqslant n$ for all n, which is impossible. The assertion about $\|f\|$ follows from the estimate $|f(x)| \leqslant |f(|x|)|$. The case of a positive operator is completely analogous. If the measure μ is not concentrated at finitely many atoms, for example, is Lebesgue measure on an interval, then every neighborhood of every function from $L^p(\mu)$ with $p < \infty$ contains a function that is strictly negative on a set of a nonzero measure. Hence the positive cone has no inner points. $\qquad\square$

It follows from this proposition that a positive functional $\varphi \mapsto \varphi(0)$ on the linear subspace of continuous functions in the space $L^2[0,1]$ has no positive extensions to $L^2[0,1]$.

Let us mention the Kantorovich theorem on extensions of positive operators, a particular case of which is Corollary 6.10.52 (for a proof, see, e.g., [14, p. 29]).

6.10.56. Theorem. *Let X be an ordered vector space with a positive cone K, let $X_0 \subset X$ be a linear subspace such that $X_0 + K = X$, and let Y be a complete vector lattice. Then every positive linear operator $T\colon X_0 \to Y$ extends to a positive linear operator on all of X.*

6.10.57. Proposition. *Let E be an ordered vector space, let f be a linear functional on E, and let p be a positively homogeneous convex functional. Suppose also that on a linear subspace $E_0 \subset E$ we are given a linear functional g. Then the following conditions are equivalent:*

(i) *there is a linear extension $\widetilde{g}$ of the functional g to E such that*

$$f(z) \leqslant \widetilde{g}(z) \quad \text{for all } z \geqslant 0 \text{ and} \quad \widetilde{g}(x) \leqslant p(x) \quad \text{for all } x \in E; \qquad (6.10.4)$$

(ii) *there is a linear extension $\widetilde{g}$ of the functional g to E such that*

$$\widetilde{g}(x) + f(z) \leqslant p(x + z) \quad \text{for all } z \geqslant 0 \text{ and } x \in E; \qquad (6.10.5)$$

(iii) *the functional g satisfies the condition*

$$g(y) + f(z) \leqslant p(y + z) \quad \text{for all } z \geqslant 0 \text{ and } y \in E_0. \qquad (6.10.6)$$

PROOF. Condition (i) implies (ii), because we have

$$p(x + z) \geqslant \widetilde{g}(x + z) = \widetilde{g}(x) + \widetilde{g}(z) \geqslant \widetilde{g}(x) + f(z).$$

Condition (ii) implies (iii). Let (iii) be fulfilled. Set $p_1(x) := p(x) - f(x)$ if $x \in E$ and $g_1(y) := g(y) - f(y)$ if $y \in E_0$. Then by (6.10.5) we obtain

$g_1(y) \leqslant p_1(y + z)$ if $y \in E_0$, $z \geqslant 0$. By Theorem 6.10.49 there is a linear extension $\widetilde{g}_1$ of the functional g_1 to E with $\widetilde{g}_1(x) \leqslant p_1(x + z)$ for all $x \in E$, $z \geqslant 0$. Set $\widetilde{g} := \widetilde{g}_1 + f$. Then $\widetilde{g}_1$ is a linear extension of g. For all $z \geqslant 0$ and $x \in E$ we have $\widetilde{g}(x) + f(z) = \widetilde{g}_1(x) + f(x) + f(z) \leqslant p(x + z)$. For $x = -z$ we obtain $f(z) \leqslant g(z)$ and for $z = 0$ we have $\widetilde{g}(x) \leqslant p(x)$. $\qquad\square$

6.10.58. Corollary. *Let E be an ordered vector space, let f be a linear functional on E, and let p be a positively homogeneous convex functional. Then a necessary and sufficient condition for the existence of a linear functional g on E such that $g(x) \leqslant p(x)$ for $x \in E$ and $f(z) \leqslant g(z)$ for all $z \geqslant 0$ is this: $f(z) \leqslant p(z)$ for all vectors $z \geqslant 0$.*

PROOF. The necessity is clear. For the proof of sufficiency we take $E_0 = \{0\}$ with $g = 0$ on E_0. Then condition (6.10.6) is fulfilled, so g extends linearly to E such that (6.10.4) is fulfilled. $\qquad\square$

We now turn to ordered normed spaces. Suppose that an ordered vector space E is equipped with a norm $\|\cdot\|$ such that for some $c > 0$ we have

$$\|x\| \leqslant c\|z\| \quad \text{whenever } 0 \leqslant x \leqslant z. \qquad (6.10.7)$$

Then the positive cone is called *normal*. For example, in the Banach spaces $B(\Omega)$ and $L^p(\mu)$ (more generally, in all Banach lattices) this condition is fulfilled, but in the space $C^1[0, 1]$ of continuously differentiable functions with its natural norm and the pointwise partial order this condition fails. First we note the following fact which can be easily derived from Theorem 6.10.49.

6.10.59. Proposition. *If E_0 is a Banach sublattice of a Banach lattice E, then every positive functional on E_0 extends to a positive functional on E with the same norm.*

Let us discuss decompositions of functionals in differences of positive functionals.

6.10.60. Proposition. *Let E be an ordered normed space with a normal positive cone satisfying condition (6.10.7). Then for every continuous linear functional f there exist positive continuous functionals f_1 and f_2 such that*

$$f = f_1 - f_2, \quad \|f_1\| \leqslant c\|f\|, \quad \|f_2\| \leqslant (1 + c)\|f\|.$$

Conversely, if this is true for every $f \in E^*$, then the positive cone is normal and (6.10.7) holds with $2(1 + c)$ in place of c.

PROOF. Let $\|f\| = 1$, $U := \{x \in E : \|x\| < 1/c\}$, $W := U - K$, where $K = \{z \in E : z \geqslant 0\}$. It is readily seen that W is an open convex set and $0 \in W$. For all $z \in K \cap W$ we have $\|z\| \leqslant 1$, since $0 \leqslant z = u - z_1$, where $\|u\| < 1/c$ and $z_1 \geqslant 0$, whence $0 \leqslant z \leqslant u$. By (6.10.7) this gives $\|z\| \leqslant c\|u\| \leqslant 1$. Denote by p the Minkowski functional of the set W. We observe that

$$f(z) \leqslant p(z) \text{ and } p(-z) < -1 \text{ for all } z \in K.$$

Indeed, setting $a := p(z)$, for each $\varepsilon > 0$ we have $p\big(z/(a + \varepsilon)\big) < 1$, hence $(a + \varepsilon)^{-1}z \in W \cap K$. Then $\|z/(a + \varepsilon)\| < 1$ as shown above. Hence we have

$f(z) \leqslant \|z\| \leqslant a$. Next, $-z \leqslant 0 \in U$, whence $-z \in W$ and hence $p(-z) < 1$. The previous corollary gives a linear functional $g \leqslant p$ such that $f(z) \leqslant g(z)$ for all $z \in K$. The functional g is positive, since for all $z \in K$ and $\lambda > 0$ we have $g(-\lambda z) \leqslant p(-\lambda z) < 1$, whence $g(z) \geqslant -1/\lambda$. Finally, if $\|x\| < 1$, we obtain $x/c \in U \subset W$, which gives $g(x/c) \leqslant p(x/c) \leqslant 1$. Thus, $\|g\| \leqslant c$. Now we can let $f_1 := g$, $f_2 := g - f$.

Suppose now that for every $f \in E^*$ there exist positive functionals f_1 and f_2 with $f = f_1 - f_2$ and $\|f_1\| \leqslant (1+c)\|f\|$, $\|f_2\| \leqslant (1+c)\|f\|$. Let $0 \leqslant x \leqslant z$. A corollary of the Hahn–Banach theorem gives $f \in E^*$ with $f(x) = \|x\|$ and $\|f\| = 1$ (we can assume that $x \neq 0$). Taking f_1 and f_2 as above, we obtain $0 \leqslant f_i(x) \leqslant f_i(z) \leqslant (1+c)\|z\|$. Hence $\|x\| = f(x) = f_1(x) - f_2(x)$ does not exceed $2(1+c)\|z\|$. $\square$

The following result about decompositions of functionals in differences of positive functionals gives a bit more than Proposition 6.10.60.

6.10.61. Theorem. *Let $\mathcal{F}$ be a vector lattice of bounded functions on a set Ω containing 1. Suppose that on $\mathcal{F}$ we are given a linear functional L continuous with respect to the norm $\|f\| = \sup_{\Omega} |f(x)|$. Then L can be represented in the form $L = L^+ - L^-$, where $L^+ \geqslant 0$, $L^- \geqslant 0$ and for all nonnegative $f \in \mathcal{F}$ we have*

$$L^+(f) = \sup_{0 \leqslant g \leqslant f} L(g), \quad L^-(f) = - \inf_{0 \leqslant g \leqslant f} L(g). \qquad (6.10.8)$$

In addition, letting $|L| := L^+ + L^-$, for all $f \geqslant 0$ we have

$$|L|(f) = \sup_{0 \leqslant |g| \leqslant f} |L(g)|, \quad \|L\| = L^+(1) + L^-(1).$$

A similar assertion with the exception of the equality is true for all continuous linear functionals on normed vector lattices.

PROOF. For nonnegative functions $f, g \in \mathcal{F}$ and every function $h \in \mathcal{F}$ such that $0 \leqslant h \leqslant f + g$ we can write $h = h_1 + h_2$, where $h_1, h_2 \in \mathcal{F}$, $0 \leqslant h_1 \leqslant f$, $0 \leqslant h_2 \leqslant g$. Indeed, let $h_1 = \min(f, h)$, $h_2 = h - h_1$. Then $h_1, h_2 \in \mathcal{F}$, $0 \leqslant h_1 \leqslant f$ and $h_2 \geqslant 0$. Finally, $h_2 \leqslant g$. Indeed, if $h_1(x) = h(x)$, then $h_2(x) = 0$ if $h_1(x) = f(x)$, then $h_2(x) = h(x) - f(x) \leqslant g(x)$, since $h \leqslant g + f$.

Let L^+ be defined by equality (6.10.8). We observe that the quantity $L^+(f)$ is finite, since $|L(h)| \leqslant \|L\| \|h\| \leqslant \|L\| \|f\|$. It is clear that $L^+(tf) = tL^+(f)$ for all nonnegative numbers t and $f \geqslant 0$. Let $f \geqslant 0$ and $g \geqslant 0$ belong to $\mathcal{F}$. Using the notation above, we obtain

$$L^+(f + g) = \sup\{L(h): 0 \leqslant h \leqslant f + g\}$$
$$= \sup\{L(h_1) + L(h_2): 0 \leqslant h_1 \leqslant f, 0 \leqslant h_2 \leqslant g\} = L^+(f) + L^+(g).$$

Now for all $f \in \mathcal{F}$ we set $L^+(f) = L^+(f^+) - L^+(f^-)$, where $f^+ = \max(f, 0)$, $f^- = -\min(f, 0)$. Note that if $f = f_1 - f_2$, where $f_1, f_2 \geqslant 0$, then we have $L^+(f) = L^+(f_1) - L^+(f_2)$. Indeed, obviously $f_1 + f^- = f_2 + f^+$ and hence $L^+(f_1) + L^+(f^-) = L^+(f_2) + L^+(f^+)$. It is clear that $L^+(tf) = tL^+(f)$ for all $t \in \mathbb{R}^1$ and $f \in \mathcal{F}$. The additivity of the functional L^+ follows from its additivity on nonnegative functions. Indeed, for any f and g we have $f = f^+ - f^-$ and $g = g^+ - g^-$, whence $f + g = (f^+ + g^+) - (f^- + g^-)$ and according to what has been said above we obtain

$$L^+(f + g) = L^+(f^+ + g^+) - L^+(f^- + g^-) = L^+(f) + L^+(g).$$

By definition, $L^+(f) \geqslant L(f)$ for nonnegative f, so the functional $L^- := L^+ - L$ is nonnegative. It is readily seen that L^- is given by the announced formula.

Finally, $\|L\| \leqslant \|L^+\| + \|L^-\| = L^+(1) + L^-(1)$. On the other hand,

$$L^+(1) + L^-(1) = 2L^+(1) - L(1) = \sup\{L(2\varphi - 1) \colon 0 \leqslant \varphi \leqslant 1\}$$
$$\leqslant \sup\{L(h) \colon -1 \leqslant h \leqslant 1\} \leqslant \|L\|.$$

The theorem is proved for the lattice $\mathcal{F}$. The case of a general vector lattice is similar, see [**14**, p. 14]. □

Let us mention a remarkable result due to P. P. Korovkin on convergence of positive operators (for a proof, see [**339**, Chapter 1]).

6.10.62. Theorem. *Let T_n be positive linear operators on $C[0, 1]$ in the sense that $T_n x \geqslant 0$ if $x \geqslant 0$ such that for the three functions $x_k(t) = t^k$, $k = 0, 1, 2$, we have $\|T_n x_k - x_k\| \to 0$ as $n \to \infty$. Then $\|T_n x - x\| \to 0$ for every $x \in C[0, 1]$.*

The same assertion is true for the space $C_{2\pi}$ of continuous 2π-periodic functions if $x_1(t) = 1$, $x_2(t) = \sin t$ and $x_2(t) = \cos t$.

6.10(vi). Vector integration

Here we briefly discuss the Lebesgue integral for vector mappings, which is called in this case the Bochner integral (see [**147**], [**149**], [**164**], [**171**], [**397**], and [**614**]). Let $(\Omega, \mathcal{B})$ be a measurable space and let μ be a nonnegative measure on $\mathcal{B}$. First we suppose that the measure μ is bounded and then indicate the necessary changes for the general case.

Let E be a real or complex Banach space. Let $\mathcal{B}(E)$ denote the σ-algebra generated by open sets in E. This σ-algebra is called *Borel*.

6.10.63. Definition. *A mapping $f \colon \Omega \to E$ that is defined μ-almost everywhere is called μ-measurable if, for every $B \in \mathcal{B}(E)$, the set $f^{-1}(B)$ is μ-measurable.*

It is clear that if f is a μ-measurable mapping with values in a Banach space E, then the function $\omega \mapsto \|f(\omega)\|$ is also μ-measurable by the Borel measurability of open balls in E.

6.10.64. Proposition. *Let E be a separable Banach space and let $f \colon \Omega \to E$ be a mapping defined μ-almost everywhere. This mapping is μ-measurable precisely when for every $l \in E^*$ the scalar function $l \circ f$ is μ-measurable.*

A sufficient condition for the μ-measurability of f is the μ-measurability of all functions $l_n \circ f$, where $\{l_n\} \subset E^*$ is a countable set such that every element in E^* is the limit of a subsequence from $\{l_n\}$ in the weak-$*$ topology (such a set exists by the separability of E, see below).

PROOF. Let us consider the real case. The necessity of the indicated condition is clear from the fact that for every $l \in E^*$ and every open set $V \subset \mathbb{R}$ the set $f^{-1}(l^{-1}(V))$ is μ-measurable, because $l^{-1}(V)$ is open. Let us show that the σ-algebra $\mathcal{B}_0$ generated by the halfspaces of the form $\{x : l(x) \leqslant r\}$ coincides with $\mathcal{B}(E)$. By the separability of the space E every open set in it equals the union of a countable collection of open balls with centers at points from $\{x_n\}$. Let U be such a ball. We show that $U \in \mathcal{B}_0$. Since U is the union of a sequence of closed balls with the same centers, we can pass to the closed ball U. It remains to represent it as the intersection of a countable collection of closed halfspaces. To this end, for every point $x_n \in E \backslash U$ and every closed ball $B(x_n, r_k)$ of rational radius r_k disjoint with U we find a halfspace $\Pi_{n,k}$ such that $U \subset \Pi_n$ and $B(x_n, r_k) \subset E \backslash \Pi_{n,k}$. We have $U = \bigcap_{n,k} \Pi_{n,k}$. Indeed, if $x \notin U$, then there exist x_n and r_k such that $x \in B(x_n, r_k)$ and $U \cap B(x_n, r_k) = \varnothing$, whence $x \notin \Pi_{n,k}$. Thus, $U \in \mathcal{B}_0$, which gives the equality $\mathcal{B}_0 = \mathcal{B}(E)$.

Let $\{l_n\} \subset E^*$ be a countable set such that every element l in E^* is the limit of a subsequence in $\{l_n\}$ in the weak-$*$ topology. The existence of such a set is obvious from the fact that E^* is the union of closed balls of radius n each of which is a metrizable compact space in the weak-$*$ topology (see Theorem 6.10.23). Then the function l is measurable with respect to the σ-algebra generated by $\{l_n\}$. Since the function l_n is measurable with respect to $\mathcal{B}_0$, so is the function l. $\qquad\square$

Note that the measurability of f is equivalent to the measurability of the functions $g_n \circ f$, where $\{g_n\} \subset E^*$ is a countable set separating points in E, since for $\{l_n\}$ we can take the set of finite linear combinations of g_i with rational coefficients (the intersection of this set with the ball in E^* is dense in this ball in the weak-$*$ topology, which is easily verified).

For a broad class of measure spaces (in particular, for intervals with Borel measures) every measurable mapping with values in a Banach space automatically takes values in a separable subspace after redefinition on a set of measure zero (see Corollary 6.10.16 and Theorem 7.14.25 in [73]).

As in the case of scalar functions, μ-measurable vector valued mappings with finite sets of values will be called simple. For a such mapping ψ with values $y_1, \ldots, y_n$ on disjoint sets $\Omega_1, \ldots, \Omega_n$, the *Bochner integral* is defined by

$$\int_{\Omega} \psi(\omega)\, \mu(d\omega) := \sum_{i=1}^{n} y_i \mu(\Omega_i).$$

By the additivity of μ this integral is well-defined, i.e., does not depend on the partition of Ω into disjoint parts on which ψ is constant.

A sequence of simple mappings ψ_n is called *Cauchy or fundamental in mean* if for every $\varepsilon > 0$ there is N such that

$$\int_{\Omega} \|\psi_n(\omega) - \psi_k(\omega)\|\, \mu(d\omega) < \varepsilon \quad \text{if } n, k \geqslant N.$$

The sequence of integrals of ψ_n is Cauchy in E. Indeed, let $\Omega_1, \ldots, \Omega_N$ be disjoint measurable sets picked for fixed n and k such that ψ_n and ψ_k are constant on them. Such sets can be easily obtained by refining the sets on which ψ_n and

ψ_k are constant. Then we have the estimate which proves our claim:

$$\left\| \int_\Omega [\psi_n(\omega) - \psi_k(\omega)] \, \mu(d\omega) \right\| \leqslant \sum_{i=1}^{N} \| y_i^n - y_i^k \| \mu(\Omega_i)$$

$$= \int_\Omega \| \psi_n(\omega) - \psi_k(\omega) \| \, \mu(d\omega).$$

6.10.65. Definition. *Let E be a Banach space. A mapping $f \colon \Omega \to E$ is called Bochner integrable if there exists a sequence of simple E-valued mappings ψ_n that converges to f μ-almost everywhere and is Cauchy in mean. The Bochner integral of f is defined as the limit of the integrals of ψ_n and denoted by the symbol $\int_\Omega f(\omega) \, \mu(d\omega)$.*

It follows from the scalar case that this definition is not ambiguous, since for every $l \in E^*$ the sequence of functions $l \circ \psi_n$ converges almost everywhere and is Cauchy in mean. Any integrable mapping f is measurable with respect to μ. Indeed, it follows from the definition that there exists a separable subspace $X_0 \subset X$ such that $f(\omega) \in X_0$ for μ-a.e. ω. Now the measurability of f follows from the proposition proved above and the measurability of the limit of a sequence of scalar measurable functions.

6.10.66. Example. Every bounded measurable mapping f with values in a separable Banach space E is Bochner integrable.

PROOF. Let us fix $n \in \mathbb{N}$. Let $\{x_i\}$ be a countable everywhere dense set in E. The space E is covered by the sequence of balls $B_{i,n} := B(x_i, 2^{-n})$. Let us find N_n such that

$$\mu\left(\Omega \backslash \bigcup_{i=1}^{N_n} f^{-1}(B_{i,n}) \right) < 2^{-n}.$$

Set $\Omega_{n,1} := f^{-1}(B_{1,n})$, $\Omega_{n,k} := f^{-1}(B_{k,n}) \backslash \Omega_{n,k-1}$, $k \leqslant N_n$. We define simple mappings ψ_n as follows. Let $\psi_n = x_k$ on $\Omega_{n,k}$ and let $\psi_n = 0$ outside $\Omega_n := \bigcup_{k=1}^{N_n} \Omega_{n,k}$. For every $\omega \in \Omega_{n,k}$ we have $\| \psi_n(\omega) - f(\omega) \| \leqslant 2^{-n}$. Hence this estimate is fulfilled on a set Ω_n with $\mu(\Omega \backslash \Omega_n) < 2^{-n}$. Therefore, $\lim_{n \to \infty} \psi_n(\omega) = f(\omega)$ for μ-almost all ω. This is true for every ω from the set $\Omega' := \bigcap_{m=1}^{\infty} \bigcup_{n=m}^{\infty} \Omega_n$ of full μ-measure. By hypothesis $\| f(\omega) \| \leqslant M$ for some M. Then $\| \psi_n(\omega) \| \leqslant M + 1$. For all $n > k$ and $\omega \in \Omega_n \cap \Omega_k$ we have $\| \psi_n(\omega) - \psi_k(\omega) \| \leqslant 2^{1-k}$, whence it follows that $\{\psi_n\}$ is Cauchy in mean. $\square$

The estimate above implies that the function $\| f(\omega) \|$ is μ-integrable and

$$\left\| \int_\Omega f(\omega) \, \mu(d\omega) \right\| \leqslant \int_\Omega \| f(\omega) \| \, \mu(d\omega)$$

for any μ-integrable mapping f. The converse is also true.

6.10.67. Theorem. *Let f be a μ-measurable mapping with values in a separable Banach space E such that the function $\omega \mapsto \| f(\omega) \|$ is integrable with respect to μ. Then the mapping f is Bochner μ-integrable.*

PROOF. For every $n \in \mathbb{N}$ we find a measurable set Ω_n with $\mu(\Omega \backslash \Omega_n) < 2^{-n}$ and

$$\int_{\Omega \backslash \Omega_n} \|f(\omega)\| \, \mu(d\omega) < 2^{-n}.$$

Slightly decreasing the set Ω_n with the preservation of these bounds, we find simple mappings ψ_n for which $\|\psi_n(\omega) - f(\omega)\| \leqslant 2^{-n}$ for all $\omega \in \Omega_n$ and $\|\psi_n(\omega)\| \leqslant \|f(\omega)\| + 1$ for almost all ω. This is clear from the proof of the previous example. As above, we obtain convergence of the sequence $\{\psi_n\}$ to f almost everywhere and its fundamentality in mean. □

It follows from the definition that for every Bochner integrable mapping f with values in a Banach space E and every continuous linear operator T from E to a Banach space Y the mapping $T \circ f$ is also Bochner integrable and

$$\int_{\Omega} T\big(f(\omega)\big) \, \mu(d\omega) = T\bigg(\int_{\Omega} f(\omega) \, \mu(d\omega) \bigg).$$

Similarly to the scalar case the following assertion can be proved (Exercise 7.10.98).

6.10.68. Theorem. *The class $\mathcal{L}^1(\mu; E)$ of all Bochner μ-integrable everywhere defined mappings with values in a Banach space E is a linear space and the Bochner integral is linear on it. The set of μ-equivalence classes $L^1(\mu; E)$ with the norm*

$$\|f\|_{L^1(\mu; E)} := \int_{\Omega} \|f(\omega)\| \, \mu(d\omega),$$

given by means of a representative of the equivalence class is a Banach space. In addition, for every $p \in [1, +\infty)$ the subspace $L^p(\mu; E)$ in $L^1(\mu; E)$ corresponding to mappings with a finite norm

$$\|f\|_{L^p(\mu; E)} := \bigg(\int_{\Omega} \|f(\omega)\|^p \, \mu(d\omega) \bigg)^{1/p}$$

is also a Banach space.

Similarly to the scalar case one can include in $\mathcal{L}^1(\mu; E)$ mappings that are not defined on a measure zero set and coincide outside a measure zero set with a mapping from the previously defined class. For the same reasons as for real functions, this set will not be a linear space, although one can take sums of such mappings and multiply them by constants (on a set of full measure). This is a matter of convenience, but the difference disappears after passage to the factor-space $L^1(\mu; E)$.

In case of an infinite measure the construction of the Bochner integral is similar, but in the definition of a simple mapping it is required in addition that it must be zero outside a set of finite measure.

In applications it is often useful to work with weaker notions of the scalar integrability or the Pettis integrability.

A mapping f with values in a Banach space E is called *scalarly integrable* if for every $l \in E^*$ the function $l \circ f$ is integrable with respect to μ and there is an element $h \in E$, called the *scalar integral* of f, such that the integral of the function $l \circ f$ equals $l(h)$. If f is scalarly integrable on every measurable set, then it is called *Pettis integrable* and its scalar integral is called the *Pettis integral* of f.

If E is separable, then the scalar integrability implies the measurability of f, but in the general case it does not yield the Bochner integrability. For example, let the measure μ on l^2 be concentrated at the points of the form $n e_n$ and $-n e_n$, where $\{e_n\}$ is the standard basis in l^2, and let the measure of such point be n^{-2}. The mapping $f(x) = x$ has zero Pettis integral, since for every $y = (y_n) \in l^2$ the series with the general term $y_n n^{-1}$ is absolutely convergent. However, the function $\|x\|$ is not integrable with respect to μ due to divergence of the series with the general term n^{-1}.

6.10(vii). The Daniell integral

The construction of the Lebesgue integral presented in this book is based on a preliminary study of measure. It is possible, however, to act in the opposite order: to define measure by means of integral. At the basis of this approach there is the following result due to Daniell. Its formulation employs the notion of a vector lattice of functions, i.e., a linear space $\mathcal{F}$ of real functions on a nonempty set Ω such that $\max(f, g) \in \mathcal{F}$ for all $f, g \in \mathcal{F}$ (this is equivalent to the closedness of $\mathcal{F}$ with respect to taking absolute values). Vector lattices of functions considered here are a particular case of abstract vector lattices mentioned in §5.6(iii). Suppose that on $\mathcal{F}$ we are given a nonnegative linear functional L, i.e., $L(f) \geq 0$ for all $f \geq 0$, and that $L(f_n) \to 0$ for every sequence $\{f_n\} \subset \mathcal{F}$ pointwise decreasing to zero. Such a functional is called the *Daniell integral*. Our nearest goal is to extend L to a larger domain of definition $\mathcal{L}$ such that the extension will possess the main properties of the integral, i.e., admit analogs of monotone and dominated convergence theorems and $\mathcal{L}$ will be complete. An example which can be taken as model is an extension of the Riemann integral from the set of step or continuous functions. Then we clarify the connections between the Daniell integral and the true integral with respect to a measure.

6.10.69. Definition. *A set $S \subset \Omega$ will be called L-zero if there exists a nondecreasing sequence of nonnegative functions $f_n \in \mathcal{F}$ for which $\sup_n L(f_n) < \infty$ and $\sup_n f_n(x) = +\infty$ on S.*

This definition is inspired by the monotone convergence theorem.

6.10.70. Lemma. (i) *The union of countably many L-zero sets S_k is an L-zero set.*

(ii) *A set S is L-zero precisely when for every $\varepsilon > 0$ there exists a nondecreasing sequence of nonnegative functions $f_n \in \mathcal{F}$ with $L(f_n) < \varepsilon$ and $\sup_n f_n(x) \geq 1$ on S.*

(iii) *If $f \in \mathcal{F}$ and $f \geq 0$ outside an L-zero set, then $L(f) \geq 0$. If $f, g \in \mathcal{F}$ and $f \leq g$ outside an L-zero set, then $L(f) \leq L(g)$.*

PROOF. (i) For any fixed k we take functions $f_{k,n} \in \mathcal{F}$, $n \in \mathbb{N}$, such that $L(f_{k,n}) \leqslant 2^{-k}$ and $\sup_n f_{k,n}(x) = \infty$ on S_k. Let $g_n = f_{1,n} + \cdots + f_{n,n}$. Then $g_n \in \mathcal{F}$, $L(g_n) \leqslant 1$ and $\sup_n g_n(x) = +\infty$ on $\bigcup_{k=1}^{\infty} S_k$, since the latter is true on every S_k.

(ii) If S is L-zero and $\{f_n\} \subset \mathcal{F}$ is an increasing sequence with the properties indicated in the definition and $L(f_n) \leqslant M$, then, for any given $\varepsilon > 0$, we take the function $\varepsilon M^{-1} f_n$. Conversely, if the condition in (i) is fulfilled, then for every k there exists an increasing sequence $\{f_{k,n}\} \subset \mathcal{F}$ such that $L(f_{k,n}) \leqslant 2^{-k}$ for all n and $\sup_n f_{k,n} \geqslant k$ on S. Let $g_n = f_{1,1} + \cdots + f_{n,n}$. Then $g_n \in \mathcal{F}$, $0 \leqslant g_n \leqslant g_{n+1}$, $L(g_n) \leqslant 1$, and $\sup_n g_n = +\infty$ on S. Hence S is L-zero.

(iii) Let $S = \{f < 0\}$ and $c = L(f) < 0$. Let us take an increasing sequence of nonnegative functions $f_n \in \mathcal{F}$ with $L(f_n) \leqslant |c|/2$ and $\sup_n f_n(x) = \infty$ for all $x \in S$. Then the sequence of functions $f + f_n$ is increasing and its finite or infinite limit is everywhere nonnegative, since the function f is finite on S and nonnegative outside S. In addition, $L(f + f_n) \leqslant -|c|/2$. Set $\varphi_n = (f + f_n)^-$. Then $\varphi_n \in \mathcal{F}$ and the functions φ_n are pointwise increasing to zero. Hence $L(\varphi_n) \to 0$ contrary to that $L(\varphi_n) \leqslant L(f + f_n) \leqslant -|c|/2$. Thus, $c = 0$. □

This yields the following stronger continuity property of L.

6.10.71. Corollary. *Let $\{f_n\} \subset \mathcal{F}$ and $f_n \downarrow 0$ outside some L-zero set S. Then $L(f_n) \downarrow 0$.*

PROOF. Set $g_n = \min(f_1, \ldots, f_n)$. Then $g_n \in \mathcal{F}$. Outside S we have $g_n = f_n$. By the lemma $L(f_n) = L(g_n)$. In addition, $\{g_n\}$ is everywhere decreasing. If $L(g_n) \geqslant c > 0$ for all n, then we take an increasing sequence of nonnegative functions $\varphi_n \in \mathcal{F}$ with $L(\varphi_n) < c/2$ and $\sup_n \varphi_n(x) = \infty$ for all $x \in S$. Then the functions $g_n - \varphi_n \in \mathcal{F}$ decrease everywhere to a nonpositive limit, because outside S the functions g_n decrease to zero and on S we have $g_n - \varphi_n \leqslant g_1 - \varphi_n \to -\infty$. The functions $(g_n - \varphi_n)^+ \in \mathcal{F}$ pointwise decrease to zero, whence we obtain $L\big((g_n - \varphi_n)^+\big) \to 0$. This contradicts the estimate $L\big((g_n - \varphi_n)^+\big) \geqslant L(g_n - \varphi_n) \geqslant c/2$. □

Denote by $\mathcal{L}^{\uparrow}$ the class of all functions $f \colon \Omega \to (-\infty, +\infty]$ for which there exists a sequence of nonnegative functions $f_n \in \mathcal{F}$ such that outside some L-zero set we have $f(x) = \lim\limits_{n\to\infty} f_n(x)$ and $f_n(x) \leqslant f_{n+1}(x)$ for all n and the sequence $\{L(f_n)\}$ is bounded. Set

$$L(f) := \lim_{n\to\infty} L(f_n).$$

The limit exists, since by assertion (iii) of the lemma above the sequence $\{L(f_n)\}$ increases. The next lemma shows that L is well-defined on $\mathcal{L}^{\uparrow}$.

6.10.72. Lemma. (i) *Suppose that $\{f_n\}$ and $\{g_n\}$ are two sequences from $\mathcal{F}$ such that outside some L-zero set S they increase and satisfy the condition*

$$\lim_{n\to\infty} f_n(x) \leqslant \lim_{n\to\infty} g_n(x),$$

where also infinite limits are allowed. Then we have $\lim\limits_{n\to\infty} L(f_n) \leqslant \lim\limits_{n\to\infty} L(g_n)$.
In particular, these limits coincide if outside S the sequences $\{f_n\}$ and $\{g_n\}$ increase to a common limit.

(ii) *Every function from $\mathcal{L}^{\uparrow}$ is finite outside some L-zero set.*

PROOF. (i) For any fixed n the functions $f_n - g_k$ outside S decrease to a nonpositive limit. Hence $(f_n - g_k)^+ \downarrow 0$ outside S, which by the corollary above gives $L\big((f_n - g_k)^+\big) \downarrow 0$. Hence $\lim\limits_{k\to\infty} L(f_n - g_k) \leqslant 0$, i.e., $L(f_n) \leqslant \lim\limits_{k\to\infty} L(g_k)$, which gives the desired inequality.

(ii) Suppose that functions $f_n \in \mathcal{F}$ increase to f outside an L-zero set S and $L(f_n) \leqslant C$. Set $g_n = \max(f_1, \ldots, f_n)$. Then $g_n \in \mathcal{F}$, $g_n \leqslant g_{n+1}$, and $g_n = f_n$ outside S. The set $Z = \{x\colon \sup_n g_n(x) = \infty\}$ is L-zero by definition. The function f is finite outside the set $S \cup Z$. $\square$

It follows from the established facts that if $f \in \mathcal{L}^{\uparrow}$ and a function $g \geqslant 0$ equals f on the complement of an L-zero set, then $g \in \mathcal{L}^{\uparrow}$ and $L(g) = L(f)$. Hence every function from $\mathcal{L}^{\uparrow}$ can be made everywhere finite without changing $L(f)$.

Let $\mathcal{L}$ denote the set of all real functions f representable as the difference $f = f_1 - f_2$ of two *everywhere finite* functions $f_1, f_2 \in \mathcal{L}^{\uparrow}$. For such functions we set $L(f) := L(f_1) - L(f_2)$. This value is well-defined, which is verified in the next theorem.

On $\mathcal{L}$ we can introduce an equivalence relation by declaring to be equivalent those functions which coincide outside an L-zero set. Then the set $\widetilde{\mathcal{L}}$ of equivalence classes becomes a metric space with the metric $d_L(f,g) := L(|f - g|)$. In addition, $\widetilde{\mathcal{L}}$ is a linear space and L is naturally defined on $\widetilde{\mathcal{L}}$. It is clear that $\mathcal{F}$ is everywhere dense in $\widetilde{\mathcal{L}}$.

6.10.73. Theorem. (i) *The functional L on $\mathcal{L}$ (or $\widetilde{\mathcal{L}}$) is well-defined and linear.*

(ii) *If $f \in \mathcal{L}$, then $|f| \in \mathcal{L}$ and $|L(f)| \leqslant L(|f|)$.*

(iii) *The assertions of the monotone convergence, Lebesgue dominated convergence and Fatou theorems hold for L with L in place of the integral: if $f_n \in \mathcal{L}$, $f_n \to f$ outside an L-zero set and either there is a function $\Phi \in \mathcal{L}$ such that $|f_n| \leqslant \Phi$ outside an L-zero set or $\{f_n\}$ is increasing outside an L-zero set and $\sup_n L(f_n) < \infty$, then $f \in \mathcal{L}$ and $\lim\limits_{n\to\infty} L(f_n) = L(f)$; if $f_n \geqslant 0$ and $\sup_n L(f_n) < \infty$, then $f \in \mathcal{L}$ and $L(f_n) \leqslant \liminf_n L(f_n)$.*

In addition, the space $\widetilde{\mathcal{L}}$ is complete with respect to the metric d_L.

PROOF. (i) It is readily seen that $f + g \in \mathcal{L}^{\uparrow}$ for all $f, g \in \mathcal{L}^{\uparrow}$, moreover, $L(f + g) = L(f) + L(g)$. If $f_1, f_2, g_1, g_2 \in \mathcal{L}^{\uparrow}$ and $f_1 - f_2 = g_1 - g_2$, then $L(f_1) + L(g_2) = L(g_1) + L(f_2)$, which shows that L is well-defined on $\mathcal{L}$. Since $L(\alpha f) = \alpha L(g)$ for all $f \in \mathcal{L}^{\uparrow}$ and all $\alpha \geqslant 0$, then this is true for all $f \in \mathcal{L}$ and $\alpha \in \mathbb{R}$. If $f, g \in \mathcal{L}$ and $f = f_1 - f_2$, $g = g_1 - g_2$, where $f_1, f_2, g_1, g_2 \in \mathcal{L}^{\uparrow}$, then by the facts already proved above we have $f + g = f_1 + g_1 - (f_2 + g_2) \in \mathcal{L}$ and $L(f + g) = L(f_1 + g_1) - L(f_2 + g_2) = L(f) + L(g)$.

The closedness of $\mathcal{F}$ with respect to the operations $(\varphi, \psi) \mapsto \min(\varphi, \psi)$ and $(\varphi, \psi) \mapsto \max(\varphi, \psi)$ yields the closedness of $\mathcal{L}^\uparrow$ with respect to these operations. Hence $\min(f_1, f_2) \in \mathcal{L}^\uparrow$ and $\max(f_1, f_2) \in \mathcal{L}^\uparrow$, whence

$$|f| = |f_1 - f_2| = \max(f_1, f_2) - \min(f_1, f_2) \in \mathcal{L}.$$

Finally, for the proof of the estimate $|L(f)| \leqslant L(|f|)$ it suffices to verify that $L\varphi \geqslant 0$ if $\varphi \in \mathcal{L}$ and $\varphi \geqslant 0$. Then $\varphi = \psi_1 - \psi_2$, where $\psi_1, \psi_2 \in \mathcal{L}^\uparrow$ and $\psi_2 \leqslant \psi_1$. Hence we can apply assertion (i) of the lemma above.

(ii) Let $\{f_n\} \subset \mathcal{L}$ be increasing outside an L-zero set and let the sequence $\{L(f_n)\}$ be bounded. For every n we find functions $f_{n,k} \in \mathcal{F}$ increasing to f_n outside some L-zero set S_n. Let $g_n = \max_{k,m \leqslant n} f_{m,k}$. Then $g_n \in \mathcal{F}$, the sequence $\{g_n\}$ is increasing and $\{L(g_n)\}$ is bounded. Hence $f = \lim\limits_{n\to\infty} g_n \in \mathcal{L}^+$ and $L(f) = \lim\limits_{n\to\infty} L(g_n)$. It is clear that $f_n(x) \to f(x)$ outside an L-zero set and $L(f) = \lim\limits_{n\to\infty} L(f_n)$, since $L(g_n) \leqslant L(f_n)$ and $L(f_n) = \lim\limits_{k\to\infty} L(f_{n,k})$. Fatou's theorem is deduced precisely as in the case of the Lebesgue integral.

Let $f_n(x) \to f(x)$ and $|f_n(x)| \leqslant \Phi(x)$ outside an L-zero set, where $f_n, \Phi \in \mathcal{L}$. Set $\varphi_n(x) := \inf_{k \geqslant n} f_k(x)$, $\psi_n(x) := \sup_{k \geqslant n} f_k(x)$. Then outside an L-zero set we have $\varphi_n \leqslant f_n \leqslant \psi_n$, $\varphi_n \geqslant -\Phi$, $\psi_n \leqslant \Phi$, $\varphi_n \uparrow f$, $\psi_n \downarrow f$. Hence $f \in \mathcal{L}$ and $L(f) = \lim\limits_{n\to\infty} L(\varphi_n) = \lim\limits_{n\to\infty} L(\psi_n)$, which gives $L(f) = \lim\limits_{n\to\infty} L(f_n)$.

Let now a sequence $\{f_n\} \subset \mathcal{L}$ be Cauchy in the metric d_L. Passing to a subsequence, we can assume that $d_L(f_n, f_{n+1}) \leqslant 2^{-n}$. As shown above, the series of $|f_n - f_{n-1}|$, where $f_0 := 0$, converges outside some L-zero set S to an element $\Phi \in \mathcal{L}$. Then the sums $f_n = \sum_{k=1}^n (f_k - f_{k-1})$ converge to a finite limit f outside S. Since $|f_n| \leqslant \Phi$, we conclude that $\{f_n\}$ converges to f in $\widetilde{\mathcal{L}}$. $\square$

The Daniell integral possesses the most important properties of the Lebesgue integral, so the question arises whether it can be defined as the integral with respect to some countably additive measure. Moreover, some measure arises automatically. Indeed, denote by $\mathcal{R}_L$ the class of all sets $E \in \Omega$ for which $I_E \in \mathcal{L}$ and let $\nu(E) := L(I_E)$. It follows from the previous theorem that $\mathcal{R}_L$ is a δ-ring and that the measure ν on it is countably additive. However, in the general case the integral with respect to this measure can fail to coincide with L. As the following example shows, without additional assumptions it is not always possible to represent L as the Lebesgue integral with respect to a countably additive measure.

6.10.74. Example. Let $\mathcal{F}$ be the set of all finite real functions f on $[0, 1]$ with the following property: for some number $\alpha = \alpha(f)$, the set $\{t: f(t) \neq \alpha(1+t)\}$ is a first category set. Let $L(f) := \alpha$. Then $\mathcal{F}$ is a vector lattice of functions with the natural order relation from $\mathbb{R}^{[0,1]}$, L is a nonnegative linear functional on $\mathcal{F}$ and $L(f_n) \to 0$ for every sequence of functions $f_n \in \mathcal{F}$ pointwise decreasing to zero, but even on the subspace of all bounded functions in $\mathcal{F}$ the functional L cannot be defined as the integral with respect to a countably additive measure.

PROOF. We observe that for every function $f \in \mathcal{F}$ there is only one number α with the indicated property, since the interval is not a first category set. Hence the function L is well-defined. For any $f \in \mathcal{F}$ let $E_f := \{t: f(t) \neq \alpha(1+t)\}$

for the number α corresponding to f. If $f, g \in \mathcal{F}$ and $\alpha = \alpha(f)$, $\beta = \alpha(g)$ are the corresponding numbers, then the set $E_f \cup E_g$ has the first category and outside this set we have $f(t) + g(t) = (\alpha + \beta)(1 + t)$. For every scalar c we have $cf(t) = c\alpha(1 + t)$ outside the set E_f. Thus, $\mathcal{F}$ is a linear space. It is readily seen that $|f| \in \mathcal{F}$ for all $f \in \mathcal{F}$. It is clear from what has been said that the function L is linear. For all $f \geqslant 0$ we have $L(f) \geqslant 0$. If functions $f_n \in \mathcal{F}$ pointwise decrease to zero, then the union of the sets E_{f_n} is a first category set. Hence there exists a point t such that $L(f_n) = f_n(t)/(1 + t)$ for all n at once, whence $\lim_{n \to \infty} L(f_n) = 0$.

Suppose now that there exists a measure μ on $\sigma(\mathcal{F})$ with values in $[0, +\infty]$ such that every bounded function f from $\mathcal{F}$ belongs to $\mathcal{L}^1(\mu)$ and $L(f)$ coincides with the integral of f with respect to the measure μ. The function $\psi \colon t \mapsto 1 + t$ belongs to $\mathcal{F}$, whence it follows that all open sets from $[0, 1]$ belong to $\sigma(\mathcal{F})$. By the estimate $\psi \geqslant 1$ we obtain $\mu([0, 1]) \leqslant L(\psi) = 1$. Thus, the restriction of μ to $\mathcal{B}([0, 1])$ is a finite measure. Hence there exists a Borel first category set E such that $\mu([0, 1] \backslash E) = 0$. Indeed, we can take the union of nowhere dense compact sets K_n with $\mu([0, 1] \backslash K_n) < 1/n$, which are constructed by means of deleting sufficiently small intervals with centers at the points of an everywhere dense countable set of μ-measure zero. Let us consider the following function f: $f(t) = 0$ if $t \in E$, $f(t) = 1 + t$ if $t \notin E$. It is clear that $f \in \mathcal{F}$ and $L(f) = 1$. On the other hand, the integral of f with respect to the measure μ is zero, which gives a contradiction. $\qquad\square$

In this example the measure ν generated by L on the δ-ring $\mathcal{R}_L$ is identically zero. Indeed, here L-zero sets are first category sets, since if $\alpha(f_n) \leqslant 1/3$, then $f_n(t) \leqslant 2/3$ outside a first category set. Hence the class $\mathcal{L}$ coincides with $\mathcal{F}$. The indicator function of a set can belong to $\mathcal{F}$ only for $\alpha = 0$, i.e., $\mathcal{R}_L$ consists of first category sets and they are zero sets.

One should bear in mind that the measure ν can be zero also in the case where L is given as the integral with respect to some nonzero measure μ. For example, let us take for $\mathcal{F}$ the one-dimensional linear space generated by the function $f(t) = t$ on $(0, 1)$ and for L take the Riemann integral, i.e., $L(\alpha f) = \alpha/2$. Then $\mathcal{R}_L$ consists of the empty set and $\nu = 0$.

We now prove that adding one simply stated condition, fulfilled in all applications, leads to the effect that the Daniell integral is given by the Lebesgue integral with respect to some measure. This is the so-called *Stone condition*:

$$\min(f, 1) \in \mathcal{F} \quad \text{for all } f \in \mathcal{F}.$$

This condition is trivially fulfilled if the lattice of functions $\mathcal{F}$ contains 1. A nontrivial example of a lattice with the Stone condition, but without 1, is the space of continuous functions with compact support on $\mathbb{R}^n$. The space $\mathcal{F}$ from the previous example and its subspace consisting of bounded functions obviously do not satisfy the Stone condition.

6.10.75. Theorem. *Let $\mathcal{F}$ satisfy the Stone condition and let L be a nonnegative linear functional on $\mathcal{F}$ such that $L(f_n) \to 0$ for every sequence of functions $f_n \in \mathcal{F}$ pointwise decreasing to zero. Then there exists a countably additive* *measure μ on $\sigma(\mathcal{F})$ with values in $[0, +\infty]$ such that $\mathcal{F} \subset L^1(\mu)$ and*

$$L(f) = \int_\Omega f(\omega)\,\mu(d\omega), \quad f \in \mathcal{F}. \tag{6.10.9}$$

In addition, $\widetilde{\mathcal{L}} = L^1(\mu)$ and such a measure μ is unique on the σ-ring generated by the sets of the form $\{f > c\}$, where $f \in \mathcal{F}$ and $c > 0$. Finally, if $1 \in \mathcal{F}$, then the measure μ is finite.

PROOF. The measure ν constructed above on the δ-ring $\mathcal{R}_L$ uniquely extends to a countably additive measure μ (with values in $[0, +\infty]$) on the σ-ring $\mathcal{R}_L^\sigma$ generated by $\mathcal{R}_L$. In addition, it extends (but not necessarily uniquely) to a countably additive measure μ on the σ-algebra generated by $\mathcal{R}_L$. Let us show that the latter coincides with the σ-algebra $\sigma(\mathcal{L})$ generated by $\mathcal{L}$. To this end we observe that one can easily derive from the Stone condition that $\min(f, 1) \in \mathcal{L}$ for all $f \in \mathcal{L}$. This gives the inclusion $E_c := f^{-1}(c, +\infty) \in \mathcal{R}_L$ for all $c > 0$ and $f \in \mathcal{L}$. Indeed, it suffices to verify this for $c = 1$. Then the functions $\varphi_n := \min(1, nf - n\min(1, f)) \in \mathcal{L}$ increase pointwise to I_{E_1}, $0 \leqslant \varphi_n \leqslant 1$ and $\varphi_n \leqslant |f|$, whence $E(\varphi_n) \leqslant L(|f|)$. Therefore, $f^{-1}((\alpha, \beta]) \in \mathcal{R}_L$ whenever $0 < \alpha < \beta$ and $f \in \mathcal{L}$. It follows from this that all functions from $\mathcal{L}$ are measurable with respect to the σ-algebra generated by $\mathcal{R}_L$. Let $f \in \mathcal{L}$ and $0 \leqslant f \leqslant 1$. Set $f_k := \sum_{n=1}^{2^k - 1} n 2^{-k} I_{(n2^{-k}, (n+1)2^{-k}]}$. Then $f_k \in \mathcal{L}$, $f_k \to f$ and $f_k \leqslant f$. Hence

$$\int_\Omega f_k\,d\mu = L(f_k) \to L(f).$$

Therefore, the function f is integrable with respect to μ and its integral is $L(f)$. Now for every nonnegative function $f \in \mathcal{L}$ we obtain $\min(f, n) \in L^1(\mu)$ and the integral of $\min(f, n)$ equals $L(\min(f, n))$, which gives the μ-integrability of f and the equality of its integral to $L(f)$ by the monotone convergence theorem for the integral and for L. Then this equality remains true for all $f \in \mathcal{L}$. Clearly, $\sigma(\mathcal{F}) \subset \sigma(\mathcal{L})$ and $\mathcal{F} \subset \mathcal{L}$. The uniqueness of a representing measure on the σ-ring generated by the sets of the form $\{f > c\}$, where $f \in \mathcal{F}$ and $c > 0$, is clear from the proof. Finally, if $1 \in \mathcal{F}$, then $\mu(\Omega) = L(1) < \infty$. □

On $\sigma(\mathcal{F})$ the measure μ is not always unique. For example, if $\mathcal{F}$ is the space of all finite Lebesgue integrable functions f on $[0, +\infty)$ with $f(0) = 0$ and L is the Lebesgue integral, then for μ we take any measure $\lambda + c\delta_0$, where λ is Lebesgue measure, δ_0 is Dirac's measure at zero and $c \geqslant 0$.

Typical applications of the Daniell–Stone method are connected with extensions of functionals on spaces of continuous functions.

6.10.76. Theorem. (THE RIESZ THEOREM) *Let K be a compact space and let L be a linear functional on $C(K)$ such that $L(f) \geqslant 0$ if $f \geqslant 0$. Then on the σ-algebra generated by all continuous functions on K there exists a nonnegative finite measure μ such that*

$$L(f) = \int_K f(x)\,\mu(dx), \quad f \in C(K).$$

PROOF. By assumption we have $|L(f)| \leqslant L(1)\max_x |f(x)|$. It remains to observe that if functions $f_n \in C(K)$ decrease pointwise to zero, then by Dini's theorem $\max_x |f_n(x)| \to 0$ and hence $L(f_n) \to 0$. $\qquad\square$

Note that the measure μ uniquely extends to a measure on the Borel σ-algebra of $\mathcal{B}(K)$ (in case of nonmetrizable K the latter can be larger than the σ-algebra $\mathcal{B}a(K)$ generated by $C(K)$) with the following regularity property: $\mu(B) = \sup\{\mu(C)\colon C \subset B \text{ is compact}\}$ for all Borel sets $B \subset K$. A proof can be found in [73, Chapter 7]. If K is metrizable, then $\mathcal{B}(K) = \mathcal{B}a(K)$.

A more general example (with a possibly infinite measure) is obtained if we consider a positive functional on the space $C_0(T)$ of all continuous functions with compact support on a locally compact space T. For example, in this way we can extend the Riemann integral to the Lebesgue integral on $\mathrm{I\!R}^n$ or on a manifold. By the way, it is clear from this why in the definition of the integral on $\mathcal{L}^\uparrow$ we used L-zero sets: if we take only pointwise limits of continuous functions, then not every Lebesgue integrable function will be equal almost everywhere to a function from the obtained class. Finally, we observe that if $1 \in \mathcal{F}$, then representation (6.10.9) takes place without the assumption that L is nonnegative. For this the functional L satisfying the condition $L(f_n) \to 0$ as $f_n \downarrow 0$ can be decomposed into the difference of two nonnegative functionals satisfying the same condition (see Theorem 6.10.61 and [73, §7.8]).

6.10(viii). Interpolation theorems

Here we prove the M. Riesz and Thorin interpolation theorem, which is one of the most important results in the theory of interpolation of linear operators. Let μ and ν be nonnegative measures on measurable spaces $(\Omega_1, \mathcal{A}_1)$ and $(\Omega_2, \mathcal{A}_2)$.

6.10.77. Theorem. *Let $p_0, q_0, p_1, q_1 \in [1, +\infty]$, where $p_0 \neq p_1$ and $q_0 \neq q_1$. Suppose we are given a linear mapping*

$$T\colon L^{p_0}(\mu) \cap L^{p_1}(\mu) \to L^{q_0}(\nu) \cap L^{q_1}(\nu),$$

where we consider complex spaces, such that

$$\|Tf\|_{L^{q_0}(\nu)} \leqslant M_0\|f\|_{L^{p_0}(\mu)} \quad and \quad \|Tf\|_{L^{q_1}(\nu)} \leqslant M_1\|f\|_{L^{p_1}(\mu)}.$$

Then T extends to an operator between the spaces $L^p(\mu)$ and $L^q(\nu)$ with the norm $M \leqslant M_0^{1-\theta}M_1^\theta$ provided that $0 < \theta < 1$ and

$$\frac{1}{p} = \frac{1-\theta}{p_0} + \frac{\theta}{p_1}, \qquad \frac{1}{q} = \frac{1-\theta}{q_0} + \frac{\theta}{q_1}.$$

PROOF. It is clear that T extends to an operator from $L^{p_0}(\mu)$ to $L^{q_0}(\nu)$ and from $L^{p_1}(\mu)$ to $L^{q_1}(\nu)$ with norms not exceeding M_0 and M_1. The point (p^{-1}, q^{-1}) belongs to the interval joining the points (p_0^{-1}, q_0^{-1}) and (p_1^{-1}, q_1^{-1}) in the plane. This explains the term "interpolation". If the measure μ is finite and $p_0 < p_1$, then p^{-1} is between p_1^{-1} and p_0^{-1} and hence $L^p(\mu)$ is between $L^{p_1}(\mu)$ and $L^{p_0}(\mu)$, i.e., T can be restricted from $L^{p_0}(\mu)$ to $L^p(\mu)$. However, it is not obvious at all that $L^p(\mu)$ will take values in $L^q(\nu)$. We observe that $1 < p < \infty$

and $1 < q < \infty$. Hence out further considerations can be conducted for simple integrable functions. Whenever $0 \leqslant \operatorname{Re} z \leqslant 1$, let

$$\frac{1}{p(z)} = \frac{1-z}{p_0} + \frac{z}{p_1}, \quad \frac{1}{q'(z)} = \frac{1-z}{q_0'} + \frac{z}{q_1'},$$

$$\varphi(z) = |f|^{p/p(z)} f/|f|, \quad \psi(z) = |g|^{q'/q'(z)} g/|g|,$$

where f and g are simple integrable functions on Ω_1 and Ω_2, respectively, and $\|f\|_{L^p(\mu)} = \|g\|_{L^{q'}(\nu)} = 1$. Then the function

$$F(z) = \int_{\Omega_2} T\varphi(z)\psi(z)\,d\nu$$

is analytic in the open strip $0 < \operatorname{Re} z < 1$ and continuous and bounded in its closure. Straightforward calculations show that

$$\|\varphi(it)\|_{L^{p_0}(\mu)} = \|\varphi(1+it)\|_{L^{p_1}(\mu)} = \|\psi(it)\|_{L^{q_0'}(\nu)} = \|\psi(1+it)\|_{L^{q_1'}(\nu)} = 1.$$

By assumption, $|F(it)| \leqslant M_0$ and $|F(1+it)| \leqslant M_1$. Note that $F(\theta)$ is the integral of $(Tf)g$ with respect to the measure ν, since $\varphi(\theta) = f$, $\psi(\theta) = g$. Since the norm of T as an operator from $L^p(\mu)$ to $L^q(\nu)$ is the supremum of the values $|F(\theta)|$ over f and g of the indicated form, for obtaining the desired estimate it suffices to apply the Hadamard three lines theorem from complex analysis, which gives the estimate $|F(\theta + it)| \leqslant M_0^{1-\theta} M_1^{\theta}$ for all $t \in \mathbb{R}$. $\qquad\square$

In the real case the same is true with the estimate $M \leqslant 2M_0^{1-\theta} M_1^{\theta}$.

6.10.78. Example. If an operator T belongs to the spaces $\mathcal{L}\big(L^p(\mu), L^p(\nu)\big)$ and $\mathcal{L}\big(L^q(\mu), L^q(\nu)\big)$ with $p < q$, then its norms are finite in $\mathcal{L}\big(L^r(\mu), L^r(\nu)\big)$ with $r \in [p, q]$. If T is bounded as a mapping between $L^1(\mu)$ and $L^1(\nu)$ and also between $L^\infty(\mu)$ and $L^\infty(\nu)$, then T has a finite norm in $\mathcal{L}\big(L^p(\mu), L^p(\nu)\big)$ for all $p \in (1, \infty)$. This norm does not exceed 1 if this is true for the norms on L^1 and on L^∞ (over $\mathbb{C}$).

For other results, including the Marcinkiewicz interpolation theorem, see Bennett, Sharpley [**54**], Bergh, Löfström [**61**], Krein, Petunin, Semenov [**353**], Lunardi [**398**], and Triebel [**607**].

Exercises

6.10.79.° Give an example of a discontinuous function $f\colon \mathbb{R} \to \mathbb{R}$ with a closed graph.

HINT: consider the function $f(x) = 1/x$, $x \neq 0$, $f(0) = 0$.

6.10.80.° Let X, Y be normed spaces and let $\lim\limits_{n\to\infty} A_n x = Ax$ for all $x \in X$, where $A, A_n \in \mathcal{L}(X, Y)$. Prove that $\|A\| \leqslant \liminf_{n\to\infty} \|A_n\|$, and give an example where the equality fails.

HINT: consider the one-dimensional projections generated by the standard orthonormal basis in l^2.

6.10.81.° Let X be a normed space and $f \in X^*$, $\|f\| = 1$. Prove that for every x the equality $\operatorname{dist}(x, \operatorname{Ker} f) = |f(x)|$ holds.

6.10.82. Let Y be a closed subspace in a Banach space X and $x_0 \in X\backslash Y$. Prove that there exists a functional $l \in X^*$ such that $\|l\| = 1$, $l|_Y = 0$ and $l(x_0) = \text{dist}\,(x_0, Y)$.

HINT: let $d = \text{dist}\,(x_0, Y)$; on the linear span of Y and x_0 consider the functional $l(y + tx_0) = dt$. Clearly, $l|_Y = 0$. Since $\|y + tx_0\| \geqslant d|t|$, one has $\|l\| \leqslant 1$. Take $\{y_n\} \subset Y$ with $\|y_n - x_0\| \to d$, which gives $d = |l(x_0) - l(y_n)| \leqslant \|l\|\,\|x_0 - y_n\| \to d\|l\|$, whence $\|l\| = 1$. Now extend the functional l to X with the same norm.

6.10.83. Show that the cardinality of an infinite-dimensional Banach space is always strictly smaller than that of its algebraic dual.

HINT: use a Hamel basis and Exercise 5.6.55 along with Cantor's theorem.

6.10.84. Suppose that a linear space E is equipped with two non-equivalent norms. Prove that the duals to E with respect to these norms are different as sets of linear functions.

HINT: if the unit ball U with respect to the first norm p_1 is not bounded with respect to the second norm p_2, then it is not weakly bounded in (E, p_2) and hence there exists a functional l continuous in the norm p_2 such that $\sup_{u \in U} |l(u)| = \infty$.

6.10.85. Let Y be a linear subspace in a normed space X with separable dual X^*. Prove that Y^* is separable.

HINT: apply the Hahn–Banach theorem.

6.10.86. Let X be a normed space such that X^* is norm separable. Prove that X is separable.

HINT: take a countable set $\{l_n\}$ dense in the unit sphere of X^*; for every n find x_n with $\|x_n\| = 1$ and $l_n(x_n) > 1/2$. Let Y be the closure of the linear span of $\{x_n\}$. Then Y is separable. If $Y \neq X$, then there exists $l \in X^*$ with $\|l\| = 1$ and $l|_Y = 0$. Let us take l_n with $\|l - l_n\| < 1/2$. Then we obtain $|l_n(x_n)| = |(l_n - l)(x_n)| \leqslant \|l_n - l\| < 1/2$, which is a contradiction.

6.10.87. Let S_n be the operator on $L^1[0, 2\pi]$ that to a function f associates the nth partial sum $S_n(f)$ of its Fourier series. Prove that $\sup_n \|S_n\| = \infty$. Deduce from this that there exist functions in $L^1[0, 2\pi]$ whose Fourier series do not converge in L^1.

6.10.88. Let X and Y be Banach spaces, $A \in \mathcal{L}(X, Y)$, and let $\widetilde{A} \colon X/\text{Ker}\,A \to Y$, $\widetilde{A}[x] := Ax$, be the operator on the quotient space induced by A.
(i) Prove that $\|\widetilde{A}\| = \|A\|$. (ii) Prove that $\|(\widetilde{A})^* y^*\| = \|A^* y^*\|$, $y^* \in Y^*$.

6.10.89. Let X and Y be normed spaces and let $J_X \colon X \to X^{**}$ and $J_Y \colon Y \to Y^{**}$ be the canonical embeddings. Prove that $T^{**} J_X = J_Y T$ for every operator $T \in \mathcal{L}(X, Y)$. Hence $T^{**} J_X(X) \subset J_Y(Y)$.

6.10.90. Prove that every element $l \in C[0, 1]^*$ is the limit in the weak-$*$ topology of a sequence of functionals of the form $x \mapsto \sum_{i=1}^n c_i x(t_i)$, where $t \in [0, 1]$ and c_i are scalars.

HINT: l is given by a bounded Borel measure μ; one can partition $[0, 1]$ into intervals $I_1 = [0, 1/n)$, $I_2 = [1/n, 2/n)$ and so on and take $t_i = i/n$, $c_i = \mu(I_i)$.

6.10.91. Let C be a convex balanced set in a normed space X and let f be a linear function on X such that the restriction of f to C is continuous at the origin. Prove that f is uniformly continuous on C.

6.10.92. Let X and Y be normed spaces, let X be complete, and let $T \in \mathcal{L}(X, Y)$ be an open mapping. Prove that Y is complete.

6.10.93. (i) Prove that there exists a discontinuous linear mapping that maps a normed space one-to-one onto a Banach space and has a closed graph. (ii) Prove that there exists a discontinuous linear mapping that maps a Banach space one-to-one onto a normed space and has a closed graph.

HINT: (i) consider an injective compact diagonal operator on l^2 and its inverse defined on its range. (ii) Take a Hamel basis $\{h_\alpha\}$ in l^2 with $\|h_\alpha\| = 1$ and consider the norm $\|x\| = \sum_\alpha |x_\alpha|$, $x = \sum_\alpha x_\alpha h_\alpha$ and the identity mapping to this norm from the standard norm on l^2.

6.10.94. Can a continuous linear operator map an incomplete normed space one-to-one onto a complete one?

HINT: use the previous exercise.

6.10.95. Let X and Y be Banach spaces and $T \in \mathcal{L}(X, Y)$. Show that if T takes every bounded closed set to a closed set, then $T(X)$ is closed. Construct an example showing that the closedness of the images of closed balls can be insufficient for this.

HINT: considering the quotient by the kernel reduce the general case to an injective operator and observe that if $Tx_n \to y$, then either $\{x_n\}$ contains a bounded subsequence of $\|x_n\| \to \infty$, so the vectors $x_n/\|x_n\|$ belong to the unit sphere whose image is closed, but $T(x_n/\|x_n\|) \to 0$.

6.10.96. Suppose that a sequence $\{x_n\}$ in a normed space is Cauchy in norm and converges weakly to some vector x. Prove that $\{x_n\}$ converges to x in norm.

HINT: observe that this is true in complete spaces and consider the completion.

6.10.97. Let X be a normed space, $f \in X^*$, $\|f\| = 1$. Prove that f attains its maximum on the unit sphere precisely when $f^{-1}(1)$ has a vector of minimal norm. This is also equivalent to the property that $\operatorname{Ker} f$ has a nearest element to some vector outside $\operatorname{Ker} f$.

HINT: observe that if e is a vector of minimal norm in $f^{-1}(1)$, then $\|e\| = 1$, since in case $\|e\| > 1$ we can find a unit vector u with $f(u) > 1/\|e\|$ and take $v = u/f(u)$ with $f(v) = 1$ and $\|v\| < \|e\|$. On the other hand, if $f(e) = 1$ and $\|e\| = 1$, then there is no element in $f^{-1}(1)$ with a smaller norm.

6.10.98. (i) Find an example of two closed linear subspaces H_1 and H_2 in a Hilbert space for which $H_1 \cap H_2 = \{0\}$, but the algebraic sum of H_1 and H_2 is not closed (see also Exercise 5.6.57).

(ii)* Prove that such an example exists in every infinite-dimensional Banach space.

HINT: (i) consider the operator T on l^2 given by the formula $Tx = (2^{-n}x_n)$, let H_1 be the graph of T in $l^2 \oplus l^2$, $H_2 := l^2 \oplus \{0\} \subset l^2 \oplus l^2$; then H_1 and H_2 possess the required properties. (ii) Consider a separable infinite-dimensional Banach space X and observe that X^* contains a linearly independent weak-$*$ dense sequence $\{f_n\}$; let L_n be the intersection of the hyperplanes $f_i^{-1}(0)$, $i \leqslant 2n$; take linearly independent vectors x_n and y_n in the algebraic complement of L_n to L_{n+1} with $\|x_n\| \geqslant 1$, $\|y_n\| \geqslant 1$, $\|x_n - y_n\| \leqslant 1/n$ and consider the closed linear spans of $\{x_n\}$ and $\{y_n\}$. To see that their intersection is zero, for each n, find linear combinations φ_n and ψ_n of f_{2n+1} and f_{2n+2} such that $\varphi_n(x_n) = \psi_n(y_n) = 0$, $\varphi_n(y_n) = \psi_n(x_n) = 1$ and observe that the closed linear span of $x_n, x_{n+1}, \ldots$ is contained in the intersection of $\varphi_n^{-1}(0)$ and $f_i^{-1}(0)$ with $i > n$.

6.10.99. Let Y and Z be closed subspaces in a Banach space X and let $Y \cap Z = 0$. Prove that the sum $Y + Z$ is closed precisely when $\operatorname{dist}(S_Y, S_Z) > 0$, where S_Y and S_Z are the unit spheres in Y and Z.

6.10.100. Apply the Hahn–Banach theorem to construct a continuous linear functional l on the set of all bounded functions on the interval $[0, 1]$ such that the action of l on continuous functions coincides with the Riemann integral, but there exists a Borel set B for which $l(I_B)$ does not coincide with the Lebesgue measure of B.

6.10.101. Let X and Y be Banach spaces and $A \in \mathcal{L}(X, Y)$. Prove that the equality $A^*(Y^*) = X^*$ holds precisely when A has the zero kernel and a closed range.

HINT: use that A has the zero kernel if and only if the range of A^* is dense and apply Corollary 6.8.6.

6.10.102. Let X and Y be Banach spaces and let X be reflexive. Prove that for every operator $A \in \mathcal{L}(X, Y)$ the image of any closed ball is closed.

HINT: use the weak compactness of balls in X.

6.10.103. Suppose that $C[0, 1]$ is equipped with a Banach norm such that convergence in this norm implies the pointwise convergence. Prove that this norm is equivalent to the usual sup-norm.

HINT: use the closed graph theorem.

6.10.104. (Schur's theorem) Prove that in the space l^1 every weakly convergent sequence converges in norm. Deduce from this that l^1 and $L^1[0, 1]$ are not linearly homeomorphic.

HINT: consider the case of weak convergence to zero and argue from the opposite.

6.10.105. (i) Prove that l^1 has no subspaces linearly homeomorphic to l^2. (ii) Let $A: l^1 \to l^2$ be a continuous linear surjection from Theorem 6.10.8. Prove that l^1 has no subspaces isomorphic to $l^1/\mathrm{Ker}\, A$. In particular, l^1 is not isomorphic to $l^2 \times (l^1/\mathrm{Ker}\, A)$.

HINT: in (i) use Schur's theorem and weak convergence to zero of vectors of the standard basis in l^2; derive (ii) from (i) and the fact that l^2 is isomorphic to $l^1/\mathrm{Ker}\, A$.

6.10.106. Show that the unit ball in $L^1[0, 1]$ has no extreme points (see §5.4).

6.10.107. Prove that the spaces $c_0 \oplus c_0$ (the sum is equipped with the norm equal the sum of the norms of components) and c_0 are isomorphic (i.e., are linearly homeomorphic), but are not isometric.

HINT: if ψ is a linear isometry between these spaces, then each element $\psi(e_n, 0)$, where $\{e_n\}$ is the standard basis in c_0, has 1 and -1 at positions of some finite set S_n. Clearly, $S_n \cap S_m$ if $n \neq k$, since $\|e_n + e_k\| = 1$, $\|e_n - e_k\| = 1$. Hence there is n such that the components of $\psi(0, e_1)$ with indices in S_n are strictly less than 1 in absolute value. This leads to a contradiction, since the norm of (e_n, e_1) equals 2.

6.10.108. Let X and Y be Banach spaces and let $T \in \mathcal{L}(X, Y)$ be an operator such that T^* is an isometry between Y^* and X^*. Is it true that T is an isometry?

6.10.109. Let $f: l^\infty \to l^\infty$, $f: (x_n) \mapsto \left(\tan\left((2\pi^{-1} \arctan x_n)^{2n+1}\right)\right)$. Show that f is continuous and maps l^∞ one-to-one onto l^∞, but f^{-1} is discontinuous at the origin.

6.10.110. Let L be a linear subspace of a normed space X. Prove that every norm on L equivalent to the original one can be extended to a norm on X also equivalent to the original one.

HINT: let U be the unit ball with respect to the original norm and let V be the unit ball in L with respect to the equivalent norm. We can assume that $U \cap L \subset V$. Now take the Minkowski functional of the convex envelope of $U \cup V$.

6.10.111. Show that on the spaces $L^p[0, 1]$, $1 \leqslant p < \infty$, there exists no linear lifting, i.e., one cannot choose a representative Lf in every equivalence class $f \in L^p[0, 1]$ such that $L(f + g)(t) = Lf(t) + Lg(t)$ and $L(cf)(t) = cf(t)$ for all $f, g \in L^p[0, 1]$, all $c \in \mathbb{R}^1$ and all $t \in [0, 1]$, and $Lf(t) \geqslant 0$ for all t if $f \geqslant 0$ a.e.

HINT: if L is a linear lifting on $L^p[0, 1]$, $1 \leqslant p < \infty$, then for every t the functional $l_t(f) = L(f)(t)$ on $L^p[0, 1]$ is linear and nonnegative on nonnegative functions, whence by Proposition 6.10.55 its continuity follows. Hence the functional l_t is given by a function $g_t \in L^{p'}[0,1]$. For every n we partition $[0,1]$ into n intervals $J_{n,1}, \ldots, J_{n,n}$ by the points k/n. Let $E_{n,k} := \{x\colon L(I_{J_{n,k}})(x) = 1\}$ and $E_n := \bigcup_{k=1}^n E_{n,k}$. Then $\lambda(E_n) = 1$ by properties of liftings. There is a point $t \in \bigcap_{n=1}^\infty E_n$. For every n there is $j(n)$ with $t \in E_{n,j(n)}$, i.e., $L(I_{J_{n,j(n)}})(t) = 1$. Since $L(I_{J_{n,k}}) = I_{J_{n,k}}$ a.e., we have

$$L(I_{J_{n,k}})(t) = \int_0^1 I_{J_{n,k}}(s)g_t(s)\,ds \leqslant n^{-1/p}\|g_t\|_{L^{p'}}$$

for all k, which leads to a contradiction.

6.10.112. Extend Goldstein's Theorem 6.7.6 to the case of the sphere: the unit sphere of a Banach space X is dense in the unit sphere of X^{**} with the topology $\sigma(X^{**}, X^*)$.

6.10.113. Prove that every separable Banach space is linearly isometric to a closed subspace in l^∞.

HINT: find a sequence of functionals f_n of unit norm on the given space X such that $\|x\| = \sup_n |f_n(x)|$ for all $x \in X$; observe that it suffices to have the latter for all vectors from a countable dense set.

6.10.114. Prove that on every separable Banach space there is an equivalent strictly convex norm.

HINT: embed this space as a closed subspace in the space $C[0,1]$ with its sup-norm and take the norm $\|x\|_0 = \|x\|_{C[0,1]} + \|x\|_{L^2[0,1]}$.

6.10.115.° Let $\mathcal{K}_n \in C([a,b]^2)$, $n \in \mathbb{N}$, and let K_n be the operator with the integral kernel $\mathcal{K}_n$ on $C[a,b]$. Prove that $K_n f \to f$ for every $f \in C[a,b]$ precisely when

1) $K_n f \to f$ for all f from a dense set, 2) $\displaystyle\sup_n \max_x \int_a^b |\mathcal{K}_n(x,t)|\,dt < \infty$.

6.10.116. Let X be a Banach space and let Y be its n-dimensional subspace. Prove that there exists a linear projection $P\colon X \to Y$ for which $\|P\| \leqslant n$.

HINT: use Proposition 6.10.45.

6.10.117. (R. Phillips) Prove that c_0 is not complemented in l^∞.

HINT: see [9, §2.5].

6.10.118. Let X be a reflexive Banach space. Prove that every bounded sequence in X contains a weakly convergent subsequence.

6.10.119. Let K be a weakly compact set in a Banach space X such that X^* contains a countable set $\{f_n\}$ separating points of K. Show that $(K, \sigma(X, X^*))$ is a metrizable compact space.

HINT: we can assume that $\|f_n\| \leqslant 1$; take the metric $d(x,y) := \sum_{n=1}^\infty 2^{-n}|f_n(x-y)|$ and verify that the identity mapping from $(K, \sigma(X, X^*))$ to (K, d) is continuous, so it is a homeomorphism.

6.10.120. Let X be a Banach space. Prove that the unit ball in X^* is metrizable in the weak-∗ topology if and only if X is separable.

HINT: use that if the ball is metrizable in the weak-∗ topology, then there is a countable collection of vectors in X determining a basis of neighborhoods of zero in the weak-∗ topology.

6.10.121. Let X be an infinite-dimensional Banach space such that X^* is norm separable. (i) Prove that there exists a sequence of vectors $x_n \in X$ with $\|x_n\| = 1$ weakly converging to zero. (ii) Prove that there exists a sequence of vectors $x_n \in X$ with $\|x_n\| = 1$

weakly converging to zero and a sequence of functionals $l_n \in X^*$ with $l_n(x_n) = 1$ that is weak-∗ convergent to zero. Consider l^1 to see that (ii) can fail for spaces with nonseparable duals.

HINT: (i) use that X is also separable (Exercise 6.10.86) and the unit ball in X^* is metrizable in the weak-∗ topology, hence one can pick points in the intersection of the unit sphere with elements of a countable basis of zero in the weak-∗ topology. (ii) See [75, Exercise 3.12.127].

6.10.122.∗ Let X be a Banach space and let $C \subset X^*$ be a weak-∗ compact set. (i) Show that if C is norm separable, then the norm closure of the convex envelope of C is also weak-∗ compact. (ii) Show that without the separability of C the conclusion in (i) can be false by taking for C the set of Dirac measures δ_t in $C[0,1]^*$.

HINT: (i) see [185, Exercise 106, p. 104]. (ii) Show that the norm closure of the set of Dirac measures consists of all probability measures concentrated on countable sets, which is not closed in the weak-∗ topology.

6.10.123. Let X be a metrizable compact set and let $f\colon X \to Y$ be a continuous mapping, where Y is a Hausdorff space. Prove that the compact set $f(X)$ is also metrizable.

HINT: the space $C(X)$ is separable and $C\big(f(X)\big)$ is isometrically embedded into it by the mapping $\varphi \mapsto \varphi \circ f$. Hence $C\big(f(X)\big)$ is also separable, which yields the metrizability of $f(X)$.

6.10.124. Let X and Y be Banach spaces and let $A, B \in \mathcal{L}(X, Y)$. Prove that the following conditions are equivalent: (i) $A^*(Y^*) \subset B^*(Y^*)$,

(ii) there is a number k such that $\|Ax\| \leqslant k\|Bx\|$ for all $x \in X$.

HINT: if such k exists, then for every $y^* \in Y^*$ the functional f on $B(X)$ given by the formula $f(Bx) := y^*(Ax)$ is well-defined and bounded, since

$$|y^*(Ax)| \leqslant \|y^*\|\,\|Ax\| \leqslant k\|y^*\|\,\|Bx\|.$$

By the Hahn–Banach theorem it extends to an element $z^* \in Y^*$. Then $B^*z^* = A^*y^*$, whence $A^*(Y^*) \subset B^*(Y^*)$. Conversely, if this inclusion holds, then $A^* = \widetilde{B^*}C$, where the operator $\widetilde{B^*}\colon Y^*/\mathrm{Ker}\,B^* \to X^*$ is generated by the operator B^* and C is a continuous operator (see Theorem 6.10.5). Then $A^{**} = C^*(\widetilde{B^*})^*$, which gives the desired estimate by the known equalities $\|A^{**}\| = \|A^*\| = \|A\|$, $\|(\widetilde{B^*})^*\| = \|\widetilde{B^*}\| = \|B^*\| = \|B\|$ and Exercise 6.10.88.

6.10.125. Let X and Y be Banach spaces and let $A, B \in \mathcal{L}(X, Y)$ be such that $A(X) \subset B(X)$. (i) Prove that there exists a number k such that

$$\|A^*y^*\|_{X^*} \leqslant k\|B^*y^*\|_{X^*}, \quad y^* \in Y^*.$$

(ii) Show that $A^{**}(X^{**}) \subset B^{**}(X^{**})$.

(iii) Show that if X is reflexive, then the estimate in (i) is equivalent to the inclusion $A(X) \subset B(X)$.

HINT: (i) under the same notation as in the hint to the previous exercise we have $A = \widetilde{B}C$, whence

$$\|A^*y^*\|_{X^*} = \|C^*\widetilde{B}^*y^*\|_{X^*} \leqslant \|C^*\|\,\|B^*y^*\|_{X^*}$$

by Exercise 6.10.88. Now (ii) and (iii) follow from the previous exercise.

6.10.126. Let X and Y be Banach spaces, $A \in \mathcal{L}(X, Y)$, $A(X) = Y$, and let Y be separable. Prove that X contains a closed separable subspace Z such that $A(Z) = Y$.

HINT: take a countable set $\{y_n\}$ dense in the unit ball of Y, use Remark 6.2.4 to pick a bounded countable set $\{x_n\} \subset X$ with $Ax_n = y_n$ and take for Z the closure of the linear span of $\{x_n\}$; apply Lemma 6.2.1.

6.10.127.* Prove that the space $C[0,1]$ can be mapped linearly and continuously onto c_0, but not onto l^1. Deduce from this that l^1 embedded isometrically into $C[0,1]$ is not complemented.

HINT: the first assertion is clear from Corollary 6.10.48; for the second assertion, see [**185**, p. 274]. Note that every infinite-dimensional complemented closed subspace in $C[0,1]$ contains a complemented subspace isomorphic to c_0 (see [**185**, Proposition 5.6.4]). There is an unproved conjecture that every infinite-dimensional complemented subspace X in $C[0,1]$ is isomorphic to the space $C(K)$ for some metric compact space K; H. Rosenthal proved that if X^* is nonseparable, then this true with $K = [0,1]$.

6.10.128. (i) Prove that $L^1[0,1]$ contains a complemented subspace isometric to l^1. (ii) Prove that $L^1[0,1]$ embedded isometrically into $C[0,1]$ as a closed subspaces is not complemented.

HINT: (i) take functions constant on $[2^{-1-n}, 2^{-n})$; (ii) apply (i) and the previous exercise. According to an unproved conjecture, every infinite-dimensional complemented subspace X in $L^1[0,1]$ is isomorphic either to l^1 or to $L^1[0,1]$.

6.10.129.* (i) (I. Kaplansky) Let A be a set in a Banach space X and let a point x belong to the closure of A in the weak topology. Prove that x belongs to the weak closure of some countable subset in A.

(ii) Let A be a subset of a weakly compact set in a Banach space X and let a point x belong to the closure of A in the weak topology. Prove that x is the limit of some sequence $\{a_n\} \subset A$ in the weak topology.

HINT: see [**185**, Theorems 4.49 and 4.50, p. 129–130].

6.10.130. (E. A. Lifshits [**695**]) A set W in a Banach space X is called *ideally convex* if the series $\sum_{n=1}^{\infty} \alpha_n x_n$ converges in X for every bounded sequence $\{x_n\} \subset W$ and every sequence of numbers $\alpha_n \geqslant 0$ with $\sum_{n=1}^{\infty} \alpha_n = 1$.

(i) Prove that in a finite-dimensional space any convex set is ideally convex.

(ii) Give an example of a convex set that is not ideally convex.

(iii) Prove that if a convex set is closed or open, then it is ideally convex.

(iv) Prove that if a set W is ideally convex and $T\colon Z \to X$ is a continuous linear operator from a Banach space Z, then $T^{-1}(W)$ is ideally convex.

(v) Prove that if a set W is ideally convex and bounded and $T\colon X \to Y$ is a continuous linear operator to a Banach space Y, then $T(W)$ is ideally convex.

(vi) Let W be an ideally convex set. Prove that the interior of W coincides with the interior of the closure of W and also with the algebraic kernel of W and the algebraic kernel of the closure of W.

(vii) From the previous results deduce the Banach–Steinhaus theorem and the open mapping theorem.

6.10.131. Let E be a closed linear subspace in $C[0,1]$ such that $E \subset C^1[0,1]$. Prove that $\dim E < \infty$.

HINT: observe that $C^1[0,1]$ belongs to the range of a compact operator and hence cannot contain infinite-dimensional closed subspaces.

6.10.132.° Let X and Y be normed spaces and let $A\colon X \to Y$ be a linear mapping continuous from the weak topology to the norm topology. Prove that $A(X)$ is finite-dimensional.

HINT: use that a weak neighborhood of zero in an infinite-dimensional space contains a subspace of finite codimension.

6.10.133. Let X and Y be Banach spaces and let $J_X : X \to X^{**}$ and $J_Y : Y \to Y^{**}$ be the canonical embeddings. Let $S \in \mathcal{L}(Y^*, X^*)$ and $S^* J_X(X) \subset J_Y(Y)$. Prove that $S = T^*$, where $T \in \mathcal{L}(X, Y)$ is defined as follows: $T = J_Y^{-1} S^* J_X$.

6.10.134. Let X be a normed space, $M > 0$, $\{x_n\} \subset X$, $\{c_n\} \subset \mathbb{R}^1$. Prove that the existence of a functional $f \in X^*$ with $\|f\| \leq M$ and $f(x_n) = c_n$ for all n is equivalent to the condition that $\left| \sum_{i=1}^n \lambda_i c_i \right| \leq M \left\| \sum_{i=1}^n \lambda_i x_i \right\|$ for all n and all $\lambda_i \in \mathbb{R}^1$.

6.10.135.° Suppose that we are given two sequences of numbers $a_n > 0$ and $b_n > 0$, where $\{b_n\}$ decreases to zero and $\{a_n b_n\}$ has a finite limit. Let us define an operator $T : C[0,1] \to c$ by the formula
$$(Tx)_n := a_n \int_0^{b_n} x(t)\, dt.$$
Prove that K is compact precisely when $a_n b_n \to 0$.

HINT: show that $\|K\| = \sup_n |a_n b_n|$.

6.10.136. (Holmgren's theorem) Let μ and ν be probability measures on spaces Ω_1 and let Ω_2 and let $\mathcal{K}$ be a $\mu \otimes \nu$-measurable function such that
$$C_1 = \operatorname{ess\,sup}_s \int_{\Omega_2} |\mathcal{K}(s,t)|\,\nu(dt) < \infty, \quad C_2 = \operatorname{ess\,sup}_t \int_{\Omega_1} |\mathcal{K}(s,t)|\,\mu(ds) < \infty.$$
Prove that the operator
$$Kx(t) = \int_{\Omega_2} \mathcal{K}(s,t) x(s)\,\mu(ds), \quad K : L^2(\mu) \to L^2(\nu),$$
is bounded and $\|K\| \leq C_1^{1/2} C_2^{1/2}$.

HINT: let $C_2 > 0$ and $c = C_1^{1/2} C_2^{-1/2}$; for $x \in \mathcal{L}^\infty(\mu)$ and $y \in \mathcal{L}^\infty(\nu)$ such that $\|x\|_{L^2(\mu)} \leq 1$ and $\|y\|_{L^2(\nu)} \leq 1$, we have
$$\left| \int_{\Omega_2} \int_{\Omega_1} \mathcal{K}(s,t) x(s) y(t)\, \mu(ds)\, \nu(dt) \right| \leq \int_{\Omega_2} \int_{\Omega_1} |\mathcal{K}(s,t)| \left[2^{-1} c |y(t)|^2 \right.$$
$$\left. + 2^{-1} c^{-1} |x(s)|^2 \right] \mu(ds)\, \nu(dt) \leq 2^{-1} c C_2 + 2^{-1} c^{-1} C_1 = C_1^{1/2} C_2^{1/2},$$
whence the desired bound follows.

6.10.137. (Schur's test) Let μ and ν be nonnegative measures on measurable spaces T and S, $\mathcal{K} \geq 0$ a measurable function on $T \times S$, and let $\varphi > 0$ and $\psi > 0$ be measurable functions on T and S, respectively, such that
$$\int_S \mathcal{K}(t,s) \psi(s)\, \nu(ds) \leq \alpha \varphi(t) \quad \mu\text{-a.e.},$$
$$\int_T \mathcal{K}(t,s) \varphi(t)\, \mu(dt) \leq \beta \psi(s) \quad \nu\text{-a.e.},$$
where α and β are numbers. Prove that the operator
$$K : L^2(\nu) \to L^2(\mu), \quad Kx(t) := \int_S \mathcal{K}(t,s) x(s)\, \nu(ds),$$
is bounded and $\|K\|^2 \leq \alpha\beta$. Deduce from this the assertion of the previous exercise.

HINT: observe that

$$\|Kx\|_{L^2(\mu)}^2 \leqslant \int_T \left(\int_S \mathcal{K}(t,s)\psi(s)\, \nu(ds) \right) \left(\int_S \mathcal{K}(t,s)\psi(s)^{-1}|x(s)|^2\, \nu(ds) \right) \mu(dt)$$

$$\leqslant \alpha \int_T \varphi(t) \left(\int_S \mathcal{K}(t,s)\psi(s)^{-1}|x(s)|^2\, \nu(ds) \right) \mu(dt) \leqslant \alpha\beta \|x\|_{L^2(\nu)}^2.$$

6.10.138. Prove that the formula

$$Ax(t) = \frac{1}{t} \int_0^t x(s)\, ds$$

defines a bounded operator on $L^2[0,1]$.

HINT: apply Schur's test to the kernel $\mathcal{K}(t,s) = t^{-1}$ if $s < t$ and $\mathcal{K}(t,s) = 0$ if $s \geqslant t$ and the function $\varphi(t) = \psi(t) = t^{-1/2}$.

6.10.139. Let X and Y be Banach spaces and $A \in \mathcal{L}(X,Y)$. Prove that the compactness of the operator A is equivalent to the following condition: there exist functionals $l_n \in Y^*$ with $\|l_n\| \to 0$ such that $\|Ax\| \leqslant \sup_n |l_n(x)|$ for all $x \in X$.

Deduce from this assertion that the compactness of A is equivalent to the existence of a bounded sequence $\{f_n\}$ in X^* and a sequence of numbers λ_n such that $\lambda_n \to 0$ and $\|Ax\| \leqslant \sup_n |\lambda_n f_n(x)|$ for all $x \in X$.

HINT: if $A \in \mathcal{K}(X,Y)$, then $A^* \in \mathcal{K}(Y^*, X^*)$. Let W be the unit ball in Y^*. Then $A^*(W)$ is contained in the closure of the convex envelope of the sequence $l_n \to 0$ (Proposition 5.5.7). Hence for every $x \in X$ we have

$$\|Ax\| = \sup_{f \in W} |f(Ax)| = \sup_{l \in A^*(W)} |l(x)| \leqslant \sup_n |l_n(x)|.$$

Conversely, if the indicated condition is fulfilled, then $S = \{l_n\} \cup \{0\}$ is compact in Y^*. The image of the unit ball U in X is totally bounded as a set in $C(S)$ by the Arzelà–Ascoli theorem, since $|l(Ax) - l'(Ax)| \leqslant \|A\| \|l - l'\|$ whenever $\|x\| \leqslant 1$, which means the uniform Lipschitzness of the elements Ax as functions on S. The estimate from our condition shows that $\|Ax\|$ is estimated by the norm of Ax as an element of $C(S)$. Hence the set $A(U)$ is totally bounded in Y. The second assertion follows from the first one.

6.10.140. Let X and Y be Banach spaces and let Y be separable. Prove that the compactness of an operator $A \in \mathcal{L}(X,Y)$ is equivalent to the following condition: for every sequence $\{y_n^*\} \subset Y^*$ that is weak-$*$ convergent to zero we have $\|A^*y_n^*\| \to 0$.

HINT: in one direction, use that the adjoint of a compact operator is compact. In the opposite direction, use that for a bounded sequence of vectors $x_n \in X$, the sequence of vectors $Ax_n \in Y$ regarded as functions on the unit ball in Y^* with a metric defining the weak-$*$ topology is uniformly equicontinuous due to the given condition.

6.10.141. Let X be a reflexive Banach space. Prove that every operator $A \in \mathcal{L}(X, l^1)$ is compact.

HINT: use that weakly convergent sequences in l^1 are norm convergent.

6.10.142. (i) Let μ be a nonnegative measure on a measurable space $(\Omega, \mathcal{A})$ and let $(t,s) \mapsto \mathcal{K}(t,s)$ be a measurable function such that the function $s \mapsto \mathcal{K}(t,s)$ belongs to $\mathcal{L}^2(\mu)$ for μ-a.e. t and the function

$$Kx(t) = \int_\Omega \mathcal{K}(t,s)x(s)\, \mu(ds)$$

belongs to $\mathcal{L}^2(\mu)$ for all $x \in \mathcal{L}^2(\mu)$. Prove that K is a bounded operator on $L^2(\mu)$.

(ii) Generalize (i) to the case where it is only known that, for every function x from $\mathcal{L}^2(\mu)$, the function $s \mapsto \mathcal{K}(t,s)x(s)$ is integrable and $Kx \in \mathcal{L}^2(\mu)$.

(iii) Show that if in (i) or (ii) the operator K is zero, then $\mathcal{K}(t,s) = 0$ $\mu \otimes \mu$-a.e.

(iv) Let μ be Lebesgue measure on $[0,1]$. Prove that the unit operator I cannot be represented in the form indicated in (ii).

(v) Let the measure μ in (ii) be finite. Show that the operator $K \colon L^2(\mu) \to L^1(\mu)$ is compact, although it need not be compact as an operator with values in $L^2(\mu)$.

(vi) Let μ be Lebesgue measure on $[0,1]$, $1/2 \leqslant \alpha < 1$, and let $\mathcal{K}_\alpha(t,s) = |t-s|^{-\alpha}$ if $t > s$ and $\mathcal{K}_\alpha(t,s) = 0$ if $t \leqslant s$. Prove that the integral kernel $\mathcal{K}_\alpha$ generates a bounded operator on $L^2[0,1]$.

HINT: in (i) and (ii) apply the closed graph theorem (see [**252**, Theorem 3.10]). For assertions (iii)–(vi), see [**252**, Theorem 8.1, Theorem 8.5, Theorem 13.8, Example 11.1].

6.10.143. (i) Let $\Omega \subset \mathrm{I\!R}^d$ be a bounded measurable set, let $\mathcal{K}_0$ be a bounded measurable function, and let $\mathcal{K}(t,s) = \mathcal{K}_0(t,s)|t-s|^{-\alpha}$, where $\alpha < d$. Prove that the operator K defined by the kernel $\mathcal{K}$ on $L^2(\Omega)$ is compact.

(ii) Prove that if the function $\mathcal{K}_0$ is continuous in t, then the operator K is compact also on $C(\Omega)$.

6.10.144.° Is the operator $Ax(t) = x(\sqrt{t})$ compact on $C[0,1]$? On $L^2[0,1]$?

6.10.145.° Suppose we are given a sequence of disjoint intervals $[a_n, b_n]$ in $[0,1]$. Prove that the operator on $L^1[0,1]$ defined by the integral kernel

$$\mathcal{K}(t,s) = \sum_{n=1}^{\infty} (b_n - a_n)^{-1} I_{[a_n,b_n]}(t) I_{[a_n,b_n]}(s),$$

is not compact.

6.10.146. Prove that every compact operator T on $L^1[0,1]$ can be expressed in the form $Tx(t) = \displaystyle\int_0^1 \mathcal{K}(t,s)x(s)\,ds$ with a measurable kernel $\mathcal{K}(\,\cdot\,,\,\cdot\,)$ such that the condition $\sup_s \|\mathcal{K}(\,\cdot\,,s)\|_{L^1} < \infty$ is fulfilled.

HINT: see [**164**, p. 508].

6.10.147. Let X and Y be Banach spaces. An operator $T \colon X \to Y$ is called *completely continuous* if it takes weakly compact sets to norm compact sets.

(i) Prove that an operator $T \in \mathcal{L}(X,Y)$ is completely continuous precisely when it takes weakly convergent sequences to norm convergent sequences.

(ii) Prove that the set of completely continuous operators is a closed linear subspace in $\mathcal{L}(X,Y)$.

(iii) Prove that every operator $T \in \mathcal{L}(l^1, Y)$ is completely continuous. In particular, the identity operator on l^1 is completely continuous, but not compact.

6.10.148. Let X be an infinite-dimensional Banach space. Prove that X^* with the weak-$*$ topology is not metrizable. Show that the completeness is important considering the subspace E of finite sequences in c_0 (here $E^* = l^1$, but $\sigma(l^1, E) \neq \sigma(l^1, c_0)$).

HINT: observe that if such a metric exists, then the corresponding ball of radius $1/n$ centered at zero in X^* must contain a weak-$*$ neighborhood of zero, hence must contain a functional f_n with $\|f_n\| = n$, which is impossible, since the sequence $\{f_n\}$ converges to zero pointwise, but is not norm bounded.

6.10.149. Give an example of a sequence of continuous linear functionals f_n on a Banach space X that is weak-$*$ convergent to zero, but the convex envelope of $\{f_n\}$ is contained in the sphere $\{f \colon \|f\| = 1\}$.

HINT: consider $X = c_0$ and coordinate functions; use that $c_0^* = l^1$ and that convex combinations of coordinate functions have unit norms in l^1.

6.10.150°. Let H be a Hilbert space and let $A_n \in \mathcal{L}(H)$ be such that $A_n x \to 0$ for every $x \in H$. Is it true that $A_n^* x \to 0$ for every x?

6.10.151. Let X and Y be normed spaces and let $S \colon Y^* \to X^*$ be a linear mapping. Prove that the existence of an operator $T \in \mathcal{L}(X, Y)$ for which $S = T^*$ is equivalent to the continuity of S with respect to the topologies $\sigma(Y^*, Y)$ and $\sigma(X^*, X)$. In particular, the continuity of S with respect to the weak-$*$ topologies yields the norm continuity of S.

HINT: if the operator S is continuous with respect to the weak-$*$ topologies, then, for every $x \in X$, the functional $y^* \mapsto Sy^*(x)$ on Y^* is continuous in the topology $\sigma(Y^*, Y)$, hence there exists an element $Tx \in X$ for which $Sy^*(x) = y^*(Tx)$; verify that T is the required operator.

6.10.152. Let $X = c_0$. The formula $Sy = \left(\sum_{n=1}^{\infty} y_n, y_2, y_3, \ldots \right)$, $y = (y_n)$, defines a bounded operator on $X^* = l^1$. Prove that S maps X^* one-to-one onto X^*, i.e., is a linear homeomorphism of X^*, but cannot be the adjoint for an operator $T \in \mathcal{L}(X)$.

HINT: one has the weak-$*$ convergence $e_n \to 0$, where $\{e_n\}$ is the standard basis in l^1, but $(Se_n)_1 = 1$ for all n, i.e., S is not continuous in the weak-$*$ topology.

6.10.153. Let X and Y be Banach spaces and let Y be separable. Let $S \in \mathcal{L}(Y^*, X^*)$. Prove that the existence of an operator $T \in \mathcal{L}(X, Y)$ such that $T^* = S$ is equivalent to the following condition: if $y_n^* \to 0$ in the weak-$*$ topology Y^*, then $Sy_n^* \to 0$ in the weak-$*$ topology X^*.

6.10.154. Prove Theorem 6.10.44 by an inductive construction without using the embedding into $C[0, 1]$ in the following stronger formulation: if we are given two sequences $\{y_n\} \subset X$ and $\{f_n\} \subset X^*$ and the latter separates points in X, then a Markushevich basis $\{x_n\}$ in the linear span of $\{y_n\}$ can be chosen in such a way that the corresponding sequence $\{l_n\} \subset X^*$ exists in the linear span of $\{f_n\}$.

HINT: see [**185**, p. 188].

6.10.155. Let X be an infinite-dimensional separable Banach space. Show that there exist sequences $\{x_n\} \subset X$ and $\{l_n\} \subset X^*$ such that $l_i(x_j) = \delta_{ij}$, the linear span of $\{x_n\}$ is dense in X, but there exists a nonzero element $x \in X$ with $l_n(x) = 0$ for all n.

HINT: take a sequence of linearly independent vectors a_i with $\|a_i\| = 1$ whose linear span L is dense in X; take $x \in X \backslash L$. By the Hahn–Banach theorem there exist $f_i \in X^*$ with $\|f_i\| = 1$, $f_i(x) = 0$, $f_i(a_i) = 1$ and $f_i(a_j) = 0$ for $j = 1, \ldots, i - 1$. One can find vectors $x_n \in L$ and functionals l_n in the linear span of $\{f_i\}$ with $l_i(x_j) = \delta_{ij}$ (see Exercise 6.10.154); one has $l_n(x) = 0$, since $f_i(x) = 0$ for all i.

6.10.156. Let X be a real normed space with the closed unit ball U and let $f, g \in X^*$ be such that $\|f\| = \|g\| = 1$ and $f^{-1}(0) \cap U \subset g^{-1}([-\varepsilon, \varepsilon])$, where $0 < \varepsilon < 1/2$. Prove that either $\|f - g\| \leqslant 2\varepsilon$ or $\|f + g\| \leqslant 2\varepsilon$.

HINT: see [**77**, p. 128].

6.10.157. Prove that the functions $\varphi_n = (n + 1/\pi)^{1/2} z^n$ form an orthonormal basis in the Bergman space $A^2(U)$ (see Example 5.2.2), where U is the unit disc in $\mathbb{C}^1$.

HINT: verify that these functions are mutually orthogonal and that for every function $f(z) = \sum_{n=0}^{\infty} c_n z^n$ from $A^2(U)$ we have Parseval's relation $\|f\|_{L^2}^2 = \sum_{n=0}^{\infty} |(f, \varphi_n)|^2$. To this end, it suffices to show that $\|f\|_{L^2}^2 = \pi \sum_{n=0}^{\infty} |c_n|^2 / (n + 1)$, which is verified by evaluating the integral of $|f(z)|^2$ over the disc of radius $r < 1$, where the series converges uniformly.

6.10.158. Let $\mathcal{A}(U)$ be the space of analytic functions on the open unit disc U in $\mathbb{C}^1$ that are continuous on the closure of U. Equip $\mathcal{A}(U)$ with the norm $\|\varphi\| = \max_z |\varphi(z)|$. Prove that any continuous linear functional on $\mathcal{A}(U)$ has the form

$$l(\varphi) = \lim_{q \to 1-} \sum_{k=0}^{\infty} q^k \frac{\varphi^{(k)}(0)}{k!} l(z^k), \quad \sup_k |l(z^k)| < \infty.$$

6.10.159. Let V be a nonempty convex closed set in a reflexive Banach space X. Prove that every point $x \in X$ has a nearest point in V.

6.10.160. Let X be a Banach space. Prove that the following conditions are equivalent: (i) X is reflexive, (ii) every nonempty closed convex set in X has a point nearest to the origin, (iii) for every closed separable linear subspace $Y \subset X$ and every functional $f \in Y^*$ with $\|f\| = 1$ the set $f^{-1}(1)$ contains a point nearest to the origin.

HINT: use Theorem 6.10.10 and Exercise 6.10.97.

6.10.161. Let X be a Banach space that is not reflexive. Prove that X^* contains a norm closed linear subspace that is not closed in the weak-$*$ topology.

HINT: take an element $F \in X^{**}$ not belonging to the image of X under the canonical embedding $X \subset X^{**}$ and consider the kernel of F.

6.10.162. Prove that every Efimov–Stechkin normed space (see Exercise 5.6.66) is complete and reflexive.

6.10.163. Prove that for any infinite-dimensional separable Banach spaces X_1 and X_2 there is a compact operator $T\colon X_1 \to X_2$ with the zero kernel and a dense range containing an a priori given sequence from X_2.

HINT: if $X_2 = l^2$, then one can take a sequence $\{l_n\} \subset X_1^*$ separating points in X_1 with $\|l_n\| \leqslant 1$ and set $(Tx)_n = 2^{-n} l_n(x)$; if $X_1 = l^2$, then one can take a sequence of unit vectors $y_n \in X_2$ with a dense linear span, set $T_0 x = \sum_{n=1}^{\infty} 2^{-n} x_n y_n$ and take the operator $T_0/\mathrm{Ker}\, T_0$.

6.10.164. (i) Let E_1 and E_2 be Hilbert spaces and let $A \in \mathcal{L}(E_1, E_2)$. Suppose that E_2 is separable and the operator A is injective. Prove that E_1 is also separable.

(ii) Extend (i) to the case where E_1 and E_2 are Banach spaces and E_1 is reflexive. Give an example showing that this can be false if E_1 is not reflexive.

(iii) Show that if in (i) we do not assume the separability of E_2, then one can assert that the cardinality of an orthonormal basis of the space E_1 does not exceed the cardinality of an orthonormal basis in the space E_2 and that there exists an operator $B \in \mathcal{L}(E_2)$ such that $B(E_2) = E_1$.

HINT: (i) the set $A^*(E_2^*)$ is dense in E_1^* by the injectivity of A; the separability of E_1^* implies the separability of E_1. (ii) Embed E_2 injectively into $L^2[0,1]$. (iii) Use the density of $A^*(E_2^*)$ in E_1^* and Exercise 5.6.53.

6.10.165. Let H be a separable Hilbert space. (i) Give an example of bounded operators A and B on H for which the sets $A(H)$ and $B(H)$ are dense in H, but $A(H) \cap B(H) = 0$.

(ii) Let $A \in \mathcal{L}(H)$ and let $A(H)$ be non-closed. Prove that there exists an operator $B \in \mathcal{L}(H)$ such that $B(H)$ is dense and $A(H) \cap B(H) = 0$.

HINT: see Theorem 7.10.18 in the next chapter.

6.10.166. Let X be a Banach space. A sequence of vectors $x_n \in X$ is called ω-independent if the relation $\sum_{n=1}^{\infty} c_n x_n = 0$ implies that $c_n = 0$ for all n.

(i) Give an example of a linearly independent sequence that is not ω-independent.

(ii) (V. I. Gurarii) Let $\{x_n\} \subset X$ and a nonzero $x_0 \in X$ be such that for some $C > 0$ and all n one has

$$\sum_{k=n+1}^{\infty} \|x_k - x_0\| < C\varrho_n, \quad \varrho_n := \inf\{\|x_0 - \sum_{k=1}^{n} \alpha_k x_k\| : \max_{k \leqslant n} |\alpha_k| \geqslant 1\}.$$

Prove that $\{x_n\}$ is ω-independent.

(iii) Prove that every linearly independent sequence contains an infinite ω-independent subsequence.

HINT: see [**690**].

6.10.167. (i) Prove that the norm on any infinite-dimensional normed space is not continuous in the weak topology.

(ii) Prove that the norm on the space l^1 is sequentially continuous in the weak topology, although is not continuous.

(iii) Prove that the function $f(x) = \sum_{n=1}^{\infty} n^{-2} x_n^2$ on l^2 with the weak topology is sequentially continuous, but is discontinuous at every point.

HINT: (i) observe that the norm is not bounded on weak neighborhoods of zero; (ii) apply Schur's theorem from Exercise 6.10.104; (iii) use the compactness of the operator $A: (x_n) \mapsto (n^{-1}x_n)$ to verify the weak sequential continuity (or use the uniform convergence of the series on balls); to prove the discontinuity of f in the weak topology show that the operator A cannot be bounded on a weak neighborhood of zero.

6.10.168. Let X be a Banach space. Prove that any two closed subspaces in X of codimension 1 are linearly homeomorphic. Deduce from this that any two closed subspaces in X of the same finite codimension are linearly homeomorphic.

HINT: see [**185**, Exercise 2.7, p. 53].

6.10.169. Prove that a Banach space X is linearly homeomorphic to $X \oplus \mathbb{R}^1$ precisely when X is linearly homeomorphic to every closed hyperplane in X.

HINT: use the previous exercise and the fact that the space X is linearly homeomorphic to $H \oplus \mathbb{R}^1$, where H is a closed hyperplane in X. Note that there exists an infinite-dimensional separable Banach space X that cannot be linearly homeomorphic to its closed hyperplane (see [**689**]).

6.10.170. Prove that every closed hyperplane in $C[0,1]$ is linearly homeomorphic to the whole space $C[0,1]$.

HINT: see [**185**, Exercise 5.33, p. 153].

6.10.171. Let l be a discontinuous linear function on a Banach space. (i) Can $l^{-1}(0)$ be a second category set?

(ii)** Can $l^{-1}(0)$ be a first category set? (See [**667**].)

6.10.172. Prove that every bounded closed convex set in a Hilbert space is the intersection of some family of closed balls.

HINT: observe that every point in the complement of this set is outside some ball containing this set. A more general result is mentioned in Theorem 6.10.36.

6.10.173. (i) Show that the closure of the convex envelope of an orthonormal basis in l^2 has no interior points. (ii)* Show that the closure of the convex envelope of a weakly convergent sequence in an infinite-dimensional Banach space has no interior points.

HINT: (ii) see [**185**, p. 87].

6.10.174. Prove that there exists a continuous linear surjection $T: C[0,1] \to L^2[0,1]$.

HINT: see [**185**, p. 195] or [**75**, Exercise 3.12.188, p. 241].

6.10.175. Let K be a compact space. Show that the extreme points (see §4.6) of the unit ball of $C(K)$ are the functions with values in $\{1, -1\}$, and the extreme points of the unit ball of $C(K)^*$ are Dirac's measures δ_k and the measures $-\delta_k$, $k \in K$.

6.10.176. Prove Theorem 5.6.5: if K_1 and K_2 are compact spaces, then $C(K_1)$ and $C(K_2)$ are linearly isometric precisely when K_1 and K_2 are homeomorphic.

HINT: if $h\colon K_1 \to K_2$ is a homeomorphism, then $J(f) := f \circ h$ is a linear isometry between $C(K_2)$ and $C(K_1)$. Conversely, if $J\colon C(K_1) \to C(K_2)$ is a linear isometry, then $J^*\colon C(K_2)^* \to C(K_2)^*$ is an isometry. For every $k \in K_2$, the measure $J^*\delta_k$ is a point of the unit ball in $C(K_1)^*$. By Exercise 6.10.175 one has $J^*\delta_k = \varepsilon_k \delta_{h(k)}$, where $h(k) \in K_1$, $\varepsilon_k = 1$ or $\varepsilon_k = -1$. It is readily seen that the mapping $k \mapsto \varepsilon_k \delta_{h(k)}$ is continuous, since J^* is continuous with respect to the weak-$*$ topologies. Moreover, the function $k \mapsto \varepsilon_k$ is also continuous, because $\varepsilon_k = \varepsilon_k \delta_{h(k)}(1) = J^*\delta_k(1) = J(1)(k)$. Therefore, the mapping $h\colon k \mapsto h(k)$ is continuous. It gives the required homeomorphism. About recovering of K by $C(K)$ see [**304**, § 18.2.1].

6.10.177. (The Alekhno–Zabreiko theorem) Suppose that functions $f_n \in L^\infty[0,1]$ converge to zero in the topology $\sigma(L^\infty, (L^\infty)^*)$. Prove that $f_n(t) \to 0$ a.e.

HINT: if Λ is a lifting on $L^\infty[0,1]$, then $\Lambda(f_n)(t) \to 0$ for all t.

6.10.178. Prove that $L^1[0,1]$ possesses the Dunford–Pettis property.

HINT: apply Exercise 6.10.177.

6.10.179. Let X be a Banach space. Prove that X^* is complemented in X^{***}.

HINT: set $P\colon X^{***} \to X^*$, $P(f) := f|_{X^*}$ (the Dixmier projection).

6.10.180. (*The James space*) Let J be the linear subspace in c_0 consisting of all elements with finite norm

$$\|x\|_J := \sup\left((x_{j_2} - x_{j_1})^2 + \cdots + (x_{j_{2m}} - x_{j_{2m-1}})^2 + (x_{j_{2m+1}})^2 \right)^{1/2},$$

where sup is taken over all finite collections $1 \leqslant j_1 < j_2 < \cdots < j_{2m+1}$. Prove that the space J has codimension 1 under the canonical embedding into J^{**} and hence is not reflexive, however, it is linearly isometric to J^{**}. Deduce from this that J cannot be isomorphic to $X \oplus X$ for a Banach space X. In particular, J is not isomorphic to $J \oplus J$.

6.10.181. Let $\{h_n\}$ be a Schauder basis in a Hilbert space H and let $\{l_n\}$ be a sequence of functionals on H such that $l_i(h_j) = \delta_{ij}$ and $\|l_n\| = \|h_n\| = 1$. Prove that $\{h_n\}$ is an orthonormal basis.

HINT: if $(h_1, h_2) \neq 0$, then the linear span of h_1 and h_2 contains a unit vector $v \perp h_1$. Then $h_2 = (h_2, v)v + (h_2, h_1)h_1$, whence we obtain $1 = |(h_2, v)|^2 + |(h_2, h_1)|^2$ and $|(h_2, v)| < 1$. However, $1 = |l_2(h_2)| = |(h_2, v)| \, |l_2(v)|$, which gives $|l_2(v)| > 1$. Hence $\|l_2\| > 1$, a contradiction.

6.10.182. Let $\{e_n\}$ be a Schauder basis in a Banach space X. Prove that $K \subset X$ is totally bounded precisely when for each $\varepsilon > 0$ there is n_ε with $\sup_{x \in K} \left\| \sum_{k=n_\varepsilon}^\infty c_k(x)e_k \right\| \leqslant \varepsilon$, where $x = \sum_{k=1}^\infty c_k(x)e_k$.

6.10.183. Let X be a Banach space with a Schauder basis $\{h_n\}$ and let $\{l_n\}$ be the corresponding coordinate functionals. Prove that $\{l_n\}$ is a Schauder basis in X^* precisely when the linear span of $\{l_n\}$ is dense in X^*.

HINT: see [**417**, p. 405].

6.10.184. Let H be a Hilbert space, $A \in \mathcal{L}(H)$ and $\|A\| \leqslant 1$. Prove that the operators $S_n := n^{-1}(I + A + \cdots + A^{n-1})$ converge pointwise to the projection onto $\mathrm{Ker}\,(A - I)$. For more general results, see [**354**].

6.10.185. Let H_1 and H_2 be mutually orthogonal infinite-dimensional closed subspaces in a separable Hilbert space. Show that there is an infinite-dimensional closed subspace H_3 such that $H_1 \cap H_3 = H_2 \cap H_3 = 0$.

6.10.186. (Hardy's inequality) Let $p \in (1, +\infty)$. Prove that the operator

$$A \colon (x_n) \mapsto \left(x_1, \frac{x_1 + x_2}{2}, \ldots, \frac{x_1 + x_2 + \cdots + x_n}{n}, \ldots\right)$$

is bounded on l^p and its norm is $p/(p-1)$.
HINT: see [**254**, §9.8].

6.10.187. (i) Let H be an infinite-dimensional Hilbert space. Consider the mapping $(A, B) \mapsto AB$, $\mathcal{L}(H) \times \mathcal{L}(H) \to \mathcal{L}(H)$. Investigate the continuity and sequential continuity of this mapping equipping $\mathcal{L}(H)$ with one of the following operator topologies: (a) the norm topology, (b) the strong operator topology, (c) the weak operator topology. Consider also different combinations of topologies on the factors and on the range.

(ii) Show that if H is separable, then on norm bounded sets in $\mathcal{L}(H)$ the weak and strong operator topologies are metrizable.

HINT: investigating the sequential continuity use the fact that a pointwise convergent sequence of operators is norm bounded. To disprove the continuity of multiplication in the weak operator topology use the left and right shifts L and R on the space $l^2(\mathbb{Z})$ of two-sided sequence, for which $L^n \to 0$ and $R^n \to 0$ in the weak operator topology, but $L^n R^n = I$. To disprove the continuity of multiplication in the strong operator topology observe that, given vectors $v_1, \ldots, v_n$ and u and a number $\varepsilon > 0$, one can find a bounded operator A such that $\|Av_i\| < \varepsilon$, $i = 1, \ldots, n$, but $\|A^2 u\| > 1$.

6.10.188. Let A be a closed set in a Hilbert space contained in the open unit ball. Prove that the closure of A in the weak topology is also contained in the open unit ball.

HINT: show that every point in the unit sphere can be separated from A by a closed hyperplane.

6.10.189. Show that there is no sequence of closed subsets F_n in l^2 contained in the open unit ball and having the property that every closed set in l^2 contained in the open unit ball belongs to some F_n.

HINT: take the standard orthonormal basis $\{e_n\}$ and observe that for any numbers $\varepsilon_n \in (0, 1/2)$ the set of vectors $(1 - \varepsilon_n)e_n$ is closed.

6.10.190. Let X be a Banach space and let $L \subset X^*$ be a linear subspace separating points in X. Is it true that for every $x \in X$ one has $\|x\| = \sup\{l(x)\colon l \in L, \|l\| \leqslant 1\}$? Consider $X = c_0$, $L = \{y = (y_n) \in l^1 \colon y_k = k^{-1}\sum_{n \in I_k} y_n \ \forall\, k \in I\}$, where $I, I_1, I_2, \ldots$ are infinite disjoint sets whose union is $\mathbb{N}$.

6.10.191. Construct an example of a Banach space X and a linear subspace $L \subset X^*$ with the following property: L is dense in X^* in the topology $\sigma(X^*, X)$, but the weak-$*$ closure of the intersection of L with the unit ball of X^* contains no ball of a positive radius.
HINT: see [**85**, p. 275].

6.10.192. (i) Prove that there exists a linear mapping $A \colon l^2 \to l^1$ discontinuous on every infinite-dimensional linear subspace in l^2 (not necessarily closed).

(ii) Prove that for every linear mapping A from l^1 to a Banach space E, there exists an infinite-dimensional linear subspace L in l^1 such that the restriction of A to L is continuous.

(iii) Prove that for every linear mapping $A \colon l^2 \to l^2$, there exists an infinite-dimensional linear subspace L in l^2 such that the restriction of the mapping A to L is continuous. For more general facts, see [**676**], [**674**], and [**706**].

CHAPTER 7

Spectral Theory

This chapter is devoted to a branch of the theory of operators very important for applications — spectral theory. More than any other chapter of the present book spectral theory owes its creation and intensive development to problems in natural sciences, in particular, in mechanics, physics, and chemistry.

7.1. The Spectrum of an Operator

The main object of spectral theory is the spectrum of a linear operator. Let X be a Banach space. A bounded operator $A\colon X \to X$ is called *invertible* if it maps X one-to-one onto X. By the Banach theorem the inverse mapping A^{-1} is automatically continuous. As in linear algebra, an important role is played by the property of invertibility of the operator $A - \lambda I$ for various scalars λ, where $I\colon x \mapsto x$ is the unit operator (the identity mapping).

7.1.1. Definition. *The spectrum $\sigma(A)$ of a bounded linear operator A on a complex Banach space X consists of all $\lambda \in \mathbb{C}$ such that the operator $A - \lambda I$ is not invertible.*

For an operator on a real space similarly one defines the real spectrum. If $X = \{0\}$, then the only operator is zero; it has also zero inverse, hence its spectrum is empty. Usually this case is excluded from consideration; in the sequel we also do not always explicitly state that a nonzero space is in question.

The complement of the spectrum is called the *resolvent set* of the operator A and denoted by $\varrho(A)$. The points of the resolvent set are called *regular* points. For every $\lambda \in \varrho(A)$ the operator

$$R_\lambda(A) := (A - \lambda I)^{-1}$$

is called the *resolvent* of A (one should bear in mind that sometimes the resolvent is defined as the inverse to $\lambda I - A$). For $\lambda, \mu \in \varrho(A)$ we have the *Hilbert identity*

$$R_\lambda(A) - R_\mu(A) = (\lambda - \mu)R_\mu(A)R_\lambda(A),$$

which is easily verified by multiplying both sides by $(A - \lambda I)$ from the right and then multiplying by $(A - \mu I)$ from the left.

By Banach's inverse mapping theorem a point λ belongs to the spectrum if and only if either $\mathrm{Ker}(A - \lambda I) \neq 0$ or $(A - \lambda I)(X) \neq X$, where

$$\mathrm{Ker}(A - \lambda I) := \{x\colon Ax - \lambda x = 0\}.$$

© Springer Nature Switzerland AG 2020

V. I. Bogachev, O. G. Smolyanov, *Real and Functional Analysis*,

Moscow Lectures 4, https://doi.org/10.1007/978-3-030-38219-3_7

In the first case λ is an *eigenvalue*, i.e., $Av = \lambda v$ for some vector $v \neq 0$ (called an *eigenvector*). In the finite-dimensional space both cases can happen only simultaneously, but in infinite-dimensional spaces the situation is different.

7.1.2. Example. The operator $Ax = (0, x_1, x_2, \ldots)$ on l^2 is injective, but not surjective. The operator $Bx = (x_2, x_3, \ldots)$ on l^2 is surjective, but not injective. In both cases 0 belongs to the spectrum, but for different reasons. In addition, A has no eigenvalues, but one can show that $\sigma(A) = \{\lambda \in \mathbb{C}\colon |\lambda| \leqslant 1\}$. The Volterra operator on $L^2[0,1]$ or $C[0,1]$ (Example 6.9.4(iv)) also has no eigenvalues (if $Vx = \lambda x$, then $x(t) = \lambda x'(t)$, $x(0) = 0$).

As we shall see, the spectrum of every bounded operator (on a nonzero complex space) is a nonempty compact set. First we establish the following important fact.

7.1.3. Theorem. *The set of invertible operators on a Banach space X (complex or real) is open in the space $\mathcal{L}(X)$ with the operator norm. Moreover, if an operator $A \in \mathcal{L}(X)$ is invertible and $D \in \mathcal{L}(X)$ is an operator such that $\|D\| < 1/\|A^{-1}\|$, then the operator $A + D$ is invertible.*

PROOF. By Banach's theorem it suffices to show that for every $y \in X$ the equation $Ax + Dx = y$ is uniquely solvable. This equation is equivalent to the equation
$$A^{-1}(A + D)x = A^{-1}y,$$
which can be written as $A^{-1}y - A^{-1}Dx = x$. Set $F(x) = A^{-1}y - A^{-1}Dx$ and observe that F is a contracting mapping, since
$$\|F(x) - F(z)\| = \|A^{-1}D(x - z)\| \leqslant \|A^{-1}\|\|D\|\|x - z\|,$$
where $\|A^{-1}\|\|D\| < 1$. An alternative proof: if $\|D\| < 1$, then $\sum_{k=0}^{\infty} D^k$ converges in the operator norm and gives $(I - D)^{-1}$. Now $A + D = (I + DA^{-1})A$, hence $(A + D)^{-1}$ is given by $A^{-1}\sum_{k=0}^{\infty}(-DA^{-1})^k$. $\qquad\square$

It follows from the theorem that the resolvent set is open. This assertion can be specified as follows.

7.1.4. Corollary. (i) *Let $A \in \mathcal{L}(X)$. Then, whenever $|\lambda| > \|A\|$, we have $\lambda \in \varrho(A)$ and*
$$R_\lambda(A) = -\sum_{k=0}^{\infty} \frac{A^k}{\lambda^{1+k}},$$
where the series converges in the operator norm.

(ii) *For every point $\lambda_0 \in \varrho(A)$, whenever $|\lambda - \lambda_0| < \|R_{\lambda_0}(A)\|^{-1}$, we have $\lambda \in \varrho(A)$ and*
$$R_\lambda(A) = \sum_{k=0}^{\infty} (\lambda - \lambda_0)^k R_{\lambda_0}(A)^{k+1},$$
where the series converges in the operator norm.

PROOF. (i) We have $A - \lambda I = -\lambda I + A$, where $\|A\| < |\lambda| = 1/\|(\lambda I)^{-1}\|$. Convergence of the series of $-\lambda^{-1-k}A^k$ in the operator norm is obvious from the estimate $\|\lambda^{-k}A^k\| \leqslant |\lambda|^{-k}\|A\|^k$. It is straightforward to show that for its sum S_λ we have $S_\lambda(A - \lambda I) = (A - \lambda I)S_\lambda = I$.

(ii) Convergence of the series with respect to the norm is justified similarly. For its sum S_λ we have

$$S_\lambda(A - \lambda I) = \sum_{k=0}^{\infty}(\lambda - \lambda_0)^k R_{\lambda_0}(A)^{k+1}\big(A - \lambda_0 I - (\lambda - \lambda_0)I\big)$$

$$= \sum_{k=0}^{\infty}\big[(\lambda - \lambda_0)^k R_{\lambda_0}(A)^k - (\lambda - \lambda_0)^{k+1}R_{\lambda_0}(A)^{k+1}\big] = I.$$

Similarly, $(A - \lambda I)S_\lambda = I$. □

7.1.5. Remark. If $\dim X < \infty$, then the set of invertible operators is not only open but is dense in $\mathcal{L}(X)$. In case of l^2 this is not true: the shift operator $A\colon (x_n) \mapsto (0, x_1, x_2, \ldots)$ cannot be approximated by invertible operators. Indeed, if $\|A - B\| < 1$, then B cannot be invertible, since by the equality $A^*A = I$ we have $\|I - A^*B\| \leqslant \|A^*\|\,\|A - B\| < 1$, which by the theorem above gives the invertibility of A^*B. If B were invertible, then also A^* would be invertible, hence also A. In this relation see also Exercises 7.10.112 and 7.10.113.

7.1.6. Theorem. *The spectrum of every operator $A \in \mathcal{L}(X)$ on a complex Banach space $X \neq 0$ is a nonempty compact set in the disc of radius $\|A\|$ centered at the origin in the complex plane.*

PROOF. The inclusion $\sigma(A) \subset \{z \in \mathbb{C}\colon |z| \leqslant \|A\|\}$ and the closedness of $\sigma(A)$ are already known. Let us verify that $\sigma(A)$ is not empty. Suppose that $R_\lambda(A)$ exists for all $\lambda \in \mathbb{C}$. Let $\psi \in \mathcal{L}(X)^*$ and $F(\lambda) = \psi(R_\lambda(A))$. By assertion (ii) of the previous corollary F is an entire function, and by assertion (i) whenever $|\lambda| \to \infty$ we have $|F(\lambda)| \to 0$. By the Liouville theorem $F \equiv 0$, whence we obtain $R_\lambda(A) = 0$, which is impossible if $X \neq 0$. □

The obtained estimate of the radius of a disc containing the spectrum can be sharpened.

The *spectral radius* of the operator A is defined by the formula

$$r(A) := \inf\{\|A^n\|^{1/n}\colon n \in \mathbb{N}\}.$$

It is clear that $r(A) \leqslant \|A\|$, since $\|A^n\| \leqslant \|A\|^n$.

7.1.7. Proposition. *We have*

$$r(A) = \lim_{n\to\infty} \|A^n\|^{1/n}.$$

In addition, $r(A) = \max\{|z|\colon z \in \sigma(A)\}$.

PROOF. Let $\varepsilon > 0$. Let $p \in \mathbb{N}$ be such that $\|A^p\|^{1/p} \leqslant r(A) + \varepsilon$. If $n \geqslant p$, we have $n = kp + m$, where $0 \leqslant m \leqslant p - 1$. Then

$$\|A^n\| \leqslant \|A^p\|^k\|A^m\| \leqslant M\|A^p\|^k, \quad M := 1 + \|A\| + \cdots + \|A^{p-1}\|.$$

Therefore,

$$r(A) \leqslant \|A^n\|^{1/n} \leqslant M^{1/n}\|A^p\|^{k/n} \leqslant M^{1/n}\big(r(A) + \varepsilon\big)^{kp/n}.$$

Since $M^{1/n} \to 1$ and $kp/n \to 1$ as $n \to \infty$, we have
$$r(A) \leqslant \limsup_{n \to \infty} \|A^n\|^{1/n} \leqslant r(A) + \varepsilon.$$
Since ε was arbitrary, this proves the first assertion.

Let us show that, whenever $|\lambda| > r(A)$, the operator $A - \lambda I$ is invertible. Dividing by λ, we arrive at the case $\lambda = 1$ and $r(A) < 1$. In this case the series $\sum_{n=0}^{\infty} A^n$ converges in the operator norm, since for all sufficiently large n we have $\|A^n\| \leqslant (r(A) + \varepsilon)^n$, where $\varepsilon > 0$ is such that $r(A) + \varepsilon < 1$. A straightforward verification shows that the sum of the indicated series serves as the inverse operator to $I - A$. It remains to show that the disc of radius $r(A)$ contains at least one point of the spectrum. Otherwise by the compactness of the spectrum we could find $r < r(A)$ such that all λ with $|\lambda| > r$ would belong to the resolvent set. According to the corollary proved above this means that for every continuous linear functional ψ on $\mathcal{L}(X)$ the function $f(\lambda) := \psi(R_\lambda(A))$ is holomorphic on the set $|\lambda| > r$. Outside the disc of radius $\|A\|$ this function is represented by the Laurent series $-\sum_{k=0}^{\infty} \lambda^{-1-k} \psi(A^k)$. By the uniqueness of expansion the same series represents the function f if $|\lambda| > r$. Let us fix such $\lambda \in (r, r(A))$. Convergence of the indicated Laurent series for every ψ gives the estimate $\sup_k \|\lambda^{-1-k} A^k\| \leqslant C < \infty$ by the Banach–Steinhaus theorem. Thus, $\|A^k\|^{1/k} \leqslant C^{1/k} \lambda^{1+1/k}$, whence $r(A) \leqslant \lambda$, which is a contradiction. $\quad\square$

In the infinite-dimensional case very different operators can have equal spectra. Let us consider examples.

7.1.8. Example. Let $\{r_n\}$ be all rational numbers in $[0, 1]$, let A be the operator on l^2 given by the formula $Ax = (r_1 x_1, r_2 x_2, \ldots)$, and let B be the operator on $L^2[0, 1]$ given by the formula $Bx(t) = tx(t)$. Then both operators have the spectrum $[0, 1]$, although for A all numbers r_n are eigenvalues, while B has no eigenvalues. Indeed, $\{r_n\} \subset \sigma(A)$, whence $[0, 1] \subset \sigma(A)$ by the closedness of the spectrum. If $\lambda \notin [0, 1]$, then there exists the inverse operator $R_\lambda(A)x = ((r_1 - \lambda)^{-1} x_1, (r_2 - \lambda)^{-1} x_2, \ldots)$.

Every point $\lambda \in [0, 1]$ belongs to $\sigma(B)$, since there is no function $x \in L^2[0, 1]$ such that $(t - \lambda)x(t) = 1$ a.e. If $\lambda \notin [0, 1]$, then the inverse operator for $B - \lambda I$ is the operator by multiplication by the bounded function $\varphi(t) = (t - \lambda)^{-1}$. The operator B has no eigenvalues: the equality $\lambda x(t) = tx(t)$ a.e. is only possible if $x(t) = 0$ a.e.

For an arbitrary linear operator A on a complex linear space and a polynomial $P(z) = \sum_{k=0}^{n} c_k z^k$ with complex coefficients, the operator $P(A)$ is defined by
$$P(A) = \sum_{k=0}^{n} c_k A^k, \quad A^0 := I.$$

7.1.9. Theorem. (THE SPECTRAL MAPPING THEOREM) *Let A be a bounded linear operator on a complex Banach space X. Then, for every polynomial P of complex variable, one has*
$$\sigma(P(A)) = P(\sigma(A)),$$
i.e., the spectrum of $P(A)$ is the image of the spectrum of A under the mapping P.

PROOF. Let us fix $\lambda \in \mathbb{C}$. Let $\lambda_1, \ldots, \lambda_n$ be the roots of the polynomial $P - \lambda$. Then $\lambda = P(\lambda_i)$ for all $i = 1, \ldots, n$, $P(z) - \lambda = c(z - \lambda_1) \cdots (z - \lambda_n)$ and
$$P(A) - \lambda I = c(A - \lambda_1 I) \cdots (A - \lambda_n I).$$
Let $c \neq 0$ (otherwise the assertion is obvious). We observe that the invertibility of $P(A) - \lambda I$ is equivalent to the invertibility of all operators $A - \lambda_i I$, since they commute. Indeed, if all these operators are invertible, then their product is also invertible. If some operator $A - \lambda_{i_0} I$ is not invertible, then either $\mathrm{Ker}(A - \lambda_{i_0} I) \neq 0$ or $(A - \lambda_{i_0} I)(X) \neq X$. Since by the commutativity of the factors we can put $A - \lambda_{i_0} I$ either on the first place or on the last one, the same relation is fulfilled for the whole product. Thus, λ belongs to $\sigma(P(A))$ precisely when there exists a number i with $\lambda_i \in \sigma(A)$. The latter is equivalent to the property that $\lambda \in P(\sigma(A))$. Indeed, if such i exists, then $\lambda = P(\lambda_i) \in P(\sigma(A))$. If $\lambda = P(z)$, where $z \in \sigma(A)$, then z is one of the numbers λ_i, i.e., λ_i belongs to $\sigma(A)$. $\square$

7.1.10. Remark. If $A \in \mathcal{L}(X)$, where X is a complex Banach space, then $\sigma(A) = \sigma(A^*)$ by Corollary 6.8.6(ii) and the equality $(A - \lambda I)^* = A^* - \lambda I$. However, for a Hilbert space X the spectrum $\sigma(A^*)$ is the set $\{\overline{z}: z \in \sigma(A)\}$ complex-conjugate to $\sigma(A)$, since here $(A - \lambda I)^* = A^* - \overline{\lambda} I$, because A^* acts on X. Note that in case of a Hilbert space the equivalence of the invertibility of the operators B and B^* is obvious from the equality $(BC)^* = C^* B^*$ and here Corollary 6.8.6 is not needed.

Let us consider the following important example.

7.1.11. Example. (THE OPERATOR OF MULTIPLICATION BY A FUNCTION) Let $\mu \neq 0$ be a finite nonnegative measure on a space Ω and let φ be a bounded complex μ-measurable function. Let us define the operator A_φ of multiplication by φ on $L^2(\mu)$ by the formula
$$A_\varphi x(\omega) = \varphi(\omega) x(\omega).$$
Then (i) A_φ^* is the operator of multiplication by the conjugate function $\overline{\varphi}$, the spectrum of A_φ is the *set of essential values* of φ, i.e., the set of all numbers $\lambda \in \mathbb{C}$ such that $\mu(\omega: |\varphi(\omega) - \lambda| \leqslant \varepsilon) > 0$ for all $\varepsilon > 0$;

(ii) $\varphi(\omega) \in \sigma(A_\varphi)$ for μ-a.e. ω;

(iii) $\|A_\varphi\| = \|\varphi\|_{L^\infty(\mu)}$.

In addition, the operator A_φ is selfadjoint precisely when $\varphi(\omega) \in \mathbb{R}^1$ for μ-a.e. ω.

PROOF. (i) The operator A_φ is bounded and $\|A_\varphi\| \leqslant \|\varphi\|_{L^\infty(\mu)}$. The expression for the adjoint to A_φ is clear from the equality
$$\int_\Omega \varphi(\omega) x(\omega) \overline{y(\omega)} \, \mu(d\omega) = \int_\Omega x(\omega) \overline{\overline{\varphi(\omega)} y(\omega)} \, \mu(d\omega)$$
for all $x, y \in L^2(\mu)$. Let us evaluate the spectrum of A_φ. If λ is not an essential value of φ, then for some $\varepsilon > 0$ for μ-almost all ω we have $|\varphi(\omega) - \lambda| \geqslant \varepsilon$. Redefining φ on a set of μ-measure zero, we can assume that this inequality is true for all $\omega \in \Omega$. Then the operator of multiplication by the bounded function $1/(\varphi - \lambda)$ is inverse to $A_\varphi - \lambda \cdot I$. Conversely, let λ be a regular value. If λ is

an essential value of φ, then the sets $B_n = \{\omega\colon |\varphi(\omega) - \lambda| \leqslant 1/n\}$ have positive measures and hence the functions $x_n = I_{B_n}/\sqrt{\mu(B_n)}$ have unit norm. Moreover, $(A_\varphi - \lambda I)x_n \to 0$, since

$$\frac{1}{\mu(B_n)} \int_{B_n} |\varphi(\omega) - \lambda|^2 \, \mu(d\omega) \leqslant n^{-2}\frac{1}{\mu(B_n)}\mu(B_n) = n^{-2}.$$

Then we obtain $x_n = (A_\varphi - \lambda I)^{-1}(A_\varphi - \lambda I)x_n \to 0$ contrary to the equality $\|x_n\| = 1$. (ii) The set S_φ of essential values of the function φ can differ from the set of its actual values. For example, let us define a function φ on the interval $(0,1)$ with Lebesgue measure as follows: $\varphi(t) = t$ if $t \neq 1/2$, $\varphi(1/2) = 2$. Then the value 2 is assumed, but is not essential, while the number 1 not belonging to the actual range of the function is essential. However, one can replace φ by a μ-almost everywhere equal function $\widetilde{\varphi}$ with values in the set of essential values of φ. Indeed, for every point $z \in \mathbb{C}$ that is not an essential value of φ we find an open disc $U(z,r)$ with $\mu\big(\varphi^{-1}(U(z,r))\big) = 0$. The obtained cover of $\mathbb{C}^1\backslash S_\varphi$ contains a countable subcover by discs $U(z_j, r_j)$. The set $E = \bigcup_j \varphi^{-1}(U(z_j, r_j))$ has μ-measure zero. In particular, there are essential values (otherwise $\mu(\Omega) = 0$). Outside E we make the function $\widetilde{\varphi}$ equal to φ and on E we make $\widetilde{\varphi}$ equal to some essential value. This modification takes values in S_φ. (iii) According to (i) it remains to show that $\|A_\varphi\| \geqslant \|\varphi\|_{L^\infty(\mu)}$. This is clear from the fact that the largest essential value of the function $|\varphi|$ is $\|\varphi\|_{L^\infty(\mu)}$. The equality $A_\varphi = A_\varphi^*$ is equivalent to the property that the operators of multiplication by φ and $\overline{\varphi}$ coincide, i.e., to the property that $\varphi(\omega) = \overline{\varphi(\omega)}$ for μ-a.e. ω. $\qquad\square$

Note that the established fact does not extend to arbitrary infinite measures (Exercise 7.10.89). The importance of the considered example will be clear from the fact, which we prove later, that in such a form one can represent any selfadjoint operator and any unitary operator on a separable space.

If μ is Lebesgue measure on $[0,1]$ and $\varphi(\omega) = \omega$, then $\sigma(A_\varphi) = [0,1]$ and A_φ has no eigenvalues (Example 7.1.8). For a general Borel measure μ on $[a,b]$ and $\varphi(\omega) = \omega$, the spectrum of A_φ is the support of μ, i.e., $[a,b]$ without all intervals of μ-measure zero, and eigenvalues are points of positive μ-measure.

7.2. The Quadratic Form and Spectrum of a Selfadjoint Operator

For a continuous linear operator A on a complex Hilbert space H we define two functions

$$\Phi_A(x,y) = (Ax, y), \quad Q_A(x) = (Ax, x).$$

The function Q_A is called the *quadratic form* of the operator A. The identity

$$4\Phi_A(x,y) = Q_A(x+y) - Q_A(x-y) + iQ_A(x+iy) - iQ_A(x-iy)$$

yields that the function Φ_A and hence the operator A are uniquely determined by the quadratic form Q_A (in the real case this is false!). The function Φ_A, called the *bilinear form of the operator A*, is linear in the first argument, conjugate-linear in the second argument and continuous. Conversely, with the aid of such a function one can construct an operator generating this form.

7.2.1. Lemma. *Let H be a complex Hilbert space and let Φ be a complex function on $H \times H$ that is linear in the first argument, conjugate-linear in the second argument and continuous in every argument separately. Then there exists a continuous linear operator A on the space H such that $\Phi(u, v) = (Au, v)$ for all $u, v \in H$.*

PROOF. The mapping $v \mapsto \overline{\Phi(u, v)}$ is linear and continuous for any fixed vector u. The Riesz theorem gives a uniquely defined vector Au for which $(v, Au) = \overline{\Phi(u, v)}$, whence $(Au, v) = \Phi(u, v)$. The linearity of $\Phi(u, v)$ in u yields the linearity of the mapping $u \mapsto Au$. If $u_n \to 0$, then $\Phi(u_n, v) \to 0$ for all vectors $v \in H$, i.e., $Au_n \to 0$ weakly and hence $\{Au_n\}$ is bounded. Hence operator A is continuous (see Theorem 6.1.3). $\square$

For a selfadjoint operator A the following important identity is valid:

$$4\mathrm{Re}(Ax, y) = Q_A(x + y) - Q_A(x - y).$$

For the proof it suffices to rewrite the expression

$$\big(A(x + y), x + y\big) - \big(A(x - y), x - y\big)$$

taking into account the equality $(Ay, x) = (y, Ax)$.

7.2.2. Lemma. *An operator A on a complex Hilbert space is selfadjoint precisely when its quadratic form Q_A is real.*

PROOF. If $A = A^*$, then

$$(Ax, x) = (x, Ax) = \overline{(Ax, x)}.$$

Conversely, if Q_A is a real function, then $Q_{A^*} = Q_A$, whence $A = A^*$, since the operator is uniquely determined by its quadratic form (recall that we consider complex spaces). $\square$

7.2.3. Theorem. (WEYL'S CRITERION) *A number λ belongs to the spectrum of a selfadjoint operator A precisely when there exists a sequence of vectors x_n such that*

$$\|x_n\| = 1 \quad and \quad \|Ax_n - \lambda x_n\| \to 0.$$

PROOF. If such a sequence exists, then $\lambda \in \sigma(A)$, since otherwise

$$x_n = (A - \lambda I)^{-1}(A - \lambda I)x_n \to 0.$$

Suppose that there are no such sequences. Then

$$\inf_{\|x\|=1} \|Ax - \lambda x\| = \alpha > 0,$$

whence

$$\|Ax - \lambda x\| \geqslant \alpha \|x\| \quad \text{for all } x. \tag{7.2.1}$$

In particular, $\mathrm{Ker}(A - \lambda I) = 0$. Let us set $Y = (A - \lambda I)(H)$ and show that $Y = H$. Let $a \perp Y$, i.e., $(Ax - \lambda x, a) = 0$ for all x. Then $(x, Aa - \overline{\lambda}a) = 0$ and hence $Aa = \overline{\lambda}a$. If $a \neq 0$, then λ must be real, because Q_A is real. Hence $Aa = \lambda a$ contrary to the injectivity of $A - \lambda I$. Thus, the closure of Y coincides with H. Let $y \in H$. Pick $y_n = Ax_n - \lambda x_n \to y$. Using that $\{y_n\}$ is a Cauchy sequence and applying (7.2.1) we conclude that $\{x_n\}$ is a Cauchy sequence. By the completeness of H there exists $x = \lim\limits_{n\to\infty} x_n$, whence $y = Ax - \lambda x$. Thus, the operator $A - \lambda I$ is invertible .   □

7.2.4. Corollary. *If A is a selfadjoint operator and λ is a complex number such that* $\inf\limits_{\|x\|=1} \|Ax - \lambda x\| > 0$, *then λ is a regular number.*

We observe that Weyl's criterion and the previous corollary are also true in the real case (see also Exercise 7.10.97).

7.2.5. Corollary. *The spectrum of a selfadjoint operator is real.*

PROOF. If $\|x\| = 1$, for all real numbers α and β we have

$$(Ax - \alpha x - i\beta x, Ax - \alpha x - i\beta x)$$
$$= (Ax - \alpha x, Ax - \alpha x) - (Ax - \alpha x, i\beta x) - i\beta(x, Ax - \alpha x) + i\beta(x, i\beta x)$$
$$= \|Ax - \alpha x\|^2 + \beta^2\|x\|^2 \geqslant \beta^2.$$

If $\beta \neq 0$, then we apply the previous corollary.   □

7.2.6. Theorem. *For every selfadjoint operator A (on a nonzero complex or real Hilbert space) one has*

$$\|A\| = \sup\{|(Ax, x)|\colon \|x\| \leqslant 1\} = \sup\{|\lambda|\colon \lambda \text{ is a point of the spectrum } A\}.$$

In addition, the spectrum of A contains the points

$$m_A = \inf\limits_{\|x\|=1} (Ax, x), \quad M_A = \sup\limits_{\|x\|=1} (Ax, x).$$

PROOF. Set $M = \sup\limits_{\|x\|\leqslant 1} |(Ax, x)|$. It is clear that $M = M_A$ or $M = -m_A$. We have $|(Ax, x)| \leqslant M\|x\|^2$ and $M \leqslant \|A\|$, since $|(Ax, x)| \leqslant \|A\|\|x\|^2$. On the other hand,

$$\|A\| = \sup\limits_{\|x\|\leqslant 1} \|Ax\| = \sup\limits_{\|x\|,\|y\|\leqslant 1} \mathrm{Re}(Ax, y)$$

$$= \frac{1}{4} \sup\limits_{\|x\|,\|y\|\leqslant 1} \left[(A(x+y), x+y) - (A(x-y), x-y)\right]$$

$$\leqslant \frac{1}{4} \sup\limits_{\|x\|,\|y\|\leqslant 1} \left[M\|x+y\|^2 + M\|x-y\|^2\right]$$

$$= \frac{1}{2} \sup\limits_{\|x\|,\|y\|\leqslant 1} \left[M\|x\|^2 + M\|y\|^2\right] = M.$$

Thus, $M = \|A\|$. We can assume that $M = M_A$, because in the case $M = -m_A$ one can pass to the operator $-A$. Then there exist vectors x_n such that $\|x_n\| = 1$ and $(Ax_n, x_n) \to M$. This gives

$$\|Ax_n - Mx_n\|^2 = (Ax_n, Ax_n) - 2M(Ax_n, x_n) + M^2(x_n, x_n)$$
$$\leqslant \|A\|^2 + M^2 - 2M(Ax_n, x_n) = 2M^2 - 2M(Ax_n, x_n) \to 0.$$

By Weyl's criterion we obtain $M \in \sigma(A)$. Taking into account that the spectrum is contained in the disc of radius $\|A\|$, this completes the proof of the equality indicated in the formulation and the inclusion $M_A \in \sigma(A)$ in case $M_A = M$.

For the proof of the inclusion $m_A, M_A \in \sigma(A)$ we observe that

$$M_{A+cI} = M_A + c, \quad m_{A+cI} = m_A + c, \quad \sigma(A + cI) = \sigma(A) + c$$

for every $c \in \mathbb{R}^1$. Let us take $c = \|A\|$. Since

$$|(Ax, x)| \leqslant \|A\|\|x\|^2,$$

we have $0 \leqslant m_{A+cI} \leqslant M_{A+cI}$, whence $M_{A+cI} \in \sigma(A + cI)$, i.e., $M_A \in \sigma(A)$ in any case. Finally, taking $c = -\|A\|$, on account of the equality $\sigma(-A) = -\sigma(A)$ we similarly obtain $m_A \in \sigma(A)$. $\square$

7.2.7. Remark. It is obvious from our reasoning that

$$\sigma(A) \subset [m_A, M_A].$$

Of course, this follows at once from Weyl's criterion, since if $\|Ax_n - \lambda x_n\| \to 0$ and $\|x_n\| = 1$, then $(Ax_n, x_n) \to \lambda$.

For a selfadjoint operator A on a real Hilbert space H its complexification $A_{\mathbb{C}}$ on the complexification $H_{\mathbb{C}}$ of the space H acts by the natural formula $A_{\mathbb{C}}(x, iy) = (Ax, iAy)$ and, as one can easily see, is also a selfadjoint operator. The realification of this operator (passage to the field $\mathbb{R}$ forgetting the complex structure) is the direct sum of two copies of the operator A. It is readily verified that the spectra of A and $A_{\mathbb{C}}$ coincide.

If A, B are selfadjoint operators with $(Ax, x) \leqslant (Bx, x)$, then we write $A \leqslant B$ and $B \geqslant A$. In particular, $A \geqslant 0$ if $(Ax, x) \geqslant 0$ (as we know, in the complex case this estimate gives the selfadjointness of A, but in the real case the selfadjointness is required additionally). Such an operator is called *nonnegative* or *nonnegative definite*. It follows from what we have proved above that $A \geqslant 0$ precisely when $A = A^*$ and $\sigma(A) \subset [0, +\infty)$.

7.3. The Spectrum of a Compact Operator

Spectra of compact operators possess peculiar properties. Let X be a complex or real Banach space. Let us consider the operator $I - K$, where K is a compact operator.

7.3.1. Lemma. *Let K be a compact operator on X.*
(i) *The kernel of the operator $I - K$ is finite-dimensional.*
(ii) *The range of the operator $I - K$ is closed.*

PROOF. (i) On the kernel of the operator $I - K$ the operator I equals K and hence is compact, which is only possible if this kernel is finite-dimensional.

(ii) Let $y_n = x_n - Kx_n \to y$. We show that $y \in (I - K)(X)$. Suppose first that $\sup_n \|x_n\| < \infty$. By the compactness of K we can extract from $\{Kx_n\}$ a convergent subsequence $\{Kx_{n_i}\}$. Since $x_{n_i} = y_{n_i} + Kx_{n_i}$, the sequence $\{x_{n_i}\}$ converges as well. Denoting its limit by x, we obtain $y = x - Kx$.

We now consider the case where the sequence $\{x_n\}$ is not bounded. Set $Z = \mathrm{Ker}(I - K)$ and

$$d_n = \inf\{\|x_n - z\| \colon z \in Z\}.$$

Since Z is finite-dimensional, there exist vectors $z_n \in Z$ with $\|x_n - z_n\| = d_n$. We show that the sequence $\{d_n\}$ is bounded. Suppose the contrary. We can assume that $d_n \to +\infty$. Set

$$v_n = (x_n - z_n)/\|x_n - z_n\|.$$

Since $(I - K)z_n = 0$ and $\sup_n \|y_n\| < \infty$, we have

$$\|v_n\| = 1, \quad v_n - Kv_n = (I - K)x_n/\|x_n - z_n\| = y_n/d_n \to 0.$$

The sequence $\{Kv_n\}$ contains a convergent subsequence $\{Kv_{n_i}\}$. Then $\{v_{n_i}\}$ converges to some vector $v \in X$. Moreover,

$$v - Kv = \lim_{i \to \infty} (v_{n_i} - Kv_{n_i}) = 0,$$

i.e., $v \in Z$. However, this is impossible, since $\mathrm{dist}(v, Z) \geqslant 1$, because

$$\|v_n - z\| = \frac{1}{d_n}\|x_n - z_n - d_n z\| \geqslant \frac{d_n}{d_n} = 1 \quad \text{for all } z \in Z, n \in \mathbb{N}.$$

Thus, the sequence $\{d_n\}$ is bounded. Now everything reduces to the first case, since $(I - K)(x_n - z_n) = (I - K)x_n = y_n$. $\qquad\square$

Clearly, the lemma is also true for the operator $I + K$, since the operator $-K$ is compact too.

The next theorem is the main result of this section.

7.3.2. Theorem. *Let K be a compact operator on a complex or real infinite-dimensional Banach space X. Then the spectrum of K either coincides with the point 0 or has the form*

$$\sigma(K) = \{0\} \cup \{k_n\},$$

where all numbers k_n are eigenvalues of K of finite multiplicity, which means that $\dim \mathrm{Ker}\,(K - k_n I) < \infty$, and the collection $\{k_n\}$ is either finite or is a sequence converging to zero.

PROOF. By the noncompactness of I the operator K is not invertible and hence $0 \in \sigma(K)$. Let $\lambda \in \sigma(K)$ and $\lambda \neq 0$. We show that λ is an eigenvalue. Suppose the contrary. Passing to the operator $\lambda^{-1}K$, we can assume that $\lambda = 1$. By the lemma the subspace $X_1 = (K - I)(X)$ is closed in X. In addition, we have $X_1 \neq X$, since otherwise $K - I$ would be invertible. Set

$$X_n = (K - I)^n(X) = (K - I)(X_{n-1}), \quad n \geqslant 2.$$

It is clear that $X_{n+1} \subset X_n$, since $X_1 \subset X$, whence $X_2 \subset X_1$ and so on. By the lemma we obtain that all subspaces X_n are closed. They are all different by the injectivity of $K - I$, since if

$$(K - I)(X_n) = (K - I)(X_{n-1}),$$

then $X_n = X_{n-1}$, whence we obtain $X_n = \cdots = X_1 = X$.

According to Theorem 5.3.4 there exist vectors $x_n \in X_n$ such that $\|x_n\| = 1$ and $\text{dist}(x_n, X_{n+1}) \geqslant 1/2$. If $n < m$, we have

$$Kx_n - Kx_m = x_n - x_m + (K - I)x_n - (K - I)x_m,$$

where

$$-x_m + (K - I)x_n - (K - I)x_m \in X_m + X_{n+1} + X_{m+1} \subset X_{n+1}.$$

Hence $\|Kx_n - Kx_m\| \geqslant 1/2$, i.e., $\{Kx_n\}$ contains no Cauchy subsequence contrary to the compactness of K. The obtained contradiction means that λ is an eigenvalue of K. By the lemma $\dim \text{Ker}(K - \lambda I) < \infty$, i.e., λ has a finite multiplicity.

We now show that $\sigma(K)$ has no nonzero limit points. Suppose that $\lambda_n \to \lambda$, where λ_n are eigenvalues and $\lambda \neq 0$. We can assume that λ_n are distinct and $|\lambda_n| \geqslant \sigma > 0$. Let us take $x_n \neq 0$ with $Kx_n = \lambda_n x_n$. It is readily seen that the vectors x_n are linearly independent. Denote by X_n the linear span of $x_1, \ldots, x_n$. It is clear that $K(X_n) \subset X_n$. By Theorem 5.3.4 there exist $y_n \in X_n$ with $\|y_n\| = 1$ and $\text{dist}(y_n, X_{n-1}) \geqslant 1/2$, $n > 1$. We have

$$y_n = \alpha_n x_n + z_n, \quad z_n \in X_{n-1}.$$

Then for $n > m$ we have

$$Ky_n - Ky_m = K(\alpha_n x_n) + Kz_n - Ky_m = \alpha_n \lambda_n x_n + Kz_n - Ky_m$$
$$= \lambda_n(y_n - z_n + \lambda_n^{-1} Kz_n - \lambda_n^{-1} Ky_m),$$

where $-z_n + \lambda_n^{-1} Kz_n - \lambda_n^{-1} Ky_m \in X_{n-1}$, because $z_n \in X_{n-1}$, $Kz_n \in X_{n-1}$, $Ky_m \in X_m \subset X_{n-1}$. Since $|\lambda_n| \geqslant \sigma$ and $\text{dist}(y_n, X_{n-1}) \geqslant 1/2$, we have $\|Ky_n - Ky_m\| \geqslant \sigma/2$. Hence $\{Ky_n\}$ contains no Cauchy subsequence, which is a contradiction. $\qquad\square$

7.3.3. Example. *The Volterra operator V on $L^2[0,1]$ or on $C[0,1]$ (see Example 6.9.4(iv)) has no eigenvalues, i.e., $\sigma(V) = \{0\}$.*

7.3.4. Corollary. *Let K be a compact operator on X. Then $(I - K)(X)$ is a closed subspace of finite codimension, i.e., $X = (I - K)(X) \oplus E$, where E is a finite-dimensional linear subspace.*

PROOF. The closedness of $(I - K)(X)$ is already established. According to Lemma 6.8.1 this subspace is the intersection of the kernels of the functionals in the kernel of $I - K^*$. Since $\dim \text{Ker}(I - K^*) < \infty$ by the compactness of K^* (see Theorem 6.9.3), there are linearly independent functionals $l_1, \ldots, l_n \in \text{Ker}(I - K^*)$ such that $(I - K)(X) = \bigcap_{i=1}^{n} \text{Ker}\, l_i$. Let us take vectors $x_i \in X$ with $l_i(x_j) = \delta_{ij}$. Then X is the sum of $(I - K)(X)$ and the linear span of $x_1, \ldots, x_n$. Indeed, for every $x \in X$ we set $z = x - \sum_{i=1}^{n} l_i(x)x_i$. This gives $l_j(z) = l_j(x) - l_j(x)l_j(x_j) = 0$ for all $j = 1, \ldots, n$. $\qquad\square$

It is clear that this corollary remains in force for $\lambda I - K$ with $\lambda \neq 0$, since $\lambda^{-1}K$ is a compact operator. In the next section we use this observation.

7.4. The Fredholm Alternative

We already know that if a nonzero number λ is not an eigenvalue for a compact operator K on a Banach space X, then the operator $K - \lambda I$ is invertible and hence the equation

$$Kx - \lambda x = y \qquad (7.4.1)$$

is uniquely solvable for every $y \in X$. Here we sharpen this assertion and show that the solvability of equation (7.4.1) for all y yields its unique solvability. In other words, the nontriviality of the kernel of $K - \lambda I$ means that $(K - \lambda I)(X) \neq X$, exactly as in the finite-dimensional case.

7.4.1. Theorem. (THE FREDHOLM ALTERNATIVE) *Let K be a compact operator on a complex or real Banach space X. Then*

$$\mathrm{Ker}(K - I) = 0 \iff (K - I)(X) = X,$$

i.e., either the equation

$$Kx - x = y$$

is uniquely solvable for all $y \in X$ or for some vector $y \in X$ it has no solutions and then the homogeneous equation

$$Kx - x = 0$$

has nonzero solutions.

PROOF. If $\mathrm{Ker}(K - I) = 0$, then by Theorem 7.3.2 we have $1 \notin \sigma(K)$. Hence $(K - I)(X) = X$. Conversely, suppose that

$$(K - I)(X) = X, \quad \text{but} \quad \mathrm{Ker}(K - I) \neq 0.$$

As we know, the operator K^* on X^* is also compact (Theorem 6.9.3). We observe that $\mathrm{Ker}\,(K^* - I) = 0$. Indeed, if $f \in X^*$ and $(K^* - I)f = 0$, then

$$f\big((K - I)x\big) = (K^* - I)f(x) = 0 \quad \text{for all } x \in X.$$

Since $(K - I)(X) = X$, we have $f = 0$. By Theorem 7.3.2 the operator $K^* - I$ is invertible. We now take a nonzero element $a \in \mathrm{Ker}(K - I)$. By the Hahn–Banach theorem there is a functional $f \in X^*$ with $f(a) = 1$. Let $g = (K^* - I)^{-1}f$. Then $(K^* - I)g(a) = f(a) = 1$. On the other hand, $(K^* - I)g(a) = g\big((K - I)a\big) = 0$, which is a contradiction. Part of this reasoning could be replaced by a reference to Corollary 6.8.6. $\qquad \square$

7.4.2. Corollary. *The Fredholm alternative remains in force also for an operator $K \in \mathcal{L}(X)$ such that for some $n \in \mathbb{N}$ the operator K^n is compact.*

PROOF. Let $1 \in \sigma(K)$. Since K^n is compact and $\sigma(K^n)$ is the image of $\sigma(K)$ under the mapping $z \mapsto z$, the unit circumference can contain only finitely many points $\lambda_1, \ldots, \lambda_m$ from $\sigma(K)$. Increasing n, we can assume that n is a simple number and $\exp(2k\pi i/n) \neq \lambda_j$ for $k = 1, \ldots, n-1, j = 1, \ldots, m$. Let

$\theta := \exp(2\pi i/n)$. Then θ^k differs from all λ_j with $k = 1, \ldots, n - 1$, i.e., the operators $I - \theta^k K$ are invertible. Hence the operator $V = (I - \theta K) \cdots (I - \theta^{n-1} K)$ is also invertible. Since

$$I - K^n = (I - K)(I - \theta K) \cdots (I - \theta^{n-1} K) = (I - K)V,$$

where V is invertible and commutes with K, we conclude that $K - I$ and $K^n - I$ have equal kernels and equal ranges. □

Clearly, the Fredholm alternative remains in force for $K-\lambda I$ for all $\lambda \neq 0$, i.e., the solvability of (7.4.1) with every right-hand side is equivalent to the absence of nontrivial solutions to the equation

$$Kx - \lambda x = 0. \qquad (7.4.2)$$

As an application of the Fredholm alternative we prove the following important result due to Weyl on the behavior of spectra under compact perturbations.

7.4.3. Theorem. *Let X be a Banach space and let A be a bounded operator on X. Then for every compact operator K on X, the spectra of the operators A and $A + K$ coincide up to the sets of eigenvalues, i.e.,*

$$\sigma(A) \backslash \sigma_p(A) \subset \sigma(A + K) \quad and \quad \sigma(A + K) \backslash \sigma_p(A + K) \subset \sigma(A),$$

where $\sigma_p(A)$ denotes the so-called point spectrum of A, i.e., the set of all eigenvalues.

PROOF. Let $\lambda \in \sigma(A)$. We have to show that if the operator $C := A + K - \lambda I$ is invertible, then λ is an eigenvalue of A. Let us consider the equality

$$A - \lambda I = C + (A - \lambda I - C) = C - K = C(I - C^{-1} K).$$

Since $\lambda \in \sigma(A)$, the operator $I - C^{-1} K$ cannot be invertible. By the Fredholm theorem (which can be applied due to the compactness of $C^{-1} K$), it has a nonzero kernel: there exists a nonzero vector v such that $C^{-1} K v = v$. Then $Kv = Cv$, whence $Av = \lambda v$, as required. Applying this to the operators $A + K$ and $-K$, we obtain the second relation in the theorem. □

It is worth noting that the full spectra of A and $A + K$ can be still very different (Exercise 7.10.64).

The classic results of Fredholm were obtained in terms of integral equations. Before turning to their discussion, we include yet another abstract result, also belonging to the so-called Fredholm theorems and dealing with the connection between the solvability of equations of the form (7.4.1) and (7.4.2) and analogous equations with the adjoint operator. We recall that for any compact operator K on X the adjoint operator K^* on X^* is also compact. In addition, $\sigma(K^*) = \sigma(K)$, but if the space X is Hilbert and the adjoint operator K^* is considered on X, then we have $\sigma(K^*) = \overline{\sigma(K)}$.

We recall that by Lemma 6.8.1 the closure of the range $A(X)$ of a bounded operator is the intersection of kernels of functionals in the kernel of its adjoint $A(X)$. This gives the first assertion in the next theorem, since the range of $K - I$ is closed for a compact operator K by Lemma 7.3.1. But more can be said in this case.

7.4.4. Theorem. *Let* $K \in \mathcal{K}(X)$, $\lambda \neq 0$. *Equation* (7.4.1) *is solvable for those and only those* y *which belong to the set*

$$\{z \in X: f(z) = 0 \ \forall f \in \text{Ker}(K^* - \lambda I)\} = \bigcap_{f \in \text{Ker}(K^* - \lambda I)} \text{Ker} f,$$

called the annihilator of $\text{Ker}(K^* - \lambda I)$ *in* X. *In addition,*

$$\dim \text{Ker}\,(K - \lambda I) = \dim \text{Ker}\,(K^* - \lambda I) = \text{codim}\,(K - \lambda I)(X). \qquad (7.4.3)$$

If X *is Hilbert, then in place of* $K^* - \lambda I$ *we take* $K^* - \overline{\lambda} I$.

PROOF. It suffices to consider $\lambda = 1$. The first assertion was explained in the proof of Corollary 7.3.4. It was shown there that there are vectors $x_1, \ldots, x_n$ in X and functionals $l_1, \ldots, l_n \in \text{Ker}\,(K^* - I)$ such that the kernel $\text{Ker}\,(K^* - I)$ coincides with the linear span of the functionals $l_1, \ldots, l_n$, $l_i(x_j) = \delta_{ij}$ for all i, j, $(K - I)(X) = \bigcap_{i=1}^{n} \text{Ker}\,l_i$, and the n-dimensional subspace E generated by $x_1, \ldots, x_n$ complements the closed subspace $(K - I)(X)$ to X. Thus,

$$\dim \text{Ker}\,(K^* - I) = \dim E = \text{codim}\,(K - I)(X).$$

We now show that $\dim \text{Ker}\,(K - I) = n$. The finite-dimensional subspace $X_0 := \text{Ker}\,(K - I)$ can be complemented to X by a closed linear subspace X_1 (Corollary 6.4.2). If $\dim X_0 < n$, then we can find an injective, but not surjective operator $K_0: X_0 \to E$. Writing x in the form $x = x_0 \oplus x_1$, $x_0 \in X_0$, $x_1 \in X_1$, we obtain the operator $K_1: X \to X$, $x \mapsto K_0 x_0 + Kx$. This operator is compact by the compactness of K and the aforementioned corollary. The kernel of $K_1 - I$ is trivial: if $K_1 x = x$, then $x - Kx = K_0 x_0 \in E$, whence $K_0 x_0 = 0$ and $x_0 = 0$, since $E \cap (K - I)(X) = 0$ and $\text{Ker}\,K_0 = 0$. This gives $x_1 - Kx_1 = 0$ and $x_1 = 0$ by the injectivity of $K - I$ on X_1. In addition, the range of $K_1 - I$ differs from X (not all vectors from E belong to it), which is impossible by the Fredholm alternative. Similarly, if $\dim X_0 > n$, then there is a surjective operator $K_0: X_0 \to E$ with a nonzero kernel. This gives a surjective operator $K_1 - I$ with a nonzero kernel, which is also impossible. Thus, $\dim X_0 = n$. By the already established facts the codimension of the range of $K^* - I$ is also n. $\qquad \square$

Let us apply the established abstract results to the objects from which Fredholm's theory was beginning — integral operators. Suppose we are given a complex square integrable function (an integral kernel) $\mathcal{K}$ on $[a, b]^2$ or, more generally, on $\Omega \times \Omega$, where $(\Omega, \mathcal{A}, \mu)$ is a space with a nonnegative measure. This kernel defines a compact operator by the formula $Kx(t) = \int_a^b \mathcal{K}(t, s)x(s)\,ds$ on the complex space $L^2[a, b]$ or a similarly defined operator on $L^2(\mu)$. A straightforward calculation shows that the adjoint operator K^* is defined by the formula

$$K^* u(t) = \int_a^b \overline{\mathcal{K}(s, t)} u(s)\,ds,$$

i.e., corresponds to the integral kernel $\mathcal{K}^*(t, s) := \overline{\mathcal{K}(s, t)}$. If the kernel is real and symmetric, then $\mathcal{K} = \mathcal{K}^*$. The first question of the theory of integral equations concerns the solvability of the equation

$$x(t) - \int_a^b \mathcal{K}(t,s)x(s)\,ds = y(t) \qquad (7.4.4)$$

with a given right-hand side. The results obtained above lead to the following conclusions concerning (7.4.4).

(1) The set of solutions to equation (7.4.4) with $y = 0$ has a finite dimension n, and the same dimension has the set of solutions to the homogeneous equation corresponding to the kernel $\mathcal{K}^*$;

(2) if $u_1, \ldots, u_n$ are linearly independent solutions to the homogeneous equation corresponding to $\mathcal{K}^*$, then the set of all those y for which equation (7.4.4) is solvable consists of the functions y for which $\int_a^b y(t)\overline{u_i(t)}\,dt = 0$, $i = 1, \ldots, n$.

The same is also true for operators on real spaces given by real kernels.

Let us consider analogous questions for the integral operator K given by a continuous real kernel $\mathcal{K}$ on $C[a,b]$. This operator is also compact. Hence we can apply Theorem 7.4.4. However, this theorem employs the adjoint operator on the adjoint space $C[a,b]^*$, which is the space of measures. It is natural to ask whether in the study of the solvability problem for the equation $x - Kx = y$ it suffices to consider the adjoint operator

$$K'x(t) = \int_a^b \mathcal{K}(s,t)x(s)\,ds$$

only on functions from $C[a,b]$ (the operator K' is the restriction of K^* to the subspace in $C[a,b]^*$ corresponding to measures given by continuous densities). This turns out to be possible. Indeed, by the general theorem we have to investigate the equation $\sigma - K^*\sigma = 0$ in $C[a,b]^*$, where $K^*\sigma$ is the measure acting on functions $x \in C[a,b]$ by the formula

$$K^*\sigma(x) = \int_a^b Kx(t)\,\sigma(dt) = \int_a^b \int_a^b \mathcal{K}(t,s)x(s)\,ds\,\sigma(dt).$$

This means that the measure $K^*\sigma$ is given by the continuous density

$$\varrho(t) = \int_{[a,b]} \mathcal{K}(t,s)\,\sigma(ds).$$

Hence the existence of nontrivial solutions to the equation $\sigma - K^*\sigma = 0$ in $C[a,b]^*$ is equivalent to the existence of nonzero solutions to the equation $\varrho - K'\varrho = 0$ in $C[a,b]$. It should be noted that here we have used the continuity of $\mathcal{K}$ in both variables (more precisely, it is important here that K^* takes $C[a,b]^*$ to $C[a,b]$). Let us consider the kernel $\mathcal{K}(t,s) = (3/2)ts^{-1/2}$, which is continuous only with respect to one variable, but obviously generates a compact operator on $C[0,1]$ with a one-dimensional range. The operator K^* on $C[0,1]^*$ also has a one-dimensional range and the measure $\sigma = s^{-1/2}\,ds$ satisfies the equation $K^*\sigma = \sigma$ and spans the subspace $\mathrm{Ker}\,(K^* - I)$. By Theorem 7.4.4 the condition for the solvability of the equation $x - Kx = y$ is given by the equality

$$\int_0^1 y(s)s^{-1/2}\,ds = 0.$$

If we act here by a formal analogy with the previous case and search eigenvectors of the adjoint kernel only in $C[0, 1]$ or $L^2[0, 1]$ (and not in $C[0, 1]^*$), then, having seen that they are absent, we could arrive at the wrong conclusion about the solvability of the equation $x - Kx = y$ for all y.

Note that the continuity of the kernel $\mathcal{K}$ in the case of an interval or its square integrability in the general case were only needed to verify the compactness of K. Theorem 7.4.4 applies also to the operators on $L^2[0, 1]$ or $C[0, 1]$ given by singular kernels $\mathcal{K}(t, s) = \mathcal{K}_0(t, s)|t - s|^{-\alpha}$, where $\alpha < 1$ and a measurable function $\mathcal{K}_0$ is bounded in the case of $L^2[0, 1]$ and bounded and continuous in t in the case of $C[0, 1]$ (see Exercise 6.10.143). If the function $\mathcal{K}_0$ is continuous and $\alpha < 1/2$, then in the study of the solvability of the equation $x - Kx = y$ in $C[0, 1]$ it is also sufficient to analyze the equation $z - K'z = 0$ corresponding to the adjoint kernel only in $C[0, 1]$ and not in $C[0, 1]^*$. Indeed, for any $y \in C[0, 1]$ the solvability of the equation $x - Kx = y$ in $C[0, 1]$ is equivalent to its solvability in $L^2[0, 1]$, since $Kx \in C[0, 1]$ for all $x \in L^2[0, 1]$, which is easily verified by using the square integrability of $|s|^{-\alpha}$ with $\alpha < 1/2$. In addition, all solutions to the equation $z - K'z = 0$ in $L^2[0, 1]$ also belong to $C[0, 1]$.

7.5. The Hilbert–Schmidt Theorem

In a finite-dimensional space every selfadjoint operator is diagonal in some orthonormal al basis. In an infinite-dimensional case a selfadjoint operator can fail to have eigenvectors (for example, the operator $Ax(t) = tx(t)$ on $L^2[0, 1]$). However, for compact selfadjoint operators there is a full analogy with the finite-dimensional case. This is asserted in the following remarkable classic result.

7.5.1. Theorem. (THE HILBERT–SCHMIDT THEOREM) *Suppose that A is a compact selfadjoint operator on a real or complex separable Hilbert space $H \neq 0$. Then A has an orthonormal eigenbasis $\{e_n\}$, i.e., $Ae_n = \alpha_n e_n$, where the numbers α_n are real and converge to zero if H is infinite-dimensional.*

PROOF. We observe that A has eigenvectors. Indeed, by Theorem 7.2.6 the infimum and supremum of the function $Q_A(x) = (Ax, x)$ on the unit sphere belong to the spectrum. If they are zero, then $A = 0$. If at least one of these numbers is not zero, by the compactness of A it is an eigenvalue. All eigenvalues of A are real. The eigenvectors corresponding to different eigenvalues are mutually orthogonal. Indeed, if $Aa = \alpha a$ and $Ab = \beta b$, then

$$\beta(a, b) = (a, Ab) = (Aa, b) = \alpha(a, b),$$

whence $(a, b) = 0$ if $\alpha \neq \beta$. Therefore, by the separability of H there are at most countably many eigenvalues. Every nonzero eigenvalue has a finite multiplicity by the compactness of A. Let $\{\alpha_n\}$ be all eigenvalues of A. In every subspace $H_n := \mathrm{Ker}(A - \alpha_n I)$ we can choose an orthonormal basis (for $\alpha_n \neq 0$ such bases are finite). The union of all these bases gives an orthonormal system $\{e_n\}$ in H. It remains to show that $\{e_n\}$ is a basis. Denote by H' the closed linear span of $\{e_n\}$. It is readily seen that $A(H') \subset H'$, since $A(H_n) \subset H_n$ for all n. Let H'' be the orthogonal complement of H'. We observe that $A(H'') \subset H''$.

Indeed, if $u \in H''$, then for every $v \in H'$ we have $(Au, v) = (u, Av) = 0$, since $Av \in H'$. As shown above, in H'' we have an eigenvector of A, which leads to a contradiction if $H'' \neq 0$. Theorem 7.3.2 gives $\alpha_n \to 0$. $\square$

7.5.2. Remark. A similar assertion is true for nonseparable spaces H. Here there exists a separable closed subspace $H_0 \subset H$ such that $A(H_0) \subset H_0$ and $A(H_0^\perp) = 0$. For H_0 we can take the closure of $A(H)$, which is separable due to the compactness of A.

There is another (variational) proof of this important theorem, which does not use the spectral theory. The reasoning above shows that the main problem is to establish the existence of at least one eigenvector. Such a vector can be found by solving the maximization problem for the function $Q(x) = |(Ax, x)|$ on the closed unit ball (of course, if we already have the Hilbert–Schmidt theorem, then it is clear that the maximum is attained at the eigenvectors corresponding to the eigenvalues with the maximal absolute value). Denote by q the supremum of this function and take unit vectors h_n with $Q(h_n) \to q$. Pick a subsequence $\{h_{n_i}\}$ weakly converging to some vector h. The sequence $\{Ah_{n_i}\}$ converges in norm to Ah, whence $(Ah_{n_i}, h_{n_i}) \to (Ah, h)$ and $Q(h) = q$. It is clear that $\|h\| = 1$ if $\|A\| > 0$. It is now easy to verify that h is an eigenvector. For this it suffices to show that $Ah \perp h^\perp$, since in that case $Ah = \lambda h$. Let $e \perp h$, $\|e\| = 1$. We can assume that $(Ah, h) > 0$. Then for all real t we have $(1 + t^2)^{-1} Q(h + te) \leqslant q$, that is,

$$q + 2t\mathrm{Re}\,(Ah, e) + t^2(Ae, e) \leqslant q + qt^2.$$

This is only possible if $\mathrm{Re}\,(Ah, e) = 0$. Since this is true for all $e \in h^\perp$, we obtain that $Ah \perp h^\perp$.

Note also that in place of $|(Ax, x)|$ we could search the maximum of the function (Ax, x), provided that we assume additionally that it assumes a positive value. If $(Ax, x) \leqslant 0$, then we can pass to $-A$.

The reasoning presented above not only gives another proof, but also leads to the following useful variational principle. Denote by $\mathcal{L}_n$ the collection of all n-dimensional linear subspaces in a Hilbert space H.

7.5.3. Theorem. *Let A be a compact selfadjoint operator on a real or complex Hilbert space and let $\alpha_1 \geqslant \alpha_2 \geqslant \cdots > 0$ be all positive eigenvalues of A written in the order of decreasing taking into account their multiplicities. Then the following equalities are valid for all n for which there exists $\alpha_n > 0$.*

(i) *The Courant variational principle:*

$$\alpha_n = \min_{L \in \mathcal{L}_{n-1}} \max_{x \in L^\perp, \|x\| = 1} (Ax, x). \tag{7.5.1}$$

(ii) *The Fischer variational principle:*

$$\alpha_n = \max_{L \in \mathcal{L}_n} \min_{x \in L, \|x\| = 1} (Ax, x). \tag{7.5.2}$$

PROOF. (i) Let $L \in \mathcal{L}_{n-1}$. In the space H_n generated by the orthonormal eigenvectors $e_1, \ldots, e_n$ corresponding to the eigenvalues $\alpha_1, \ldots, \alpha_n$, there is a

unit vector $h \perp L$. Since for any $x \in H_n$ we have $(Ax, x) \geqslant \alpha_n(x, x)$, it follows that
$$\max_{x \in L^\perp, \|x\|=1} (Ax, x) \geqslant \alpha_n.$$
If we take $L = H_{n-1}$, then the equality is attained.

(ii) Let $L \in \mathcal{L}_n$. Then L contains a vector $h \perp H_{n-1}$, where H_n is the same as above. It is clear that $(Ah, h) \leqslant \alpha_n(h, h)$. Hence $\min_{x \in L, \|x\|=1}(Ax, x) \leqslant \alpha_n$. If we take $L = H_n$, then the equality is attained. $\qquad \square$

Similar variational characterizations can be written for negative eigenvalues (one can simply pass to $-A$).

Note that in the infinite-dimensional case it is necessary to separate positive and negative eigenvalues. In the finite-dimensional case the stated equalities are fulfilled for all eigenvalues written in the order of decreasing.

The next rather non-obvious result is clear from our discussion.

7.5.4. Corollary. *Let A and B be compact selfadjoint operators on a real or complex Hilbert space H such that $(Ax, x) \leqslant (Bx, x)$. Let $\{\alpha_n\}$ and $\{\beta_n\}$ be the sequences of their positive eigenvalues written in the order of decreasing taking into account their multiplicities. Then for every n we have $\alpha_n \leqslant \beta_n$.*

7.6. Unitary Operators

Let H be a complex Hilbert space and let A be a bounded operator on H. We recall that $\sigma(A^*)$ is the set complex-conjugate to $\sigma(A)$ (Remark 7.1.10).

7.6.1. Definition. *A linear operator U on a Hilbert space $H \neq 0$ is called unitary if it maps H onto H and preserves the inner product, i.e.,*
$$(Ux, Uy) = (x, y) \quad \text{for all } x, y \in H.$$

A unitary isomorphism of nonzero Hilbert spaces H_1 and H_2 is a one-to-one linear operator $U \colon H_1 \to H_2$ preserving the inner product.

An equivalent definition of a unitary operator is the equality
$$U^{-1} = U^*$$
or two equalities
$$UU^* = U^*U = I.$$

Note that to ensure that the operator U is unitary it is not enough to have only one of these equalities if we do not require its invertibility. For example, let $Ux = (0, x_1, x_2, \ldots)$ on l^2. Then U preserves the inner product and $U^*U = I$, but $UU^* \neq I$.

7.6.2. Example. Let A_φ be the operator of multiplication on $L^2(\mu)$ by a bounded μ-measurable function φ considered in Example 7.1.11. Then the operator A_φ is unitary if and only if $|\varphi(\omega)| = 1$ for μ-a.e. ω. Indeed, the equality $A_\varphi A_\varphi^* = I$ gives the equality $\varphi(\omega)\overline{\varphi(\omega)} = 1$ for μ-a.e. ω. Conversely, the latter equality yields that $A_\varphi A_\varphi^* = I = A_\varphi^* A_\varphi$.

7.6.3. Lemma. *The spectrum of a unitary operator belongs to the unit circumference.*

PROOF. Let U be a unitary operator. Then $\|U\| = 1$ and hence the spectrum of U is contained in the unit disc. The same is true for U^*, but $U^* = U^{-1}$, which excludes from the spectrum all inner points of the disc: if the operator $U - \lambda I$ is not invertible, then so is $I - \lambda U^*$, hence also $\lambda^{-1} I - U^*$. $\square$

7.6.4. Definition. *An operator A_1 on a Hilbert space H_1 is called unitarily equivalent to an operator A_2 on a Hilbert space H_2 if there exists a unitary isomorphism $J\colon H_1 \to H_2$ such that $A_1 = J^{-1} A_2 J$, i.e., we have a commutative diagram*

$$
\begin{array}{ccc}
H_1 & \xrightarrow{\ A_1\ } & H_1 \\
\downarrow{\scriptstyle J} & & \uparrow{\scriptstyle J^{-1}} \\
H_2 & \xrightarrow{\ A_2\ } & H_2
\end{array}
$$

7.6.5. Lemma. *A linear isometry of Hilbert spaces is a unitary isomorphism.*

PROOF. Let $J\colon H_1 \to H_2$ be a linear isometry. Then $\|Jx + Jy\|^2 = \|x + y\|^2$, $\|Jx\|^2 = \|x\|^2$, $\|Jy\|^2 = \|y\|^2$, whence $(Jx, Jy) + (Jy, Jx) = (x, y) + (y, x)$, i.e., $\mathrm{Re}(Jx, Jy) = \mathrm{Re}(x, y)$. Replacing x by ix in the complex case, we obtain the equality for the imaginary parts. Any isometry is surjective by definition. $\square$

Characteristics of an operator that do not change under unitary equivalence are called *unitary invariants*. For example, the spectrum and norm are unitary invariants.

7.6.6. Lemma. *Let H_1 and H_2 be Hilbert spaces, $H_0 \subset H_1$ an everywhere dense linear subspace, and $U\colon H_0 \to H_2$ a linear mapping preserving the inner product. Then U uniquely extends to a linear mapping from H_1 to H_2 preserving the inner product and having a closed range. If $U(H_0) \neq 0$ is dense in H_2, then the extension is a unitary isomorphism.*

PROOF. Let $x \in H_1$. Let us take $x_n \in H_0$ with $x_n \to x$. Then the sequence $\{U x_n\}$ is Cauchy in H_2 and hence converges to some element $y \in H_2$. Set $Ux := y$. It is clear from our assumption that y is independent of our choice of a sequence converging to x. If $z_n \to z$, then

$$\alpha x_n + \beta z_n \to \alpha x + \beta z \quad \text{and} \quad \alpha U x_n + \beta U z_n \to \alpha U x + \beta U z,$$

whence $U(\alpha x + \beta z) = \alpha U x + \beta U z$. Thus, the extension is linear. In addition,

$$(Ux, Uz) = \lim_{n \to \infty} (U x_n, U z_n) = \lim_{n \to \infty} (x_n, z_n) = (x, z).$$

Since U preserves the norm, the set $U(H_1)$ is closed. If it is dense, then it coincides with H_2. $\square$

A generalization of a unitary operator is a *partial isometry*. This is an operator J that is defined on a closed linear subspace H_1 in a Hilbert space H and maps it with the preservation of norm onto a closed subspace $H_2 \subset H$. It is clear that such an operator preserves also the inner product. However, it does not always extend to a unitary operator. For example, the operator $J\colon (0, x_1, x_2, \ldots) \mapsto (x_1, x_2, \ldots)$ isometrically maps a closed hyperplane in l^2 onto the whole space and cannot be extended to an isometry operator on all of l^2. Every partial isometry V has a maximal partially isometric extension $\widetilde{V}$. Indeed, if H_1 does not coincide with H and $V(H_1) \neq H$, then we take the orthogonal complements $E_1 := H_1^\perp$ and $E_2 := V(H_1)^\perp$. The following cases are possible: 1) the spaces E_1 and E_2 are isometric and hence V can be extended to a unitary isomorphism, 2) E_2 is larger than E_1, i.e., E_2 is not isometric to E_1, but has a closed subspace E_2' isometric to E_1, which gives an isometry between H and $V(H_1) \oplus E_2'$, 3) E_1 is larger than E_2, i.e., E_1 is not isometric to E_2, but has a closed subspace E_1' isometric to E_2, which gives an isometry between $H_1 \oplus E_1'$ and H.

Let us note some properties of partial isometries. We recall that an orthogonal projection is an operator of the orthogonal projecting onto a closed subspace (see Corollary 5.4.6).

7.6.7. Proposition. *Let $V \in \mathcal{L}(H)$. The following conditions are equivalent:*

(i) *the operator V is a partial isometry on the orthogonal complement of its kernel;*

(ii) *V^*V is an orthogonal projection onto some closed subspace;*

(iii) *$V = VV^*V$.*

In this case $\operatorname{Ker} V^\perp = V^*V(H)$ and V^* is also a partial isometry from $V(H)$ onto $\operatorname{Ker} V^\perp$ vanishing on $V(H)^\perp$.

PROOF. If V isometrically maps H_1 onto $V(H_1)$ and vanishes on $H_1^\perp$, then V^* vanishes on $V(H_1)^\perp$ and isometrically maps $V(H_1)$ onto H_1. Indeed, the equality $(Vx, y) = (x, V^*y)$ yields that $V^*y = 0$ if $y \perp V(H_1) = V(H)$. Whenever $y \in V(H_1)$, i.e., $y = Vz$ with $z \in H_1$, we have $V^*y \perp \operatorname{Ker} V$, i.e., $V^*y \in H_1$. Then $(Vx, y) = (x, z) = (x, V^*y)$ for all $x \in H_1$, whence $V^*y = z$. Hence V^*V is the projection onto H_1, whence $V = VV^*V$.

Let V^*V be the orthogonal projection P onto a closed subspace H_1. Then $V = 0$ on $H_1^\perp$ and V is an isometry on H_1, since

$$(Vx, Vx) = (V^*Vx, x) = (Px, x) = (Px, Px).$$

It is clear that we also have the equality $V = VV^*V$. Finally, suppose that V satisfies the latter equality. Then $V^*V = (V^*V)^2$, i.e., the selfadjoint operator $A = V^*V$ satisfies the identity $A = A^2$. It is readily seen (this is done below in Lemma 7.9.1) that A is an orthogonal projection. $\qquad\square$

It follows that an operator V is a maximal partial isometry defined by zero on the orthogonal complement of the subspace on which it is an isometry if and only if either $V^*V = I$ or $VV^* = I$ (i.e., one of the operators V or V^* is isometric on all of H).

Extensions of isometric operators will be considered in Chapter 10 in connection with extensions of symmetric (unbounded) operators. In the next section partial isometries are used for obtaining the so-called polar decompositions of operators.

7.7. Continuous Functions of Selfadjoint Operators

Any linear operator A on a space X can be substituted as an argument of a polynomial f of one variable to obtain the operator $f(A)$. Here we establish a highly nontrivial fact that selfadjoint operators can be substituted in arbitrary continuous functions of one variable.

7.7.1. Lemma. *Let A be a selfadjoint operator on a nonzero complex Hilbert space and let P be a polynomial with complex coefficients. Then*

$$\|P(A)\| = \max_{t\in\sigma(A)} |P(t)| \leqslant \max_{t\in[-\|A\|,\|A\|]} |P(t)|. \qquad (7.7.1)$$

PROOF. Let $P(t) = \sum_{k=0}^n c_k t^k$. Then

$$P(A)^* P(A) = \sum_{k=0}^n \overline{c_k} A^k \sum_{k=0}^n c_k A^k = Q(A),$$

$Q\colon t \mapsto \overline{P(t)}P(t)$ is a polynomial, $P(A)^*P(A)$ is selfadjoint. Hence

$$\begin{aligned}
\|P(A)\|^2 &= \sup_{\|x\|\leqslant 1} \big(P(A)x, P(A)x\big) \\
&= \sup_{\|x\|\leqslant 1} \big(P(A)^*P(A)x, x\big) = \|P(A)^*P(A)\| \\
&= \sup_{\lambda\in\sigma(P(A)^*P(A))} |\lambda| = \sup_{t\in\sigma(A)} |\overline{P(t)}P(t)| = \sup_{t\in\sigma(A)} |P(t)|^2,
\end{aligned}$$

where the third and forth equalities follow from Theorem 7.2.6, and the last but one equality is obtained from Theorem 7.1.9. □

In the real case the same is true for real polynomials.

With the aid of this lemma it is easy to define continuous functions of a selfadjoint operator. We recall that an *algebra* is a linear space $\mathcal{L}$ equipped with an associative multiplication $(a, b) \mapsto ab$ for which $(\lambda a)b = a(\lambda b) = \lambda ab$, $(a+b)(c+d) = ac + bc + ad + bd$ for all $a, b, c, d \in \mathcal{L}$ and all scalars λ (see Chapter 11). The most important examples for us is the algebra $\mathcal{L}(H)$ of bounded operators on a Hilbert space H and the algebra $C(K)$ of continuous complex functions on a compact space K. A homomorphism of an algebra is a linear operator J with $J(ab) = J(a)J(b)$.

7.7.2. Theorem. *Let A be a selfadjoint operator on a complex Hilbert space $H \neq 0$. There is a unique homomorphism J of the algebra $C(\sigma(A))$ to the algebra $\mathcal{L}(H)$ such that*
1) $J(P) = P(A)$ *for every polynomial $P\colon \mathbb{R}^1 \to \mathbb{C}$,*
2) $\|J(f)\| = \sup_{t\in\sigma(A)} |f(t)|$ *for all $f \in C(\sigma(A))$,*
3) $J(f)^* = J(\overline{f})$ *for all $f \in C(\sigma(A))$.*
A similar assertion is true in case of real spaces H and $C(\sigma(A))$.

PROOF. For every polynomial f we set

$$J(f) := f(A).$$

For every $f \in C(\sigma(A))$ there exists a sequence of polynomials f_n uniformly converging to f on the compact set $\sigma(A)$ on the real line. By the lemma the sequence of operators $f_n(A)$ is Cauchy in $\mathcal{L}(H)$ and hence converges in the operator norm to some operator $J(f) \in \mathcal{L}(H)$. It is important that this operator does not depend on the approximating sequence: if polynomials g_n also converge uniformly to f on $\sigma(A)$, then the polynomials $f_1, g_1, f_2, g_2, \ldots$ converge as well, which proves our assertion. If polynomials φ_n converge in $C(\sigma(A))$ to a function φ and polynomials ψ_n converge to a function ψ, then $\varphi(A)\psi(A)$ equals

$$\lim_{n\to\infty} \varphi_n(A) \lim_{n\to\infty} \psi_n(A) = \lim_{n\to\infty} \varphi_n(A)\psi_n(A) = (\varphi \cdot \psi)(A),$$

since the polynomials $\varphi_n\psi_n$ converge to the function $\varphi\psi$. Thus, we have constructed a homomorphism. Moreover, $\|f(A)\| = \lim\limits_{n\to\infty} \|f_n(A)\|$, which proves 2). In addition, $J(\overline{f}) = J(f)^*$. The proof also shows the uniqueness of a homomorphism with the indicated properties. □

Letting $f(A) := J(f)$ for each continuous function f on the whole real line, we obtain the equality $\|f(A)\| = \sup\limits_{t \in \sigma(A)} |f(t)|$.

7.7.3. Corollary. Let $f \in C(\sigma(A))$ and $f(t) \geqslant 0$ for all $t \in \sigma(A)$. Then the operator $f(A)$ is selfadjoint and $f(A) \geqslant 0$.

PROOF. The function $\sqrt{f} \geqslant 0$ is continuous on $\sigma(A)$. Hence the operator $B = \sqrt{f}(A)$ is selfadjoint. Since we have $f(A) = B^2$ and $(Ax, x) = (Bx, Bx) \geqslant 0$, the operator $f(A)$ is selfadjoint and nonnegative. □

Taking the function $f(t) = \sqrt{t}$ on $[0, +\infty)$ in case of an operator $A \geqslant 0$, we obtain the operator $\sqrt{A}$.

7.7.4. Corollary. If $A \geqslant 0$, then the operator $\sqrt{A}$ is selfadjoint, nonnegative and $A = \sqrt{A}\sqrt{A}$.

Note that the operator $\sqrt{A}$ is the only nonnegative operator the square of which equals A (Exercise 7.10.73).

7.7.5. Corollary. For every bounded operator A on a Hilbert space H, the operator

$$|A| := (A^*A)^{1/2}$$

is well-defined and nonnegative. The operator $|A|$ is called the absolute value of A.

PROOF. We have $(A^*Ax, x) = (Ax, Ax) \geqslant 0$. □

7.7.6. Example. Let A_φ be the operator of multiplication on $L^2(\mu)$ by a bounded μ-measurable function φ considered in Example 7.1.11. The operator A_φ can be substituted in an arbitrary bounded Borel function f on the complex plane (not necessarily continuous), by defining $f(A_\varphi)$ as the operator $A_{f\circ\varphi}$ of multiplication by the bounded μ-measurable function $f\circ\varphi$. Clearly, for any polynomial f this gives the operator $f(A_\varphi)$. Moreover, it is easy to see from Theorem 7.7.2 that in case of a real function φ the operator $f(A_\varphi)$ is the operator of multiplication by the function $f \circ \varphi$. Below, when we establish a unitary equivalence of every selfadjoint operator to some operator of multiplication A_φ, this will enable us to define easily Borel functions of selfadjoint operators.

The next useful result represents general operators by means of selfadjoint operators and partial isometries and is called the *polar decomposition* of an operator.

7.7.7. Theorem. *Let A be a bounded operator on a complex or real Hilbert space H. Then there exists a partial isometry U on the closure of the range of $|A|$, equal the orthogonal complement of the kernel of A, for which*

$$A = U|A|.$$

If $A \neq 0$ has a zero kernel and dense range, then U is a unitary operator.

PROOF. Set

$$Uy := Ax, \quad y = |A|x \in |A|(H).$$

Since

$$(Ax, Ax) = (A^*Ax, x) = (|A|^2x, x) = (|A|x, |A|x),$$

whenever $|A|x = 0$ we have $Ax = 0$, which proves that U is well-defined. From the equalities above we obtain $(Uy, Uy) = (y, y)$, i.e., U is an isometry on $|A|(H)$. Hence U extends to a partial isometry on the closure of $|A|(H)$. The same equalities show that $\mathrm{Ker}\, A = \mathrm{Ker}\, |A|$. Since $|A|$ is a selfadjoint operator, the closure of its range is the orthogonal complement of the kernel (see Lemma 6.8.4). If the operator A is injective and has a dense range, then the operator U on the closure of $|A|(H)$ is everywhere defined and has a dense range. Hence it is unitary. □

Usually it is convenient to extend U to an operator on the whole space H by setting $U|_{\mathrm{Ker}\, A} = 0$. Below, when it becomes necessary to consider U on all of H, we shall have in mind this extension. The way of extending does not influence $U|A|$, however, the chosen extension makes the adjoint operator U^* also a partial isometry. Indeed, the extended operator U is zero on $\mathrm{Ker}\, A$ and linearly and isometrically maps the closed subspace

$$H_1 := (\mathrm{Ker}\, A)^\perp = \overline{A^*(H)}$$

onto the closed subspace

$$H_2 := \overline{A(H)} = (\mathrm{Ker}\, A^*)^\perp.$$

The orthogonal decomposition

$$H = \mathrm{Ker}\, A \oplus \overline{A^*(H)} = \mathrm{Ker}\, A^* \oplus \overline{A(H)}$$

shows that the operator U^* is zero on $\mathrm{Ker}\, A^*$ and maps H_2 isometrically onto H_1 by means of the inverse to $U|_{H_1}$ (here we could also refer to Proposition 7.6.7). Hence U^*U is the orthogonal projection onto H_1, and UU^* is the orthogonal projection onto H_2. This gives (under the indicated extension of U, of course) the equality

$$|A| = U^*A.$$

Concerning the uniqueness of the polar decomposition, see Exercise 7.10.86.

The polar decomposition resembles the representation of a complex number z in the form $z = e^{i\theta}|z|$. This analogy is limited, though. Say, it is not always true that $|A + B| \leqslant |A| + |B|$. In addition, one does not always have $|A| = |A^*|$. For example, if the operator A on l^2 is the right shift $(x_1, x_2, \ldots) \mapsto (0, x_1, x_2, \ldots)$, then $A^*A = I$ and $|A| = I$, but $AA^*x = (0, x_2, x_3, \ldots)$ and $|A^*|^2 = AA^* \neq I$.

If the operator U is unitary (i.e., A is injective and has a dense range), then $W = U^*$ is also unitary, so $A^* = |A|W$, where W is unitary. Since A^* is also injective and has a dense range, we have similarly $A = |A^*|V$, where V is unitary. However, in the general case it is not always possible to decompose A into the product $A = ST$ of a selfadjoint operator S and a unitary operator T. As an example take the shift considered above: the operator S would have the zero kernel and a range that is not dense, which is impossible for a selfadjoint operator.

By using the polar decomposition and the Hilbert–Schmidt theorem we obtain the following representation of an arbitrary compact operator on a Hilbert space.

7.7.8. Proposition. *Let K be a compact operator on a complex or real separable Hilbert space $H \neq 0$. Then there exist two orthonormal sequences $\{\varphi_n\}$ and $\{\psi_n\}$ and a sequence of real numbers λ_n tending to zero for which*

$$Kx = \sum_{n=1}^{\infty} \lambda_n(x, \varphi_n)\psi_n, \quad x \in H.$$

In addition, every operator of the indicated form is compact.

PROOF. Let $K = U|K|$ be the polar decomposition. Let $\{\varphi_n\}$ be an eigenbasis of the selfadjoint compact operator $|K|$ and $|K|\varphi_n = \lambda_n \varphi_n$. Set $\psi_n := U\varphi_n$ for all n such that $U\varphi_n \neq 0$. Since $|K|x = \sum_{n=1}^{\infty} \lambda_n(x, \varphi_n)\varphi_n$, applying U to both parts of this equality we arrive at the desired representation (of course, excluding numbers n with $U\varphi_n = 0$ we must renumber φ_n and ψ_n). $\quad\square$

7.8. The Functional Model

The main result of this section shows that up to an isomorphism any selfadjoint operator is the operator of multiplication by a real function. This assertion is a continual analog of the fact known from linear algebra about representing a symmetric matrix in a diagonal form. The nontriviality of the generalization is, in particular, that in the infinite-dimensional case a selfadjoint operator can fail to have eigenvectors. The representation of an operator in the form of a multiplication by a function is called the functional model of this operator. In the next section we discuss a representation of a selfadjoint operator in the form of an integral with respect to a projection-valued measure. First we consider operators analogous to finite-dimensional operators without multiple eigenvalues. This analog is the following concept.

7.8.1. Definition. *We shall say that an operator A on a normed space X has a cyclic vector h if the linear span of the vectors $h, Ah, A^2h, \ldots$ is everywhere dense in X.*

7.8.2. Example. The unit operator on any space of dimension greater than 1 has no cyclic vectors. The operator A_φ of multiplication by the argument on $L^2(\mu)$, where μ is a bounded Borel measure on an interval, has cyclic vectors. For example, for a cyclic vector we can take the function $h(t) = 1$, because the linear span of the functions $1, t, t^2, \ldots$ consists of all polynomials and hence is dense in the space $L^2(\mu)$.

We shall say that a Hilbert space H is an *orthogonal sum* of its closed subspaces H_n if $H_n \perp H_k$ whenever $n \neq k$ and every vector $h \in H$ is the sum of the series $\sum_{n=1}^\infty P_n h$, where P_n is the operator of orthogonal projection onto H_n.

7.8.3. Lemma. *Let A be a selfadjoint operator on a separable Hilbert space H. Then H is the orthogonal sum of closed subspaces H_n such that $A(H_n) \subset H_n$ and $A|_{H_n}$ possesses a cyclic vector.*

PROOF. Let us consider the family S the elements of which are collections of pairwise orthogonal closed subspaces $E \subset H$ such that $A(E) \subset E$ and $A|_E$ has a cyclic vector. This family is partially ordered by inclusion. Every chain in S has a majorant: the union of its elements. Hence S contains a maximal element S. By the separability of H this element S consists of a finite or countable collection of pairwise orthogonal subspaces H_n with the indicated property. The linear span of all H_n is everywhere dense in H, because otherwise we can find a nonzero vector $h \perp H_n$ for all n. Then the closure E of the linear span of the sequence of vectors $A^k h$, $k = 0, 1, \ldots$, is orthogonal to all H_n and is mapped by A into itself, while the vector h is cyclic for $A|_E$. However, this is impossible by the maximality of S. It remains to observe that $h = \sum_{n=1}^\infty P_n h$ for every $h \in H$ (P_n is the projection onto H_n), since otherwise the vector $a = h - \sum_{n=1}^\infty P_n h$ is orthogonal to all H_n and cannot be approximated by vectors from the linear span of H_n. □

7.8.4. Remark. Similarly one can define a decomposition $H = \bigoplus_\gamma H_\gamma$ with an uncountable number of pairwise orthogonal separable closed subspaces H_γ. Then the previous lemma remains in force also in the case of nonseparable H, which is proved by the same reasoning.

7.8.5. Theorem. *Let A be a selfadjoint operator with a cyclic vector on a separable Hilbert space $H \neq 0$. Then there exists a nonnegative Borel measure μ on the compact set $\sigma(A)$ such that the operator A is unitarily equivalent to the operator of multiplication by the argument in $L^2(\mu)$, i.e., to the operator A_φ with the function $\varphi(t) = t$.*

PROOF. Let h be a cyclic vector for A. We define a functional l on $C(\sigma(A))$ by the formula $l(f) = (f(A)h, h)$. Clearly, we have obtained a continuous linear functional. If $f \geqslant 0$, then $l(f) \geqslant 0$ according to the results in the previous section. By the Riesz theorem there exists a nonnegative Borel measure μ on the compact set $\sigma(A)$ for which

$$l(f) = \int_{\sigma(A)} f(t)\, \mu(dt), \quad f \in C(\sigma(A)).$$

For any nonnegative integer k we set

$$U(A^k h) := p_k, \quad \text{where } p_k(t) = t^k.$$

We extend U by linearity to finite linear combinations of vectors $A^k h$. Since

$$\left(\sum \alpha_k A^k h, \sum \beta_m A^m h \right) = \left(\left(\sum \overline{\beta_m} A^m \right) \left(\sum \alpha_k A^k \right) h, h \right)$$

$$= \int_{\sigma(A)} \left(\sum \overline{\beta_m} t^m \right) \left(\sum \alpha_k t^k \right) \mu(dt)$$

$$= \left(\sum \alpha_k p_k, \sum \beta_m p_m \right)_{L^2(\mu)},$$

the mapping U is well-defined when extended by linearity (i.e., equal linear combinations are taken to the same element of $L^2(\mu)$) and preserves the inner product. The range of U is everywhere dense in $L^2(\mu)$, because the set of polynomials (the linear span of $\{p_k\}$) is everywhere dense in $L^2(\mu)$. By Lemma 7.6.6 we can extend U to a unitary operator from H to $L^2(\mu)$. By construction we have

$$UA\left(\sum \alpha_k A^k h \right) = U\left(\sum \alpha_k A^{k+1} h \right)$$

$$= \sum \alpha_k p_{k+1} = A_\varphi \sum \alpha_k p_k = A_\varphi U\left(\sum \alpha_k A^k h \right),$$

where $\varphi(t) = t$. Thus, $UA = A_\varphi U$, i.e., $A_\varphi = UAU^{-1}$. $\square$

Let us now consider the general case of the *spectral theorem*.

7.8.6. Theorem. (THE FUNCTIONAL MODEL OF A SELFADJOINT OPERATOR) *Let A be a selfadjoint operator on a separable Hilbert space $H \neq 0$. Then there exists a finite nonnegative Borel measure μ along with a bounded Borel function φ on the real line such that the operator A is unitarily equivalent to the operator of multiplication by φ in $L^2(\mu)$. Moreover, the function φ can be taken with values in $\sigma(A)$.*

PROOF. Let us decompose H in the orthogonal sum of closed subspaces H_n invariant under A such that the operators $A|_{H_n}$ have cyclic vectors. The orthogonal projection of h onto H_n is denoted by h_n. For every n, we find a probability Borel measure μ_n on the compact set $\sigma(A) \subset [-\|A\|, \|A\|]$ for which the operator $A|_{H_n}$ is unitarily equivalent to multiplication by the argument in $L^2(\mu_n)$. We can assume that $\|A\| < 1$, i.e., $\sigma(A) \subset [a, b] \subset (-1, 1)$. Let us translate the measure μ_n to the interval $\Omega_n = (2n - 3, 2n - 1)$ and denote the obtained measure by ν_n (it is concentrated on $\sigma(A) + 2n - 2$). Clearly, the operator $A|_{H_n}$ is unitarily equivalent to the operator of multiplication by the function $t \mapsto t - 2n + 2$ in $L^2(\nu_n)$. Let $J_n \colon H_n \to L^2(\nu_n)$ be the corresponding unitary equivalence. Let

$$\mu = \sum_{n=1}^{\infty} 2^{-n} \nu_n$$

and $\varphi(t) = t - 2n + 2$ if $t \in \sigma(A) + 2n - 2$. Outside these compact sets we define φ in an arbitrary way to make it a Borel function; for example, we can make it equal to a fixed number in $\sigma(A)$, which gives a function with values in $\sigma(A)$, but it can also be made a continuous periodic piece-wise linear function with $\varphi(t) = t$ if $t \in [a, b]$ (however, it is not always possible to obtain a continuous function with values in $\sigma(A)$). On the space $L^2(\mu)$ we have the operator A_φ of multiplication by φ. We now define an operator $J\colon H \to L^2(\mu)$ by the equality

$$J\colon h = \sum_{n=1}^\infty h_n \mapsto \sum_{n=1}^\infty 2^{n/2} J_n h_n.$$

The series in the right-hand side converges in $L^2(\mu)$, since the functions $J_n h_n$ have supports in pairwise disjoint sets $\sigma(A) + 2n - 2$ and the integral of a function with respect to the measure μ equals the sum of its integrals with respect to the measures $2^{-n}\nu_n$. Let us verify that we have obtained the desired objects. For every $h \in H$ we have

$$\|Jh\|_{L^2(\mu)}^2 = \sum_{n=1}^\infty 2^{-n} 2^n \|J_n h_n\|_{L^2(\nu_n)}^2 = \sum_{n=1}^\infty \|h_n\|^2 = \|h\|^2.$$

In addition,

$$A_\varphi Jh = \sum_{n=1}^\infty \varphi 2^{n/2} J_n h_n = \sum_{n=1}^\infty 2^{n/2} J_n A h_n = JAh.$$

Thus, J is a unitary equivalence of the operators A and A_φ. $\qquad\square$

7.8.7. Remark. (i) In the proof we have constructed a measure on the real line and the function φ has been taken either with values in $\sigma(A)$ or continuous and piece-wise linear, but it is easy to modify our construction in order to obtain a measure on an interval (it is also possible to transform the real line into an interval by using arctg). A continuous version of φ with values in the spectrum of A can fail to exist (say, if the spectrum is not an interval).

(ii) Remark 7.8.4 enables us to obtain an analog of this theorem in case of a nonseparable Hilbert space. To this end, we represent H as the orthogonal sum of pairwise orthogonal separable closed subspaces H_γ, $\gamma \in \Gamma$, invariant with respect to A. The restriction of A to H_γ is unitarily equivalent to the multiplication by a measurable function $\varphi_\gamma\colon \mathbb{R}^1 \to \sigma(A)$ in $L^2(\mu_\gamma)$, where μ_γ is some Borel measure on the real line. Now we can take Γ disjoint copies of the real line and take for μ the sum of the measures μ_γ on these copies (which gives a countably additive measure with values in $[0, +\infty]$; such a measure need not be σ-finite).

(iii) The choice of H_n is not unique. For example, for the operator of multiplication by the argument on $L^2[-1, 1]$ we can take the subspace H_1 of functions equal to zero on $(0, 1]$ and the subspace H_2 of functions equal to zero on $[-1, 0)$. The operators $A|_{H_1}$ and $A|_{H_2}$ have cyclic vectors. The same is possible when there is no cyclic vector in H. In §10.4 we discuss some canonical decompositions invariant under unitary isomorphisms. Note also that the spectrum of the restriction of A to H_n can be strictly smaller than $\sigma(A)$.

Representations of an operator as the multiplication by a function enable us to substitute it into Borel functions.

7.8.8. Definition. *Let A be a selfadjoint operator on a separable Hilbert space $H \neq 0$ and let f be a bounded Borel function on the real line. We know that A is unitarily equivalent by means of an isomorphism J to the operator of multiplication by a bounded Borel function φ in $L^2(\mu)$. Let us define the operator $f(A)$ by the formula*

$$f(A) := J^{-1}A_{f \circ \varphi}J.$$

This definition extends to the nonseparable case by using a decomposition of the space on which the given selfadjoint operator acts into the direct sum of invariant separable subspaces. It is readily seen that this definition agrees with the earlier introduced construction of a continuous function of a selfadjoint operator. In the next section we represent Borel functions of a selfadjoint operator by integrals with respect to projection-valued measures, whence it will follow that our definition of Borel functions of a selfadjoint operator does not depend on our choice of the functional model.

7.8.9. Corollary. *To every bounded complex Borel function f on the real line this definition associates the operator $f(A) \in \mathcal{L}(H)$ with the following properties: if $f(t) = g(t)$ for $t \in \sigma(A)$, then $f(A) = g(A)$, and $f_n(A)x \to f(A)x$ for all vectors $x \in H$ if $f_n(t) \to f(t)$ for all points $t \in \sigma(A)$ and $|f_n(t)| \leqslant C < \infty$ for all $n \in \mathbb{N}$ and $t \in \sigma(A)$.*

PROOF. Let us write A as the operator of multiplication by a function φ with values in $\sigma(A)$. Then the first assertion is obvious and the second one follows from the Lebesgue dominated convergence theorem. □

On account of Example 7.1.11 we obtain the following assertion.

7.8.10. Corollary. *Let A be a selfadjoint operator and let f be a continuous complex function on $\sigma(A)$. Then*

$$\sigma\big(f(A)\big) = f\big(\sigma(A)\big).$$

PROOF. We can assume that A is the operator of multiplication by a bounded Borel function φ on $L^2(\mu)$, where μ is a bounded nonnegative Borel measure on the real line. For μ-a.e. t, the number $\varphi(t)$ belongs to the spectrum of A. Then $f(A)$ is the multiplication by the function $f \circ \varphi$, which is defined μ-almost everywhere. If λ is an essential value of φ, then $f(\lambda)$ is an essential value of $f \circ \varphi$, since the set $f^{-1}\big(\{z \colon |f(\lambda) - z| < r\}\big)$ contains a neighborhood of λ for every $r > 0$. If the point η does not belong to the compact set $f\big(\sigma(A)\big)$, then it is at some positive distance ε from this set, hence $\big|f\big(\varphi(t)\big) - \eta\big| \geqslant \varepsilon$ for μ-a.e. t, which gives the invertibility of the operator $f(A) - \eta I$. □

An important example of a function of a selfadjoint operator A is its *Caley transform*

$$U = (A - iI)(A + iI)^{-1} = \varphi(A),$$

where $\varphi(t) = (t - i)/(t + i)$. Since $|\varphi(t)| = 1$, the operator U is unitary. The operator A is reconstructed by U by the formula

$$A = i(I + U)(I - U)^{-1},$$

since the spectrum of U does not contain 1 by the previous corollary. Conversely, for every unitary operator U the spectrum of which does not contain 1 the operator $i(I + U)(I - U)^{-1}$ is selfadjoint, which is readily verified. Indeed, the adjoint of this operator is $-i(I + U^*)(I - U^*)^{-1}$. In addition,

$$-(I + U^*)(I - U^*)^{-1} = (I + U)(I - U)^{-1},$$

since multiplying both sides by the invertible operator $(I - U^*)(I - U)$ and using the fact that the operators U and U^* commute, we arrive at the obvious equality $-(I + U^*)(I - U) = (I + U)(I - U^*)$ (on both sides we have $U - U^*$).

If the operator A has no cyclic vectors, it cannot be unitarily equivalent to the operator of multiplication by the argument on $L^2(\mu)$ for a measure μ on an interval, since this operator of multiplication possesses a cyclic vector and the property to have cyclic vectors is obviously preserved by unitary isomorphisms. However, even in the case of the absence of cyclic vectors any selfadjoint operator is represented in the form of multiplication by the argument in the space of vector functions (see § 7.10(viii)).

7.8.11. Remark. It is instructive to compare the Hilbert–Schmidt theorem about diagonalization of a compact selfadjoint operator A on a separable Hilbert space H with the theorem about representation of A in the form of multiplication by a bounded real measurable function φ on $L^2(\mu)$, where μ is a bounded Borel measure on the real line. If the operator A has no multiple eigenvalues (including the zero eigenvalue) and the space H is infinite-dimensional, then A is unitarily equivalent to the operator of multiplication by the argument on $L^2(\mu)$, where μ is a probability measure concentrated on the set $\{\alpha_n\}$ of all eigenvalues of A; we can assume that the value of μ at α_n is 2^{-n}. In case of multiple eigenvalues one has to decompose A in a direct sum of operators without multiple eigenvalues and apply the same construction as in the general theorem. On the other hand, the Hilbert–Schmidt theorem can be derived from the general spectral theorem. For this it suffices to verify that if the operator of multiplication by φ on $L^2(\mu)$ is compact, then the restriction of the measure μ to the set $\{t\colon \varphi(t) \neq 0\}$ is concentrated at countably many points (then their indicator functions will be eigenfunctions). This verification is Exercise 7.10.87.

7.8.12. Proposition. *Let A be a bounded operator on a Hilbert space H.*

(i) *The operator A is not compact precisely when there exists an infinite-dimensional closed subspace $H_0 \subset H$ such that the restriction of A to H_0 has a bounded inverse operator, i.e., the mapping $A\colon H_0 \to A(H_0)$ is one-to-one and the inverse is continuous.*

(ii) *The operator A is compact precisely when*

$$\lim_{n\to\infty} \|Ae_n\| = 0$$

for every infinite orthonormal sequence $\{e_n\}$ in H. This is equivalent to the property that

$$\lim_{n\to\infty} (A\psi_n, \varphi_n) = 0$$

for all infinite orthonormal sequences $\{\psi_n\}$ and $\{\varphi_n\}$ in the space H (see, however, Exercise 7.10.114).

PROOF. (i) We know that a compact operator cannot be invertible on an infinite-dimensional space. Suppose that A is not compact. It suffices to consider separable H. We first consider the case where A is selfadjoint and nonnegative. We can assume that A is the multiplication by a bounded nonnegative Borel function φ on $L^2(\mu)$, where μ is a Borel probability measure on the real line. Then for some $\varepsilon > 0$ the Borel set $E_\varepsilon := \{t\colon \varphi(t) \geqslant \varepsilon\}$ possesses the property that the subspace H_0 in $L^2(\mu)$ consisting of functions equal to zero outside E_ε is infinite-dimensional. In the opposite case the operator A would be compact, because it is the limit (in the operator norm) of the operators of multiplication by $\varphi I_{E_\varepsilon}$ as $\varepsilon \to 0$. It is clear that on H_0 our operator is invertible.

In the general case we take the polar decomposition $A = U|A|$. Then $|A|$ is not compact. As we have shown, there is an infinite-dimensional closed subspace H_0 on which $|A|$ is invertible, i.e., there is $c > 0$ such that $\|x\| \leqslant c\||A|x\|$ whenever $x \in H_0$. Then we have $\|x\| \leqslant c\|Ax\|$ for all $x \in H_0$.

(ii) Let $A \in \mathcal{K}(H)$ and let $\{e_n\}$ be an orthonormal sequence. Then $Ae_n \to 0$ in the weak topology, since for every $y \in H$ we have $(Ae_n, y) = (e_n, A^*y) \to 0$. Hence $\|Ae_n\| \to 0$. If $A \notin \mathcal{K}(H)$, then by (i) there exists an infinite orthonormal sequence $\{e_n\}$ with $\|Ae_n\| \geqslant c > 0$. One can also find two orthonormal sequences $\{\psi_n\}$ and $\{\varphi_n\}$ with $|(A\psi_n, \varphi_n)| \geqslant c > 0$. For this we observe that the operator $\sqrt{|A|}$ is not compact as well. Hence we can take an infinite orthonormal sequence $\{\psi_n\}$ such that $\|\sqrt{|A|}\psi_n\| \geqslant c > 0$. Such a sequence can be picked in the subspace

$$H_1 := \operatorname{Ker} \sqrt{|A|}^{\perp} = \operatorname{Ker} |A|^{\perp} = \operatorname{Ker} A^{\perp}.$$

Set $\varphi_n = U\psi_n$. This gives an orthonormal sequence, because $\psi_n \in H_1$ and the operator U is an isometry on H_1. In addition, $|A|\psi_n \in |A|(H) \subset H_1$. We obtain

$$(A\psi_n, \varphi_n) = (U|A|\psi_n, U\psi_n) = (|A|\psi_n, \psi_n) = \left\|\sqrt{|A|}\psi_n\right\|^2 \geqslant c^2,$$

which completes the proof. $\qquad\square$

Closing this section we observe that the obtained representations of selfadjoint operators in the form of operators of multiplication by functions leave open the following question: given two operators of multiplication, how can one decide whether they are equivalent? This question will be addressed in Chapter 10.

7.9. Projections and Projection-Valued Measures

An *orthogonal projection* in a Hilbert space H is the operator of the orthogonal projecting onto a closed subspace (see Corollary 5.4.6).

7.9.1. Lemma. *A bounded operator P is an orthogonal projection precisely when $P^* = P = P^2$.*

PROOF. If P is the projection onto a closed subspace H_0, then $P^2 = P$ and $(Px, y) = (x, Py)$, i.e., $P^* = P$, which has already been noted in Example 6.8.3. Conversely, if the indicated equalities are fulfilled, then we set

$$H_0 := \operatorname{Ker}(I - P), \quad H_1 := \operatorname{Ker} P.$$

It is clear that H_0 and H_1 are closed and $H_0 \perp H_1$, since for all $x \in H_0$ and $y \in H_1$ we have $(x, y) = (Px, y) = (x, Py) = 0$. Since

$$x - Px \in \operatorname{Ker} P, \quad Px \in \operatorname{Ker}(I - P),$$

for any vector x we obtain $x = (x - Px) + Px$, where $Px \in H_0$ and $x - Px \in H_1$, i.e., H is the sum of H_0 and H_1. On the subspace H_0 the operator P coincides with the identity, on H_1 it vanishes. Hence P is the projection onto H_0. $\square$

We now discuss a representation of a selfadjoint operator in the form of an integral with respect to a projection-valued measure.

7.9.2. Definition. *Let $(\Omega, \mathcal{B})$ be a measurable space and let H be a separable Hilbert space. A mapping Π from $\mathcal{B}$ to the space $\mathcal{P}(H)$ of orthogonal projections in H is called a projection-valued measure if, for every $a, b \in H$, the complex function*

$$\Pi_{a,b} \colon B \mapsto \big(\Pi(B)a, b\big)$$

is a bounded countably additive measure on $\mathcal{B}$.

Here are the simplest properties of the projection-valued measure Π:

1) the mapping Π is additive, i.e., $\Pi(B_1 \cup B_2) = \Pi(B_1) + \Pi(B_2)$ for any disjoint sets $B_1, B_2 \in \mathcal{B}$;
2) $\Pi(B_1) \leqslant \Pi(B_2)$ if $B_1, B_2 \in \mathcal{B}$ and $B_1 \subset B_2$;
3) $\Pi(B_1)\Pi(B_2) = \Pi(B_2)\Pi(B_1) = 0$ for any disjoint sets $B_1, B_2 \in \mathcal{B}$;
4) $\Pi(B_1)\Pi(B_2) = \Pi(B_2)\Pi(B_1) = \Pi(B_1 \cap B_2)$ for all $B_1, B_2 \in \mathcal{B}$.

Property 1) follows from the additivity of the measures $\Pi_{a,b}$. Property 2) follows from the equality $\Pi(B_2) = \Pi(B_1) + \Pi(B_2 \backslash B_1)$ taking into account that Π takes values in the set of nonnegative operators. For the proof of 3) we observe that 1) and the equality $\Pi(B_1 \cup B_2) = \Pi(B_1 \cup B_2)^2$ yield the equality

$$\Pi(B_1)\Pi(B_2) + \Pi(B_2)\Pi(B_1) = 0.$$

Multiplying it from the left by $\Pi(B_1)$ we obtain

$$\Pi(B_1)\Pi(B_2) + \Pi(B_1)\Pi(B_2)\Pi(B_1) = 0,$$

i.e., $\Pi(B_1)\Pi(B_2)\big(I + \Pi(B_1)\big) = 0$. Since the operator $I + \Pi(B_1) \geqslant I$ is invertible, we have $\Pi(B_1)\Pi(B_2) = 0$. The second equality in 3) follows from the first one. For any $B_1, B_2 \in \mathcal{B}$ we can write $B_1 = C_1 \cup D$, $B_2 = C_2 \cup D$, where

$$C_1 = B_1 \backslash (B_1 \cap B_2), \ C_2 = B_2 \backslash (B_1 \cap B_2), \ D = B_1 \cap B_2.$$

Since C_1, C_2, D are disjoint, the projections $\Pi(C_1)$, $\Pi(C_2)$ and $\Pi(D)$ commute as shown above and

$$\Pi(C_1)\Pi(C_2) = \Pi(C_1)\Pi(D) = \Pi(C_2)\Pi(D) = 0.$$

This gives the equality $\Pi(B_1)\Pi(B_2) = \Pi(D)^2 = \Pi(D)$.

Since each $\Pi(B)$ is a projection, we have

$$\Pi_{a,a} \geqslant 0, \quad \Pi_{a,a}(\Omega) \leqslant \|a\|^2.$$

The equality $\operatorname{Re}\Pi_{a,b} = \frac{1}{2}(\Pi_{a+b,a+b} - \Pi_{a,a} - \Pi_{b,b})$ gives the following estimate for the variation of the measure $\Pi_{a,b}$:

$$\|\Pi_{a,b}\| \leqslant 2(\|a+b\|^2 + \|a\|^2 + \|b\|^2) \leqslant 6\|a\|^2 + 6\|b\|^2.$$

For an arbitrary bounded complex $\mathcal{B}$-measurable function f, we define the integral

$$\int_\Omega f(\omega)\, d\Pi(\omega)$$

to be a bounded operator T such that

$$(Ta, b) = \int_\Omega f(\omega)\, d\Pi_{a,b}(\omega)$$

for all $a, b \in H$. Indeed, the right-hand side of this equality is linear in a, conjugate-linear in b and separately continuous in a and b, which follows from the above estimate for the variation (it is also sufficient to use the continuity of the functions $\Pi_{a,b}(B)$). One can go further and introduce the Lebesgue integral with respect to an H-valued measure in order to define Ta. Finally, one can define operator-valued integrals with the aid of partial sums converging in the operator norm. We do not develop here these approaches and give only the following assertion.

7.9.3. Proposition. *For every n, one can partition Ω into disjoint parts $\Omega_{n,1}, \ldots, \Omega_{n,n} \in \mathcal{B}$ such that for any choice of points $\omega_{n,k} \in \Omega_{n,k}$ the sums $\sum_{k=1}^n f(\omega_{n,k})\Pi(\Omega_{n,k})$ will converge to T in the operator norm.*

PROOF. It suffices to prove our assertion for real functions. Then the operator T and the aforementioned sums are selfadjoint operators. We can assume that $0 \leqslant f(\omega) < 1$. Let us divide the interval $[0, 1)$ into equal subintervals of the form $J_{n,k} = [(k-1)/n, k/n)$ and set $\Omega_{n,k} = f^{-1}(J_{n,k})$. For any choice of points $\omega_{n,k} \in \Omega_{n,k}$ for every $a \in H$ with $\|a\| \leqslant 1$ we have

$$\left|(Ta, a) - \sum_{k=1}^n f(\omega_{n,k})\Pi_{a,a}(\Omega_{n,k})\right|$$

$$= \left|\sum_{k=1}^n \int_{\Omega_{n,k}} f(\omega)\, d\Pi_{a,a}(\omega) - \sum_{k=1}^n f(\omega_{n,k})\Pi_{a,a}(\Omega_{n,k})\right|$$

$$\leqslant \sum_{k=1}^n \sup_{\omega \in \Omega_{n,k}} |f(\omega) - f(\omega_{n,k})|\Pi_{a,a}(\Omega_{n,k}) \leqslant \frac{1}{n}.$$

Since we deal with selfadjoint operators, by Theorem 7.2.6 we obtain the estimate $\left\|T - \sum_{k=1}^n f(\omega_{n,k})\Pi(\Omega_{n,k})\right\| \leqslant 1/n$. $\qquad\square$

Note, however, that the integral with respect to a projection-valued measure is not a Bochner integral with respect to the operator norm.

7.9.4. Proposition. *Let Π be a projection-valued measure on a σ-algebra $\mathcal{B}$ in a space Ω, φ and ψ bounded $\mathcal{B}$-measurable complex functions, and let A and B be the integrals of φ and ψ with respect to the measure Π in the sense defined* *above. Then for all $u, v \in H$ we have*

$$(ABu, v) = \int_\Omega \varphi(\omega)\psi(\omega)\, d\Pi_{u,v}(\omega). \tag{7.9.1}$$

PROOF. It is clear from the previous proposition that $\Pi(S)B = B\Pi(S)$ for all $S \in \mathcal{B}$. It suffices to verify (7.9.1) for functions with finitely many values, which reduces to indicator functions of sets. In the case where $\varphi = I_{S_1}$, $\psi = I_{S_2}$, we first assume that $S_1 \cap S_2 = \varnothing$. Then the right-hand side of (7.9.1) equals zero and the left-hand side is

$$\Pi_{Bu,v}(S_1) = \big(\Pi(S_1)Bu, v\big) = \big(B\Pi(S_1)u, v\big) = \big(\Pi(S_2)\Pi(S_1)u, v\big) = 0.$$

The equality to be proved remains also valid in the case where $S_1 = S_2$. The general case follows from this, since

$$S_1 = M_1 \cup (S_1 \cap S_2), \quad S_2 = M_2 \cup (S_1 \cap S_2),$$

where M_1, M_2 and $S_1 \cap S_2$ are pairwise disjoint. $\qquad\square$

Let us single out the case where $\Omega = K$ is a compact set on the real line (for example, $[a, b]$) and $\mathcal{B}$ is the Borel σ-algebra of K. In this case we obtain a selfadjoint operator

$$A := \int_K \lambda\, d\Pi(\lambda). \tag{7.9.2}$$

The previous proposition yields that for every $k \in \mathbb{N}$ the operator A^k is written as the integral of λ^k with respect to $d\Pi(\lambda)$. Hence for every polynomial f we obtain

$$f(A) = \int_K f(\lambda)\, d\Pi(\lambda).$$

Since two Borel measures on a compact set with equal integrals of polynomials coincide (see Lemma 3.8.9), we arrive at the following conclusion.

7.9.5. Corollary. *If A is represented in the form (7.9.2) with respect to a projection-valued measure Π on K, then such a measure is unique.*

We now show that every selfadjoint operator can be represented in the form of an integral with respect to a projection-valued measure.

7.9.6. Theorem. (THE SPECTRAL DECOMPOSITION OF A SELFADJOINT OPERATOR) *Let A be a selfadjoint operator on a separable Hilbert space $H \neq 0$. Then there exists a unique projection-valued measure Π on $\mathcal{B}(\mathbb{R}^1)$ with $\Pi(\mathbb{R}^1) = I$ vanishing outside some interval such that for every bounded Borel function f we have*

$$f(A) = \int_{\sigma(A)} f(\lambda)\, d\Pi(\lambda) = \int_{\mathbb{R}^1} f(\lambda)\, d\Pi(\lambda). \tag{7.9.3}$$

In particular,

$$A = \int_{\sigma(A)} \lambda\, d\Pi(\lambda) = \int_{\mathbb{R}^1} \lambda\, d\Pi(\lambda). \tag{7.9.4}$$

The measure Π is concentrated on $\sigma(A)$, i.e., $\Pi\big(\mathbb{R}^1 \backslash \sigma(A)\big) = 0$.

PROOF. Set $\Pi(B) := I_B(A)$. Then $\Pi(B)$ is an orthogonal projection, since $\Pi(B) \geqslant 0$ and $\Pi(B)^2 = \Pi(B)$. If A is realized as the operator of multiplication by a Borel function φ, then $\Pi(B)$ is the operator of multiplication by $I_{B \circ \varphi} = I_{\varphi^{-1}(B)}$. The measure Π is concentrated on $\sigma(A)$, i.e., $\Pi(B) = 0$ if $B \cap \sigma(A) = \varnothing$, since by Theorem 7.8.6 we can choose φ with values in $\sigma(A)$. If $f = I_M$, where $M \in \mathcal{B}(\mathbb{R}^1)$, then for every $a, b \in H$ we have

$$\int_{\sigma(A)} I_M(t) \, d\Pi_{a,b}(t) = \int_{\mathbb{R}^1} I_M(t) \, d\Pi_{a,b}(t) = \Pi_{a,b}(M) = \big(I_M(A)a, b\big),$$

i.e., equality (7.9.3) is true for simple functions f. By means of uniform approximations it is easily extended to bounded functions f. The uniqueness of the measure Π follows by the corollary above. $\square$

Obviously, in (7.9.3) or (7.9.4) in place of $\sigma(A)$ we can take any interval $[a, b] \supset \sigma(A)$, for example, the interval $[-\|A\|, \|A\|]$.

For the operator of multiplication by the argument the spectral measure is explicitly calculated in the proof: $\Pi(B)$ is the multiplication by I_B. It is also easy to obtain from the proof explicit expressions for the measures $\Pi_{a,b}$ for the operator of multiplication by φ in $L^2(\mu)$: $\Pi_{a,b} = (a\overline{b} \cdot \mu) \circ \varphi^{-1}$ for all $a, b \in L^2(\mu)$. For example, for the operator of multiplication by the argument the measure $\Pi_{a,b}$ is given by the density $a\overline{b}$ with respect to μ.

If A is an orthogonal projection, then in case $A \neq 0$ and $A \neq I$ we have $\Pi = (I - A)\delta_0 + A\delta_1$, where δ_0 and δ_1 are the Dirac measures at the points 0 and 1. Finally, if A is a selfadjoint operator with an eigenbasis $\{e_n\}$ and eigenvalues $\{\alpha_n\}$, i.e., $Ae_n = \alpha_n e_n$, then $\Pi = \sum_{n=1}^{\infty} P_n \delta_{\alpha_n}$, where P_n is the projection onto the linear span of e_n.

It follows from this theorem and Proposition 7.9.3 that any selfadjoint operator A is the limit in the operator norm of a sequence of finite linear combinations of orthogonal projections. Of course, this can be also seen from the theorem about representation of A as the multiplication by a bounded real function φ: it suffices to uniformly approximate φ by simple functions.

The projection-valued function $\Pi_0(\lambda) := \Pi\big((-\infty, \lambda)\big)$ is called a *resolution of the identity*. This is a very important characteristic of the operator A. Similarly to the distribution function of a scalar measure, it possesses the following properties:

(i) $\Pi_0(\lambda) \leqslant \Pi_0(\mu)$ if $\lambda \leqslant \mu$,
(ii) $\Pi_0(\lambda_n)x \to \Pi_0(\lambda)x$ for all $x \in H$ if $\lambda_n \uparrow \lambda$.
In addition, $\Pi_0(a) = 0$ if $a < -\|A\|$ and $\Pi_0(b) = I$ if $b > \|A\|$.

Projection-valued measures with equal resolutions of the identity coincide, since scalar measures are uniquely determined by their distribution functions. Moreover, the function Π_0 can be determined by the operator A without recurring to its functional model (but using the functional calculus). To this end, set

$$\Pi_0(\lambda)h = \lim_{n \to \infty} \psi_n(A)h,$$

where ψ_n is a continuous function on the real line defined as follows: $\psi_n(t) = 1$ if $t \leqslant \lambda - 1/n$, $\psi_n(t) = 0$ if $t \geqslant \lambda$, and on $(\lambda - 1/n, \lambda)$ the function ψ_n is linearly interpolated. This shows that bounded Borel functions of A constructed above are uniquely determined by the operator A itself (through the functional calculus) and do not depend on our choice of the functional model (used in our definition of these objects). Independence of the model is also seen from Corollary 7.9.5.

Any function Π_0 with the properties indicated above generates a projection-valued measure similarly to the case where a scalar measure is constructed by its distribution function. Before proving this we observe that for any orthogonal projections P_1 and P_2 in H the condition $P_1 \leqslant P_2$ is equivalent to the inclusion $P_1(H) \subset P_2(H)$. Indeed, if $(P_1 h, h) \leqslant (P_2 h, h)$, then for any h with $P_1 h = h$ we have $\|h\|^2 = (P_1 h, h) \leqslant (P_2 h, h)$, whence $P_2 h = h$, because P_2 is an orthogonal projection. The converse is obvious.

7.9.7. Proposition. *Let* $\Pi_0 \colon \mathbb{R}^1 \to \mathcal{P}(H)$ *have properties* (i) *and* (ii) *above and* $\Pi_0(a) = 0$ *and* $\Pi_0(b) = I$ *for some* $a < b$. *Then there is a selfadjoint operator* A *with* $\sigma(A) \subset [a, b]$ *for which* Π_0 *is the resolution of the identity.*

PROOF. The function $\Pi_{x,x} \colon \lambda \mapsto \big(\Pi_0(\lambda)x, x\big)$ for every $x \in H$ is the distribution function of some nonnegative Borel measure μ_x on the real line concentrated on $[a, b]$, moreover, $\|\mu_x\| \leqslant \|x\|^2$. The function $\Pi_{x,y} \colon \lambda \mapsto \big(\Pi_0(\lambda)x, y\big)$ for every $x, y \in H$ is the distribution function of the complex Borel measure $\mu_{x,y}$ on the real line generated by the measure μ_x as follows:

$$4\mu_{x,y} := \mu_{x+y} - \mu_{x-y} + i\mu_{x+iy} - i\mu_{x-iy}.$$

Then $\mu_{x,x} = \mu_x$. For all λ we have

$$\mu_{x,y}\big((-\infty, \lambda)\big) = \big(\Pi_0(\lambda)x, y\big).$$

Since measures on the real line with equal distribution functions coincide, we have $\mu_{x+z,y} = \mu_{x,y} + \mu_{z,y}$, $\mu_{\alpha x, y} = \alpha\mu_{x,y}$ and $\mu_{x,y} = \overline{\mu_{y,x}}$. It is clear that $\|\mu_{x,y}\| \leqslant 4$ if $\|x\| \leqslant 1$, $\|y\| \leqslant 1$. Hence for every Borel set B on the real line the function $(x, y) \mapsto \mu_{x,y}(B)$ is linear in x, conjugate-linear in y and continuous in every argument separately. Hence there exists an operator $P(B) \in \mathcal{L}(H)$ such that $\mu_{x,y}(B) = \big(P(B)x, y\big)$. Since $\mu_{x,x} \geqslant 0$, the operator $P(B)$ is nonnegative selfadjoint. Let us show that $P(B)$ is an orthogonal projection. The operator $P(B) = \Pi_0(\beta) - \Pi_0(\alpha)$ for $B = [\alpha, \beta)$ is an orthogonal projection, which is easily verified with the aid of the equality $\Pi_0(\alpha)\Pi_0(\beta) = \Pi_0(\beta)\Pi_0(\alpha) = \Pi_0(\alpha)$, following from the condition $\Pi_0(\alpha) \leqslant \Pi_0(\beta)$. This yields the equality

$$P(B \cap E) = P(B)P(E) = P(E)P(B)$$

for semi-intervals. Let $x, y \in H$. Let us consider two complex Borel measures $E \mapsto \big(P(B \cap E)x, P(E)y\big)$ and $E \mapsto \big(P(B)x, P(E)y\big)$. If B is a semi-interval $[\alpha, \beta)$, then these two measures have equal values on semi-intervals, hence coincide, i.e., $P(B \cap E) = P(B)P(E) = P(E)P(B)$ for all Borel sets E. Repeating this reasoning for a fixed Borel set E, we conclude that the equality remains valid for all Borel sets B. In particular, $P(B) = P(B)^2$. It is clear that $\Pi_0(\lambda) = P\big((-\infty, \lambda)\big)$. Now set

$$A := \int_{[a,b]} \lambda \, dP(\lambda).$$

By Corollary 7.9.5 the operator A generates the projection-valued measure P and has the resolution of the identity Π_0. □

7.10. Complements and Exercises

(i) The structure of the spectrum (314). (ii) Commuting selfadjoint operators (316). (iii) Operator ranges in a Hilbert space (320). (iv) Hilbert–Schmidt operators and nuclear operators (323). (v) Integral operators and Mercer's theorem (337). (vi) Tensor products (339). (vii) Fredholm operators (341). (viii) The vector form of the spectral theorem (345). (ix) Invariant subspaces (346). Exercises (347).

7.10(i). The structure of the spectrum

Let us discuss the structure of the spectrum of a bounded operator A on an infinite-dimensional separable Hilbert space H. The set $\sigma(A)$ is nonempty and compact in $\mathbb{C}$. On the other hand, every nonempty compact set $K \subset \mathbb{C}$ is the spectrum of some operator $A \in \mathcal{L}(H)$, since we can find a finite or countable everywhere dense set of points λ_n in K, take an orthonormal basis $\{e_n\}$ in H and define a bounded diagonal operator on H by the formula $Ae_n = \lambda_n e_n$. Its spectrum is the closure of $\{\lambda_n\}$ (Exercise 7.10.57), which is exactly K. It has been proved by Gowers and Maurey, there exist infinite-dimensional separable Banach spaces in which the spectrum of every bounded operator is finite or countable.

If the set $(A - \lambda I)(H)$ is dense and the operator $(A - \lambda I)^{-1}$ is continuous on this set, then it extends to a bounded operator, which will serve as the inverse to $A - \lambda I$. An important part of the spectrum of A is the *point spectrum*, i.e., the set $\sigma_p(A)$ of eigenvalues (which, as we know, can be absent in the infinite-dimensional case). The remaining points $\lambda \in \sigma(A)$ belong to the spectrum because the mapping $(A - \lambda I)^{-1} \colon (A - \lambda I)(H) \to H$ is either discontinuous or defined on a set that is not dense (although is continuous). The *continuous spectrum* $\sigma_c(A)$ is usually defined as the set of all numbers $\lambda \in \sigma(A)\backslash\sigma_p(A)$ for which $A - \lambda I$ has a dense range, but the inverse operator is discontinuous on it. Then the *residual spectrum* $\sigma_r(A)$ is $\sigma(A)\backslash(\sigma_p(A) \cup \sigma_c(A))$, i.e., the collection of numbers λ for which the range of the operator $A - \lambda I$ is not dense (in this case $(A - \lambda I)^{-1}$ can be continuous on this range or discontinuous). However, one encounters different partitions of the spectrum in the literature. For example, sometimes the residual spectrum is defined to consist of those λ for which the range of $A - \lambda I$ is not dense, but the inverse operator is bounded.

Let us describe the structure of the set of eigenvalues.

7.10.1. Theorem. *Let A be a bounded operator on an infinite-dimensional separable Hilbert space H. Then the set $\sigma_p(A)$ of all eigenvalues of A is a countable union of compact sets.*

Conversely, every bounded set that is a countable union of compact sets serves as the set of all eigenvalues of some bounded operator on H.

PROOF. The set of eigenvalues is bounded as a subset of the spectrum. The closed unit ball U in H with the weak topology is a metrizable compact space. Hence its open subset $U\backslash\{0\}$ can be represented as a countable union of closed (in the weak topology) parts $K_n \subset U$. The sets K_n are compact in the weak topology.

A point $\lambda \in \mathbb{C}$ is an eigenvalue of A precisely when $\|Ax - \lambda x\| = 0$ for some $x \in U \backslash \{0\}$. Hence $\sigma_p(A)$ is the union of the projections to $\mathbb{C}$ of the sets

$$M_n := \{(x, \lambda) \in K_n \times D : \|Ax - \lambda x\| = 0\},$$

where D is the closed disc in $\mathbb{C}$ with the center at the origin and radius $\|A\|$. We observe that M_n is compact if K_n is equipped with the weak topology. This is seen from the fact that $K_n \times D$ is compact when we equip K_n with the weak topology and M_n is closed in this product, since it is specified by the conditions $\lambda(x, e_i) = (x, A^* e_i)$, where $\{e_i\}$ is an orthonormal basis in H. Therefore, the projection of M_n to $\mathbb{C}$ is also compact.

We now show that every bounded set P equal to the union of compact sets $S_n \subset \mathbb{C}$ coincides with the point spectrum of some bounded operator on a separable Hilbert space. It suffices to show this for every S_n separately, since the direct sum of uniformly bounded operators A_n in Hilbert spaces H_n has for the family of eigenvalues the union of the point spectra of A_n. Thus, we shall deal with a single compact set S. We can assume that it is not empty and is contained in $D := \{z : |z| < 1\}$. Let us consider the Bergman space $A^2(D)$ of all functions $f \in L^2(D)$ holomorphic in D, where D is equipped with Lebesgue measure (Example 5.2.2). We know that $A^2(D)$ is a separable Hilbert space with the norm from $L^2(D)$, i.e.,

$$\|f\|^2_{A^2(D)} := \int_D |f(x + iy)|^2 \, dx \, dy.$$

Let us consider the operator T on the dual to the space $A^2(D)$ (so that T acts on distributions, not on functions) given by the formula $T\psi(f) := \psi(zf)$, where $(zf)(z) = zf(z)$. Now we do not identify $A^2(D)$ with its dual. Actually, T is the operator adjoint to the multiplication by the argument on $A^2(D)$. However, the operator T has too many eigenvalues: every point λ in D turns out to be an eigenvalue, since the functional $\psi_\lambda : f \mapsto f(\lambda)$ satisfies the equality $T\psi_\lambda = \lambda \psi_\lambda$. The continuity of this functional is clear from estimate (5.2.1) in Example 5.2.2. Hence it is natural to take for H the closed linear subspace in $A^2(D)^*$ generated by the functionals ψ_λ with $\lambda \in S$. It is clear that $T(H) \subset H$ and all elements of S remain eigenvalues of $T|_H$. Let us verify that there are no other eigenvalues. It suffices to make sure that the operator T on H has no eigenvalues on the unit circumference and for every $\lambda \in D$ the kernel of $T - \lambda I$ in the whole $A^2(D)^*$ is one-dimensional. Since T is the adjoint to the operator T_1 of multiplication by the argument on $A^2(D)$, we have to verify that the range of $T_1 - \lambda I$ is dense if $|\lambda| = 1$ and has a one-dimensional orthogonal complement if $|\lambda| < 1$. Suppose that there exists λ with $|\lambda| = 1$ and a unit vector $g \in A^2(D)$ such that $(T_1 f - \lambda f, g) = 0$ for all $f \in A^2(D)$, i.e.,

$$\int_D (x + iy) f(x + iy) \overline{g(x + iy)} \, dx \, dy = \lambda \int_D f(x + iy) \overline{g(x + iy)} \, dx \, dy$$

for all $f \in A^2(D)$. In particular, for $f = g$ we obtain in the right-hand side a number the absolute value of which equals 1. The left-hand side is strictly less, since $|x + iy| < 1$ if $x + iy \in D$. Let now $|\lambda| < 1$. Since the functional $f \mapsto f(\lambda)$ is continuous, its kernel H_λ is closed and has codimension 1. In addition, every function in H_λ belongs to the range of the operator $T_1 - \lambda I$. Indeed, for every function $f \in H_\lambda$ the function $g(z) = (z - \lambda)^{-1}f(z)$ belongs to $A^2(D)$ (it is holomorphic and square integrable in D, since $f(\lambda) = 0$) and satisfies the equality $f = (T_1 - \lambda I)g$. $\square$

Close considerations give the following result (see [**708**]).

7.10.2. Theorem. (i) *For every operator* $A \in \mathcal{L}(l^2)$, *the set* $\sigma_c(A)$ *is a countable intersection of open sets.*

(ii) *Let* $K \subset \mathbb{C}$ *be a nonempty compact set such that* $K = P \cup C \cup R$, *where* P, C *and* R *are pairwise disjoint,* P *is a countable union of compact sets, and* C *is a countable intersection of open sets. Then there exists an operator* $A \in \mathcal{L}(l^2)$ *such that* $\sigma_p(A) = P$, $\sigma_c(A) = C$, *and* $\sigma_r(A) = R$.

7.10(ii). Commuting selfadjoint operators

The class of selfadjoint operators is contained in the class of *normal* operators: bounded operators B on a Hilbert space such that $BB^* = B^*B$. The latter class contains also all unitary operators. It turns out that normal operators are also unitarily isomorphic to operators of multiplication. This fact is derived below from a more general assertion about simultaneous representation of commuting selfadjoint operators in the form of multiplication by a function. We first prove an auxiliary result on projection-valued measures. To projection-valued measures one can extend some (but not all!) results of the usual measure theory.

7.10.3. Proposition. *Suppose that on an algebra* $\mathcal{R}$ *in a space* Ω *we are given an additive set function* Π *with values in the set of orthogonal projections in a Hilbert space* H. *Suppose that for every* $a, b \in H$ *the complex function*

$$R \mapsto \Pi_{a,b}(R) := \big(\Pi(R)a, b\big)$$

is countably additive on $\mathcal{R}$. *Then the function* Π *has a unique extension to a projection-valued measure on the* σ-*algebra* $\sigma(\mathcal{R})$ *generated by* $\mathcal{R}$.

PROOF. For any fixed $a, b \in H$, the function $\Pi_{a,b}$ has a unique extension to a countably additive complex measure on $\sigma(\mathcal{R})$ denoted also by $\Pi_{a,b}$. Indeed, if $a = b$, then the function $\Pi_{a,a}$ on $\mathcal{R}$ is countably additive, nonnegative and bounded, since $\big(\Pi(R)a, a\big) \leqslant \big(\Pi(\Omega)a, a\big)$. By Theorem 2.4.6 it uniquely extends to a bounded measure on $\sigma(\mathcal{R})$. The formulas

$$2\operatorname{Re}\Pi_{a,b} = \Pi_{a+b,a+b} - \Pi_{a,a} - \Pi_{b,b}, \quad 2\operatorname{Im}\Pi_{a,b} = \Pi_{a+ib,a+ib} - \Pi_{a,a} + \Pi_{b,b}$$

give extensions of $\Pi_{a,b}$ to $\sigma(\mathcal{R})$. Note that $\Pi_{a,b} = \overline{\Pi_{b,a}}$ on $\sigma(\mathcal{R})$, since this is true on $\mathcal{R}$. Therefore, for every $S \in \sigma(\mathcal{R})$ there exists a bounded selfadjoint operator $\Pi(S)$ with $\big(\Pi(S)a, b\big) = \Pi_{a,b}(S)$ (see Lemma 7.2.2). It follows from our construction that $0 \leqslant \Pi(S) \leqslant I$. Let us show that $\Pi(S)$ is an orthogonal projection. Denote by $\mathcal{M}$ the class of sets $S \in \sigma(\mathcal{R})$ with this property. Then $\mathcal{M}$ contains the algebra $\mathcal{R}$. In addition, $\mathcal{M}$ is a monotone class: if sets M_n in $\mathcal{M}$ either increase to M or decrease to M, then $M \in \mathcal{M}$. Indeed, in the first case $\Pi(M_{n+1}) = \Pi(M_n) + \Pi(M_{n+1}\backslash M_n) \geqslant \Pi(M_n)$, i.e., $\Pi(M_n)$ are projections onto increasing closed subspaces H_n and hence M is the projection onto the closure of the union of H_n. In the second case the reasoning is similar.   □

7.10.4. Corollary. *Let* Π' *and* Π'' *be projection-valued measures on* σ-*algebras* $\mathcal{A}'$ *and* $\mathcal{A}''$ *in spaces* Ω' *and* Ω''. *Suppose that* $\Pi'(S')$ *and* $\Pi''(S'')$ *commute for all* $S' \in \mathcal{A}'$ *and* $S'' \in \mathcal{A}''$. *Then on the* σ-*algebra* $\mathcal{A} = \mathcal{A}' \otimes \mathcal{A}''$ *in the space* $\Omega = \Omega' \times \Omega''$ *there is a projection-valued measure* Π *such that* $\Pi(S' \times S'') = \Pi'(S')\Pi''(S'')$ *for all* $S' \in \mathcal{A}'$ *and* $S'' \in \mathcal{A}''$.

PROOF. Products of the form $S' \times S''$, where $S' \in \mathcal{A}'$, $S'' \in \mathcal{A}''$, constitute a semi-algebra $\mathcal{R}_0$, on which Π can be defined by $\Pi(S' \times S'') = \Pi'(S')\Pi''(S'')$. It is readily seen that we have obtained an additive projection-valued function. The sets in the algebra $\mathcal{R}$ generated by $\mathcal{R}_0$ have the form $R = R_1 \cup \cdots \cup R_n$, where $R_i \in \mathcal{R}_0$ are disjoint. Let us extend Π to $\mathcal{R}$ by the natural formula $\Pi(R) = \Pi(R_1) + \cdots + \Pi(R_n)$. We observe that $\Pi(R_i)\Pi(R_j) = 0$ if $i \neq j$. Hence $\Pi(R)$ is an orthogonal projection. Proposition 2.3.7 yields that the nonnegative scalar measures $R \mapsto (\Pi(R)a, a)$ are countably additive on $\mathcal{R}$ for all $a \in H$. This gives countably additivity complex measures $R \mapsto (\Pi(R)a, b)$, $a, b \in H$.   □

7.10.5. Lemma. *Let* A *be a selfadjoint operator with a cyclic vector and let* B *be a selfadjoint operator commuting with* A. *Then* B *is a Borel function of* A.

PROOF. We can assume that A is the operator of multiplication by the argument on $L^2(\mu)$ for some measure μ on an interval. Set $\psi = B(1)$, where we choose a Borel version of the function $B(1) \in L^2(\mu)$. We show that B coincides with the operator of multiplication by ψ. First we verify that for every function $p_k \colon t \mapsto t^k$ one has the equality $Bp_k = \psi p_k$. For $k = 0$ this is true. If the equality is true for some $k \geqslant 0$, then it remains valid for $k + 1$, because we have $Bp_{k+1} = BAp_k = ABp_k = A(\psi p_k) = \psi p_{k+1}$. Thus, $Bp = \psi \cdot p$ for every polynomial p. This yields that $Bf = \psi \cdot f$ for every function $f \in L^2(\mu)$. Indeed, there exists a sequence of polynomials f_k converging to f in $L^2(\mu)$, which gives convergence of Bf_k to Bf in $L^2(\mu)$ and convergence of $\psi \cdot f_k$ to $\psi \cdot f$ in $L^1(\mu)$. From the equality $Bf_k = \psi \cdot f_k$ we obtain the equality $Bf = \psi \cdot f$ a.e. Since this is true for all $f \in L^2(\mu)$, one has $\psi \in L^\infty(\mu)$. This can be easily verified without using the property that the operator B is bounded, but this property gives at once the estimate $|\psi(t)| \leqslant \|B\|$ for μ-a.e. t, since if the set $M = \{t \colon |\psi(t)| > \|B\|\}$ has positive measure, then $\|B\| \cdot \|I_M\|_{L^2} < \|\psi I_M\|_{L^2} = \|BI_M\|_{L^2} \leqslant \|B\| \cdot \|I_M\|_{L^2}$.   □

7.10.6. Lemma. *Suppose that selfadjoint operators* A *and* B *on a separable Hilbert space commute. Then for any bounded Borel functions* φ *and* ψ *the operators* $\varphi(A)$ *and* $\psi(B)$ *commute.*

PROOF. According to Exercise 7.10.78 there are polynomials p_n such that the operators $p_n(A)$ converge to $\varphi(A)$ on every vector. Then

$$B\varphi(A)x = \lim_{n\to\infty} Bp_n(A)x = \lim_{n\to\infty} p_n(A)Bx = \varphi(A)Bx.$$

Thus, B commutes with $\varphi(A)$. Applying the proved assertion once again, we obtain the equality $\psi(B)\varphi(A) = \varphi(A)\psi(B)$.   □

7.10.7. Theorem. *Suppose that selfadjoint operators $A_1, \ldots, A_n$ on a separable Hilbert space $H \neq 0$ commute. Then there exists a bounded nonnegative Borel measure μ on $\mathbb{R}^n$ along with a unitary isomorphism $J \colon H \to L^2(\mu)$ and bounded Borel functions $\varphi_1, \ldots, \varphi_n$ on $\mathbb{R}^n$ such that the operators $J A_i J^{-1}$ are the operators of multiplication by φ_i for all $i = 1, \ldots, n$.*

PROOF. Suppose first that there is a unit vector h such that the set of finite linear combinations of all vectors of the form $A_1^{k_1} \cdots A_n^{k_n} h$, $k_i = 0, 1, \ldots$, is dense in H. Let us write A_i in the form of integrals with respect to projection-valued measures Π_i on the real line. On $\mathbb{R}^n$ we can define a projection-valued measure Π with the aid of Corollary 7.10.4 and induction. We observe that

$$A_i = \int_{\mathbb{R}^1} t_i \, d\Pi_i(t) = \int_{\mathbb{R}^n} t_i \, d\Pi(t), \quad i = 1, \ldots, n.$$

Let $\mu = \Pi_{h,h}$. The measure μ is concentrated on the set $\prod_{i=1}^n [-\|A_i\|, \|A_i\|]$. Let us define a mapping $J \colon H \to L^2(\mu)$ by the formula

$$J(A_1^{k_1} \cdots A_n^{k_n} h) := p_{k_1, \ldots, k_n}, \quad \text{where} \quad p_{k_1, \ldots, k_n}(t_1, \ldots, t_n) = t_1^{k_1} \cdots t_n^{k_n},$$

then we extend it by linearity to the linear span of such vectors. We observe that for any finite linear combination of the indicated vectors we have the equality

$$\left| \sum c_{k_1, \ldots, k_n} A_1^{k_1} \cdots A_n^{k_n} h \right|^2 = \int \left| \sum c_{k_1, \ldots, k_n} p_{k_1, \ldots, k_n} \right|^2 d\mu,$$

since by relation (7.9.1) we have

$$(A_1^{k_1} \cdots A_n^{k_n} h, A_1^{l_1} \cdots A_n^{l_n} h) = (A_1^{k_1 + l_1} \cdots A_n^{k_n + l_n} h, h)$$

$$= \int_{\mathbb{R}^n} t_1^{k_1 + l_1} \cdots t_n^{k_n + l_n} \, \mu(dt).$$

The obtained equality means that the mapping J is well-defined and preserves the inner product. The range of J is everywhere dense in $L^2(\mu)$, since it contains all polynomials. Hence J extends to a unitary isomorphism. In the general case we can decompose H into the orthogonal sum of closed subspaces invariant with respect to all operators A_i and possessing the aforementioned property. Similarly to the case of a single operator, this is done with the aid of Zorn's lemma. $\square$

7.10.8. Corollary. *Every normal operator S on a separable Hilbert space $H \neq 0$ is unitarily equivalent to the operator of multiplication by a bounded complex Borel function on the space $L^2(\mu)$ for some bounded nonnegative Borel measure μ on $\mathbb{C}$.*

PROOF. The operators $A = S + S^*$ and $B = i^{-1}(S - S^*)$ are selfadjoint and commute. In addition, $S = A/2 + iB/2$. It remains to represent simultaneously A and B in the form of multiplication. $\square$

7.10.9. Corollary. *Every unitary operator U on a separable Hilbert space $H \neq 0$ is unitarily equivalent to the operator of multiplication by a Borel function ζ on the space $L^2(\mu)$ for some bounded Borel measure μ on $\mathbb{C}$ or on $\mathbb{R}^1$ such that*

$|\zeta(t)| = 1$ *for μ-a.e. t. In addition, there is a representation $U = \exp(iB)$, where B is a selfadjoint operator.*

PROOF. By the previous corollary we can assume that U is the multiplication by a bounded function ζ on $L^2(\mu)$ for some Borel measure μ on $\mathbb{C}$. Then $|\zeta(t)| = 1$ μ-a.e. Hence ζ can be written in the form $\zeta = \exp(ig)$, where g is a Borel function with values in $[0, 2\pi]$. For B we take the operator of multiplication by g. Now we write B in the form of multiplication by a function on an interval. □

By using the projection-valued measure Π of the selfadjoint operator B from this corollary, we write the operator U in the form

$$U = \int_{\sigma(B)} \exp(i\lambda) \, d\Pi(\lambda).$$

We now extend the previous theorem to infinite families of commuting operators.

7.10.10. Theorem. *Suppose we are given a countable set of commuting selfadjoint operators A_n on a separable Hilbert space $H \neq 0$. Then there exists a bounded nonnegative Borel measure μ on $[0,1]^\infty$ along with a unitary isomorphism $J: H \to L^2(\mu)$ and bounded Borel functions φ_n on $[0,1]^\infty$ such that the operators JA_nJ^{-1} are the operators of multiplication by φ_n for all n.*

PROOF. Without loss of generality we can assume that $\sigma(A_n) \subset [0,1]$. As in the previous theorem, on $[0,1]^\infty$ we construct a Borel projection-valued measure Π with respect to which the integrals of the coordinate functions give our operators A_n. The only difference is that for the initial algebra on which the measure Π is defined is constituted by the unions of finite powers of $\mathcal{B}([0,1])$ (the measure Π is defined on them by the previous theorem). As in the case of a finite collection of operators, the general case reduces to the situation where there is a unit vector h with the property that the linear span of all possible vectors of the form $A_1^{k_1} \cdots A_n^{k_n} h$ is dense in H. In this situation we employ the same isomorphism J as in the proof for a finite collection. Its range consists of polynomials in finitely many variables, which is everywhere dense in $L^2(\mu)$. Hence the proof is completed as above. □

Let us consider arbitrary collections of operators.

7.10.11. Theorem. *Suppose we are given a collection $\mathcal{T}$ of commuting selfadjoint operators on a separable Hilbert space H. Then there exists a selfadjoint operator A on H and, for every $T \in \mathcal{T}$, there is a bounded Borel function φ_T on the real line such that $T = \varphi_T(A)$ for all $T \in \mathcal{T}$.*

PROOF. We first consider the case of a countable collection of commuting selfadjoint operators A_n. As shown above, we can assume that the operators A_n are the operators of multiplication by bounded Borel functions φ_n on $L^2(\mu)$, where μ is a Borel measure on $[0,1]^\infty$. We now use the following fact (see [73, Corollary 6.8.8]): there exists a Borel isomorphism G between $[0,1]^\infty$ and the interval $[0,1]$. Let ν be the image of the measure μ under this isomorphism. Then the operators A_n become the operators of multiplication by the Borel functions $\psi_n = \varphi_n \circ G^{-1}$ in $L^2(\nu)$ with the measure ν on the interval, hence they become functions of the operator of multiplication by the argument on $L^2(\nu)$.

In the general case we find in $\mathcal{T}$ a countable family of operators $\mathcal{T}_0$ with the following property: for every $T \in \mathcal{T}$, there exists a sequence of operators $A_n \in \mathcal{T}_0$ converging to T in the weak operator topology, i.e., $(A_n x, y) \to (Tx, y)$ for all $x, y \in H$. This is possible, since, as is readily seen, the weak operator topology is metrizable on balls. As shown above, we can assume that all operators in $\mathcal{T}_0$ have the form of multiplication by some functions $\psi \circ \varphi$, where ψ is a bounded Borel function and φ is the Borel function on the interval determining the operator on $L^2(\nu)$ functions of which are the operators in $\mathcal{T}_0$. We can choose a Borel version of φ such that the norm of every operator in $\mathcal{T}_0$ will equal $\sup_x |\psi \circ \varphi(x)|$ for the corresponding function ψ. Let $T \in \mathcal{T}$. Let us find a sequence $\{T_n\} \subset \mathcal{T}_0$ converging to T in the weak operator topology. Then $\{\psi_n \circ \varphi\}$ converges weakly in $L^2(\nu)$ to some limit g. The sequence $\{T_n\}$ is bounded in the operator norm, which gives the boundedness of $\{\psi_n \circ \varphi\}$ in $L^\infty(\nu)$. Then $g \in L^\infty(\nu)$, since the sequence $S_n \circ \varphi$ of the arithmetic means of some subsequence in $\{\psi_n \circ \varphi\}$ converges in the norm of $L^2(\nu)$ (see Example 6.10.33). It is clear that the operator T is given by the multiplication by g, but we have to verify that $g = \psi \circ \varphi$ for some bounded Borel function ψ. Passing to a subsequence, we can assume that $S_n \circ \varphi \to g$ ν-a.e. By Luzin's theorem, the set of convergence contains compact sets K_j with $\nu(K_j) \to \nu([0,1])$ on which the function φ is continuous. The sets $\varphi(K_j)$ are compact, $E = \bigcup_{j=1}^\infty \varphi(K_j)$ is a Borel set, and for every $y \in E$ the sequence $\{S_n(y)\}$ converges. Let us denote the limit by $\psi(y)$. Outside E we define ψ by zero. We have obtained a bounded Borel function. For any $x \in \bigcup_{j=1}^\infty K_j$ we have
$$\psi\big(\varphi(x)\big) = \lim_{n\to\infty} \psi_n\big(\varphi(x)\big) = g(x), \text{ i.e., } \psi\big(\varphi(x)\big) = g(x) \text{ for } \nu\text{-a.e. } x. \qquad \square$$

7.10.12. Corollary. *Every normal operator T on a separable Hilbert space H has the form $T = f(A)$, where A is a selfadjoint operator on H and f is a bounded complex Borel function.*

PROOF. Since $T = A_1 + iA_2$, where A_1 and A_2 are commuting selfadjoint operators, we can apply the theorem above (of course, here we need its simplest case). $\qquad \square$

Hence T can be written in the form of an integral (7.9.3) of f with respect to a projection-valued measure Π. Consequently, T can be written as the integral
$$T = \int_{\sigma(T)} z \, dP(z)$$
with respect to the projection-valued measure P on $\sigma(T)$ defined as follows: $P(B) := \Pi\big(f^{-1}(B)\big)$.

7.10(iii). Operator ranges in a Hilbert space

Let us apply the polar decomposition to show that the range of every bounded operator on a Hilbert space coincides with the range of some selfadjoint operator.

7.10.13. Lemma. *For every bounded operator A on a Hilbert space H we have*

$$A(H) = |A^*|(H).$$

PROOF. We have $A^* = V|A^*|$, where the operator V isometrically maps the subspace $E_1 := (\operatorname{Ker} A^*)^\perp$ onto $E_2 := \overline{A^*(H)}$ and equals zero on $\operatorname{Ker} A^*$. Then $A = |A^*|V^*$, whence $A(H) \subset |A^*|(H)$. On the other hand, we have $\operatorname{Ker}|A^*| = \operatorname{Ker} A^*$, hence $|A^*|(H) = |A^*|(E_1)$. Since V^* isometrically maps E_2 onto E_1, we have $|A^*|(H) = |A^*|V^*(E_2) = A(E_2) \subset A(H)$. $\qquad\square$

In accordance with a general result for Banach spaces from §6.10(i), a linear subspace L of a Hilbert space H is the range of a bounded operator on H precisely when it is a Hilbert space continuously embedded into H. Indeed, the quotient space of H with respect to the closed subspace H_0 is also Hilbert (it can be identified with $H_0^\perp$). In addition, if a Hilbert space E is continuously embedded into H, then there is an operator $A \in \mathcal{L}(H)$ with $A(H) = E$ (Exercise 6.10.164). Our next result gives a more constructive condition.

7.10.14. Proposition. *A linear subspace L in a Hilbert space H is the range of a bounded operator on H precisely when there exists a sequence of pairwise orthogonal closed subspaces $H_n \subset H$ such that*

$$L = \Big\{ \sum_{n=1}^\infty x_n \colon x_n \in H_n, \ \sum_{n=1}^\infty 4^n \|x_n\|^2 < \infty \Big\}.$$

PROOF. If L has the indicated form, then $L = A(H)$, where $A = \sum_{n=1}^\infty 2^{-n} P_n$ and P_n is the projection onto H_n.

Let L be the range of a bounded operator. As noted above, we can assume that $L = A(H)$, where A is a nonnegative selfadjoint operator. The general case easily reduces to the case of separable H and an operator A with a cyclic vector. Hence we can assume that A is the operator of multiplication by the argument on the space $L^2(\mu)$ for some bounded Borel measure on $[0,1]$. Let us take for H_n the subspace in $L^2(\mu)$ consisting of the functions vanishing outside $(2^{-n}, 2^{1-n}]$. Then $Ax = \sum_{n=1}^\infty P_n Ax$, $P_n Ax \in H_n$, and $\|P_n Ax\| \leqslant 2^{1-n}\|P_n x\|$. Hence $4^n\|P_n Ax\|^2 \leqslant 4\|P_n x\|^2$, which gives a convergent series. On the other hand, if $y = \sum_{n=1}^\infty y_n$, where $y_n \in H_n$ and $\sum_{n=1}^\infty 4^n\|y_n\|^2 < \infty$, then $y_n = Ax_n$, where $x_n \in H_n$ and $\|x_n\| \leqslant 2^n\|y_n\|$, since $x_n(t) = t^{-1}y_n(t)$. Therefore, we have the bound $\sum_{n=1}^\infty \|x_n\|^2 \leqslant \sum_{n=1}^\infty 4^n\|y_n\|^2 < \infty$. Hence the series of the pairwise orthogonal vectors x_n converges to some vector x and $Ax = y$. $\qquad\square$

In case of a Hilbert space the factorization Theorem 6.10.5 can be sharpened in the following way (this result was obtained in [**686**]).

7.10.15. Theorem. *Let A and B be two bounded operators on a Hilbert space H. The following conditions are equivalent:*

(i) $A(H) \subset B(H)$,

(ii) $A = BC$ *for some* $C \in \mathcal{L}(H)$,

(iii) *there exists a number* $\lambda \geqslant 0$ *such that*

$$(A^*x, A^*x) \leqslant \lambda^2(B^*x, B^*x) \quad \text{for all } x \in H.$$

PROOF. The equivalence of (i) and (ii) is seen from Theorem 6.10.5, since in case of a Hilbert space it gives a continuous linear operator $C = B_0^{-1}A$ from H onto $(\operatorname{Ker} B)^\perp$, where B_0 is the restriction of B to $(\operatorname{Ker} B)^\perp$ and B_0^{-1} is the algebraic inverse. Next, (ii) yields (iii) at once. Let (iii) be fulfilled. On the linear space $B^*(H)$ we have the linear mapping $D\colon B^*x \mapsto A^*x$. This mapping is well-defined, because $A^*x = 0$ if $B^*x = 0$. In addition, $\|Dy\| \leqslant \lambda\|y\|$ if $y \in B^*(H)$, hence D can be extended to a bounded operator on the closure of $B^*(H)$. Next we extend D to a bounded operator on all of H, defining D by zero on $B^*(H)^\perp = \operatorname{Ker} B$. Then $DB^* = A^*$, whence $A = BD^*$. In part of this reasoning we could refer to results from §6.10(i). $\square$

Note that in (iii) it is important to have an estimate for the adjoint operator, but not for the original one (to see this, it suffices to take $A = I$ and an isometry B with $B(H) \neq H$). From the previous theorem and Corollary 7.5.4 we obtain the following result efficient in estimating the rate of decreasing of eigenvalues of compact selfadjoint operators.

7.10.16. Corollary. *Suppose that A and B are compact operators on a Hilbert space H such that $A(H) \subset B(H)$. Let $\alpha_1^+ \geqslant \alpha_2^+ \geqslant \cdots > 0$ and $\beta_1^+ \geqslant \beta_2^+ \geqslant \cdots > 0$ be positive eigenvalues of the operators $|A|$ and $|B|$, respectively, written in the decreasing order taking into account their multiplicities. Then there exists $C > 0$ such that $\alpha_n^+ \leqslant C\beta_n^+$.*

PROOF. The theorem gives the estimate $AA^* \leqslant \lambda^2 BB^*$. Now it is important that the operators AA^* and A^*A have the same nonzero eigenvalues (Exercise 7.10.65), i.e., $|A^*|$ and $|A|$ have common nonzero eigenvalues; the same is true for the pair of operators BB^* and B^*B. According to Corollary 7.5.4, the indicated estimate yields the inequalities $\lambda_n(AA^*) \leqslant \lambda^2\lambda_n(BB^*)$ for positive eigenvalues of the operators AA^* and BB^* written in the order of decreasing. $\square$

7.10.17. Example. Let $W_{2\pi}^{2,1}[0, 2\pi]$ be the class of all absolutely continuous complex functions f on $[0, 2\pi]$ with $f' \in L^2[0, 1]$ and $f(0) = f(2\pi)$. If the range of a bounded operator A on $L^2[0, 2\pi]$ belongs to $W_{2\pi}^{2,1}[0, 2\pi]$, then A is a compact operator and for the positive eigenvalues α_n^+ of the operator $|A|$ written in the order of decreasing taking into account their multiplicities, for some $C > 0$ one has the estimate $\alpha_n^+ \leqslant Cn^{-1}$.

For this we observe that $W_{2\pi}^{2,1}[0, 2\pi]$ is the range of the operator B with eigenfunctions $\exp(ikt)$ and eigenvalues $\beta_k = k^{-1}$, $k \neq 0$, $\beta_0 = 1$ (Exercise 9.10.32).

Let us now prove the following theorem due to von Neumann [**701**].

7.10.18. Theorem. *Suppose that a bounded operator A on a separable Hilbert space has a non-closed range. Then there exists a unitary operator U such that $A(H) \cap UA(H) = 0$.*

PROOF. We start with a simple explicit example of such an operator with a dense range in $L^2[-\pi, \pi]$. Let $e_n(t) = e^{int}$ and let A be given by $Ae_n = e^{-n^2} e_n$. It is readily seen that all functions in $A(H)$ are real-analytic on $[-\pi, \pi]$. Let U be the operator of multiplication by $\text{sign } t$. Then the range of A and UA intersect only by zero. This example can be easily modified to make the range of A not only dense, but also containing an infinite-dimensional closed subspace. For this it suffices take the direct sum of countably many copies of A.

The main step of the proof consists in verification of the following interesting fact: if the range of A is dense, non-closed and contains an infinite-dimensional closed subspace, then for every bounded operator B with a non-closed range there exists a unitary operator W such that $WB(H) \subset A(H)$. For this we apply Proposition 7.10.14 to $L = B(H)$ and take the corresponding pairwise orthogonal closed subspaces H_n. Since $B(H)$ is not closed, there are infinitely many nonzero subspaces H_n. Let $\mathbb{N} = \bigcup_{i=1}^{\infty} \Omega_i$ be a partition into countably many countable parts and let S_n be the set Ω_i which contains n. Set $H_n' := \bigoplus_{i \in S_n} H_i$. Applying the cited proposition to the pairwise orthogonal subspaces H_n', we obtain an operator B' with $B(H) \subset B'(H)$, but now all subspaces H_n' are infinite-dimensional.

We now apply the same proposition to $A(H)$ and take the corresponding nonzero pairwise orthogonal closed subspaces E_n. We observe that among E_n there is at least one infinite-dimensional. Indeed, if all E_n were finite-dimensional, we would obtain a compact operator $C := \sum_{n=1}^{\infty} 2^{-n} P_{E_n}$ with $C(H) = A(H)$, as shown in the cited proposition. However, the range of a compact operator cannot contain an infinite-dimensional closed subspace, because it is covered by countably many compact sets. We can assume that E_1 is infinite-dimensional. Then we can take pairwise orthogonal infinite-dimensional closed subspaces $L_i \subset E_1$ such that $E_1 = \bigoplus_{i=1}^{\infty} L_i$. Set $E_i' := E_i \bigoplus L_i$ if $i > 2$ and $E_1' := L_1$. We have obtained pairwise orthogonal infinite-dimensional closed subspaces. It is easy to see that for the corresponding operator A' in the cited proposition we have $A'(H) \subset A(H)$. A unitary operator W can be defined by means of unitary isomorphisms between H_i' and E_i'. Then $B(H) \subset B'(H) \subset A'(H) \subset A(H)$. For completing the proof it remains to transform $A(H)$ by the unitary operator W to $A_0(H)$, where A_0 is an operator with a dense range such that $A_0(H) \cap U_0 A_0(H) = 0$ for some unitary operator U_0. For U we take $W^{-1} U_0 W$. $\square$

This result of von Neumann was further developed by Dixmier [682], [683] (see also Exercise 7.10.105). The proof above follows Fillmore, Williams [687].

Let us observe that in real spaces the same fact is true with orthogonal operators in place of unitary operators. To see this, we can write a real space as the sum of two infinite-dimensional closed subspaces and introduce the corresponding complex structure.

7.10(iv). Hilbert–Schmidt operators and nuclear operators

In this subsection we discuss two classes of compact operators on Hilbert spaces, both important for applications and interesting. One of several equivalent definitions of these classes is connected with the behavior of eigenvalues.

7.10.19. Definition. *Let A be a compact operator on a Hilbert space H (real or complex) and let $\{s_n(A)\}$ be all eigenvalues of the operator $|A|$. We shall call A a Hilbert–Schmidt operator if*

$$\sum_{n=1}^{\infty} s_n(A)^2 < \infty.$$

The operator A will be called a nuclear or trace class operator if

$$\sum_{n=1}^{\infty} s_n(A) < \infty.$$

The class of all Hilbert–Schmidt operators on H is denoted by $\mathcal{H}(H)$ or $\mathcal{L}_{(2)}(H)$. The class of all nuclear operators on H is denoted by $\mathcal{N}(H)$ or $\mathcal{L}_{(1)}(H)$.

We recall that $s_n(A) \geqslant 0$. It is clear that $\mathcal{N}(H) \subset \mathcal{H}(H) \subset \mathcal{K}(H)$.

These two classes are the most important special cases of the Schatten classes $\mathcal{S}_p(H)$ defined by the condition $\{s_j(A)\} \subset l^p$.

There is an equivalent characterization of Hilbert–Schmidt operators that is often taken for the definition.

7.10.20. Theorem. (i) *An operator $A \in \mathcal{L}(H)$ on a separable Hilbert space H is a Hilbert–Schmidt operator precisely when for some orthonormal basis $\{e_n\}$ we have*

$$\sum_{n=1}^{\infty} \|Ae_n\|^2 < \infty. \tag{7.10.1}$$

In this case such a series converges for every orthonormal basis and its sum does not depend on the basis.

(ii) *A bounded operator A on a separable Hilbert space H is a Hilbert–Schmidt operator precisely when A^* is a Hilbert–Schmidt operator. In addition,*

$$\sum_{n=1}^{\infty} \|Ae_n\|^2 = \sum_{n=1}^{\infty} \|A^*e_n\|^2. \tag{7.10.2}$$

PROOF. Let us take the polar decomposition $A = U|A|$. If $A \in \mathcal{H}(H)$, then we have (7.10.1) for the eigenbasis of $|A|$. If (7.10.1) holds, then the operator A is compact. Indeed, the estimate $\left\| \sum_{n=1}^{\infty}(x, e_n)Ae_n \right\|^2 \leqslant \sum_{n=1}^{\infty} \|Ae_n\|^2 \|x\|^2$ yields that the finite-dimensional operators $x \mapsto \sum_{n=1}^{N}(x, e_n)Ae_n$ converge to A with respect to the operator norm. We verify that the sum (7.10.1) does not depend on the basis. To this end we take an arbitrary orthonormal basis $\{\varphi_n\}$ and write the following equality:

$$\sum_{n=1}^{\infty} \|Ae_n\|^2 = \sum_{n=1}^{\infty}\sum_{k=1}^{\infty} |(Ae_n, \varphi_k)|^2 = \sum_{n=1}^{\infty}\sum_{k=1}^{\infty} |(e_n, A^*\varphi_k)|^2 = \sum_{k=1}^{\infty} \|A^*\varphi_k\|^2.$$

If $\{\psi_n\}$ is yet another basis, then the right-hand side equals $\sum_{n=1}^{\infty} \|A\psi_n\|^2$, since $A^{**} = A$. Hence the sum is independent of the basis. Applying this to the eigenbasis $\{\psi_n\}$ of the operator $|A|$, we obtain that $A \in \mathcal{H}(H)$. Assertion (i) is proved. On the way we have also proved (ii). $\qquad\square$

7.10.21. Proposition. *The class $\mathcal{H}(H)$ of all Hilbert–Schmidt operators on a separable Hilbert space H equipped with the inner product*

$$(A, B)_{\mathcal{H}} := \sum_{n=1}^{\infty} (Ae_n, Be_n) = \sum_{n=1}^{\infty} (B^*Ae_n, e_n)$$

is a separable Hilbert space. The corresponding Hilbert–Schmidt norm has the form

$$\|A\|_{\mathcal{H}} = \left(\sum_{n=1}^{\infty} \|Ae_n\|^2 \right)^{1/2}.$$

If $A \in \mathcal{H}(H)$ and $B \in \mathcal{L}(H)$, then $AB \in \mathcal{H}(H)$, $BA \in \mathcal{H}(H)$ and

$$\|AB\|_{\mathcal{H}} \leqslant \|B\|_{\mathcal{L}(H)} \|A\|_{\mathcal{H}}, \quad \|BA\|_{\mathcal{H}} \leqslant \|B\|_{\mathcal{L}(H)} \|A\|_{\mathcal{H}}.$$

PROOF. It is clear from (7.10.1) that $\mathcal{H}(H)$ is a linear space. In addition, the series defining the inner product in $\mathcal{H}(H)$ converges absolutely, since $|(Ae_n, Be_n)| \leqslant \|Ae_n\|^2 + \|Be_n\|^2$. Let us verify the completeness of $\mathcal{H}(H)$. If a sequence of operators A_j is Cauchy in $\mathcal{H}(H)$, then it converges in norm to some operator $A \in \mathcal{L}(H)$, because $\|T\|_{\mathcal{L}(H)} \leqslant \|T\|_{\mathcal{H}}$. Let $\{e_n\}$ be an orthonormal basis. It is clear that $\sum_{n=1}^{N} \|Ae_n\|^2 \leqslant \sup_j \sum_{n=1}^{N} \|A_j e_n\|^2 \leqslant \sup_j \|A_j\|_{\mathcal{H}}^2$ for all N, i.e., $A \in \mathcal{H}(H)$. Let us verify that $\|A - A_j\|_{\mathcal{H}} \to 0$. Let $\varepsilon > 0$. Find a number j_0 with $\|A_j - A_i\|_{\mathcal{H}}^2 \leqslant \varepsilon$ for all $i, j \geqslant j_0$. Let $m \geqslant j_0$. There is N with $\sum_{n=N+1}^{\infty} \|(A - A_m)e_n\|^2 \leqslant \varepsilon$, since $A, A_m \in \mathcal{H}(H)$. Finally, $\sum_{n=1}^{N} \|(A - A_m)e_n\|^2 \leqslant \varepsilon$, which is obtained by letting $i \to \infty$ in the estimate $\sum_{n=1}^{N} \|(A_i - A_m)e_n\|^2 \leqslant \varepsilon$ that holds for all $i \geqslant j_0$. The separability of $\mathcal{H}(H)$ follows from the fact that finite-dimensional operators are dense in $\mathcal{H}(H)$, since the operator $|A|$ with an eigenbasis $\{e_n\}$ is the limit with respect to the norm in $\mathcal{H}(H)$ of the finite-dimensional operators $x \mapsto (x, e_1)Ae_1 + \cdots + (x, e_n)Ae_n$, and finite-dimensional operators can be approximated by rational linear combinations of operators $x \mapsto (x, v_i)v_j$, where $\{v_i\}$ is a countable everywhere dense set. If $A \in \mathcal{H}(H)$ and $B \in \mathcal{L}(H)$, then $BA \in \mathcal{H}(H)$ and $\|BA\|_{\mathcal{H}} \leqslant \|B\|_{\mathcal{L}(H)} \|A\|_{\mathcal{H}}$, since $\|BAe_n\| \leqslant \|B\|_{\mathcal{L}(H)} \|Ae_n\|$. In addition, $AB \in \mathcal{H}(H)$, since we have the equalities $(AB)^* = B^*A^*$, $\|B^*\|_{\mathcal{L}(H)} = \|B\|_{\mathcal{L}(H)}$, $\|B^*\|_{\mathcal{H}} = \|B\|_{\mathcal{H}}$. $\square$

It is useful to define Hilbert–Schmidt mappings between arbitrary Hilbert spaces.

7.10.22. Definition. *Let E_1 and E_2 be two Hilbert spaces. An operator $A \in \mathcal{L}(E_1, E_2)$ is called a Hilbert–Schmidt operator if the series $\sum_{\alpha} \|Ae_{\alpha}\|_{E_2}^2$ converges for some orthonormal basis $\{e_{\alpha}\}$ in E_1.*

The class of all Hilbert–Schmidt operators acting from E_1 to E_2 is denoted by $\mathcal{H}(E_1, E_2)$ or by $\mathcal{L}_{(2)}(E_1, E_2)$.

As above, it is easy to verify that the composition of two continuous operators between Hilbert spaces is a Hilbert–Schmidt operator if at least one of these operators has this property.

If the space E_1 is nonseparable, then the inclusion $A \in \mathcal{H}(E_1, E_2)$ means that A is zero on the orthogonal complement to some separable closed subspace

$E_0 \subset E_1$ and $\sum_{n=1}^{\infty} \|Ae_n\|_{E_2}^2 < \infty$ for some orthonormal basis $\{e_n\}$ in E_0. As above, one verifies that if the series of $\|Ae_\alpha\|_{E_2}^2$ converges for some orthonormal basis in E_1, then it converges for every orthonormal basis and the sum does not depend on the basis. The square root of the sum is called the Hilbert–Schmidt norm of the operator A and is denoted by $\|A\|_{\mathcal{H}}$. The Hilbert–Schmidt norm is generated by the inner product

$$(A, B) = \sum_\alpha (Ae_\alpha, Be_\alpha)_{E_2}.$$

This quantity does not depend on the orthonormal basis $\{e_\alpha\}$ in E_1.

The next result shows that abstract Hilbert–Schmidt operators are precisely integral operators on spaces L^2 with quadratically integrable kernels.

7.10.23. Proposition. (i) *Let μ be a nonnegative measure with values in $[0, +\infty]$ on a measurable space $(\Omega, \mathcal{A})$ and let $\mathcal{K}$ be a measurable function on $\Omega \times \Omega$ belonging to $L^2(\mu \otimes \mu)$. Then the operator defined on $L^2(\mu)$ by the formula*

$$Tx(t) = \int_\Omega \mathcal{K}(t, s)x(s)\, \mu(ds),$$

is a Hilbert–Schmidt operator. In addition,

$$\|T\|_{\mathcal{H}}^2 = \int_\Omega \int_\Omega |\mathcal{K}(t, s)|^2\, \mu(dt)\, \mu(ds) = \|\mathcal{K}\|_{L^2(\mu \otimes \mu)}^2.$$

(ii) *Every Hilbert–Schmidt operator on $L^2(\mu)$ has such a form. Hence every Hilbert–Schmidt operator is unitarily equivalent to an integral operator of the indicated form.*

PROOF. (i) The boundedness of the operator T with values in $L^2(\mu)$ is easily verified similarly to Example 6.9.4(iii). Let $\{e_n\}$ be an orthonormal sequence in $L^2(\mu)$. By Bessel's inequality applied to the functions $s \mapsto \mathcal{K}(t, s)$ we obtain

$$\sum_{n=1}^{\infty} \|Te_n\|^2 = \sum_{n=1}^{\infty} \int_\Omega \left| \int_\Omega \mathcal{K}(t, s)e_n(s)\, \mu(ds) \right|^2 \mu(dt)$$

$$\leqslant \int_\Omega \int_\Omega |\mathcal{K}(t, s)|^2\, \mu(ds)\, \mu(dt).$$

Hence T is a Hilbert–Schmidt operator (in the separable case this is obvious, for the general case see Exercise 7.10.102). Let us take an orthonormal sequence $\{e_n\}$ such that the function $\mathcal{K}$ belongs to the closed linear span of the functions $\varphi_{n,m}(t, s) := e_n(t)e_m(s)$. Then for μ-a.e. t the function $s \mapsto \mathcal{K}(t, s)$ has an expansion in a series with respect to the elements e_n (Exercise 5.6.59). For such s in the relation above, in place of the inequality we have the Parseval equality, which gives $\sum_{n=1}^{\infty} \|Te_n\|^2 = \|\mathcal{K}\|_{L^2(\mu \otimes \mu)}^2$.

(ii) Let T be a Hilbert–Schmidt operator on $L^2(\mu)$ and let $\{e_\alpha\}$ be an orthonormal basis in $L^2(\mu)$. Then there is an at most countable part $\{e_n\}$ with $\|Te_n\| > 0$. We define an integral kernel $\mathcal{K}$ by the formula

$$\mathcal{K}(t, s) = \sum_{n,m \geqslant 1} (Te_n, e_m)e_n(t)e_m(s).$$

This double series converges in the space $L^2(\mu \otimes \mu)$, since the sequence of functions $e_n(t)e_m(s)$ is orthonormal and $\sum_{n,m \geqslant 1} |(Te_n, e_m)|^2 = \sum_{n=1}^{\infty} \|Te_n\|^2 < \infty$. The kernel $\mathcal{K}$ defines the operator $T_{\mathcal{K}}$ that coincides with T, because for every vector e_n in the chosen sequence we have $Te_n = T_{\mathcal{K}}e_n$ and for all other e_α we have $T_{\mathcal{K}}e_\alpha = 0 = Te_\alpha$. Finally, an abstract Hilbert–Schmidt operator is unitarily equivalent to some Hilbert–Schmidt operator on a suitable space $L^2(\mu)$ unitarily isomorphic to the original space (see Exercise 5.6.42).   □

From Theorem 7.10.15 we obtain at once the following result.

7.10.24. Proposition. *Let H_1 and H_2 be Hilbert spaces, $A, B \in \mathcal{L}(H_1, H_2)$, and $A(H_1) \subset B(H_1)$. If B is a Hilbert–Schmidt operator, then so is the operator A. If $H_1 = H_2$ and B is a nuclear operator, then the operator A is also nuclear.*

Yet another important characterization of Hilbert–Schmidt operators describes them as absolutely summing and 2-summing operators.

A series $\sum_{n=1}^{\infty} x_n$ in a Banach space is called *unconditionally convergent* if it converges for all permutations of indices. If $\sum_{n=1}^{\infty} \|x_n\| < \infty$, then this series is called *absolutely convergent*.

7.10.25. Definition. *Let X and Y be two Banach spaces. An operator $T \in \mathcal{L}(X, Y)$ is called absolutely p-summing if for every weakly p-summable sequence $\{x_n\} \subset X$, i.e., satisfying the condition $\sum_{n=1}^{\infty} |l(x_n)|^p < \infty$ for all $l \in X^*$, we have $\sum_{n=1}^{\infty} \|Tx_n\|_Y^p < \infty$.*
The operator T is called absolutely summing if it takes unconditionally convergent series to absolutely convergent ones.

For any absolutely p-summing operator there is $C > 0$ such that

$$\sum_{n=1}^{\infty} \|Tx_n\|_Y^p \leqslant C \sup_{\|l\| \leqslant 1} \sum_{n=1}^{\infty} |l(x_n)|^p \qquad (7.10.3)$$

for every sequence $\{x_n\} \subset X$. Indeed, otherwise for each m there is a finite set $x_{m,1}, \ldots, x_{m,k}$ with $\sup_{\|l\| \leqslant 1} \sum_{i=1}^{k} |l(x_{m,i})|^p \leqslant 2^{-m}$ and $\sum_{i=1}^{k} \|Tx_{m,i}\|_Y^p \geqslant 1$, which leads to a contradiction. The smallest possible C is denoted by $\pi_p(T)$.

These classes are stable under compositions from the right and left with bounded operators. According to the Dvoretzky–Rogers theorem, in every infinite-dimensional Banach space X there is a conditionally convergent series that is not absolutely convergent. If X has no subspaces isomorphic to c_0 (and only in this case), then the unconditional convergence series of x_n is equivalent to the weak 1-summability of $\{x_n\}$ (see [303, Chapters 3, 4]).

7.10.26. Proposition. *Let E_1 and E_2 be two Hilbert spaces. An operator $T \in \mathcal{L}(E_1, E_2)$ is a Hilbert–Schmidt operator precisely when is it absolutely 2-summing or absolutely summing.*

PROOF. Let T be an absolutely 2-summing operator and $\{e_\alpha\}$ an orthonormal basis in E_1. For every $y \in E_1$ we have $\sum_\alpha |(y, e_\alpha)|^2 = \|y\|^2 < \infty$, whence

$\sum_\alpha \|Te_\alpha\|^2 < \infty$. Suppose now that $T \in \mathcal{H}(E_1, E_2)$ and $\{h_n\}$ is a sequence in E_1 with $\sum_{n=1}^\infty |(h_n, y)|^2 < \infty$ for all $y \in E_1$. We can assume that E_1 is the closure of the linear span of the vectors h_n. If E_1 is finite-dimensional, then the assertion is trivial. Hence we assume that E_1 is infinite-dimensional and take a basis $\{e_n\}$. Let us define an operator S on the space E_1 by $Se_n = h_n$, i.e., $Sx = \sum_{n=1}^\infty (x, e_n) h_n$. This series converges weakly in E_1, since for every $y \in E_1$ the series $\sum_{n=1}^\infty (x, e_n)(h_n, y)$ converges absolutely, which follows from convergence of the series of $|(x, e_n)|^2$ and $|(h_n, y)|^2$. Hence the operator S is continuous. This shows that TS is a Hilbert–Schmidt operator. Therefore, the series of $\|Th_n\|^2 = \|TSe_n\|^2$ converges. If T is absolutely summing and $\{e_n\}$ is an orthonormal sequence, then for any $(x_n) \in l^2$ the series of $x_n e_n$ converges unconditionally. This yields convergence of the series of $|x_n| \|Te_n\|$, whence $\{\|Te_n\|\} \in l^2$. Hence $T \in \mathcal{H}(E_1, E_2)$. The converse follows from Exercise 7.10.130. We observe that this proposition is true for all $p > 0$ in place of 2 or 1. $\square$

7.10.27. Theorem. *Suppose that μ is a Borel probability measure on a topological space Ω. Then the identity embedding $j\colon C_b(\Omega) \to L^2(\mu)$ is absolutely 2-summing.*

The same is true if for Ω we take an arbitrary probability space and replace $C_b(\Omega)$ by $L^\infty(\mu)$.

PROOF. Let $\{x_n\} \subset C_b(\Omega)$ and $\sum_{n=1}^\infty |l(x_n)|^2 < \infty$ for all $l \in C_b(\Omega)^*$. Then the sequence of operators $T_n\colon C_b(\Omega)^* \to l^2$, $l \mapsto (l(x_1), \ldots, l(x_n), 0, 0, \ldots)$, is pointwise bounded. By the Banach–Steinhaus theorem

$$\sup_{\|l\| \leqslant 1} \sum_{n=1}^\infty |l(x_n)|^2 \leqslant C < \infty.$$

Taking for l the functionals $x \mapsto x(t)$, we obtain $\sum_{n=1}^\infty |x_n(t)|^2 \leqslant C$. Thus, $\sum_{n=1}^\infty \|j(x_n)\|_2^2 \leqslant C$. The case of $L^\infty(\mu)$ for a general measure space reduces to the case of $C(K)$ with a compact space K. For this we need the following fact from §11.7(i): there exists a compact space K such that the algebra $L^\infty(\mu)$ is linearly isometric to the algebra $C(K)$ and on K there is a probability Borel measure ν such that the indicated isomorphism between $L^\infty(\mu)$ and $C(K)$ defines an isomorphism between $L^2(\mu)$ and $L^2(\nu)$. $\square$

On account of Proposition 7.10.26 we obtain the following interesting fact.

7.10.28. Corollary. *Let H be a Hilbert space, $(\Omega, \mathcal{A}, \mu)$ a probability space, and $A\colon H \to L^2(\mu)$ a continuous operator such that $A(H) \subset L^\infty(\mu)$. Then A is a Hilbert–Schmidt operator.*

This corollary admits an important reinforcement that is a particular case of a more general result due to V. B. Korotkov.

7.10.29. Theorem. *Let μ be a probability measure on a measurable space $(\Omega, \mathcal{A})$. A bounded operator T on $L^2(\mu)$ is a Hilbert–Schmidt operator precisely when there exists a nonnegative function $\Phi \in \mathcal{L}^2(\mu)$ such that, for every function*

$x \in L^2(\mu)$, *the element* Tx *has a modification which satisfies the estimate*

$$|Tx(\omega)| \leqslant C_x \Phi(\omega) \qquad (7.10.4)$$

with some number $C_x \geqslant 0$.

PROOF. The necessity of this condition is clear from the facts established above and the estimate

$$|Tx(t)| \leqslant \|x\| \left(\int_\Omega |\mathcal{K}(t,s)|^2 \, \mu(ds) \right)^{1/2}$$

for the operator T defined by the kernel $\mathcal{K} \in L^2(\mu \otimes \mu)$, since on the right we have a function from $\mathcal{L}^2(\mu)$. Suppose that (7.10.4) is fulfilled. Replacing Φ by $\Phi + 1$, we can assume that $\Phi \geqslant 1$. We observe that for C_x we can take $C\|x\|$ with some $C > 0$. Indeed, the bounded operator $S = \Phi^{-1}T$ takes values in $L^\infty(\mu)$ and hence is bounded as an operator from $L^2(\mu)$ to $L^\infty(\mu)$ (see Corollary 6.2.8), which gives the desired number C. Set

$$T_\varepsilon x(\omega) := |\varepsilon\Phi(\omega) + 1|^{-1} Tx(\omega), \quad \varepsilon > 0.$$

Then $|T_\varepsilon x(\omega)| \leqslant C\varepsilon^{-1}\|x\|$, i.e., the range of the operator T_ε is contained in $L^\infty(\mu)$. According to the previous corollary the operator T_ε is a Hilbert–Schmidt operator. For every $x \in L^2(\mu)$ we have $\lim_{\varepsilon \to 0} T_\varepsilon x = Tx$. Hence it suffices to establish the uniform boundedness of the Hilbert–Schmidt norms of the operators T_ε. Since by the Lebesgue dominated convergence theorem

$$\left\| \Phi(\varepsilon\Phi + 1)^{-1} \right\|_{L^2(\mu)} \to \|\Phi\|_{L^2(\mu)} \quad \text{as } \varepsilon \to 0,$$

it suffices to show that for a bounded function Φ one has $\|T\|_{\mathcal{H}} \leqslant C\|\Phi\|_{L^2(\mu)}$. We already know that in case of a bounded function Φ the operator T is a Hilbert–Schmidt operator, hence is defined by some kernel $\mathcal{K} \in L^2(\mu \otimes \mu)$. Hence for every function $x \in \mathcal{L}^2(\mu)$ with $\|x\|_{L^2(\mu)} \leqslant 1$ we have

$$|Tx(t)| = \left| \int_\Omega \mathcal{K}(t,s)x(s) \, \mu(ds) \right| \leqslant C\Phi(t) \qquad (7.10.5)$$

for μ-a.e. t. However, the corresponding measure zero set can depend on x. By the compactness of T there is a separable closed subspace $H \subset L^2(\mu)$ with $T(H) \subset H$ and $T(H^\perp) = 0$. Let us take a countable set $\{x_n\}$ dense in the unit ball of H. This enables us to pass to the case of a separable space $L^2(\mu)$ (for which it suffices to consider the measure μ on the σ-algebra generated by all functions x_n and Φ). Then (7.10.5) is fulfilled almost everywhere simultaneously for all x_n. By our choice of $\{x_n\}$ the quantity $\sup_n \left| \int_\Omega \mathcal{K}(t,s)x_n(s) \, \mu(ds) \right|$ is the L^2-norm of the function $s \mapsto \mathcal{K}(t,s)$ for points t such that this function belongs to $L^2(\mu)$. Therefore, for μ-a.e. t we have

$$\int_\Omega |\mathcal{K}(t,s)|^2 \, \mu(ds) \leqslant C^2 \Phi(t)^2.$$

Hence $\|T\|_{\mathcal{H}} \leqslant \|\mathcal{K}\|_{L^2(\mu \otimes \mu)} \leqslant C^2 \|\Phi\|_{L^2(\mu)}$, as required. $\qquad \square$

The following theorem due to Pietch shows that in case of a general Banach space absolutely 2-summing operators are also connected with the space L^2.

7.10.30. Theorem. *Let X and Y be Banach spaces, B^* the closed unit ball in the space X^* equipped with the $*$-weak topology, $\sigma(C(B^*))$ the σ-algebra generated by continuous functions on the compact space B^*. An operator $T \in \mathcal{L}(X, Y)$ is absolutely 2-summing precisely when there exists a bounded nonnegative measure ν on $\sigma(C(B^*))$ such that*

$$\|Tx\|_Y^2 \leqslant \int_{B^*} |\xi(x)|^2 \, \nu(d\xi).$$

PROOF. If such a measure exists, then for all $x_1, \ldots, x_n \in X$ we have

$$\sum_{i=1}^n \|Tx_i\|_Y^2 \leqslant \int_{B^*} \sum_{i=1}^n |\xi(x_i)|^2 \, \nu(d\xi) \leqslant \nu(B^*) \sup_{\|\xi\| \leqslant 1} \sum_{i=1}^n |\xi(x_i)|^2,$$

whence $\pi_2(T) \leqslant \nu(B^*)$. Conversely, suppose that $\pi_2(T) = 1$. Let us consider the following subsets in the Banach space $C(B^*)$:

$$F_1 := \{f \in C(B^*) \colon \sup_{\|\xi\| \leqslant 1} f(\xi) < 1\}$$

and F_2 equal to the convex envelope of the set of functions $f \in C(B^*)$ of the form $f(\xi) = |\xi(x)|^2$, where $\|Tx\|_Y = 1$. These sets are convex, F_1 is open and $F_1 \cap F_2 = \varnothing$, since $\pi_2(T) = 1$. By the Hahn–Banach theorem and the Riesz theorem on the representation of functionals on $C(B^*)$ by measures, there exists a measure ν on the aforementioned σ-algebra in B^* such that

$$\sup_{f \in F_1} \int_{B^*} f(\xi) \, \nu(d\xi) \leqslant \inf_{f \in F_2} \int_{B^*} f(\xi) \, \nu(d\xi).$$

Since F_1 contains all negative functions, the measure ν is nonnegative. We can assume that $\nu(B^*) = 1$. Then the left-hand side of the previous estimate equals 1. Whenever $\|Tx\|_Y = 1$, for the function $f_x(\xi) := |\xi(x)|^2$ we obtain $f_x \in F_2$, hence the integral of f_x with respect to the measure ν is not less than 1, which yields the desired inequality. $\qquad\square$

For example, from this theorem one can derive Theorem 7.10.27, taking for ν the image of the measure μ under the embedding $\Omega \subset C_b(\Omega)^*$, $\omega \mapsto \delta_\omega$. Then the integral of $|x(\omega)|^2$ with respect to the measure μ will equal the integral of $|\delta_\omega(x)|^2$ with respect to the measure μ, i.e., the integral of $|\xi(x)|^2$ with respect to the measure ν.

Diagonal operators are defined in Example 6.1.5(vii). Let us mention an interesting theorem due to von Neumann (see [**252**, Theorem 14.13]).

7.10.31. Theorem. *For every selfadjoint operator A on a separable Hilbert space H and every number $\varepsilon > 0$, there exists a diagonal selfadjoint operator D and a selfadjoint Hilbert–Schmidt operator S_ε on the space H such that $\|S_\varepsilon\|_\mathcal{H} \leqslant \varepsilon$ and $A = D + S_\varepsilon$.*

We now return to nuclear operators and their useful connections with Hilbert–Schmidt operators.

7.10.32. Remark. Let $A \in \mathcal{L}(H)$ be such that for some C one has

$$\sum_{n=1}^{\infty} |(A\psi_n, \varphi_n)| \leqslant C$$

for all pairs of orthonormal bases $\{\psi_n\}$ and $\{\varphi_n\}$. Then $A \in \mathcal{N}(H)$. Indeed, Proposition 7.8.12 yields that the operator A is compact. Let us take its polar decomposition $A = U|A|$ and the eigenbasis $\{\psi_n\}$ for the operator $|A|$. For $\{\varphi_n\}$ we take the orthonormal sequence of all nonzero vectors $U\psi_n$ complemented to an orthonormal basis. This gives convergence of the series of $s_n(A)$.

7.10.33. Theorem. *Let H be a separable Hilbert space and $A \in \mathcal{L}(H)$. The following conditions are equivalent:*
(i) $A \in \mathcal{N}(H)$; (ii) $A^* \in \mathcal{N}(H)$; (iii) $A = A_1 A_2$, where $A_1, A_2 \in \mathcal{H}(H)$;
(iv) there exist two sequences of vectors $\{v_k\}$ and $\{u_k\}$ with $\|u_k\| = \|v_k\| = 1$ and a scalar sequence $\{\lambda_k\} \in l^1$ such that

$$Ax = \sum_{k=1}^{\infty} \lambda_k (x, u_k) v_k; \qquad (7.10.6)$$

(v) there is an orthonormal basis $\{e_n\}$ with $\sum_{n=1}^{\infty} \|Ae_n\| < \infty$.

PROOF. The equivalence of (i) and (ii) follows from the fact that according to Exercise 7.10.65 the operators A^*A and AA^* have the same nonzero eigenvalues (for zero this can be false). Hence for all nonzero eigenvalues $s_j(A) = s_j(A^*)$. If A is nuclear, then $|A|^{1/2}$ is a Hilbert–Schmidt operator, hence the polar decomposition $A = U|A|$ gives the representation $A = U|A|^{1/2}|A|^{1/2}$, where $U|A|^{1/2} \in \mathcal{H}(H)$, i.e., we obtain (iii).

Let us derive (iv) from (iii). Suppose we have a representation $A = A_1 A_2$, where $A_1, A_2 \in \mathcal{H}(H)$. Let us take the polar decomposition $A_2 = V|A_2|$. The operator $|A_2|$ has an eigenbasis $\{e_n\}$ and eigenvalues $\{\alpha_n\}$. Hence we have $Ax = \sum_{n=1}^{\infty} \alpha_n (x, e_n) A_1 V e_n$. If $\beta_n := \|A_1 V e_n\| > 0$, we set $v_n := \beta_n^{-1} A_1 V e_n$. Then for the numbers $\lambda_n := \alpha_n \beta_n$ we obtain $\{\lambda_n\} \in l^1$, which gives (7.10.6) with $u_n = e_n$. Suppose now that (iv) holds. Then A is obviously compact. If $\{\psi_n\}$ and $\{\varphi_n\}$ are two orthonormal bases, then the series of $|(A\psi_n, \varphi_n)|$ is estimated by the double series of $|\lambda_k(\psi_n, u_k)(v_k, \varphi_n)|$, which is dominated by the series of $|\lambda_k|$, since $|(\psi_n, u_k)(v_k, \varphi_n)| \leqslant [|(\psi_n, u_k)|^2 + |(v_k, \varphi_n)|^2]/2$ and the sum over n in the right-hand side equals 1. According to Remark 7.10.32 the operator A is nuclear. Moreover, (v) is fulfilled for the eigenbasis of $|A|$. Finally, (v) implies (iv), because we take $\lambda_k = \|Ae_k\|$, $u_k = e_k$ and $v_k = \|Ae_k\|^{-1} Ae_k$ if $Ae_k \neq 0$. $\qquad \square$

Note that the series in (v) converges for some, but in general not for every orthonormal basis (Exercise 7.10.115). For a basis for which convergence holds we can take the eigenbasis of $|A|$. Then for the expansion (7.10.6) we can take $Ax = \sum_{n=1}^{\infty} s_n(A)(x, e_n) U e_n$, where only e_n with $U e_n \neq 0$ are taken into account.

The next result enables us to introduce the trace of any nuclear operator.

7.10.34. Lemma. *Let $A \in \mathcal{N}(H)$ and let $\{\varphi_n\}$ be an orthonormal basis in H. If A has the form (7.10.6), then*

$$\sum_{n=1}^{\infty} \lambda_n(v_n, u_n) = \sum_{n=1}^{\infty} (A\varphi_n, \varphi_n), \qquad (7.10.7)$$

the series in the right-hand side converges absolutely and its sum does not depend on our choice of the basis. In particular,

$$\sum_{n=1}^{\infty} (A\varphi_n, \varphi_n) = \sum_{n=1}^{\infty} s_n(A)(Ue_n, e_n),$$

where $\{e_n\}$ is the eigenbasis of the operator $|A|$.

PROOF. The series in the left-hand side of (7.10.7) converges absolutely and

$$\sum_{n=1}^{\infty} \lambda_n(v_n, u_n) = \sum_{n=1}^{\infty} \lambda_n \sum_{k=1}^{\infty} (v_n, \varphi_k)(\varphi_k, u_n)$$

$$= \sum_{k=1}^{\infty} \sum_{n=1}^{\infty} \lambda_n(v_n, \varphi_k)(\varphi_k, u_n) = \sum_{k=1}^{\infty} (A\varphi_k, \varphi_k),$$

which shows convergence of the series in the right-hand side. Its absolute convergence follows from the fact that the sum does not change under permutations of terms in this series (note that any permutation of $\{\varphi_k\}$ remains an orthonormal basis). □

The *trace of an operator $A \in \mathcal{N}(H)$* is defined by

$$\operatorname{tr} A := \sum_{k=1}^{\infty} (A\varphi_k, \varphi_k),$$

where $\{\varphi_k\}$ is an orthonormal basis. The lemma shows that the trace is well-defined. The following fact is true.

7.10.35. Proposition. *Let H be a complex Hilbert space. An operator $A \in \mathcal{L}(H)$ is nuclear precisely when for every orthonormal basis $\{e_n\}$ in H the series $\sum_{n=1}^{\infty}(Ae_n, e_n)$ converges.*

PROOF. Let us show that $A \in \mathcal{N}(H)$ if all such series converge. If $A \geqslant 0$, then it suffices to have convergence for some basis, since in that case $A^{1/2}$ is a Hilbert–Schmidt operator. If A is selfadjoint, then there exist closed subspaces H_1 and H_2 such that $H_1 \perp H_2$, $H = H_1 \oplus H_2$, $A(H_i) \subset H_i$. This follows from the theorem about representation of A as the multiplication by a real function φ on $L^2(\mu)$: for H_1 and H_2 we can take the subspaces of functions vanishing outside $\{\varphi > 0\}$ and $\{\varphi \leqslant 0\}$. The operators $A|_{H_1}$ and $-A|_{H_2}$ are nonnegative. Let us take orthonormal bases $\{\varphi_n\}$ and $\{\psi_n\}$ in H_1 and H_2. Let $e_n = \varphi_n$ for odd n and $e_n = \psi_n$ for even n. Then both series $\sum_{n=1}^{\infty}(A\varphi_n, \varphi_n)$ and $\sum_{n=1}^{\infty}(A\psi_n, \psi_n)$ must converge separately, since the series of (Ae_n, e_n) converges for every permutation of its terms. Therefore, the restrictions of A to H_1 and H_2 are nuclear operators, whence $A \in \mathcal{N}(H)$. Finally, in the general case we observe that the operator A^* satisfies the same condition as A. Then this condition is fulfilled for the selfadjoint operators $B_1 = A + A^*$ and $B_2 = i(A - A^*)$. As we have shown, $B_1, B_2 \in \mathcal{N}(H)$. Hence $A = (B_1 - iB_2)/2 \in \mathcal{N}(H)$.   □

Note that here it is not enough to have convergence of only one such series (even absolute): see Exercise 7.10.116. For real spaces this proposition is false (it suffices to take a noncompact operator A with $(Ax, x) = 0$).

7.10.36. Theorem. *Let H be a complex or real Hilbert space.*
(i) *Let $A \in \mathcal{N}(H)$. Then*

$$\operatorname{tr}|A| = \sup \sum_{n=1}^{\infty} |(A\psi_n, \varphi_n)|, \qquad (7.10.8)$$

where sup *is taken over all pairs of orthonormal bases $\{\psi_n\}$ and $\{\varphi_n\}$.*
(ii) *Let $A \in \mathcal{N}(H)$. Then*

$$\operatorname{tr}|A| = \inf \sum_{n=1}^{\infty} |\lambda_n|, \qquad (7.10.9)$$

where inf *is taken over all representations of the form (7.10.6).*
(iii) *The space of nuclear operators on a separable Hilbert space H is a separable Banach space with respect to the norm $\|A\|_{(1)} := \operatorname{tr}|A|$.*
(iv) *An operator $A \in \mathcal{L}(H)$ is nuclear precisely when so is A^*. In addition, $\|A\|_{(1)} = \|A^*\|_{(1)}$ and $\operatorname{tr} A = \overline{\operatorname{tr} A^*}$.*
(v) *If $A \in \mathcal{N}(H)$ and $T \in \mathcal{L}(H)$, then $AT, TA \in \mathcal{N}(H)$ and*

$$\operatorname{tr} AT = \operatorname{tr} TA \quad and \quad \|AT\|_{(1)} \leqslant \|T\|_{\mathcal{L}(H)} \|A\|_{(1)}.$$

PROOF. We shall assume that H is complex (the real case is similar). (i) Let us take the polar decomposition $A = U|A|$ and the eigenbasis $\{e_n\}$ of the operator $|A|$. Then $A\psi_n = \sum_{j=1}^{\infty} s_j(A)(\psi_n, e_j)Ue_j$, whence

$$\sum_{n=1}^{\infty} (A\psi_n, \varphi_n) = \sum_{n=1}^{\infty} \sum_{j=1}^{\infty} s_j(A)(\psi_n, e_j)(Ue_j, \varphi_n),$$

which is estimated by $\operatorname{tr}|A|$ in absolute value, as in the proof of Theorem 7.10.33. Since the same is true when we replace e_n by $e^{i\theta_n}e_n$ with arbitrary real θ_n, the right-hand side of (7.10.8) is not greater than the left-hand side. On the other hand, all nonzero vectors Ue_n form an orthonormal system. Complementing it to a basis $\{\varphi_n\}$, we obtain a pair of bases for which the equality is achieved.
(ii) We have seen in the proof of Theorem 7.10.33 that the sum of the series of $s_n(A)$ does not exceed the sum of the series of $|\lambda_n|$ for every representation (7.10.6). The equality is achieved for $Ax = \sum_{n=1}^{\infty} s_n(A)(x, e_n)Ue_n$, where $A = U|A|$ is the polar decomposition and $\{e_n\}$ is the eigenbasis for $|A|$.
(iii) It is clear from assertion (iv) of Theorem 7.10.33 that $\mathcal{N}(H)$ is a linear space. Estimate (7.10.8) along with the obvious inequality $\|A\| \leqslant \|A\|_{(1)}$ shows that $\| \cdot \|_{(1)}$ is a norm. If a sequence $\{A_n\}$ is Cauchy with respect to this norm, then it converges to some operator $A \in \mathcal{K}(H)$ in the operator norm. Let $A = U|A|$ be the polar decomposition of A and $\{e_n\}$ the eigenbasis of $|A|$. Set $\psi_n = e_n$

and $\varphi_n = Ue_n$ if $Ue_n \neq 0$. Let us complement $\{\varphi_n\}$ to an orthonormal basis. We obtain convergence of the series of $s_n(A)$, since the series of $|(A\psi_n, \varphi_n)|$ converges by (7.10.8) and the uniform boundedness of the norms $\|A_n\|_{(1)}$. We show that $\|A - A_n\|_{(1)} \to 0$. Let $\varepsilon > 0$. Find a number n_0 with $\|A_m - A_n\|_{(1)} \leqslant \varepsilon$ for all $n, m \geqslant n_0$. Suppose that we have two orthonormal bases $\{\psi_k\}$ and $\{\varphi_k\}$ in the space H. Let $m \geqslant n_0$. Find N with $\sum_{k=N+1}^{\infty} |(A\psi_k - A_m\psi_k, \varphi_k)| \leqslant \varepsilon$. The bound $\sum_{k=1}^{N} |(A_j\psi_k - A_m\psi_k, \varphi_k)| \leqslant \varepsilon$ for $j \geqslant n_0$ yields in the limit the same bound for A in place of A_j. Thus, by (7.10.8) we obtain that $\|A - A_m\|_{(1)} \leqslant 2\varepsilon$. It remains to observe that every operator $A \in \mathcal{N}(H)$ is approximated with respect to the norm $\| \cdot \|_{(1)}$ by finite-dimensional operators, since this is true for $|A|$ (it suffices to use the eigenbasis of $|A|$).

The first assertion in (iv) follows from the fact that $|A|$ and $|A^*|$ have the same nonzero eigenvalues, which has already been noted above. The same reasoning gives the equality $\|A\|_{(1)} = \|A^*\|_{(1)}$. The last equality in (iv) is obvious.

In assertion (v) the nuclearity of the operators AT and TA is obvious from assertion (iv) of Theorem 7.10.33 (one can also use the connection with Hilbert–Schmidt operators). The estimate $\|AT\|_{(1)} \leqslant \|T\|_{\mathcal{L}(H)}\|A\|_{(1)}$ follows from assertion (ii) and representation (7.10.6). Let us verify the equality of the traces of the operators AT and TA. Writing the operator A in the form (7.10.6), we arrive at the following two equalities:

$$TAx = \sum_{n=1}^{\infty} \lambda_n(x, u_n)Tv_n, \quad ATx = \sum_{n=1}^{\infty} \lambda_n(Tx, u_n)v_n = \sum_{n=1}^{\infty} \lambda_n(x, T^*u_n)v_n.$$

Equality (7.10.7) shows that the trace of TA equals $\sum_{n=1}^{\infty} \lambda_n(Tv_n, u_n)$ and the trace of AT equals $\sum_{n=1}^{\infty} \lambda_n(v_n, T^*u_n)$, i.e., the same number. $\qquad \square$

Let us give a sufficient condition for an operator on $L^2[0, 1]$ to be nuclear.

7.10.37. Example. Let T be a bounded operator on $L^2[0, 1]$ such that its range is contained in the set of Lipschitz functions (or, more generally, there exists a function $\Phi \in L^2[0, 1]$ such that the range of T is contained in the class of absolutely continuous functions x for which $|x'(t)| \leqslant C_x|\Phi(t)|$ a.e.). Then T is a nuclear operator.

PROOF. First, we suppose additionally that all functions from the range of T equal zero at 0. The operator $Sx = (Tx)'$ takes values in $L^\infty[0, 1]$. According to Corollary 7.10.28 the operator S is a Hilbert–Schmidt operator. In the case of the second more general condition we apply Theorem 7.10.29. The Volterra operator V of indefinite integration is also a Hilbert–Schmidt operator. Hence $T = VS$ is nuclear. In the general case we observe that the operator T is continuous as an operator with values in $C[0, 1]$. Hence the functional $l: x \mapsto Tx(0)$ is continuous. This enables us to write T in the form $Tx = l(x)1 + T_0x$, where T_0 has the form considered above. Then T is nuclear, since so are T_0 and the one-dimensional operator $x \mapsto l(x)1$. $\qquad \square$

An important particular case is the following integral operator with an integral kernel satisfying the Lipschitz condition in the first variable.

7.10.38. Example. Suppose that a measurable function $\mathcal{K}$ belongs to $\mathcal{L}^2$ on the square $[0,1] \times [0,1]$ and that for almost every s the function $t \mapsto \mathcal{K}(t,s)$ satisfies the Lipschitz condition with a constant $\Phi(s)$ such that the function Φ belongs to $\mathcal{L}^2[0,1]$. Then the operator

$$Tx(t) = \int_0^1 \mathcal{K}(t,s)x(s)\,ds$$

is nuclear. The same is true under the following weaker condition: for almost every point s the function $t \mapsto \mathcal{K}(t,s)$ is absolutely continuous on the interval $[0,1]$ and $\partial \mathcal{K}(t,s)/\partial t \in \mathcal{L}^2([0,1] \times [0,1])$.

Note that the continuity of the kernel $\mathcal{K}$ is not sufficient for the nuclearity of the integral operator with the kernel $\mathcal{K}$. T. Carleman gave the following example: $\mathcal{K}(t,s) = F(t-s)$, where F is a continuous function with a period 1 and the Fourier expansion $F(t) = \sum_{k \in \mathbb{Z}} c_k e^{2\pi i k t}$ such that $\sum_k |c_k| = \infty$. Then the functions $e_k(t) = e^{2\pi i k t}$ constitute an orthonormal eigenbasis for the integral operator with the kernel $\mathcal{K}$ and $\{c_k\}$ is the sequence of the corresponding eigenvalues. However, if it is known in addition that the integral operator with the kernel $\mathcal{K}$ is nonnegative (its quadratic form is nonnegative), then it must be nuclear.

7.10.39. Example. Let $\mathcal{K}$ be a continuous real function on $[0,1] \times [0,1]$ with $\mathcal{K}(t,s) = \mathcal{K}(s,t)$ such that the integral operator T given on $L^2[0,1]$ by the kernel $\mathcal{K}$ is nonnegative. Then T is a nuclear operator and

$$\operatorname{tr} T = \int_0^1 \mathcal{K}(t,t)\,dt.$$

Indeed, let us take the eigenbasis $\{e_n\}$ of the operator T with eigenvalues λ_n. By Mercer's theorem 7.10.43 proved below the series $\sum_{n=1}^{\infty} \lambda_n e_n(t) e_n(s)$ converges to $\mathcal{K}(t,s)$ uniformly on the square. In particular, it converges uniformly on the diagonal, which after integration gives the indicated equality. Note that T is nuclear if in place of the continuity of $\mathcal{K}$ we assume only the measurability and boundedness (of course, keeping the condition $T \geqslant 0$). This follows from the result above, since one can take a sequence of continuous kernels $\mathcal{K}_n$ with $\sup |\mathcal{K}_n(t,s)| \leqslant \sup |\mathcal{K}(t,s)|$ converging in measure to $\mathcal{K}$ and generating nonnegative operators (see Gokhberg, Krein [**226**, p. 149–151]).

The next theorem describes an interesting connection between nuclear operators and functionals on the space of operators.

7.10.40. Theorem. (i) *For every $S \in \mathcal{N}(H)$ the functional*

$$K \mapsto \operatorname{tr} SK$$

on $\mathcal{K}(H)$ has the norm $\|S\|_{(1)}$. Conversely, every continuous functional on $\mathcal{K}(H)$ admits such a representation, i.e., the space $\mathcal{K}(H)^*$ is naturally isomorphic to the space $\mathcal{N}(H)$.

(ii) *For every $T \in \mathcal{L}(H)$ the functional*

$$S \mapsto \operatorname{tr} TS$$

on $\mathcal{N}(H)$ has the norm $\|T\|_{\mathcal{L}(H)}$. Conversely, every continuous functional on $\mathcal{N}(H)$ admits such a representation, i.e., the space $\mathcal{N}(H)^*$ is naturally isomorphic to the space $\mathcal{L}(H)$.

PROOF. (i) We already know that the norm $\|\Lambda\|$ of the indicated functional Λ does not exceed $\|S\|_{(1)}$. We show that $\|\Lambda\| \geqslant \|S\|_{(1)}$. Let $\varepsilon > 0$. Let us take the polar decomposition $S = U|S|$ and the eigenbasis $\{e_n\}$ of the operator $|S|$ with eigenvalues s_n. Then

$$\|S\|_{(1)} \leqslant \sum_{n=1}^{N} s_n + \varepsilon$$

for some N, where $s_n > 0$. Let us take a finite-dimensional operator K_N with $K_N U e_n = e_n$, $i = 1, \ldots, N$ and extend it by zero on the orthogonal complement of the linear span of $e_1, \ldots, e_N$. Then $\|K_N\|_{\mathcal{L}(H)} = 1$ and

$$\operatorname{tr} K_N S = \sum_{n=1}^{N} s_n (K_N U e_n, e_n) = \sum_{n=1}^{N} s_n \geqslant \|S\|_{(1)} - \varepsilon,$$

which gives the desired bound. Thus, we have an isometric embedding of the space $\mathcal{N}(H)$ into $\mathcal{K}(H)^*$. Let $\Lambda \in \mathcal{K}(H)^*$. For every $u, v \in H$, we have a one-dimensional operator $K_{u,v}(x) := (x, u)v$. Set

$$B(u, v) := \Lambda(K_{u,v}).$$

The function B is linear in u and conjugate-linear in v. The continuity of B is clear from the equality $\|K_{u,v}\|_{(1)} = \|u\| \, \|v\|$ (see Exercise 7.10.63). According to Exercise 7.2.1 there exists an operator $S \in \mathcal{L}(H)$ with $(Su, v) = B(u, v)$. We show that this is the required operator. For any two infinite-dimensional orthonormal sequences $\{\psi_n\}$ and $\{\varphi_n\}$ and every element $(\xi_n) \in c_0$, the series $\sum_{n=1}^{\infty} \xi_n K_{\psi_n, \varphi_n}$ converges with respect to the operator norm to some operator $K \in \mathcal{K}(H)$. By the continuity of Λ the series of $\xi_n \Lambda(K_{\psi_n, \varphi_n}) = \xi_n (S\psi_n, \varphi_n)$ converges and the absolute value of the sum does not exceed $\|\Lambda\| \, \|K\|_{\mathcal{L}(H)}$. It is readily seen that $\|K\|_{\mathcal{L}(H)} = \sup_n |\xi_n|$. Therefore, $\{(S\psi_n, \varphi_n)\} \in l^1$ and $\sum_{n=1}^{\infty} |(S\psi_n, \varphi_n)| \leqslant \|\Lambda\|$, which gives the estimate $\|S\|_{(1)} \leqslant \|\Lambda\|$ by the previous remark. Assertion (ii) is proved similarly. $\square$

This theorem yields the equality $\mathcal{K}(H)^{**} = \mathcal{L}(H)$.

One of the deepest results on operator traces is the following theorem due to V. B. Lidskii. Its rather difficult proof can be read in several books, for example, see Gokhberg, Krein [**226**, Chapter III, §8] (the shortest proof is presented in the book Simon [**554**]).

7.10.41. Theorem. *For every $A \in \mathcal{N}(H)$ we have*

$$\operatorname{tr} A = \sum_{n=1}^{\infty} \lambda_n,$$

where λ_n are all eigenvalues of A counted with multiplicities and in the case of absence of eigenvalues the right-hand side is defined to be zero.

The most difficult part of the proof concerns precisely the case of absence of eigenvalues. It is far from being obvious that in this case the trace vanishes. Nonzero eigenvalues are considered in Exercise 7.10.106. About eigenvalues of compact operators, see also König [**333**] and Pietsch [**484**].

7.10(v). Integral operators and Mercer's theorem

Let us mention several interesting facts connected with the Hilbert–Schmidt theorem and useful in the study of integral equations. Let $\mathcal{K} \in \mathcal{L}^2([0,1]\times[0,1])$ be a complex function with $\mathcal{K}(t,s) = \overline{\mathcal{K}(s,t)}$. Let us consider the compact selfadjoint operator

$$T_{\mathcal{K}} x(t) = \int_0^1 \mathcal{K}(t,s)x(s)\,ds$$

on $L^2[0,1]$. By the Hilbert–Schmidt theorem there exists an orthonormal basis $\{e_n\}$ with $T_{\mathcal{K}} e_n = \lambda_n e_n$, where $\lambda_n \in \mathbb{R}^1$. Hence for every n we have

$$\lambda_n e_n(t) = \int_0^1 \mathcal{K}(t,s)e_n(s)\,ds$$

for almost all t. Hence for almost all t this equality is true simultaneously for all n. In addition, for almost all t the function $s \mapsto \mathcal{K}(s,t)$ belongs to $\mathcal{L}^2[0,1]$. Let t possess both properties (the set of such points also has full measure). Then the previous equality means that

$$\lambda_n \overline{e_n(t)} = (\mathcal{K}(\,\cdot\,,t), e_n),$$

which gives the equality

$$\sum_{n=1}^{\infty} \lambda_n^2 |e_n(t)|^2 = \int_0^1 |\mathcal{K}(s,t)|^2\,ds. \qquad (7.10.10)$$

After integration in t we obtain

$$\sum_{n=1}^{\infty} \lambda_n^2 = \int_0^1 \int_0^1 |\mathcal{K}(s,t)|^2\,ds\,dt < \infty.$$

Since for almost every t we have the orthogonal expansion

$$\mathcal{K}(\,\cdot\,,t) = \sum_{n=1}^{\infty}(\mathcal{K}(\,\cdot\,,t), e_n)e_n = \sum_{n=1}^{\infty} \lambda_n \overline{e_n(t)}e_n,$$

we have

$$\mathcal{K}(s,t) = \sum_{n=1}^{\infty} \lambda_n e_n(s)\overline{e_n(t)}, \qquad (7.10.11)$$

where the series converges in $L^2([0,1] \times [0,1])$. If we impose some additional conditions on $\mathcal{K}$, then a stronger conclusion can be obtained.

7.10.42. Example. Suppose that there is a number C such that

$$\int_0^1 |\mathcal{K}(t,s)|^2\,ds \leqslant C \quad \text{almost everywhere on } [0,1].$$

Then $\sum_{n=1}^{\infty} |\lambda_n e_n(t)|^2 \leqslant C$ a.e., moreover, for every $x \in L^2[0,1]$ the series

$$T_{\mathcal{K}} x(t) = \sum_{n=1}^{\infty} \lambda_n (x, e_n) e_n(t) \qquad (7.10.12)$$

converges absolutely and uniformly on some set E of full measure.

PROOF. Since we have $|\mathcal{K}(t,s)| = |\mathcal{K}(s,t)|$, the first assertion is obvious from (7.10.10). For E we take the set of those points t for which (7.10.10) and the inequality from the hypotheses are fulfilled. Now the estimate

$$\sum_{n=m}^{\infty} |\lambda_n (x, e_n) e_n(t)| \leqslant \left(\sum_{n=m}^{\infty} |\lambda_n e_n(t)|^2 \right)^{1/2} \left(\sum_{n=m}^{\infty} |(x, e_n)|^2 \right)^{1/2}$$

$$\leqslant C^{1/2} \left(\sum_{n=m}^{\infty} |(x, e_n)|^2 \right)^{1/2}$$

proves the second assertion. □

7.10.43. Theorem. (MERCER'S THEOREM) *Suppose that $\mathcal{K}$ is a continuous function such that the operator $T_{\mathcal{K}}$ is nonnegative, i.e., $(T_{\mathcal{K}} x, x) \geqslant 0$. Then the functions e_n corresponding to eigenvalues $\lambda_n \neq 0$ have continuous modifications and the series (7.10.11) and (7.10.12) converge absolutely and uniformly. In addition,*

$$\operatorname{tr} T_{\mathcal{K}} = \int_0^1 \mathcal{K}(t,t)\, dt.$$

PROOF. For $\lambda_n \neq 0$ the continuous version of e_n is given by the formula

$$e_n(t) = \lambda_n^{-1} \int_0^1 \mathcal{K}(t,s) e_n(s)\, ds.$$

We observe that $\mathcal{K}(t,t) \geqslant 0$. Indeed, for a fixed $\tau \in [0,1)$ we take the functions $x_n = n I_{[\tau, \tau+1/n]} I_{[0,1]}$. By assumption

$$0 \leqslant (T_{\mathcal{K}} x_n, x_n) = n^2 \int_{\tau}^{\min(\tau+1/n, 1)} \int_{\tau}^{\min(\tau+1/n, 1)} \mathcal{K}(t,s)\, dt\, ds.$$

As $n \to \infty$, the integral in the right-hand side tends to $\mathcal{K}(\tau, \tau)$, hence $\mathcal{K}(\tau, \tau) \geqslant 0$. Let us apply this observation to the continuous kernels $\mathcal{K} - \mathcal{K}_n$, where

$$\mathcal{K}_n(t,s) = \sum_{j=1}^{n} \lambda_j e_j(t) \overline{e_j(s)},$$

defining nonnegative operators, since $\lambda_j \geqslant 0$ and

$$(T_{\mathcal{K}} x, x) - (T_{\mathcal{K}_n} x, x) = \sum_{j=n+1}^{\infty} \lambda_j |(x, e_j)|^2 \geqslant 0.$$

Thus, $\mathcal{K}_n(t,t) \leqslant \mathcal{K}(t,t)$. Therefore,

$$\sum_{j=1}^{n} \lambda_j |e_j(t)|^2 \leqslant \mathcal{K}(t,t) \leqslant M := \sup_{t,s} |\mathcal{K}(t,s)|.$$

Hence
$$\sum_{j=1}^{\infty} \lambda_j |e_j(t)e_j(s)| \leqslant \sum_{j=1}^{\infty} \lambda_j \frac{|e_j(t)|^2 + |e_j(s)|^2}{2} \leqslant M.$$

In addition, for every fixed s the series $\sum_{j=1}^{\infty} \lambda_j e_j(t)\overline{e_j(s)}$ converges uniformly in t. Indeed, for any given $\varepsilon > 0$ we can find m with $\sum_{j=m}^{\infty} \lambda_j |e_j(s)|^2 \leqslant \varepsilon$ and by the Cauchy–Bunyakovskii inequality we obtain

$$\sum_{j=m}^{\infty} \lambda_j |e_j(t)e_j(s)| \leqslant \left(\sum_{j=m}^{\infty} \lambda_j |e_j(t)|^2\right)^{1/2} \left(\sum_{j=m}^{\infty} \lambda_j |e_j(s)|^2\right)^{1/2} \leqslant (M\varepsilon)^{1/2}.$$

Now we can conclude that the series $\sum_{n=1}^{\infty} \lambda_n e_n(t)\overline{e_n(s)}$ converges to $\mathcal{K}(t,s)$ at every point, not only almost everywhere. Indeed, let us denote the sum of this pointwise convergent series by $Q(t,s)$. It follows from what we have proved that the function Q is bounded and continuous in every variable separately. Let us fix s and show that $Q(t,s) = \mathcal{K}(t,s)$ for all t. By the continuity of both functions in t it suffices to verify that they have equal inner products with all functions e_k. Since the series defining Q converges uniformly in t (which is shown above), the integral of $Q(t,s)\overline{e_k(t)}$ in t equals $\lambda_k \overline{e_k(s)}$. The same value has the integral of $\mathcal{K}(t,s)\overline{e_k(t)}$ in t according to our choice of versions of e_k. The pointwise equality $\mathcal{K}(t,s) = Q(t,s)$ yields that the series $\sum_{n=1}^{\infty} \lambda_n |e_n(t)|^2$ converges pointwise to the continuous function $\mathcal{K}(t,t)$. By Dini's theorem this convergence is uniform. From this with the aid of the Cauchy–Bunyakovskii inequality one can easily derive the uniform convergence of series (7.10.11). The equality for the trace is obtained by integration of this series for $t = s$. $\qquad\square$

For a generalization, see Exercise 7.10.128.

It is readily verified that for a continuous real symmetric kernel $\mathcal{K}$ on $[0,1]^2$ the corresponding operator $T_{\mathcal{K}}$ is nonnegative (in the sense of quadratic forms) precisely when the matrices $\big(\mathcal{K}(t_i,t_j)\big)_{i,j\leqslant n}$ are nonnegative definite. Hence the condition $T_{\mathcal{K}} \geqslant 0$ does not follow from the condition $\mathcal{K}(t,s) \geqslant 0$. As an example let us consider the kernel $\mathcal{K}(t,s) = |t - s|$ that generates an operator which is not nonnegative (of course, this can be easily seen from the formula for the trace in §7.10(iv)).

Note also that the nonnegativity of an operator T on $L^2(\mu)$ in the sense of quadratic forms should not be confused with the nonnegativity in the sense of ordered spaces, i.e., with the condition that $Tx \geqslant 0$ whenever $x \geqslant 0$ (none of these two conditions follows from the other).

7.10(vi). Tensor products

Let X and Y be Banach spaces. For every pair (x,y) in $X \times Y$, the formula $l \mapsto l(x)y$ defines a one-dimensional operator from X^* to Y; this operator will be denoted by the symbol $x \otimes y$. Denote by $X \otimes Y$ the linear space in $\mathcal{L}(X^*, Y)$ generated by all operators $x \otimes y$. Note that a representation of an operator in the form $x_1 \otimes y_1 + \cdots + x_n \otimes y_n$ is not unique. For example, $x \otimes (y+z) = x \otimes y + x \otimes z$. The linear space $X \otimes Y$ is called the *algebraic tensor product* of the spaces X and Y.

It can be completed with respect to any norm on the space of finite-dimensional operators. A norm p on $X \otimes Y$ is called a cross-norm if $p(x \otimes y) = \|x\| \, \|y\|$ for all $x \in X$, $y \in Y$. It turns out that among cross-norms there are two extreme ones that correspond to the usual operator norm on $\mathcal{L}(X^*, Y)$ and the nuclear norm. These two cross-norms are given by the equalities

$$\|u\|_\infty := \varepsilon(u) := \sup\Big\{\sum_i f(x_i)g(y_i) \colon \ u = \sum_i x_i \otimes y_i, \|f\|_{X^*} \leqslant 1, \|g\|_{Y^*} \leqslant 1\Big\},$$

$$\|u\|_{\mathcal{N}} := \pi(u) := \inf\Big\{\sum_i \|x_i\| \, \|y_i\| \colon \ u = \sum_i x_i \otimes y_i\Big\}.$$

They are called the injective and projective norms, respectively.

For every cross-norm p we have $\|u\|_\infty \leqslant p(u) \leqslant \|u\|_{\mathcal{N}}$ (Exercise 7.10.119). The term the "nuclear norm" is closely connected with the fact that an operator $T \in \mathcal{L}(X, Y)$ is called nuclear if it is representable in the form

$$Tx = \sum_{i=1}^\infty u_i(x)v_i, \quad \text{where } u_i \in X^*, \ v_i \in Y, \ \sum_{i=1}^\infty \|u_i\| \, \|v_i\| < \infty.$$

The infimum of sums $\sum_{i=1}^\infty \|u_i\| \, \|v_i\|$ over all possible representations of T is called the nuclear norm of T and is denoted by the symbol $\|T\|_{\mathcal{N}}$.

The completions of $X \otimes Y$ with respect to the norms $\| \cdot \|_{\mathcal{N}}$ and $\| \cdot \|_\infty$ are denoted by the symbols $X \widehat{\otimes} Y$ and $X \widetilde{\otimes} Y$, respectively. Alternative symbols are $X \widehat{\otimes}_\pi Y$ and $X \widehat{\otimes}_\varepsilon Y$.

The space $X \widehat{\otimes} Y = X \widehat{\otimes}_\pi Y$ is called the *Banach tensor product* of X and Y. Every element $u \in X \widehat{\otimes} Y$ can be represented as a series

$$u = \sum_{i=1}^\infty x_i \otimes y_i, \quad \text{where } \sum_{i=1}^\infty \|x_i\| \, \|y_i\| < \infty,$$

and $\|u\|_{\mathcal{N}}$ is the infimum of sums of the indicated form (this is clear from Exercise 5.6.51).

If X and Y are Hilbert spaces, then $X \otimes Y$ can be equipped with the Hilbert–Schmidt norm, which leads to the *Hilbert tensor product* $X \otimes_2 Y$. Thus, in the case where $X = Y = H$ is a Hilbert space, the Banach tensor product $H \widehat{\otimes} H$ is the space $\mathcal{N}(H)$ of nuclear operators, the Hilbert tensor product $H \otimes_2 H$ is the space $\mathcal{H}(H)$ of Hilbert–Schmidt operators, and the tensor product $H \widetilde{\otimes} H$ is the space $\mathcal{K}(H)$ of compact operators, moreover, the tensor norms introduced above are exactly the corresponding operator norms.

If A is a bounded operator on X and B is a bounded operator on Y, then $A \otimes B$ is defined as a bounded operator on $X \otimes Y$ by setting

$$A \otimes B(u \otimes v) = Au \otimes Bv.$$

If p is a cross-norm on $X \otimes Y$, then for $z = \sum_i x_i \otimes y_i$ we have

$$p(A \otimes Bz) \leqslant \|A\| \, \|B\| \sum_i \|x_i\| \, \|y_i\|,$$

hence the left-hand side is estimated by $\|A\| \, \|B\| \, \|z\|_{\mathcal{N}}$. In particular, $A \otimes B$ is bounded on the Banach tensor product and its norm is $\|A\| \, \|B\|$. It is readily seen that the same is true for the injective norm. Cross-norms with this property are called uniform in Ryan [**523**, p. 128]. If $X = Y = H$ is a Hilbert space, then $A \otimes B$ is also bounded on the Hilbert tensor product and its norm is $\|A\| \|B\|$. Indeed, it suffices to prove that the norm is bounded by $\|A\| \|B\|$. Moreover, it suffices to do that in the case where one of the operators is the identity, since

$$A \otimes B = (A \otimes I)(I \otimes B).$$

Now let $B = I$. If $v = \sum_i u_i \otimes v_i$, where v_i are mutually orthogonal, then

$$\|v\|_{\mathcal{H}}^2 = \sum_i \|u_i \otimes v_i\|^2 = \sum_i \|u_i\|^2.$$

Hence

$$\|(A \otimes I)v\|_{\mathcal{H}}^2 = \left\| \sum_i Au_i \otimes v_i \right\|_{\mathcal{H}}^2 = \sum_i \|Au_i\|^2 \leqslant \|A\|^2 \sum_i \|u_i\|^2 = \|A\|^2 \|v\|_{\mathcal{H}}^2.$$

A useful application of this construction can be found in Exercise 7.10.135.

7.10(vii). Fredholm operators

Let X and Y be Banach spaces.

7.10.44. Definition. *A bounded operator* $A \colon X \to Y$ *is called Fredholm if its kernel* Ker A *has a finite dimension and its range* Ran $A = A(X)$ *has a finite codimension.*

The number dim Ker A − codim Ran A *is called the index of the operator* A *and denoted by the symbol* Ind A.

According to Proposition 6.2.12 the range of any Fredholm operator is closed.

Sometimes Fredholm operators are called Noether operators, while the term Fredholm is reserved for Noether operators with zero index.

The index of a Fredholm operator turns out to be a more important characteristic than the dimension of its kernel or the codimension of its range. This is connected with the behavior of the index under compositions and perturbations of operators, which will be demonstrated below.

7.10.45. Example. (i) Every linear operator A on a finite-dimensional space E is Fredholm with zero index, since

$$\dim \operatorname{Ker} A + \dim A(E) = \dim E.$$

(ii) A compact operator $K \colon X \to Y$ can be Fredholm only in the case where both spaces X and Y are finite-dimensional. Indeed, the closed space $K(X)$ must be finite-dimensional by the open mapping theorem (the image of an open ball is open in $K(X)$, but in this case it is totally bounded). Then the kernel of K has a finite codimension, i.e., X is also finite-dimensional.

(iii) Let $K \colon X \to X$ be a compact operator. Then the operator $I + K$ is Fredholm. In addition, Ind $(I + K) = 0$. This was proved in §7.4. As we shall see below, this example is very typical. The term a "Fredholm operator" gives credit to Fredholm, who investigated integral equations. His results concerned with integral operators paved the way to subsequent research of Riesz and Schauder about abstract compact operators. The Nikolskii theorem proved below exhibits a close connection between Fredholm operators and compact perturbations of the identity operator.

(iv) The operator $A_\varphi\colon L^2[0,1] \to L^2[0,1]$ of multiplication by a bounded measurable function φ is Fredholm precisely when it is invertible, i.e., $|\varphi(t)| \geq c$ almost everywhere on $[0,1]$ for some number $c > 0$. In this case the index of A_φ is zero. Indeed, the set $Z := \varphi^{-1}(0)$ has measure zero, since otherwise the space of functions from $L^2[0,1]$ vanishing outside Z is infinite-dimensional. This space coincides with the kernel of A. Thus, the operator A is injective. In addition, its range is everywhere dense in $L^2[0,1]$, because for every function f in $L^2[0,1]$ the functions $f I_{E_n}$, where $E_n := \{t\colon |\varphi|(t) \geq n^{-1}\}$, converge to f in $L^2[0,1]$, but such functions belong to the range of A. Hence the operator A is one-to-one.

(v) Let U be the open unit disc in $\mathbb{C}$ and let $X = H(U)$ be the Banach space of analytic functions on U continuous on the closure of U with the norm $\|f\| = \max_{z \in U} |f(z)|$. Set $Af(z) = z^n f(z)$. Then A is a Fredholm operator with index n. Indeed, this operator has the zero kernel and a range of finite codimension. The functions $1, z, \ldots, z^{n-1}$ along with the range of A generate $H(U)$, which is clear from the expansion $f(z) = \sum_{k=0}^\infty a_k z^k$.

Let us consider in greater detail the structure of a Fredholm operator T between two Banach spaces X and Y. Let us take a closed linear complement X_1 to the kernel of T in X (see Corollary 6.4.2). There is also a finite-dimensional complement Y_0 to the range Y_1 of the operator T. The operator T maps X_1 one-to-one to Y_1 and equals zero on $X_0 := \operatorname{Ker} T$. Since X_1 and Y_1 are Banach spaces, the operator $T\colon X_1 \to Y_1$ is a linear homeomorphism. This yields the following assertion.

7.10.46. Proposition. *Let X, Y, Z be Banach spaces and let $T \in \mathcal{L}(X, Y)$ be a Fredholm operator. Then the image $T(Z)$ of every closed linear subspace $Z \subset X$ is closed in Y.*

PROOF. The linear subspace $Z_1 := Z \cap X_1$ is closed and has a finite-dimensional linear complement Z_2 to Z. Hence $T(Z) = T(Z_1) + T(Z_2)$, where the subspace $T(Z_1)$ is closed by the property that T is a homeomorphism on X_1 and the subspace $T(Z_2)$ is finite-dimensional. According to Proposition 5.3.7 the set $T(Z)$ is closed. $\qquad\square$

Now with the aid of the above considerations we prove the following theorem due to S. M. Nikolskii, which connects Fredholm operators with compact perturbations of the identity.

7.10.47. Theorem. *Let X and Y be Banach spaces and $T \in \mathcal{L}(X, Y)$. The operator T is Fredholm precisely when there exists an operator $S \in \mathcal{L}(Y, X)$ such that the operators $ST - I_X$ and $TS - I_Y$ are compact (on X and Y, respectively). In this case the operator S can be chosen such that the indicated operators will be even finite-dimensional.*

PROOF. Let T be Fredholm. As above, we take a closed linear complement X_1 to $X_0 = \operatorname{Ker} T$ in X. We know that the projection operator $P_1\colon X \to X_1$ is continuous. The projection $P = I_X - P_1$ onto $\operatorname{Ker} T$ is also continuous. There is a finite-dimensional complement Y_0 to the range Y_1 of the operator T and the continuous projection operator $Q_1 \colon Y \to Y_1$. Then $Q := I_Y - Q_1$ is the continuous projection operator onto Y_0. The operator T maps X_1 one-to-one onto Y_1. The Banach inverse mapping theorem gives a mapping $S_0 \in \mathcal{L}(Y_1, X_1)$ with $S_0 T x = x$ for all $x \in X_1$ and $T S_0 y = y$ for all $y \in Y_1$. Set $S \colon Y \to X$, $Sy = S_0 Q_1 y$. Then $S \in \mathcal{L}(Y, X)$ and $TSy = TS_0 Q_1 y = Q_1 y = y - Qy$ for all vectors $y \in Y$. In addition, we have

$$STx = S_0 P_0 T x = S_0 T x = S_0 T P_0 x = P_0 x = x - Px \quad \text{for all } x \in X.$$

Thus, the operators $ST - I_X$ and $TS - I_Y$ are finite-dimensional.

Conversely, suppose that there exists an operator $S \in \mathcal{L}(Y, X)$ such that the operators $ST - I_X$ and $TS - I_Y$ are compact. This gives the Fredholm property of the operators ST and TS. Since $\operatorname{Ker} T \subset \operatorname{Ker} ST$, the kernel of T is finite-dimensional. In addition, the range of TS is contained in the range of T, hence the latter has a finite codimension. $\qquad \square$

Note that the operator S ("almost inverse" to the Fredholm operator T) is also Fredholm, since it satisfies the Nikolskii condition.

7.10.48. Theorem. *Let X, Y and Z be Banach spaces and let $S \in \mathcal{L}(X, Y)$ and $T \in \mathcal{L}(Y, Z)$ be Fredholm operators. Then the operator $TS \in \mathcal{L}(X, Z)$ is also Fredholm and $\operatorname{Ind}(TS) = \operatorname{Ind} T + \operatorname{Ind} S$.*

PROOF. Let us consider the finite-dimensional subspace $Y_0 := S(X) \cap \operatorname{Ker} T$ in Y. It is clear that $\operatorname{Ker}(TS) = S^{-1}(\operatorname{Ker} T)$. The operator S maps $\operatorname{Ker}(TS)$ onto Y_0 and the kernel of this mapping is the set $\operatorname{Ker} S$. Thus,

$$\dim \operatorname{Ker}(TS) = \dim \operatorname{Ker} S + \dim Y_0.$$

Since the range of S is finite-dimensional, the subspace $S(X) + \operatorname{Ker} T$ possesses a finite-dimensional algebraic complement Y_1. It is clear that $\operatorname{Ran} T$ is the algebraic sum of $\operatorname{Ran}(TS)$ and $T(Y_1)$. Let us verify that this is a direct sum. Indeed, if $TSx = Ty$, where $x \in X$ and $y \in Y_1$, then $Sx - y \in \operatorname{Ker} T$. Then we have $y \in (S(X) + \operatorname{Ker} T) \cap Y_1$. Hence $y = 0$, since the intersection above consists only of zero. Thus, the operator TS is Fredholm, and we arrive at the equality

$$\operatorname{codim} \operatorname{Ran}(TS) = \operatorname{codim} \operatorname{Ran} T + \dim Y_1.$$

On the ground of the already proved facts we obtain the following relation:

$$\operatorname{Ind}(TS) = (\dim \operatorname{Ker} S + \dim Y_0) - (\operatorname{codim} \operatorname{Ran} T + \dim Y_1).$$

The subspace Y_0 in the finite-dimensional subspace $\operatorname{Ker} T$ has a linear complement Y_2. Hence $\dim \operatorname{Ker} T = \dim Y_0 + \dim Y_2$. For obtaining the desired relation it remains to observe that

$$\operatorname{codim} \operatorname{Ran} S = \dim Y_1 + \dim Y_2.$$

This equality follows from the fact that by construction Y is the direct sum of $\operatorname{Ran} S + \operatorname{Ker} T$ and Y_1, moreover, $\operatorname{Ran} S + \operatorname{Ker} T$ is a direct sum of $\operatorname{Ran} S$ and Y_2 due to the fact that Y_2 is a complement of $\operatorname{Ran} S \cap \operatorname{Ker} T$ in $\operatorname{Ker} T$. $\qquad \square$

7.10.49. Corollary. *Let X and Y be Banach spaces, let $T \in \mathcal{L}(X, Y)$ be a Fredholm operator, and let $K \in \mathcal{K}(X, Y)$ be a compact operator. Then the operator $T + K$ is also Fredholm, moreover,*

$$\operatorname{Ind}(T + K) = \operatorname{Ind} T.$$

PROOF. By the Nikolskii theorem there is an operator $S \in \mathcal{L}(Y, X)$ such that the operators $P_1 := ST - I_X$ and $P_2 := TS - I_Y$ are finite-dimensional. Hence $S(T + K) = I_X + P_1 + SK$ and $(T + K)S = I_Y + P_2 + KS$, where the operators $P_1 + SK$ and $P_2 + KS$ are compact. By the same theorem the operator $T + K$ is Fredholm. It has been noted in Example 7.10.45(iii) that compact perturbations of the identity operator have zero index. By the theorem proved above this gives the equality

$$\operatorname{Ind} T + \operatorname{Ind} S = 0 = \operatorname{Ind}(T + K) + \operatorname{Ind} S,$$

which yields the equality $\operatorname{Ind}(T + K) = \operatorname{Ind} T$. $\qquad\square$

7.10.50. Theorem. *Let X and Y be two Banach spaces. An operator $T \in \mathcal{L}(X, Y)$ is Fredholm precisely when its adjoint operator T^* is Fredholm. Moreover, $\operatorname{Ind} T = -\operatorname{Ind} T^*$.*

PROOF. Let T be Fredholm. Then

$$\dim \operatorname{Ker} T^* = \operatorname{codim} \operatorname{Ran} T, \quad \operatorname{codim} \operatorname{Ran} T^* = \dim \operatorname{Ker} T.$$

Indeed, let us represent X as $X_0 \oplus X_1$, where $X_0 = \operatorname{Ker} T$ and X_1 is a closed linear complement to X_0. In addition, let us represent Y as $Y = Y_0 \oplus Y_1$, where Y_1 is the closed range of T and Y_0 is a finite-dimensional complement to Y_1. In this representation the operator T is written as $(x_0, x_1) \mapsto (0, T_1 x_1)$, where $T_1 \colon X_1 \to Y_1$ is an invertible operator. Then the operator $T^* \colon Y_0^* \oplus Y_1^* \to X_0^* \oplus X_1^*$ acts as follows: $T^*(y_0^*, y_1^*) = (0, T_1^* y_1^*)$, $y_0^* \in Y_0^*$, $y_1^* \in Y_1^*$. The operator T_1^* is an isomorphism between the spaces Y_1^* and X_1^*. For finite-dimensional spaces we have the equalities $\dim X_0 = \dim X_0^*$ and $\dim Y_0 = \dim Y_0^*$. This shows that T^* is Fredholm and gives the desired equality. Note that the Fredholm property of T^* by itself is obvious from the Nikolskii theorem and the equalities $(ST)^* = T^* S^*$, $(TS)^* = S^* T^*$. Conversely, let the operator $T^* \colon Y^* \to X^*$ be Fredholm. Then the operator $T^{**} \colon X^{**} \to Y^{**}$ is also Fredholm. Hence $\dim \operatorname{Ker} T < \infty$. The range of T has a finite codimension. Indeed, this range is closed, since it coincides with the image of the closed subspace X in X^{**} under the action of the Fredholm operator T^{**}. Therefore, the range of T coincides with the annihilator of the finite-dimensional kernel of T^* (see Lemma 6.8.1). $\qquad\square$

It is suggested in Exercise 7.10.120 to prove that a Fredholm operator T has zero index precisely when $T = S + K$, where the operator S is invertible and the operator K is finite-dimensional.

In many infinite-dimensional Banach spaces one can easily construct operators that are neither compact nor Fredholm. For a long time the following problem remained open: does there exist an infinite-dimensional Banach space in which every bounded operator has the form $\lambda I + K$, where K is compact? Only recently S. Argyros and R. Haydon [**666**] have constructed such a space.

7.10(viii). The vector form of the spectral theorem

Here we obtain two more functional representations of selfadjoint operators, but we employ spaces of vector functions. The next assertion follows directly from the proof of Theorem 7.8.6.

7.10.51. Theorem. *Let A be a selfadjoint operator on a separable Hilbert space $H \neq 0$. Then there exists a finite or countable family of nonnegative Borel measures μ_n on $[-\|A\|, \|A\|]$ such that the operator A is unitarily equivalent to the operator B on the space $\bigoplus_{n=1}^{\infty} L^2(\mu_n)$ acting by the formula*

$$Bf(t) = \big(tf_1(t), tf_2(t), \ldots, tf_n(t), \ldots\big), \quad f = (f_1, f_2, \ldots, f_n, \ldots).$$

This theorem can be reformulated in a different form, where the operator acts not in the sum of the spaces $L^2(\mu_n)$, but in a subspace of some common space $L^2(\mu, H)$ of square integrable vector functions. Let us first introduce the space $L^2(\mu, H)$, where μ is a finite nonnegative measure on a σ-algebra $\mathcal{A}$ in a space Ω and H is a separable Hilbert space. A mapping x with values in H will be called μ-measurable if it is defined μ-a.e. and for some orthonormal basis $\{e_n\}$ in H the scalar functions $\omega \mapsto \big(x(\omega), e_n\big)$ are μ-measurable. This property does not depend on the choice of a basis: if $\{\varphi_k\}$ is another orthonormal basis, then

$$\big(x(\omega), \varphi_k\big) = \sum_{n=1}^{\infty} \big(x(\omega), e_n\big)_H (e_n, \varphi_k)_H.$$

Similarly to the scalar case one introduces the class $\mathcal{L}^2(\mu, H)$ of all μ-measurable mappings x with values in H such that $\| \cdot \|_H^2 \in L^1(\mu)$, i.e.,

$$\sum_{n=1}^{\infty} \int_{\Omega} \big|\big(x(\omega), e_n\big)_H\big|^2 \, \mu(d\omega) < \infty.$$

Finally, let $L^2(\mu, H)$ denote the space of equivalence classes in $\mathcal{L}^2(\mu, H)$, where equivalent mappings are μ-a.e. equal mappings (see §6.10(vi)). As in the scalar case, the space $L^2(\mu, H)$ is equipped with a structure of a linear space by means of operations on representatives of classes and the inner product is defined by

$$(x, y) := \int_{\Omega} \big(x(\omega), y(\omega)\big)_H \, \mu(d\omega) = \sum_{n=1}^{\infty} \int_{\Omega} \big(x(\omega), e_n\big)_H \overline{\big(y(\omega), e_n\big)_H} \, \mu(d\omega),$$

where in the right-hand side one takes representatives of equivalence classes. In particular, the mapping x has the norm

$$\|x\| := \left(\int_{\Omega} \|x(\omega)\|_H^2 \, \mu(d\omega) \right)^{1/2}.$$

Exercise 7.10.98 suggests to verify the completeness of $L^2(\mu, H)$. Choosing orthonormal bases $\{e_n\}$ in H and $\{f_k\}$ in $L^2(\mu)$, we obtain an orthonormal basis in $L^2(\mu, H)$ consisting of all mappings $\omega \mapsto f_k(\omega)e_n$.

7.10.52. Theorem. *Let A be a selfadjoint operator on a separable Hilbert space $H \neq 0$. Then there exists a nonnegative Borel measure μ on $[-\|A\|, \|A\|]$ such that the operator A is unitarily equivalent to the operator A_0 defined by*

$$A_0 f(t) = t f(t)$$

on some closed linear subspace of the Hilbert space $L^2(\mu, l^2)$.

PROOF. Let $\|A\| \leqslant 1$. We know that the operator A is unitarily equivalent to the direct sum of operators of multiplication by the argument on the spaces $L^2(\mu_n)$ with some nonnegative Borel measures μ_n on $\Omega := \sigma(A)$, where $\mu_n(\Omega) \leqslant 1$. For μ we take the measure $\mu := \sum_{n=1}^{\infty} 2^{-n} \mu_n$ on Ω. Hence it suffices to prove our assertion for this direct sum. Since every measure μ_n is obviously absolutely continuous with respect to the measure μ, by the Radon–Nikodym theorem it possesses a density ϱ_n with respect to μ, i.e., there exists a nonnegative μ-integrable function ϱ_n such that

$$\mu_n(B) = \int_B \varrho_n(t)\, \mu(dt)$$

for every Borel set $B \subset \Omega$. This equality yields (see §3.9) that for every μ_n-integrable function φ we have

$$\int_\Omega \varphi\, d\mu_n = \int_\Omega \varphi \varrho_n\, d\mu.$$

In particular, this is true for all $\varphi \in L^2(\mu_n)$. Let us define an embedding J of the space $\bigoplus_{n=1}^{\infty} L^2(\mu_n)$ into $L^2(\mu, l^2)$ by the formula

$$J\big(\{\varphi_n\}_{n=1}^{\infty}\big)(t) = \{\varphi_n(t)\sqrt{\varrho_n}(t)\}_{n=1}^{\infty}.$$

It is clear that J is linear and preserves the inner product, since for every function $\varphi = \{\varphi_n\}_{n=1}^{\infty}$ in $\bigoplus_{n=1}^{\infty} L^2(\mu_n)$ we have

$$\|J\varphi\|^2 = \sum_{n=1}^{\infty} \int_\Omega |\varphi_n(t)|^2 \varrho_n(t)\, \mu(dt) = \sum_{n=1}^{\infty} \int_\Omega |\varphi_n(t)|^2 \mu_n(dt) = \|\varphi\|^2.$$

Therefore, the range of the isometry J is a closed linear subspace E in $L^2(\mu, l^2)$. In the general case this subspace does not coincide with $L^2(\mu, l^2)$. However, the operator of multiplication by the argument acts on E by the natural formula $A_0(\{x_n\})(t) = \{tx_n(t)\}$ and corresponds to our operator on the space $\bigoplus_{n=1}^{\infty} L^2(\mu_n)$ under the isomorphism J. $\square$

7.10(ix). Invariant subspaces

In our study of selfadjoint operators we have occasionally made use of their invariant subspaces, i.e., closed subspaces H_0 such that $A(H_0) \subset H_0$. Every selfadjoint operator has many such subspaces (Exercise 7.10.91). However, already for several decades the following question remains open: does every bounded operator on a separable complex Hilbert space H have invariant closed subspaces different from 0 and H? The same question for general separable Banach spaces also remained open for a long time, but in 1981 P. Enflo constructed a counter-example (later C. Read constructed a counter-example in l^1). Here we present a remarkable result of V. I. Lomonosov obtained before the discovery of these counter-examples and strengthening some results due to von Neumann, Aronszajn and Smith. The proof gives an unexpected and beautiful application of the nonlinear Schauder theorem to linear operators.

7.10.53. Theorem. *Let $K \neq 0$ be a compact linear operator on an infinite-dimensional Banach space X. Then all bounded operators commuting with K (including K itself) have a common nontrivial closed invariant subspace.*

PROOF. Suppose the contrary and take an open ball $U \subset X$ such that 0 does not belong to the convex compact set $S := \overline{K(U)}$. Let us consider the subalgebra $\mathcal{M} := \{A \in \mathcal{L}(X) : AK = KA\}$ and the linear subspace $M(x) := \{Ax : A \in \mathcal{M}\}$ for every $x \in X$. For $x \neq 0$ we have $\overline{M(x)} = X$, since the closed subspace $\overline{M(x)}$ is invariant with respect to all $A \in \mathcal{M}$ for all $x \in M(x)$, because $I \in \mathcal{M}$. Hence for every $s \in S$ there exists $A_s \in \mathcal{M}$ with $A_s(s) \in U$. This gives an open ball V_s centered at s with $A_s(V_s) \subset U$. By the compactness of S we obtain a finite collection $s_1, \ldots, s_n \in S$ such that $S \subset V_{s_1} \cup \ldots \cup V_{s_n}$. As shown in §1.9(iv), there exist functions $\varphi_i \in C(S)$, $i = 1, \ldots, n$, such that $\varphi_i \geqslant 0$, $\sum_{i=1}^{n} \varphi_i(x) = 1$ and $\varphi_i(x) = 0$ whenever $x \notin V_{s_i}$. Let us consider the mapping

$$F \colon S \to X, \quad F(x) = K\Big(\sum_{i=1}^{n} \varphi_i(x) A_{s_i}(x)\Big).$$

This mapping is continuous and $F(S) \subset S$, since for any $x \in S$ we have $\varphi_i(x) \neq 0$ only for $x \in V_{s_i}$, but then $A_{s_i}(x) \in U$ and the set U is convex. By Schauder's theorem there is a point $x_0 \in S$ with $F(x_0) = x_0$. Finally, let us take the compact operator $T = \sum_{i=1}^{n} \varphi_i(x_0) K A_{s_i}$. It is clear that $T \in \mathcal{M}$ and $T(x_0) = x_0$. The closed subspace $L = \operatorname{Ker}(T - I)$ is finite-dimensional, $L \neq 0$. Since $KT = TK$, we have $K(L) \subset L$. Hence in L the operator K has an eigenvector with some eigenvalue λ. It remains to observe that the subspace $E = \operatorname{Ker}(K - \lambda I)$ is invariant with respect to all operators commuting with K. □

If X is a real space, the operator K itself has an invariant subspace. A similar reasoning gives the following result due to V. I. Lomonosov.

7.10.54. Theorem. *Let X be an infinite-dimensional Banach space. Suppose that an operator $T \in \mathcal{L}(X)$ commutes with some nonzero compact operator and is not a multiple of the unit operator. Then all bounded operators commuting with T have a common nontrivial closed invariant subspace.*

The class of operators T covered by this theorem is so large that it took several years to construct an operator that does not belong to it. About invariant subspaces, see [52] and [662].

Exercises

7.10.55.° Find the spectra of the following operators on l^2: (i) $Ax = (0, x_1, x_2, \ldots)$; (ii) $Ax = (x_2, x_3, \ldots)$; (iii) $Ax = (x_2/2, x_3/3, \ldots, x_n/n, \ldots)$.

HINT: use that the operators in (i) and (ii) are adjoint to each other and that the operator in (iii) is compact.

7.10.56.° Find the spectra of the following operators on the space $l^2(\mathbb{Z})$ of two-sided sequences: (i) $(Ax)_n = x_{n+1}$; (ii) $(Ax)_n = 0$ if n is odd, $(Ax)_n = x_{n+1}$ if n is even; (iii) $(Ax)_n = x_{n+1}/(|n| + 1)$.

7.10.57.° Let $\{e_n\}$ be an orthonormal basis in a separable Hilbert space H and let a bounded operator A be given by the formula $Ae_n = \alpha_n e_n$, where $\{\alpha_n\}$ is a bounded sequence in $\mathbb{C}$. Prove that the spectrum of A is the closure of $\{\alpha_n\}$.

7.10.58.° The operator A on l^2 is defined by $Ax = (r_1 x_1, r_2 x_2, \ldots, r_n x_n, \ldots)$, where $\{r_n\}$ is some enumeration of the set of rational numbers in $[0,1]$. Prove that A has a cyclic vector and the spectrum $[0,1]$, but is not unitarily equivalent to the operator of multiplication by the argument on $L^2[0,1]$.

7.10.59.° Investigate whether there are unitarily equivalent operators among the following operators of multiplication by a function φ on $L^2[a,b]$, where $[a,b]$ is equipped with Lebesgue measure: (a) $\varphi(t) = t$, $[a,b] = [0,1]$, (b) $\varphi(t) = |t|$, $[a,b] = [-1,1]$, (c) $\varphi(t) = t^2$, $[a,b] = [0,1]$, (d) $\varphi(t) = t^3$, $[a,b] = [0,1]$, (e) $\varphi(t) = t^{1/2}$, $[a,b] = [0,1]$, (f) $\varphi(t) = \sin t$, $[a,b] = [0,1]$.

7.10.60. (i) Prove that a selfadjoint operator on $\mathbb{C}^n$ possesses a cyclic vector precisely when it has no multiple eigenvalues. (ii) Prove that a compact selfadjoint operator on a Hilbert space possesses a cyclic vector precisely when it has no multiple eigenvalues.

HINT: write the operator in the diagonal form and observe that in case of an eigenvalue of multiplicity at least two for every vector h there is a nonzero vector u orthogonal to the orbit of h.

7.10.61. Suppose that μ is a Borel probability measure on $[0,1]$ mutually singular with Lebesgue measure λ, having no points of positive measure and satisfying the condition $\mu((a,b)) > 0$ whenever $0 \leqslant a < b \leqslant 1$. Prove that the operators of multiplication by the argument on $L^2(\mu)$ and $L^2(\lambda)$ have cyclic vectors, equal spectra and have no eigenvalues, but are not unitarily equivalent.

7.10.62.° Let A be a bounded operator on a complex Banach space X such that one has $\|A\| \in \sigma(A)$. Prove that $\|I + A\| = 1 + \|A\|$.

HINT: use that otherwise the norm of the operator $(1 + \|A\|)^{-1}(I + A)$ is less than 1, hence the difference with the unit operator must be invertible.

7.10.63.° Let H be a Hilbert space, $u, v \in H$, and let the operator $K_{u,v}$ be given by the formula $K_{u,v} x := (x, u)v$. Show that $K_{u,v}^* = K_{v,u}$ and $|K_{u,v}| x = \|v\|(x, u)u$.

7.10.64.° Let U be the shift in the space l^2 of two-sided sequences $(z_n)_{n \in \mathbb{Z}}$ defined by $Ue_n = e_{n+1}$. Set $Kz := (z, e_{-1})e_0$. Prove that the spectrum of $U - K$ coincides with the unit disc.

7.10.65.° Let A and B be linear operators on a linear space X. Show that the operators AB and BA have the same nonzero eigenvalues.

7.10.66.° Let X be a complex Banach space and $A \in \mathcal{L}(X)$. Prove that if A^2 has an eigenvalue, then A also does.

HINT: use that $A^2 - \lambda = (A + \lambda^{1/2})(A - \lambda^{1/2})$.

7.10.67.° Construct a bounded linear operator on the complex space l^2 such that its spectrum consists of the two points 0 and 1 that are not its eigenvalues.

HINT: consider first a compact operator without eigenvalues.

7.10.68.° Find the eigenvalues of the operator V^*V for the Volterra operator on $L^2[0,1]$ defined by $Vx(t) = \int_0^t x(s)\, ds$.

7.10.69. Let X be a Banach space and let $t \mapsto A_t$ be a mapping from $[0, 1]$ to $\mathcal{L}(X)$ continuous in the operator norm such that there exists a number $C > 0$ for which $\|x\| \leqslant C\|A_t x\|$ for all $x \in X$ and $t \in [0, 1]$. Prove that the invertibility of A_0 is equivalent to the invertibility of A_1. In particular, the operators $A_0, A_1 \in \mathcal{L}(X)$ are simultaneously invertible or non-invertible if the condition is fulfilled for $A_t := tA_1 + (1-t)A_0$, $t \in [0, 1]$.

HINT: observe that if A_t is invertible for some t, then $\|A_t^{-1}\| \leqslant C$, hence its perturbations by operators with norm less that C^{-1} are invertible.

7.10.70. Let X be a complex Banach space and let $A \in \mathcal{L}(X)$. Suppose that $f(z) = \sum_{n=0}^{\infty} c_n z^n$ is a function analytic in the disc of radius $r > \|A\|$ centered at zero. Prove that $\sigma(f(A)) = f(\sigma(A))$, where $f(A) := \sum_{n=0}^{\infty} c_n A^n$.

HINT: observe that $f(A) - f(\lambda)I = (A - \lambda I)g(A) = g(A)(A - \lambda I)$ is not invertible if $\lambda \in \sigma(A)$. Conversely, if $f(A) - \mu I$ is not invertible, then on a compact disc of some radius larger than $\|A\|$ we have $f(z) - \mu = (z - \mu_1) \cdots (z - \mu_k)h(z)$, where h has no zeros on this disc. Hence $h(A)$ is invertible, which means that some μ_i belongs to $\sigma(A)$, so $\mu = f(\mu_i) \in f(\sigma(A))$.

7.10.71. Let A be a selfadjoint operator and $A \geqslant 0$. Prove that $\|Ax\|^2 \leqslant \|A\|(Ax, x)$.
HINT: use that $A = A^{1/2}A^{1/2}$.

7.10.72. Let A, B be selfadjoint operators such that $A, B \geqslant 0$ and $AB = BA$. Prove that $AB \geqslant 0$. Give an example showing that this is not always true if A and B do not commute.
HINT: observe that $\sqrt{A}\sqrt{B} = \sqrt{B}\sqrt{A}$, so $(ABx, x) = \|\sqrt{A}\sqrt{B}x\|^2$.

7.10.73. Let B be a selfadjoint operator, $B \geqslant 0$ and $A = B^2$. Prove that $B = \sqrt{A}$.
HINT: write B as the multiplication by a function.

7.10.74. Let A be a selfadjoint operator on a nonzero separable Hilbert space such that $\sigma(A) = K_1 \cup K_2$, where K_1 and K_2 are compact and $K_1 \cap K_2 = \varnothing$. Prove that A can be written as the direct sum of operators A_1 and A_2 with $\sigma(A_1) = K_1$ and $\sigma(A_2) = K_2$.
HINT: represent A as an operator of multiplication.

7.10.75. Let P_1 and P_2 be the orthogonal projections on subspaces H_1 and H_2. Prove that $P_1 P_2$ is the orthogonal projection if and only if $P_1 P_2 = P_2 P_1$, and in this case $P_1 P_2$ is the projection onto $H_1 \cap H_2$.

7.10.76. Let P_j, $j \in \mathbb{N}$, be orthogonal projections in a complex Hilbert space.
(i) Let P be an orthogonal projection such that $(P_j x, x) \to (Px, x)$ for all x. Prove that $\|P_j x - Px\| \to 0$ for all x, using that $(P_j x, y) \to (Px, y)$ for all x, y. (ii) Suppose that for every x the sequence $(P_j x, x)$ is increasing (or, for every x, is decreasing). Prove that there exists an orthogonal projection P for which $\|P_j x - Px\| \to 0$ for all x. (iii) Show that if P is an operator such that $\|P_j x - Px\| \to 0$ for all x, then P is a projection. (iv) Give an example of an operator P that is not a projection, but for which $(P_j x, y) \to (Px, y)$ for all x, y. For this consider the indicator functions of the sets $B_n \subset [0, 1]$ such that B_n consists of the left halves of the intervals obtained by partitioning $[0, 1]$ into 2^n equal pieces.

7.10.77. Let H be a separable Hilbert space and P a projection-valued measure on the Borel σ-algebra of the real line with values in $\mathcal{P}(H)$. (i) Prove that there exists a finite nonnegative Borel measure μ on $\mathbb{R}$ such that all measures $\mu_x(B) = (P(B)x, x)$ are absolutely continuous with respect to μ. (ii) Prove that for every Borel set $B \subset \mathbb{R}^1$ one can find a sequence of sets B_j written as finite unions of intervals such that $P(B_j)x \to P(B)x$ for every $x \in H$.

7.10.78. Let H be a separable Hilbert space, $A \in \mathcal{L}(H)$ a selfadjoint operator, and ψ a bounded Borel function. Prove that there exist polynomials p_n such that the operators $p_n(A)$ converge to $\psi(A)$ on every vector.

HINT: let $A = A_f$ be the operator of multiplication on $L^2(\mu)$ by a bounded function f; the measure $\nu := \mu \circ f^{-1}$ is concentrated on a compact interval, hence one can find a uniformly bounded sequence of polynomials p_n that converges to ψ ν-a.e.; then the functions $p_n \circ f$ are uniformly bounded and μ-a.e. converge to $\psi \circ f$, which by the dominated convergence theorem gives convergence of $(p_n \circ f)x$ to $(\psi \circ f)x$ in $L^2(\mu)$ for every element $x \in L^2(\mu)$. One can also use Theorem 7.9.6 and the measure μ from (i) in the previous exercise.

7.10.79. Let A_n be selfadjoint operators on a Hilbert space H such that

$$(A_1 x, x) \leqslant (A_2 x, x) \leqslant \cdots \leqslant (A_n x, x) \leqslant \cdots$$

and $\sup_n \|A_n\| < \infty$. Prove that there exists a selfadjoint operator A on H for which $Ax = \lim_{n \to \infty} A_n x$ for all $x \in H$.

HINT: define A through its quadratic form and apply Exercise 7.10.71 to $A - A_n$.

7.10.80. Let H be a Hilbert space and $A \in \mathcal{L}(H)$ a selfadjoint operator. Prove that there exist unique selfadjoint operators $A^+, A^- \geqslant 0$ such that $A^+ A^- = A^- A^+ = 0$ and $A = A^+ - A^-$, and if $A_1, A_2 \in \mathcal{L}(H)$ are selfadjoint operators for which $0 \leqslant A_1 \leqslant A^+$, $0 \leqslant A_2 \leqslant A^-$ and $A = A_1 - A_2$, then $A_1 = A^+$ and $A_2 = A^-$.

HINT: take $A^+ = f_1(A)$, $A^- = f_2(A)$, $f_1(t) = \max(t, 0)$, $f_2(t) = -\min(t, 0)$.

7.10.81. Construct an example of a continuous quadratic form on a Banach space which cannot be decomposed into the difference of two nonnegative continuous quadratic forms.

HINT: construct a sequence of two-dimensional Banach spaces $(X_n, \| \cdot \|_n)$ with quadratic forms Q_n such that $|Q_n| \leqslant 1$ on the unit ball U_n in X_n, but the positive part of Q_n assumes the value 2^n on U_n; consider the space of bounded sequences $x = (x_n)$, where $x_n \in X_n$, with the norm $\|x\| = \sup_n \|x_n\|_n$ (or its separable subspace of a sequences (x_n) with $\|x_n\|_n \to 0$), and the form $\sum_{n=1}^{\infty} n^{-2} Q_n$.

7.10.82. Prove that for every two given Hilbert spaces at least one is linearly isometric to a closed subspace of the other.

HINT: take orthonormal bases and compare their cardinalities.

7.10.83. Let H be a Hilbert space and let $A \in \mathcal{L}(H)$ have a dense range. Prove that the operator A is invertible precisely when $|A|$ is invertible. Give an example showing that this can be false if the range of A is not dense.

HINT: use the polar decomposition; consider the operator $x \mapsto (0, x_1, x_2, \ldots)$ on l^2.

7.10.84. (i) Let A be the operator on $C[0, 1]$ defined by the formula

$$Ax(t) = \frac{1}{t} \int_0^t x(s)\, ds, \quad Ax(0) = x(0).$$

Prove that for the spectral radius we have $r(A) = 1$.

(ii) Find eigenvalues of the operator defined by the formula above on the spaces $C[0, 1]$ and $L^2[0, 1]$ (see Exercise 6.10.138). Prove that in both cases it is not compact.

7.10.85. Let X be a Banach space, $A \in \mathcal{L}(X)$ and λ a boundary point of the spectrum of A. Prove that there exists a sequence of vectors x_n in the space X such that $\|x_n\| = 1$ and $\|Ax_n - \lambda x_n\| \to 0$.

HINT: if $\|Ax - \lambda x\| \geqslant \varepsilon > 0$ whenever $\|x\| = 1$, then $\|Ax - \lambda_n x\| \geqslant \varepsilon/2$ on the unit sphere for a sequence of points $\lambda_n \to \lambda$ from the resolvent set, so $\|(A - \lambda_n I)^{-1}\| \leqslant 2/\varepsilon$, hence $A - \lambda I$ must be invertible.

7.10.86. Let H be a Hilbert space, $A \in \mathcal{L}(H)$, and let $A = U|A|$ be the polar decomposition of A. Suppose that $A = WB$ and $B = W^*A$, where $B \in \mathcal{L}(H)$ is a nonnegative selfadjoint operator and $W \in \mathcal{L}(H)$ is an isometry on the closure of $B(H)$. Prove that $B = |A|$ and $W = U$ on the closure of $|A|(H)$.

7.10.87. Let μ be a bounded Borel measure on the real line and φ a bounded μ-measurable function. Prove that the operator A_φ of multiplication by φ on $L^2(\mu)$ is compact precisely when the restriction of μ to the set $\{t \colon \varphi(t) \neq 0\}$ is concentrated on a finite or countable set of points α_n and $\varphi(\alpha_n) \to 0$ if the set of these points is infinite.

HINT: use that the operator of multiplication by a function separated from zero is invertible and that in case of an atomless measure the L^2-space over a set of positive measure is infinite-dimensional.

7.10.88. Let μ be a bounded Borel measure on the real line and φ a bounded μ-measurable function. When does the operator A_φ of multiplication by φ on $L^2(\mu)$ have a closed range?

7.10.89. Construct an infinite measure μ and a bounded μ-measurable function φ for which the spectrum of the operator of multiplication by φ on $L^2(\mu)$ does not coincide with the set of essential values of φ.

7.10.90. Let A be a selfadjoint operator on a separable Hilbert space and Π_0 the corresponding resolution of the identity. Prove that $\sigma(A)$ coincides with the complement to the union of all intervals on which Π_0 is constant and also with the set of points λ such that $\Pi_0(\lambda) \neq \lim_{\lambda_n \downarrow \lambda} \Pi_0(\lambda_n)$.

HINT: it suffices to consider the case where A is the multiplication by the argument on the space $L^2(\mu)$ for some measure μ with support $\sigma(A)$ in some interval.

7.10.91.° Let A be a selfadjoint operator on an infinite-dimensional Hilbert space. Prove that A has nontrivial closed invariant subspaces.

HINT: consider the closed linear span of the orbit of a nonzero vector.

7.10.92. Consider the shift operator $Ux = (0, x_1, x_2, \ldots)$ on l^2. Prove that there is no compact operator $K \neq 0$ such that $UK = KU$.

HINT: use that $U^n K = KU^n$ and that the sequence $U^n x$ converges to zero weakly, hence $\|KU^n x\| \to 0$.

7.10.93. Prove that a selfadjoint operator $A \geqslant 0$ on a Hilbert space is compact precisely when for some $\alpha > 0$ (then for every $\alpha > 0$) the operator A^α is compact.

7.10.94. Let A be a selfadjoint operator on a Hilbert space H such that $A \geqslant 0$ and the range $A(H)$ is closed. Prove that the range $A^\alpha(H)$ is closed for all $\alpha > 0$. In case $\alpha \in \mathbb{N}$ prove the same without the assumption that A is nonnegative.

HINT: here A is a linear isomorphism of $A(H)$.

7.10.95. Let A be a selfadjoint operator with $A > 0$. Show that for any vector y the function $F(x) = (Ax, x) - 2\mathrm{Re}\,(x, y)$ attains its minimum $-(A^{-1}y, y)$ at $x_0 = A^{-1}y$.

HINT: use that $F(x) = (A(x - x_0), x - x_0) - (Ax_0, x_0)$.

7.10.96. Let A and B be selfadjoint operators on a Hilbert space H and $A \leqslant B$.
(i) Prove that $TAT^* \leqslant TBT^*$ for every $T \in \mathcal{L}(H)$.

(ii) Prove that if the operator A is invertible and $A \geqslant 0$, then the operator B is invertible and $B^{-1} \leqslant A^{-1}$.

(iii) Let $A \geqslant 0$. Show that for all $\alpha \in (0, 1)$ one has

$$A^\alpha = c_\alpha \int_0^{+\infty} t^{\alpha-1} A(A + tI)^{-1} \, dt.$$

(iv) Prove that if $A \geqslant 0$, then $A^\alpha \leqslant B^\alpha$ for all $\alpha \in (0, 1]$. Give an example where it is not true that $A^2 \leqslant B^2$.

HINT: to prove (ii) use the previous exercise. Check (iii) for operators of multiplication. Deduce (iv) from (iii) and (ii).

7.10.97. Prove that Theorem 7.2.3 and Corollary 7.2.4 are true for normal operators.

7.10.98. Prove the completeness of the space $L^2(\mu, H)$ introduced before Theorem 7.10.52. Prove the more general assertion from Theorem 6.10.68.

7.10.99. Let L be a linear subspace in a Hilbert space containing no infinite-dimensional closed subspaces. Prove that every operator $A \in \mathcal{L}(H)$ with $A(H) \subset L$ is compact.

HINT: reduce the assertion to the case of a separable space and a selfadjoint operator, then write A as a multiplication operator.

7.10.100. Let H be a Hilbert space and $A \in \mathcal{L}(H)$. Prove that a vector y belongs to the range $A(H)$ precisely when $\sup_x |(x, y)|/\|A^*x\| < \infty$, where $0/0 := 1$.

HINT: apply Theorem 7.10.15.

7.10.101. Let A_n and A be selfadjoint operators on a separable Hilbert space H and $f \in C_b(\mathbb{R}^1)$. (i) Suppose that $A_n \to A$ in the operator norm. Prove that $f(A_n) \to f(A)$ in the operator norm. In particular, if A_n and A are nonnegative, then $\sqrt{A_n} \to \sqrt{A}$ in the operator norm.

(ii) Suppose that $A_n x \to A x$ for all $x \in H$. Prove that $f(A_n)x \to f(A)x$ for all vectors $x \in H$. In particular, if A_n and A are nonnegative, then $\sqrt{A_n} \to \sqrt{A}$ in the strong operator topology.

HINT: observe that in both cases $\{A_n\}$ is norm bounded and verify the assertions for polynomials; in (ii) also observe that $A_n x_n \to A x$ if $x_n \to x$.

7.10.102. Let $A \in \mathcal{L}(H)$, where H is a Hilbert space. Show that A is a Hilbert–Schmidt operator precisely when there is a number C such that $\sum_{i=1}^n \|Ae_i\|^2 \leqslant C$ for every finite orthonormal collection $e_1, \ldots, e_n$. Moreover, $\|A\|_{\mathcal{H}}^2 \leqslant C$.

7.10.103. A Hilbert–Schmidt ellipsoid in a separable Hilbert space is the image of the closed unit ball with respect to a Hilbert–Schmidt operator.

(i) Prove that a set V is a Hilbert–Schmidt ellipsoid precisely when one can find an orthonormal sequence $\{\varphi_n\}$ and numbers $\alpha_n > 0$ with $\sum_{n=1}^\infty \alpha_n^2 < \infty$ such that

$$V = \{x \colon x = \textstyle\sum_{n=1}^\infty x_n \varphi_n, \ \sum_{n=1}^\infty |x_n/\alpha_n|^2 \leqslant 1\}.$$

(ii) (V. N. Sudakov) Prove that if a bounded set W is contained in no Hilbert–Schmidt ellipsoid, then there exists an orthonormal sequence $\{\varphi_n\}$ such that

$$\textstyle\sum_{n=1}^\infty \sup_{w \in W} |(w, \varphi_n)|^2 = \infty.$$

7.10.104. Let H be a Hilbert space and $A, B \in \mathcal{L}(H)$.

(i) Prove that $A(H) + B(H) = \sqrt{AA^* + BB^*}(H)$. In particular, if the operators A and B are selfadjoint and nonnegative, then $\sqrt{A + B}(H) = \sqrt{A}(H) + \sqrt{B}(H)$.

(ii) Let the operators A and B be selfadjoint and nonnegative and have closed ranges. Prove that the range of $A + B$ is closed precisely when the linear subspace $A(H) + B(H)$ is closed.

HINT: Apply Theorem 7.10.15.

7.10.105. Let H be a Hilbert space and let $A, B \in \mathcal{L}(H)$ be two operators such that $H = A(H) + B(H)$. (i) Prove that there exist closed subspaces $H_1 \subset A(H)$ and $H_2 \subset B(H)$ such that $H_1 + H_2 = H$, $H_1 \cap H_2 = 0$ and $H_1^\perp \cap H_2^\perp = 0$.

(ii) Let $A(H)$ and $B(H)$ be dense. Prove that the set $A(H) \cap B(H)$ is also dense.

HINT: see [**682**], [**683**].

7.10.106. Let K be a compact operator on a separable complex Hilbert space H and let $\{\lambda_n(K)\}$ be all nonzero eigenvalues of K written in the order of decreasing of their absolute values taking into account their multiplicities. (i) Prove that there is an orthonormal system $\{e_n\}$ such that $(Ke_n, e_n) = \lambda_n(K)$ for all n.

(ii) Let $\{s_n(K)\}$ be all eigenvalues of $|K|$. Prove that for every $p \in [1, \infty)$ Weyl's inequality $\sum_{n=1}^{\infty} |\lambda_n(K)|^p \leqslant \sum_{n=1}^{\infty} |s_n(K)|^p$ holds. Note that the inequality is also true for summing from 1 to any finite N (see [**554**, §1.6]).

7.10.107. Let H be a complex Hilbert space and $A \in \mathcal{L}(H)$. Prove that the operator A is normal (see §7.10(ii)) precisely when $\|Ax\| = \|A^*x\|$ for all $x \in H$.

HINT: observe that $(AA^*x - A^*Ax, x) = 0$.

7.10.108. Let T be a bounded operator on a Hilbert space H. Prove that it can be represented in the form $T = UA$, where U is a unitary operator and $A \geqslant 0$ is a selfadjoint operator, precisely when the subspaces $\operatorname{Ker} T$ and $\operatorname{Ker} T^*$ are isometric.

HINT: observe that $\operatorname{Ker} T = |T|(H)^\perp$ and $\operatorname{Ker} T^* = T(H)^\perp$.

7.10.109. Prove that a bounded operator T on a Hilbert space H is normal precisely when it is representable in the form $T = UA$, where U is a unitary operator, A is a nonnegative selfadjoint operator and $UA = AU$.

HINT: show that T admits a polar decomposition with a unitary operator.

7.10.110. Suppose that a bounded operator T on a Hilbert space H has the form $T = UA$, where U is unitary, A is selfadjoint, $A \geqslant 0$, and $UA \neq AU$. Prove that T is not normal.

7.10.111. Show that any bounded linear operator $A\colon L^2[0, 1] \to L^2[0, 1]$ has the form

$$Ax(t) = \frac{d}{dt} \int_0^1 \mathcal{K}(t, s)x(s)\, ds, \quad \text{where } \mathcal{K} \in L^2([0, 1] \times [0, 1]).$$

HINT: the composition of A and the Volterra operator is a Hilbert–Schmidt operator.

7.10.112. Let H be a separable Hilbert space. Prove that the set of operators having a left or right inverse is everywhere dense in $\mathcal{L}(H)$.

HINT: use the polar decomposition, observe that a nonnegative selfadjoint operator can be approximated by positive operators and consider partial isometries.

7.10.113. Let K be a compact operator on a complex Banach space X. Let us consider the operator $A = \lambda I + K$, where $\lambda \in \mathbb{C}$. Prove that for every $\varepsilon > 0$ there exists an invertible operator A_ε with $\|A - A_\varepsilon\| \leqslant \varepsilon$.

HINT: use that the spectrum of K is at most countable.

7.10.114. Give an example of a noncompact nonnegative selfadjoint operator A on a separable Hilbert space such that there exists an orthonormal basis $\{e_n\}$ for which $\lim_{n \to \infty} \|Ae_n\| = 0$.

HINT: consider the operator on l^2 defined by the infinite matrix of the following form: its nonzero elements are diagonal blocks, the nth of which is of size $n \times$ with all entries $1/n$; see also [251, p. 273].

7.10.115. (i) Let $e_1, \ldots, e_n$ be the standard basis in $\mathbb{C}^n$ or $\mathbb{R}^n$. Then the function $\|A\|_{\{e_i\}} = \sum_{i=1}^n \|Ae_i\|$ is a norm on the space of operators. Show that for the nuclear norm (the sum of eigenvalues of $\sqrt{A^*A}$) one has $\|A\|_{\{e_i\}} \leqslant \sqrt{n}\|A\|_1$ and there is a symmetric nonzero operator A for which there holds the equality.

(ii) Give an example of a nuclear operator $A \geqslant 0$ on a complex separable Hilbert space such that there exists an orthonormal basis $\{e_n\}$ for which $\sum_{n=1}^\infty \|Ae_n\| = +\infty$.

HINT: (i) take any orthonomal basis $\{\varphi_i\}$ such that φ_1 has equal coordinates $n^{-1/2}$ and the operator A defined by $A\varphi_1 = \varphi_1$, $A\varphi_j = 0$ if $j > 1$; then $\|A\|_1 = 1$ and $\|Ae_i\| = n^{-1/2}$ for all i, hence $\|A\|_{\{e_i\}} = n^{1/2}$. On the other hand, if α_i are eigenvalues of $\sqrt{A^*A}$, then $\|A\|_{\{e_i\}} \leqslant n^{1/2}\left(\sum_{i=1}^n \alpha_i^2\right)^{1/2}$, which is estimated by $n^{1/2}\sum_{i=1}^n \alpha_i$. An example in (ii) is easily constructed by using (i).

7.10.116. Give an example of a bounded operator A on the *real* space l^2 that is not nuclear, although there exists an orthonormal basis $\{e_n\}$ such that $(Ae_n, e_n) = 0$ and hence $\sum_{n=1}^\infty |(Ae_n, e_n)| < \infty$.

HINT: write l^2 as a countable sum of two-dimensional planes in which A acts as rotations by $\pi/2$.

7.10.117. (i) Let E be a closed linear subspace in $L^2[0,1]$ such that $E \subset L^\infty[0,1]$. Prove that $\dim E < \infty$.

HINT: apply Corollary 7.10.28 to obtain the compactness of the projection onto E.

7.10.118. Let H be an infinite-dimensional complex separable Hilbert space and let $\mathcal{U}(H)$ be the set of all unitary operators on H. (i) Investigate whether the space $\mathcal{U}(H)$ with the operator norm is connected (a space is connected if it cannot be decomposed into nonempty disjoint open parts). (ii) Find the closure of $\mathcal{U}(H)$ in the weak operator topology and in the strong operator topology.

HINT: (i) represent a given unitary operator U as $\exp(iA)$ with a bounded selfadjoint operator A and consider the family $\exp(itA)$. (ii) Observe that for every operator A with $\|A\| \leqslant 1$ and any two finite orthonormal collections $e_1, \ldots, e_n$ and $\varphi_1, \ldots, \varphi_n$, one can find a unitary operator U such that $(Ue_i, \varphi_j) = (Ae_i, \varphi_j)$, $i, j = 1, \ldots, n$. Observe also that if $\|Tx\| = \|x\|$ for all x, then on every finite-dimensional subspace T coincides with a unitary operator.

7.10.119. Let X and Y be Banach spaces and let p be a cross-norm on the tensor product $X \otimes Y$, i.e., $p(x \otimes y) = \|x\| \|y\|$. Prove the inequality $\|u\|_\infty \leqslant p(u) \leqslant \|u\|_\mathcal{N}$.

7.10.120. Show that a Fredholm operator $T \in \mathcal{L}(X, Y)$ between Banach spaces X and Y has zero index precisely when $T = S + K$, where S is invertible and K is compact, moreover, in this case K can be chosen finite-dimensional.

7.10.121. Suppose that X and Y are Banach spaces, $T_n \in \mathcal{L}(X, Y)$, $S_n \in \mathcal{L}(Y, X)$, the operators T_n converge pointwise to T, and the operators S_n converge pointwise to S. Suppose that $S_n T_n = I + K_n$, $T_n S_n = I + H_n$, where the operators K_n and H_n are uniformly compact in the sense that if U_X is the unit ball of X and U_Y is the unit ball of Y, then the sets $K_n(U_X)$ are contained in a common compact set and similarly for the sets $H_n(U_Y)$. Prove that $\operatorname{Ind} T = \lim_{n \to \infty} \operatorname{Ind} T_n$.

HINT: see [281, Theorem 19.1.10].

7.10.122. Let T and S be nuclear operators on a separable Hilbert space H and let $A, B \in \mathcal{L}(H)$ be such that $AB = I - T$, $BA = I - S$. Prove that $\operatorname{Ind} A = \operatorname{tr} S - \operatorname{tr} T$.
HINT: see [**281**, Proposition 19.1.14].

7.10.123. Suppose that A is a nuclear operator on a separable Hilbert space H and operators $P_n, Q_n \in \mathcal{L}(H)$ converge pointwise to I. Prove that $\operatorname{tr}(P_n A Q_n^*) \to \operatorname{tr} A$.

7.10.124.* Let V be the Volterra operator of indefinite integration on $L^2[0,1]$. Let H_a be the subspace of functions vanishing on $[0,a]$, $a \in [0,1]$. Prove that V has no closed invariant subspaces different from H_a.
HINT: see [**378**, p. 280].

7.10.125. (Gleason's theorem) Let H be a separable complex Hilbert space, $\mathcal{P}$ the set of all orthogonal projections in H and $\mu \colon \mathcal{P} \to [0,1]$ a function such that $\mu(P) \leqslant \mu(Q)$ whenever $P \leqslant Q$ and $\mu\left(\sum_{i=1}^\infty P_i\right) = \sum_{i=1}^\infty \mu(P_i)$ if orthogonal projections P_i are pairwise orthogonal. Prove that $\mu(P) = \operatorname{tr} SP$, where S is a nonnegative trace class operator.

7.10.126. (Wigner's theorem) Let H be a separable complex Hilbert space, $\mathcal{P}$ the set of all orthogonal projections in H and $\xi \colon \mathcal{P} \to \mathcal{P}$ a mapping such that $\xi(I) = I$, $\xi(P) \leqslant \xi(Q)$ whenever $P \leqslant Q$ and for every sequence of pairwise orthogonal projections P_i the projections $\xi(P_i)$ are pairwise orthogonal and $\xi\left(\sum_{i=1}^\infty P_i\right) = \sum_{i=1}^\infty \xi(P_i)$. Prove that $\xi(P) = UPU^{-1}$, where U is a real-linear isometry that is either complex-linear or conjugate-linear.

7.10.127. Let X be a complex Banach space, $A \in \mathcal{L}(X)$ and let λ_0 be a pole of order m of the operator function $\lambda \mapsto R_\lambda(A) = (A - \lambda I)^{-1}$. Prove that λ_0 is an eigenvalue and for all $n \geqslant m$ one has the decomposition $X = \operatorname{Ker}(A - \lambda_0 I)^n \oplus (A - \lambda_0 I)^n(X)$.
HINT: see [**640**, Chapter VIII, §8].

7.10.128. Let H be a separable Hilbert space that is a dense linear subspace in a Banach space E such that the identity mapping $H \to E$ is continuous. Then the continuous embedding $E^* \to H^* = H$ defines the so-called triple of spaces $E^* \subset H \subset E$.

(i) Let A be a compact selfadjoint operator on H such that $A(H) \subset E^*$ and A is compact as an operator with values in E^*. Prove that the expansion $Ax = \sum_{n=1}^\infty \alpha_n(x, e_n)e_n$ in the eigenbasis $\{e_n\}$ of A converges with respect to the operator norm on $\mathcal{L}(H, E^*)$.

(ii) Suppose in addition that A extends to a compact operator from the space E to E^*. Prove that the indicated expansion also converges with respect to the operator norm on $\mathcal{L}(E, E^*)$.

(iii) Let (Ω, μ) be a probability space and let $\mathcal{K} \colon \Omega \times \Omega \to \mathbb{R}^1$ be a bounded measurable function such that $\mathcal{K}(t,s) = \mathcal{K}(s,t)$. Let $\{\lambda_i\}$ and $\{e_i\}$ be the eigenvalues and eigenfunctions of the operator T on $L^2(\mu)$ given by the integral kernel $\mathcal{K}$. Suppose that T is compact also as an operator from $L^1(\mu)$ to $L^\infty(\mu)$. Prove that

$$\lim_{n \to \infty} \operatorname{esssup}_{t \in \Omega} \operatorname{esssup}_{s \in \Omega} \left| \mathcal{K}(t,s) - \sum_{i=1}^n \lambda_i e_i(t) e_i(s) \right| = 0.$$

HINT: see [**349**, Chapter III, §9].

7.10.129. Let A be a selfadjoint operator on a separable Hilbert space H with the resolution of the identity Π_0, $T \in \mathcal{L}(H)$.

(i) Prove that $AT = TA$ precisely when $T\Pi_0(\lambda) = \Pi_0(\lambda)T$ for all $\lambda \in \mathbb{R}^1$.

(ii) Prove that T commutes with all bounded operators commuting with A precisely when $T = f(A)$, where f is a bounded Borel function.

HINT: (i) use that in the strong operator topology A is a limit of linear combinations of operators $\Pi_0(\lambda)$ and that $\Pi_0(\lambda)$ is a limit of polynomials of A. (ii) Use Lemma 7.10.5 and observe that A is a function of a selfadjoint operator possessing a cyclic vector.

7.10.130. Prove that the identity embedding $l^1 \to l^2$ is absolutely summing.

7.10.131. Let S be a countable union of compact sets in l^2. Prove that there is a unitary operator U (orthogonal in the real case) such that $U(S) \cap S$ is either empty or the zero element.

HINT: observe that for compact sets K_n there are numbers $\varepsilon_n > 0$ such that the union of $\varepsilon_n K_n$ has compact closure, hence is contained in a compact ellipsoid and consequently in the range of a compact operator, then apply Theorem 7.10.18.

7.10.132. (Kalisch [**691**]) Let us consider the operator

$$Kx(t) = tx(t) - \int_0^t x(s) \, ds$$

on the real space $L^2[0,1]$. Prove that for every nonempty compact set S in $\mathbb{R}^1$ there exists a closed subspace $H \subset L^2[0,1]$ such that $K(H) \subset H$ and the spectrum of K on H coincides with S and consists of eigenvalues.

7.10.133. Let E be a Banach lattice, $A \in \mathcal{L}(E)$ and $A \geqslant 0$ (see §6.10(v)). (i) Show that the spectral radius $r(A)$ is contained in $\sigma(A)$. (ii) Prove the Krein–Rutman theorem: if the operator A is compact and its spectral radius $r(A)$ is positive, then $r(A)$ is an eigenvalue and there exists an eigenvector $v \geqslant 0$.

HINT: see [**141**, Theorem 19.2], [**533**, p. 265], [**534**, Chapter V].

7.10.134. Let H be a complex Hilbert space and let $A \in \mathcal{L}(H)$. Prove that the set $\{(Ax, x): \|x\| = 1\}$ is convex in $\mathbb{C}$.

HINT: see [**251**].

7.10.135. (F. A. Sukochev) Let A and B be two nonnegative nuclear operators on a Hilbert space H. Then the operator $\sqrt{A^2 + B^2}$ is nuclear. More generally, if $A_1, \ldots, A_n$ are nonnegative nuclear operators, then the operator $\sqrt{A_1^2 + \cdots + A_n^2}$ is nuclear.

HINT: It suffices to prove the second assertion and apply induction on n. It is readily seen that for any 2×2-matrix M the operators $M \otimes A$ and $M \otimes B$ are nuclear on $\mathbb{C}^2 \otimes H$. Let e_{ij} denote the 2×2-matrix with zero entries except for 1 at the intersection of the array number i and the column number j. Then the operator $T = e_{11} \otimes A + e_{21} \otimes B$ is nuclear. Since $T^* = e_{11} \otimes A + e_{12} \otimes B$, we have $T^*T = e_{11} \otimes (A^2 + B^2)$. Hence $|T| = e_{11} \otimes \sqrt{A^2 + B^2}$ is nuclear. It follows that $\sqrt{A^2 + B^2}$ is nuclear as well.

7.10.136. (Brown, Pearcy [**678**]) Let A and B be bounded operators on a Hilbert space H. Then the spectrum of $A \otimes B$ on the Hilbert tensor product is $\sigma(A)\sigma(B)$.

7.10.137. (Brown, Pearcy [**677**]) Let H be a complex infinite-dimensional separable Hilbert space. An operator $A \in \mathcal{L}(H)$ is a commutator, i.e., has the form $A = ST - TS$ with $S, T \in \mathcal{L}(H)$, if and only if it is not of the form $\lambda I + K$, where $\lambda \neq 0$ and K is a compact operator.

7.10.138. (Brown, Pearcy, Salinas [**680**]) If T is a noncompact bounded linear operator on a separable Hilbert space H, then there exists a bounded nilpotent operator N on H (even with $N^3 = 0$) such that $N + T$ is invertible.

7.10.139. (Brown, Pearcy [**679**]) Every bounded operator T on a Hilbert space H with $\dim H > 1$ has the form $T = PAQ - QAP$, where P, A and Q are invertible operators.

7.10.140. Let A and B be positive operators on a Hilbert space H. Show that there is a positive operator T such that $B = TAT$.

HINT: consider $T = A^{-1/2}\big(A^{1/2}BA^{1/2}\big)^{1/2}A^{-1/2}$.

CHAPTER 8

Locally Convex Spaces and Distributions

In this chapter we study linear spaces more general than normed spaces. For many problems in applications the framework of normed spaces turns out to be too stringent; a typical example is the space of infinitely differentiable functions that possesses its natural convergence, which, however, cannot be described by means of a norm. Some elements of this theory have already been encountered in our discussion of weak topologies. In this chapter we also consider generalized functions or distributions, an important tool in the theory of partial differential equations and mathematical physics.

8.1. Locally Convex Spaces

Let E be a real or complex linear space and let $\{p_\alpha\}$ be some (nonempty) collection of seminorms on E such that for every $x \neq 0$ there exists $\alpha = \alpha(x)$ with $p_\alpha(x) > 0$. The simplest example is a normed space with the family of seminorms consisting of its single norm.

Let us consider sets of the form

$$U_{\alpha_1,\ldots,\alpha_n;\varepsilon}(a) = \{x \in E\colon p_{\alpha_i}(x - a) < \varepsilon,\ i = 1,\ldots,n\}, \qquad (8.1.1)$$

where $\varepsilon > 0$, $a \in E$, $p_{\alpha_i} \in \{p_\alpha\}$. In the case of a normed space with norm p and $p_\alpha = p$ such sets are open balls. Arbitrary unions of such sets (unions in an arbitrary number, where one can vary points a, numbers n, indices α_i and numbers ε) and the empty set will be called open in E. We have obtained a Hausdorff topology in E. The verification of this is similar to the case of the topology $\sigma(E, F)$ considered earlier. Indeed, by definition arbitrary unions of these open sets are open. In order to prove that the intersection of two open sets is open it suffices to show that the intersection of any two sets $U := U_{\alpha_1,\ldots,\alpha_n;\varepsilon}(a)$ and $U' := U_{\beta_1,\ldots,\beta_k;\delta}(b)$ is open. To this end it suffices to verify that every point $x \in U \cap U'$ enters this intersection with a neighborhood

$$V := U_{\alpha_1,\ldots,\alpha_n,\beta_1,\ldots,\beta_k;\varepsilon_1}(x) \subset U \cap U'$$

for some $\varepsilon_1 > 0$. The inclusion $x \in U \cap U'$ gives the estimates

$$p_{\alpha_i}(x - a) < \varepsilon \quad \text{and} \quad p_{\beta_j}(x - b) < \delta.$$

V. I. Bogachev, O. G. Smolyanov, *Real and Functional Analysis*,
Moscow Lectures 4, https://doi.org/10.1007/978-3-030-38219-3_8

Set $\varepsilon_1 := \min\limits_{i \leqslant n, j \leqslant k} \big(\varepsilon - p_{\alpha_i}(x - a), \delta - p_{\beta_j}(x - b)\big)$. Let $y \in V$. The triangle inequality yields

$$p_{\alpha_i}(y - a) \leqslant p_{\alpha_i}(y - x) + p_{\alpha_i}(x - a) < \varepsilon_1 + p_{\alpha_i}(x - a) < \varepsilon.$$

A similar estimate holds for $p_{\beta_j}(y - b)$. The Hausdorff property of the topology obtained follows from the fact that for every pair of different points x and y there exists p_α with $p_\alpha(x - y) = c > 0$. By the triangle inequality this gives the desired equality $U_{\alpha, c/4}(x) \cap U_{\alpha, c/4}(y) = \varnothing$.

8.1.1. Definition. *The space E is called a locally convex space with the topology generated by the family of seminorms p_α.*

It is seen directly from the definition that for every neighborhood U of any point a the set $U - a$ is a neighborhood of zero and the set $U - a + b$ is a neighborhood of the point b. The neighborhoods of zero (8.1.1) are convex and balanced.

It is sometimes convenient to speak of a basis $\mathcal{W}$ of closed convex balanced neighborhoods of zero: here it is required that every set in $\mathcal{W}$ must contain an open neighborhood of zero and that every open neighborhood of zero must contain a set from $\mathcal{W}$. For example, for $\mathcal{W}$ one can take the closures of the sets from a basis of open convex balanced neighborhoods of zero (see Exercise 8.6.38).

8.1.2. Remark. In a normed space two equivalent norms define the same topology. An analog of this in a locally convex space X with the topology given by seminorms $\{p_\alpha\}$ is the following assertion: a collection of seminorms $\{q_\beta\}$ on X defines the same topology precisely when for every seminorm q_β there exists a number C_β and seminorms $p_{\alpha_1}, \ldots, p_{\alpha_n}$ with $q_\beta \leqslant C_\beta[p_{\alpha_1} + \cdots + p_{\alpha_n}]$ and, conversely, for every seminorm p_α there exist a number C_α and seminorms $q_{\beta_1}, \ldots, q_{\beta_k}$ with $p_\alpha \leqslant C_\alpha[q_{\beta_1} + \cdots + q_{\beta_k}]$. Indeed, the coincidence of two topologies is equivalent to the property that every neighborhood of zero of the form (8.1.1) in one of the two systems of seminorms must contain a neighborhood from the other system. This is exactly the indicated condition.

Let us consider some typical examples. In addition to normed spaces, we have already encountered an important class of examples: spaces E with the topology $\sigma(E, F)$. For seminorms we take here functions $|f|$, where $f \in F$. Generally speaking, one of the purposes of introducing seminorms is to introduce quantitative characteristics of a natural convergence in a given functional space.

8.1.3. Example. (i) The space $C_b^\infty(0, 1)$ of all real infinitely differentiable functions on $(0, 1)$ with finite seminorms

$$p_k(\varphi) := \sup_{t \in (0,1)} |\varphi^{(k)}(t)|.$$

Similarly one introduces the space $C_b^\infty(U)$ of all real infinitely differentiable functions on an open set $U \subset \mathbb{R}^n$ with finite seminorms

$$p_k(\varphi) := \sup_{x \in U} \|\varphi^{(k)}(x)\|.$$

Convergence in $C_b^\infty(U)$ is the uniform convergence on U of the derivatives of every fixed order. The broader space $C^\infty(U)$ of all infinitely differentiable functions on an open set $U \subset \mathbb{R}^n$ is equipped with seminorms

$$p_{k,S}(\varphi) := \sup_{x \in S} \|\varphi^{(k)}(x)\|,$$

where S is a compact set in U. Convergence in $C^\infty(U)$ is the uniform convergence of the derivatives of every fixed order on every compact set in U. On all of U these functions need not be even bounded.

(ii) Let U be an open set in $\mathbb{C}$. The complex space $H(U)$ of all functions holomorphic in U is equipped with the system of seminorms

$$p_S(\varphi) := \sup_{z \in S} |\varphi(z)|,$$

where S is a compact set in U. Convergence in $H(U)$ is the uniform convergence on compact subsets of U. It is known from complex analysis that this convergence yields also the uniform convergence of the derivatives on all compact sets. Moreover, if we add to the seminorms p_S also the seminorm $p_{k,S}(\varphi) = \sup_{x \in S} \|\varphi^{(k)}(x)\|$, then the topology does not change, because for every compact set $S \subset U$ and every $k \in \mathbb{N}$ there exist a compact set $S' \subset U$ and a number $C = C(S, U, k) > 0$ such that $p_{k,S}(\varphi) \leqslant C p_{S'}(\varphi)$ for all $\varphi \in H(U)$. This is easily verified with the aid of the Cauchy formula.

(iii) The space $\mathcal{S}(\mathbb{R}^1)$ consists of all infinitely differentiable functions φ on $\mathbb{R}^1$ with finite seminorms

$$p_{k,m}(\varphi) := \sup_{x \in \mathbb{R}^1} (1 + |x|^2)^m |\varphi^{(k)}(x)|,$$

where m, k are nonnegative integer numbers. Thus, the space $\mathcal{S}(\mathbb{R}^1)$ consists of smooth functions all derivatives of which decrease at infinity faster than every negative power. The same topology can be defined by means of the seminorms (which are norms, as $p_{k,m}$)

$$p_m(\varphi) := \sup_{k \leqslant m} \sup_{x \in \mathbb{R}^1} (1 + |x|^2)^m |\varphi^{(k)}(x)|.$$

Similarly one introduces the space $\mathcal{S}(\mathbb{R}^n)$ of functions on $\mathbb{R}^n$. Here in place of $|\varphi^{(k)}(x)|$ we take either the norm of the k-linear mapping $\varphi^{(k)}(x)$, where $\|\varphi^{(k)}(x)\| := \sup_{|h_i| \leqslant 1} |\varphi^{(k)}(x)(h_1, \ldots, h_k)|$, or $\max_{|\alpha| \leqslant k} |\partial^{(\alpha)} \varphi(x)|$, where

$$\partial^{(\alpha)} \varphi := \partial_{x_1}^{(\alpha_1)} \cdots \partial_{x_n}^{(\alpha_n)} \varphi, \quad \alpha = (\alpha_1, \ldots, \alpha_n), \quad |\alpha| := \alpha_1 + \cdots + \alpha_n.$$

Since $|\partial^{(\alpha)} \varphi(x)| \leqslant \|\varphi^{(|\alpha|)}(x)\|$ and $\|\varphi^{(k)}(x)\|$ is estimated by the sum of the functions $|\partial^{(\alpha)} \varphi(x)|$ over all $|\alpha| = k$, we obtain equivalent seminorms.

(iv) The space $\mathcal{D}_m(\mathbb{R}^1)$, $m \in \mathbb{N}$, consists of all infinitely differentiable functions vanishing outside $[-m, m]$. It is equipped with the countable system of seminorms (actually norms)

$$p_k(\varphi) := \max_t |\varphi^{(k)}(t)|.$$

In place of all norms p_k one can take an arbitrary countable part of them, since $|\varphi^{(k)}(t)| \leqslant 2m \max_t |\varphi^{(k+1)}(t)|$ by the mean value theorem. Similarly one defines the space $\mathcal{D}([a,b])$ of all smooth functions on the real line vanishing outside $[a,b]$, equipped with the same norms p_k. The space $\mathcal{D}_m(\mathbb{R}^n)$ is introduced precisely in the same way: in place of $[-m,m]$ we take the ball $|x| \leqslant m$. In the multidimensional case, as in (iii), in place of the derivatives $\varphi^{(k)}(x)$ as multilinear functions one can use the partial derivatives, by setting

$$p_k(\varphi) := \max_{k_1 + \cdots + k_n = k} \max_{|x| \leqslant m} |\partial_{x_1}^{k_1} \cdots \partial_{x_n}^{k_n} \varphi(x)|.$$

Moreover, one can take an even smaller collection of norms

$$q_k(\varphi) := \sup_{|x| \leqslant m} |\partial_{x_1}^k \cdots \partial_{x_n}^k \varphi(x)|.$$

Indeed, by the mean value theorem we have

$$\sup_x |\partial_{x_1}^{k_1} \cdots \partial_{x_n}^{k_n} \varphi(x)| \leqslant 2m \sup_x |\partial_{x_i} \partial_{x_1}^{k_1} \cdots \partial_{x_n}^{k_n} \varphi(x)|.$$

Hence the left-hand side can be estimated by $(2m)^d q_l$, where l is the maximal number in $k_1, \ldots, k_n$ and $d = \sum_{i=1}^n (l - k_i)$.

(v) The space $\mathcal{D}(\mathbb{R}^1)$ consists of all infinitely differentiable functions on the real line with bounded support, i.e., $\mathcal{D}(\mathbb{R}^1) = \bigcup_{m=1}^\infty \mathcal{D}_m(\mathbb{R}^1)$. It is equipped with the seminorms

$$p_{\{a_k\}}(\varphi) = \sum_{k=-\infty}^\infty a_k \max_{m=0,\ldots,a_k} \max_{x \in [k, k+1]} |\varphi^{(m)}(x)|, \tag{8.1.2}$$

where for $\{a_k\}$ we take all possible collections of nonnegative integer numbers a_k, $k \in \mathbb{Z}$. Similarly we define the space $\mathcal{D}(\mathbb{R}^n)$: in place of $|\varphi^{(m)}(x)|$ we take either $\|\varphi^{(m)}(x)\|$ or the maximum of numbers $|\partial^{(\alpha)}\varphi(x)|$ with $|\alpha| = m$. Of course, $x \in [k, k+1]$ is replaced by $k \leqslant |x| \leqslant k+1$.

As one can see from §8.6(i), the topologies of the spaces in (i)–(iv) are generated by metrics, but not by norms, and the topology of $\mathcal{D}$ is not metrizable (see also Exercise 8.6.39). In the sequel we shall only need convergence of sequences in $\mathcal{D}$, so we define it without topologies.

8.1.4. Definition. *A sequence $\{\varphi_j\} \subset \mathcal{D}(\mathbb{R}^1)$ converges to $\varphi \in \mathcal{D}(\mathbb{R}^1)$ in $\mathcal{D}(\mathbb{R}^1)$ if all functions φ_j and φ vanish outside some common compact set and for every nonnegative integer k the uniform convergence $\varphi_j^{(k)} \rightrightarrows \varphi^{(k)}$ holds. Similarly we introduce convergence in $\mathcal{D}(\mathbb{R}^n)$.*

8.1.5. Remark. A sequence $\{\varphi_j\}$ converges to φ in $\mathcal{D}(\mathbb{R}^1)$ in the sense of this definition if and only if it converges in the locally convex topology τ on $\mathcal{D}(\mathbb{R}^1)$ generated by the seminorms indicated in (v). Indeed, convergence in the topology τ obviously follows from the convergence defined above, since by the condition that all φ_j and φ vanish outside some interval $[-n,n]$ the series (8.1.2) with $\varphi - \varphi_j$ in place of φ becomes a finite sum. The converse assertion will be proved once we establish the existence of a common interval outside of which all functions from a sequence $\{\varphi_j\}$ converging in the topology τ vanish. We can assume that the limit function is zero. Suppose that for every n there exists j_n with $\varphi_{j_n}(t_n) \neq 0$, where $t_n \in [k_n, k_n + 1]$, $k_n \in \mathbb{Z}$ and either $k_n \geqslant n$ or $k_n \leqslant -n-1$.

Then there is $a_{k_n} \in \mathbb{N}$ such that $a_{k_n} |\varphi_{j_n}(t_n)| > 1$, which gives a contradiction. The reader should be warned that the locally convex topology introduced above is not the only locally convex topology generating the indicated convergence of sequences; moreover, there are also not locally convex topologies with the same supply of convergent sequences. All these subtleties, which are not needed for the sequel, are discussed in Exercises 8.6.45 and 8.6.83.

In §8.6(i) we give simple criteria of the metrizability and normability of a locally convex space. The class of all seminorms on X defines the strongest locally convex topology.

There is another, more geometric, way of introducing locally convex topologies on linear spaces.

8.1.6. Definition. (i) *A real linear space E with a Hausdorff topology τ is called a topological vector space if the operations $E \times E \to E$, $(x, y) \mapsto x + y$ and $\mathbb{R}^1 \times E \to E$, $(t, x) \mapsto tx$ are continuous with respect to the natural topology on $E \times E$ and $\mathbb{R}^1 \times E$. A complex topological vector space is defined similarly.*

(ii) *A topological vector space E is called locally convex if every neighborhood of zero contains a convex neighborhood of zero.*

Every neighborhood of zero U in a topological vector space contains a balanced neighborhood of zero W. Indeed, there exist $\varepsilon > 0$ and a neighborhood of zero V such that $\lambda v \in U$ whenever $|\lambda| \leqslant \varepsilon$ and $v \in V$. The set $W_0 = \bigcap_{|\mu| \geqslant 1} \mu U$ is obviously balanced and contains a neighborhood of zero εV. According to Exercise 8.6.37 the interior W of the set W_0 is balanced. It is clear that $\varepsilon V \subset W \subset U$.

A locally convex space in the sense of our first definition is locally convex in the sense of part (ii) of the definition above. Indeed, sets of the form (8.1.1) with $a = 0$ are convex neighborhoods of zero, moreover, every neighborhood of zero contains a set of such a form. The continuity of the addition operation follows from the inclusion

$$U_{\alpha_1,\ldots,\alpha_n;\varepsilon/2}(0) + U_{\alpha_1,\ldots,\alpha_n;\varepsilon/2}(0) \subset U_{\alpha_1,\ldots,\alpha_n;\varepsilon}(0),$$

following in turn from the triangle inequality. Similarly one verifies the continuity of multiplication by scalars. It is less obvious that the topology of a locally convex space E in the sense of the second definition can be generated by means of a family of seminorms. To see this, we first observe that every convex neighborhood of zero V contains a convex balanced neighborhood of zero W. Indeed, as shown above, every convex neighborhood of zero U contains a balanced neighborhood of zero V. The convex envelope of the set V is convex, open and belongs to U. It is readily seen that it is balanced. It is easy to show that the Minkowski functionals of such neighborhoods W form a family of seminorms generating the original locally convex topology.

In topological vector spaces there is a special definition of boundedness (coinciding with the usual one for normed spaces, but not for general metric spaces!).

8.1.7. Definition. *A set A in a topological vector space E is called bounded if for every neighborhood of zero U there is $\lambda > 0$ such that $A \subset \lambda U$.*

We emphasize that if E is metrizable, then the boundedness with respect to the metric does not always yield the boundedness in the indicated sense. For example, the real line can be metrized by a bounded metric, which does not make it a bounded topological vector space. If E has a bounded neighborhood of zero, then E is metrizable (see [**533**, Chapter I, §6.2]) and if there exists a convex bounded neighborhood of zero, then E is normable (see Theorem 8.6.3). If a topological vector space E is metrizable, then E is called a *metric linear space*. In this case a metric can be made invariant with respect to translations, i.e., $d(x, y) = d(x - y, 0)$ (see §8.6(i)). A complete metrizable locally convex space is called a *Fréchet space*.

A linear subspace of a locally convex space is locally convex with the same seminorms.

8.2. Linear Mappings

The basic properties of linear functionals and operators on locally convex spaces are similar to their properties on normed spaces.

8.2.1. Lemma. *If a linear mapping* $F\colon X \to Y$ *between topological vector spaces is continuous at some point, then it is continuous at every point.*

PROOF. Let $a, b \in X$ and let F be continuous at the point a. If V is a neighborhood of $F(b)$, then $W = V - F(b) + F(a)$ is a neighborhood of $F(a)$. By our condition there exists a neighborhood U of the point a such that $F(U) \subset W$. Then $U - a + b$ is a neighborhood of the point b and

$$F(U - a + b) = F(U) - F(a) + F(b) \subset W - F(a) + F(b) = V,$$

which shows the continuity at the point b. $\square$

8.2.2. Theorem. *Let X be a locally convex space with the topology generated by a family of seminorms* $\{p_\alpha\}$. *A linear function f on X is continuous precisely when there exist a finite subfamily* $p_{\alpha_1}, \ldots, p_{\alpha_n}$ *and a number C for which*

$$|f(x)| \leqslant C[p_{\alpha_1}(x) + \cdots + p_{\alpha_n}(x)]$$

for all $x \in X$.

PROOF. The indicated estimate yields the continuity of f at zero, hence at every other point. If the function f is continuous, then the set $\{x\colon |f(x)| < 1\}$ contains a neighborhood of zero of the form $\{x\colon p_{\alpha_1}(x) < \varepsilon, \ldots, p_{\alpha_n}(x) < \varepsilon\}$. Therefore, the condition $p_{\alpha_1}(x) + \cdots + p_{\alpha_n}(x) < \varepsilon$ yields the inequality $|f(x)| < 1$. Hence for C we can take $C = \varepsilon^{-1}$. $\square$

Exercise 8.6.40 contains the following generalization of this theorem.

8.2.3. Theorem. *Let X and Y be locally convex spaces whose topologies are generated by collections of seminorms* $\{p_\alpha\}$ *and* $\{q_\beta\}$, *respectively. A linear mapping* $F\colon X \to Y$ *is continuous if and only if for every β there exist a finite family* $p_{\beta,\alpha_1}, \ldots, p_{\beta,\alpha_n}$ *in* $\{p_\alpha\}$ *and a number C_β for which*

$$q_\beta(F(x)) \leqslant C_\beta[p_{\beta,\alpha_1}(x) + \cdots + p_{\beta,\alpha_n}(x)], \quad x \in X.$$

As in the case of normed spaces, the Hahn–Banach theorem implies that the dual to a locally convex space is not trivial.

8.2.4. Theorem. *Let X be a locally convex space and let X_0 be its linear subspace. Every continuous linear functional on X_0 extends to a continuous linear functional on all of X.*

PROOF. For any continuous functional f on X_0 Theorem 8.2.2 gives seminorms $p_{\alpha_1}, \ldots, p_{\alpha_n}$ from a family generating the topology and a number C such that $|f| \leqslant C[p_{\alpha_1} + \cdots + p_{\alpha_n}]$ on X_0. By the Hahn–Banach theorem f possesses a linear extension to X satisfying the same inequality on X, which means the continuity of the extension. $\square$

8.2.5. Corollary. *For any linearly independent vectors $x_1, \ldots, x_n$ in a locally convex space X, there exist continuous linear functions $l_1, \ldots, l_n$ on X such that one has $l_i(x_j) = \delta_{ij}$.*

It is easy to verify that the space $L^0(\lambda)$, where λ is Lebesgue measure on $[0, 1]$, with the metric from Exercise 3.12.29 is a topological vector space, but it has the zero dual.

The following frequently used result says that in a locally convex space bounded sets in the original topology and in the weak topology are the same.

8.2.6. Theorem. *A set A in a locally convex space E is bounded precisely when it is weakly bounded, i.e., $\sup_{x \in A} |f(x)| < \infty \quad \forall f \in E^*$.*

PROOF. It is clear that the boundedness in the original topology yields the weak boundedness. Conversely, let A be weakly bounded and let p be a continuous seminorm on E. Let us consider the normed factor-space E/E_0, $E_0 = p^{-1}(0)$, equipped with the norm generated by p (see §5.1). We have to show the boundedness of the image of A under the natural mapping from E onto E/E_0. By assumption, for every $c \geqslant 0$ every linear functional f for which $|f| \leqslant cp$ is bounded on A. This means the weak boundedness in E/E_0 of the image of A. It remains to apply the result already known for normed spaces. $\square$

For some classes of topological vector spaces, the most important theorems about linear mappings such as the Banach–Steinhaus theorem, the closed graph theorem, and the open mapping theorem still hold. A detailed discussion of these questions is beyond our book (see Bogachev, Smolyanov [75]), but we mention one very elegant result due to P. P. Zabreiko [712], which enables one to extend the aforementioned theorems in a simple and unified manner to complete metric linear spaces (of course, it is also possible to use suitable modifications of the reasoning given in the case of a Banach space); see also §8.6(iv).

8.2.7. Theorem. (ZABREIKO'S THEOREM) *Let X be a second category metric linear space (for example, complete) and let $\psi \colon X \to [0, +\infty)$ be a function such that (i) $\lim_{t \to 0} \psi(tx) = 0$ for every $x \in X$; (ii) $\psi(x + y) \leqslant \psi(x) + \psi(y)$ for all $x, y \in X$ and $\psi\left(\sum_{n=1}^{\infty} x_n\right) \leqslant \sum_{n=1}^{\infty} \psi(x_n)$ for every series of x_n converging in the space X. Then the function ψ is continuous.*

PROOF. The reasoning is based on a modification of the proof of the Banach open mapping theorem. The inequality $\psi(x+y) \leqslant \psi(x)+\psi(y)$ yields the estimate

$$|\psi(x + h) - \psi(x)| \leqslant \psi(h) + \psi(-h).$$

Hence it suffices to verify the continuity of ψ at zero. Let

$$M(\varepsilon) := \{x \in X : \psi(x) + \psi(-x) \leqslant \varepsilon\}, \quad \varepsilon > 0.$$

Let us fix $\varepsilon > 0$. By condition (i) we have $X = \bigcup_{n=1}^{\infty} nM(\varepsilon)$. Since X is a second category set, the set $\overline{nM(\varepsilon)}$ for some n contains a ball of positive radius. Hence $\overline{M(\varepsilon)}$ contains an open ball $B(x_0, r)$ centered at some $x_0 \in M(\varepsilon)$ of radius $r = r(\varepsilon) > 0$. Then $-x_0 \in M(\varepsilon)$. There exists $\delta(\varepsilon) > 0$ such that $x_0 + B(0, \delta(\varepsilon)) \subset B(x_0, r)$ (if we use a translation invariant metric, then $\delta(\varepsilon) = r$ works). The ball $B(0, \delta(\varepsilon))$ belongs to $\overline{M(2\varepsilon)}$. Indeed, for every $v \in B(0, \delta(\varepsilon))$ there is $\{x_n\} \subset M(\varepsilon)$ with $x_n \to x_0 + v$, whence $x_n - x_0 \to v$ and $x_n - x_0 \in M(2\varepsilon)$. We show that $\psi(v) \leqslant 4\varepsilon$ whenever $v \in B(0, \delta(\varepsilon))$. As shown above, there exists $v_1 \in M(2\varepsilon)$ with $v - v_1 \in B(0, \delta(\varepsilon/2))$. Next we find $v_2 \in M(\varepsilon)$ with $v - v_1 - v_2 \in B(0, \delta(\varepsilon/4))$. By induction we obtain a sequence of elements $v_n \in M(2^{2-n}\varepsilon)$ with $v - v_1 - \cdots - v_n \in B(0, \delta(2^{-n}\varepsilon))$. Hence $v = \sum_{n=1}^{\infty} v_n$ and $\psi(v) \leqslant \sum_{n=1}^{\infty} \psi(v_n) \leqslant \varepsilon \sum_{n=1}^{\infty} 2^{2-n}$. $\square$

8.2.8. Theorem. *Suppose that X and Y are metric linear spaces and X is a second category set. Let $A_n \colon X \to Y$ be a sequence of continuous linear mappings such that for every $x \in X$ the sequence $\{A_n x\}$ is bounded. Then for every neighborhood of zero $V \subset Y$ there exists a neighborhood of zero $U \subset X$ such that $A_n(U) \subset V$ for all n.*

PROOF. Let d be a metric on Y invariant with respect to translations (its existence is proved in §8.6(i)). Set

$$\psi(x) := \sup_n d(0, A_n x), \quad x \in X.$$

Since we have $A_n(tx) = tA_n x$, the condition of the theorem gives the equality $\lim_{t \to 0} \psi(tx) = 0$ for all $x \in X$. The estimate

$$d(0, u + v) \leqslant d(0, u) + d(u, u + v) = d(0, u) + d(0, v)$$

yields the inequality $\psi(x + y) \leqslant \psi(x) + \psi(y)$. Finally, if a series $\sum_{i=1}^{\infty} x_i$ converges in X to x, then for every n we have

$$d(0, A_n x) = \lim_{k \to \infty} d\left(0, \sum_{i=1}^{k} A_n x_i\right) \leqslant \sum_{i=1}^{\infty} d(0, A_n x_i).$$

Hence $\psi(x) \leqslant \sum_{i=1}^{\infty} \psi(x_i)$. Thus, all conditions of the previous theorem are fulfilled, which gives the continuity of ψ, whence our claim follows. $\square$

Let us single out an important particular case.

8.2.9. Corollary. *If in the theorem above for all $x \in X$ there exists a limit $Ax = \lim_{n \to \infty} A_n x$, then the operator A is continuous.*

We now prove a generalization of the open mapping theorem.

8.2.10. Theorem. *Suppose that X and Y are metric linear spaces, where X is complete and Y is a second category set. Let $A\colon X \to Y$ be a continuous linear mapping such that $A(X) = Y$. Then for every open set $U \subset X$ the set $A(U)$ is open in Y.*

PROOF. Let d be a metric on X invariant with respect to translations. Set

$$\psi(y) := \inf\{d(0,x)\colon Ax = y\}, \quad y \in Y.$$

Condition (i) in Theorem 8.2.7 is obviously fulfilled, since if $Ax = y$, then $A(tx) = ty$. The inequality $\psi(y_1 + y_2) \leqslant \psi(y_1) + \psi(y_2)$ follows from the estimate $d(0, x_1 + x_2) \leqslant d(0, x_1) + d(0, x_2)$. Suppose that a series $\sum_{n=1}^{\infty} y_n$ converges in Y to a vector y. We verify that $\psi(y) \leqslant \sum_{n=1}^{\infty} \psi(y_n)$. Otherwise there exists $\varepsilon > 0$ such that $\sum_{n=1}^{\infty} \psi(y_n) < \psi(y) - \varepsilon$. Pick vectors $x_n \in A^{-1}(y_n)$ such that $d(0, x_n) \leqslant \psi(y_n) + \varepsilon 4^{-n}$. Then $\sum_{n=1}^{\infty} d(0, x_n) \leqslant \psi(y) - \varepsilon/2$. By the completeness of X the series of x_n converges to some element $x \in X$, moreover, $d(0, x) \leqslant \psi(y) - \varepsilon/2$, which contradicts the definition of $\psi(y)$. $\square$

8.2.11. Corollary. *Suppose that X and Y are metric linear spaces such that X is complete and Y is a second category set. Let $A\colon X \to Y$ be a continuous linear mapping such that $\operatorname{Ker} A = 0$ and $A(X) = Y$. Then the mapping $A^{-1}\colon Y \to X$ is continuous.*

As in the case of Banach spaces, for complete linear metric spaces the inverse mapping theorem yields the closed graph theorem (since the projection of the graph $A\colon X \to Y$ onto X is a bijection). However, we also derive it from Theorem 8.2.7.

8.2.12. Theorem. *Suppose that X and Y are metric linear spaces such that Y is complete and X is a second category set. Let $A\colon X \to Y$ be a linear mapping with a closed graph. Then it is continuous.*

PROOF. Let d be a metric on Y invariant with respect to translations. Set

$$\psi(x) := d(0, Ax), \quad x \in X.$$

Condition (i) of Theorem 8.2.7 is obviously fulfilled. In addition, we have

$$d(0, Ax_1 + Ax_2) \leqslant d(0, Ax_1) + d(0, Ax_2).$$

If a series $\sum_{n=1}^{\infty} x_n$ converges to $x \in X$, then $d(0, Ax) \leqslant \sum_{n=1}^{\infty} d(0, Ax_n)$ holds if the sum is infinite. If the sum is finite, then by the completeness of Y the series of Ax_n converges to some element $y \in Y$. The condition that the graph of A is closed gives the equality $Ax = y$, whence we obtain the desired bound $d(0, Ax) = d\big(0, \sum_{n=1}^{\infty} Ax_n\big) \leqslant \sum_{n=1}^{\infty} d(0, Ax_n)$. $\square$

For additional results in this direction and related references, see [75].

8.3. Separation of Convex Sets

Here we discuss a number of geometric corollaries of the analytic form of the Hahn–Banach theorem generalizing and complementing the results obtained earlier for normed spaces.

A *hyperplane* in a linear space E is defined as a set of the form $a + L$, where $a \in E$ and L is a linear subspace of codimension 1. In other words, a hyperplane is a set of the form $\{x \in E : l(x) = \alpha\}$, where l is a linear function on E and α is a number. The following lemma shows that a hyperplane is closed precisely when the generating linear function l is continuous.

8.3.1. Lemma. *A linear function l on a topological vector space E is continuous precisely when the set $l^{-1}(0)$ is closed, which is equivalent to the closedness of some of the sets $l^{-1}(\alpha)$. If the linear function l is discontinuous, then the set $l^{-1}(0)$ is everywhere dense in E.*

PROOF. If the function l is continuous, then all sets $l^{-1}(\alpha)$ are closed. Suppose that at least one of these sets $l^{-1}(\alpha)$ is closed. We show that the set $l^{-1}(0)$ is closed as well. Indeed, either $l = 0$ or there exists $v \in E$ with $l(v) = \alpha$. Then $l^{-1}(0) = l^{-1}(\alpha) - v$ is closed as a shift of a closed set. For the proof of continuity of l it suffices to establish the continuity of l at zero. This will be done if we verify that every set $U_\varepsilon = \{x : |l(x)| < \varepsilon\}$ with $\varepsilon > 0$ contains a neighborhood of zero. If $l \neq 0$, then there exists an element $e \in E$ with $l(e) = \varepsilon$. By the closedness of $l^{-1}(0)$ there exists a balanced neighborhood of zero W such that $(e + W) \cap l^{-1}(0) = \varnothing$. It follows from this that $W \subset U_\varepsilon$. Otherwise there is a vector $w \in W$ with $|l(w)| \geqslant \varepsilon$. Then $v := -\varepsilon w f(w)^{-1} \in W$, whence $l(e + v) = 0$, i.e., $e + v \in (e + W) \cap l^{-1}(0)$, which is impossible. Finally, if the linear function l is discontinuous, then its kernel is not closed as shown above. Hence its closure is a larger linear subspace (Exercise 8.6.43). Since the kernel has codimension 1, then the closure is the whole space. $\qquad\square$

We shall say that two sets A and B in a real topological vector space E are *separated* by the closed hyperplane H of the form $l^{-1}(\alpha)$, where l is a continuous linear function and $\alpha \in \mathbb{R}^1$, if either $l(x) \leqslant \alpha$ for all $x \in A$ and $l(x) \geqslant \alpha$ for all $x \in B$, or $l(x) \leqslant \alpha$ for all $x \in B$ and $l(x) \geqslant \alpha$ for all $x \in A$, i.e., A and B are contained in different closed halfspaces generated by the hyperplane H. If A and B are contained in different open halfspaces generated by the hyperplane H, then we say that A and B are *strictly separated* by the hyperplane H. For example, the closed halfspaces $\{x : l(x) \geqslant 0\}$ and $\{x : l(x) \leqslant 0\}$ are separated by the hyperplane $l^{-1}(0)$, but not strictly, while the corresponding open halfspaces are strictly separated. A closed hyperplane H passing through a point of a set A is called *supporting* if A is contained in one of the two halfspaces generated by H.

8.3.2. Theorem. *Let E be a real topological vector space and let A and B be nonempty convex sets in E such that the interior of A is nonempty and does not intersect B. Then there is a closed hyperplane H separating A and B.*

If, in addition, A and B are open, then they are strictly separated by a closed hyperplane.

PROOF. It is not difficult to verify that the set A° of all inner points of A is also convex. By Theorem 6.3.7 there exists a nonzero linear function l separating A° and B, i.e., for some c we have $l(x) \leqslant c$ for all $x \in A^\circ$ and $l(x) \geqslant c$ for all $x \in B$. The boundedness from above on an open set yields the continuity of l. Clearly, $l(x) \leqslant c$ for all $x \in A$.

If the sets A and B are open, then the hyperplane $H = l^{-1}(c)$ strictly separates them. Indeed, if, say, $l(b) = c$ for some $b \in B$, then there exists a neighborhood of zero U such that $b + U \subset B$, but U contains a vector u with $l(u) < 0$ (because we have $l \neq 0$), whence $l(b + u) < 0$, which is impossible. □

8.3.3. Corollary. *Let E be a real topological vector space, V a nonempty open convex set in E, and let L be an affine subspace such that $L \cap V = \varnothing$. Then E contains a closed hyperplane H containing L and not intersecting V.*

PROOF. By a shift we can pass to the case where L is a linear subspace. By the theorem proved above V and L are separated by a hyperplane H of the form $H = l^{-1}(c)$, where l is a nonzero continuous linear function and $c \in \mathbb{R}^1$, i.e., $l(x) \leqslant c$ if $x \in V$ and $l(x) \geqslant c$ if $x \in L$. Since we have passed to the case of a linear subspace, we have $c = 0$. As in the proof of the theorem, we have $V \cap H = \varnothing$. Finally, $L \subset H$, since if $x \in L$ and $l(x) \neq 0$, then by multiplying by a scalar we obtain a vector $y \in L$ for which $l(y) < 0$. □

8.3.4. Corollary. *Let V be a closed convex set with a nonempty interior in a real topological vector space E. Then, for every boundary point of the set V, there is at least one closed supporting hyperplane passing through this point. Moreover, V is the intersection of closed halfspaces defined by supporting hyperplanes and containing V.*

PROOF. By the theorem proved above every boundary point x of the set V is separated from V by a closed hyperplane. It is clear that this hyperplane is supporting. For the proof of the second assertion we denote by W the intersection of all closed halfspaces defined by supporting hyperplanes and containing V. Then $V \subset W$. Suppose that there is a point $z \in W \backslash V$. Let us take a point x in the interior of V. The line segment $[x, z]$ contains some boundary point y of the set V. We already know that there is a closed supporting hyperplane H for V passing through y. Then H does not contain z, since otherwise the inner point x would also belong to H, which is impossible, as shown in the proof of the theorem. Therefore, z and V belong to different halfspaces defined by H. Thus, $z \notin W$, which is a contradiction. □

8.3.5. Corollary. *Let A and B be nonempty convex sets in a real locally convex space E such that A is closed and B is compact. Then there exists a closed hyperplane strictly separating A and B.*

PROOF. We show that there exists a convex neighborhood of zero U such that $(A + U) \cap (B + U) = \varnothing$. Since $A + U$ and $B + U$ are open and convex, the theorem above will apply. It suffices to find a convex balanced neighborhood of zero W such that $A \cap (B + W) = \varnothing$; then we can take $U = W/2$. For every point $b \in B$, there exists a convex balanced neighborhood of zero W_b such that $A \cap (b + W_b) = \varnothing$. The open cover of the compact set B by the sets $b + 4^{-1}W_b$ contains a finite subcover $b_1 + 4^{-1}W_{b_1}, \ldots, b_n + 4^{-1}W_{b_n}$. The neighborhood $W := 4^{-1} \bigcap_{i=1}^n W_{b_i}$ is the desired one. $\square$

8.3.6. Corollary. *Every nonempty closed convex set in a real locally convex space is the intersection of closed halfspaces containing it.*

PROOF. By the previous corollary the given set can be separated by a closed hyperplane from every point of the exterior. $\square$

8.3.7. Corollary. *In any real locally convex space E, the closure of every convex set coincides with its closure in the weak topology $\sigma(E, E^*)$.*

PROOF. The closure of a convex set V in the original topology is convex and closed. By the previous corollary this closure is closed in the weak topology. On the other hand, the closure in the weak topology is not smaller than the closure in the original topology. $\square$

8.3.8. Corollary. *Let A be a nonempty closed convex balanced set in a real locally convex space E. Let $v \in E \backslash A$. Then there exists a continuous linear function f on E such that $|f(x)| \leqslant 1$ for all $x \in A$ and $f(v) > 1$.*

PROOF. By Corollary 8.3.5 there is a functional $f \in E^*$ such that for some number c we have $f(x) \leqslant c$ for all $x \in A$ and $f(v) > c$. Since A is balanced, we have $|f(x)| \leqslant c$ for all $x \in A$. Hence $c \geqslant 0$. We can assume that $c > 0$. Now we replace f by f/c. $\square$

In case of a locally convex space Corollary 8.3.4 extends to convex compact sets (in the infinite-dimensional case they have no inner points).

8.3.9. Corollary. *Let C be a nonempty convex compact set in a real locally convex space E. Then for every closed hyperplane H in E there exists at least one and not more than two supporting hyperplanes that are shifts of H. Moreover, the set C is the intersection of closed halfspaces defined by closed supporting hyperplanes and containing C.*

PROOF. Let H have the form $H = l^{-1}(c)$, where $l \in E^*$ and $c \in \mathbb{R}^1$. By the compactness of C there exist points $a \in C$ and $b \in C$ such that we have $l(a) = \alpha := \inf_{x \in C} l(x)$ and $l(b) = \beta := \sup_{x \in C} l(x)$. It is clear that $l^{-1}(\alpha)$ and $l^{-1}(\beta)$ are closed supporting hyperplanes (they can coincide) that are shifts of H and there are no other supporting hyperplanes parallel to H.

Let us show that C is the intersection of closed halfspaces generated by supporting hyperplanes containing C. According to Corollary 8.3.5 every point $z \notin C$ is strictly separated from C by a closed hyperplane H. It is then clear that there exists a supporting hyperplane parallel to H such that C and z are in the different halfspaces defined by it. This proves our assertion. $\square$

8.3.10. Corollary. *Let C be a convex compact set in a real locally convex space E. Then*

$$C = \bigcap_{l \in E^*} l^{-1}(l(C)).$$

A characteristic feature of the results above on separation of closed convex sets by hyperplanes is the requirement of certain additional properties such as the existence of inner points or compactness for one of these sets. Let us show that in the general case one cannot omit additional conditions.

8.3.11. Example. In the Banach space l^1 we consider the straight line L given by the conditions $x_n = 0$, $n \geqslant 2$, and also the set

$$A := \{x = (x_n) \colon |n^3 x_n - n| \leqslant x_1 \; \forall n \geqslant 2\}.$$

It is readily seen that A is closed and convex and that $A \cap L = \varnothing$. However, A and L cannot be separated by a closed hyperplane, since the set $A - L$ cannot be separated from zero by a closed hyperplane, because this set is everywhere dense. Indeed, every point $x \in l^1$ such that $x_n = n^{-2}$ for all $n \geqslant n_0$ with some n_0 belongs to $A - L$.

Suppose we are given a linear space E and some linear space F of linear functions on E separating points in E. Then we shall say that we are given a *dual pair* $\langle E, F \rangle$. We already know from results in §6.6 that once the space E is equipped with the topology $\sigma(E, F)$, for the space of all continuous linear functions we obtain precisely the space F. If this construction is applied to a locally convex space E (equipped with some topology τ) and we take for F the space E^* of all linear functionals on E continuous in the topology τ, then we obtain a weaker topology on E, but the dual space does not change. The question arises about all locally convex topologies on E for which the dual remains E^*. Such topologies are called *consistent with duality*. It is clear that $\sigma(E, E^*)$ is the weakest among them. It turns out that among these topologies there is the strongest one: the Mackey topology τ_M. The Mackey–Arens theorem proved in §8.6(ii) gives a complete description of topologies consistent with duality as topologies intermediate between $\sigma(E, E^*)$ and τ_M. The main idea of the characterization of topologies consistent with duality is employment of topologies of uniform convergence on certain classes of sets. It is useful to introduce polars of sets.

Let $\langle E, F \rangle$ be a dual pair. The *polar* of a set $A \subset E$ with respect to the dual pair $\langle E, F \rangle$ is defined as the set

$$A^\circ := \{f \in F \colon |f(a)| \leqslant 1 \; \forall a \in A\}.$$

The following properties of polars are obvious from the definition.

8.3.12. Lemma. *The set A° is convex, balanced and closed in the topology $\sigma(F, E)$.*

The same reasoning as in the case of normed spaces gives the following version of the Banach–Alaoglu–Bourbaki theorem for locally convex spaces.

8.3.13. Theorem. *Let X be a locally convex space. Then the polar of every neighborhood of zero U in X is compact in the topology $\sigma(X^*, X)$.*

The following bipolar theorem is of great value.

8.3.14. Theorem. *The bipolar $A^{\circ\circ}$ of any set $A \subset E$, i.e., the polar of A° with respect to the pair $\langle F, E \rangle$, coincides with the $\sigma(E, F)$-closed convex balanced envelope of A.*

In particular, if X is a locally convex space, the polar A° is taken with respect to $\langle X, X^* \rangle$, and the bipolar is taken with respect to $\langle X^*, X \rangle$, then $A^{\circ\circ}$ coincides with the closed convex balanced envelope of A.

PROOF. It is clear that $A \subset A^{\circ\circ}$. Let B be the $\sigma(E, F)$-closed convex balanced envelope of A. If an element $v \in E$ does not belong to B, then by a corollary of the Hahn–Banach theorem there exists a $\sigma(E, F)$-continuous linear functional f such that $|f(v)| > 1$ and $|f(b)| \leqslant 1$ for all $b \in B$. We know that $f \in F$ (see §6.6). Hence $f \in A^\circ$. Therefore, $v \notin A^{\circ\circ}$. Thus, $B \subset A^{\circ\circ}$. $\qquad\square$

8.4. Distributions

Generalized functions (distributions or Schwartz distributions) are introduced as continuous linear functionals on various spaces of smooth functions. The main idea of this construction (originated in physics) is that certain usual highly discontinuous functions are meaningful not due to their values at individual points, but in the sense of their integration with smooth functions. In the theory of integration there is a conceptually close identification of equivalent functions, but here we deal with a further development of these motives.

8.4.1. Definition. (i) *The space of all continuous linear functions on $\mathcal{S}(\mathbb{R}^n)$ is denoted by $\mathcal{S}'(\mathbb{R}^n)$ and called the class of tempered distributions.*

(ii) *The space of all linear functions F on $\mathcal{D}(\mathbb{R}^n)$ such that $F(\varphi_j) \to 0$ for every sequence $\{\varphi_j\}$ converging to zero in $\mathcal{D}(\mathbb{R}^n)$ is denoted by $\mathcal{D}'(\mathbb{R}^n)$. Functionals from $\mathcal{D}'(\mathbb{R}^n)$ are called distributions, or generalized functions, on $\mathbb{R}^n$.*

The spaces $\mathcal{S}'$ and $\mathcal{D}'$ are linear. Note that in (ii) we require the sequential continuity of F. It was mentioned above that $\mathcal{D}$ is not metrizable. Hence the question arises about the coincidence of $\mathcal{D}'$ with the space of all continuous linear functions on $\mathcal{D}$. It turns out that this coincidence holds (although for nonlinear functions this is no longer true: see Exercise 8.6.45!). We give this definition in such a form with the only purpose to avoid the use of topologies (which are not needed for our subsequent discussion). However, for completeness of presentation in Proposition 8.4.8 below we give a proof of the aforementioned coincidence.

Note that a linear function F on $\mathcal{S}$ is in $\mathcal{S}'$ precisely when $|F(\varphi)| \leqslant C p_m(\varphi)$ for some C and m, where $\{p_m\}$ are the norms from Example 8.1.3(iii).

8.4.2. Example. (i) Let F be a locally integrable function on $\mathbb{R}^n$. Then

$$\varphi \mapsto \int_{\mathbb{R}^n} \varphi(x) F(x)\, dx$$

is an element of $\mathcal{D}'(\mathbb{R}^n)$ denoted also by F.

(ii) Every Borel measure μ on $\mathbb{R}^n$ bounded on balls defines a distribution of class $\mathcal{D}'(\mathbb{R}^n)$ by the formula

$$\varphi \mapsto \int_{\mathbb{R}^n} \varphi(x)\, \mu(dx).$$

(iii) The formula $\varphi \mapsto \varphi(0)$ defines the distribution called the δ-function (or Dirac's δ-function). Similarly, for every $a \in \mathbb{R}^n$ we can define the distribution

$$\delta_a(\varphi) := \varphi(a).$$

It is clear that δ_a corresponds to the Dirac measure δ_a at the point a. We observe that δ cannot be defined by a locally integrable function F: it suffices to take a sequence of functions $\varphi_j \in \mathcal{D}(\mathbb{R}^n)$ such that $\varphi_j(0) = 1$, $0 \leqslant \varphi_j \leqslant 1$ and $\varphi_j(x) \to 0$ for every $x \neq 0$. Then $\delta(\varphi_j) = 1$, but the integral of $\varphi_j F$ must tend to zero by the Lebesgue dominated convergence theorem.

(iv) The distribution $\varphi \mapsto -\varphi'(0)$, $\varphi \in \mathcal{D}(\mathbb{R}^1)$ is not defined even by a measure, since it is easy to construct functions $\varphi_j \in \mathcal{D}(\mathbb{R}^1)$ with $\varphi_j'(0) = 1$ that are uniformly bounded and converge pointwise to zero.

Functions in the classes $\mathcal{D}$ and $\mathcal{S}$ are called *test functions*. The spaces $\mathcal{S}$ and $\mathcal{D}_m$ are complete metrizable, which gives the following analog of a corollary of the Banach–Steinhaus theorem (see Corollary 8.2.9).

8.4.3. Theorem. *Let $F_j \in \mathcal{S}'(\mathbb{R}^n)$ be such that for every φ in $\mathcal{S}(\mathbb{R}^n)$ there exists a finite limit $F(\varphi) := \lim\limits_{j \to \infty} F_j(\varphi)$. Then $F \in \mathcal{S}'(\mathbb{R}^n)$. A similar assertion is true for $\mathcal{D}'(\mathbb{R}^n)$.*

With the aid of this result it is easy to construct several other interesting examples of distributions.

8.4.4. Example. For every $\varphi \in \mathcal{S}(\mathbb{R}^1)$ there exists a finite limit

$$\text{V.P.}\frac{1}{x}(\varphi) = \lim_{\varepsilon \to 0} \int_{|x|>\varepsilon} \frac{\varphi(x)}{x}\, dx = \lim_{R \to \infty} \int_{-R}^{R} \frac{\varphi(x) - \varphi(0)}{x}\, dx, \qquad (8.4.1)$$

defining the distribution $\text{V.P.}\frac{1}{x}$ ($1/x$ is not integrable near zero). Similarly, the distributions $(x+i0)^{-1}$ and $(x-i0)^{-1}$ are defined by the equalities

$$(x+i0)^{-1}(\varphi) := \lim_{\varepsilon \to 0+} \int_{-\infty}^{+\infty} \frac{\varphi(x)}{x+i\varepsilon}\, dx,$$

$$(x-i0)^{-1}(\varphi) := \lim_{\varepsilon \to 0+} \int_{-\infty}^{+\infty} \frac{\varphi(x)}{x-i\varepsilon}\, dx.$$

For justification of (8.4.1) we observe that our assertion is obvious for functions φ equal to zero in a neighborhood of zero. Hence it suffices to consider functions $\varphi \in \mathcal{D}(\mathbb{R}^1)$. Let $\varphi(x) = 0$ if $|x| \geqslant m$. Then the function $[\varphi(x) - \varphi(0)]/x$ is integrable on $[-m, m]$, since it is estimated by $\sup_t |\varphi'(t)|$. Hence

$$\int_{-m}^{m} \frac{\varphi(x) - \varphi(0)}{x}\, dx = \lim_{\varepsilon \to 0} \int_{\varepsilon < |x| \leqslant m} \frac{\varphi(x) - \varphi(0)}{x}\, dx,$$

which proves our assertion, since the integral of $\varphi(0)/x$ over $[-m, -\varepsilon] \cup [\varepsilon, m]$ is zero. The remaining assertions are proved similarly.

Convergence of distributions F_j to F is understood as convergence of values $F_j(\varphi) \to F(\varphi)$ for all test functions φ. Exercise 8.6.60 shows that any distribution F is the limit of a sequence of distributions F_j generated by smooth functions.

Although distributions are not functions of a point on the real line, to them one can apply many constructions and notions known for usual functions. Here we consider multiplication by functions and the notions of support and singular support, and in the next section we discuss differentiation.

If $F \in \mathcal{D}'(\mathbb{R}^n)$ and $f \in C^\infty(\mathbb{R}^n)$, then the distribution fF is defined by the natural formula $(fF)(\varphi) := F(f\varphi)$. It is clear that $fF \in \mathcal{D}'(\mathbb{R}^n)$. Similarly, $fF \in \mathcal{S}'(\mathbb{R}^n)$ if $F \in \mathcal{S}'(\mathbb{R}^n)$ and $f \in C^\infty(\mathbb{R}^n)$ are such that for every k for some $m = m(k)$ one has the estimate
$$\|f^{(k)}(x)\| \leqslant c_{k,m}(1 + |x|^2)^m.$$
The support $\operatorname{supp} f$ of a usual function f is the closure of the set $\{f \neq 0\}$.

8.4.5. Definition. *We shall say that a distribution $F \in \mathcal{D}'(\mathbb{R}^n)$ is smooth on an open set $U \subset \mathbb{R}^n$ if there exists a function $g \in C^\infty(U)$ such that*
$$F(\varphi) = \int_U \varphi(x)g(x)\,dx$$
for all $\varphi \in \mathcal{D}(\mathbb{R}^n)$ with $\operatorname{supp}\varphi \subset U$. If $g = 0$, then we say that F is zero on U.

The complement to the union of all open sets on which F is smooth is called the singular support of F and denoted by the symbol $\operatorname{singsupp} F$.

The complement to the union of all open sets on which F equals zero is called the support of F and denoted by the symbol $\operatorname{supp} F$.

8.4.6. Example. The following equalities hold
$$\operatorname{supp}\delta = \{0\}, \ \operatorname{supp}\mathrm{V.P.}\frac{1}{x} = \mathbb{R}^1, \ \operatorname{singsupp}\mathrm{V.P.}\frac{1}{x} = \{0\}.$$

It is clear from the definition that if $F \in \mathcal{D}'(\mathbb{R}^n)$ and $\varphi \in \mathcal{D}(\mathbb{R}^n)$ are such that $\operatorname{supp} F \cap \operatorname{supp}\varphi = \varnothing$, then $F(\varphi) = 0$.

8.4.7. Example. If $F \in \mathcal{D}'(\mathbb{R}^n)$, $f \in C^\infty(\mathbb{R}^n)$ and $f = 1$ on an open set $U \supset \operatorname{supp} F$, then $fF = F$, since the support of $f\varphi - \varphi$ is disjoint with $\operatorname{supp} F$. However, it is not enough to have the equality $f = 1$ on $\operatorname{supp} F$. For example, if $F(\varphi) = \varphi'(0)$ and $f(x) = 1 + x$, then $\operatorname{supp} F = \{0\}$ and $f(0) = 1$, but we have $fF = F + \delta$, which is seen from the following obvious equalities: $(fF)(\varphi) = F(f\varphi) = (f\varphi)'(0) = \varphi(0) + \varphi'(0)$.

8.4.8. Proposition. *The space $\mathcal{D}'(\mathbb{R}^1)$ coincides with the space of continuous linear functions on $\mathcal{D}(\mathbb{R}^1)$ with the topology introduced above. The same is true for $\mathcal{D}'(\mathbb{R}^n)$.*

PROOF. We have to show that every sequentially continuous linear function F on $\mathcal{D}(\mathbb{R}^1)$ is actually continuous in the usual topological sense. It suffices to verify that there exists a seminorm $p_{\{\alpha_n\}}$ of the form (8.1.2) such that $|F(\varphi)| \leqslant p_{\{\alpha_n\}}(\varphi)$ for all $\varphi \in \mathcal{D}(\mathbb{R}^1)$. The sequential continuity gives the continuity of the restriction of F to every metrizable locally convex space $\mathcal{D}_m(\mathbb{R}^1)$.

It is easy to construct a sequence of functions $\zeta_k \in \mathcal{D}(\mathbb{R}^1)$, $k \in \mathbb{Z}$, with the following properties: $0 \leqslant \zeta_k \leqslant 1$, $\zeta_k(x) = 1$ if $x \in [k - 1/3, k + 1/3]$, $\zeta_k(x) = 0$

if $x \notin [k - 2/3, k + 2/3]$, and $\zeta_k(x) + \zeta_{k+1}(x) = 1$ for all points x. Then we have the equality $\sum_{k=-\infty}^{+\infty} \zeta_k(x) = 1$.

For every k, we consider the functional $F_k(\varphi) := F(\zeta_k \varphi)$ on $\mathcal{D}(\mathbb{R}^1)$. The restriction of F_k to $\mathcal{D}([k-1, k+1])$ is continuous and the value $F_k(\varphi)$ depends only on the restriction of φ to $[k - 2/3, k + 2/3]$ and does not change if we replace φ by $\theta_k \varphi$, where $\theta_k \in \mathcal{D}(\mathbb{R}^1)$, $\theta_k(x) = 1$ if $x \in [k - 2/3, k + 2/3]$ and $\theta_k(x) = 0$ if $x \notin [k-1, k+1]$. Applying Theorem 8.2.2 for $\mathcal{D}([k-1, k+1])$, we find $C_k, r_k \in \mathbb{N}$ such that

$$|F_k(\varphi)| \leqslant C_k \sup_{t \in [k-1, k+1]} [|\varphi(t)| + \cdots + |\varphi^{(r_k)}(t)|] \quad \forall \varphi \in \mathcal{D}(\mathbb{R}^1).$$

Set $\alpha_k := \sum_{l=k-1}^{k+2}(C_l + r_l)$. For all $\varphi \in \mathcal{D}(\mathbb{R}^1)$ we have

$$|F(\varphi)| = \left| F\left(\sum_{k=-m}^{m} \zeta_k \varphi \right) \right| \leqslant \sum_{k=-\infty}^{+\infty} |F(\zeta_k \varphi)|$$

$$\leqslant \sum_{k=-\infty}^{+\infty} C_k \sup_{t \in [k-1, k+1]} [|\varphi(t)| + \cdots + |\varphi^{(r_k)}(t)|],$$

whence $|F(\varphi)| \leqslant p_{\{\alpha_k\}}(\varphi)$. The case $n > 1$ is similar. $\qquad\square$

Some other notions known for usual functions are meaningful for distributions. The derivative will be considered in the next section and here we only observe that the direct product of distributions $F \in \mathcal{D}'(\mathbb{R}^n)$ and $G \in \mathcal{D}'(\mathbb{R}^k)$ is defined as an element $F \otimes G$ from $\mathcal{D}'(\mathbb{R}^{n+k})$ acting on test functions φ from the space $\mathcal{D}(\mathbb{R}^{n+k}) = \mathcal{D}(\mathbb{R}^n \times \mathbb{R}^k)$ by the formula

$$F \otimes G(\varphi) := F(\psi), \quad \psi(x) := G\big(\varphi(x, \cdot)\big).$$

It is clear that the function ψ has compact support. Verification of its smoothness and the proof of the inclusion $F \otimes G \in \mathcal{D}'(\mathbb{R}^{n+k})$ are delegated to Exercise 8.6.53. Similarly, for an arbitrary diffeomorphism Λ of $\mathbb{R}^n$ and any $F \in \mathcal{D}'(\mathbb{R}^n)$ we set $F \circ \Lambda(\varphi) := F(\varphi \circ \Lambda)$, which gives $F \circ \Lambda \in \mathcal{D}'(\mathbb{R}^n)$.

In applications also other classes of test and generalized functions are employed. For example, taking the class $\mathcal{E}(\mathbb{R}^n)$ of all infinitely differentiable functions on $\mathbb{R}^n$ with the topology of uniform convergence of all derivatives on bounded sets (it is defined by the seminorms $p_{k,m}(\varphi) = \sup_{|x| \leqslant m} \|\varphi^{(k)}(x)\|$), we obtain the class $\mathcal{E}'$ of distributions contained in $\mathcal{S}'$. Distributions from $\mathcal{E}'$ are precisely the elements of $\mathcal{D}'$ with compact support.

8.4.9. Proposition. *A functional $F \in \mathcal{D}'(\mathbb{R}^n)$ extends to an element of the space $\mathcal{E}'(\mathbb{R}^n)$ (i.e., is continuous with respect to the topology from $\mathcal{E}(\mathbb{R}^n)$) precisely when it has compact support.*

PROOF. If F has support in a ball U, then an extension to $\mathcal{E}(\mathbb{R}^n)$ is given by the formula $F(\varphi) := F(\theta \varphi)$, where θ is a fixed function from $C_0^\infty(\mathbb{R}^n)$ equal to 1 in a neighborhood of U. Conversely, if F is continuous in the topology from $\mathcal{E}(\mathbb{R}^n)$, then there exist k, m and C such that $|F(\varphi)| \leqslant C p_{k,m}(\varphi)$ for all $\varphi \in \mathcal{D}(\mathbb{R}^n)$. This means that the support of F is contained in the ball of radius m (for every point a outside this ball there is a function $\varphi \in \mathcal{D}(\mathbb{R}^n)$ such that $\varphi(a) = 1$ and $\varphi(x) = 0$ whenever $|x| \leqslant m$). $\qquad\square$

8.5. Derivatives of Distributions

It is known that there are continuous functions having a derivative at no point. In the framework of the theory of distributions derivatives can be defined not only for such functions, but even for everywhere discontinuous locally integrable functions. However, these derivatives will be not be usual function, but distributions.

8.5.1. Definition. *Let $F \in \mathcal{D}'(\mathbb{R}^n)$. The generalized partial derivative $\partial_{x_i} F$ of the distribution F is defined as the element of $\mathcal{D}'(\mathbb{R}^n)$ given by the formula $\partial_{x_i} F(\varphi) := -F(\partial_{x_i}\varphi)$. Similarly, we set $\partial^{(\alpha)} F(\varphi) := (-1)^{|\alpha|} F(\partial^{(\alpha)}\varphi)$.*

For $n = 1$ we obtain $F'(\varphi) = -F(\varphi')$.

The functional $\varphi \mapsto -F(\partial_{x_i}\varphi)$ is indeed a distribution, since if $\varphi_j \to 0$ in $\mathcal{D}(\mathbb{R}^n)$, then $\partial_{x_i}\varphi_j \to 0$ in $\mathcal{D}(\mathbb{R}^n)$.

Thus, every distribution has generalized partial derivatives of every order in the indicated sense. In particular, any usual locally integrable function, even without usual derivatives at any point, is infinitely differentiable in the sense of distributions. However, its derivatives are not usual functions, but distributions.

8.5.2. Example. (i) Let us consider the distribution F defined by the usual function $I_{[0,+\infty)}$ (called the *Heaviside function*). Then $F' = \delta$ in the sense of distributions, since

$$F'(\varphi) = -F(\varphi') = -\int_0^\infty \varphi'(x)\,dx = \varphi(0) = \delta(0), \quad \varphi \in \mathcal{D}(\mathbb{R}^1).$$

(ii) Let us consider the distribution F in $\mathcal{D}'$ defined by the locally integrable function $\ln|x|$. Then $F' = \text{V.P.}\frac{1}{x}$. Indeed,

$$F'(\varphi) = -F(\varphi') = -\int_{-\infty}^{+\infty} \varphi'(x)\ln|x|\,dx$$

$$= -\lim_{\varepsilon \to 0}\int_{|x|>\varepsilon} \varphi'(x)\ln|x|\,dx = \lim_{\varepsilon \to 0}\int_{|x|>\varepsilon} \frac{\varphi(x)}{x}\,dx.$$

As in the case of usual functions, distributions with zero derivative are constants, i.e., are defined by constant usual functions (of course, this does not mean that such a distribution is a constant linear functional: a linear function is constant only if it is identically zero).

8.5.3. Proposition. *Let $F \in \mathcal{D}'(\mathbb{R}^1)$ and $F' = 0$. Then F is given by some constant c, i.e.,*

$$F(\varphi) = c\int_{\mathbb{R}^1} \varphi(x)\,dx.$$

If $F' = 0$ on an interval U, then there exists a constant c for which the equality above is true for all $\varphi \in C_0^\infty(U)$.

PROOF. Let $\varphi_0 \in \mathcal{D}(\mathbb{R}^1)$ have the integral 1 over the whole real line. Set $c := F(\varphi_0)$ and show that we have obtained the required constant. For all functions $\varphi \in \mathcal{D}(\mathbb{R}^1)$ we have $\varphi = \varphi - \theta\varphi_0 + \theta\varphi_0$, where θ is the integral of φ. Set

$$\psi(x) := \int_{-\infty}^{x} [\varphi(t) - \theta\varphi_0(t)] \, dt.$$

We observe that $\psi \in \mathcal{D}(\mathbb{R}^1)$, since if both functions φ and φ_0 vanish outside some interval $[a, b]$, then ψ also vanishes outside $[a, b]$, because the integral of $\varphi - \theta\varphi_0$ over the whole real line is zero. Since $\psi' = \varphi - \theta\varphi_0$, we obtain

$$F(\varphi) = F(\psi') + \theta F(\varphi_0) = -F'(\psi) + c\theta = c\theta,$$

as required. The case of an interval is similar. $\qquad\square$

For the Cantor function C_0 from Proposition 3.1.12 we have $C_0'(x) = 0$ a.e., however, C_0' in $\mathcal{D}'$ is not zero, but the Borel probability measure ν for which $\nu\big([0, x)\big) = C_0(x)$. More generally, if μ is a Borel measure on the real line and $F(x) = \mu\big((-\infty, x)\big)$, then $F' = \mu$ in $\mathcal{D}'$ (see §4.5(i)).

Similarly to locally integrable functions, distributions possess primitives.

8.5.4. Proposition. *For every distribution F in $\mathcal{D}'(\mathbb{R}^1)$ there exists a distribution $G \in \mathcal{D}'(\mathbb{R}^1)$ such that $G' = F$. If we fix one such function, then every other one differs by a distribution defined by a constant.*

PROOF. For any $\varphi \in \mathcal{D}(\mathbb{R}^1)$ we take the same function ψ as in the previous proof. Set

$$G(\varphi) := -F(\psi).$$

The functional G is linear on $\mathcal{D}(\mathbb{R}^1)$. We show that it is continuous. It suffices to verify that the linear mapping $\varphi \mapsto \psi$ is continuous on $\mathcal{D}(\mathbb{R}^1)$. Let $\varphi_j \to 0$ in $\mathcal{D}(\mathbb{R}^1)$. Then the functions φ_j vanish outside some interval and their integrals θ_j tend to zero. The functions $\varphi_j - \theta_j\varphi_0$ vanish outside some interval $[a, b]$ and have zero integrals. Hence the corresponding functions ψ_j vanish outside $[a, b]$. Since $\varphi_j - \theta_j\varphi_0 \to 0$ in $\mathcal{D}(\mathbb{R}^1)$, the sequence of functions ψ_j also tends to zero in $\mathcal{D}(\mathbb{R}^1)$, which completes our proof. $\qquad\square$

The distribution G also has a primitive and so on. It turns out that if we fix the interval $[a, b]$ and restrict F to functions with support in $[a, b]$, then in finitely many steps we arrive at a usual function (continuous or even continuously differentiable).

8.5.5. Theorem. *If $F \in \mathcal{D}'(\mathbb{R}^n)$ has support in an open ball U, then there exists a continuous function Ψ on U and a number $k \in \mathbb{N}$ for which $F = \partial_{x_1}^k \cdots \partial_{x_n}^k \Psi$. Every element $F \in \mathcal{S}'(\mathbb{R}^n)$ has the form $F = \partial_{x_1}^k \cdots \partial_{x_n}^k \Psi$, where $k \in \mathbb{N}$, $\Psi \in C(\mathbb{R}^n)$, $|\Psi(x)| \leqslant C + C|x|^m$, $C, m > 0$. In the one-dimensional case $F = \Psi^{(k)}$.*

PROOF. Let $n = 1$ and $U = (0, 1)$. By Theorem 8.2.2 there are numbers $m, C \in \mathbb{N}$ such that $|F(\varphi)| \leqslant C \sup_{x \in U} |\varphi^{(m)}(x)|$, $\varphi \in \mathcal{D}(U)$. By the mean value theorem $|\varphi^{(m)}(x)| \leqslant \|\varphi^{(m+1)}\|_{L^1} \leqslant \|\varphi^{(m+1)}\|_{L^2}$. Thus, we have

$|F(\varphi)| \leqslant C\|\varphi^{(m+1)}\|_{L^2}$. The Riesz theorem (see Example 6.5.3) gives a function $g \in L^2[0,1]$ for which $F(\varphi) = (\varphi^{(m+1)}, g)_{L^2(U)}$, $\varphi \in \mathcal{D}(U)$. Hence $F = (-1)^{m+1} g^{(m+1)}$. For Ψ we can take the primitive of g. In the multidimensional case the reasoning is similar. Let $U \subset (0,1)^n$. As above, there exist numbers C and k for which $|F(\varphi)| \leqslant C \sup_{x \in U} |\partial_{x_1}^k \cdots \partial_{x_n}^k \varphi(x)|$. We observe that the right-hand side is estimated by $C\|\partial_{x_1}^{k+1} \cdots \partial_{x_n}^{k+1} \varphi\|_{L^2}$. Indeed, for every function $\psi \in \mathcal{D}(U)$ we have $|\psi(x)| \leqslant \|\partial_{x_1} \cdots \partial_{x_n} \psi\|_{L^1}$, since by the n-fold application of the Newton–Leibniz formula $\psi(x)$ is the integral of the function $\partial_{x_1} \cdots \partial_{x_n} \psi$ over the parallelepiped $\{0 \leqslant y_i \leqslant x_i, i = 1, \ldots, n\}$. As above, we obtain a function $g \in L^2$ for which $F = (-1)^{n(k+1)} \partial_{x_1}^{k+1} \cdots \partial_{x_n}^{k+1} g$. For Ψ we take the function $\displaystyle\int_0^{x_n} \cdots \int_0^{x_1} g(y)\, dy_1 \cdots dy_n$. The case of $\mathcal{S}'$ is similar, but we have to use the estimate with the L^1-norm and Example 6.4.12 for $Y = L^1$ (see Vladimirov [620, pp. 80, 81]). Note that the function $-e^x \sin x = (\cos e^x)'$ belongs to $\mathcal{S}'$, but has no power bounds. $\qquad\square$

8.5.6. Proposition. *In the previous theorem $F(\varphi) = 0$ if $\partial_{x_1}^{k_1} \cdots \partial_{x_n}^{k_n} \varphi = 0$ on* supp F *for all $k_i \leqslant k$.*

PROOF. Let U_ε be the open ε-neighborhood of the compact set supp F and $\varrho_\varepsilon(x) = \varepsilon^{-n} \varrho(x/\varepsilon)$, where $\varrho \in C_0^\infty(\mathbb{R}^n)$ has the unit integral and support in the closed unit ball U. The smooth function $\zeta_\varepsilon := \varrho_\varepsilon * I_{K_{2\varepsilon}}$ has support in $U_{3\varepsilon}$ and equals 1 on U_ε. In addition, $|\partial^{(\alpha)} \zeta_\varepsilon(x)| \leqslant C_\alpha \varepsilon^{-|\alpha|}$. The Taylor formula and our condition on φ yield the estimate $|\partial_{x_1}^{k_1} \cdots \partial_{x_n}^{k_n} \varphi(x)| \leqslant C_1 \varepsilon^{kn - k_1 - \cdots - k_n}$. Hence $|\partial_{x_1}^k \cdots \partial_{x_n}^k (\zeta_\varepsilon \varphi)| \leqslant C\varepsilon$, where C does not depend on ε. We can assume that supp $F \subset U$, supp $\varphi \subset U$ and the function Ψ is continuous on U. Thus, according to Example 8.4.7 we obtain

$$|\langle F, \varphi \rangle| = |\langle F, \zeta_\varepsilon \varphi \rangle| = |\langle \Psi, \partial_{x_1}^k \cdots \partial_{x_n}^k (\zeta_\varepsilon \varphi) \rangle| \leqslant C\varepsilon \sup_U |\Psi(x)| \lambda_n(U),$$

where the right-hand side tends to zero as $\varepsilon \to 0$. $\qquad\square$

8.5.7. Example. Let $F \in \mathcal{D}'(\mathbb{R}^n)$ and supp $F = \{0\}$. Then

$$F = \sum_{|\alpha| \leqslant N} c_\alpha \partial^{(\alpha)} \delta$$

for some $c_\alpha \in \mathbb{R}^1$ and $N \in \mathbb{N} \cup \{0\}$. If $n = 1$, then $F = c_0 \delta + c_1 \delta' + \cdots + c_N \delta^{(N)}$.

PROOF. Let us fix a function $\zeta \in \mathcal{D}(\mathbb{R}^n)$ equal to 1 on the unit ball and let $c_\alpha := F(\zeta x_1^{\alpha_1} \cdots x_n^{\alpha_n})$, $\alpha = (\alpha_1, \ldots, \alpha_n)$. Set

$$\psi := \varphi - \zeta \sum_{|\alpha| \leqslant kn} \partial^{(\alpha)} \varphi(0) x_1^{\alpha_1} \cdots x_n^{\alpha_n}.$$

The previous proposition gives $F(\psi) = 0$, since we have $\partial^{(\alpha)} \psi(0) = 0$ whenever $|\alpha| \leqslant kn$. In the case $n = 1$ there is another justification: since $F = \Psi^{(k)}$, where Ψ is a continuous function, there exist polynomials P_1 and P_2 of degree $k - 1$ such that $\Psi(x) = P_1(x)$ if $x > 0$ and $\Psi(x) = P_2(x)$ if $x < 0$. $\qquad\square$

8.6. Complements and Exercises

(i) Metrizability and normability (377). (ii) The Mackey topology (379). (iii) Inductive and projective limits (381). (iv) Barrelled and bornological spaces (382). (v) Banach spaces generated by Minkowski functionals (383). (vi) The Krein–Milman theorem (389). (vii) The measurable graph theorem (391). (viii) Fixed point theorems in locally convex spaces (391). Exercises (392).

8.6(i). Metrizability and normability

8.6.1. Proposition. *A locally convex space E is metrizable precisely when its topology is generated by a countable family of seminorms. In this case the original family of seminorms $\{p_\alpha\}$ generating the topology contains a finite or countable subfamily $\{p_{\alpha_n}\}$ with the following property: for every α there exist a number C_α and an index $n = n(\alpha)$ such that $p_\alpha \leqslant C_\alpha(p_{\alpha_1} + \cdots + p_{\alpha_n})$.*

PROOF. If the topology of the space E is generated by a countable family of seminorms p_n, then we set

$$d(x,y) := \sum_{n=1}^{\infty} 2^{-n} \min(p_n(x-y), 1).$$

It is readily seen that d is a metric. Let us show that d defines the topology generated by the seminorms $\{p_n\}$. For this it suffices to verify that every ball $K(a,r) := \{x\colon d(x,a) < r\}$ contains a neighborhood of the form (8.1.1) and every such neighborhood contains some ball $K(a,r')$. We find N with $2^{-N} < r/2$. Set $\varepsilon = r/2$. Then $U_{p_1,\ldots,p_N,\varepsilon}(a) \subset K(a,r)$. Conversely, the neighborhood (8.1.1) defined by seminorms $p_1,\ldots,p_n$ contains the ball $K(a,r')$ of radius $r' = 2^{-n-1}\varepsilon$.

Suppose now that the locally convex topology E is metrizable by some metric d. Then every ball of radius $1/k$ centered at zero contains a neighborhood of the form (8.1.1) with $a = 0$ and some n, ε and α_i. Hence for every k we take a finite collection of seminorms. Denote by $\{p_{\alpha_n}\}$ the countable family that is the union of all these finite collections. The sets of the form (8.1.1) generated by p_{α_n} are part of all such sets. In order to see the coincidence of the corresponding topologies (i.e., arbitrary unions of sets of such a form), it suffices to show that every set of the form (8.1.1) for the original family contains a set of this form for our countable family $\{p_{\alpha_n}\}$. In turn, for this it suffices to verify the last assertion of the proposition. Now let α be fixed. By our condition the open set $U := \{x\colon p_\alpha(x) < 1\}$ contains some ball V of radius $1/k$ in the metric d centered at zero. By construction this ball contains a set of the form (8.1.1) with α_i from the countable set we picked. Let $C = \varepsilon^{-1}$. Then the bound $Cp_{\alpha_i}(x) < 1$ for all $i = 1,\ldots,n$ gives the inclusion $x \in V \subset U$, i.e., the estimate $p_\alpha(x) < 1$. This means that $p_\alpha \leqslant C(p_{\alpha_1} + \cdots + p_{\alpha_n})$. $\square$

Metrizability of a topological vector space is equivalent to the existence of a countable base of neighborhoods of zero. If the space is separable, then there is also a countable base of the whole topology.

8.6.2. Theorem. *A topological vector space X with a countable base of neighborhoods of zero can be equipped with a translation invariant metric d (i.e., $d(x,y) = d(x+z, y+z)$) defining the same topology and such that the balls centered at zero will be balanced. In particular, this is true if X is metrizable.*

PROOF. It is clear from our discussion in §8.1 that there exists a base of neighborhoods of zero $\{V_n\}$ with the following properties: the sets V_n are balanced and $V_{n+1} + V_{n+1} \subset V_n$. Let D be the set of all dyadic rationals in $[0,1)$. Such a number r can be uniquely written in the form $r = c_1(r)2^{-1} + \cdots + c_n(r)2^{-n}$, where $c_i(r) \in \{0,1\}$. For any $r \in D$ let $A(r) := c_1(r)V_1 + \cdots + c_n(r)V_n$; let $A(r) = X$ if $r \geqslant 1$. Finally, let

$$f(x) := \inf\{r\colon x \in A(r)\}, \quad d(x,y) := f(x-y), \quad x,y \in X.$$

We show that the following relation is fulfilled:

$$A(r) + A(s) \subset A(r+s), \quad r,s \in D. \tag{8.6.1}$$

By induction we verify that if $r,s \in D$ are such that $r+s < 1$ and $c_n(r) = c_n(s) = 0$ for all $n > N$, then (8.6.1) is true. For $N = 1$ this is true. Suppose that this is true for some $N - 1 > 0$. Let $r,s \in D$, $r+s < 1$ and $c_n(r) = c_n(s) = 0$ if $n > N$. Then $r = r' + c_N(r)2^{-N}$, $s = s' + c_N(s)2^{-N}$, i.e., $A(r) = A(r') + c_N(r)V_N$, $A(s) = A(s') + c_N(s)V_N$. By the inductive assumption $A(r') + A(s') \subset A(r'+s')$. Hence $A(r) + A(s) \subset A(r'+s') + c_N(r)V_N + c_N(s)V_N$. If $c_N(r) = c_N(s) = 0$, then (8.6.1) is true by the inductive assumption. Hence it suffices to consider the cases $c_N(r) = 0$, $c_N(s) = 1$ and $c_N(r) = c_N(s) = 1$. In the first one $A(r'+s') + c_N(r)V_N + c_N(s)V_N = A(r'+s'+2^{-N}) = A(r+s)$, which gives inclusion (8.6.1). In the second case the inductive assumption gives

$$A(r'+s') + c_N(r)V_N + c_N(s)V_N \subset A(r'+s') + V_{N-1}$$
$$= A(r'+s') + A(2^{-N+1}) \subset A(r'+s'+2^{-N+1}) = A(r+s),$$

i.e., (8.6.1) is fulfilled. Note that (8.6.1) yields the inequality

$$f(x+y) \leqslant f(x) + f(y).$$

Indeed, since $f \leqslant 1$, we can assume that $f(x) + f(y) < 1$. For every $\varepsilon > 0$ there exist $r,s \in D$ such that $f(x) < r$, $f(y) < s$, $r + s < f(x) + f(y) + \varepsilon$. From (8.6.1) we have $x + y \in A(r+s)$, whence $f(x+y) \leqslant r+s < f(x) + f(y) + \varepsilon$. Note also that $f(x) = f(-x)$ because $A(r)$ is balanced, $f(0) = 0$ and $f(x) > 0$ if $x \neq 0$, since whenever $x \notin V_n = A(2^{-n})$ we have $f(x) \geqslant 2^{-n}$. Therefore, the function d is a metric invariant with respect to translations. Note also that the sets $B(0,\delta) = \{x\colon f(x) < \delta\} = \bigcup_{r<\delta} A(r)$ are open and balanced. In addition, $B(\delta,0) \subset V_n$ if $\delta < 2^{-n}$. Hence the balls $B(\delta,0)$ form a base of neighborhoods of zero, which completes the proof. $\qquad\square$

The question naturally arises about conditions for normability of a locally convex space, i.e., the existence of a single norm defining its topology. It is easy to see from the results above that such a condition is the existence of a norm p_{α_0} in $\{p_\alpha\}$ such that every other seminorm of this family is estimated by it in the form $p_\alpha \leqslant C_\alpha p_{\alpha_0}$. A. N. Kolmogorov found the following geometric criterion for normability.

8.6.3. Theorem. *A locally convex space is normable precisely when it has a bounded neighborhood of zero.*

PROOF. The necessity of the indicated condition is obvious, since balls with respect to a norm possess the required property. If such neighborhood exists, then it contains a convex balanced neighborhood of zero W. The Minkowski functional p_W gives the desired norm. $\qquad\square$

8.6(ii). The Mackey topology

Here we show that any locally convex topology is the topology of uniform convergence on a certain class of sets, consider the Mackey topology and prove the Mackey–Arens theorem about topologies consistent with the given duality.

First we describe a general construction of a topology of uniform convergence on a class of sets. Suppose we are given a dual pair $\langle X, Y \rangle$. Suppose also that $\mathcal{K}$ is some class of $\sigma(Y, X)$-bounded sets in Y (i.e., $\sup_{f \in K} |f(x)| < \infty$ for every $K \in \mathcal{K}$ and $x \in X$), having the property that the union of all sets in $\mathcal{K}$ separates points in X, which means that whenever $x \neq y$, there exists a functional l from this union with $l(x) \neq l(y)$. Then the family of seminorms

$$p_K(x) := \sup_{f \in K} |f(x)|, \quad K \in \mathcal{K},$$

defines a locally convex topology on X. Convergence of nets in this topology is uniform convergence on every set in $\mathcal{K}$. The locally convex topology we obtained is called the topology of uniform convergence on sets from $\mathcal{K}$. If the class $\mathcal{K}$ admits multiplication by scalars and the union of two sets from $\mathcal{K}$ is contained in a set from $\mathcal{K}$, then the polars of sets from $\mathcal{K}$ form a base of closed convex balanced neighborhoods of zero in X.

A particular role is played by the *Mackey topology* $\tau_M(X, Y)$ — the topology of uniform convergence on $\sigma(Y, X)$-compact convex balanced sets from Y.

8.6.4. Definition. *A set $\mathcal{T}$ of linear mappings from a locally convex space X to a locally convex space Y is called equicontinuous if for every neighborhood of zero $V \subset Y$ there exists a neighborhood of zero $U \subset X$ such that $T(U) \subset V$ for all $T \in \mathcal{T}$.*

In particular, a set $\mathcal{F} \subset X^*$ is equicontinuous if for every $\varepsilon > 0$ there exists a neighborhood of zero $U \subset X$ such that $|f(x)| < \varepsilon$ for all $f \in \mathcal{F}$ and $x \in U$.

8.6.5. Theorem. *Every locally convex topology is the topology of uniform convergence on equicontinuous subsets of the dual space.*

PROOF. We know that every locally convex space X possesses a base of zero consisting of closed convex balanced sets. The polar U° of such a set U is equicontinuous. In addition, $U^{\circ\circ} = U$ by the bipolar theorem, which proves the theorem. $\qquad\square$

We now prove the following Mackey–Arens theorem.

8.6.6. Theorem. *Let $\langle X, Y \rangle$ be a dual pair. A locally convex topology τ on the space X agrees with the given duality, i.e., $(X, \tau)^* = Y$, if and only if $\sigma(X, Y) \leqslant \tau \leqslant \tau_M(X, Y)$. In this case τ is the topology of uniform convergence on some family of $\sigma(Y, X)$-compact convex balanced sets in Y.*

PROOF. If $(X,\tau)^* = Y$, then τ is the topology of uniform convergence on the polars of neighborhoods of zero in X. Such polars are convex, balanced and $\sigma(Y,X)$-compact by the Banach–Alaoglu–Bourbaki theorem. Hence we have $\tau \leqslant \tau_M(X,Y)$. It is also clear that $\sigma(X,Y) \leqslant \tau$, since $\sigma(X,Y)$ is the weakest locally convex topology that agrees with duality.

We now show that the Mackey topology agrees with the given duality. The inclusion $Y \subset \big(X, \tau_M(X,Y)\big)^*$ is obvious. Let us prove the inverse inclusion. Let f be an element of $F := \big(X, \tau_M(X,Y)\big)^*$. By the definition of the Mackey topology there exists a $\sigma(Y,X)$-compact convex balanced set $S \subset Y$ such that

$$S^\circ \subset V := \{x \in X: \ |f(x)| \leqslant 1\}.$$

Then $V^\circ \subset S^{\circ\circ}$, where the polar V is taken in F and the polar of the set S° is taken in X. The set S is compact not only in the topology $\sigma(Y,X)$, but also in the topology $\sigma(F,X)$, since $Y \subset F$. By the bipolar theorem $S = S^{\circ\circ}$. Thus, we have $f \in V^\circ \subset S \subset Y$, which means that $F = Y$. $\qquad\square$

A locally convex space is called a *Mackey space* if its topology is the Mackey topology.

Note that sometimes one has to deal with natural topologies of uniform convergence stronger than the Mackey topology (hence not necessarily agrees with duality). So is the so-called *strong topology* $\beta(X,Y)$, defined as the topology of uniform convergence on all $\sigma(Y,X)$-bounded sets. For example, let $X = l^1$ and $Y = c_0$ be equipped with their natural duality $\langle x, y \rangle = \sum_{n=1}^\infty x_n y_n$. Then $\sigma(Y,X)$-bounded sets in Y are precisely the sets bounded in the usual norm of the space c_0. Hence the dual to X with the topology $\beta(X,Y)$ is l^∞, but not Y.

Topologies of convergence on sets are used for introducing topologies on spaces of linear mappings. If X and Y are locally convex spaces, then, for every class $\mathcal{K}$ of bounded sets in X, the space $\mathcal{L}(X,Y)$ of continuous linear mappings can be equipped with seminorms

$$p_{K,\beta}(T) := \sup\nolimits_{x \in K} p_\beta(Tx),$$

where $\{p_\beta\}$ is the family of seminorms defining the topology of Y. The obtained locally convex topology on $\mathcal{L}(X,Y)$ is called the topology of uniform convergence on sets from $\mathcal{K}$. For example, if for the class $\mathcal{K}$ we take the class of all finite sets, then we obtain the topology of pointwise convergence.

By analogy with the case of normed spaces, a set A in a topological vector space is called *totally bounded* if, for every neighborhood of zero U, there exists a finite set of points $a_1, \ldots, a_n \in A$ with $A \subset \bigcup_{i=1}^n (U + a_i)$.

8.6.7. Theorem. *Let $\mathcal{T}$ be an equicontinuous family of linear mappings between locally convex spaces X and Y. The following topologies coincide on $\mathcal{T}$:*

(i) *the topology of pointwise convergence on an everywhere dense set,*

(ii) *the topology of pointwise convergence,*

(iii) *the topology of uniform convergence on the class of totally bounded sets.*

PROOF. It suffices to verify that the topology from (i) on $\mathcal{T}$ is stronger than the topology from (iii). Let A be an everywhere dense set in X, let $T_\alpha \in \mathcal{T}$ be a net of elements converging to an element $T \in \mathcal{T}$ on every $a \in A$, and let C be a totally bounded set in X. Let us show that convergence is uniform on C, i.e., for every convex balanced neighborhood of zero V in Y there exists an index α_0 such that $(T_\alpha - T)(C) \subset V$ for all $\alpha \geqslant \alpha_0$. By assumption there exists a neighborhood of zero $U \subset X$ such that $T_\alpha(U) \subset V/4$ and $T(U) \subset V/4$ for all α. Since A is everywhere dense and C is totally bounded, there exists a finite set $a_1, \ldots, a_n \in A$ such that $C \subset \bigcup_{i=1}^{n}(a_i + U)$. By the pointwise convergence on A there is an index α_0 such that $(T_\alpha - T)(a_i) \subset V/2$ for all $\alpha \geqslant \alpha_0$ and $i \leqslant n$. Then $(T_\alpha - T)(a_i + U) \subset V$, whence $(T_\alpha - T)(C) \subset V$ for all indices $\alpha \geqslant \alpha_0$. $\square$

The Mackey–Arens theorem gives a hint how to construct the *completion of a locally convex space* by analogy with the completion of a normed space.

8.6.8. Definition. *A net $\{x_\alpha\}$ in a topological vector space E is called Cauchy or fundamental if, for every neighborhood of zero V, there exists an index α_0 such that $x_\alpha - x_\beta \in V$ for all $\alpha, \beta \geqslant \alpha_0$.*

A set A in E is called complete if every Cauchy net of elements of A has a limit in this set. If every fundamental countable sequence in A has a limit in A, then A is called sequentially complete.

If E is linearly homeomorphic to a dense linear subspace in a complete topological vector space $\widehat{E}$, then $\widehat{E}$ is called a completion of E.

8.6.9. Theorem. *Every locally convex space E has a completion, moreover, a completion is unique up to a linear homeomorphism.*

A proof can be read in Robertson, Robertson [512, Chapter 6] or in Bogachev, Smolyanov [75, § 1.7]. We only note that one can use the Mackey–Arens theorem to represent X as the space of functionals on X^* with the topology of uniform convergence on some family $\mathcal{A}$ of convex balanced $\sigma(X^*, X)$-compact sets, then for the completion one can take the subspace

$$\widehat{E} := \bigcap_{A \in \mathcal{A}} (E + A^\circ)$$

of the space $(E^*)'$ (the algebraic dual to E^*), where polars are taken in $(E^*)'$ and $\widehat{E}$ is equipped with the topology of uniform convergence on sets from $\mathcal{A}$.

A weaker property than completeness is quasi-completeness: a locally space is called *quasi-complete* if every bounded set is contained in a complete set. Hence such a space is sequentially complete. There is a version of the James theorem (see Theorem 6.10.12) for locally convex spaces that characterizes weakly compact sets. See [147, Chapter 1, Theorem 5] for a proof.

8.6.10. Theorem. *Let B be a weakly closed bounded set in a quasi-complete locally convex space. Then B is weakly compact if and only if every continuous linear functional attains its maximum on B.*

8.6(iii). Inductive and projective limits

Suppose we are given a linear space X and a family of locally convex spaces X_α with certain linear mappings $f_\alpha \colon X_\alpha \to X$ such that $X = \bigcup_\alpha f_\alpha(X_\alpha)$. Then on X there is the strongest locally convex topology with respect to which all mappings f_α are continuous. A base of neighborhoods of zero in this topology consists of convex balanced sets U such that $f_\alpha^{-1}(U)$ is a neighborhood of zero in X_α for all α. The space X with the indicated topology is called the *inductive limit* of the spaces X_α.

The most important special case is the situation where we are given a sequence of locally convex spaces X_n with topologies τ_n and X_n is a linear subspace in X_{n+1} such that the restriction of τ_{n+1} to X_n coincides with τ_n. The space $X = \bigcup_{n=1}^\infty X_n$ with the topology of the inductive limit of the spaces X_n with natural embeddings $f_n \colon X_n \to X$ is called the *strict inductive limit* of X_n. The following assertions are true (proofs can be found in [75], [512], and [533]).

8.6.11. Theorem. *The restriction of the inductive topology from X to X_n is τ_n. A linear mapping from X to a locally convex space Y is continuous precisely when its restrictions to all X_n are continuous.*

8.6.12. Theorem. *Let X be the strict inductive limit of a sequence of increasing spaces X_n such that X_n is closed in X_{n+1}. A set in X is bounded precisely when it is contained and bounded in some X_n.*

8.6.13. Example. The space $\mathcal{D}(\mathbb{R}^n)$ is the strict inductive limit of the spaces $\mathcal{D}_m$ (a proof is Exercise 8.6.80), $\mathcal{D}_m$ is closed in $\mathcal{D}_{m+1}$.

One should bear in mind that in order a set A in the strict inductive limit of spaces X_n be open or closed it is not enough that all intersections $A \cap X_n$ be open or closed in X_n, respectively (see Exercise 8.6.45). The space $\mathcal{D}(\mathbb{R}^1)$ possesses even a non-closed linear subspace the intersections of which with all $\mathcal{D}_m$ are closed (Exercise 8.6.83).

Suppose now that in place of mappings $f_\alpha \colon X_\alpha \to X$ we are given linear mappings $g_\alpha \colon X \to X_\alpha$ such that $\bigcap_\alpha g_\alpha^{-1}(0) = 0$. The *projective topology* on X is defined as the weakest locally convex topology with respect to which all g_α are continuous; the space X is called the *projective limit* of X_α. If the topology of X_α is given by seminorms $p_{\alpha,\beta}$, then the topology of X is defined by the seminorms $p_{\alpha,\beta} \circ g_\alpha$.

8.6.14. Example. The space $\mathcal{S}(\mathbb{R}^n)$ is the projective limit of the sequence of the spaces of k-fold continuously differentiable functions with finite norms $p_{k,k}$ from Example 8.1.3(iii).

8.6(iv). Barrelled and bornological spaces

Let X be a locally convex space. A set $A \subset X$ is called *absorbing* or *absorbent* if, for every $x \in X$, there exists a number $\lambda > 0$ such that $x \in \mu A$ whenever $|\mu| \geqslant \lambda$. A *barrel* in X is an absorbing balanced convex closed set.

The space X is called *barrelled* if every barrel in X contains a neighborhood of zero.

8.6.15. Example. Every Baire locally convex space (i.e., a space that is not a first category set) is barrelled. For example, Banach spaces are barrelled and complete metrizable locally convex spaces are barrelled. Indeed, if A is a barrel, then X is the union of closed sets nA. Hence some nA has interior points, whence by the absolute convexity of A we obtain that A contains a neighborhood of zero.

8.6.16. Proposition. *Let X be a barrelled locally convex space and let $\{p_\alpha\}$ be a family of continuous seminorms on X such that $p(x) := \sup_\alpha p_\alpha(x) < \infty$ for all $x \in X$. Then p is a continuous seminorm.*

PROOF. The set $\{x \colon p(x) \leqslant 1\} = \bigcap_\alpha \{x \colon p_\alpha(x) \leqslant 1\}$, is, as one can easily see, closed, convex and balanced. In addition, it is absorbing and hence contains a neighborhood of zero. This gives the continuity of p (Exercise 8.6.42). $\quad\square$

8.6.17. Corollary. *Let X be a barralled space and let $f_n \in X^*$ be functionals such that for every x there exists a finite limit $f(x) = \lim\limits_{n\to\infty} f_n(x)$. Then the functional f is continuous.*

A space X is barrelled precisely when for every Fréchet space Y (see p. 362) every linear mapping from X to Y with a closed graph is continuous (see [75, § 3.5] or [533, Chapter IV, §8.6]).

Let us now consider *bornological* spaces. This class consists of locally convex spaces in which every convex balanced set A with the property that for every bounded set B there exists a number $\lambda > 0$ such that $B \subset \lambda A$ contains a neighborhood of zero. The space X is bornological precisely when the continuity of a seminorm on X is equivalent to its boundedness on all bounded sets.

8.6.18. Example. Every metrizable locally convex space is bornological. Indeed, suppose that a seminorm p is bounded on bounded sets. If p is discontinuous, then there exists a sequence $x_n \to 0$ such that $p(x_n) \to +\infty$. This is impossible by the boundedness of $\{x_n\}$.

8.6.19. Proposition. *Let X be a bornological space. A linear function f on the space X is continuous precisely when $f(x_n) \to 0$ as $x_n \to 0$.*

PROOF. We have to verify the continuity of every sequentially continuous linear function f. If the function f is discontinuous, then the seminorm $p = |f|$ is discontinuous as well. Hence this seminorm is unbounded on some bounded set B and there are vectors $x_n \in B$ such that $c_n := |f(x_n)| \to \infty$. Then $c_n^{-1/2} x_n \to 0$, whence $f(c_n^{-1/2} x_n) \to 0$ contrary to the equality $|f(c_n^{-1/2} x_n)| = c_n^{1/2}$. $\quad\square$

8.6.20. Proposition. *Let X be the inductive limit of locally convex spaces X_α. If all X_α are barrelled, then X is barrelled as well, and if all X_α are bornological, then so is X.*

The proof is delegated to Exercise 8.6.75.

8.6.21. Example. The space $\mathcal{D}(\mathbb{R}^n)$ is barrelled and bornological.

8.6(v). Banach spaces generated by Minkowski functionals

We have seen in Theorem 6.3.6 that any convex balanced set V generates a seminorm p_V (the Minkowski functional of the set V) if zero belongs to its algebraic kernel (defined before that theorem). The latter condition is always fulfilled on the linear subspace E_V, i.e., on the set $\bigcup_{n=1}^{\infty} nV$. Indeed, the algebraic kernel of V in this subspace is the set $U = \bigcup_{\lambda \in [0,1)} \lambda V$, since for every $u \in U$ and every $v \in E_V$ there exists $\varepsilon > 0$ such that $u + tv \in V$ if $|t| < \varepsilon$. For ε we can take the number $n^{-1}(1 - \lambda)$, where n and $\lambda \in [0,1)$ are such that $u = \lambda v_0$, $v = nv_1$, $v_0, v_1 \in V$. In addition, if we assume that V contains no straight lines, then p_V is a norm on E_V. Here we give several useful results about properties of the space (E_V, p_V).

8.6.22. Theorem. *Let V be a bounded convex balanced set in a Hausdorff topological vector space E and let V be sequentially complete. Then the space E_V with the norm p_V is Banach, the set V is the closed unit ball in it and the norm topology on E_V is stronger than the original topology.*

PROOF. It follows from our discussion above that p_V is a norm on E_V, since in a Hausdorff topological vector space a bounded set contains no straight lines. Let us show that $V = U_1 := \{x \in E_V : p_V(x) \leqslant 1\}$. It is clear that $V \subset U_1$. Let $p_V(u) = 1$. Then there exist numbers $r_n \in (0, 1/n)$ such that $u \in (1 + r_n)V$, i.e., $r_n(r_n + 1)^{-1}u \in V$. By the sequential closedness of V (following from the sequential completeness and the Hausdorff property) we obtain that $u \in V$. Hence $U_1 \subset V$. It also follows that the norm topology in E_V is stronger than the original one. Let us prove the completeness of E_V. Let $\{x_n\}$ be a Cauchy sequence in E_V. Then $\sup_n p(x_n) = k < \infty$, i.e., $x_n \in kV$ for all n. The set kV is also sequentially complete. Since the norm topology is stronger than the original one, the sequence $\{x_n\}$ is fundamental in the topology of E and hence converges to some element $x \in kV$. We have $\varepsilon_{n,m} := p_V(x_n - x_m) \to 0$ as $n, m \to \infty$. Then $(x_n - x_m)/\varepsilon_{n,m} \in V$. Set $\varepsilon_n := n^{-1} + \limsup_{m \to \infty} \varepsilon_{n,m}$. Then $\varepsilon_n \to 0$. We show that $(x - x_n)/\varepsilon_n \in V$. Indeed, for any fixed n the bounded sequence of numbers $\varepsilon_{n,m}$ contains a subsequence ε_{n,m_j} converging to some $\delta_n \leqslant \varepsilon_n$. If $\delta_n > 0$, then the elements $(x_n - x_{m_j})(\delta_n + \varepsilon_{n,m_j})^{-1}$ also belong to V and converge in the original topology to $(x_n - x)(n^{-1} + \delta_n)$. By the sequential closedness of V we obtain $(x_n - x)(n^{-1} + \delta_n) \in V$. Then $(x - x_n)/\varepsilon_n \in V$, because $\varepsilon_n \geqslant n^{-1} + \delta_n$. Thus, $p_V(x - x_n) \leqslant \varepsilon_n$, i.e., $x_n \to x$ in E_V. $\square$

A typical example of a set V satisfying the conditions of this theorem is any convex balanced compact set. The norm topology on V is usually strictly stronger than the original topology. For example, if V is compact in E, then it can be compact in E_V only when E_V is finite-dimensional. It is sometimes useful to have a larger normed space p_W which induces on V the original topology. This becomes possible in metrizable locally convex spaces.

8.6.23. Theorem. *Let V be a bounded set in a metrizable locally convex space E. Then in X there exists a bounded closed convex balanced set W containing V such that the norm p_W induces on V the same topology as E. In particular, if the set V is compact, then it is also compact in the normed space E_W. If E is complete and V is compact, then W can be found compact.*

PROOF. By the metrizability of E there is a sequence of convex balanced neighborhoods of zero U_n in E such that every neighborhood of zero contains at least one of the sets U_n. We can assume that V is convex and balanced, since the convex balanced envelope of V is bounded. By the boundedness of V there exist numbers $\alpha_n \geqslant n$ such that $V \subset \alpha_n U_n$. Set $W := \bigcap_{n=1}^{\infty} n\alpha_n \overline{U}_n$. It is clear that we have obtained a balanced convex closed set. This set is bounded, since for every n we have $W \subset n\alpha_n U_n$. For every $\varepsilon > 0$, there exists n_1 such that $\varepsilon < 1/n_1$. Hence $V \subset \varepsilon n\alpha_n U_n$ if $n \geqslant n_1$. In addition, for some m we have $U_m \subset \bigcap_{k=1}^{n_1} \varepsilon k\alpha_k U_k$. Therefore, $V \cap U_m \subset \varepsilon n\alpha_n U_n$ for all n, i.e., $V \cap U_m \subset \varepsilon W$. It follows that every neighborhood of every point $v \in V$ in the norm topology p_W contains a neighborhood of this point in the topology induced in V by the original topology of E. Indeed, let $v \in V$. For every $\varepsilon > 0$ we can find m such that $V \cap U_m \subset \varepsilon W/2$. Then $(x + U_m) \cap V \subset (x + \varepsilon W) \cap V$. This is seen from the fact that if $u \in U_m$ and $v + u \in V$, then $u \in V - V \subset 2V$, because V is convex and balanced. Then $u/2 \in V \cap U_m \subset \varepsilon W/2$, i.e., $u \in \varepsilon W$. Thus, both topologies coincide on V.

If V is compact, then it is compact in E_W. Let us show that if E is complete, then W can be chosen compact. Passing to E_W, we reduce this assertion to the case of a Banach space E. According to Proposition 5.5.7, the compact set V is contained in the closed convex envelope of some sequence $\{x_n\}$ converging to zero. Set $\alpha_n := 1 + \|x_n\|^{-1/2}$ if $x_n \neq 0$ and $\alpha_n = 0$ if $x_n = 0$. The sequence $\{\alpha_n x_n\}$ also tends to zero. Let B be its closed balanced convex envelope. It is clear that B is compact and $V \subset B$. Let us show that V is compact in the Banach space (E_B, p_B). We have $p_B(x_n) \to 0$, since $p_B(\alpha_n x_n) \leqslant 1$. Hence the closed convex envelope C of the sequence $\{x_n\}$ in the space E_B is compact. Then C is compact in E, whence the equality $C = V$ follows, since the set C contains the convex envelope of $\{x_n\}$, which is dense in V. $\qquad\square$

Even if V is a convex balanced compact set in a separable Hilbert space, the Banach space (E_V, p_V) is not always separable. For example, let us take the set $V = \{(x_n): \sup_n |x_n| \leqslant 1\}$ in the weighted Hilbert space

$$E = \Big\{ x = (x_n): \|x\|^2 := \sum_{n=1}^{\infty} n^2 x_n^2 < \infty \Big\}.$$

In this case E_V coincides with l^∞. However, the following assertion is valid.

8.6.24. Theorem. *Let E be a complete metrizable locally convex space and let K be a compact set in E. Then there exists a convex balanced compact set V, containing K, such that the Banach space (E_V, p_V) is separable and K is compact in it.*

PROOF. On account of the completeness of the space E and the two previous theorems there exists a balanced convex compact set W_0, containing K, such that K is compact as a subset of the Banach space E_{W_0}. Then the linear span of K in E_{W_0} is separable with respect to the norm p_{W_0}. The closure of E_0 in this linear span by the norm p_{W_0} on E_{W_0} gives the desired separable Banach space E_W, the unit ball of which is the set $W := W_0 \cap E_0$. $\qquad\square$

One can go further and obtain the following useful assertion — the Davies–Fiegel–Johnson–Pełczyński theorem.

8.6.25. Theorem. *Let X be a Banach space and let K be a convex balanced weakly compact set in X. Then there exists a bounded closed balanced convex set W, containing K, such that the Banach space (E_W, p_W) is reflexive. If K is compact, then W can be chosen also compact and E_W can be made separable.*

PROOF. The set K is bounded and closed, being weakly compact and convex. For every $n \in \mathbb{N}$, we set $U_n := 2^n K + 2^{-n} U$, where U is the closed unit ball in X. It is not difficult to verify that the sets U_n are closed. Every norm p_{U_n} is equivalent to the norm of X. By means of these norms we construct p_W. Let

$$W := \Big\{ x \in X : \ \textstyle\sum_{n=1}^{\infty} p_{U_n}(x)^2 \leqslant 1 \Big\}.$$

It is readily seen that W is closed. Hence the space (E_W, p_W) is complete. It is clear that $p_W(x)^2 = \sum_{n=1}^{\infty} p_{U_n}(x)^2$.

Let $v \in K$. Then $2^n v \in 2^n K$ and hence $(2^n + \varepsilon)v \in U_n$ for all sufficiently small $\varepsilon > 0$ (if $K \subset U$, then for all $\varepsilon < 2^{-n}$). Hence $p_{U_n}(v) \leqslant 2^{-n}$, whence $p_W(v) < 1$, i.e., $v \in W$. It is clear that W is a bounded set in X, so the natural embedding $j \colon (E_W, p_W) \to X$ is continuous.

We now show that the mapping $j^{**} \colon E_W^{**} \to X^{**}$ is injective and that $(j^{**})^{-1}(X) = E_W$. Indeed, let us set $X_n := (E_{U_n}, p_{U_n})$ and consider the l^2-sum of the spaces X_n, i.e., the Banach space of all sequences

$$Z := \Big(\textstyle\sum_n \oplus X_n \Big)_{l^2} := \Big\{ z = (z_n) : \ z_n \in X, \ \textstyle\sum_{n=1}^{\infty} \|z_n\|_X^2 < \infty \Big\}$$

with the norm $\|z\|_Z = \big(\sum_{n=1}^{\infty} \|z_n\|_X^2 \big)^{1/2}$. There is an obvious linear isometry ψ between E_W and a closed subspace in Z defined by the formula $\psi(x) = (x, x, \ldots)$. It is not difficult to observe that the mapping

$$\psi^{**} \colon E_W^{**} \to Z^{**}$$

takes any $F \in E_W^{**}$ to the element $(j^{**}F, j^{**}F, \ldots) \in \big(\sum_n \oplus X_n^{**} \big)_{l^2}$. Since the mapping ψ preserves the norm, so is ψ^{**}. This yields the injectivity of j^{**}. In addition, $(\psi^{**})^{-1}\big(\psi(E_W) \big) = E_W$, which gives the equality $(j^{**})^{-1}(X) = E_W$.

Let us verify the reflexivity of E_W. Denote by B the closed unit ball in E_W^{**}. We observe that the $\sigma(X^{**}, X^*)$-closure of the set W in E_W^{**} coincides with $j^{**}(B)$. Indeed, the set B is compact in the topology $\sigma(E_W^{**}, E_W^*)$ and W is dense in B in this topology. The mapping $j^{**} \colon E_W^{**} \to X^{**}$ is continuous with respect to the topologies $\sigma(E_W^{**}, E_W^*)$ and $\sigma(X^{**}, X^*)$. Hence $j^{**}(B)$ is compact in the topology $\sigma(X^{**}, X^*)$. Since $j^{**}(W) = W$, the set W is also dense in the compact set $j^{**}(B)$ in the topology $\sigma(X^{**}, X^*)$, i.e., $j^{**}(B)$ is the closure of W in the indicated topology.

All sets $2^n K + 2^{-n} B$ in X^{**} are closed in the topology $\sigma(X^{**}, X^*)$ by the compactness of K in the topology $\sigma(X, X^*)$. Each of them contains W, which is clear from the definition of p_W. Therefore, each of these sets contains the $\sigma(X^{**}, X^*)$-closure of W, i.e., the set $j^{**}(B)$. Thus,

$$j^{**}(B) \subset \textstyle\bigcap_{n=1}^{\infty} (2^n K + 2^{-n} B) \subset \textstyle\bigcap_{n=1}^{\infty} (X + 2^{-n} B) = X.$$

According to the facts proved above, we obtain $E_W^{**} \subset (j^{**})^{-1}(X) \subset E_W$, which shows the reflexivity of (E_W, p_W).

If K is compact, then we first use the previous theorem to find an intermediate Banach space E_{W_1} in which K is compact and whose closed unit ball W_1 is compact in E. Repeating this construction for W_1, we find a Banach space E_{W_2} continuously embedded into X in which W_1 is compact and the unit ball W_2 of which is compact in X. We now apply the assertion proved above to the compact set W_1 in E_{W_2} and find a reflexive Banach space E_W continuously embedded into E_{W_2} in which W_1 is bounded. Then K is compact in E_W. The closed unit ball W of the space E_W is contained in the set W_2 that is compact in X. Hence the set W is totally bounded. Moreover, it is closed, since it is convex and weakly compact by the weak compactness of the closed balls in a reflexive space and the weak continuity of the embedding of E_W into X (see Theorem 6.6.8). Finally, the space E_W can be made separable by replacing it with the closure of the linear span of K in the norm p_W. □

8.6.26. Corollary. *In the situation of Theorem 8.6.24, the Banach space (E_V, p_V) can be chosen separable reflexive.*

PROOF. By Theorem 8.6.24 we can find a separable space E_V in which K is compact. Next, by the previous theorem we find a reflexive Banach space E_W, continuously embedded into E_V, in which K is also compact. The closure of the linear span of K is the desired separable reflexive space (actually, the space E_W itself turns out to be separable by Exercise 6.10.164). □

Can one go yet further and obtain for E_W a Hilbert space or a space with a Schauder basis? The answer is negative (see Fonf, Johnson, Pisier, Preiss [**688**]).

In case of a compact set V, the topology of the space E_V is much stronger than the original topology of E. Nevertheless, as the following result shows, on E_V there are sufficiently many linear functionals continuous with respect to this original topology.

8.6.27. Proposition. *Let V be a compact convex balanced set in a locally convex space E. Let B^* be the unit ball in the dual to the Banach space E_V. Then the set of all functionals in B^* continuous with respect to the topology induced from E is dense in B^* in the topology of uniform convergence on compact sets from E_V.*

PROOF. According to Theorem 8.6.7, it suffices to verify that the indicated set F is dense in the topology of pointwise convergence. Let $f \in B^*$, let $\varepsilon > 0$, and let $v_1, \ldots, v_n \in V$ be a finite collection of vectors. We have to find a functional $g \in E^*$ such that $g|_{E_V} \in B^*$ and $|f(v_i) - g(v_i)| < \varepsilon$ for all $i = 1, \ldots, n$. We can assume that $f(v_1) \neq 0$ (if $f(v_i) = 0$ for all i, then we take $g = 0$). Denote by L the linear subspace generated by $v_1, \ldots, v_n$. Let us consider the closed affine subspace $H = L \cap f^{-1}(1 + \varepsilon)$ in L. It is also closed in E, since L is finite-dimensional. By the compactness of V there exists an open convex set U in E such that $V \subset U$ and $U \cap H = \varnothing$. By the Hahn–Banach theorem there exists a closed hyperplane H_1 in E with $H \subset H_1$ and $H_1 \cap U = \varnothing$. This hyperplane has the form $H_1 = g^{-1}(0)$, where $g \in E^*$. Since $V \cap H_1 = \varnothing$, we have $|g(v)| \leqslant 1$ for all $v \in V$, i.e., $g|_{E_V} \in B^*$. In addition, $H_1 \cap L = H$, since H is a hyperplane in L, $H \subset H_1$ and $H_1 \cap L$ differs from L due to the fact that $v_i \notin H_1$. Therefore, for all $x \in L$ we have $f(x) = (1+\varepsilon)g(x)$, whence $|f(v_i) - g(v_i)| \leqslant \varepsilon|g(v_i)| \leqslant \varepsilon$, because $v_i \in V$, $g|_{E_V} \in B^*$. $\qquad\square$

The convex envelope of a compact set can fail to be compact even in the Hilbert space if it is infinite-dimensional (compactness of the convex envelope of any compact set in a finite-dimensional space is the subject of Exercise 5.6.32). This happens for the sequence $n^{-1}e_n \to 0$, where $\{e_n\}$ is an orthonormal basis, with the added limit 0. However, the following useful assertion is valid.

8.6.28. Proposition. *Suppose we are given a finite collection of convex compact sets in a topological vector space X. Then the convex envelope and the convex balanced envelope of their union are compact. In particular, the convex balanced envelope of a convex compact set is also compact.*

PROOF. It is clear that it suffices to prove our assertion for two sets. The set $V := K_1 \times K_2 \times S$, where $S := \{(t_1, t_2) \in \mathbb{R}^2 : t_i \geqslant 0, t_1 + t_2 = 1\}$, is compact. Its image W under the continuous mapping $(x_1, x_2, t_1, t_2) \mapsto t_1 x_1 + t_2 x_2$ is compact as well. It is clear that $W \subset \operatorname{conv}(K_1 \cup K_2)$. For the proof of the inverse inclusion it suffices to verify the convexity of W. Let $u, v \in W$ and $0 < \lambda < 1$. Then $u = t_1 x_1 + t_2 x_2$, $v = s_1 y_1 + s_2 y_2$, where $x_i, y_i \in K_i$, $t_i, s_i \geqslant 0$, $t_1 + t_2 = s_1 + s_2 = 1$. We show that $\lambda u + (1 - \lambda)v \in W$. If one of the numbers $t_1 + s_1$ or $t_2 + s_2$ is zero, then this is obvious. If both numbers are nonzero, then $\lambda u + (1 - \lambda)v = \tau z_1 + (1 - \tau)z_2$, where $\tau = \lambda t_1 + (1 - \lambda)s_1$,

$$z_1 = \lambda t_1 \big(\lambda t_1 + (1 - \lambda)s_1\big)^{-1} x_1 + (1 - \lambda)s_1 \big(\lambda t_1 + (1 - \lambda)s_1\big)^{-1} y_1 \in K_1,$$

$$z_2 = \lambda t_2 \big(\lambda t_2 + (1 - \lambda)s_2\big)^{-1} x_2 + (1 - \lambda)s_2 \big(\lambda t_1 + (1 - \lambda)s_1\big)^{-1} y_2 \in K_2.$$

This is clear from the equality $\lambda t_1 + (1 - \lambda)s_1 + \lambda t_2 + (1 - \lambda)s_2 = 1$. The case of the convex balanced envelope is similar. $\qquad\square$

Let K be a compact set in a locally convex space X and let μ be a bounded measure on the σ-algebra in K generated by all functionals $l \in X^*$. Then we can define a functional on X^* by the formula

$$I(\mu)(l) := \int_K l(x)\, \mu(dx).$$

If there is a vector $h_\mu \in X$ with $l(h)(\mu) = I(\mu)(l)$, then h_μ is called the *mean* or *barycenter* of μ. If the means exist for the restrictions of μ to all measurable sets, then $I(\mu)$ is called the *Pettis integral*.

8.6.29. Lemma. *If X is sequentially complete and the compact set K is metrizable, then every Borel measure μ on K has a mean $I(\mu)$ in X. If μ is a probability measure, then $I(\mu)$ belongs to the closed convex envelope of K.*

PROOF. According to Exercise 1.9.68 there exist functionals $l_j \in X^*$ such that the topology of K is defined by the metric $d(x, y) := \sum_{j=1}^{\infty} 2^{-j}|l_j(x - y)|$. Let us take refining finite partitions of K into Borel parts $K_{n,i}$, $i \leqslant N_n$ of diameter less than 2^{-n} (in the metric d). Pick $x_{n,i} \in K_{n,i}$ and set $I_n := \sum_{i=1}^{N_n} x_{n,i} \mu(K_{n,i})$. Let us show that the vectors I_n converge; then their limit will be the mean of μ. By condition, it suffices to verify that $\{I_n\}$ is a Cauchy sequence. Let p be a continuous seminorm on X and $\varepsilon > 0$. The compactness of K yields that there exists $\delta > 0$ such that $p(x - y) \leqslant \varepsilon$ if $x, y \in K$ and $d(x, y) \leqslant \delta$. Let $2^{-n_0} < \delta$. Let $m > n \geqslant n_0$. Then I_n and I_m can be written as

$$I_n = \sum_{i=1}^{k} a_{n,i} \mu(E_i), \quad I_m = \sum_{i=1}^{k} a_{m,i} \mu(E_i),$$

where $\{E_i\}$ is the partition obtained by refining the partitions $\{K_{n,i}\}$ and $\{K_{m,i}\}$, $a_{n,i}, b_{n,i} \in E_i$. Since $d(a_{n,i}, a_{m,i}) < \delta$, we have $p(a_{n,i} - a_{m,i}) \leqslant \varepsilon$. Hence $p(I_n - I_m) \leqslant \sum_{i=1}^{k} p(a_{n,i} - a_{m,i}) |\mu(E_i)| \leqslant \varepsilon |\mu|(K)$. Note that $I(\mu)$ is the Pettis integral, which is proved by the same reasoning. □

8.6.30. Theorem. *The closed absolutely convex envelope A of a metrizable compact set K in a locally convex space X is metrizable, and if X is sequentially complete, then it is compact.*

PROOF. The first assertion follows from the second one, since X has a completion (which is certainly sequentially complete). Let X be sequentially complete. By the Riesz theorem $C(K)^*$ is the space of Borel measures on K. The closed unit ball U in $C(K)^*$ is compact in the weak-$*$ topology. Since $C(K)$ is separable (Exercise 1.9.69), the set U is a metrizable compact space in the weak topology (Exercise 6.10.23). Let us consider the mapping $I: U \to X$ from the previous lemma. It is readily seen that this mapping is continuous if U is considered with the weak-$*$ topology and X with the weak topology. Hence the absolutely convex set $I(U)$ is weakly compact in X. By the metrizability of U this set is metrizable (Exercise 6.10.123). It is clear that $A \subset I(U)$, since $K \subset I(U)$ by the equality $k = I(\delta_k)$, where δ_k is the probability measure at the point k. Hence A is a metrizable compact set as a closed subset of a metrizable compact space (actually, as one can easily verify, $I(U)$ coincides with A). □

Related to integral representations of the considered form is an important theorem due to Choquet, which gives a representation of points of a convex compact set in the form of the means of probability measures on the set of extreme points. About this theorem, see Akilov, Kutateladze [8], Alfsen [11], and Phelps [479]; extreme points are also discussed in the next subsection.

8.6(vi). The Krein–Milman theorem

A point v of a set V in a linear space E is called an *extreme point* of V if it cannot be represented in the form $ta + (1 - t)b$, where $a, b \in V$ and $t \in (0, 1)$. In other words, the point v is not an inner point of an interval with endpoints in V.

A *supporting closed affine subspace* for the convex set V is any set M of the form $M = L + v$, where $v \in V$, L is a closed linear subspace in X, satisfying the following condition: if points $a, b \in V$ are such that $ta + (1 - t)b \in M$ for some $t \in (0, 1)$, then the whole interval $[a, b]$ belongs to M. If L is of codimension 1, then we arrive at the previously introduced notion of a supporting hyperplane.

8.6.31. Theorem. (THE KREIN–MILMAN THEOREM) *Every nonempty convex compact set in a locally convex space is the closed convex envelope of its extreme points.*

PROOF. We can assume that the space is real, since the properties considered in the theorem do not change when we pass to real spaces. We prove the following auxiliary assertion of independent interest:

Every closed hyperplane supporting for a convex compact set C in a real locally convex space contains at least one extreme point of C.

For the proof we take any supporting closed hyperplane H for C, which exists by Corollary 8.3.9. Let us consider the set $\mathcal{M}$ of all supporting closed affine subspaces for C contained in H and partially ordered by the inverse inclusion, i.e., $M_1 \leqslant M_2$ if $M_1 \supset M_2$. Let $\{M_\alpha\}$ be some chain in $\mathcal{M}$. The set $M := \bigcap_\alpha M_\alpha$ is nonempty, since the intersection of the compact sets $M_\alpha \cap V$ is nonempty by the compactness of V and the fact that every finite subfamily of $M_\alpha \cap V$ contains one of the elements of this subfamily by the definition of a chain. It is readily seen that the set M is a supporting closed affine subspace. Moreover, $M_\alpha \leqslant M$ for all α, i.e., M is a majorant for this chain. By Zorn's lemma $\mathcal{M}$ has a maximal element M_0 (from the viewpoint of the direct inclusion this element is minimal). We show that M_0 consists of a single point. Then the definition of a supporting affine subspace yields that this point is extreme. Let M_0 be different from a singleton. The set $C_0 := C \cap M_0$ is nonempty convex compact in the affine subspace M_0. Since M_0 is at least one-dimensional, it contains a closed hyperplane M_1 supporting for C_0. Let us show that $M_1 \subset \mathcal{M}$. Indeed, if $a, b \in C$ and $at + (1-t)b \in M_1$ for some $t \in (0,1)$, then $at + (1-t)b \in M_0$, whence $[a,b] \subset M_0$, i.e., $[a,b] \subset C_0$ and hence $[a,b] \subset M_1$. Thus, $M_1 \subset \mathcal{M}$ contrary to our choice of M_0. The contradiction we obtained proves our auxiliary assertion.

Let us proceed to the main assertion. Denote by D the closed convex envelope of the set of all extreme points of the given convex compact set C. It is clear that $D \subset C$. Suppose that there exists a point $z \in C \backslash D$. By a corollary of the Hahn–Banach theorem there exists a continuous linear functional f on X and a point α such that $f(z) > \alpha$ and $f(x) \leqslant \alpha$ for all $x \in D$. Let $\beta := \max_{x \in C} f(x)$. Since $z \in C$ and $D \subset C$, we have $\beta > \alpha$. Hence the hyperplane $H := f^{-1}(\beta)$ does not intersect D. This leads to a contradiction, because according to the auxiliary assertion above H contains an extreme point of C, since H is a supporting hyperplane: C has a point x with $f(x) = \beta$. □

The compactness of V cannot be replaced by the closedness and boundedness even in the case of a Banach space. For example, the closed unit ball of the Banach space c_0 has no extreme points. Of course, the theorem applies to weakly compact convex sets in Banach spaces (which need not be norm compact).

The following result due to D. P. Milman is also of interest.

8.6.32. Theorem. *Let K be a compact set in a locally convex space X such that its closed convex envelope C is compact (which is automatically the case if X is complete). Then every extreme point of the set C belongs to K.*

PROOF. Let x be an extreme point of C and U a convex neighborhood of zero. There are points $x_1, \ldots, x_n$ in K such that $K \subset \bigcup_{i=1}^{n}(x_i + U)$. Let V_i be the closed convex envelope of $K \cap (x_i + U)$. The sets V_i belong to C and hence are compact. Hence the convex envelope of their union is also compact on (see Proposition 8.6.28) and so equals C. Thus, $x = \sum_{i=1}^{n} \lambda_i v_i$, where $v_i \in V_i$, $\lambda_i \geqslant 0$ and $\sum_{i=1}^{n} \lambda_i = 1$. Since x is an extreme point of C, we have $x = v_i$ for some i, whence $x \in x_i + U \subset K + U$. Since U was arbitrary and K is closed, we obtain that $x \in K$. $\square$

8.6(vii). The measurable graph theorem

Here we give without proof an interesting generalization of the closed graph theorem: the measurable graph theorem. A *Souslin set* is the image of a complete separable metric space under a continuous mapping. Every Borel set in a complete separable metric space is Souslin (see [73], v. 2, Chapter 6).

8.6.33. Theorem. (i) *Let X and Y be separable complete metrizable topological vector spaces and let $A \colon X \to Y$ be a linear mapping. If the graph of A is a Souslin set (for example, a Borel set), then A is continuous.*

(ii) *If Z is a Souslin (for example, Borel) linear subspace in a separable complete metrizable topological vector space X and the codimension of Z is finite or countable, then Z is closed and has a finite codimension.*

For the proofs, see [**75**, § 3.12], [**116**, Chapter 5]. It follows from this theorem that the kernel of a discontinuous linear functional on a Banach space cannot be a Borel or Souslin set. Indeed, the codimension of the kernel is one, hence it is closed, but this implies the continuity of the functional.

8.6(viii). Fix point theorems in locally convex spaces

Schauder's theorem (see Theorem 5.5.3) was extended to locally convex spaces by Leray and Tychonoff. The formulation is the same and the proof is also similar, so we do not include it.

8.6.34. Theorem. *Let K be a convex compact set in a locally convex space and let $f \colon K \to K$ be a continuous mapping. Then there exists a point $x \in K$ with $f(x) = x$.*

There are two other useful fixed point theorems for families of affine mappings. The first one is due to Markov and Kakutani and deals with commuting affine mappings.

8.6.35. Theorem. *Let K be a convex compact set in a locally convex space and let G be a family of pairwise commuting continuous mappings from K to K such that $g(\lambda x + (1 - \lambda)y) = \lambda g(x) + (1 - \lambda)g(y)$ for all $x, y \in K$, $g \in G$. Then, there exists a point $x_0 \in K$ with $g(x_0) = x_0$ for all $g \in G$.*

Kakutani proved that if G is a group of equicontinuous affine mappings, then their commutativity is not needed.

8.6.36. Theorem. *Let K be a convex compact set in a locally convex space and let G be a group of affine mappings from K to K equicontinuous on K. Then, there exists $x_0 \in K$ with $g(x_0) = x_0$ for all $g \in G$.*

Proofs of these theorems can be found in [**522**, Chapter 5]. In Chapter 11 we apply Kakutani's theorem for constructing Haar measures.

Exercises

8.6.37.° Prove that the set of inner points of a convex set in a topological vector space is convex and the set of inner points of a balanced set is balanced.

8.6.38.° Prove that the closure of the neighborhood $U_{\alpha_1,\ldots,\alpha_n;\varepsilon}(0)$ of the form (8.1.1) belongs to $U_{\alpha_1,\ldots,\alpha_n;\delta}(0)$ for all $\delta > \varepsilon$.

8.6.39. Prove that convergence in $\mathcal{D}$ is not generated by a metric.

HINT: take a nonzero function $f \in \mathcal{D}$ and let $f_{n,k}(t) = k^{-1}f(t/n)$; observe that $f_{n,k} \to 0$ in $\mathcal{D}$ for each n, but there is no convergent sequence of the form f_{n,k_n} with indices $k_n \to \infty$.

8.6.40.° Prove Theorem 8.2.3.

8.6.41.° Let X and Y be topological vector spaces and let $T\colon X \to Y$ be a linear mapping taking some neighborhood of zero to a bounded set. Prove that T is continuous.

8.6.42.° Let p be a seminorm on a locally convex space bounded on a nonempty open set. Prove that p is continuous.

8.6.43.° Prove that the closure of a linear subspace in a topological vector space is a linear subspace.

8.6.44. Prove that a linear mapping A from a locally convex space to $\mathbb{R}^n$ is continuous if its kernel $\operatorname{Ker} A$ is closed.

HINT: use induction and consider the case of a linear functional f; if this functional is not zero, there is a vector v with $f(v) = 1$, hence there is a balanced neighborhood of zero W such that $v + W$ does not intersect $f^{-1}(0)$, but then f must be bounded on W.

8.6.45. (i) Prove that the topology τ on the space $\mathcal{D}(\mathbb{R}^1)$ introduced above is strictly weaker than the topology τ_1 on $\mathcal{D}(\mathbb{R}^1)$ in which open sets are all sets that have open intersections with all $\mathcal{D}_n$. For this show that the quadratic function $F(\varphi) = \sum_{n=1}^{\infty} \varphi(n)\varphi^{(n)}(0)$ is discontinuous in the topology τ, but continuous in τ_1.

(ii) Prove that the topology τ strictly stronger than the topology τ_2 on $\mathcal{D}(\mathbb{R}^1)$ generated by all norms $p_\psi(\varphi) = \sup_x |\psi(x)\varphi^{(m)}(x)|$, where one takes all possible nonnegative integer numbers m and positive locally bounded functions ψ. For this verify that the linear function $F(\varphi) = \sum_{n=1}^{\infty} \varphi^{(n)}(n)$ is continuous in the topology τ, but discontinuous in the topology τ_2.

8.6.46. Prove that the topology of the spaces $\mathcal{S}(\mathbb{R}^n)$, $H(U)$ and $C^\infty(U)$ from Example 8.1.3 cannot be generated by a norm.

HINT: if the topology of a locally convex space is generated by a norm, then this norm is estimated with a constant by a finite sum of seminorms from the family of seminorms generating the topology. Hence all other seminorms in this family are estimated with constants by the same finite sum, which is not the case for the indicated spaces.

8.6.47. (i) Let X be a normed space and let τ be some locally convex topology on X intermediate between the weak topology and the norm topology. Show that if a linear mapping $A\colon X \to X$ is continuous as a mapping from (X, τ) to (X, τ), then it is a bounded operator.

(ii) Show that the assertion converse to (i) is not always true by constructing a unitary operator A on a Hilbert space H which is discontinuous as a mapping from (H, τ) to (H, τ) for some locally convex topology on H intermediate between the weak topology and the norm topology.

HINT: (i) follows from the proof of Theorem 6.6.8. In (ii) one can take $H = L^2(\mathbb{R}^1)$ and the Fourier transform for A (see §9.2), defining the topology τ by adding to the seminorms of the weak topology $x \mapsto |(x, y)|$ one more seminorm $p(x) := \|I_{[0,1]} x\|$. On the subspace H_0 of functions with support in $[0, 1]$, for no $C > 0$ and $y_1, \ldots, y_n \in L^2(\mathbb{R}^1)$ one can have the estimate $p(Ax) \leqslant C[p(x) + |(x, y_1)| + \cdots + |(x, y_n)|]$, since $p(x) = 0$ on H_0 and $p(Ax) > 0$ for nonzero $x \in H_0$ due to the fact that Ax is an analytic function, moreover, for x one can take a function orthogonal to $y_1, \ldots, y_n$.

8.6.48. Let X be an infinite-dimensional Banach space. Show that X with the weak topology and X^* with the weak-$*$ topology are not complete, i.e., the indicated topologies have Cauchy nets that do not converge.

HINT: show that every linear functional f on X (not necessarily continuous) is the limit of a net of continuous functionals that is Cauchy in the weak topology; to this end, observe that for any finite set $x_1, \ldots, x_n$, there is a continuous functional g such that $f(x_i) = g(x_i)$ for each i.

8.6.49. Prove that the product of a test function and a distribution (for pairs $(\mathcal{S}, \mathcal{S}')$ and $(\mathcal{D}, \mathcal{D}')$) cannot be extended to a product of all distributions in such a way that we have $(fg)h = f(gh)$ for all distributions and $fg = gf$ is the previously defined product if f or g is a smooth function.

HINT: consider the product of δ, x and V.P.x^{-1}.

8.6.50.° Find the first three derivatives in $\mathcal{D}'$ of the following function F: $F(x) = -x$ if $x < 0$, $F(x) = x^2$ if $x \geqslant 0$.

8.6.51.° Prove that the derivative of $\ln|x|$ in $\mathcal{D}'$ is V.P.$\frac{1}{x}$.

8.6.52.° Prove that the formula
$$\text{V.P.}\frac{1}{|x|}(\varphi) := \int_{|x|<1} \frac{\varphi(x) - \varphi(0)}{|x|}\, dx + \int_{|x| \geqslant 1} \frac{\varphi(x)}{|x|}\, dx$$
defines a distribution from the classes $\mathcal{D}'(\mathbb{R}^1)$ and $\mathcal{S}'(\mathbb{R}^1)$.

8.6.53.° Let $G \in \mathcal{D}'(\mathbb{R}^k)$, $\varphi \in \mathcal{D}(\mathbb{R}^n \times \mathbb{R}^k)$ and $\psi(x) := G(\varphi(x, \cdot))$, $x \in \mathbb{R}^n$. Prove that $\psi \in \mathcal{D}(\mathbb{R}^n)$ and $F \otimes G \in \mathcal{D}'(\mathbb{R}^{n+k})$, where $F \otimes G(\varphi) := F(\psi)$.

8.6.54. Investigate whether $e^{ixt}(x + i0)^{-1}$ has a limit in $\mathcal{D}'(\mathbb{R}^1)$ as $t \to +\infty$.

8.6.55.° Prove that if $|c_k| \leqslant \alpha k^r + \beta$, then the series $\sum_{k=1}^{\infty} c_k \sin kx$ converges in the space $\mathcal{D}'(\mathbb{R}^1)$.

HINT: use that the Fourier coefficients of every function of class $\mathcal{D}(\mathbb{R}^1)$ converge to zero faster than k^{-r-2}.

8.6.56. (i) Set $f(x) = x$ for $x \in [0, 2\pi)$ and extend f to the real line with the period 2π. Represent f in the form $f = \sum_{k \in \mathbb{Z}} c_k \exp(ikx)$ in $\mathcal{D}'$ and find c_k.

(ii) Prove the Poisson formula $\sum_{k \in \mathbb{Z}} \exp(ikx) = 2\pi \sum_{k \in \mathbb{Z}} \delta_{2k\pi}$ in $\mathcal{D}'$.

HINT: to find c_k integrate by parts in the integral of $x \exp(ikx)$ over $[0, 2\pi)$; observe that the distribution in the right-hand side in (ii) is the generalized derivative of the function f in (i) and $\exp(ikx)$ is the derivative of $\exp(ikx)/(ik)$ for $k \neq 0$.

8.6.57.° (i) Find the first and second derivatives in $\mathcal{D}'$ of the usual functions $\sin|x|$ and $|\sin x|$. (ii) Prove the equality $|\sin x|'' + |\sin x| = 2\sum_{k\in\mathbb{Z}} \delta_{k\pi}$.

8.6.58.° Without using general theorems prove that every solution in $\mathcal{D}'$ of the equation $F'' - F' + F = 0$ is generated by a smooth function.

8.6.59.° Solve in $\mathcal{D}'$ the following differential equations: (i) $xF = 0$, (ii) $x^n F = 0$, (iii) $(\sin x)F = 0$, (iv) $xF' = 1$, (v) $F'' = \delta$, (vi) $F'' - F = \delta$.

8.6.60. Let $F \in \mathcal{D}'(\mathbb{R}^1)$. Prove that there exist $F_j \in \mathcal{D}(\mathbb{R}^1)$ such that $F_j \to F$ in $\mathcal{D}'(\mathbb{R}^1)$, i.e., $F_j(\varphi) \to F(\varphi)$ for all $\varphi \in \mathcal{D}(\mathbb{R}^1)$. Prove the same for $\mathcal{S}'(\mathbb{R}^1)$.

HINT: use approximations of the form $\psi_k * (\theta_k F)$, where $\psi_k(t) = k\psi(kt)$, the function $\psi \in \mathcal{D}(\mathbb{R}^1)$ has unit integral, $\theta_k(t) = 1$ if $t \in [-k, k]$, $\theta_k(t) = 0$ if $|t| \geq k + 1$, and for every m the functions $\theta_k^{(m)}$ are uniformly bounded.

8.6.61. Let $F \in \mathcal{D}'(\mathbb{R}^d)$ and $F(\varphi) \geq 0$ if $\varphi \in \mathcal{D}(\mathbb{R}^d)$ and $\varphi \geq 0$. Prove that F is given by a nonnegative locally bounded Borel measure.

HINT: consider first the case of F with bounded support and show that F is continuous with respect to the sup-norm (see [620, p. 17]); one can also use Theorem 8.5.5 and convolutions of F with nonnegative functions of class $\mathcal{D}(\mathbb{R}^d)$.

8.6.62. Let $\mu \geq 0$ be a locally bounded Borel measure on $\mathbb{R}^d$ and let $F(\varphi)$ be the integral of $\varphi \in \mathcal{D}$ against the measure μ. Prove that F extends to an element of $\mathcal{S}'$ if and only if $\mu(x: \|x\| \leq R) \leq C + CR^k$, where $C, k > 0$.

8.6.63. Prove that a locally convex space is sequentially complete if and only if all Cauchy sequences with respect to every seminorm from a family generating the topology converge in this space. This is also equivalent to convergence of all Cauchy sequences with respect to every continuous seminorm.

Show also that a sequentially complete metrizable locally convex space can be metrized by a complete translation invariant metric.

8.6.64. Let E be a sequentially complete locally convex space. Prove that if a sequence of continuous linear functionals f_n on E is pointwise bounded, then it is uniformly bounded on every bounded set.

HINT: it suffices to consider absolutely convex closed bounded sets B and apply the Banach–Steinhaus theorem for the Banach space E_B.

8.6.65. Show that the spaces $C[0, 1]$, c_0 and $L^1[0, 1]$ are not linearly homeomorphic to duals of normed spaces.

HINT: otherwise their closed unit balls must be convex and compact in the corresponding weak-$*$ topologies and by the Krein–Milman theorem possess sufficient supplies of extreme points.

8.6.66. Let B be a nonempty closed convex set in a locally convex space X, $K \subset X$ a nonempty compact set, and let $A \subset X$ be such that $A + K \subset B + K$. Prove that $A \subset B$. In particular, if A is also nonempty, closed and convex and $A + K = B + K$, then $A = B$.

HINT: it suffices to consider the case $A = \{0\}$; then one can take $k_1 \in K$ and by induction find points $k_n \in K$ and $b_n \in B$ such that $k_n = b_n + k_{n+1}$; observe that $n^{-1}(b_1 + \cdots + b_n) = n^{-1}(k_1 - k_{n+1}) \to 0$, so $0 \in B$, since B is convex and closed.

8.6.67. Prove that any convex compact set in a metrizable locally convex space is the closure of the convex envelope of some sequence converging to zero.
HINT: see [**75**, Proposition 1.8.14].

8.6.68. Let A be a convex compact set and let B be a closed bounded convex set in a locally convex space. Prove that the convex envelope of $A \cup B$ is closed.
HINT: see [**75**, Exercise 1.12.54].

8.6.69. Prove that in a topological vector space every finite-dimensional linear subspace is closed.

8.6.70. Prove that a topological vector space is finite-dimensional if it has a neighborhood of zero with compact closure.
HINT: let V be a neighborhood of zero in a space X with compact closure $\overline{V}$; there are $x_1, \ldots, x_n \in \overline{V}$ such that $\overline{V} \subset (x_1 + V/2) \cup \cdots \cup (x_n + V/2)$. Let L be the linear span of $x_1, \ldots, x_n$. Then $V \subset L + V/2$, which by induction gives $V \subset L + 2^{-k}V$ for all k. For every neighborhood of zero U there is n with $\overline{V} \subset nU$. Hence $V \subset \overline{L} = L$, so $X = L$.

8.6.71. Consider the compact ellipsoid $K := \{(x_n): \sum_{n=1}^{\infty} \sqrt{n}x_n^2 \leqslant 1\}$ in l^2 and the point $a \in K$, where $a_n = cn^{-1}$ and $c > 0$ is such that $\sum_{n=1}^{\infty} c^2 n^{-3/2} = 1$. Prove that a is an extreme point of K, but no supporting hyperplane to K passes through a.
HINT: if the supporting hyperplane is defined by the equation $(x, y) = 1$, where $y \in l^2$, then $y_n = c\sqrt{n}a_n$, whence $\sum_{n=1}^{\infty} y_n^2 = \infty$.

8.6.72. Let K be a convex compact set in a real locally convex space. Prove that every continuous linear functional attains its minimal and maximal values on K at some extreme points of K.
HINT: use Theorem 8.6.31.

8.6.73. Prove that there are no continuous norms on $\mathbb{R}^{\infty}$ (according to the Bessaga–Pełczyński theorem, every complete metrizable locally convex space on which there are no continuous norms contains a subspace isomorphic to $\mathbb{R}^{\infty}$).
HINT: use that any neighborhood of zero in $\mathbb{R}^{\infty}$ contains a nontrivial linear subspace.

8.6.74. Let X be a normed space. In relation to Theorem 6.10.26 consider the topology bw^* on X^* defined as follows: a set is bw^*-closed in X^* if its intersection with every closed ball is closed in the weak-$*$ topology.
(i) Prove that the topology bw^* is generated by a basis of sets of the form
$$B(f_0, \{a_n\}) = \{f \in X^*: |(f - f_0)(a_n)| < 1 \ \forall n \in \mathbb{N}\},$$
where $\{a_n\} \subset X$ and $\|a_n\| \to 0$. In particular, this topology is locally convex.
(ii) Prove that if X is infinite-dimensional, then the topology bw^* is strictly stronger than the topology $\sigma(X^*, X)$ and strictly weaker than the norm topology. Prove that if the space X is complete, then bw^* is not generated by a metric.
(iii) Give an example of a locally convex topology τ on X^* such that it is not stronger than $\sigma(X^*, X)$, but on all bounded sets induces the same topology as $\sigma(X^*, X)$ (hence the same one as bw^*); for this consider c_0^* with the topology $\sigma(c_0^*, L)$, where L is the space of finite sequences. (iv) Let X, Y be Banach spaces and $T \in \mathcal{L}(X, Y)$. Prove that the operator T is compact precisely when the operator T^* is continuous from the topology bw^* on Y^* to the norm topology on X^*.
HINT: see [**417**, §2.7, §3.4].

8.6.75. Prove Proposition 8.6.20.
HINT: see [**75**, Theorem 3.5.10] and [**171**, Theorem 7.3.3].

8.6.76. (The Schwartz kernel theorem) Suppose that $T\colon \mathcal{D}(\mathbb{R}^n) \to \mathcal{D}'(\mathbb{R}^n)$ is a continuous linear mapping. Then there exists a distribution $\Phi \in \mathcal{D}'(\mathbb{R}^n \times \mathbb{R}^n)$ such that $\langle T\varphi, \psi \rangle = \Phi(\varphi \otimes \psi)$ for all $\varphi, \psi \in \mathcal{D}(\mathbb{R}^n)$.

8.6.77. Prove that every locally convex space is completely regular.
HINT: see [**75**, §1.6].

8.6.78. (Peetre's theorem) Prove that every linear mapping $L\colon C^\infty(\mathbb{R}^n) \to C^\infty(\mathbb{R}^n)$ with $\operatorname{supp} Lf \subset \operatorname{supp} f$ is a differential operator of a locally finite order with smooth coefficients.
HINT: see [**452**, Theorem 3.3.3].

8.6.79. Prove that the nonnegative cones in l^2 and $C[0,1]^*$ have empty algebraic kernels.
HINT: if $x = (x_n) \in l^2$ and $x_n \geqslant 0$, then find $y = (y_n) \in l^2$ such that for no $\varepsilon > 0$ one has $x_n + \varepsilon y_n \geqslant 0$ for all n. Any nonnegative functional $l \in C[0,1]^*$ is represented by a nonnegative measure μ on $[0,1]$; take a point x of μ-measure zero and observe that $\mu - \varepsilon \delta_x$ cannot be nonnegative for $\varepsilon > 0$.

8.6.80. Prove that $\mathcal{D}(\mathbb{R}^n)$ is the strict inductive limit of the spaces $\mathcal{D}_m$ and every bounded set in $\mathcal{D}$ is contained in some $\mathcal{D}_m$.
HINT: see [**75**, §2.6].

8.6.81. A locally convex space X is called a Montel space if all closed bounded sets are compact in this space. Prove that $\mathbb{R}^\infty$, $\mathcal{D}(\mathbb{R}^n)$, $\mathcal{S}(\mathbb{R}^n)$, and the space $H(U)$ of holomorphic functions on an open set $U \subset \mathbb{C}^1$ with the topology of uniform convergence on compact sets are Montel spaces.
HINT: use that in $\mathcal{D}(\mathbb{R}^n)$ a bounded set is contained and is bounded in some $\mathcal{D}_m$; use also that a sequence bounded with all derivatives on a ball contains a subsequence converging uniformly with all derivatives; for holomorphic functions the uniform convergence on compact sets implies the uniform convergence of all derivatives.

8.6.82. Let $F_j \in \mathcal{D}'(\mathbb{R}^1)$ and $F_j(\varphi) \to 0$ for every function $\varphi \in \mathcal{D}(\mathbb{R}^1)$. Prove that there exists a seminorm $p_{\{a_k\}}$ of the form (8.1.2) along with numbers $\varepsilon_j \to 0$ for which $|F_j(\varphi)| \leqslant \varepsilon_j p_{\{a_k\}}(\varphi)$ for all j.
HINT: see [**75**, Exercise 2.10.48].

8.6.83.** (i) Prove that in the space $\mathcal{D}(\mathbb{R}^1)$ there is a nonclosed countable set the intersection of which with every subspace $\mathcal{D}_m$ is closed. (ii) Prove that in the space $\mathcal{D}(\mathbb{R}^1)$ there is a nonclosed linear subspace the intersection of which with every $\mathcal{D}_m$ is closed.
HINT: see [**707**], [**75**, §2.10].

8.6.84.* Let L be a linear subspace in $\mathcal{D}(\mathbb{R}^1)$ of finite codimension such that all intersections $L \cap \mathcal{D}_m$ are closed. Prove that L is closed.
HINT: see [**75**, Exercise 2.10.50].

8.6.85.* Suppose that f is a measurable function on $\mathbb{R}^1$ such that the linear subspace $L := \{\varphi \in \mathcal{D}(\mathbb{R}^1)\colon f\varphi \in L^1(\mathbb{R}^1)\}$ has a finite codimension. Prove that there exists a distribution $F \in \mathcal{D}'(\mathbb{R}^1)$, the action of which on $\varphi \in L$ is the integral of $f\varphi$.
HINT: considering separately the functions f^+ and f^-, we can assume that $f \geqslant 0$; let $f_n = \min(f, n)$ and let E be the set of all functions $\varphi \in \mathcal{D}(\mathbb{R}^1)$ for which there exists a finite limit $l(\varphi)$ of the integrals of $f_n\varphi$ as $n \to \infty$; then E is a linear subspace of finite codimension, since $L \subset E$. Observe that the intersections $E \cap \mathcal{D}_m$ are Borel sets in $\mathcal{D}_m$; apply Theorem 8.6.33, which gives their closedness and the continuity of l.

CHAPTER 9

The Fourier Transform and Sobolev Spaces

In this chapter we discuss one of the most classical objects of analysis — the Fourier transform, and also give an elementary introduction to the theory of Sobolev spaces, which plays a very important role in modern analysis. It will not be an exaggeration to say that Sobolev spaces are applied everywhere where the derivative and integral are used. Along with Lebesgue's integral and Banach spaces, Sobolev spaces belong to the greatest achievements of analysis in the XX century, determining its modern appearance. Of course, this introduction cannot pretend even to be a brief course of the theory of Sobolev spaces. It actually contains only a discussion of main definitions and examples. As we shall see, the theory of Sobolev spaces is closely connected with the theory of Fourier transform, and both have direct relations to the theory of operators and the theory of distributions.

9.1. The Fourier Transform in L^1

In this section we consider Fourier transforms of functions: one of the most important tools in analysis.

9.1.1. Definition. *The Fourier transform of a function f in $\mathcal{L}^1(\mathbb{R}^n)$ (possibly, complex) is the complex function*

$$\widehat{f}(y) = \frac{1}{(2\pi)^{n/2}} \int_{\mathbb{R}^n} e^{-i(y,x)} f(x)\, dx.$$

The Fourier transform of an element $f \in L^1(\mathbb{R}^n)$ is the function $\widehat{f}$ for an arbitrary representative of the equivalence class of f.

The necessity of distinguishing between versions of an integrable function when dealing with its Fourier transform will be clear from the sequel, when we discuss recovering of the values of f at individual points through the function $\widehat{f}$. The numerical factor in front of the Fourier transform of functions leads to a unitary operator on $L^2(\mathbb{R}^n)$. Finally, our choice of the minus sign in the exponent is just a tradition. As we shall see below, changing the minus by plus gives the definition of the inverse Fourier transform.

In some cases one can explicitly calculate Fourier transforms. Let us consider one of the most important examples.

© Springer Nature Switzerland AG 2020   397
V. I. Bogachev, O. G. Smolyanov, *Real and Functional Analysis*,
Moscow Lectures 4, https://doi.org/10.1007/978-3-030-38219-3_9

9.1.2. Example. Let $\alpha > 0$. Then

$$\frac{1}{(2\pi)^{n/2}} \int_{\mathbb{R}^n} \exp[-i(y,x)]\exp[-\alpha|x|^2]\,dx = \frac{1}{(2\alpha)^{n/2}} \exp\left[-\frac{1}{4\alpha}|y|^2\right].$$

PROOF. Evaluating this integral by Fubini's theorem reduces everything to the one-dimensional case. After an obvious change of variable it suffices to consider the case $\alpha = 1/2$. In this case both parts of the equality to be proved are analytic functions of y and coincide at $y = it$, $t \in \mathbb{R}$, which is known from calculus (the reader may recall the details). Hence these functions coincide at all points y. $\quad\square$

9.1.3. Proposition. *Let $f \in \mathcal{L}^1(\mathbb{R}^n)$. Then the function $\widehat{f}$ is uniformly continuous and*

$$|\widehat{f}(y)| \leqslant (2\pi)^{-n/2}\|f\|_{L^1} \quad and \quad \lim_{|y|\to\infty} \widehat{f}(y) = 0. \tag{9.1.1}$$

PROOF. The first relation in (9.1.1) is obvious. If f is the indicator function of a cube with edges parallel to the coordinate axes, then $\widehat{f}$ is easily evaluated explicitly with the aid of Fubini's theorem (Exercise 9.10.13), which shows the second relation. Hence it is valid for linear combinations of indicators functions of such cubes. Now it remains to take a sequence $\{f_j\}$ of such linear combinations converging to f in $L^1(\mathbb{R}^n)$ and observe that the functions $\widehat{f}_j$ converge uniformly to the function $\widehat{f}$ by the first relation in (9.1.1). $\quad\square$

Let us consider some additional useful properties of the Fourier transform.

9.1.4. Proposition. (i) *If a continuously differentiable and integrable function f possesses an integrable partial derivative $\partial_{x_j} f$, then*

$$\widehat{\partial_{x_j} f}(y) = iy_j\widehat{f}(y).$$

In the one-dimensional case $\widehat{f'}(y) = iy\widehat{f}(y)$.

(ii) *If $f \in \mathcal{L}^1(\mathbb{R}^n)$ is a function such that for some $j \in \{1,\ldots,n\}$ the function $g_j\colon y \mapsto y_j f(y)$ is integrable, then the function $\widehat{f}$ is continuously differentiable in the variable x_j and*

$$\partial_{x_j}\widehat{f}(x) = -i\widehat{g_j}(x).$$

In the one-dimensional case $\widehat{f}'(x) = -i\widehat{yf(y)}(x)$.

PROOF. (i) If f has bounded support, then the desired equality follows from the integration by parts formula. In order to reduce to this the general case it suffices to take a sequence of smooth functions ζ_k on $\mathbb{R}^n$ with the following properties: $0 \leqslant \zeta_k \leqslant 1$, $\sup_k |\partial_{x_j}\zeta_k| \leqslant C$, $\zeta_k(x) = 1$ if $|x| \leqslant k$. Then the functions $\zeta_k f$ converge in $L^1(\mathbb{R}^n)$ to f and the functions $\partial_{x_j}(\zeta_k f)$ converge to $\partial_{x_j} f$, since $f\partial_{x_j}\zeta_k \to 0$ in $L^1(\mathbb{R}^n)$ by the Lebesgue dominated convergence theorem. Actually the assertion is true without continuity of $\partial_{x_j} f$. Assertion (ii) follows from the theorem on the differentiability of the integral with respect to a parameter, which is applicable due to the relation $|\partial_{x_j}\exp[i(x,y)]f(y)| = |y_j f(y)|$. $\quad\square$

The complexification of $\mathcal{S}(\mathbb{R}^n)$ will be denoted by the same symbol and we shall deal with it when considering the Fourier transform.

9.1.5. Corollary. *If $f \in \mathcal{S}(\mathbb{R}^n)$, then $\widehat{f} \in \mathcal{S}(\mathbb{R}^n)$.*

PROOF. As shown above, for each $j = 1, \ldots, n$ and $k \in \mathbb{N}$ the function $\partial_{x_j}^k \widehat{f} = (-i)^k \widehat{h_{j,k}}$, where $h_{j,k}(y) = y_j^k f(y)$, decreases at infinity faster than $|x|^{-m}$ for all m. $\qquad\square$

Below we show that the Fourier transform is an isomorphism of the complex space $\mathcal{S}(\mathbb{R}^n)$. Next we extend the Fourier transform to L^2 and obtain a unitary operator.

Naturally the question arises of how to recover the function f by its Fourier transform defining this function up to a modification. To this end the *inverse Fourier transform* is used. For an integrable function f the inverse Fourier transform is given by the formula

$$\check{f}(x) = (2\pi)^{-n/2} \int_{\mathbb{R}^n} e^{i(y,x)} f(y) \, dy.$$

We shall see that if the direct Fourier transform of f is integrable, then its inverse transform gives the original function f. Actually this is true even without the assumption of the integrability of $\widehat{f}$ if one defines the inverse Fourier transform for distributions. We now postpone this question, but give a sufficient condition for recovering a function at a fixed point by its Fourier transform and then prove the Parseval equality serving as a basis of the definition of the Fourier transform of distributions.

9.1.6. Theorem. *Let f be a function integrable on the real line and satisfying Dini's condition at a point x: the function $t \mapsto [f(x+t) - 2f(x) + f(x-t)]/t$ is integrable in a neighborhood of zero. Then the following inversion formula is valid:*

$$f(x) = \lim_{R \to +\infty} \frac{1}{\sqrt{2\pi}} \int_{-R}^{R} e^{ixy} \widehat{f}(y) \, dy. \tag{9.1.2}$$

In particular, (9.1.2) is true at all points of differentiability of the function f.

PROOF. Set

$$J_R := \frac{1}{\sqrt{2\pi}} \int_{-R}^{R} e^{ixy} \widehat{f}(y) \, dy,$$

where $R > 0$. By Fubini's theorem we obtain

$$J_R = \frac{1}{2\pi} \int_{-\infty}^{+\infty} f(z) \int_{-R}^{R} e^{iy(x-z)} \, dy \, dz$$

$$= \frac{1}{2\pi} \int_{-\infty}^{+\infty} f(z) \frac{2 \sin(R(x-z))}{x - z} \, dz = \frac{1}{\pi} \int_{-\infty}^{+\infty} f(x-t) \frac{\sin(Rt)}{t} \, dt$$

$$= \frac{1}{2\pi} \int_{-\infty}^{+\infty} [f(x+t) + f(x-t)] \frac{\sin(Rt)}{t} \, dt.$$

It is well-known that

$$\lim_{T \to +\infty} \int_{-T}^{T} \frac{\sin t}{t} \, dt = \pi.$$

Let $\varepsilon > 0$. Since the integral of $\sin(Rt)/t$ over $[-T,T]$ equals the integral of $\sin t/t$ over $[-RT, RT]$, there exists $T_1 > 0$ such that for all $T > T_1$ and $R > 1$ we have

$$\left| \frac{f(x)}{\pi} \int_{-T}^{T} \frac{\sin(Rt)}{t}\, dt - f(x) \right| < \varepsilon.$$

By the integrability of f there exists $T_2 > T_1$ such that

$$\int_{|t| \geqslant T_2} \frac{|f(x+t) + f(x-t)|}{|t|}\, dt < \varepsilon.$$

By the assumption of the theorem the function $[f(x+t) - 2f(x) + f(x-t)]/t$ is integrable in t on $[-T_2, T_2]$. According to (9.1.1) there exists a number $R_1 > 1$ such that for all $R > R_1$ one has the estimate

$$\left| \int_{-T_2}^{T_2} \frac{f(x+t) - 2f(x) + f(x-t)}{t} \sin(Rt)\, dt \right| < \varepsilon.$$

On account of the three estimates above for all $R > R_1$ we obtain

$$|J_R - f(x)|$$

$$\leqslant \left| J_R - \frac{f(x)}{\pi} \int_{-T_2}^{T_2} \frac{\sin(Rt)}{t}\, dt \right| + \left| \frac{f(x)}{\pi} \int_{-T_2}^{T_2} \frac{\sin(Rt)}{t}\, dt - f(x) \right|$$

$$\leqslant \left| J_R - \frac{f(x)}{\pi} \int_{-T_2}^{T_2} \frac{\sin(Rt)}{t}\, dt \right| + \varepsilon$$

$$= \frac{1}{2\pi} \left| \int_{-\infty}^{+\infty} [f(x+t) + f(x-t)] \frac{\sin(Rt)}{t}\, dt - 2 \int_{-T_2}^{T_2} f(x) \frac{\sin(Rt)}{t}\, dt \right| + \varepsilon$$

$$\leqslant \frac{1}{2\pi} \left| \int_{-T_2}^{T_2} [f(x+t) - 2f(x) + f(x-t)] \frac{\sin(Rt)}{t}\, dt \right|$$

$$+ \frac{1}{2\pi} \left| \int_{|t| \geqslant T_2} [f(x+t) + f(x-t)] \frac{\sin(Rt)}{t}\, dt \right| + \varepsilon < 3\varepsilon.$$

Note that there exists a function $f \in \mathcal{L}^1$ for which the limit in (9.1.2) exists at no point (Kolmogorov's example), but for $f \in \mathcal{L}^2$ Carleson's theorem gives (9.1.2) almost everywhere (see [235]). $\square$

9.1.7. Corollary. *Let $f \in \mathcal{S}(\mathbb{R}^n)$. Then*

$$f(x) = (2\pi)^{-n/2} \int_{\mathbb{R}^n} e^{i(y,x)} \widehat{f}(y)\, dy. \tag{9.1.3}$$

PROOF. We already know that $\widehat{f} \in \mathcal{S}(\mathbb{R}^n)$. Hence in the case $n = 1$ equality (9.1.3) follows from (9.1.2). By Fubini's theorem this equality extends to functions on $\mathbb{R}^n$. We shall assume that our assertion is already proved for $n - 1$ in place of n. For a vector $u = (u_1, \ldots, u_n)$ set $\overline{u} = (u_1, \ldots, u_{n-1})$. Then

$$f(x) = \frac{1}{\sqrt{2\pi}} \int_{-\infty}^{+\infty} e^{iy_n z_n} g(\overline{x}, y_n)\, dy_n,$$

where the function $y_n \mapsto g(\overline{x}, y_n)$ is the Fourier transform of the function of one variable $x_n \mapsto f(\overline{x}, x_n)$. For every fixed y_n the function $\overline{x} \mapsto g(\overline{x}, y_n)$ belongs to the class $\mathcal{S}(\mathrm{I\!R}^{n-1})$, which is easily verified. Hence

$$g(\overline{x}, y_n) = (2\pi)^{-(n-1)} \int_{\mathrm{I\!R}^{n-1}} e^{i(\overline{x}, \overline{y})} \int_{\mathrm{I\!R}^{n-1}} e^{-i(\overline{z}, \overline{y})} g(\overline{z}, y_n) \, d\overline{z} \, d\overline{y}.$$

Substituting

$$g(\overline{z}, y_n) = \frac{1}{\sqrt{2\pi}} \int_{-\infty}^{+\infty} e^{iy_n z_n} f(\overline{z}, z_n) \, dz_n,$$

we arrive at formula (9.1.3). $\qquad\qquad\square$

9.1.8. Corollary. *The Fourier transform is a linear homeomorphism of the complex space $\mathcal{S}(\mathrm{I\!R}^n)$.*

PROOF. It was shown above that $\widehat{f} \in \mathcal{S}(\mathrm{I\!R}^n)$ for all $f \in \mathcal{S}(\mathrm{I\!R}^n)$ and that $f(-\cdot)$ is the Fourier transform of the function $\widehat{f}$. Hence the Fourier transform is a linear isomorphism of $\mathcal{S}(\mathrm{I\!R}^n)$. Let us verify the continuity of $\mathcal{F}$ on $\mathcal{S}(\mathrm{I\!R}^n)$. For this we use the relation

$$y_j^{2m} \widehat{f}^{(k)}(y) = (-i)^k y_j^{2m} \mathcal{F}(x_j^k f)(y) = (-i)^{k+2m} \mathcal{F}[\partial_{x_j}^{2m}(x_j^k f)](y).$$

The function $\partial_{x_j}^{2m}(x_j^k f)$ is a finite sum of functions of the form $x_j^l \partial_{x_j}^r f$. Their L^1-norms do not exceed $c_n \sup_x (1 + |x|^2)^{n+l} |\partial_{x_j}^r f(x)|$, where c_n is the integral of $(1 + |x|^2)^{-n}$ over $\mathrm{I\!R}^n$. According to Theorem 8.2.3 the mapping $\mathcal{F}$ is continuous on $\mathcal{S}(\mathrm{I\!R}^n)$. $\qquad\square$

The following simple fact is frequently used.

9.1.9. Theorem. *For all $\varphi, \psi \in L^1(\mathrm{I\!R}^n)$ one has*

$$\int_{\mathrm{I\!R}^n} \widehat{\varphi} \psi \, dx = \int_{\mathrm{I\!R}^n} \varphi \widehat{\psi} \, dx, \qquad \int_{\mathrm{I\!R}^n} \widehat{\varphi} \overline{\psi} \, dx = \int_{\mathrm{I\!R}^n} \varphi \overline{\widecheck{\psi}} \, dx. \qquad (9.1.4)$$

PROOF. Applying Fubini's theorem to the equality

$$\int_{\mathrm{I\!R}^n} \widehat{\varphi}(x) \psi(x) \, dx = \frac{1}{(2\pi)^{n/2}} \int_{\mathrm{I\!R}^n} \int_{\mathrm{I\!R}^n} e^{-i(x,y)} \varphi(y) \psi(x) \, dy \, dx,$$

we obtain the first formula, and the second one follows by the identity $\overline{\widecheck{\psi}} = \widehat{\overline{\psi}}$. $\qquad\square$

Applying the complex conjugation to the second equality, we obtain one more useful formula:

$$\int_{\mathrm{I\!R}^n} \widecheck{\psi} \overline{\varphi} \, dx = \int_{\mathrm{I\!R}^n} \psi \overline{\widehat{\varphi}} \, dx.$$

The following fact is one of the most important results in this chapter.

9.1.10. Corollary. *Let $\varphi \in L^1(\mathrm{I\!R}^n)$. Then for every function $\psi \in \mathcal{S}(\mathrm{I\!R}^n)$ the following Parseval equality holds*

$$\int_{\mathrm{I\!R}^n} \varphi(x) \overline{\psi(x)} \, dx = \int_{\mathrm{I\!R}^n} \widehat{\varphi}(y) \overline{\widehat{\psi}(y)} \, dy. \qquad (9.1.5)$$

PROOF. The function $f := \widehat{\psi}$ belongs to $S(\mathbb{R}^n)$. It remains to apply the inversion formula $\psi = \check{f}$. □

9.1.11. Corollary. *Let $f \in \mathcal{L}^1(\mathbb{R}^n)$ and $\widehat{f} \in \mathcal{L}^1(\mathbb{R}^n)$. Then the function f has a continuous modification f_0 and*

$$f_0(x) = (2\pi)^{-n/2} \int_{\mathbb{R}^n} e^{i(y,x)} \widehat{f}(y)\, dy \quad \forall\, x \in \mathbb{R}^n.$$

In particular, if $\widehat{f} = 0$, then $f(x) = 0$ almost everywhere.

PROOF. By our condition the function $g := \widehat{f}$ is integrable. Hence its inverse Fourier transform f_0 is continuous. We verify that $f = f_0$ almost everywhere. To this end it suffices to show that for every smooth real function φ with bounded support

$$\int_{\mathbb{R}^n} f\varphi\, dx = \int_{\mathbb{R}^n} f_0\varphi\, dx.$$

By the Parseval equality we have $\displaystyle\int f\varphi\, dx = \int \widehat{f}\,\overline{\widehat{\varphi}}\, dx$. On the other hand, we have $\displaystyle\int g\overline{\widehat{\varphi}}\, dx = \int f_0\varphi\, dx$, whence our claim follows. □

Let us apply the Fourier transform for the proof of completeness of the system of Hermite functions (see Example 5.4.12(iv)).

9.1.12. Theorem. *Let $f \in \mathcal{L}^2(\mathbb{R}^1)$ be a function such that*

$$\int_{-\infty}^{+\infty} t^k f(t) \exp(-t^2/2)\, dt = 0 \quad \text{for all } k = 0, 1, \ldots.$$

Then $f(t) = 0$ a.e.

PROOF. The function

$$g(z) = \int_{-\infty}^{+\infty} f(t) \exp(itz - t^2/2)\, dt$$

is defined and analytic in the complex plane, since for every $R > 0$ the function $f(t) \exp(R|t| - t^2/2)$ is integrable due to the fact that the function $\exp(R|t| - t^2/2)$ belongs to $\mathcal{L}^2(\mathbb{R}^1)$. It follows from our condition that all derivatives of g at zero are zero (including the value of g at zero). Hence $g(z) = 0$. Hence the integrable function $f(t) \exp(-t^2/2)$ has the identically zero Fourier transform and hence equals zero almost everywhere. □

9.2. The Fourier Transform in L^2

With the aid of the Parseval equality for Fourier integrals the Fourier transform extends to functions from $L^2(\mathbb{R}^n)$.

9.2.1. Definition. *The Fourier transform in the complex space $L^2(\mathbb{R}^n)$ is the unitary operator on $L^2(\mathbb{R}^n)$ that extends the Fourier transform on $S(\mathbb{R}^n)$ by means of equality (9.1.5) and Lemma 7.6.6. The Fourier transform of $f \in L^2(\mathbb{R}^n)$ is denoted by $\mathcal{F}f$ or $\widehat{f}$.*

Note that the cited lemma is applicable, since the space $\mathcal{S}(\mathbb{R}^n)$ is dense in $L^2(\mathbb{R}^n)$ and is mapped onto itself under the action of the Fourier transform preserving on $\mathcal{S}(\mathbb{R}^n)$ the inner product by equality (9.1.5). This definition means that for finding the Fourier transform of an element $f \in L^2(\mathbb{R}^n)$ we should take a sequence of functions $f_j \in \mathcal{S}(\mathbb{R}^n)$ converging to f in $L^2(\mathbb{R}^n)$ and set $\widehat{f} := \lim_{j\to\infty} \widehat{f_j}$, where the limit is taken in L^2 (which exists because $\{f_j\}$ is Cauchy in L^2 and $\|\widehat{f_j} - \widehat{f_i}\|_2 = \|f_j - f_i\|_2$).

It is obvious from the definition that on $\mathcal{S}(\mathbb{R}^n)$ the operator $\mathcal{F}$ is given by the usual Fourier transform. However, if a function f from $L^2(\mathbb{R}^n)$ does not belong to $L^1(\mathbb{R}^n)$, then the function $\mathcal{F}f$ cannot be defined as the Lebesgue integral of $(2\pi)^{-n/2}\exp[-i(x,y)]f(y)$. On the other hand, for all $f \in L^2 \cap L^1$ the given definition is consistent with the one for integrable functions.

9.2.2. Lemma. *If $f \in L^2(\mathbb{R}^n) \cap L^1(\mathbb{R}^n)$, then the Fourier transform of the function f in $L^2(\mathbb{R}^n)$ is given by its Fourier transform $\widehat{f}$ in $L^1(\mathbb{R}^n)$.*

PROOF. Let $\mathcal{F}f$ be the Fourier transform of f in $L^2(\mathbb{R}^n)$ and let $\widehat{f}$ be its Fourier transform in $L^1(\mathbb{R}^n)$. We show that $\mathcal{F}f = \widehat{f}$ a.e. It suffices to verify that for every real function $\varphi \in \mathcal{D}(\mathbb{R}^n)$ the functions $(\mathcal{F}f)\varphi$ and $\widehat{f}\varphi$ have equal integrals. Let ψ be the inverse Fourier transform of φ. Then $\psi \in \mathcal{S}(\mathbb{R}^n)$ and $\widehat{\psi} = \varphi$. By the Parseval equality we have

$$\int_{\mathbb{R}^n} \widehat{f}(x)\varphi(x)\,dx = \int_{\mathbb{R}^n} f(x)\overline{\psi(x)}\,dx.$$

On the other side, taking a sequence of functions $f_j \in \mathcal{S}(\mathbb{R}^n)$ converging to f in $L^2(\mathbb{R}^n)$ we obtain

$$\int_{\mathbb{R}^n} \mathcal{F}f(x)\varphi(x)\,dx = \lim_{j\to\infty} \int_{\mathbb{R}^n} \mathcal{F}f_j(x)\varphi(x)\,dx$$
$$= \lim_{j\to\infty} \int_{\mathbb{R}^n} \widehat{f_j}(x)\varphi(x)\,dx = \lim_{j\to\infty} \int_{\mathbb{R}^n} f_j(x)\overline{\psi(x)}\,dx = \int_{\mathbb{R}^n} f(x)\overline{\psi(x)}\,dx,$$

which proves our assertion. $\square$

Note that the class of Fourier transforms of functions from L^1 has no constructive description (see, for example, Exercise 9.10.31).

It is now easy to prove the following Plancherel theorem, which is frequently used (especially in old textbooks) as a way of defining the Fourier transform in L^2.

9.2.3. Theorem. *Let $f \in L^2(\mathbb{R}^n)$. Then the functions*

$$g_R(x) := \frac{1}{(2\pi)^{n/2}} \int_{|y|\leqslant R} \exp\bigl(-i(x,y)\bigr) f(y)\,dy$$

converge in $L^2(\mathbb{R}^n)$ as $R \to +\infty$ to the Fourier transform of the function f in $L^2(\mathbb{R}^n)$ defined above.

PROOF. Let $f_R := fI_{\{|x|\leqslant R\}}$. Then $f_R \to f$ in $L^2(\mathbb{R}^n)$ as $R \to +\infty$. Hence $\mathcal{F}f_R \to \mathcal{F}f$ in $L^2(\mathbb{R}^n)$. Finally, by the lemma above $\mathcal{F}f_R = \widehat{f_R} = g_R$. $\square$

Thus, there holds the representation (the inversion formula)

$$\widehat{f}(x) = \lim_{R\to+\infty} \frac{1}{(2\pi)^{n/2}} \int_{|y|\leqslant R} \exp\big(-i(x,y)\big) f(y)\, dy$$

in the sense of convergence in L^2. Now the question about convergence almost everywhere arises. Of course, convergence in L^2 enables us to find a sequence g_{R_j} with $R_j \to \infty$ converging almost everywhere. For $n = 1$ it is known that convergence holds almost everywhere, but the proof of this fact is very difficult (see Grafakos [**235**, Chapter 10]). For $n > 1$ the question remains open (if the balls are replaced by the cubes $[-R, R]^n$, then also here the answer is positive). For additional information about Fourier transform, see Bochner [**72**], Dym, McKean [**169**], Duoandikoetxea [**167**], Folland [**194**], Grafakos [**235**], Kammler [**309**], Katznelson [**318**], Petersen [**477**], Stein [**571**], Strichartz [**577**], Titchmarsh [**596**], and Wiener [**630**].

9.3. The Fourier Transform in $\mathcal{S}'$

The Parseval equality used above in the definition of the Fourier transform in L^2 and also formula (9.1.4) suggest a way of extending it to distributions.

9.3.1. Definition. *Let $F \in \mathcal{S}'(\mathbb{R}^n)$. Set*

$$\mathcal{F}F(\varphi) := \widehat{F}(\varphi) := F(\widehat{\varphi}), \quad \text{where } \varphi \in \mathcal{S}(\mathbb{R}^n),$$

and call the functional $\mathcal{F}F = \widehat{F}$ on $\mathcal{S}(\mathbb{R}^n)$ the Fourier transform of the distribution F; here $\mathcal{S}$ and $\mathcal{S}'$ are complex spaces.

Since the Fourier transform $\varphi \mapsto \widehat{\varphi}$ is a linear homeomorphism of the complex space $\mathcal{S}(\mathbb{R}^n)$, we have $\widehat{F} \in \mathcal{S}'(\mathbb{R}^n)$ and the Fourier transform is a one-to-one linear operator on the complex space $\mathcal{S}'(\mathbb{R}^n)$.

If a distribution F is given by a usual integrable function F_0, i.e.,

$$F(\varphi) = \int_{\mathbb{R}^n} F_0(x)\varphi(x)\, dx,$$

then $\widehat{F}$ is given by the usual function $\widehat{F_0}$, since by formula (9.1.4) for all functions $\varphi \in \mathcal{S}(\mathbb{R}^n)$ we have

$$\widehat{F}(\varphi) = F(\widehat{\varphi}) = \int_{\mathbb{R}^n} F_0(x)\widehat{\varphi}(x)\, dx = \int_{\mathbb{R}^n} \widehat{F_0}(x)\varphi(x)\, dx.$$

Note that if the action of usual functions as distributions is defined by means of the inner product in L^2, i.e., with $\overline{\varphi}$ in place of φ in the integral, then for consistency of the generalized and usual Fourier transforms on integrable functions one has to set $\mathcal{F}F(\varphi) = F(\mathcal{F}^{-1}\varphi)$.

Let us evaluate Fourier transforms of some typical distributions.

9.3.2. Example. (i) We have the equalities

$$\widehat{1} = (2\pi)^{n/2}\delta, \quad \widehat{\delta} = (2\pi)^{-n/2}.$$

Indeed, for every function $\varphi \in \mathcal{S}(\mathbb{R}^n)$ we have

$$\widehat{1}(\varphi) = \int_{\mathbb{R}^n} \widehat{\varphi}(x)\,dx = (2\pi)^{n/2}\varphi(0) = (2\pi)^{n/2}\delta(\varphi).$$

From the same formula for $\widehat{\varphi}$ in place of φ we obtain the second equality, since $\mathcal{F}^2\varphi(x) = \varphi(-x)$.

(ii) For the Heaviside function $\chi := I_{[0,+\infty)}$ on the real line we have

$$\widehat{\chi}(x) = \frac{-i}{\sqrt{2\pi}}\frac{1}{x - i0}.$$

Indeed, let $\chi_\varepsilon := e^{-\varepsilon x}I_{[0,+\infty)}$, $\varepsilon > 0$. For all functions $\varphi \in \mathcal{S}(\mathbb{R}^1)$ we have $\widehat{\chi}(\varphi) = \lim_{\varepsilon \to 0} \widehat{\chi_\varepsilon}(\varphi)$. Since $\widehat{\chi_\varepsilon}(x) = -i(2\pi)^{-1/2}(x - i\varepsilon)^{-1}$, the indicated equality is true by the definition of $(x - i0)^{-1}$.

(iii) Let μ be a bounded Borel measure on $\mathbb{R}^n$. Then

$$\widehat{\mu}(x) = (2\pi)^{-n/2}\int_{\mathbb{R}^n} \exp[-i(x,y)]\,\mu(dy).$$

Indeed, for all $\varphi \in \mathcal{S}(\mathbb{R}^n)$ Fubini's theorem gives the equality

$$\widehat{\mu}(\varphi) = \int_{\mathbb{R}^n} \widehat{\varphi}(y)\,\mu(dy) = (2\pi)^{-n/2}\int_{\mathbb{R}^n}\left(\int_{\mathbb{R}^n}\varphi(x)\exp[-i(x,y)]\,dx\right)\mu(dy)$$

$$= (2\pi)^{-n/2}\int_{\mathbb{R}^n}\varphi(x)\left(\int_{\mathbb{R}^n}\exp[-i(x,y)]\,\mu(dy)\right)dx,$$

which proves our assertion.

(iv) For the function $f(x) = \exp(-\alpha|x|)$ on the real line, where $\alpha > 0$, a straightforward calculation gives

$$\widehat{f}(y) = \frac{1}{\sqrt{2\pi}}\left(\frac{1}{\alpha - iy} + \frac{1}{\alpha + iy}\right) = \frac{\sqrt{2}}{\sqrt{\pi}}\frac{\alpha}{y^2 + \alpha^2}.$$

Hence for the function $g(y) = (y^2 + \alpha^2)^{-1}$ we obtain

$$\widehat{g}(x) = \sqrt{\pi/2}\,\alpha^{-1}e^{-\alpha|x|}.$$

If in case (iii) we take for μ Dirac's measure δ_a at the point a, then we obtain

$$\widehat{\delta_a}(x) = (2\pi)^{-n/2}\exp[-i(x,a)].$$

The same equality can be obtained directly similarly to case (i). In addition, for the function $e_a(x) = \exp[i(x,a)]$ we have

$$\widehat{e_a} = (2\pi)^{n/2}\delta_a.$$

With the aid of this equality it is easy to find the Fourier transforms of the generalized functions given by the usual functions $\sin x$ and $\cos x$. By using differentiation and multiplication by polynomials the supply of explicitly calculated Fourier transforms can be enlarged, since for all $F \in \mathcal{S}'(\mathbb{R}^1)$ we have

$$\widehat{\partial_{y_j}F} = ix_j\widehat{F}, \quad \partial_{x_j}\widehat{F} = -i\widehat{y_jF} \tag{9.3.1}$$

due to analogous relations for functions from $\mathcal{S}(\mathbb{R}^1)$.

9.3.3. Example. Suppose that a functional $F \in \mathcal{S}'(\mathbb{R}^n)$ extends to an element of $\mathcal{E}'(\mathbb{R}^n)$, i.e., a continuous functional on the space $\mathcal{E}(\mathbb{R}^n)$ of all infinitely differentiable functions with the topology of uniform convergence of all derivatives on bounded sets. Then $\widehat{F}$ is given by a smooth function ψ, where $\psi(x) = F(e_x)$, $e_x(y) := (2\pi)^{-n/2} \exp[-i(x, y)]$. For example, if F is given by a measure ν with compact support, then $\psi(x)$ is the integral of $e_x(y)$ in y with respect to ν.

PROOF. For simplicity we consider the case $n = 1$. Let $\varphi \in \mathcal{D}(\mathbb{R}^1)$ have support in $[a, b]$. We show that $\widehat{F}(\varphi)$ is the integral of $\varphi\psi$. It is easy to see that ψ is smooth; for example, the continuity is clear from the fact that $e_{x_j} \to e_x$ in $\mathcal{E}(\mathbb{R}^1)$ as $x_j \to x$. The integral of $\varphi\psi$ is the limit of the Riemann sums $S_k = \sum_{j=1}^{k} \varphi(t_j)\psi(t_j)\Delta_j$ corresponding to the partitions of $[a, b]$ into 2^k equal intervals. Let $g_k(y) = \sum_{j=1}^{k} \varphi(t_j)e_{t_j}(y)\Delta_j$. Then $S_k = F(g_k)$ and $g_k \to \widehat{\varphi}$ in $\mathcal{E}(\mathbb{R}^1)$. The pointwise convergence $g_k(y) \to \widehat{\varphi}(y)$ is convergence of Riemann sums to the integral; this convergence is locally uniform in y and the same is true for all derivatives. Hence $F(g_k) \to F(\widehat{\varphi}) = \widehat{F}(\varphi)$. Another proof follows from Proposition 8.4.9, Theorem 8.5.5 and equalities (9.3.1). □

There are other classes of distributions for which one can define the Fourier transform. For example, the class $\mathcal{S}'$ does not contain the function $F(x) = \exp x^2$, hence there is no its Fourier transform in $\mathcal{S}'$, but if we take for the space of test functions the class of Fourier transforms of functions from $\mathcal{D}$, then $\widehat{F}$ can be defined as a functional on this space by the equality $\widehat{F}(\widehat{\varphi}) := F(\varphi)$. This scheme leads to a quadruple of spaces in place of the pair "test functions and generalized functions": a class $\mathcal{P}$ of test functions, the class $\mathcal{FP}$ of Fourier transforms of test functions and the spaces of functionals $\mathcal{P}'$ and $\mathcal{FP}'$ on these two classes. Then the Fourier transform of an element $F \in \mathcal{P}'$ is the element $\widehat{F} \in \mathcal{FP}'$ defined by the formula $\widehat{F}(\widehat{\varphi}) = F(\varphi)$. The Fourier transform of distributions plays an important role in the theory of partial differential equations.

9.4. Convolution

In this section the convolution of integrable functions known from §3.11 is extended to the case where one of the two functions is generalized, i.e., a distribution. The convolution is connected with the Fourier transform; it has numerous applications in analysis, algebra, and differential equations.

9.4.1. Theorem. If $f, g \in \mathcal{L}^1(\mathbb{R}^n)$, then

$$\widehat{f * g}(y) = (2\pi)^{n/2}\widehat{f}(y)\widehat{g}(y). \tag{9.4.1}$$

PROOF. We know that $f * g \in L^1(\mathbb{R}^n)$. By Fubini's theorem we have

$$(2\pi)^{-n/2} \int_{\mathbb{R}^n} \int_{\mathbb{R}^n} e^{-i(y,x)} f(x - z)g(z)\, dz dx$$

$$= (2\pi)^{-n/2} \int_{\mathbb{R}^n} \int_{\mathbb{R}^n} e^{-i(y,u)} e^{-i(y,z)} f(u)g(z)\, dz du,$$

which gives (9.4.1). See also Exercise 9.10.18. □

9.4.2. Example. (The Titchmarsh theorem) Let $f, g \in \mathcal{L}^1(\mathbb{R}^1)$ vanish on the ray $(-\infty, 0]$ and let $f * g = 0$. Then either $f = 0$ a.e. or $g = 0$ a.e. Indeed, $\widehat{f}$ and $\widehat{g}$ extend holomorphically to the lower halfplane Π, are bounded there and continuous on the real axis. Since $\widehat{fg} = 0$ on $\mathbb{R}^1$, we have $\widehat{fg} = 0$ in Π, whence either $\widehat{f} \equiv 0$ or $\widehat{g} \equiv 0$ in Π, whence our claim follows.

Let us now discuss the convolution $f * F$ of a usual function and a distribution. We consider only the following two cases: $f \in \mathcal{D}(\mathbb{R}^n)$, $F \in \mathcal{D}'(\mathbb{R}^n)$ and also $f \in \mathcal{S}(\mathbb{R}^n)$, $F \in \mathcal{S}'(\mathbb{R}^n)$. Our goal is to define the convolution $f * F$ in such a way that it be consistent with the classical one for distributions F given by integrable functions. To this end we write the action of the convolution of integrable functions f and F on a function $\varphi \in \mathcal{S}(\mathbb{R}^n)$ as follows:

$$\int_{\mathbb{R}^n} f * F(x)\varphi(x)\,dx = \int_{\mathbb{R}^n}\int_{\mathbb{R}^n} f(x-y)F(y)\varphi(x)\,dy\,dx$$
$$= \int_{\mathbb{R}^n}\int_{\mathbb{R}^n} f(x-y)\varphi(x)F(y)\,dx\,dy = \int_{\mathbb{R}^n} f(-\cdot) * \varphi(y)F(y)\,dy,$$

where $f(-\cdot)$ denotes the function $x \mapsto f(-x)$.

This suggests the following definition. Suppose that either $f \in \mathcal{D}(\mathbb{R}^n)$ and $F \in \mathcal{D}'(\mathbb{R}^n)$ or $f \in \mathcal{S}(\mathbb{R}^n)$ and $F \in \mathcal{S}'(\mathbb{R}^n)$. Set

$$f * F(\varphi) := F\big(f(-\cdot) * \varphi\big).$$

We observe that in the first case $f(-\cdot) * \varphi \in \mathcal{D}(\mathbb{R}^n)$ for all $\varphi \in \mathcal{D}(\mathbb{R}^n)$, and in the second case $f(-\cdot) * \varphi \in \mathcal{S}(\mathbb{R}^n)$ for all $\varphi \in \mathcal{S}(\mathbb{R}^n)$. The latter is easily seen from the fact that the Fourier transform of the convolution of two functions from $\mathcal{S}(\mathbb{R}^n)$ is the product of two elements of $\mathcal{S}(\mathbb{R}^n)$, while the Fourier transform maps $\mathcal{S}(\mathbb{R}^n)$ to $\mathcal{S}(\mathbb{R}^n)$. Thus, the convolution with the desired consistency property is well-defined. It turns out that it is given by a smooth function. Below some other cases of existence of convolution are mentioned.

9.4.3. Proposition. *If either $f \in \mathcal{D}(\mathbb{R}^n)$ and $F \in \mathcal{D}'(\mathbb{R}^n)$ or $f \in \mathcal{S}(\mathbb{R}^n)$ and $F \in \mathcal{S}'(\mathbb{R}^n)$, then $f * F$ is given by an infinitely differentiable function.*

PROOF. As suggested by the above expression for the action of convolution, we set

$$g(x) = F(f_x), \quad f_x(y) := f(x-y).$$

The definition is meaningful, since f_x belongs to the same class as f. We show that g is the required function.

We verify that the function g has partial derivatives and $\partial_{x_j} g = F\big((\partial_{x_j} f)_x\big)$. By induction this will yield the infinite differentiability of g. We fix a point x. By the continuity of F it suffices to show that the difference quotient $t^{-1}(f_{x+te_j} - f_x)$ converges to $(\partial_{x_j} f)_x$ as $t \to 0$ in the sense of convergence in the space $\mathcal{D}(\mathbb{R}^n)$ or $\mathcal{S}(\mathbb{R}^n)$, respectively. In the case of $\mathcal{D}$ it is clear that these difference quotients vanish outside some common cube and converge uniformly by the mean value theorem. Similarly one verifies uniform convergence of their derivatives of every order.

Let us consider the case of S. We observe that

$$t^{-1}\big(\widehat{f_{x+te_j}}(y) - \widehat{f_x}(y)\big) = t^{-1}\big(\exp(-ity_j) - 1\big)\widehat{f_x}(y).$$

The right-hand side converges in S to $-iy_j\widehat{f_x}(y) = -\widehat{\partial_{y_j}f_x}(y) = \widehat{(\partial_{x_j}f)_x}(y)$ by the estimate $|\exp(-ity_j) - 1| \leqslant |ty_j|$. Therefore, $t^{-1}(f_{x+te_j} - f_x) \to (\partial_{x_j}f)_x$ in S as $t \to 0$. Let us show that g defines the action of $f * F$. We have to verify that

$$\int_{\mathbb{R}^n} g(x)\varphi(x)\,dx = F\big(f(-\cdot) * \varphi\big), \tag{9.4.2}$$

i.e., that F can be interchanged with the integral. If φ has support in a cube Q, then the integral of $g\varphi$ is the limit of the Riemann sums of the form

$$\sum_{k=1}^{m} g(x_k)\varphi(x_k)\lambda_n(Q_k) = \sum_{k=1}^{m} F(f_{x_k})\varphi(x_k)\lambda_n(Q_k),$$

which can be written as $F\big(\sum_{k=1}^{m} f_{x_k}\varphi(x_k)\lambda_n(Q_k)\big)$, corresponding to refining partitions of the cube Q into cubes Q_k. We observe that the Riemann sums $\Psi_m(y) := \sum_{k=1}^{m} f_{x_k}(y)\varphi(x_k)\lambda_n(Q_k)$ for every fixed y tend to the integral of the function $x \mapsto f(x - y)\varphi(x)$ over Q, i.e., to the function $f(-\cdot) * \varphi(y)$. Moreover, this convergence takes place in $\mathcal{D}$ or in S for the respective f. Indeed, for any $f \in \mathcal{D}$ we have uniform convergence of Ψ_m, moreover, the derivatives of Ψ_m have a similar form (equal analogous sums with derivatives of f in place of f). The case of $f \in S$ is analogous, since the Fourier transform of Ψ_m is $\widehat{f(-\cdot)}(y)\sum_{k=1}^{m} \exp[-i(x_k,y)]\varphi(x_k)\lambda_n(Q_k)$, where the right-hand side converges in S to the function $(2\pi)^{n/2}\widehat{f(-\cdot)}\widehat{\varphi}$. By the continuity of F we arrive at equality (9.4.2) for φ with support in Q. In case of $f \in S$ the function g is growing not faster than some function $C(1 + |x|^2)^m$. Indeed, by the continuity of F there exist numbers C, m and k such that

$$|F(\varphi)| \leqslant C\sup_y(1 + |y|^2)^m\big(|\varphi(y)| + |\partial_{y_1}^k\varphi(y)| + \cdots + |\partial_{y_n}^k\varphi(y)|\big).$$

If we substitute $\varphi(y) = (1 + |x|^2)^{-m}f_x(y)$, then the right-hand side will be estimated by $C'\sup_y\big[(1+|x|^2)^{-m}(1+|x-y|^2)^{-m}(1+|y|^2)^m\big]$, where C' does not depend on x. We obtain an expression uniformly bounded in $x \in \mathbb{R}^n$. Therefore, the function $(1 + |x|^2)^{-m}|F(f_x)|$ is bounded. Now for justifying (9.4.2) with $\varphi \in S$ it suffices to take a sequence from $\mathcal{D}$ converging to φ in S. $\square$

Let us note the easily verified equality

$$\partial_{x_i}(f * F) = (\partial_{x_i}f) * F = f * (\partial_{x_i}F).$$

There are also other cases when one can define the convolution of a usual function and a distribution or even of two distributions (see [620, Chapter 1, §4]). For example, if $F, G \in \mathcal{D}'$ and G has compact support, then the convolution $F * G$ is well-defined: as above, we set

$$F * G(\varphi) := G\big(F(-\cdot) * \varphi\big),$$

where $F(-\cdot) * \varphi(x) = F\big(\varphi(\cdot - x)\big)$ is a smooth function, as shown above. If a function $F \in S'$ is such that its Fourier transform $\widetilde{F}$ is a usual function and

$\widetilde{F}\varphi \in S$ for all $\varphi \in S$, then for every distribution $G \in S'$ one can define the distribution $\widetilde{F} \cdot \widetilde{G} \in S'$, which by means of the inverse Fourier transform enables us to define $F * G$. One can also apply Theorem 8.5.5: if $F = \partial_{x_1}^k \cdots \partial_{x_n}^k \Psi$, $G = \partial_{x_1}^m \cdots \partial_{x_n}^m \Phi$, then we set $F * G := \partial_{x_1}^{k+m} \cdots \partial_{x_n}^{k+m} (\Psi * \Phi)$ if the convolution $F * G$ is meaningful in $\mathcal{D}'$ (for example, if G has compact support). However, the convolution $f * F$ on $S \times S'$ cannot be extended to a commutative associative convolution on $S' \times S'$ (see Exercise 8.6.49).

9.5. The Spectrum of the Fourier Transform and Convolution

Since the Fourier transform is a unitary operator on L^2, the question arises about its spectrum. We observe that the square of the Fourier transform is the operator of inverting the argument, i.e., $\mathcal{F}^2 f(x) = f(-x)$. This equality follows from the inversion formula for $f \in S(\mathbb{R}^n)$ and hence remains valid for all $f \in L^2(\mathbb{R}^n)$. Hence $\mathcal{F}^4$ is the unit operator. Therefore, taking the spectrum of $\mathcal{F}$ to power 4 we obtain 1. This shows at once that $\mathcal{F}$ has a finite spectrum belonging to the set $\{1, -1, i, -i\}$ and that all points in the spectrum are eigenvalues. In particular, the operator $\mathcal{F}$ has an eigenbasis. But how can we find eigenfunctions and decide which points in the indicated set belong to the spectrum and what are their multiplicities? Let $n = 1$. So far we know only one eigenfunction of the operator $\mathcal{F}$: the Gaussian function $\exp(-|x|^2/2)$ coincides with its Fourier transform. It turns out that starting from this eigenfunction one can find the whole eigenbasis. For this we observe that the linear span of the functions $\exp(-|x|^2/2), x\exp(-|x|^2/2), \ldots, x^k \exp(-|x|^2/2)$ is taken by the operator $\mathcal{F}$ to the same linear span. Indeed, let $f_0(x) = \exp(-|x|^2/2)$, $f_m(x) = x^m \exp(-|x|^2/2)$. Then

$$\mathcal{F}f_m(x) = i^m \frac{d^m}{dx^m} \mathcal{F}f_0(x) = P_m(x)e_0(x),$$

where P_m is a polynomial of degree m. A closer look at this formula shows that the linear span L_k of the four functions $f_{4k}, f_{4k+1}, f_{4k+2}, f_{4k+3}$ is invariant for every $k = 0, 1, \ldots$, which by the mutual orthogonality of eigenfunctions of a unitary operator hints to take the second eigenfunction in the form $e_1(x) = xe_0(x)$, since $e_1 \perp e_0$. We have $\mathcal{F}e_1 = -ie_1$. Next we find an eigenfunction e_2 in the form $e_2(x) = (x^2 - c)e_0(x)$, where c is chosen such that $e_2 \perp e_0$. Clearly, $e_2 \perp e_1$ and $\mathcal{F}e_2 = -e_2$. Finally, e_3 is found in the form $e_3(x) = (x^3 - \alpha x)e_0(x)$, where α is chosen such that $e_3 \perp e_1$. We have $\mathcal{F}e_3 = ie_3$. Note that $c = 1/2$, $\alpha = 3/2$. Thus, each of the four numbers indicated above is an eigenvalue. If we have decided to look for eigenfunctions among Hermite functions

$$E_k(t) = c_k \exp(t^2/2) \frac{d^k}{dt^k} \exp(-t^2),$$

forming a basis in $L^2(\mathbb{R}^1)$ according to Example 5.4.12(iv) with suitable norming numbers c_k, then we first verify the relationship

$$E_k(t) = \alpha_k G_k, \quad G_k(t) := \left(t - \frac{d}{dt}\right)^k f_0(t), \tag{9.5.1}$$

which after taking the Fourier transform gives

$$\mathcal{F}E_k(s) = \alpha_k\left(i\frac{d}{ds} - is\right)^k \mathcal{F}f_0(s) = (-i)^k\alpha_k\left(s - \frac{d}{ds}\right)^k f_0(s) = (-i)^k E_k(s).$$

For justification of formula (9.5.1) we use induction on k. For $k = 0$ this formula is true. Suppose it is true for some natural $k > 0$. It is readily seen that the function G_k has the form $G_k(t) = P_k(t)f_0(t)$, where P_k is a polynomial of degree k, and for odd k this polynomial includes only odd powers of t, while for even k it includes only even powers. In addition, the coefficient at t^k in $P_k(t)$ equals 2^k. In order to make the inductive step, we have to verify that the function G_{k+1} is orthogonal to all G_j with $j \leqslant k$. By the integration by parts formula we have

$$\int_{-\infty}^{+\infty}\left(t - \frac{d}{dt}\right)G_k(t)G_j(t)\,dt$$

$$= \int_{-\infty}^{+\infty} tG_k(t)G_j(t)\,dt + \int_{-\infty}^{+\infty} G_k(t)G_j'(t)\,dt$$

$$= \int_{-\infty}^{+\infty} 2tG_k(t)G_j(t)\,dt - \int_{-\infty}^{+\infty} G_k(t)[tG_j(t) - G_j'(t)]\,dt$$

$$= \int_{-\infty}^{+\infty} 2tG_k(t)G_j(t)\,dt - \int_{-\infty}^{+\infty} G_k(t)G_{j+1}(t)\,dt.$$

If $j < k - 1$, then the right-hand side of the relation above equals zero by the inductive assumption. If $j = k - 1$, then the right-hand side contains the integral of the function $G_k(t)[2tG_{k-1}(t) - G_k(t)]$, which also equals zero by the inductive assumption, since $2tG_{k-1}(t) - G_k(t) = Q(t)f_0(t)$, where Q is a polynomial of degree $k - 1$. Finally, the function $G_{k+1}G_k$ is odd and hence has zero integral. This gives the equality $(G_{k+1}, G_k)_{L^2} = 0$.

Let us now present another method of evaluating the spectrum of $\mathcal{F}$ (see Dirac [153, Chapter 2, §9]). This method is based on the following obvious algebraic relations:

$$\mathcal{F}^* = \mathcal{F}^{-1}, \qquad \mathcal{F}^{4+m} = \mathcal{F}^m.$$

Let us introduce operators

$$P_0 := \frac{1}{4}(I + \mathcal{F} + \mathcal{F}^2 + \mathcal{F}^3), \qquad P_1 := \frac{1}{4}(I - i\mathcal{F} - \mathcal{F}^2 + i\mathcal{F}^3),$$

$$P_2 := \frac{1}{4}(I - \mathcal{F} + \mathcal{F}^2 - \mathcal{F}^3), \qquad P_3 := \frac{1}{4}(I + i\mathcal{F} - \mathcal{F}^2 - i\mathcal{F}^3).$$

The general formula is this:

$$P_j := \frac{1}{4}\sum_{k=0}^{3}(-i)^{jk}\mathcal{F}^k, \qquad j = 0, 1, 2, 3.$$

Note also that these four operators are obtained by multiplying three of the four operators $\mathcal{F} - I$, $\mathcal{F} + I$, $\mathcal{F} - iI$, $\mathcal{F} + iI$. We do not discuss how one can come to the idea of using these operators. The straightforward verification shows that

$$P_j^* = P_j, \qquad P_j^2 = P_j, \qquad \mathcal{F}P_j = i^j P_j, \qquad P_j P_k = 0 \text{ if } j \neq k.$$

Thus, P_j are projections onto pairwise orthogonal subspaces on which $\mathcal{F}$ acts as the multiplication by i^j. Moreover,

$$P_0 + P_1 + P_2 + P_3 = I.$$

Set $H_j := P_j(H)$, where $H = L^2(\mathbb{R}^1)$. Then H turns out to be the direct sum of pairwise orthogonal closed subspaces H_j and $\mathcal{F}_{H_j}$ is the multiplication by i^j. In every subspace H_j every orthogonal basis is an eigenbasis for $\mathcal{F}$. Of course, for the completeness of the picture it is necessary to verify that the subspaces H_j are infinite-dimensional.

In the case $n > 1$ an eigenbasis of $\mathcal{F}$ is formed by the products of Hermite functions in different variables, i.e., by functions of the form

$$E_{k_1,\ldots,k_n}(x_1,\ldots,x_n) := E_{k_1}(x_1) \cdots E_{k_n}(x_n), \quad k_i \geqslant 0.$$

Hermite functions can be also introduced on infinite-dimensional spaces.

Let us now evaluate the spectrum of the convolution operator

$$S_\psi \varphi(x) = \psi * \varphi(x)$$

on the space $L^2(\mathbb{R}^n)$, where $\psi \in L^1(\mathbb{R}^n)$. We know (see Theorem 3.12.42) that $\psi * \varphi \in L^2(\mathbb{R}^n)$. Set $f = \mathcal{F}\psi$. Applying the Fourier transform we obtain (see Theorem 9.4.1 and Exercise 9.10.18)

$$\mathcal{F}S_\psi \varphi = (2\pi)^{n/2} \mathcal{F}\psi \cdot \mathcal{F}\varphi,$$

i.e., one has the equality

$$\mathcal{F}S_\psi \mathcal{F}^{-1} = (2\pi)^{n/2} A_f,$$

where A_f is the operator of multiplication by the bounded continuous function f. The spectrum of this operator is known: this is the set of essential values of f, i.e., the closure of the set of all values of f by the continuity of f. If the function f is real, then the spectrum of S_ψ is a closed interval of the real line. Let us observe that the convolution operator is noncompact (unlike integral operators from Example 6.9.4 on L^2 on a compact interval).

9.6. The Laplace Transform

One more important integral transform, a relative of the Fourier transform, is called the Laplace transform. It is also defined on different classes of functions. Set

$$\mathcal{L}f(x) := \frac{1}{\sqrt{\pi}} \int_0^\infty e^{-xy} f(y)\, dy \tag{9.6.1}$$

if the integral exists in the sense of Lebesgue. Note that if for some constants C and γ we have

$$|f(y)| \leqslant C \exp(\gamma y) \quad \text{for all } y \geqslant 0,$$

then the function $\mathcal{L}f$ is defined and analytic in the strip $\operatorname{Re} x > \gamma$ in the complex plane.

The Laplace transform can be also defined on $L^2[0, +\infty)$.

9.6.1. Definition. *The Laplace transform of a complex function f from the space $L^2[0, +\infty)$ is defined by the formula*

$$\mathcal{L}f(s) := \frac{1}{\sqrt{\pi}} \int_0^\infty e^{-st} f(t) \, dt, \quad s > 0.$$

9.6.2. Theorem. *The function $\mathcal{L}f$ belongs to $L^2[0, +\infty)$ and $\|\mathcal{L}f\|_2 \leqslant \|f\|_2$.*

PROOF. Note that unlike the case of the Fourier transform the indicated integral exists in the usual sense for every fixed $s > 0$, since the functions $t \mapsto f(t)$ and $t \mapsto e^{-st}$ belong to $L^2[0, +\infty)$. Suppose first that f vanishes in some neighborhood of zero. By the Cauchy–Bunyakovskii inequality we have

$$|\mathcal{L}f(s)|^2 \leqslant \frac{1}{\pi} \int_0^\infty e^{-st}|f(t)|^2 t^{1/2} \, dt \int_0^\infty e^{-st} t^{-1/2} \, dt$$

$$= \frac{1}{\pi}\sqrt{\frac{\pi}{s}} \int_0^\infty e^{-st}|f(t)|^2 t^{1/2} \, dt.$$

Integrating this inequality in s over $[0, +\infty)$, changing the order of integration and using that the integral of $e^{-st} t^{1/2} s^{-1/2}$ in s equals π, we obtain the bound

$$\|\mathcal{L}f\|_2^2 \leqslant \|f\|_2^2.$$

The general case follows by approximating f by functions of the indicated form. Indeed, if functions f_j converge to f in $L^2[0, +\infty)$ and each of them vanishes on some interval $[0, a_j]$, then by the considered case the sequence of functions $\mathcal{L}f_j$ is Cauchy in $L^2[0, +\infty)$ and hence converges to some function $h \in L^2[0, +\infty)$. Moreover, for every fixed $s > 0$ the sequence $\mathcal{L}f_j(s)$ converges to $\mathcal{L}f(s)$. Therefore, $\mathcal{L}f(s) = h(s)$ a.e. In addition, $\|\mathcal{L}f_j\|_2 \leqslant \|f_j\|_2$, whence $\|h\|_2 \leqslant \|f\|_2$. $\square$

The norm of the Laplace transform $\mathcal{L}$ on $L^2[0, +\infty)$ equals exactly 1 (Exercise 9.10.22). In addition, $\operatorname{Ker} \mathcal{L} = 0$ (Exercise 9.10.21), but $0 \in \sigma(\mathcal{L})$, since $\mathcal{L}$ is not surjective, because

$$|\mathcal{L}f(t)|^2 \leqslant (2\pi)^{-1}\|f\|_2 t^{-1}$$

by the Cauchy–Bunyakovskii inequality.

The Laplace transform can be defined for some generalized functions. Let $\mathcal{D}'_{+,\sigma}$ be the class of distributions $F \in \mathcal{D}'(\mathbb{R}^1)$ such that $\operatorname{supp} F \subset [0, +\infty)$ and $F_a := \exp(-ax)F \in \mathcal{S}'(\mathbb{R}^1)$ for all $a > \sigma$, moreover, $\widehat{F_a}$ is given by a usual function. The Laplace transform of the distribution F is defined as the usual function $\mathcal{L}F$ in the half-plane $\operatorname{Re} z > \sigma$ given by the formula

$$\mathcal{L}F(p + iq) := \sqrt{2}\widehat{F_p}(q).$$

It is clear that if distributions F_a are given by integrable functions, then we arrive at a formula of type (9.6.1). The Laplace transform is defined for certain functions (usual or generalized) for which the Fourier transform is not defined. As explained in the next section, this proves very useful for solving linear differential equations with constant coefficients.

9.7. Applications to Differential Equations

In this section we briefly discuss applications of the Fourier and Laplace transforms and convolution for solving differential equations. We consider only the general scheme and some typical examples.

Let L be a differential operator on $\mathbb{R}^n$ with constant coefficients $a_{j_1,\ldots,j_k}$, $j_i = 1, \ldots, n$, $k = 0, \ldots, m$, of the form

$$LF = \sum_{j_1,\ldots,j_k} a_{j_1,\ldots,j_k} \partial_{x_{j_1}} \cdots \partial_{x_{j_k}} F.$$

We observe that for functions F from $\mathcal{S}(\mathbb{R}^n)$ as well as for distributions F from $\mathcal{S}'(\mathbb{R}^n)$ one has the equality

$$\widehat{LF} = \sum_{j_1,\ldots,j_k} a_{j_1,\ldots,j_k} \cdot (ix_{j_1}) \cdots (ix_{j_k}) \widehat{F}.$$

The polynomial

$$P(x) = \sum_{j_1,\ldots,j_k} a_{j_1,\ldots,j_k} \cdot (ix_{j_1}) \cdots (ix_{j_k})$$

is called the *symbol of the differential operator* L; it is customary to write

$$L = P(-iD).$$

Thus, the problem of resolving the equation

$$LF = G$$

in the class $\mathcal{S}'(\mathbb{R}^n)$ (it can be also considered in other classes) reduces to the problem of resolving the equation

$$P\widehat{F} = \widehat{G}.$$

Formally the solution is given by the expression $\mathcal{F}^{-1}(\frac{1}{P}\mathcal{F}G)$. Of course, here the question arises about the validity of this operation. There is a simple, but important case where no problem arises: this is the case where $P(x) \geqslant c > 0$. Then for every element $G \in \mathcal{S}'$ we have $\mathcal{F}G \in \mathcal{S}'$, whence $\frac{1}{P}\mathcal{F}G \in \mathcal{S}'$ and hence there exists $\mathcal{F}^{-1}(\frac{1}{P}\mathcal{F}G) \in \mathcal{S}'$.

9.7.1. Example. Let $LF = \Delta F - F$. The equation $LF = G$ with respect to F is uniquely solvable in $\mathcal{S}'(\mathbb{R}^n)$ for every right-hand side $G \in \mathcal{S}'(\mathbb{R}^n)$ and the solution is given by the formula

$$F = \mathcal{F}^{-1}[(1 + |x|^2)^{-1}\mathcal{F}G].$$

If $G \in L^2(\mathbb{R}^n)$, then $F \in L^2(\mathbb{R}^n)$.

A *fundamental solution* (or a fundamental function) for the differential operator L is defined as a solution Φ of the equation $L\Phi = \delta$ in the corresponding class of distributions (for example, in $\mathcal{D}'$ or $\mathcal{S}'$). In case of $\mathcal{S}'$ the Fourier transform reduces this equation to the algebraic identity $P\widehat{\Phi} = (2\pi)^{-n/2}$, i.e., we arrive at the problem of dividing by a nonzero polynomial. There are the following two important results in this direction.

9.7.2. Theorem. *Let L be a differential operator with constant coefficients and the symbol $P \not\equiv 0$ on $\mathbb{R}^n$.*

(i) (THE MALGRANGE–EHRENPREIS THEOREM) *For every $G \in \mathcal{D}'(\mathbb{R}^n)$, the equation $LF = G$ has a solution in $\mathcal{D}'(\mathbb{R}^n)$. In particular, the operator L has a fundamental solution in $\mathcal{D}'(\mathbb{R}^n)$.*

(ii) (THE HÖRMANDER–LOJASIEWICZ THEOREM) *For every $G \in \mathcal{S}'(\mathbb{R}^n)$, the equation $LF = G$ has a solution in $\mathcal{S}'(\mathbb{R}^n)$. In particular, the operator L has a fundamental solution in $\mathcal{S}'(\mathbb{R}^n)$.*

One more close result is given in Exercise 9.10.45. Note that the equation $LF = G$ can have solutions in $\mathcal{D}'$ not belonging to $\mathcal{S}'$. For example, the equation $F' - F = 0$ has only zero solution in $\mathcal{S}'$, but has a solution e^x in $\mathcal{D}'$. If $\mathcal{E}$ is a fundamental solution in $\mathcal{D}'$ or $\mathcal{S}'$ and $G \in \mathcal{D}'$ has compact support, then the convolution $F := \mathcal{E} * G$ is defined and gives a solution to the equation $LF = G$.

If every solution in $\mathcal{D}'$ of the equation $LF = G$ with $G \in C^\infty$ is also given by a function of class C^∞, then the operator L is called *hypoelliptic*. This condition is equivalent to the property that

$$\operatorname{singsupp} LF = \operatorname{singsupp} F \quad \text{for all } F \in \mathcal{D}'(\mathbb{R}^n).$$

The latter equality is taken for the definition of hypoellipticity for general differential operators; it means that if $F \in \mathcal{D}'$ is such that LF in some open set U is given by a function of class $C^\infty(U)$, then F in U is also given by a function of this class. For operators with constant coefficients Hörmander found the following algebraic criterion for hypoellipticity: the equality $\lim_{|y|\to\infty} \nabla P(y)/P(y) = 0$; another equivalent condition is this: $\lim_{|y|\to\infty} P(y + v)/P(y) = 1$ for all $v \in \mathbb{R}^n$. For example, the Laplace operator Δ is hypoelliptic, but $\partial_{x_1}^2$ is not. For operators with constant coefficients hypoellipticity is equivalent to the existence of a fundamental solution Φ given by a function of class $C^\infty(\mathbb{R}^n \backslash 0)$ (in this case all fundamental solutions have this property). These questions are thoroughly considered in the books [281], [603, §18–20], and [548, Chapter 3].

Note that for $P \neq \text{const}$ the equation $LF = 0$ in $\mathcal{D}'$ always has a nonzero solution $\exp\big(i(x,\xi)\big)$, where $\xi \in \mathbb{C}^n$ and $P(\xi) = 0$.

We have discussed above equations on the whole space, but in applications one often encounters boundary value problems, in which also boundary values are given. For example, one can solve an equation with constant coefficients

$$f^{(n)} + c_1 f^{(n-1)} + \cdots + c_n f = g$$

on $[0, +\infty)$ with initial conditions $f^{(k)} = a_k$, $k = 0, \ldots, n - 1$. The Fourier transform does not apply here directly for two reasons: solutions are defined only on the half-line and may fail to be in $\mathcal{S}'$ even if we define then by the zero value on $(-\infty, 0)$ (for example, for the equation $f' - f = 0$, $f(0) = 1$ with a solution $f(x) = e^x$). Here it is useful to employ the Laplace transform. First we consider the case $n = 1$. Assuming that $g \in L^1[0, +\infty)$, we apply the Laplace transform to the equation and after integration by parts find

$$x\mathcal{L}f(x) + \frac{1}{\sqrt{\pi}} f(0) + c_1 \mathcal{L}f(x) = \mathcal{L}g(x).$$

Hence $\mathcal{L}f(x) = (x + c_1)^{-1}(\mathcal{L}g(x) - a_0/\sqrt{\pi})$. In principle, from here we could find f by means of the inverse Laplace transform. To this end, we have to extend $\mathcal{L}f$ to the imaginary axis and find the integral of $(2\pi^{-1/2}i)^{-1}e^{st}\mathcal{L}f(s)$ over this axis, see [154], [377], [629], where the inversion formula for the Laplace transform is discussed. In the general case $n > 1$ we similarly arrive at the equation

$$\Psi(x) + \sum_{k=0}^{n} c_{n-k}x^k \mathcal{L}f(x) = \mathcal{L}g(x),$$

where Ψ is a polynomial arising after n integrations by parts (the coefficients of Ψ are determined by the numbers a_k). Now the solution is given by the inverse Laplace transform of the function $[\mathcal{L}g(x) - \Psi(x)](\sum_{k=0}^{n} c_{n-k}x^k)^{-1}$. We do not discuss technical conditions under which this is possible. We only comment on the connection between the obtained solution with the one which is obtained by applying the Fourier transform to the function f extended by zero to $(-\infty, 0)$ if $f \in \mathcal{S}'(\mathbb{R}^1)$. Then the initial value disappears, but the equation itself changes due to the artificial jump of the solution at zero. For example, for $n = 1$ we obtain the equation

$$f' + f(0)\delta + c_1 f = g.$$

The Fourier transform gives $ix\mathcal{F}f(x) + a_0/\sqrt{2\pi} + c_1\mathcal{F}f(x) = \mathcal{F}g(x)$. Since formally $\mathcal{L}f(x) = \sqrt{2}\mathcal{F}f(ix)$, there is a full agreement between the obtained relations. These formal manipulations can be given a rigorous interpretation by continuing the Fourier and Laplace transforms to the complex domain. The picture is similar for equations of the nth order. This reasoning gives a basis of the so-called operational calculus, which is discussed in more detail in [154], [377, Chapter VI].

The Fourier transform is also useful for equations with variables coefficients, although for them there are no such explicit formulas for solutions. If we are given a function σ on the space $\mathbb{R}^n \times \mathbb{R}^n$, then we set

$$LF(x) := \mathcal{F}^{-1}[\sigma(x, \cdot)\mathcal{F}F(\cdot)](x),$$

which means that for fixed x the Fourier transform of F (as a function of argument y) is multiplied by the function $\sigma(x, y)$ and we evaluate the inverse Fourier transform at the point x. The operator L is called the *pseudo-differential operator* with the *symbol* σ. Under appropriate assumptions about F and σ (which we do not precise here) one can write

$$LF(x) = (2\pi)^{-n} \iint e^{i(x, y-z)}\sigma(x, y)F(z)\, dz\, dy.$$

If $\sigma(x, y) = \sigma(y)$ is a polynomial in y, then we obtain a differential operator with constant coefficients. If

$$\sigma(x, y) = \sum_{j_1, \ldots, j_k} a_{j_1, \ldots, j_k}(x)\, iy_{j_1} \cdots iy_{j_k},$$

then

$$LF(x) = \sum_{j_1, \ldots, j_k} a_{j_1, \ldots, j_k}(x)\partial_{x_{j_1}} \cdots \partial_{x_{j_k}} F(x).$$

Similarly one can consider the Cauchy problem for the evolution equation

$$\frac{\partial F}{\partial t} = LF$$

with the initial condition $F(0, x) = F_0(x)$, where L is a differential operator with constant coefficients. Applying the Fourier transform reduces this equation to an ordinary differential equation

$$\frac{\partial \Psi(t, y)}{\partial t} = P(y)\Psi(t, y)$$

for $\Psi(t, y) = [\mathcal{F}F(t, \cdot)](y)$, where the Fourier transform is taken in the second argument. A solution to this equation with the initial condition $\Psi_0(y) = \mathcal{F}F_0(y)$ is given by the formula

$$\Psi_0(y) \exp(tP(y)).$$

Hence the formal expression

$$\mathcal{F}^{-1}\big[\Psi_0 \exp(tP(y))\big](t, x) = (2\pi)^{-n/2} F_0 * \mathcal{F}^{-1}\big[\exp(tP(y))\big](t, x)$$

gives a "solution" to the original equation. Of course, as above, the question arises about conditions under which one can justify these formal manipulations (for example, for $P \leqslant 0$ and $F_0 \in L^2$).

9.7.3. Example. Let $L = \Delta$. Then $P(y) = -|y|^2$. Set

$$G_t(x) = (4\pi t)^{-n/2} \exp\big[-(4t)^{-1}|x|^2\big].$$

For every $F_0 \in L^2(\mathbb{R}^n)$ the function

$$F(t, x) = F_0 * G_t(x) = (4\pi t)^{-n/2} \int_{\mathbb{R}^n} \exp\big[-(4t)^{-1}|y|^2\big] F_0(x - y)\, dy$$

is a solution to the Cauchy problem

$$\frac{\partial F(t, x)}{\partial t} = \Delta F(t, y), \quad F(0, x) = F_0(x)$$

in the following sense (besides the pointwise equality): for every test function $\varphi \in \mathcal{S}(\mathbb{R}^n)$, for all $t \geqslant 0$ we have

$$\int_{\mathbb{R}^n} F(t, x)\varphi(x)\, dx = \int_{\mathbb{R}^n} F_0(x)\varphi(x)\, dx$$

$$+ \int_0^t \int_{\mathbb{R}^n} F(s, x)\Delta\varphi(x)\, dx\, ds. \quad (9.7.1)$$

Indeed, let $\psi = \mathcal{F}^{-1}\varphi$ and let $y \mapsto \Phi(s, y)$ be the Fourier transform of the function $x \mapsto F(s, x)$ belonging to $L^2(\mathbb{R}^n)$. By the Parseval equality and the relation $\Delta\varphi(x) = \Delta\mathcal{F}\psi(x) = \mathcal{F}\big[-|\cdot|^2\psi\big](x)$ the left-hand side of (9.7.1) coincides with

$$\int_{\mathbb{R}^n} \Phi(t, y)\overline{\psi(y)}\, dy,$$

and the right-hand side equals

$$\int_{\mathbb{R}^n} \Phi(0, y)\overline{\psi(y)}\, dy - \int_0^t \int_{\mathbb{R}^n} \Phi(s, y)|y|^2\overline{\psi(y)}\, dy\, dt.$$

Since $\Phi(s, y) = \Phi(0, y) \exp(-t|y|^2)$, both sides are equal. A similar representation of solutions to the Cauchy problem is valid in the class S' for initial conditions $F_0 \in S'$ if the convolution $F_0 * G_t$ is understood in the sense of distributions.

For the Cauchy problem

$$\frac{\partial f(t, x)}{\partial t} = Lf(t, x) + g(t, x), \quad f(0, x) = f_0(x)$$

with the boundary conditions $Mf = 0$ in a domain $\Omega \subset \mathbb{R}^n$, where L and M are differential operators on Ω and $\partial\Omega$, the Laplace transform can be applied, but this time in the variable t. At a formal level this gives the following equation for $\Psi(s, x) = \mathcal{L}F(\,\cdot\,, x)(s)$, where we take the Laplace transform in the first argument:

$$s\Psi(s, x) + f(0, x) = L\Psi(s, x) + \mathcal{L}g(\,\cdot\,, x)(s).$$

We have $M\Psi(s, x) = 0$ on $\partial\Omega$. Thus, for every s we have to solve the boundary value problem with the operator $sI - L$ on Ω. This problem can be simpler than the original initial-boundary value problem.

The explicit formulas for solutions to elliptic and parabolic equations found above with the Laplace operator are very useful in the study of general elliptic and parabolic equations with variables coefficients.

9.8. Sobolev Spaces $W^{p,k}$

Here we begin a discussion of one of the most important classes of spaces of functional analysis: Sobolev spaces.

Let Ω be an open set in $\mathbb{R}^n$, $p \in [1, +\infty)$, $k \in \mathbb{N}$.

9.8.1. Definition. *The Sobolev space $W^{p,k}(\Omega)$ is the class of all functions $f \in L^p(\Omega)$ such that for every $m \leqslant k$ and every multiindex $i_1, \ldots, i_m$ with $i_j \leqslant n$ the generalized derivative $\partial_{x_{i_1}} \cdots \partial_{x_{i_m}} f$ is a function of class $L^p(\Omega)$. The space $W^{p,k}(\Omega)$ is equipped with the Sobolev norm*

$$\|f\|_{p,k} := \|f\|_{L^p} + \sum_{m \leqslant k} \sum_{i_1, \ldots, i_m \leqslant n} \|\partial_{x_{i_1}} \cdots \partial_{x_{i_m}} f\|_{L^p}.$$

As in the case of spaces $L^p(\Omega)$, almost everywhere equal functions are identified, i.e., $W^{p,k}(\Omega)$ is the space of equivalence classes. One can also say that $W^{p,k}(\Omega)$ a subspace in $L^p(\Omega)$ consisting of all elements with finite norm $\|\cdot\|_{p,k}$. It is clear that $W^{p,k}(\mathbb{R}^n)$ is a linear subspace in $L^p(\Omega)$.

9.8.2. Example. A function f belongs to the class $W^{1,1}(\mathbb{R}^1)$ precisely when it is integrable and has an absolutely continuous version with a derivative integrable on $\mathbb{R}^1$.

PROOF. If f is integrable and absolutely continuous with $f' \in L^1(\mathbb{R}^1)$, then the integration by parts formula for absolutely continuous functions shows that f' serves as a generalized derivative. Conversely, suppose that f has a generalized derivative $g \in L^1(\mathbb{R}^1)$. Set

$$f_0(x) := \int_{-\infty}^{x} g(t)\, dt.$$

For every function $\varphi \in C_0^\infty(\mathbb{R}^1)$ we have the equalities

$$\int_{-\infty}^{+\infty} \varphi'(t)f(t)\,dt = -\int_{-\infty}^{+\infty} \varphi(t)g(t)\,dt = \int_{-\infty}^{+\infty} \varphi'(t)f_0(t)\,dt.$$

As shown in §8.5, this implies that the function $f - f_0$ almost everywhere equals some constant C. This constant is zero, since $g(x) \to 0$ as $x \to -\infty$ and by the integrability of f we can find points $x_n \to -\infty$ with $f(x_n) = f_0(x_n) + C$ such that $f(x_n) \to 0$. $\qquad\square$

9.8.3. Lemma. *The space $W^{p,k}(\Omega)$ is complete with respect to the indicated norm.*

PROOF. If a sequence $\{f_j\}$ is Cauchy with respect to the norm $\|\cdot\|_{p,k}$, then all partial derivatives of the functions f_j up to order k inclusive form Cauchy sequences and hence converge in $L^p(\Omega)$. It is readily seen that the limits obtained are generalized partial derivatives of the corresponding order for the limit of $\{f_j\}$ in $L^p(\Omega)$. $\qquad\square$

9.8.4. Lemma. *Let $f \in W^{p,k}(\Omega)$ and $\zeta \in C_b^\infty(\Omega)$. Then $\zeta f \in W^{p,k}(\Omega)$ and the generalized partial derivatives ζf up to order k are calculated by the formal application of the Leibniz rule. For example, $\partial_{x_i}(\zeta f) = \zeta \partial_{x_i} f + f \partial_{x_i} \zeta$.*

PROOF. The product of a smooth function and a distribution is differentiated by the Leibniz rule. The corresponding derivatives belong to $L^p(\Omega)$. $\qquad\square$

The next result gives a characterization of $W^{p,k}(\mathbb{R}^n)$ by means of the completion with respect to the Sobolev norm.

9.8.5. Theorem. *The space $W^{p,k}(\mathbb{R}^n)$ coincides with the completion of $C_0^\infty(\mathbb{R}^n)$ with respect to the Sobolev norm $\|\cdot\|_{p,k}$.*

PROOF. By the completeness of $W^{p,k}(\mathbb{R}^n)$ the aforementioned completion belongs to $W^{p,k}(\mathbb{R}^n)$. We show that every function f from $W^{p,k}(\mathbb{R}^n)$ can be approximated by smooth compactly supported functions with respect to the Sobolev norm. We first observe that f can be approximated in $W^{p,k}(\mathbb{R}^n)$ by functions with bounded support. To this end it suffices to take a sequence of smooth functions $\zeta_j(x) = \zeta(x/j)$, where $\zeta \in C_0^\infty(\mathbb{R}^n)$, $0 \leqslant \zeta \leqslant 1$, $\zeta(x) = 1$ if $|x| \leqslant 1$, $\zeta(x) = 0$ if $|x| \geqslant 2$. Then by the lemma $\zeta_j f \in W^{p,k}(\mathbb{R}^n)$ and the derivatives are calculated by the Leibniz rule. Moreover,

$$\partial_{x_{i_1}} \cdots \partial_{x_{i_m}}(\zeta_j f) \to \partial_{x_{i_1}} \cdots \partial_{x_{i_m}} f \quad \text{in } L^p(\mathbb{R}^n),$$

which follows by the Lebesgue dominated convergence theorem and the fact that $\zeta_j(x) = 1$ and $\partial_{x_i}\zeta_j(x) = 0$ as $|x| \leqslant j$. For example, in the equality

$$\partial_{x_i}(\zeta_j f) = (\partial_{x_i}\zeta_j)f + \zeta_j \partial_{x_i} f$$

the first term tends to zero and the second one tends to $\partial_{x_i} f$.

In the case where f has support in the ball of radius R centered at zero, for approximations we take the convolutions $f_j := f * \varphi_j$, where $\varphi_j(x) = j^d \varphi(jx)$ and φ is a smooth probability density with support in the unit ball. It is clear that $f_j \in C_0^\infty(\mathbb{R}^n)$. We observe that $\partial_{x_i} f_j = \partial_{x_i} f * \varphi_j$. This follows by taking

Fourier transforms in S', but it also suffices to verify that the functions $\psi \partial_{x_i} f_j$ and $\psi \partial_{x_i} f * \varphi_j$ have equal integrals for every function $\psi \in C_0^\infty(\mathbb{R}^n)$. By the definition of a generalized derivative for every fixed x we have

$$\int_{\mathbb{R}^n} \partial_{y_i} f(y) \varphi_j(x-y) \, dy = -\int_{\mathbb{R}^n} f(y) \partial_{y_i} \varphi_j(x-y) \, dy.$$

Multiplying both sides of this equality by $\psi(x)$ and integrating in x, on account of Fubini's theorem and the equality

$$\partial_{y_i} \varphi_j(x-y) = -\partial_{x_i} \varphi_j(x-y)$$

we obtain the following relations:

$$\int_{\mathbb{R}^n} \partial_{x_i} f * \varphi_j(x) \psi(x) \, dx = -\int_{\mathbb{R}^n} \int_{\mathbb{R}^n} f(y) \partial_{y_i} \varphi_j(x-y) \psi(x) \, dy \, dx$$

$$= \int_{\mathbb{R}^n} \int_{\mathbb{R}^n} f(y) \partial_{x_i} \varphi_j(x-y) \psi(x) \, dx \, dy$$

$$= -\int_{\mathbb{R}^n} \int_{\mathbb{R}^n} f(y) \varphi_j(x-y) \partial_{x_i} \psi(x) \, dx \, dy$$

$$= -\int_{\mathbb{R}^n} f * \varphi_j(x) \partial_{x_i} \psi(x) \, dx = \int_{\mathbb{R}^n} \partial_{x_i}(f * \varphi_j)(x) \psi(x) \, dx.$$

By induction we obtain $\partial_{x_{i_1}} \cdots \partial_{x_{i_m}} (f * \varphi_j) = \partial_{x_{i_1}} \cdots \partial_{x_{i_m}} f * \varphi_j$ for all $m \leqslant k$ and $i_j \leqslant n$. By Corollary 3.11.12 this gives convergence of $f * \varphi_j$ to f in the Sobolev norm. □

Let us prove one more useful result.

9.8.6. Theorem. *Let $1 < p < \infty$ and let a sequence $\{f_j\} \subset W^{p,k}(\mathbb{R}^n)$ be bounded in the Sobolev norm $\| \cdot \|_{p,k}$. Then it has a subsequence $\{f_{j_l}\}$ that converges weakly in $L^p(\mathbb{R}^n)$ to a function from $W^{p,k}(\mathbb{R}^n)$, moreover, the generalized partial derivatives of the functions f_{j_l} up to order k converge weakly in $L^p(\mathbb{R}^n)$ to the corresponding partial derivatives of the limit function.*

PROOF. It suffices to apply Theorem 6.7.4 to extract in $\{f_j\}$ a subsequence weakly converging in $L^p(\mathbb{R}^n)$ such that its generalized partial derivatives up to order k also converge weakly in $L^p(\mathbb{R}^n)$. □

For $p = 1$ this assertion is not valid. For example, the indicator function of $[-1, 1]$ is the limit in L^1 of the sequence of functions f_j such that $f_j(t) = 1$ if $t \in [-1, 1]$, $f_j(t) = 0$ if $|t| > 1 + 1/j$ and f_j is linear on $[-1 - 1/j, -1]$ and on $[1, 1 + 1/j]$. This sequence is bounded in $W^{1,1}$.

9.8.7. Theorem. *Let $f \in W^{p,1}(\mathbb{R}^n)$ and let ψ be a function on the real line satisfying the Lipschitz condition and $\psi(0) = 0$. Then $\psi(f) \in W^{p,1}(\mathbb{R}^n)$.*

PROOF. Suppose first that $\psi \in C^\infty(\mathbb{R}^n)$. Let us take functions $f_j \in C_0^\infty(\mathbb{R}^n)$ converging to f in $W^{p,1}(\mathbb{R}^n)$. We observe that $\psi(f_j) \in C_0^\infty(\mathbb{R}^n)$. Moreover, $|\psi(f_j) - \psi(f)| \leqslant C|f_j - f|$, where C is the Lipschitz constant of the function ψ. Hence $\psi(f_j) \to \psi(f)$ in $L^p(\mathbb{R}^n)$. In addition, $\partial_{x_i} \psi(f_j) = \psi'(f_j) \partial_{x_i} f_j$. Passing to a subsequence, we can assume that $f_j \to f$ almost everywhere. Then, by

Lebesgue's dominated convergence theorem, $\psi'(f_j)\partial_{x_i}f \to \psi'(f)\partial_{x_i}f$ in $L^p(\mathbb{R}^n)$. Since also $\psi'(f_j)\partial_{x_i}f_j - \psi'(f_j)\partial_{x_i}f \to 0$ in $L^p(\mathbb{R}^n)$ by the uniform boundedness of ψ', we obtain convergence of $\psi(f_j)$ to $\psi(f)$ in $W^{p,1}(\mathbb{R}^n)$.

In the general case we can find a sequence of smooth functions ψ_j with $\psi_j(0) = 0$ that converge pointwise to ψ and are uniformly Lipschitz. Then $\psi_j(f) \to \psi(f)$ in $L^p(\mathbb{R}^n)$, since

$$|\psi_j(f) - \psi(f)| \leqslant C|f|$$

with some constant C and $|\psi_j(f) - \psi(f)| \to 0$ pointwise. If $p > 1$, then by the relations

$$|\partial_{x_i}\psi_j(f)| = |\psi_j'(f)\partial_{x_i}f| \leqslant C|\partial_{x_i}f|$$

we obtain the inclusion $f \in W^{p,1}(\mathbb{R}^n)$ by Theorem 9.8.6. The case $p = 1$ is delegated to Exercise 9.10.37. $\qquad\Box$

For $n = 1$ every function from the Sobolev class $W^{p,1}$ has a continuous modification. For $n > 1$ this is not true.

9.8.8. Example. (i) The function $f(x) = \varphi(x)\ln|x|$ on $\mathbb{R}^2$, where φ from $C_0^\infty(\mathbb{R}^2)$ has support in the disc of radius $1/2$ centered at zero, $\varphi \geqslant 0$ and $\varphi = 1$ in a neighborhood of zero, belongs to $W^{p,1}(\mathbb{R}^2)$ whenever $1 \leqslant p < 2$, but has no bounded modification. Indeed, $f \in L^p(\mathbb{R}^2)$ for all $p < \infty$. The functions

$$\partial_{x_i}f(x) = \varphi(x)x_i|x|^{-2} + \partial_{x_i}\varphi(x)\ln|x|$$

are defined outside the origin and belong to $L^p(\mathbb{R}^2)$ with any $p < 2$. By means of the integration by parts formula it is straightforward to verify that they serve as generalized partial derivatives of f.

(ii) Similarly, the function $f(x) = \varphi(x)\ln(-\ln|x|)$ belongs to $W^{2,1}(\mathbb{R}^2)$.

(iii) The function $f(x) = \varphi(x)|x|^{-1}$ on $\mathbb{R}^3$, where $\varphi \in C_0^\infty(\mathbb{R}^3)$ and $\varphi = 1$ in a neighborhood of zero, belongs to $W^{p,1}(\mathbb{R}^3)$ whenever $1 \leqslant p < 3/2$, but has no bounded modification. This is verified similarly.

(iv) With the aid of the indicated functions f one can construct a function of the form $g(x) = \sum_{j=1}^\infty 2^{-j}f(x - x_j)$, where $\{x_j\}$ is an everywhere dense set. This function belongs to the corresponding $W^{p,1}(\mathbb{R}^n)$, but has no modification bounded at least in some neighborhood.

In spite of such examples, there are important positive results about some higher integrability of Sobolev functions. For a brief discussion, see §9.10(ii).

9.9. Characterization of $W^{2,k}$ by Means of the Fourier Transform

In the case $p = 2$ the Sobolev space $W^{p,k}$ is Hilbert, which leads to considerable simplifications in many problems. As an example we give a description of the classes $W^{2,k}$ in terms of the Fourier transform.

9.9.1. Theorem. *The space $W^{2,k}(\mathbb{R}^n)$ consists of all functions $f \in L^2(\mathbb{R}^n)$ such that the function $y \mapsto |y|^k\widehat{f}(y)$ belongs to $L^2(\mathbb{R}^n)$.*

PROOF. Let $f \in W^{2,k}(\mathbb{R}^n)$. Then for generalized derivatives we have $\partial_{x_i}^k f \in L^2(\mathbb{R}^n)$, which gives the quadratic integrability of the functions $y_i^k \widehat{f}(y)$. Hence the function $|y|^k \widehat{f}(y)$ belongs to $L^2(\mathbb{R}^n)$ as well. Conversely, let f and $|y|^k \widehat{f}(y)$ be in L^2. Then for any nonnegative integer numbers $k_1, \ldots, k_n$ with $k_1 + \ldots + k_n = m \leqslant k$ the function

$$g_{k_1,\ldots,k_n}(y) := (i)^m y_1^{k_1} \cdots y_n^{k_n} \widehat{f}(y)$$

belongs to $L^2(\mathbb{R}^n)$. Denote by $f_{k_1,\ldots,k_n}$ the inverse Fourier transform of the function $g_{k_1,\ldots,k_n}$. We show that $f_{k_1,\ldots,k_n}$ is the generalized derivative $\partial_{x_1}^{k_1} \cdots \partial_{x_n}^{k_n} f$. Fix a real function $\varphi \in \mathcal{D}(\mathbb{R}^n)$ and denote by ψ its Fourier transform. Then

$$\int_{\mathbb{R}^n} f(x) \partial_{x_1}^{k_1} \cdots \partial_{x_n}^{k_n} \varphi(x)\, dx = \int_{\mathbb{R}^n} \widehat{f}(y)(-i)^m y_1^{k_1} \cdots y_n^{k_n} \overline{\psi}(y)\, dy$$

$$= (-1)^m \int_{\mathbb{R}^n} g_{k_1,\ldots,k_n}(y)\overline{\psi}(y)\, dy = (-1)^m \int_{\mathbb{R}^n} f_{k_1,\ldots,k_n}(x)\varphi(x)\, dx$$

by the Parseval equality, which proves our assertion. $\square$

9.9.2. Example. For every function $g \in L^2(\mathbb{R}^n)$, there exists a unique solution $f \in W^{2,2}(\mathbb{R}^n)$ to the equation $\Delta f - f = g$. This solution is defined by the equality $\widehat{f}(y) = (1 + |y|^2)^{-1}\widehat{g}(y)$.

The previous theorem gives the following special case of the embedding theorem (for the general case, see §9.10(ii)).

9.9.3. Example. Let $f \in W^{2,k}(\mathbb{R}^n)$, where $k > n/2$. Then f has a bounded continuous version.

PROOF. An even stronger assertion is true: $\widehat{f} \in L^1(\mathbb{R}^n)$. Indeed, by the previous theorem the function $g(y) = |y|^k \widehat{f}(y)$ belongs to $L^2(\mathbb{R}^n)$. We have $|\widehat{f}(y)| \leqslant |g(y)| |y|^{-k}$. The right-hand side is integrable over the set $|y| \geqslant 1$ by the Cauchy–Bunyakovskii inequality, since the functions $|g(y)|^2$ and $|y|^{-2k}$ are integrable over this set if $2k > n$, which is obvious in polar coordinates. On the set $|y| \leqslant 1$ the function $\widehat{f}$ is integrable as well, because it belongs to $L^2(\mathbb{R}^n)$. $\square$

9.9.4. Remark. It follows from what has been said that for any $2k > n$ there exists a number $C(k, n)$ such that

$$\|f\|_{L^\infty} \leqslant C_1(k, n)\|f\|_{W^{2,k}}.$$

Indeed, the left-hand side does not exceed $(2\pi)^{-n/2}\|\widehat{f}\|_{L^1}$, which is estimated by $C_1(k, n) \sum_{i=1}^n \|\partial_{x_i}^k f\|_{L^2}$, as one can easily see from our justifications of the theorem and example. If $k = 2m$ is even, then

$$\|f\|_{L^\infty} \leqslant C_1(k, n)\|(I - \Delta)^m f\|_{L^2}.$$

With the aid of the Fourier transform one can also introduce fractional Sobolev classes $H^{2,r}(\mathbb{R}^n)$ with $r \in (-\infty, +\infty)$ consisting of distributions $f \in \mathcal{S}'(\mathbb{R}^n)$ such that $(1 + |y|^2)^{r/2}\widehat{f}(y) \in L^2(\mathbb{R}^n)$. In applications, the fractional classes $W^{p,r}$ and other generalizations or analogs of Sobolev classes (for example, Besov classes) are also useful. These questions are discussed in the books [5], [63], [184], [415], [566], [570], [574], and [657].

9.10. Complements and Exercises

(i) Singular integrals (422). (ii) Embedding theorems (425). (iii) The Bochner and Paley–Wiener theorems (428). Exercises (429).

9.10(i). Singular integrals

The term "singular integrals" is traditionally applied to a circle of problems at the junction of real analysis, harmonic analysis, operator theory, and the theory of integral equations, where objects of the following two types are used.

1. Sequences of integral kernels $\Phi_n(x, y)$ defining integral operators

$$T_n f(x) := \int f(y) \Phi_n(x, y)\, dy,$$

which converge in a certain sense (for example, in measure or in L^p) to an operator Tf, for example, to $Tf = f$.

2. Integral kernels $\Phi(x, y)$ defining operators of the form

$$Tf(x) = \int f(y) \Phi(x, y)\, dy,$$

where, however, the integral does not exist in the usual sense, but requires for definition some additional tricks (regularization, convergence in the sense of the principal value and so on). In applications one frequently deals with kernels of the form $\Phi(x, y) = \Psi(x - y)$, where Ψ is not necessarily integrable. Below we consider an important example: the Hilbert transform, in which $\Psi(z) = 1/z$.

Integrals of the second type are often defined by means of integrals of the first kind, so the difference between the two cases is very conditional. The word "singular" emphasizes the existence of a singularity of the kernel Φ, which in the second case does not allow to define the integral operator directly and in the first case indicates that the limit object need not be given in the integral form (as is the case with the operator $T = I$). The term itself is not formalized. Say, Chapter X of Natanson's book [**453**] is concerned with singular integrals and deals with the first case, while Chapter II of Stein's book [**570**] studies the second case. The goal of this subsection is to give a number of typical results about singular integrals. Actually, meaningful examples (which in many respects motivated the creation of the modern theory) have already been encountered above: this is the Fourier transform on $L^2(\mathbb{R}^n)$ defined as a singular integral operator with the kernel $(2\pi)^{-n/2} \exp i(x, y)$ and approximations of functions by trigonometric polynomials with the aid of Fourier and Fejér sums, which leads to integral operators with the Dirichlet kernels D_n and the Fejér kernels Φ_n.

9.10.1. Example. (i) Let $\varrho \in \mathcal{L}^1(\mathbb{R}^1)$ have the integral 1 over the real line. Set

$$\varrho_n(x) = \sigma_n^{-1} \varrho(x/\sigma_n),$$

where $\sigma_n > 0$ and $\lim\limits_{n \to \infty} \sigma_n = 0$. Then for every bounded continuous function f on the real line we have

$$f(x) = \lim_{n \to \infty} \int_{-\infty}^{+\infty} f(y) \varrho_n(x - y)\, dy.$$

Indeed, by the change of variables formula we have the equalities

$$f(x) - \int_{-\infty}^{+\infty} f(y)\varrho_n(x-y)\,dy = f(x) - \int_{-\infty}^{+\infty} f(x-u)\varrho_n(u)\,du$$

$$= f(x) - \int_{-\infty}^{+\infty} f(x-\sigma_n y)\varrho(y)\,dy = \int_{-\infty}^{+\infty} [f(x) - f(x-\sigma_n y)]\varrho(y)\,dy.$$

The integral in the right-hand side tends to zero as $n \to \infty$ by the Lebesgue dominated convergence theorem, since the quantities $f(x) - f(x - \sigma_n y)$ are uniformly bounded and tend to zero.

(ii) If f is a bounded measurable function and ϱ is locally bounded, then the established equality holds at every Lebesgue point of the function f (see [73, Theorem 5.8.4]).

(iii) If $f \in L^p(\mathbb{R}^1)$, where $1 \leqslant p < \infty$, then $f * \varrho_n \to f$ in $L^p(\mathbb{R}^1)$, which is deduced from (i) with the aid of approximations by continuous functions and Hölder's inequality.

Similar questions arise for more general kernels. Let us mention the following result; for its proof, generalizations, and other versions, see, e.g., Dunford, Schwartz [164, Chapter III, §12, Theorem 10], Hewitt, Stromberg [271, §21], and Natanson [453, Chapter X].

9.10.2. Theorem. *Let Q_n be nonnegative right-continuous functions on the real line such that Q_n is increasing on $(-\infty, 0]$, decreasing on $[0, +\infty)$,*
$$\lim_{|t| \to \infty} Q_n(t) = 0, \ \lim_{n \to \infty} Q_n(t) = 0 \ \text{if}\ t \neq 0, \ \text{and} \ \lim_{n \to \infty} \int_{-1}^{1} Q_n(t)\,dt = 1. \ \text{Then, for}$$
every function $f \in \mathcal{L}^1(\mathbb{R})$, one has
$$\lim_{n \to \infty} \int_{-\infty}^{+\infty} Q_n(t-s)f(s)\,ds = f(t)$$
at every Lebesgue point t of f, hence almost everywhere.

A typical example of the second type is the *Hilbert transform* defined by the singular integral

$$\mathcal{H}f(x) := \frac{1}{\pi} \int_{-\infty}^{+\infty} \frac{f(x-y)}{y}\,dy \qquad (9.10.1)$$

for functions from $L^p(\mathbb{R}^1)$. Here the role of the kernel is played by the function $G(z) = 1/z$. Let $\operatorname{sign} y = 1$ if $y \geqslant 0$ and $\operatorname{sign} y = -1$ if $y < 0$.

9.10.3. Theorem. *The Hilbert transform (9.10.1) on $L^2(\mathbb{R}^1)$ defined as a limit in $L^2(\mathbb{R}^1)$ of integral operators*

$$\mathcal{H}_\varepsilon f(x) := \frac{1}{\pi} \int_{\{|y| \geqslant \varepsilon\}} \frac{f(x-y)}{y}\,dy$$

exists and is a bounded operator. The operator $\mathcal{H}$ is unitarily equivalent to the operator of multiplication by the function $\xi(y) := -i \operatorname{sign} y$ and hence is also a unitary operator.

PROOF. Set $\theta_\varepsilon := I_{\mathbb{R}^1 \setminus [-\varepsilon, \varepsilon]}$. The operator $\mathcal{H}_\varepsilon$ is the convolution $f * h_\varepsilon$, where the function $h_\varepsilon(y) = (\pi y)^{-1} \theta_\varepsilon(y)$ is square integrable. Let us find its Fourier transform. It suffices to find the Fourier transform in the sense of distributions. Since $\pi y h_\varepsilon(y) = \theta_\varepsilon(y)$, we have $i\pi(\mathcal{F}h_\varepsilon)' = \mathcal{F}\theta_\varepsilon$. The Fourier transform of the function θ_ε in $\mathcal{S}'(\mathbb{R}^1)$ is easily evaluated explicitly, since $\theta_\varepsilon = 1 - I_{[-\varepsilon, \varepsilon]}$, where $\mathcal{F}1 = \sqrt{2\pi}\delta$, $\mathcal{F}I_{[-\varepsilon, \varepsilon]}(x) = \sqrt{2/\pi} x^{-1} \sin(\varepsilon x)$. Thus,

$$i\pi(\mathcal{F}h_\varepsilon)' = \sqrt{2\pi}\delta - \sqrt{2/\pi} x^{-1} \sin(\varepsilon x) = \sqrt{\pi/2}(\operatorname{sign} x)' - \sqrt{2/\pi} x^{-1} \sin(\varepsilon x).$$

Therefore, there exists a constant C_ε such that

$$\sqrt{2\pi}\mathcal{F}h_\varepsilon(x) = -i \operatorname{sign} x - i\frac{2}{\pi} \int_0^x \frac{\sin \varepsilon y}{y} \, dy + C_\varepsilon.$$

Since $h_\varepsilon(-x) = -h_\varepsilon(x)$, the function $\mathcal{F}h_\varepsilon$ is odd, whence $C_\varepsilon = 0$. The function

$$\psi_\varepsilon(x) := \int_0^x \frac{\sin \varepsilon y}{y} \, dy = \int_0^{\varepsilon x} \frac{\sin u}{u} \, du$$

is uniformly bounded in x and ε, since

$$\int_1^R \frac{\sin u}{u} \, du = \int_1^R \frac{\cos u}{u^2} \, du - \frac{\cos R}{R} + \cos R.$$

Moreover, $\lim_{\varepsilon \to 0} \psi_\varepsilon(x) = 0$ for every x. Therefore, for every $\varphi \in L^2(\mathbb{R}^1)$ we have $(\sqrt{2\pi}\mathcal{F}h_\varepsilon)\varphi \to -i(\operatorname{sign} x)\varphi$ in $L^2(\mathbb{R}^1)$. Thus, the Fourier transform is a unitary isomorphism between the operator $\mathcal{H}_\varepsilon$ and the operator of multiplication by the bounded measurable function $\sqrt{2\pi}\mathcal{F}h_\varepsilon(x) = -i \operatorname{sign} x - 2i\pi^{-1}\psi_\varepsilon(x)$. Since the Fourier transform is a unitary operator, we obtain convergence of $\mathcal{H}_\varepsilon f$ to $\mathcal{F}^{-1}A_\xi \mathcal{F}f$ in $L^2(\mathbb{R}^1)$, where A_ξ is the operator of multiplication by the bounded function $\xi(x) = -i \operatorname{sign} x$. $\square$

9.10.4. Remark. There is another justification of existence of the Hilbert transform. For this let us consider the convolution

$$\mathcal{G}_\varepsilon f(x) = \frac{1}{\pi} \int_{-\infty}^{+\infty} f(x - y) \frac{y}{y^2 + \varepsilon^2} \, dy.$$

The function $g_\varepsilon(y) := (\pi)^{-1} y(y^2 + \varepsilon^2)^{-1}$ belongs to $L^2(\mathbb{R}^1)$. Its Fourier transform has the form $\mathcal{F}g_\varepsilon(x) = i[\mathcal{F}G_\varepsilon(x)]'$, where $G_\varepsilon(y) = (\pi)^{-1}(y^2 + \varepsilon^2)^{-1}$. By the equality $\mathcal{F}G_\varepsilon(x) = (2\pi)^{-1/2}\varepsilon^{-1}e^{-\varepsilon|x|}$ (see Example 9.3.2(iv)) we obtain

$$\sqrt{2\pi}\mathcal{F}g_\varepsilon(x) = -ie^{-\varepsilon|x|}\operatorname{sign} x.$$

The bounded operators of multiplication by these functions converge on every element in $L^2(\mathbb{R}^1)$ to the operator of multiplication by $-i \operatorname{sign} x$.

A closer look shows that the Hilbert transform $\mathcal{H}$ extends to a bounded operator on every space $L^p(\mathbb{R}^1)$ with $1 < p < \infty$ and $\mathcal{H}_\varepsilon f \to \mathcal{H}f$ as $\varepsilon \to 0$ with respect to the norm of $L^p(\mathbb{R}^1)$ and almost everywhere for every function $f \in L^p(\mathbb{R}^1)$. Let us present the M. Riesz proof of the boundedness of $\mathcal{H}$ on $L^p(\mathbb{R}^1)$.

9.10.5. Theorem. *For every $p \in (1, \infty)$ there exists a number C_p such that* $\|\mathcal{H}u\|_p \leqslant C_p \|u\|_p$ *for all $u \in L^p(\mathbb{R}^1)$.*

PROOF. We can consider the case $p = 2k$, where $k \in \mathbb{N}$, since by the interpolation theorem 6.10.77 this will cover all $p > 2$, and then all $p \in (1, 2)$ on account of the equality $\mathcal{H}^* = -\mathcal{H}$. It suffices to find a common number C_p for all real $u \in C_0^\infty(\mathbb{R}^1)$. Set

$$f(z) := \frac{1}{\pi i} \int_{\mathbb{R}} \frac{u(t)}{t - z} \, dt, \quad \operatorname{Re} z > 0.$$

The function f is analytic in the upper half-plane and is estimated by $C(u)|z|^{-1}$ outside some disc. Using the boundedness of the relation $|u(t) - u(\xi)|/|t - \xi|$ on the real line and that u has compact support, it is not difficult to verify that the function f is bounded and when the argument is approaching the real line it tends to $u + iv$, where $v = \mathcal{H}u$. The function f^{2k} is analytic in the upper half-plane and has the zero integral over the contour consisting of the interval $[-R, R] + i\varepsilon$ and the semicircle of radius R over this interval. As $R \to \infty$, the integral over the semicircle tends to zero by the estimate $|f(z)|^{2k} \leqslant C(u)^{2k}|z|^{-2k}$. As $\varepsilon \to 0$ we obtain that the integral of the function $(u + iv)^{2k}$ over $\mathbb{R}^1$ is zero. Since for every $\delta > 0$ we have $u^l v^{2k-l} \leqslant \delta v^{2k} + \delta^{-2k}u^{2k}$, $1 \leqslant l \leqslant 2k - 1$, the integral of v^{2k} is estimated by the integral of $2k\delta v^{2k} + M_k u^{2k}$, where M_k depends only on k. For $\delta = (4k)^{-1}$ this gives the desired estimate with the constant $C_{2k} = (2M_k)^{1/(2k)}$. $\qquad\square$

If $f \in L^1(\mathbb{R}^1)$, then a finite limit $\lim_{\varepsilon \to 0} \mathcal{H}_\varepsilon f(x)$ exists almost everywhere, but it can fail to belong to $L^1(\mathbb{R}^1)$ (it suffices to evaluate $\mathcal{H}f$ for $f = I_{[a,b]}$). We observe that the operator $\mathcal{G}$ on $L^2[0, +\infty)$ with the kernel $(t + s)^{-1}$ is also continuous, since $\mathcal{G} = \mathcal{L}^2$, where $\mathcal{L}$ is the Laplace transform. About more general kernels with singularities, see Stein [**570**] and Torchinsky [**598**].

9.10(ii). Embedding theorems

In the theory of Sobolev spaces and its applications an important role is played by the so-called embedding theorems, which assert that under suitable conditions on p, k and n the space $W^{p,k}(\mathbb{R}^n)$ consists of continuous functions and is embedded into some $L^q(\mathbb{R}^n)$. The simplest example: the space $W^{1,1}(\mathbb{R}^1)$ is embedded into the space of bounded continuous functions. In the multidimensional case the situation is more complicated, but it is remarkable that also in that case there are simply stated embedding theorems. We confine ourselves to the proof of the most typical and important result and formulations of some other embedding theorems.

9.10.6. Theorem. (i) *If $p > n$ or $p = n = 1$, then one has the embedding* $W^{p,1}(\mathbb{R}^n) \subset C_b(\mathbb{R}^n) = C(\mathbb{R}^n) \cap L^\infty(\mathbb{R}^n)$. *Moreover, there exists a number* $C(p, n) > 0$ *such that*

$$\|f\|_{L^\infty(\mathbb{R}^n)} \leqslant C(p, n)\|f\|_{W^{p,1}}, \quad f \in W^{p,1}(\mathbb{R}^n). \tag{9.10.2}$$

(ii) *If $p \in [1, n)$, then $W^{p,1}(\mathbb{R}^n) \subset L^{np/(n-p)}(\mathbb{R}^n)$. Moreover, there exists a number $C(p, n) > 0$ (where $C(1, n) = 1$) such that*

$$\|f\|_{L^{np/(n-p)}(\mathbb{R}^n)} \leqslant C(p, n) \big\| |\nabla f| \big\|_{L^p}, \quad f \in W^{p,1}(\mathbb{R}^n). \tag{9.10.3}$$

PROOF. (i) The case $p = n = 1$ is trivial. Let $p > n > 1$. Due to Theorem 9.8.5 it suffices to establish the estimate

$$\sup_x |\varphi(x)| \leqslant C(p, n) \|\varphi\|_{W^{p,1}}, \quad \varphi \in C_0^\infty(\mathbb{R}^n).$$

In turn, it suffices to show that $|\varphi(0)| \leqslant C(p, n)\|\varphi\|_{W^{p,1}}$ for all $\varphi \in C_0^\infty(\mathbb{R}^n)$. This will be done if we find a number C such that $|\varphi(0)| \leqslant C\|\nabla\varphi\|_{L^p}$ for all φ with support in the unit ball U. Indeed, let us take a function $\zeta \in C^\infty(\mathbb{R}^n)$ with $\zeta(0) = 1$ and support in U. Then for all $u \in C_0^\infty(\mathbb{R}^n)$ we obtain

$$|\zeta(0)\varphi(0)| \leqslant C\|\nabla(\zeta\varphi)\|_{L^p} \leqslant C[\max_x[|\zeta(x)| + |\nabla\zeta(x)|]\,\|\varphi\|_{W^{p,1}}.$$

For every unit vector u the Newton–Leibniz formula gives

$$-\varphi(0) = \int_0^1 \big(\nabla\varphi(tu), u\big)\,dt.$$

Integrating this equality over the unit sphere S with respect to the standard surface measure σ_{n-1} we obtain

$$-\varphi(0) = \sigma_n(S)^{-1} \int_S \int_0^1 \big(\nabla\varphi(tu), u\big)\,dt\,\sigma_{n-1}(du).$$

In the right-hand side one can easily recognize the integral over U written in spherical coordinates. By the change of variables formula we find

$$-\varphi(0) = \sigma_n(S)^{-1} \int_U |x|^{1-n}\big(\nabla\varphi(x), |x|^{-1}x\big)\,dx.$$

Hölder's inequality yields

$$|\varphi(0)| \leqslant \sigma_n(S)^{-1}\|\nabla\varphi\|_{L^p}\left(\int_U |x|^{p(1-n)/(p-1)}\,dx\right)^{(p-1)/p} = C(n, p)\|\nabla\varphi\|_{L^p},$$

since the integral of $|x|^{p(1-n)/(p-1)}$ over U is finite. This is verified again in spherical coordinates, which up to a factor gives a power of the integral of t^α over the interval $[0, 1]$, where $\alpha = (1 - n)/(p - 1) > -1$.

(ii) Again it suffices to establish estimate (9.10.3) for $\varphi \in C_0^\infty(\mathbb{R}^n)$. For every $j \leqslant n$ by the Newton–Leibniz formula we have

$$|\varphi(x)| \leqslant \int_{-\infty}^{+\infty} |\partial_{y_j}\varphi(x_1, \ldots, y_j, \ldots, x_n)|\,dy_j.$$

The function under the integral will be denoted by $|\partial_{y_j}\varphi|$. Then

$$|\varphi(x)|^{n/(n-1)} \leqslant \prod_{j=1}^n \left(\int_{-\infty}^{+\infty} |\partial_{y_j}\varphi|\,dy_j\right)^{1/(n-1)}.$$

Integrating this inequality consequently in $x_1, \ldots, x_n$ and applying after every integration in one variable the generalized Hölder inequality

$$\|\psi_1 \cdots \psi_k\|_{L^1} \leqslant \|\psi_1\|_{L^k} \cdots \|\psi_k\|_{L^k}$$

with $k = n - 1$ (the integral of $|\partial_{y_j}\varphi|$ does not depend on x_j in y_j) we arrive at the estimate

$$\|\varphi\|_{L^{n/(n-1)}} \leqslant \left\{ \prod_{j=1}^{n} \int_{\mathbb{R}^n} |\partial_{y_j}\varphi| \, dx \right\}^{1/n} \leqslant \|\nabla\varphi\|_{L^1}.$$

Thus, we have obtained (9.10.3) for $p = 1$ with $C(1, n) = 1$. It is clear that it remains in force for $|\varphi|^{\alpha}$, where $\alpha = (n-1)p/(n-p) > 1$ in case $p > 1$. Hence

$$\big\| |\varphi|^{\alpha} \big\|_{L^{n/(n-1)}} \leqslant \alpha \big\| |\varphi|^{\alpha-1} |\nabla\varphi| \big\|_{L^1} \leqslant \alpha \big\| |\varphi|^{\alpha-1} \big\|_{L^{p'}} \big\| |\nabla\varphi| \big\|_{L^p}.$$

It remains to observe that $\alpha n/(n-1) = np/(n-p)$ and $(\alpha - 1)p' = np/(n-p)$, whence

$$\big\| |\varphi|^{\alpha} \big\|_{L^{n/(n-1)}} = \|\varphi\|_{L^{np/(n-p)}}^{(n-1)p/(n-p)}, \quad \big\| |\varphi|^{\alpha-1} \big\|_{L^{p'}} = \|\varphi\|_{L^{np/(n-p)}}^{n(p-1)/(n-p)},$$

which gives (9.10.3). See [384, p. 317] for the optimal constant $C(p, n)$. $\qquad\square$

For $p = 1$ inequality (9.10.3) is called the *Gagliardo–Nirenberg inequality*; it shows that any integrable function on $\mathbb{R}^n$ with an integrable gradient actually belongs to $L^{n/(n-1)}(\mathbb{R}^n)$.

Note that a function from $W^{n,1}(\mathbb{R}^n)$ need not be locally bounded (see Example 9.8.8(ii)), but on every ball U it belongs to all $L^r(U)$.

From the established results one can derive the following assertions.

9.10.7. Corollary. *The following embeddings hold.*
(i) *If $kp < n$, then $W^{p,k}(\mathbb{R}^n) \subset L^{np/(n-kp)}(\mathbb{R}^n)$.*
(ii) *If $kp > n$, then $W^{p,k}(\mathbb{R}^n) \subset C(\mathbb{R}^n) \cap L^{\infty}(\mathbb{R}^n)$.*
(iii) *$W^{1,n}(\mathbb{R}^n) \subset C(\mathbb{R}^n) \cap L^{\infty}(\mathbb{R}^n)$.*

Analogous assertions are also valid for domains with sufficiently regular boundaries, but the constants will depend on domains. Unlike the whole space, for a bounded domain Ω there are inclusions $L^p(\Omega) \subset L^r(\Omega)$ for $p > r$. This leads to a broader spectrum of embedding theorems. Let us formulate the principal results for the ball $\Omega = B(a, r) \subset \mathbb{R}^n$. Set $W^{q,0} := L^q$.

9.10.8. Theorem. (i) *If $kp < n$, then*

$$W^{p,j+k}(\Omega) \subset W^{q,j}(\Omega), \quad j = 0, 1, \ldots, \quad q \leqslant \frac{np}{n - kp}.$$

(ii) *If $kp = n$, then $W^{p,j+k}(\Omega) \subset W^{q,j}(\Omega)$, $j = 0, 1, \ldots$, $q < \infty$. For $p = 1$ we have $W^{1,j+n}(\Omega) \subset C_b^j(\Omega)$.*
(iii) *If $kp > n$, then $W^{p,j+k}(\Omega) \subset C_b^j(\Omega)$, $j = 0, 1, \ldots$*
In addition, these embeddings, excepting case (i) with $q = np/(n - kp)$ and the embedding $W^{1,j+n}(\Omega) \subset C_b^j(\Omega)$, are compact operators.

For $k = 1$ Sobolev functions can be characterized in terms of their behavior on straight lines as follows.

9.10.9. Theorem. *A function $f \in L^p(\mathbb{R}^n)$ belongs to $W^{p,1}(\mathbb{R}^n)$ precisely when it has a modification g such that for every fixed i for almost every point $(y_1, \ldots, y_{i-1}, y_{i+1}, \cdots, y_n) \in \mathbb{R}^{n-1}$ the function*

$$x_i \mapsto g(y_1, \ldots, y_{i-1}, x_i, y_{i+1}, \cdots, y_n)$$

is absolutely continuous on all compact intervals and the function $\partial_{x_i} g$ (existing almost everywhere) belongs to the class $L^p(\mathbb{R}^n)$.

PROOF. For simplification of our formulas let us consider the case $n = 2$, $p = 1$. If such a version g exists, then with the aid of the one-dimensional integration by parts formula one can easily verify that the functions $\partial_{x_i} g$ serve as generalized partial derivatives of f. Let $f \in W^{1,1}(\mathbb{R}^2)$. Let us take functions $\varphi_j \in C_0^\infty(\mathbb{R}^2)$ converging to f in $W^{1,1}(\mathbb{R}^2)$. Passing to a subsequence, we can assume that $\|\varphi_{j+1} - \varphi_j\|_{W^{1,1}} \leqslant 2^{-j}$ and $\{\varphi_j\}$ converges almost everywhere. Let us define a version g of the function f as the limit $\{\varphi_j(x)\}$, where it exists. For any fixed x_2 we set

$$h_j(x_2) = \int_{\mathbb{R}^1} \big[|\varphi_{j+1}(x_1, x_2) - \varphi_j(x_1, x_2)|$$
$$+ |\partial_{x_1} \varphi_{j+1}(x_1, x_2) - \partial_{x_1} \varphi_j(x_1, x_2)| \big] \, dx_1.$$

Then the series of the integrals of h_j converges, so $\sum_{j=1}^\infty h_j(x_2) < \infty$ for almost all x_2. In addition, for almost every x_2 there is a point x_1 at which the functions $\psi_{x_2,j} \colon t \mapsto \varphi_j(t, x_2)$ converge. Hence for almost every x_2 the sequence of functions $\psi_{x_2,j}$ converges at some point and is Cauchy in $W^{1,1}(\mathbb{R}^1)$. By the Newton–Leibniz formula this implies its convergence on the whole real line to a locally absolutely continuous function from $W^{1,1}(\mathbb{R}^1)$. Thus, the function g possesses the required property in x_1. It is clear that the same is true for x_2. $\square$

Note that it is not always possible to find a version continuous everywhere in every variable separately (Exercise 9.10.38).

9.10(iii). The Bochner and Paley–Wiener theorems

Here we formulate Bochner's theorem describing Fourier transforms of nonnegative measures and two theorems due to Paley and Wiener, which characterize Fourier transforms of test functions and distributions with compact support. A function $\varphi \colon \mathbb{R}^n \to \mathbb{C}$ is called positive definite if $\sum_{i,j=1}^k c_i \overline{c_j} \varphi(x_i - x_j) \geqslant 0$ for all $k \in \mathbb{N}$, $c_1, \ldots, c_k \in \mathbb{C}$, $x_1, \ldots, x_k \in \mathbb{R}^n$.

9.10.10. Theorem. *A function $\varphi \colon \mathbb{R}^n \to \mathbb{C}$ is the Fourier transform of a nonnegative Borel measure on $\mathbb{R}^n$ precisely when it is continuous and positive definite.*

For a proof, see [**73**, §3.10(v), §7.13].

9.10.11. Theorem. *The Fourier transform of a function f from $\mathcal{D}(\mathbb{R}^n)$ with support in the ball $B_R = \{x \colon |x| \leqslant R\}$ extends to an entire function g on $\mathbb{C}^n$ such that for every $N \geqslant 0$ there exists a number C_N such that*

$$|g(z)| \leqslant C_N (1 + |z|)^{-N} \exp(R|\mathrm{Im}\, z|), \quad z \in \mathbb{C}^n.$$

Conversely, every entire function g on $\mathbb{C}^n$ satisfying the stated condition is a holomorphic continuation of the Fourier transform of a function from the complex space $\mathcal{D}(\mathbb{R}^n)$ with support in the ball B_R.

9.10.12. Theorem. *The Fourier transform of a distribution F from $S'(\mathbb{R}^n)$ with support in the ball $B_R = \{x: |x| \leqslant R\}$ extends to an entire function G on $\mathbb{C}^n$ such that for some numbers $C, N \geqslant 0$ we have*

$$|G(z)| \leqslant C(1 + |z|)^N \exp(R|\mathrm{Im}\, z|), \quad z \in \mathbb{C}^n.$$

Conversely, every entire function G on $\mathbb{C}^n$ satisfying the stated condition is a holomorphic continuation of the Fourier transform of a distribution from the complex space $S'(\mathbb{R}^n)$ with support in the ball B_R.

For the proofs, see Reed, Simon [**503**, §IX.3] and Rudin [**522**, Chapter 7].

Exercises

9.10.13° Find the Fourier transform of the indicator function of the parallelogram $[a_1, b_1] \times \cdots \times [a_n, b_n] \subset \mathbb{R}^n$.

9.10.14° Give an example of a function $f \in L^1(\mathbb{R}^1)$ the Fourier transform of which belongs neither to $L^1(\mathbb{R}^1)$ nor to $L^2(\mathbb{R}^1)$, and also an example of a function g in $L^2(\mathbb{R}^1)$ the Fourier transform of which does not belong to $L^1(\mathbb{R}^1)$.

9.10.15° Find the Fourier transforms of the following distributions of class $S'(\mathbb{R}^1)$: (i) $\mathrm{sign}\, x$, (ii) V.P. x^{-1}, (iii) $(x + i0)^{-1}$, (iv) $\theta(x) \sin x$, where $\theta = I_{[0,+\infty)}$ is the Heaviside function, (v) $|x|$, (vi) $F(x) = -x$ if $x < 0$, $F(x) = x^2$ if $x \geqslant 0$, (vii) $F(x) = 0$ if $x < 0$, $F(x) = \sin x$ if $x \geqslant 0$, (viii) $F(x) = x$ if $x < 0$, $F(x) = \sin x$ if $x \geqslant 0$, (ix) $F(x) = \sin(x^2)$, (x) $F(x) = x^{-1} \sin x$.

9.10.16. Prove that $\mathcal{F}(\text{V.P.}|x|^{-1})(y) = c_1 + c_2 \ln|y|$, where c_1, c_2 are numbers and the distribution V.P.$|x|^{-1}$ is defined in Exercise 8.6.52. With the aid of this equality calculate the Fourier transform of $\ln|x|$ in $S'(\mathbb{R}^1)$.

9.10.17° Prove that the equation $F' = F$ has no nonzero solutions in $S'(\mathbb{R}^1)$.

9.10.18° From Theorem 9.4.1 derive the equality $\widehat{f * g} = (2\pi)^{n/2} \widehat{f}\widehat{g}$ if $f \in L^2(\mathbb{R}^n)$ and $g \in L^2(\mathbb{R}^n)$ or $g \in L^1(\mathbb{R}^n)$.

9.10.19. With the aid of the Laplace transform solve the ordinary differential equation $f''(x) + f(x) = x$ with the initial conditions $f(0) = 1$, $f'(0) = 0$.

9.10.20. Let $f \in L^1(\mathbb{R}^1) \cap L^\infty(\mathbb{R}^1)$ and $\widehat{f} \geqslant 0$. Prove that $\widehat{f} \in L^1(\mathbb{R}^1)$ and that for every Lebesgue point x of the function f we have

$$f(x) = (2\pi)^{-1/2} \int_{-\infty}^{+\infty} e^{ixy} \widehat{f}(y)\, dy.$$

Hence f has a continuous version satisfying the last equality everywhere.

HINT: observe that the integrals of $\widehat{f}(y) \exp(-\alpha y^2)$ are uniformly bounded in $\alpha > 0$, since the integrals of $f(x)\alpha^{-1/2} \exp(-x^2/(4\alpha))$ are uniformly bounded by the boundedness of f. By Fatou's theorem $\widehat{f}$ is integrable. One can show that this assertion fails if f is not bounded, but it is easily seen that it suffices to know that f is bounded in a neighborhood of the origin.

9.10.21. Let $f, g \in \mathcal{L}^1[0, +\infty)$ be extended by zero to $(-\infty, 0)$.
(i) Prove that $\mathcal{L}(f * g) = \sqrt{\pi}\mathcal{L}f \cdot \mathcal{L}g$, where $\mathcal{L}$ is the Laplace transform.
(ii) Let $f \in \mathcal{L}^1[0, +\infty)$ and $\mathcal{L}f = 0$. Prove that $f = 0$ a.e.
(iii) Prove assertions (i) and (ii) for functions from $\mathcal{L}^2[0, +\infty)$.

9.10.22. (i) Show that the operator $\mathcal{L}^2$ on $L^2[0, +\infty)$, where $\mathcal{L}$ is the Laplace transform, is given by the integral kernel $\mathcal{K}(t, s) = \pi^{-1}(t + s)^{-1}$.
(ii) Prove that the norm of the Laplace transform on $L^2[0, +\infty)$ equals 1.
HINT: (ii) show that the operator $\mathcal{L}^2$ has the unit norm by considering the functions $\psi_{a,b}(t) = t^{-1/2}I_{[a,b]}$, where $a \to 0$, $b \to +\infty$.

9.10.23. Let $g_\varepsilon(x) = \exp(-\varepsilon|x|^2)$ or $g_\varepsilon(x) = \exp(-\varepsilon|x|)$, where $\varepsilon > 0$. Prove that for every function $f \in \mathcal{L}^1(\mathbb{R}^1)$ the functions

$$f_\varepsilon(x) = (2\pi)^{-1/2} \int_{-\infty}^{+\infty} e^{ixy}\widehat{f}(y)g_\varepsilon(y)\, dy$$

converge to f in $L^1(\mathbb{R}^1)$ as $\varepsilon \to 0$.
HINT: find the Fourier transform of g_ε, write f_ε as a convolution, observe that the assertion holds for smooth compactly support functions and approximate f by such functions.

9.10.24. Let $F(x) = \operatorname{sech}(x\sqrt{\pi/2})$, where $\operatorname{sech}(x) = 2(e^x + e^{-x})^{-1}$, $x \in \mathbb{R}$. Prove that $\widehat{F} = F$.

9.10.25. Prove the Poisson summation formula:
$$(2\pi)^{n/2} \sum_{k \in \mathbb{Z}^n} \widehat{f}(2\pi k) = \sum_{k \in \mathbb{Z}^n} f(k), \quad f \in \mathcal{S}(\mathbb{R}^n).$$
HINT: see Exercise 8.6.56.

9.10.26. Let $\mathcal{E}'(\mathbb{R}^d)$ be the space of distributions with compact support. It is an algebra with respect to the convolution. Prove that this algebra has no nonzero divisors of zero, i.e., if $F * G = 0$, where $F, G \in \mathcal{E}'(\mathbb{R}^d)$, then at least one of the elements F or G is zero.

9.10.27. Let $f \in L^2(\mathbb{R}^1)$. Prove the Heisenberg inequality:
$$\left(\int_{-\infty}^{+\infty} |f(x)|^2\, dx\right)^2 \leqslant 4 \int_{-\infty}^{+\infty} x^2|f(x)|^2\, dx \int_{-\infty}^{+\infty} y^2|\widehat{f}(y)|^2\, dy.$$
HINT: for any $f \in C_0^\infty(\mathbb{R}^1)$ we have
$$|f(x)|^2 = \int_{-\infty}^{x} f'(t)\overline{f(t)}\, dt + \int_{-\infty}^{x} f(t)\overline{f'(t)}\, dt,$$
which by means of the integration by parts formula gives
$$\|f\|_{L^2}^2 = \int_{-\infty}^{+\infty} xf'(x)\overline{f(x)}\, dx + \int_{-\infty}^{+\infty} xf(x)\overline{f'(x)}\, dx.$$
Apply to the right-hand side the Cauchy–Bunyakovskii inequality and observe that the integral of $|f'(x)|^2$ equals the integral of $|\widehat{f}'(y)|^2$, and the latter is the integral of $|y\widehat{f}(y)|^2$. Another proof: take the operators $Df = if'$ and $Af(x) = xf(x)$ and use the equality $(AD - DA)f = if$ and the symmetry of A and D.

9.10.28. Let ϱ be an absolutely continuous probability density on $\mathbb{R}^1$. Prove the inequality
$$\int_{-\infty}^{+\infty} x^2\varrho(x)\, dx \int_{-\infty}^{+\infty} \frac{|\varrho'(x)|^2}{\varrho(x)}\, dx \geqslant 1.$$

The integral of the function $x^2 \varrho(x)$ is called the second moment of ϱ, the integral of $|\varrho'(x)|^2/\varrho(x)$ is called the Fisher information of ϱ.
HINT: apply the previous exercise to $f = \sqrt{\varrho}$.

9.10.29. (i) Prove that φ is the Fourier transform of a function from $L^1(\mathbb{R}^n)$ if and only if $\varphi = \psi_1 * \psi_2$, where $\psi_1, \psi_2 \in L^2(\mathbb{R}^n)$.

(ii) Prove that a complex function φ equals the Fourier transform of a nonnegative absolutely continuous measure precisely when there exists a complex function $\psi \in L^2(\mathbb{R}^n)$ such that

$$\varphi(x) = \int_{\mathbb{R}^n} \psi(x+y)\overline{\psi(y)}\, dy.$$

HINT: if $f \in L^1(\mathbb{R}^n)$ and $f \geqslant 0$, then $h := \sqrt{f} \in L^2(\mathbb{R}^n)$, whence we obtain $\check{f} = (2\pi)^{-n/2}\check{h} * \check{h}$, where $\check{h}(-x) = \overline{\check{h}(x)}$; the converse is proved similarly taking into account that $|\widehat{g}|^2 \in L^1(\mathbb{R}^n)$ and $|\widehat{g}|^2 \geqslant 0$.

9.10.30. Prove that whenever $1 < p < 2$ the Fourier transform extends to a continuous operator from $L^p(\mathbb{R}^n)$ to $L^q(\mathbb{R}^n)$, where $1/p + 1/q = 1$.
HINT: apply Theorem 6.10.77.

9.10.31. Prove that there exists a uniformly continuous function f on $\mathbb{R}^1$ that satisfies the condition $\lim_{|x|\to\infty} f(x) = 0$, but is not the Fourier transform of a function from $L^1(\mathbb{R}^1)$.
HINT: otherwise the Fourier transform $\mathcal{F}$ would be a bijection between $L^1(\mathbb{R}^1)$ and the space E of bounded uniformly continuous complex functions tending to zero at infinity. Since E is Banach with the sup-norm and $\mathcal{F}$ is continuous with values in E with this norm, then by the Banach inverse mapping theorem $\mathcal{F}^{-1}$ would be also continuous. However, this is false: one can take smooth even functions f_j with support in $[-1,1]$ and $0 \leqslant f_j \leqslant 1$ such that $f_j(x) \to f(x) = I_{[-1,1]}(x)$ pointwise. The sequence $g_j = \widehat{f_j}$ is not bounded in L^1, since $\widehat{f} \notin L^1$. However, $g_j = \mathcal{F}^{-1}f_j$, where the sequence $\{f_j\}$ is bounded in E.

9.10.32. Prove that the space $W_{2\pi}^{2,1}[0, 2\pi]$ of all absolutely continuous complex functions f on $[0, 2\pi]$ with $f' \in L^2[0, 2\pi]$ and $f(0) = f(2\pi)$ is the image of the operator B on $L^2[0, 2\pi]$ with eigenfunctions $\exp(ikt)$ and eigenvalues $\beta_0 = 1$, $\beta_k = k^{-1}$ if $k \neq 0$.
HINT: if $f \in W_{2\pi}^{2,1}[0, 2\pi]$ and $f(t) = \sum_{k=-\infty}^{+\infty} c_k \exp(ikt)$, then

$$f'(t) = i \sum_{k=-\infty}^{+\infty} kc_k \exp(ikt).$$

Conversely, if for the Fourier coefficients c_k of the function f the series of $|kc_k|^2$ converges, then it is easy to verify that $f \in W_{2\pi}^{2,1}[0, 2\pi]$.

9.10.33. Prove that for some $C > 0$, for every odd function $f \in \mathcal{L}^1(\mathbb{R}^1)$ we have the bound $\left|\int_0^t s^{-1}\widehat{f}(s)\, ds\right| \leqslant C\|f\|_{L^1}$ for all $t \geqslant 0$.
HINT: use that the integrals of the functions $s^{-1}\sin(sx)$ in s over $[0,t]$ are uniformly bounded in x and t.

9.10.34. Let $f \in \mathcal{L}^1(\mathbb{R}^1)$. Prove that f is recovered from $\widehat{f}$ by means of Cesàro summation in the following sense:

$$f(x) = \frac{1}{\sqrt{2\pi}} \lim_{T\to+\infty} \frac{1}{T} \int_0^T \int_{-t}^t e^{ixy}\widehat{f}(y)\, dy\, dt$$

for almost all points x, including all points of continuity of the function f.

HINT: use Fubini's theorem to express the double integral on the right as the integral of $2T^{-1}f(u)u^{-2}\big(\cos(T(x-u))-1\big)$, use that $\big(\cos(T(x-u))-1\big) = -2[\sin(T^2(x-u)/2)]^2$ and apply Theorem 9.10.2.

9.10.35. Let $f \in \mathcal{L}^1(\mathbb{R}^1)$. Prove that for all ξ we have

$$\int_0^\xi f(x)\,dx = \frac{1}{\sqrt{2\pi}} \lim_{T\to+\infty} \int_{-T}^T \frac{e^{i\xi y}-1}{iy}\,\widehat{f}(y)\,dy.$$

HINT: find the limit of the integral of $(e^{iy(\xi-x)} - e^{-yx})/y$ over $[-T,T]$ as $T \to +\infty$ in the cases $x \leqslant \xi$ and $x > \xi$ and use the dominated convergence theorem.

9.10.36. Let $f \in W^{1,1}(\mathbb{R}^n)$. Prove that the gradient ∇f vanishes almost everywhere on the set of zeros $\{x\colon f(x) = 0\}$.

HINT: show that $\partial_{x_i}f$ vanishes almost everywhere on $\{f = 0\}$. To this end use a version of f absolutely continuous in x_i; observe that if a function f on $\mathbb{R}$ is differentiable at a point x, $f(x) = 0$ and x is not an isolated point of the set $\{f = 0\}$, then $f'(x) = 0$.

9.10.37. Prove Theorem 9.8.7 for $p = 1$.

9.10.38. Give an example of a function $f \in W^{1,1}(\mathbb{R}^2)$ that has no modification continuous in every argument separately.

9.10.39. Let U be the halfspace $\{x\colon x_1 > 0\}$ in $\mathbb{R}^n$. Prove that to every function $f \in W^{p,1}(U)$ one can associate a function $Tf \in W^{p,1}(\mathbb{R}^n)$ such that the operator T is linear and continuous and $Tf = f$ if f the restriction to U of a smooth function that is even with respect to the variable x_1.

HINT: extend functions by reflexion.

9.10.40. Investigate whether the embedding of $W^{1,1}\big((0,1)\big)$ to $C(0,1)$ is compact.

9.10.41. (S. N. Bernstein) For every multiindex $\alpha = (\alpha_1,\ldots,\alpha_n)$ there is a number $C(n,\alpha) > 0$ such that if $f \in \mathcal{L}^\infty(\mathbb{R}^n)$ and $\widehat{f}$ in the sense of $\mathcal{S}'(\mathbb{R}^n)$ is a function with support in the unit ball, then $\|\partial^{(\alpha)}f\|_{L^\infty} \leqslant C(n,\alpha)\|f\|_{L^\infty}$.

HINT: $f = f * g$, where $g \in \mathcal{S}(\mathbb{R}^n)$, $\widehat{g}$ has support in the ball of radius 2 and equals $(2\pi)^{-n/2}$ on the unit ball.

9.10.42. (Hardy and Littlewood, see [596, §4.8, §4.9]) (i) If a measurable function f on the real line is such that the function $|f(x)|^q|x|^{q-2}$, where $q > 2$, belongs to $\mathcal{L}^1(\mathbb{R}^1)$, then $\widehat{f} \in \mathcal{L}^q(\mathbb{R}^1)$. (ii) If $f \in \mathcal{L}^p(\mathbb{R}^1)$, where $1 < p < 2$, then the function $|x|^{p-2}|\widehat{f}(x)|^p$ belongs to $\mathcal{L}^1(\mathbb{R}^1)$.

9.10.43. (Titchmarsh [596, §4.13]) Let $0 < \alpha \leqslant 1$ and $f \in \mathcal{L}^p(\mathbb{R}^1)$, where $1 < p \leqslant 2$. If $\|f(\cdot + h) - f\|_{L^p} = O(h^\alpha)$ as $h \to 0$, then $\widehat{f} \in L^r(\mathbb{R}^1)$ if $\frac{p}{p+\alpha p-1} < r \leqslant \frac{p}{p-1}$.

9.10.44. A distribution $F \in \mathcal{D}'(\mathbb{R}^n)$ is called homogeneous of degree $d > 0$ if $\langle F, \varphi(\lambda\cdot)\rangle = \lambda^{-d-n}\langle F, \varphi\rangle$ for all $\varphi \in \mathcal{D}(\mathbb{R}^n)$ and $\lambda > 0$. Prove that in this case F extends to a functional from $\mathcal{S}'(\mathbb{R}^n)$.

HINT: see [620, p. 85, 87].

9.10.45. Let $P \not\equiv 0$ be a polynomial on $\mathbb{R}^n$ and let L be the generated differential operator. Let Ω be a bounded domain in $\mathbb{R}^n$. Prove the Hörmander inequality

$$\|f\|_{L^2(\Omega)} \leqslant C(P,\Omega)\|Lf\|_{L^2(\Omega)}, \ f \in L^2(\Omega).$$

Deduce from it that for every $G \in L^2(\Omega)$ there exists $F \in L^2(\Omega)$ with $LF = G$ in the sense of distributions.

HINT: see [620, p. 207, 208].

CHAPTER 10

Unbounded Operators and Operator Semigroups

This chapter is devoted to fundamentals of the spectral theory of unbounded selfadjoint operators and some elements of the theory of operator semigroups. Some of the principal applications of these theories are connected with partial differential equations and mathematical physics, in particular, quantum mechanics, but there are also important applications in many other areas of mathematics, for example, in the theory of random processes and geometry. The spaces considered in this chapter are complex unless if explicitly stated otherwise.

10.1. Graphs and Adjoints

We already know that on every infinite-dimensional normed space there are discontinuous linear functionals. However, the theory of unbounded selfadjoint operators deals with unboundedness of another sort: a characteristic feature of these operators is that their domain of definition does not coincide with the whole space. A typical example is a differential operator regarded as an operator on the space L^2, although its actual domain of definition is smaller.

10.1.1. Definition. *Let X and Y be Banach spaces. A linear mapping T from a dense linear subspace $\mathcal{D}(T) \subset X$, called the domain of definition of T, to the space Y is called a densely defined linear operator on X.*

An operator T with the domain of definition $\mathcal{D}(T)$ is called closed if its graph

$$\Gamma(T) := \big\{ \{x, Tx\} \in X \times Y \colon \ x \in \mathcal{D}(T) \big\}$$

is closed in $X \times Y$.

Note that the kernel $\operatorname{Ker} T$ of a closed operator T is always closed, although $\mathcal{D}(T)$ can fail to be closed. Indeed, if $x_n \in \mathcal{D}(T)$, $x_n \to x$ and $Tx_n = 0$, then $x \in \mathcal{D}(T)$ and $Tx = 0$.

The set of values of a linear mapping T defined on a linear domain of definition $\mathcal{D}$ is called the range of T and denoted by $\operatorname{Ran} T$, i.e.,

$$\operatorname{Ran} T := T(\mathcal{D}).$$

According to the closed graph theorem, every closed operator defined on the whole Banach space is bounded. Hence every unbounded everywhere defined operator serves as an example of a densely defined nonclosed linear operator. Let us give an example of an unbounded densely defined closed operator. Until §10.4 inclusive for simplification of exposition we shall discuss only operators on Hilbert

© Springer Nature Switzerland AG 2020

V. I. Bogachev, O. G. Smolyanov, *Real and Functional Analysis*,

Moscow Lectures 4, https://doi.org/10.1007/978-3-030-38219-3_10

spaces. Only in §§10.5–10.6 in connection with semigroups we return to general Banach spaces.

10.1.2. Example. Let $H = l^2$. Set

$$\mathfrak{D}(T) = \Big\{ x \in l^2 : \ \sum_{n=1}^{\infty} n^2 x_n^2 < \infty \Big\}, \quad Tx = (x_1, 2x_2, 3x_3, \ldots).$$

If vectors $x^k = (x_n^k)$ converge to x and vectors Tx^k converge to y, then it is clear that $x \in \mathfrak{D}(T)$ and $Tx = y$. So T is closed.

10.1.3. Definition. *We shall say that an operator $\big(S, \mathfrak{D}(S)\big)$ is an extension of an operator $\big(T, \mathfrak{D}(T)\big)$ if*

$$\mathfrak{D}(T) \subset \mathfrak{D}(S) \quad and \quad S|_{\mathfrak{D}(T)} = T.$$

Notation: $T \subset S$.

For an unbounded operator its domain of definition is a characteristic as important as the way of defining the operator on this domain. Operators defined by the same expression on different domains are different operators and, as we shall see below, can possess completely different properties.

10.1.4. Definition. *A densely defined linear operator T is called closable if it has a closed extension.*

If an operator T is closable, then the closure $\overline{\Gamma(T)}$ of its graph is the graph of the operator called the *closure* of T and denoted by the symbol $\overline{T}$.

The domain $\mathfrak{D}(T)$ of an operator T on a Hilbert space H can be equipped with the inner product

$$(x, y)_T := (x, y) + (Tx, Ty).$$

The Euclidean space obtained in this way is denoted by D_T. The norm

$$\|x\|_T := \sqrt{(x, x)_T}$$

is equivalent to the norm induced from H only if the operator T is bounded.

10.1.5. Lemma. *A densely defined operator T is closed precisely when the space D_T is Hilbert.*

PROOF. Obviously, the mapping $x \mapsto \{x, Tx\}$ is an isometry of the spaces D_T and $\Gamma(T)$. Hence the completeness of one of them is equivalent to the completeness of the other. $\square$

Let us give an example of an operator that has no closure.

10.1.6. Example. Let $H = L^2[0, 1]$, $\mathfrak{D}(T) = C[0, 1]$, $Tx(t) = x(1) \cdot 1(t)$, where 1 is the function identically equal to 1. The operator T has no closure. Indeed, let $x_n(t) = t^n$. Then $x_n \in \mathfrak{D}(T)$, $x_n \to 0$ in H, but $Tx_n = 1$.

For consideration of graphs of operators on a Hilbert space H let us introduce on $H \times H$ two unitary operators

$$U: H \times H \to H \times H, \quad U\{x, y\} = \{y, x\},$$
$$V: H \times H \to H \times H, \quad V\{x, y\} = \{-y, x\}.$$

The operators U and V satisfy the equalities
$$U^2 = I, \quad V^2 = -I, \quad UV = -VU.$$
Let us note the following relation valid for every set $E \subset H \times H$:
$$(VE)^\perp = V(E^\perp) = V^{-1}(E^\perp). \tag{10.1.1}$$
The first equality follows from the fact that a unitary operator preserves the orthogonality and the second one follows from the equalities $V^2 = -I$ and $E^\perp = -E^\perp$.

10.1.7. Proposition. *Suppose that a densely defined operator T on a Hilbert space H has a dense range and* $\operatorname{Ker} T = \{0\}$. *Then the closedness of T is equivalent to the closedness of the operator T^{-1} with domain* $\mathfrak{D}(T^{-1}) = \operatorname{Ran} T$.

PROOF. It suffices to observe that $\Gamma(T^{-1}) = U\Gamma(T)$. $\qquad\square$

10.1.8. Definition. *Suppose that an operator T on a Hilbert space H has a dense domain of definition* $\mathfrak{D}(T)$. *Let us define the operator T^* as follows:* $y \in \mathfrak{D}(T^*)$ *if there exists an element $T^*y := z \in H$ such that*
$$(Tx, y) = (x, z) \quad \text{for all } x \in \mathfrak{D}(T).$$

By the Riesz theorem the vector y belongs to $\mathfrak{D}(T^*)$ precisely when the functional $x \mapsto (Tx, y)$ on $\mathfrak{D}(T)$ is continuous with respect to the norm from H.

The graph of T^* is connected with the graph of T in the following way.

10.1.9. Proposition. *Let T be a densely defined operator on a Hilbert space H. Then*
$$\Gamma(T^*) = [V\Gamma(T)]^\perp,$$
where the orthogonal complement is taken in $H \times H$. In particular, the operator T^* is closed.

PROOF. By definition $\{y, z\} \in \Gamma(T^*)$ precisely when $(Tx, y) = (x, z)$ for all $x \in \mathfrak{D}(T)$, which can be written as the orthogonality of $\{-Tx, x\}$ and $\{y, z\}$ in $H \times H$. The latter is the orthogonality of $\{y, z\}$ to the subspace $V\Gamma(T)$. $\qquad\square$

10.1.10. Corollary. *If T is a closed densely defined operator on H, then $H \times H$ is the orthogonal sum of the closed subspaces $V\Gamma(T)$ and $\Gamma(T^*)$, i.e.,*
$$H \times H = V\Gamma(T) \oplus \Gamma(T^*).$$
In particular, for every pair of vectors $x, y \in H$, there exists a unique pair of vectors $u \in \mathfrak{D}(T)$ and $v \in \mathfrak{D}(T^*)$ satisfying the equalities
$$x = -Tu + v, \quad y = u + T^*v,$$
$$\|x\|^2 + \|y\|^2 = \|u\|^2 + \|Tu\|^2 + \|v\|^2 + \|T^*v\|^2.$$

It is clear that $0 \in \mathfrak{D}(T^*)$ for every densely defined operator T. However, it can happen that $\mathfrak{D}(T^*)$ contains no nonzero elements.

10.1.11. Example. It suffices to construct an operator on $H = l^2$ the graph of which is everywhere dense in $H \oplus H$. For this we take a linearly independent everywhere dense countable set $B_1 = \{b_n^1\}$ in H, add an everywhere dense countable set $B_2 = \{b_n^2\}$ linearly independent with B_1, and continue by induction. We obtain a countable collection of everywhere dense countable sets $B_k = \{b_n^k\}$ that are linearly independent in total. Let us complement the constructed countable linearly independent set to a Hamel basis and set $Ab_n^k = b_k^1$; on the additional vectors of the basis we define A by zero. We extend A to H by linearity. The graph of A is everywhere dense in $H \times H$: by the linearity of the graph it suffices to verify that every vector form $\{b_m^1, b_k^1\}$ belongs to the closure of the graph and for this it suffices to take a sequence $b_{n_j}^k \to b_m^1$.

10.1.12. Proposition. *Let T be an operator with a dense domain $\mathfrak{D}(T)$ in a Hilbert space H. Then*
(i) $(T + B)^* = T^* + B^*$ *for all $B \in \mathcal{L}(H)$,*
(ii) $(\lambda T)^* = \overline{\lambda} T^*$ *and $(T + \lambda I)^* = T^* + \overline{\lambda} I$ for all $\lambda \in \mathbb{C}$.*

PROOF. (i) The equalities $\mathfrak{D}(T + B) = \mathfrak{D}(T)$, $\mathfrak{D}(T^* + B^*) = \mathfrak{D}(T^*)$ are obvious. For any $y \in \mathfrak{D}(T^*)$, $x \in \mathfrak{D}(T)$ we have

$$(Tx + Bx, y) = (x, T^*y) + (x, B^*y) = (x, T^*y + B^*y),$$

so $T^* + B^* \subset (T + B)^*$. Applying this to $T + B$ and $-B$, we obtain the inverse inclusion. (ii) The first equality is obvious, the second one follows from (i). $\quad\square$

There is the same connection between the range of the operator and the kernel of its adjoint as in the case of bounded operators.

10.1.13. Proposition. *Let T be a linear operator with a dense domain of definition $\mathfrak{D}(T)$ in a Hilbert space H. Then the closed subspaces $\overline{\mathrm{Ran}\, T}$ and $\mathrm{Ker}\, T^*$ are mutually orthogonal and $H = \overline{\mathrm{Ran}\, T} \oplus \mathrm{Ker}\, T^*$.*
In addition, for every $\lambda \in \mathbb{C}$ we have

$$H = \overline{\mathrm{Ran}\,(T - \lambda I)} \oplus \mathrm{Ker}\,(T^* - \overline{\lambda} I).$$

PROOF. The subspace $\mathrm{Ker}\, T^*$ is closed by the closedness of T^*. The inclusion $y \in \mathrm{Ker}\, T^*$ is equivalent to the property that $(Tx, y) = 0$ for all $x \in \mathfrak{D}(T)$. The latter is precisely the condition $y \perp \mathrm{Ran}\, T$. Therefore, for every $u \in H$ we obtain an element v in $\overline{\mathrm{Ran}\, T}$ that is the orthogonal projection of u onto this closed subspace and also the element $w := u - v$ orthogonal to $\mathrm{Ran}\, T$, i.e., belonging to $\mathrm{Ker}\, T^*$ according to what has been said above. The last assertion of the proposition follows from the first one applied to the operator $T - \lambda I$. $\quad\square$

10.1.14. Proposition. (i) *If a densely defined operator T has the closure $\overline{T}$, then $(\overline{T})^* = T^*$.*
(ii) *Suppose that an operator T is densely defined. The operator T^* is densely defined precisely when T is closable. In this case $T^{**} = \overline{T}$.*
(iii) *Let T be an operator such that the sets $\mathfrak{D}(T)$ and $\mathrm{Ran}\, T$ are dense and $\mathrm{Ker}\, T = \{0\}$. Then the operator $(T^*)^{-1}$ is densely defined and $(T^*)^{-1} = (T^{-1})^*$, where $\mathfrak{D}(T^{-1}) = \mathrm{Ran}\, T$.*

PROOF. (i) We have $\Gamma(T^*) = [V\Gamma(T)]^\perp$. If T is closable, then $\Gamma(\overline{T}) = \overline{\Gamma(T)}$, hence

$$[V\Gamma(T)]^\perp = [V\Gamma(\overline{T})]^\perp = \Gamma(\overline{T}^*).$$

(ii) Let T be densely defined and closable. According to assertion (i) we have $T^* = (\overline{T})^*$. Hence we can assume that T is closed. If $y \perp \mathfrak{D}(T^*)$, then $\{y, 0\} \perp \Gamma(T^*)$, which by Corollary 10.1.10 gives the inclusion $\{y, 0\} \in \Gamma(T)$. Hence $y = 0$.

Conversely, let the operator T^* be densely defined. The operator T^{**} is closed. In addition, $T \subset T^{**}$, which yields the closability of T and the inclusion $\overline{T} \subset T^{**}$. We show that actually the equality holds. It is clear that it suffices to show the equality $T = T^{**}$ for the closed operator T. This equality follows from the relations

$$\Gamma(T^{**}) = [V\Gamma(T^*)]^{\perp}, \quad \Gamma(T) = V^{-1}([\Gamma(T^*)]^{\perp})$$

and equality (10.1.1).

(iii) The density of the range of T gives the equality $\operatorname{Ker} T^* = 0$. Hence the operator $(T^*)^{-1}$ is defined on the range of T^*. Let us compare the graphs of the operators $(T^*)^{-1}$ and $(T^{-1})^*$. We have the equalities

$$\Gamma\big((T^*)^{-1}\big) = U\Gamma(T^*) = U\big([V\Gamma(T)]^{\perp}\big),$$
$$\Gamma\big((T^{-1})^*\big) = [V\Gamma(T^{-1})]^{\perp} = [VU\Gamma(T)]^{\perp}.$$

Their right-hand sides coincide, which follows from relations (10.1.1) and the equality $UV = -VU$. $\qquad\square$

For unbounded operators it is also useful to introduce the notion of a regular point. Let T be a closed operator. The number $\lambda \in \mathbb{C}$ is called *regular* for T if $T - \lambda I$ has a bounded inverse on the domain $\operatorname{Ran}(T - \lambda I)$, i.e., for some $c > 0$ we have

$$\|(T - \lambda I)x\| \geqslant c\|x\|, \quad x \in \mathfrak{D}(T). \tag{10.1.2}$$

Set

$$d_T(\lambda) := \dim\big(\operatorname{Ran}(T - \lambda I)\big)^{\perp}, \quad \lambda \in \mathbb{C}.$$

The number $d_T(\lambda) \in [0, +\infty]$ is called the *defect* of the operator $T - \lambda I$.

10.1.15. Proposition. *Let T be a closed operator on a Hilbert space and let λ be its regular point. Then there exists an open disc centered at λ consisting of regular points and the function d_T is constant on this disc.*

PROOF. If we have (10.1.2) and $|\lambda - \mu| < c$, then

$$\|(T - \mu I)x\| \geqslant (c - |\lambda - \mu|)\|x\|, \quad x \in \mathfrak{D}(T).$$

The operator $T - \lambda I$ is closed, since if

$$x_n \in \mathfrak{D}(T), \quad x_n \to x \quad \text{and} \quad Tx_n - \lambda x_n \to y,$$

then $Tx_n \to \lambda x + y$, whence $x \in \mathfrak{D}(T)$ and $Tx = \lambda x + y$, i.e., $(T - \lambda I)x = y$. Hence we can assume that $\lambda = 0$. Then (10.1.2) yields that the ranges of T and $T - \mu I$ are closed. Let us show that $d_T(0) \leqslant d_T(\mu)$. If $d_T(0) > d_T(\mu)$, then the subspace $E_\mu := \big(\operatorname{Ran}(T - \mu I)\big)^{\perp}$ is finite-dimensional and there exists a nonzero element $z \in (\operatorname{Ran} T)^{\perp}$ orthogonal to E_μ. Hence $z \in \operatorname{Ran}(T - \mu I)$, i.e., $z = (T - \mu I)y$, $y \in \mathfrak{D}(T)$, $y \neq 0$. Since $z \perp \operatorname{Ran} T$, we have $(z, Ty) = 0$. Hence

$$(Ty, Ty) = \mu(y, Ty) \leqslant |\mu|\,\|y\|\,\|Ty\|,$$

whence $\|Ty\| \leqslant |\mu| \, \|y\|$. Therefore, $y = 0$, which is a contradiction. Hence $d_T(0) \leqslant d_T(\mu)$. Similarly we verify that $d_T(0) \geqslant d_T(\mu)$, since in the opposite case there is a nonzero vector y for which $z = Ty$ is orthogonal to $Ty - \mu y$. $\qquad \square$

10.2. Symmetric and Selfadjoint Operators

A very important class for applications is constituted by unbounded selfadjoint operators on Hilbert spaces.

10.2.1. Definition. *An operator A with a dense domain in a Hilbert space is called selfadjoint if $A^* = A$.*

The equality of operators includes the equality of their domains of definition. So for unbounded operators the selfadjointness is stronger than symmetry.

10.2.2. Definition. *An operator A with a dense domain of definition $\mathfrak{D}(A)$ in a Hilbert space is called symmetric if*
$$(Ax, y) = (x, Ay) \quad \text{for all } x, y \in \mathfrak{D}(A).$$

We have seen in Example 6.8.8 that an *everywhere* defined symmetric operator on a Hilbert space is necessarily continuous. Hence an unbounded symmetric operator cannot be extended to the whole space as a symmetric operator. The following example of the differentiation operator is classical in the theory of operators in many respects. In particular, it gives a simple example of a symmetric operator that is not selfadjoint. Let $AC[0,1]$ be the class of all absolutely continuous functions on $[0, 1]$.

10.2.3. Example. (The differentiation operator) (i) Let $H = L^2[0,1]$. On the set $\mathfrak{D}(A) = C_0^\infty(0,1)$ of all infinitely differentiable complex functions with support in the open interval $(0,1)$ we define an operator A by the formula
$$Au(t) = iu'(t).$$
This operator is densely defined. For all $u, v \in \mathfrak{D}(A)$ we have
$$(Au, v) = i \int_0^1 u'(t)\overline{v(t)} \, dt = \int_0^1 u(t)\overline{iv'(t)} \, dt = (u, Av)$$
by the integration by parts formula. Thus, the operator A is symmetric. However, it is not selfadjoint, since it is not closed. Indeed, let u be a continuously differentiable, but not infinitely differentiable function on $[0, 1]$ with $u(0) = u(1) = 0$. It is readily seen that there exists a sequence of functions $u_n \in C_0^\infty(0, 1)$ such that $u_n \to u$ and $u_n' \to u'$ in $L^2[0,1]$.

(ii) We show that
$$\mathfrak{D}(A^*) = \big\{ u \in AC[0,1] \colon \ u' \in L^2[0,1] \big\}, \quad A^*u = iu'.$$
The fact that every absolutely continuous function u with $u' \in L^2[0,1]$ belongs to $\mathfrak{D}(A^*)$ and that $A^*u = iu'$ is obvious from the integration by parts formula above with $v \in C_0^\infty(0,1)$. This also shows that $\mathfrak{D}(A^*)$ does not coincide with $\mathfrak{D}(A)$. Let $u \in \mathfrak{D}(A^*)$, which means that there exists a function $w \in L^2[0,1]$ such that for all $v \in C_0^\infty(0, 1)$ we have
$$\int_0^1 w(t)\overline{v(t)} \, dt = (w, v) = (A^*u, v) = (u, Av) = \int_0^1 u(t)\overline{iv'(t)} \, dt.$$

Set $u_0(t) = -i \int_0^t w(s)\,ds$. For real $v \in C_0^\infty(0,1)$, the integration by parts formula and the equality above give

$$\int_0^1 u_0(t)v'(t)\,dt = i\int_0^1 w(t)v(t)\,dt = \int_0^1 u(t)v'(t)\,dt.$$

For every function h from $L^2[0,1]$ with zero integral we can find a sequence of functions $v_n \in C_0^\infty(0,1)$ such that $v_n' \to h$ in $L^2[0,1]$. Hence the function $u - u_0$ is orthogonal to all such elements h. Hence it equals a constant almost everywhere (to its integral). Thus, along with u_0, the function u is also absolutely continuous and has a square integrable derivative. Finally, it is clear that $A^*u = iu'$. Note that the operator A^* is not symmetric.

(iii) Let us now describe the closure of A. We show that

$$\mathfrak{D}(A^{**}) = \{u \in AC[0,1]\colon\ u' \in L^2[0,1], u(0) = u(1) = 0\}, \quad A^{**}u = iu'.$$

The indicated domain belongs to $\mathfrak{D}(A^{**})$ by the integration by parts formula. As above, the main thing is to prove the inverse inclusion. Let $u \in \mathfrak{D}(A^{**})$. It follows from what has been proved that u is an absolutely continuous function, $u' \in L^2[0,1]$ and $A^{**}u = iu'$. Now, however, at our disposal we have all smooth functions v on $[0,1]$, not only having support in $(0,1)$. Hence the equality

$$\int_0^1 iu'(t)v(t)\,dt = (A^{**}u, v) = (u, A^*v) = -i\int_0^1 u(t)v'(t)\,dt$$

for smooth real functions v gives a new relation $u(1)v(1) = u(0)v(0)$, whence $u(0) = u(1) = 0$. The operator A^{**} is closed as any adjoint. By assertion (ii) in Proposition 10.1.14 the operator A^{**} coincides with the closure of A. In particular, A^{**} is a symmetric closed, but not selfadjoint operator, since $(A^{**})^* = \overline{A}^* = A^*$ differs from A^{**}.

(iv) The operator A has selfadjoint extensions (which are also extensions of $\overline{A}$), moreover, all these extensions have the form A_θ, where

$$\mathfrak{D}(A_\theta) = \{u \in AC[0,1]\colon\ u' \in L^2[0,1], u(1) = \theta u(0)\}, \quad |\theta| = 1,$$
$$A_\theta u = iu'.$$

Indeed, if $\widetilde{A}$ is a selfadjoint extension of A, then $\widetilde{A} = \widetilde{A}^* \subset A^*$, i.e., we have $\mathfrak{D}(\widetilde{A}) \subset \mathfrak{D}(A^*)$ and $\widetilde{A}u = iu'$. The condition that $\widetilde{A}$ is symmetric leads to the requirement $u(0)\overline{v(0)} = u(1)\overline{v(1)}$ for all $u, v \in \mathfrak{D}(\widetilde{A})$. Since $\overline{A}$ is not selfadjoint, we see that $\mathfrak{D}(\widetilde{A})$ is larger than $\mathfrak{D}(\overline{A})$, which means that there exists a function $u \in \mathfrak{D}(\widetilde{A})$ for the continuous version of which we have $u(0) \neq 0$. We can assume that $u(0) = 1$. Then we set $\theta = \overline{u(1)}$ and for $v = u$ obtain $|\theta|^2 = 1$. In addition, for all $v \in \mathfrak{D}(\widetilde{A})$ we obtain $\overline{v(0)} = \overline{\theta v(1)}$, i.e., $v(0) = \theta v(1)$. Conversely, the operator A_θ on $\mathfrak{D}(A_\theta)$ is symmetric and extends A. Hence $A_\theta^* \subset A^*$, i.e., for all $u \in \mathfrak{D}(A_\theta^*)$ we have $u \in AC[0,1]$ and $u' \in L^2[0,1]$. Hence for all $u \in \mathfrak{D}(A_\theta^*)$, $v \in \mathfrak{D}(A_\theta)$ we obtain

$$\int_0^1 iu'(t)\overline{v(t)}\,dt = (A_\theta^*u, v) = (u, A_\theta v) = \int_0^1 u(t)\overline{iv'(t)}\,dt,$$

which leads to the condition $u(0)\overline{\theta v(1)} = u(1)\overline{v(1)}$. Since we can take v with $v(1) = 1$, the equality $u(0) = \theta u(1)$ holds. Thus, we obtain $\mathfrak{D}(A_\theta^*) = \mathfrak{D}(A_\theta)$ and the operator A_θ is selfadjoint.

The next example is also important for the general theory. We shall establish below that every selfadjoint operator on a separable Hilbert space is unitarily equivalent to an operator of such a form.

10.2.4. Example. Let μ be a finite nonnegative measure on a measurable space $(\Omega, \mathcal{B})$ and let φ be a real μ-measurable function (not necessarily bounded). Let us define the operator A_φ (the same notation as in case of bounded functions) on the domain

$$\mathfrak{D}(A_\varphi) = \{x \in L^2(\mu) \colon \varphi \cdot x \in L^2(\mu)\}$$

by the formula $A_\varphi x(\omega) = \varphi(\omega)x(\omega)$. Then the operator A_φ is selfadjoint. Indeed, the set $\mathfrak{D}(A_\varphi)$ is dense in $L^2(\mu)$, since for every $x \in L^2(\mu)$ it contains all functions xI_{Ω_n}, where $\Omega_n := \{\omega \colon |\varphi(\omega)| \leqslant n\}$. The operator A_φ is symmetric on this domain. Let $u \in \mathfrak{D}(A_\varphi^*)$. Then there is $w \in L^2(\mu)$ with

$$\int_\Omega u(\omega)\varphi(\omega)v(\omega)\,\mu(d\omega) = \int_\Omega w(\omega)v(\omega)\,\mu(d\omega)$$

for all real functions $v \in \mathfrak{D}(A_\varphi)$. Taking for v indicator functions of measurable subsets of the sets Ω_n, we obtain for μ-a.e. ω that $w(\omega) = \varphi(\omega)u(\omega)$. Thus, $\mathfrak{D}(A_\varphi^*) = \mathfrak{D}(A_\varphi)$ and the operator A_φ is selfadjoint.

10.2.5. Example. Let H be a Hilbert space and let Π be a projection-valued measure on $\mathcal{B}(\mathbb{R}^1)$ with values in $\mathcal{L}(H)$. Set

$$\mathfrak{D}(A) := \left\{x \in H \colon \int_{\mathbb{R}^1} \lambda^2\,d\Pi_{x,x}(\lambda) < \infty\right\}, \quad Ax := \int_{\mathbb{R}^1} \lambda\,d\Pi_x(\lambda),$$

where the integral is understood as the limit in H of the integrals over the intervals $[-n, n]$ (which exist, as we have seen in §7.9). The operator A is selfadjoint. Indeed, it is easy to see that $\mathfrak{D}(A)$ is a linear space. It is dense, since for every $x \in H$ the vectors $x_n := \Pi([-n, n])x$ converging to x belong to $\mathfrak{D}(A)$. In addition,

$$Ax_n = A_n x = \int_{[-n,n]} \lambda\,d\Pi_x(\lambda),$$

where the bounded selfadjoint operator A_n is defined by restricting the projection-valued measure Π to $[-n, n]$ (see §7.9). If $k > n$, we have

$$\|A_n x - A_k x\|^2 = ((A_n - A_k)^2 x, x) = \int_{n < |\lambda| \leqslant k} \lambda^2\,d\Pi_{x,x}(\lambda),$$

whence it follows that $\{Ax_n\}$ is Cauchy in norm and there is $Ax = \lim_{n\to\infty} A_n x$. We show that $A^* = A$. Let $y \in \mathfrak{D}(A^*)$. Hence $|(Ax, y)| \leqslant \|A^* y\|\,\|x\|$ for all $x \in \mathfrak{D}(A)$. Therefore, $y \in \mathfrak{D}(A)$, since otherwise $c_n := \|A_n y\| \to \infty$, which is impossible by the relation

$$\|A_n y\| = c_n^{-1}(A_n y, A_n y) = c_n^{-1}(A_n A_n y, y) = c_n^{-1}(AA_n y, y) \leqslant \|A^* y\|$$

following from the estimate $\|c_n^{-1}A_n y\| \leqslant 1$. Since the operator A is symmetric and $\mathfrak{D}(A^*) = \mathfrak{D}(A)$, we have $A^* = A$.

10.3. The Spectral Theorem

Here we shall see that unbounded selfadjoint operators are also unitarily isomorphic to operators of multiplication by functions, i.e., Example 10.2.4 is universal. Unitary equivalence of unbounded selfadjoint operators is defined naturally (but now it is also required that isomorphisms must interchange the domains). The Caley transform also extends to such operators.

10.3.1. Lemma. *Let A be a selfadjoint operator on a Hilbert space $H \neq 0$. Then the operators $A + iI$ and $A - iI$ are injective and their ranges coincide with H. In addition, the operator $U = (A - iI)(A + iI)^{-1}$ is unitary.*

PROOF. We have $\mathrm{Ker}(A - iI) = \mathrm{Ker}(A - iI) = 0$, since if $Ax = ix$, then $(Ax, x) = i(x, x)$ and $x = 0$, because $(Ax, x) \in \mathbb{R}^1$. There is a similar equality for $A + iI$. The range of $A + iI$ is dense, since if $y \in H$ is such that $(Ax + ix, y) = 0$ for all $x \in \mathfrak{D}(A)$, then $y \in \mathfrak{D}(A^*) = \mathfrak{D}(A)$, whence $(x, Ay - iy) = 0$, so $Ay = iy$ and $y = 0$. We observe that

$$\|Ax + ix\|^2 = \|Ax\|^2 + \|x\|^2 = \|Ax - ix\|^2 \quad \forall x \in \mathfrak{D}(A),$$

because $(Ax, ix) = -i(Ax, x) = -i(x, Ax)$. Hence there exists a linear isometry U between the everywhere dense linear subspaces $(A + iI)(H)$ and $(A - iI)(H)$ defined by the formula

$$U(Ax + ix) = Ax - ix, \quad x \in \mathfrak{D}(A).$$

By the injectivity of $A + iI$ the operator U on $(A + iI)(H)$ can be written in the form $U = (A - iI)(A + iI)^{-1}$. This operator uniquely extends to a unitary operator, also denoted by U. We show that $(A + iI)\big(\mathfrak{D}(A)\big) = H$. Let $y \in H$. Find $x_n \in \mathfrak{D}(H)$ for which $Ax_n + ix_n \to y$. Then $Ax_n - ix_n = U(Ax_n + ix_n) \to Uy$, which gives convergence of $\{x_n\}$ to some $x \in H$. Since A is closed, we obtain $x \in \mathfrak{D}(A)$ and $Ax = y$. Similarly, $(A - iI)\big(\mathfrak{D}(A)\big) = H$. $\square$

The unitary operator U from this lemma is called the *Caley transform* of the operator A.

10.3.2. Lemma. *Suppose that selfadjoint operators A and B on H possess equal Caley transforms. Then $A = B$. In addition, if U the Caley transform of A, then the operator $U - I$ is injective and*

$$\mathfrak{D}(A) = (U - I)(H), \quad Ax = i(I + U)(I - U)^{-1}x.$$

PROOF. Let $y \in H$. As shown above, there is $x \in \mathfrak{D}(A)$ with $y = Ax + ix$. Then $Uy = Ax - ix$ and $y - Uy = 2ix$. Hence $\mathrm{Ker}(U - I) = 0$. In addition, $(U - I)(H) = \mathfrak{D}(A)$, since for every $x \in \mathfrak{D}(A)$ we can take $y = Ax + ix$. Therefore, if U is the common Caley transform of the operators A and B, then $\mathfrak{D}(A) = \mathfrak{D}(B)$ and for every x from this common domain we obtain

$$Ax = i(I + U)(I - U)^{-1}x = Bx,$$

which completes the proof. $\square$

10.3.3. Theorem. *Every selfadjoint operator A on a separable Hilbert space $H \neq 0$ is unitarily equivalent to some operator A_φ of multiplication by a function φ from Example 10.2.4 with some probability measure μ.*

PROOF. Let U be the Caley transform of the operator A. According to Corollary 7.10.9 we can assume that U is the operator of multiplication on $L^2(\mu)$ by a μ-measurable function ψ with $|\psi| = 1$, where μ is some probability measure on a space Ω. Since U is the Caley transform of A, the operator $U - I$ is injective, i.e., $\psi(\omega) \neq 1$ for μ-a.e. ω. Set $\varphi = i(1 + \psi)(1 - \psi)^{-1}$. It is straightforward to verify that the Caley transform of the operator A_φ is the operator $A_\psi = U$. By the previous lemma $A = A_\varphi$. $\square$

For μ one can take some Borel probability measure on the real line and for φ some Borel function. By using Remark 7.8.7 this theorem can be extended to nonseparable spaces. There is an analog of Theorem 7.10.11 in which the commutativity of unbounded operators means the commutativity of their projection measures.

Similarly to the case of bounded operators, by means of a functional model it is easy to define Borel functions of selfadjoint operators and construct projection-valued measures. Let A be a selfadjoint operator on a separable Hilbert space and let f be a complex Borel function on the real line. Let us define the operator $f(A)$ as follows: we represent A in the form of multiplication by a μ-measurable real function φ and set

$$\mathfrak{D}(f(A)) := \{x \in L^2(\mu): (f \circ \varphi) \cdot x \in L^2(\mu)\}, \quad f(A)x := (f \circ \varphi) \cdot x.$$

If the function f is real, then the operator $f(A)$ is selfadjoint. If the function f is bounded, then the operator $f(A)$ is bounded as well. If f is the indicator function of a Borel set B, then $\Pi(B) := I_B(A)$ is the orthogonal projection and the mapping $B \mapsto \Pi(B)$ is a projection-valued measure. This measure generates complex scalar measures $\Pi_{x,y}(B) = (\Pi(B), x, y)$. Similarly to Theorem 7.9.6 the following relations hold; they can be obtained as a corollary of the cited theorem and Example 10.2.5 if we represent A as the direct sum of the countable collection of bounded operators of multiplication by the functions $\varphi I_{\{k \leqslant \varphi < k+1\}}$.

10.3.4. Theorem. *There holds the equality*

$$A = \int_{\sigma(A)} \lambda \, d\Pi(\lambda), \quad \text{where } \mathfrak{D}(A) = \left\{x: \int_{\sigma(A)} \lambda^2 \, d\Pi_{x,x}(\lambda) < \infty\right\},$$

understood as the identity

$$(Ax, y) = \int_{\sigma(A)} \lambda \, \Pi_{x,y}(d\lambda), \quad x \in \mathfrak{D}(A), y \in H.$$

In addition, for every Borel function f we have the equality

$$f(A) = \int_{\sigma(A)} f(\lambda) \, d\Pi(\lambda),$$

understood similarly.

10.4. Unitary Invariants of Selfadjoint Operators

It was already noted in Chapter 7 that the representations of selfadjoint operators constructed there do not provide any way to establish the equivalence or non-equivalence of two such operators. Of course, in some cases non-equivalence can be seen directly from characteristics such as the spectrum or existence and absence of cyclic vectors. But even for operators with cyclic vectors the equality of spectra does not imply the equivalence of the operators. In the finite-dimensional case the eigenvalues and their multiplicities determine operators up to unitary equivalence. What can serve as an analog of this in the infinite-dimensional case? It is clear that we should somehow distinguish measures used in our representations. The first step in this direction is the following result.

10.4.1. Theorem. *Let μ and ν be two bounded nonnegative Borel measures on the real line and let A_μ and A_ν be the operators of multiplication by the argument on $L^2(\mu)$ and $L^2(\nu)$, respectively. These operators are unitarily equivalent precisely when the measures μ and ν are equivalent.*

PROOF. Let μ and ν be equivalent and $\varrho = d\nu/d\mu$. We recall (see Chapter 3) that equivalent measures possess equal supplies of measurable functions and a measurable function φ is integrable with respect to ν precisely when the function $\varphi\varrho$ is integrable with respect to μ. In addition,

$$\int \varphi(t)\,\nu(dt) = \int \varphi(t)\varrho(t)\,\mu(dt).$$

Let us define an operator $U\colon L^2(\nu) \to L^2(\mu)$ by the equality

$$Uf(t) = \sqrt{\varrho(t)}f(t).$$

Then

$$\|Uf\|_{L^2(\mu)}^2 = \int |f(t)|^2\varrho(t)\,\mu(dt) = \int |f(t)|^2\,\nu(dt) = \|f\|_{L^2(\nu)}^2,$$

i.e., U is an isometry. The operator U is surjective, since for every function $g \in L^2(\mu)$ the function $\varrho^{-1/2}g$ belongs to $L^2(\nu)$. Finally,

$$A_\mu Uf(t) = t\sqrt{\varrho(t)}f(t) = \sqrt{\varrho(t)}tf(t) = UA_\nu f(t).$$

Let A_μ and A_ν be unitarily equivalent. Let $U\colon L^2(\nu) \to L^2(\mu)$ be their isometry. We can assume that μ and ν have supports in some $S = [-n, n]$, since $g(A_\mu)$ and $g(A_\nu)$ for $g(t) = tI_S(t)$ are also equivalent. Set $\psi := U1$ and $p_k(t) := t^k$. Then

$$Up_k(t) = UA_\nu p_{k-1}(t) = tUp_{k-1}(t),$$

which gives the equality $Up_k(t) = t^k\psi(t)$. For every polynomial f we obtain $Uf(t) = f(t)\psi(t)$. Since U is unitary, we have

$$\int_S |f(t)|^2\,\nu(dt) = \int_S |\psi(t)|^2|f(t)|^2\,\mu(dt).$$

Let now B be a Borel set in S. Taking the measure $\mu + \nu$, we find a uniformly bounded sequence of polynomials f_n such that $f_n(t) \to I_B(t)$ for almost all t with respect to both measures. By the Lebesgue dominated convergence theorem

$$\nu(B) = \int_B |\psi(t)|^2 \, \mu(dt),$$

which means that $\nu \ll \mu$ and $d\nu/d\mu = |\psi|^2$. By the symmetry of the roles of both measures they are equivalent. $\square$

In the case of decompositions into subspaces with cyclic vectors the situation becomes more complicated. For example, the operator of multiplication by the argument on $L^2[0,1]$ can be decomposed into the sum of operators of multiplication by the argument on the intervals $[0,1/2]$ and $(1/2,1]$. How can we avoid such redundant terms? On the other hand, one should somehow take into account the multiplicities of several copies of the operator of multiplication by the argument. In the finite-dimensional case this reduces to counting the multiplicity of every eigenvalue, but only different eigenvalues are taken (for example, an eigenvalue of multiplicity 2 is not considered as two eigenvalues of multiplicity 1). Moreover, in case of multiplicity 1 the corresponding subspace is one-dimensional and cannot be further decomposed. In the infinite-dimensional case an analog of an operator with a simple spectrum is an operator with a cyclic vector, but, as the example above shows, such an operator can fail to have a "minimal" canonical decomposition. For this reason for obtaining unitarily invariant representations in the infinite-dimensional space one has to reject the finite-dimensional picture $\mathbb{C}^n = H = H_{\lambda_1} \oplus H_{\lambda_2} \oplus \cdots \oplus H_{\lambda_k}$, where $\lambda_1, \ldots, \lambda_k$ are all distinct eigenvalues of A and H_{λ_i} are the corresponding n_i-dimensional kernel subspaces. To this picture there corresponds the Jordan form

$$\begin{pmatrix} \lambda_1 I_{n_1} & & \\ & \ddots & \\ & & \lambda_k I_{n_k} \end{pmatrix}.$$

Here $\lambda_j I_{n_j}$ denotes the block consisting of the n_j-dimensional unit matrix multiplied by λ_j. We consider instead another picture obtained by some rearrangement of the previous one. In this new picture eigenvalues are ordered according to increasing of their multiplicities: $n_1 \leqslant n_2 \leqslant \cdots \leqslant n_k$. Suppose that there are N different multiplicities $m_1 < m_2 < \cdots < m_N$. Then we create the following blocks:

$$B_1 = \begin{pmatrix} A_1 & & & \\ & A_1 & & \\ & & \ddots & \\ & & & A_1 \end{pmatrix}, \quad \text{where} \quad A_1 = \begin{pmatrix} \lambda_1 & & & \\ & \lambda_2 & & \\ & & \ddots & \\ & & & \lambda_l \end{pmatrix},$$

λ_l is the last number of multiplicity m_1 and A_1 is taken m_1 times, next,

$$B_2 = \begin{pmatrix} A_2 & & & \\ & A_2 & & \\ & & \ddots & \\ & & & A_2 \end{pmatrix}, \quad \text{where} \quad A_2 = \begin{pmatrix} \lambda_{l+1} & & & \\ & \lambda_{l+2} & & \\ & & \ddots & \\ & & & \lambda_p \end{pmatrix},$$

here the block A_2 is taken m_2 times (it employs the eigenvalues λ_j of multiplicity m_2, i.e., λ_2 appears in it only in the case where $n_2 > n_1$), and further blocks $A_3, \ldots, A_N$ of increasing multiplicity appear. This procedure represents the operator A as the direct sum of N blocks $B_1, \ldots, B_N$, where every B_j is the direct sum of m_j copies of the operator A_j with a simple spectrum (the spectra of A_j are disjoint). Thus, the algorithm is this: we take the eigenvalue of the minimal multiplicity m_1 and separated the part of the standard Jordan form constituted by the blocks $\lambda_1 I_{m_1}, \ldots, \lambda_l I_{m_1}$, which excludes the eigenvalue of the minimal multiplicity (or several eigenvalues of the same minimal multiplicity). Next we repeat the same with the remaining part.

10.4.2. Definition. *A selfadjoint operator is called an operator of homogeneous multiplicity* m, *where* $m \in \mathbb{N} \cup \{\infty\}$, *if it is unitarily equivalent to the direct sum of* m *copies of the operator of multiplication by the argument on the space* $L^2(\mu)$ *for some nonnegative σ-finite Borel measure μ on the real line.*

It is clear that bounded operators of homogeneous multiplicity correspond to measures with bounded supports.

In the finite-dimensional example described above the operators B_j are operators of homogeneous multiplicity. It turns out that such a representation already has a reasonable infinite-dimensional analog. In the final classification we use the notions of equivalence and mutual singularity of measures. We recall that two measures μ and ν are equivalent ($\mu \sim \nu$) if they are given by densities with respect to each other; the measures μ and ν are mutually singular ($\mu \perp \nu$) if they are concentrated on disjoint sets. These notions extend to classes of measures: two classes of measures M and L are equivalent if $\mu \sim \lambda$ for all $\mu \in M$ and $\lambda \in L$; two classes of measures M and L are mutually singular if $\mu \perp \lambda$ for all $\mu \in M$ and $\lambda \in L$. Let $\langle \mu \rangle$ be the equivalence class of μ.

We first show that for operators of homogeneous multiplicity m a full unitary invariant is the equivalence type of the measure.

10.4.3. Theorem. (i) *Let μ be a nonnegative σ-finite Borel measure on the real line that is not identically zero. If the direct sum of m copies of the operator of multiplication by the argument on the space $L^2(\mu)$ is unitarily equivalent to the direct sum of n copies of this operator, where $m, n \in [1, \ldots, \infty]$, then $m = n$.*

(ii) *Two operators of homogeneous multiplicities m and n are unitarily equivalent precisely when $m = n$ and the measures generating these operators are equivalent.*

PROOF. (i) Let $U: \left(L^2(\mu)\right)^m \to \left(L^2(\mu)\right)^n$ be a unitary isomorphism of the regarded operators A_n and A_m and let $n < m \leqslant \infty$. The isomorphism of operators gives the equality

$$I_B(A_n)U = UI_B(A_m)$$

for all Borel sets B. The projections $I_B(A_n)$ and $I_B(A_m)$ are operators of multiplication by I_B. Let us take a Borel set E with $0 < \mu(E) < \infty$ and consider $n + 1$ vectors $\varphi_1, \ldots, \varphi_{n+1}$ from the direct sum of m copies of $L^2(\mu)$ defined as follows: the component with the number j of the element φ_j is I_E and the remaining components are zero. Set

$$\psi_j := U\varphi_j = (\psi_{j,1}, \ldots, \psi_{j,n}).$$

For every Borel set $B \subset E$ we have

$$\big(I_B(A_m)\varphi_i, \varphi_j\big) = \big(I_B(A_n)\psi_i, \psi_j\big) = \int_B \sum_{k=1}^{n} \psi_{i,k}(t)\overline{\psi_{j,k}(t)}\, \mu(dt).$$

The left-hand side equals $\mu(B)\delta_{ij}$. Since this is true for all $B \subset E$, for any fixed i, j we obtain

$$\sum_{k=1}^{n} \psi_{i,k}(t)\overline{\psi_{j,k}(t)} = \delta_{ij} \quad \text{for } \mu\text{-a.e. } t \in E.$$

Then this equality is true μ-almost everywhere on E for all $i, j = 1, \ldots, n+1$ at once. Hence there exists a point t at which the indicated equality is fulfilled for all $i, j = 1, \ldots, n+1$. This leads to a contradiction, since gives $n+1$ linearly independent vectors in an n-dimensional space.

(ii) Suppose that the operators $A_{\mu,n}$ and $A_{\nu,m}$ of homogeneous multiplicities n and m, generated by the measures μ and ν, are unitarily equivalent. Then $\mu \sim \nu$, since the projections $I_B(A_{\mu,n})$ and $I_B(A_{\nu,m})$ are also unitarily equivalent and $I_B(A_{\mu,n}) = 0$ precisely when $\mu(B) = 0$, and similarly for the second operator. According to the previous theorem the operators $A_{\mu,n}$ and $A_{\nu,n}$ are equivalent. Assertion (i) gives the equality $n = m$. $\qquad\square$

Let A be a selfadjoint operator on a separable Hilbert space H. In addition to the idea to consider operators of homogeneous multiplicity, the structural spectral theorem employs some analysis of the measures μ_x, $x \in H$, defined by

$$\mu_x(B) = \Pi_{x,x}(B) = \big(\Pi(B)x, x\big),$$

where Π is the projection-valued measure generated by the operator A.

We shall say that $x \in H$ is a *vector of maximal type* in H if $\mu_y \ll \mu_x$ for all $y \in H$. In a natural way we define vectors of maximal type in any subspace $H' \subset H$.

Given x, let H_x be the closure of the linear span of all vectors $\Pi(B)x$, where $B \in \mathcal{B}(\mathbb{R}^1)$. For a bounded operator A, the subspace H_x equals the closure of the linear span of the sequence $\{x, Ax, A^2x, \ldots\}$; see Exercise 10.7.35.

10.4.4. Lemma. (i) *If* $y \in H_x$, *then* $\mu_y \ll \mu_x$.

(ii) *If* $\mu_x \perp \mu_y$, *then* $H_x \perp H_y$ *and* $\mu_{x+y} = \mu_x + \mu_y$.

(iii) *There are vectors of maximal type. Moreover, for every* $v \in H$, *there exists a vector* x *of maximal type such that* $v \in H_x$.

PROOF. (i) Let B be a Borel set such that $\mu_x(B) = 0$, i.e.,

$$\big(\Pi(B)x, \Pi(B)x\big) = \big(\Pi(B)x, x\big) = 0.$$

Then $\Pi(B)z = 0$ for every vector z of the form $z = \Pi(E)x$, where $E \in \mathcal{B}(\mathbb{R}^1)$, since $\Pi(B)\Pi(E)x = \Pi(E)\Pi(B)x$. The equality $\Pi(B)z = 0$ remains valid for all vectors z from the closure of the linear span of the vectors $\Pi(E)x$, i.e., for all $z \in H_x$. Thus, $\Pi(B)y = 0$ and $\mu_y(B) = \big(\Pi(B)y, y\big) = 0$ for all $y \in H_x$. Since $\big(\Pi(B)x, y\big) = 0$, we have $\mu_{x+y} = \mu_x + \mu_y$.

(ii) There are Borel sets S_1 and S_2 such that
$$S_1 \cap S_2 = \varnothing, \quad S_1 \cup S_2 = \mathbb{R}^1, \quad \mu_x(S_2) = 0, \quad \mu_y(S_1) = 0.$$
For every set $B \in \mathcal{B}(\mathbb{R}^1)$ we have $\Pi(B \cap S_2)x = \Pi(B \cap S_1)y = 0$, since $\Pi(B \cap S_2) \leqslant \Pi(S_2)$ and $\|\Pi(S_2)x\| = 0$, and similarly for $\Pi(B \cap S_1)y$. Then for any $B_1, B_2 \in \mathcal{B}(\mathbb{R}^1)$ we obtain
$$\Pi(B_1)x = \Pi(B_1 \cap S_1)x + \Pi(B_1 \cap S_2)x = \Pi(B_1 \cap S_1)x \in \Pi(S_1)(H)$$
and $\Pi(B_2)y = \Pi(B_2 \cap S_2)y \in \Pi(S_2)(H)$. Since $S_1 \cap S_2 = \varnothing$, we have $\Pi(S_1)(H) \perp \Pi(S_2)(H)$ (by the properties of projection-valued measure).

(iii) It suffices to consider a unit vector v. Let us take the set $\mathfrak{M}$ the elements of which are all possible families of measures μ_x with $x \neq 0$ that are pairwise mutually singular (within every family) and are also mutually singular with μ_v. It is partially ordered by inclusion. Zorn's lemma gives a maximal family, which by the separability of H consists of some countable collection of measures μ_{v_n} (for every family, the corresponding subspaces H_x are mutually orthogonal by assertion (ii)). For the required vector we take $x = v_0 + \sum_{n=1}^{\infty} c_n v_n$, where $v_0 = v$, $c_n := n^{-2}\|v_n\|^{-1}$. Since the measures μ_{v_n} and μ_v are mutually singular, there exist pairwise disjoint Borel sets B_n, $n = 0, 1, \ldots$, such that each measure μ_{v_n} is concentrated on B_n. Indeed, for every pair of different numbers n, k we can find Borel sets $B_{n,k}$ with $\mu_{v_k}(B_{n,k}) = \mu_{v_n}(\mathbb{R}^1 \backslash B_{n,k}) = 0$ and then take $B_n := \bigcap_{k \neq n} B_{n,k}$. This proves the inclusion $v_0 = \Pi(B_0)v_0 = \Pi(B_0)x \in H_x$. Let us verify that x is a vector of maximal type. If this is not true, then there exists a vector $u \in H$ such that the measure μ_u is not absolutely continuous with respect to μ_x. This means that for some Borel set B we have $\mu_x(B) = 0$ and $\mu_u(B) > 0$. Set $z := \Pi(B)u$. Then $\mu_z(B) = \mu_u(B) > 0$ and
$$\mu_z(\mathbb{R}^1 \backslash B) = \big(\Pi(\mathbb{R}^1 \backslash B)\Pi(B)u, z\big) = 0,$$
since $\Pi(\mathbb{R}^1 \backslash B)\Pi(B) = 0$. The equality $\mu_x = \mu_{v_0} + \sum_{n=1}^{\infty} c_n^2 \mu_{v_n}$ yields that $\mu_z \perp \mu_{v_n}$ for all $n \geqslant 0$ contrary to the maximality of the regarded family. $\quad\square$

10.4.5. Lemma. *Let $x \in H$ and let the measure μ_x be written as the sum $\mu_x = \sum_{n=1}^{\infty} \mu_n$ of pairwise mutually singular nonnegative Borel measures. Then there exist pairwise orthogonal vectors $x_n \in H_x$ such that*
$$x = \sum_{n=1}^{\infty} x_n, \quad \mu_{x_n} = \mu_n, \quad H_x = \bigoplus_{n=1}^{\infty} H_{x_n}.$$
If a finite measure $\nu \geqslant 0$ satisfies the condition $\nu \ll \Pi_{x,x}$, then $\nu = \Pi_{y,y}$ for some vector $y \in H_x$.

PROOF. As in the previous lemma, we split the real line into pairwise disjoint Borel sets B_n such that μ_n is concentrated on B_n. Let $x_n := \Pi(B_n)x$. Since the measure μ is concentrated on the union of B_n, we have
$$x = \Pi\Big(\bigcup_{n=1}^{\infty} B_n\Big)x = \sum_{n=1}^{\infty} x_n.$$
The vectors x_n are pairwise orthogonal by the disjointness of the sets B_n. For the same reason for every n the measure μ_{x_n} is concentrated on the set B_n. In

addition, $\mu_{x_n} = \mu_n$, since for every set $B \in \mathcal{B}(\mathbb{R}^1)$ by the equality

$$\Pi(B \cap B_n) = \Pi(B)\Pi(B_n) = \Pi(B_n)\Pi(B)$$

we have

$$\mu_x(B) = (\Pi(B)x, x) = \sum_{n=1}^{\infty} (\Pi(B \cap B_n)x, x) = \sum_{n=1}^{\infty} (\Pi(B)\Pi(B_n)x, x)$$

$$= \sum_{n=1}^{\infty} (\Pi(B)\Pi(B_n)x, \Pi(B_n)x) = \sum_{n=1}^{\infty} \mu_{x_n}(B).$$

Further, by the previous lemma $H_{x_n} \perp H_{x_k}$ if $n \neq k$. For every $B \in \mathcal{B}(\mathbb{R}^1)$ we have $\Pi(B)x_n = \Pi(B)\Pi(B_n)x = P(B \cap B_n)x \in H_x$, i.e., $H_{x_n} \subset H_x$. On the other hand, $\Pi(B)x = \sum_{n=1}^{\infty} \Pi(B \cap B_n)x$ belongs to $\bigoplus_{n=1}^{\infty} H_n$. Thus, $\bigoplus_{n=1}^{\infty} H_n = H_x$. Finally, let $\nu \ll \Pi_{x,x}$, where $\|x\| = 1$, and let A be the selfadjoint operator on H_x generated by Π. Let us represent A as the multiplication by the argument on $L^2(\mu)$, $\mu = \Pi_{x,x}$. Then $\nu = f \cdot \mu$, $f \in L^1(\mu)$, hence we can take $y = f^{1/2} \in L^2(\mu)$. $\qquad \square$

The following fundamental result gives a complete classification of selfadjoint operators.

10.4.6. Theorem. *For every selfadjoint operator A on a separable Hilbert space H there is a decomposition of H into the sum $H = H_1 \oplus H_2 \oplus \cdots \oplus H_\infty$ of mutually orthogonal closed subspaces H_m (some of them can be absent) with the following properties:*

(i) $A(H_m \cap \mathfrak{D}(A)) \subset H_m$ ($A(H_m) \subset H_m$ for bounded A) and $A|_{H_m}$ is an operator of homogeneous multiplicity m for each $m \in [1, 2, \ldots, \infty]$;

(ii) *the classes of measures $\langle \mu_m \rangle$, corresponding to the operators $A|_{H_m}$, are mutually singular for different m.*

The equivalence classes of measures μ_m give a complete collection of unitary invariants of the operator, i.e., two operators with equal collections are unitarily equivalent.

PROOF. Actually we shall find unitary invariants of the projection-valued measure of the operator and use them to obtain the desired invariants of the operator itself. We establish the existence of a decomposition of the indicated form. We first show that there exists a finite or countable set of unit vectors v_n such that

$$H = \bigoplus_n H_{v_n}, \qquad \mu_{v_{n+1}} \ll \mu_{v_n}.$$

It is natural to take a unit vector v_1 of maximal type, next in the orthogonal complement $H_{v_1}^\perp$ of the subspace H_{v_1} take a unit vector v_2 of maximal type for this complement and so on: at the nth step we take a unit vector v_n of maximal type in the orthogonal complement to $H_{v_1} \oplus \cdots \oplus H_{v_{n-1}}$. We observe that $H_{v_2} \perp H_{v_1}$, since for any sets $B, C \in \mathcal{B}(\mathbb{R}^1)$ we have

$$(\Pi(B)v_1, \Pi(C)v_2) = (\Pi(C)\Pi(B)v_1, v_2) = (\Pi(C \cap B)v_1, v_2) = 0,$$

because $\Pi(C \cap B)v_1 \in H_{v_1}$. Hence we have mutually orthogonal subspaces H_{v_n} in a finite or countable number (for example, if v_1 is cyclic, then $H_{v_1} = H$).

We would like to obtain the whole space H as the sum of H_{v_n}, but this is not always true. For example, if A is the identity operator and $\{e_n\}$ is an orthonormal basis in H, then as the result of our construction we can obtain e_1, e_3, e_5 and so on. In order to avoid this unpleasant thing, we have to slightly modify the construction. Taking a basis $\{e_n\}$, we pick v_1 such that $e_1 \in H_{v_1}$, which is possible by the lemma. If e_2 has a nonzero projection φ_2 on $H_{v_1}^\perp$, then we pick v_2 such that $\varphi_2 \in H_{v_2}$. Hence $e_1, e_2 \in H_{v_1} \oplus H_{v_2}$. At the nth step we pick $v_n \in H_{v_1} \oplus \cdots \oplus H_{v_{n-1}}$ such that H_{v_n} contains the projection of e_n onto the orthogonal complement of $H_{v_1} \oplus \cdots \oplus H_{v_{n-1}}$. As a result, the direct sum of H_n contains all e_n and hence coincides with H.

We now rearrange the obtained decomposition. For every n, the measure μ_{v_n} has the form $\mu_{v_n} = \varrho_n \cdot \mu_{v_1}$, where ϱ_n is a Borel measurable Radon–Nikodym density of the measure μ_{v_n} with respect to μ_{v_1}. By the relation $\mu_{v_{n+1}} \ll \mu_{v_n}$, these densities can be chosen such that the sets $S_n := \{t \colon \varrho_n(t) > 0\}$ decrease. This decreasing need not be strict, because S_n can coincide with S_{n+1} if $\mu_{v_{n+1}}$ and μ_{v_n} are equivalent. The restriction of the measure μ_{v_1} to $S_{n-1} \backslash S_n$, where $S_0 := \mathbb{R}^1$, will be denoted by μ_n, and the restriction of μ_{v_1} to $S_\infty := \bigcap_{n=1}^\infty S_n$ will be denoted by μ_∞. It is clear that the obtained measures are pairwise orthogonal and their sum is μ_{v_1}. For every n, the measure μ_{v_n} is equivalent to the measure $\mu_n \oplus \mu_{n+1} \oplus \cdots \oplus \mu_\infty$ (note that such sums are finite measures dominated by μ_{v_1}).

Suppose now that the operator A is bounded. Then v_n is a cyclic vector for the restriction of A to H_{v_n} and this restriction is unitarily equivalent to multiplication by the argument on L^2 with respect to the measure $\mu_n \oplus \mu_{n+1} \oplus \cdots \oplus \mu_\infty$, which can be written as the direct sum of operators of multiplication by the argument on the spaces $L^2(\mu_n), L^2(\mu_{n+1}), \ldots, L^2(\mu_\infty)$. It is clear that we have obtained the desired decomposition.

If the operator A is not bounded, then we first represent it as the direct sum of bounded operators A_k each of which is already decomposed into the sum of operators of homogeneous multiplicity n by means of measures $\mu_{k,n}$ concentrated on some bounded Borel set B_k, where B_k are pairwise disjoint. We can assume that $\mu_{k,n}(\mathbb{R}^1) \leqslant 2^{-n-k}$. For every m we take the measure $\mu_m := \sum_{k=1}^\infty \mu_{k,m}$.

Let us proceed to the proof of unitary invariance of the obtained objects: the appearing multiplicities n and measure types μ_n. It is clear from Theorem 10.4.1 that two decompositions of the indicated form with $\mu_n \sim \mu_n'$ are unitarily equivalent. Suppose that two operators A_1 and A_2 of such a form corresponding to collections of measures $\{\mu_n\}$ and $\{\nu_n\}$ are unitarily equivalent. Suppose that for some $n \in \mathbb{N} \cup \{\infty\}$ the measures μ_n and ν_n are not equivalent. We can assume that there is a Borel set B with $\mu_n(B) = 0$ and $0 < \nu_n(B) < \infty$. Pairwise orthogonal measures ν_k are concentrated on some disjoint Borel sets B_k. Then $\nu_n(B \cap B_n) > 0$ and $\nu_k(B \cap B_n) = 0$ if $k \neq n$. Hence the operator $A_2 I_{B \cap B_n}(A_2)$, which is the direct sum of n copies of the operator of multiplication by the argument on the space $L^2(\nu_n|_{B \cap B_n})$, is nonzero and of homogeneous multiplicity n. This contradicts Theorem 10.4.3, since the operators $A_1 I_{B \cap B_n}(A_1)$ and $A_2 I_{B \cap B_n}(A_2)$ are unitarily equivalent and by our construction $A_1 I_{B \cap B_n}(A_1) = 0$. $\qquad\square$

Let us explain how this theorem excludes "false" components of the type mentioned above in the decomposition of $L^2[0,1]$ into the sum $L^2[0,1/2] \oplus L^2(1/2,1]$ for the operator of multiplication by the argument. In this sum the summands are generated by mutually singular measures, but have the same multiplicity 1, which is forbidden in the theorem. The theorem can be restated in terms of the unitary equivalence of projection-valued measures of operators (see [**67**, Chapter 7]).

Note also that even in the case of operators with simple spectra the equivalence of generating measures cannot be in general seen from the equality of spectra. For example, the operator of multiplication by the argument on $L^2(\lambda)$ with Lebesgue measure λ has the same spectrum $[0,1]$ as the operator of multiplication by the argument on $L^2(\mu)$ with any singular measure μ such that the interval $[0,1]$ is the minimal closed set of measure 1 (i.e., positive on all intervals in $(0,1)$).

10.5. Operator Semigroups

Let X be a real or complex Banach space. In this section we study families of operators that in the finite-dimensional case correspond to semigroups of the form $\exp(tA)$. The main feature of the infinite-dimensional case is that for A one cannot always take a bounded operator. For this reason, as in the previous sections, we deal with operators whose domains of definition do not coincide with the whole space. Generators of semigroups are among the most typical examples of unbounded operators.

10.5.1. Definition. *A family of bounded operators $\{T_t\}_{t \geqslant 0}$ on X is called a strongly continuous operator semigroup (or a C_0-semigroup) if*
   (i) $T_0 = I$, $T_{t+s} = T_t T_s$ *for all $t, s \geqslant 0$,*
   (ii) $\lim\limits_{t \to 0} \|T_t x - x\| = 0$ *for all $x \in X$.*

The definition yields the continuity of all mappings
$$t \mapsto T_t x, \quad x \in X,$$
on the half-line (not only at zero), since $\|T_{t+s}x - T_t x\| \leqslant \|T_t\| \, \|T_s x - x\|$.
   Set
$$\mathfrak{D}(L) := \Big\{ x \in X \colon \ \exists \lim_{t \to 0} \frac{T_t x - x}{t} \Big\},$$
$$Lx := \lim_{t \to 0} \frac{T_t x - x}{t}, \ x \in \mathfrak{D}(L).$$
The operator L with the indicated domain of definition $\mathfrak{D}(L)$ is called the *generator* or the *infinitesimal operator* of the semigroup $\{T_t\}_{t \geqslant 0}$.

The Banach–Steinhaus theorem yields the uniform boundedness of operators T_t with $t \in [0,1]$, i.e., $\sup_{t \in [0,1]} \|T_t\| \leqslant C < \infty$. Since $T_n = T_1^n$ and $T_t = T_{[t]} T_r$, where $[t]$ is the integer part of t and r is the fractional part of t, we arrive at the following estimate with some constant β:
$$\|T_t\| \leqslant C \|T_1\|^{[t]} \leqslant C e^{\beta t}. \tag{10.5.1}$$

We observe that the operators $S_t = e^{-\beta t} T_t$ also form a strongly continuous semigroup, but are uniformly bounded. The generator of the semigroup $\{S_t\}_{t \geqslant 0}$ equals $L - \beta I$, where L is the generator of the semigroup $\{T_t\}_{t \geqslant 0}$.

The estimate obtained above enables us to consider the operators R_λ defined for $\operatorname{Re}\lambda > \beta$ by the formula

$$R_\lambda x = \int_0^\infty e^{-\lambda s} T_s x\, ds. \tag{10.5.2}$$

This integral exists as a limit in norm of the integrals over $[0, N]$ as $N \to \infty$, and the integrals over the intervals exist as Riemann integrals of continuous vector-valued functions (i.e., as limits of usual Riemann sums).

10.5.2. Example. Let L be a bounded operator and

$$T_t = \exp(tL) := \sum_{n=0}^\infty t^n L^n / n!.$$

Then this series converges with respect to the operator norm and defines a strongly continuous semigroup the generator of which coincides with the operator L.

PROOF. Norm convergence of the series above is obvious from the estimate $\|L^n\| \leqslant \|L\|^n$. Multiplying the series for $\exp(tL)$ and $\exp(sL)$, it is readily verified that at L^m we have the coefficient $(t + s)^m / m!$, since this is true for scalar expansions. Finally, the equality $\lim_{t\to 0} t^{-1}[\exp(tL) - I] = L$ follows from the equality

$$t^{-1}[\exp(tL) - I] - L = \sum_{n=2}^\infty t^{n-1} L^n / n!$$

and convergence of the series with the general term $|t|^n \|L\|^n / n!$. $\quad\square$

10.5.3. Example. Let A be a selfadjoint (possibly, unbounded) operator on a Hilbert space H. Then the unitary operators $\exp(itA)$, $t \in \mathbb{R}^1$, form a strongly continuous group with the generator iA.

PROOF. It suffices to consider the case of a separable space H. We can assume that A is the operator of multiplication on $L^2(\mu)$ by a real measurable function φ and

$$\mathfrak{D}(A) = \{x \in L^2(\mu)\colon \varphi \cdot x \in L^2(\mu)\}.$$

Then $\exp(itA)$ is the operator of multiplication by $\exp(it\varphi)$. It is clear that this is a strongly continuous group. Let L be its generator. Then $\mathfrak{D}(L) \subset \mathfrak{D}(A)$ and $Lx = i\varphi \cdot x$ for all $x \in \mathfrak{D}(L)$, since convergence $t^{-1}[\exp(it\varphi)x - x] \to Lx$ in $L^2(\mu)$ yields convergence $t_n^{-1}[\exp(it_n\varphi(\omega))x(\omega) - x(\omega)] \to Lx(\omega)$ μ-a.e. for some sequence $t_n \to 0$ and the left-hand side converges to $i\varphi(\omega)x(\omega)$. On the other hand, for every $x \in \mathfrak{D}(A)$ we have convergence $t^{-1}[\exp(it\varphi)x - x] \to i\varphi x$ in $L^2(\mu)$. This follows from the Lebesgue dominated convergence theorem, since we have the pointwise convergence and the bound

$$\left|t^{-1}[\exp(it\varphi(\omega))x(\omega) - x(\omega)]\right| \leqslant 2|\varphi(\omega)x(\omega)|.$$

Thus, $L = iA$. In particular, $\mathfrak{D}(L) = \mathfrak{D}(A)$. $\quad\square$

The next example is justified similarly.

10.5.4. Example. Let A be a selfadjoint (possibly, unbounded) operator on a Hilbert space H such that $A \geqslant 0$. Then the operators $\exp(-tA)$, $t \geqslant 0$, form a strongly continuous semigroup with the generator $-A$.

We shall consider below complex spaces, but all results in this section remain valid for real spaces. Let β satisfy (10.5.1).

10.5.5. Proposition. *For every λ with $\operatorname{Re} \lambda > \beta$ we have*

$$R_\lambda(X) \subset \mathfrak{D}(L), \quad (\lambda I - L) R_\lambda = I,$$

$$\lim_{\lambda > 0, \, \lambda \to \infty} \lambda R_\lambda x = x \quad \text{for all } x \in X.$$

PROOF. We observe that

$$t^{-1}(T_t - I) R_\lambda x = t^{-1} \int_0^\infty e^{-\lambda s} T_t T_s x \, ds - t^{-1} \int_0^\infty e^{-\lambda s} T_s x \, ds$$

$$= t^{-1} \int_0^\infty e^{-\lambda s} T_{t+s} x \, ds - t^{-1} \int_0^\infty e^{-\lambda s} T_s x \, ds$$

$$= t^{-1} e^{\lambda t} \int_t^\infty e^{-\lambda s} T_s x \, ds - t^{-1} \int_0^\infty e^{-\lambda s} T_s x \, ds.$$

The right-hand side can be written in the form

$$t^{-1}(e^{\lambda t} - 1) \int_t^\infty e^{-\lambda s} T_s x \, ds - t^{-1} \int_0^t e^{-\lambda s} T_s x \, ds$$

$$= t^{-1}(e^{\lambda t} - 1)\left(R_\lambda x - \int_0^t e^{-\lambda s} T_s x \, ds\right) - t^{-1} \int_0^t e^{-\lambda s} T_s x \, ds.$$

By the continuity of the function $s \mapsto e^{-\lambda s} T_s x$ we conclude that, as $t \to 0+$, the obtained expression tends to $\lambda R_\lambda x - x$. Thus, we have shown that $R_\lambda x \in \mathfrak{D}(L)$ and $L R_\lambda x = \lambda R_\lambda x - x$. For the proof of the second assertion we observe that

$$\lambda R_\lambda x - x = \lambda \int_0^\infty e^{-\lambda s}(T_s x - x) \, ds.$$

In addition, whenever $\lambda > 0$, we have

$$\lambda \int_{\beta+1}^\infty e^{-\lambda s} \|T_s x - x\| \, ds \leqslant \lambda \int_{\beta+1}^\infty e^{-\lambda s}(Ce^{\beta s} + 1)\|x\| \, ds,$$

which tends to zero as $\lambda \to \infty$. Since $\|T_s x - x\| \to 0$ as $s \to 0$, for every $\varepsilon > 0$ there exists $\delta > 0$ such that

$$\lambda \int_0^\delta e^{-\lambda s} \|T_s x - x\| \, ds \leqslant \varepsilon \lambda \int_0^\infty e^{-\lambda s} \, ds = \varepsilon$$

for all $\lambda > \beta$. Finally, the integral of $\lambda e^{-\lambda s} \|T_s x - x\|$ over $[\delta, \beta + 1]$ tends to zero as $\lambda \to \infty$ by the Lebesgue dominated convergence theorem, since $\lambda e^{-\lambda s} \to 0$ for all $s > 0$ and $\lambda e^{-\lambda s} \leqslant \lambda e^{-\delta \lambda}$. $\square$

10.5.6. Theorem. *For every strongly continuous operator semigroup* $\{T_t\}_{t\geqslant 0}$ *the following assertions hold*:

(i) *the operator* L — *the generator of the semigroup* — *is densely defined and closed*;

(ii) *for every* $x \in \mathfrak{D}(L)$ *we have* $T_t x \in \mathfrak{D}(L)$ *and*

$$\frac{d}{dt}T_t x = LT_t x = T_t L x;$$

(iii) *for every* $x \in X$ *we have*

$$T_t x = \lim_{\varepsilon \to 0} \exp\left(t\frac{T_\varepsilon - I}{\varepsilon}\right)x, \quad t \geqslant 0,$$

moreover, convergence is uniform on compact intervals in $[0, +\infty)$.

PROOF. (i) Set $A_\varepsilon := \varepsilon^{-1}(T_\varepsilon - I)$. Then

$$\exp(tA_\varepsilon) = \exp(-t/\varepsilon)\exp(t\varepsilon^{-1}T_\varepsilon) = \exp(-t/\varepsilon)\sum_{n=0}^{\infty}\frac{t^n T_{n\varepsilon}}{\varepsilon^n n!},$$

since the operators T_t are bounded. The continuity of the mapping $t \mapsto T_t x$ yields the existence of the vector integral

$$B_t x := \frac{1}{t}\int_0^t T_s x\, ds, \quad t > 0.$$

We observe that (10.5.1) implies the estimate

$$\|B_t x\| \leqslant \frac{C}{\beta t}(e^{\beta t} - 1).$$

Hence for every t the operator B_t is bounded. We show that

$$A_\varepsilon B_t = A_t B_\varepsilon, \quad \varepsilon > 0, t > 0. \tag{10.5.3}$$

Indeed, we have

$$\varepsilon t A_\varepsilon B_t x = \int_0^t (T_\varepsilon - I)T_s x\, ds = \int_0^t (T_{s+\varepsilon} - T_s)x\, ds$$

$$= \int_t^{t+\varepsilon} T_s x\, ds - \int_0^\varepsilon T_s x\, ds,$$

$$\varepsilon t A_t B_\varepsilon x = \int_0^\varepsilon (T_t - I)T_s x\, ds = \int_0^\varepsilon (T_{s+t} - T_s)x\, ds$$

$$= \int_t^{t+\varepsilon} T_s x\, ds - \int_0^\varepsilon T_s x\, ds.$$

Next, the continuity of $T_t x$ in t yields that $B_t x \to x$ as $t \to 0$. By (10.5.3) this gives the equality

$$\lim_{\varepsilon \to 0} A_\varepsilon B_t x = A_t \lim_{\varepsilon \to 0} B_\varepsilon x = A_t x.$$

It follows that $B_t x \in \mathfrak{D}(L)$ for all $t > 0$ and that $LB_t x = A_t x$. The set of elements of the form $B_t x$ is everywhere dense in X, since we have $\lim_{t \to 0} B_t x = x$ for every $x \in X$.

We now show that the operator L is closed. Let $\{x_n\} \subset \mathfrak{D}(L)$, $x_n \to x$ and $Lx_n \to y$ in X as $n \to \infty$. Since the operators T_t and T_s commute, the operators A_ε and B_t commute as well. Hence (10.5.3) yields the equalities

$$B_t L u = B_t \lim_{\varepsilon \to 0} A_\varepsilon u = \lim_{\varepsilon \to 0} B_\varepsilon A_t u = A_t u, \quad u \in \mathfrak{D}(L), \varepsilon > 0, t > 0.$$

Therefore, $A_t x_n = B_t L x_n$, whence $A_t x = B_t y$. Since $\lim_{t \to 0} B_t y = y$, we obtain $x \in \mathfrak{D}(L)$ and $Lx = y$. Thus, the operator L is closed.

(ii) Let $x \in \mathfrak{D}(L)$. For any $t > 0$ we have

$$\lim_{\varepsilon \to 0} A_\varepsilon T_t x = \lim_{\varepsilon \to 0} T_t A_\varepsilon x = T_t L x.$$

Hence $T_t x \in \mathfrak{D}(L)$ and

$$L T_t x = T_t L x.$$

The equality $B_t L x = A_t x$ proved above can be written in the form

$$\int_0^t T_s L x \, ds = T_t x - x.$$

By the continuity of the integrand this completes the proof of (ii).

(iii) Suppose first that $x \in \mathfrak{D}(L)$. According to assertion (ii), whenever $0 < s < t$ we have

$$\frac{d}{ds}\left[\exp\big((t-s)A_\varepsilon\big) T_s x \right] = \exp\big((t-s)A_\varepsilon\big) T_s (Lx - A_\varepsilon x),$$

which after integration in s over the interval $[0, t]$ gives the equality

$$T_t x - \exp(t A_\varepsilon) x = \int_0^t \exp\big((t-s)A_\varepsilon\big) T_s (Lx - A_\varepsilon x) \, ds.$$

From (10.5.1), letting $\gamma := e^\beta$, we obtain the following estimates:

$$\| \exp(t A_\varepsilon) \| \leqslant \exp(-t/\varepsilon) \sum_{n=0}^\infty \frac{t^n}{\varepsilon^n n!} \| T_{n\varepsilon} \|$$

$$\leqslant C \exp(-t/\varepsilon) \sum_{n=0}^\infty \frac{t^n}{\varepsilon^n n!} \exp(\beta \varepsilon n) = C \exp\left(\frac{t}{\varepsilon}(\gamma^\varepsilon - 1) \right).$$

Whenever $0 < \varepsilon \leqslant 1$, by the bound $\gamma^\varepsilon - 1 \leqslant \varepsilon \gamma$ we obtain

$$\| \exp(t A_\varepsilon) \| \leqslant C e^{\gamma t}.$$

Therefore, we arrive at the estimate

$$\| T_t x - \exp(t A_\varepsilon) x \| \leqslant C^2 \| Lx - A_\varepsilon x \| \int_0^t e^{\gamma(t-s)} e^{\beta s} \, ds$$

$$\leqslant C_1 e^{\gamma t} \| Lx - A_\varepsilon x \|, \tag{10.5.4}$$

where C_1 is a constant. We now fix $x_0 \in X$, an interval $[0, \tau]$ and $\delta > 0$. Set $M := C e^{\beta \tau} + C e^{\gamma \tau}$. Pick $x \in \mathfrak{D}(L)$ with $\| x - x_0 \| \leqslant \delta/M$. By using (10.5.4) for $t \in [0, \tau]$ we obtain

$$\|T_t x_0 - \exp(tA_\varepsilon)x_0\| \leqslant \|T_t x - \exp(tA_\varepsilon)x\| + \|T_t - \exp(tA_\varepsilon)\| \, \|x_0 - x\|$$
$$\leqslant C_1 e^{\gamma \tau} \|Lx - A_\varepsilon x\| + \delta.$$

For all sufficiently small ε the right-hand side is estimated by 2δ, since we have $\|Lx - A_\varepsilon x\| \to 0$ as $\varepsilon \to 0$. $\square$

10.5.7. Corollary. *If* $\mathrm{Re}\,\lambda > \beta$, *where* β *satisfies inequality* (10.5.1), *then the operator* R_λ *is inverse to the operator* $\lambda I - L$ *and maps* X *one-to-one onto* $\mathfrak{D}(L)$.

PROOF. By Proposition 10.5.5 we have the equality $(\lambda I - L)R_\lambda = I$. We show that $R_\lambda(\lambda I - L)x = x$ for all $x \in \mathfrak{D}(L)$. The semigroup $S_t = e^{-\lambda t}T_t$ has generator $L - \lambda I$. This gives the equality

$$e^{-\lambda t}T_t x - x = \int_0^t e^{-\lambda s}T_s(L - \lambda I)x\,ds, \quad t \geqslant 0.$$

Letting $t \to +\infty$ we obtain $e^{-\lambda t}T_t x \to 0$, since $\|e^{-\lambda t}T_t\| \leqslant C e^{(\beta - \lambda)t}$, where $\mathrm{Re}\,(\beta - \lambda) < 0$. Hence $R_\lambda(L - \lambda I)x = -x$. $\square$

Thus, we have the following representation for all λ with a sufficiently large real part (if $\|T_t\| \leqslant 1$, then for all λ with $\mathrm{Re}\,\lambda > 0$):

$$(\lambda I - L)^{-1}x = \int_0^\infty e^{-\lambda s}T_s x\,ds. \tag{10.5.5}$$

This representation is a basis for many further results in the theory of operator semigroups. Let us give without proof (which can be found in [317, p. 622] or [471, p. 85]) the following theorem due to Trotter.

10.5.8. Theorem. *Let* L *and* L_n, *where* $n \in \mathbb{N}$, *be the generators of strongly continuous semigroups* $\{T_t\}_{t \geqslant 0}$ *and* $\{T_t^{(n)}\}_{t \geqslant 0}$ *on a Banach space* X *for which there exists a number* $C, \omega \in \mathbb{R}^1$ *such that* $\|T_t^{(n)}\| \leqslant C e^{\omega t}$, $\|T_t\| \leqslant C e^{\omega t}$ *and for some* λ *with* $\mathrm{Re}\,\lambda > \omega$ *we have* $\lim_{n\to\infty} (\lambda I - L_n)^{-1}x = (\lambda I - L)^{-1}x$ *for every* $x \in X$. *Then this is true for all* λ *with* $\mathrm{Re}\,\lambda > \omega$ *and, as* $n \to \infty$, *for every* $x \in X$ *we have* $T_t^{(n)}x \to T_t x$ *uniformly in* t *from every compact interval.*

Conversely, if $T_t^{(n)}x \to T_t x$ *for all* $x \in X$ *and* $t \geqslant 0$, *then for every* λ *with* $\mathrm{Re}\,\lambda > \omega$ *we have* $(\lambda I - L_n)^{-1}x \to (\lambda I - L)^{-1}x$ *for all* $x \in X$.

There is also a condition for convergence of semigroups in terms of their generators (see [471, p. 88]).

10.5.9. Theorem. *The pointwise convergence of semigroups in the situation of the previous theorem holds if in place of convergence of resolvents the following condition is satisfied: there exist a dense linear subspace* D *in the intersection of the domains of definition of* L *and* L_n *and a number* λ *with* $\mathrm{Re}\,\lambda > \omega$ *such that* $L_n x \to Lx$ *for all* $x \in D$ *and* $(\lambda I - L)(D)$ *is dense in* X.

In the next section we continue our discussion of semigroups.

10.6. Generators of Semigroups

Here we prove two important results about semigroup generators: Stone's theorem about generators of unitary groups on Hilbert spaces and the Hille–Yosida theorem, which gives a description of all semigroup generators. We first observe that the generator uniquely determines the semigroup.

10.6.1. Proposition. *Suppose that two strongly continuous operator semigroups* $\{T_t\}_{t \geqslant 0}$ *and* $\{S_t\}_{t \geqslant 0}$ *on a Banach space have equal generators. Then* $T_t = S_t$ *for all* $t \geqslant 0$.

PROOF. Let L be the common generator of the given semigroups. Then by the equality $(\lambda - L)^{-1} = R_\lambda$ proved above for every λ with a sufficiently large real part we have

$$\int_0^\infty e^{-\lambda t} T_t x \, dt = \int_0^\infty e^{-\lambda t} S_t x \, dt.$$

Let $l \in X^*$, $\varphi(t) = e^{-\eta t} l(T_t x)$, $\psi(t) = e^{-\eta t} l(S_t x)$, where η is a large real number and $\varphi = \psi = 0$ on $(-\infty, 0)$. The previous equality means that φ and ψ have equal Fourier transforms. Hence $\varphi(t) = \psi(t)$, which gives the equality $T_t x = S_t x$, because l was arbitrary. $\qquad \square$

We now prove the following Stone theorem.

10.6.2. Theorem. *Let* $\{U_t\}_{t \in \mathbb{R}}$ *be a strongly continuous group of unitary operators on a Hilbert space* H. *Then its generator* L *has the form* $L = iA$ *with some selfadjoint operator* A *and* $U_t = \exp(itA)$.

PROOF. For any $f \in C_0^\infty(\mathbb{R}^1)$ and $\varphi \in H$ set

$$\varphi_f := \int_{-\infty}^{+\infty} f(t) U_t \varphi \, dt,$$

where the integral is understood in the Riemann sense. Let D be the set of finite linear combinations of elements φ_f with $f \in C_0^\infty(\mathbb{R}^1)$ and $\varphi \in H$. The linear subspace D is dense in H, since for every $\varphi \in H$ we can take elements $\varphi_{\theta_\varepsilon}$ with $\theta_\varepsilon(t) = \varepsilon^{-1} \theta(t/\varepsilon)$, where θ is a smooth probability density with support in $[0,1]$. Then, letting $\varepsilon \to 0$, we have

$$\|\varphi_{\theta_\varepsilon} - \varphi\| = \left\| \int_{-\infty}^{+\infty} \theta_\varepsilon(t)[U_t \varphi - \varphi] \, dt \right\| \leqslant \sup_{t \in [0,\varepsilon]} \|U_t \varphi - \varphi\| \to 0$$

by the strong continuity of the semigroup. Letting $s \to 0$ we find that

$$s^{-1}(U_s - I)\varphi_f = s^{-1} \int_{-\infty}^{+\infty} f(t)[U_{t+s} - U_t]\varphi \, dt =$$

$$= \int_{-\infty}^{+\infty} \frac{f(\tau - s) - f(\tau)}{s} U_\tau \varphi \, d\tau \to -\int_{-\infty}^{+\infty} f'(\tau) U_\tau \varphi \, d\tau = \varphi_{-f'},$$

since the difference quotients $s^{-1}[f(\tau - s) - f(\tau)]$ converge to $-f'(\tau)$ uniformly on some interval outside of which they vanish. Set

$$A\varphi_f := -i\varphi_{-f'} = -i \lim_{s \to 0} s^{-1}(U_s - I)\varphi_f.$$

So the operator A is defined on D. We observe that $U_t(D) \subset D$, $A(D) \subset D$ and $U_t A\psi = AU_t\psi$ for all $\psi \in D$, which is verified on the vectors $\psi = \varphi_f$. In addition, if $f, g \in C_0^\infty(\mathbb{R}^1)$ and $\varphi, \psi \in H$, then

$$(A\varphi_f, \psi_g) = \lim_{s \to 0} \frac{1}{is}(U_s\varphi_f - \varphi_f, \psi_g) = \lim_{s \to 0} \frac{1}{is}(\varphi_f, U_{-s}\psi_g - \psi_g)$$
$$= (\varphi_f, i^{-1}\psi_{-g'}) = (\varphi_f, A\psi_g).$$

Thus, A is a symmetric operator on D. We show that it is essentially selfadjoint. Suppose that $u \in \mathfrak{D}(A^*)$ and $A^*u = iu$. Then for every $\varphi \in D$ we have

$$\frac{d}{dt}(U_t\varphi, u) = (iAU_t\varphi, u) = -i(U_t\varphi, A^*u) = -i(U_t\varphi, iu) = (U_t\varphi, u).$$

Thus, the complex function $\zeta(t) = (U_t\varphi, u)$ satisfies the equation $\zeta'(t) = \zeta$, so $\zeta(t) = ce^t$. Since $|\zeta(t)| \leqslant \|\varphi\| \|u\|$, we have $c = 0$, hence $(\varphi, u) = 0$. Using that D is dense, we obtain that $u = 0$. Similarly we verify that if $u \in \mathfrak{D}(A^*)$ and $A^*u = -iu$, then $u = 0$. By Corollary 10.7.11 proved below the closure $\overline{A}$ of the operator A is selfadjoint. Let $V_t = \exp(it\overline{A})$. It remains to show that $U_t = V_t$. It suffices to verify that $U_t\varphi = V_t\varphi$ for all $\varphi \in D$, since D is dense. Thus, let $\varphi \in D$. Since $D \subset \mathfrak{D}(\overline{A})$, according to Example 10.5.3 and Theorem 10.5.6 we have $V_t\varphi \in \mathfrak{D}(\overline{A})$ and $(V_t\varphi)' = i\overline{A}V_t\varphi$. On the other hand, $U_t\varphi \in D$ and $(U_t\varphi)' = iAU_t\varphi$, since $(U_t\varphi)' = U_t(U_s\varphi)'|_{s=0} = iU_tA\varphi$. Set $w(t) := U_t\varphi - V_t\varphi$. Then

$$w'(t) = iAU_t\varphi - i\overline{A}V_t\varphi = i\overline{A}w(t).$$

Therefore,

$$\frac{d}{dt}(w(t), w(t)) = -i(\overline{A}w(t), w(t)) + i(w(t), \overline{A}w(t)) = 0.$$

Since $w(0) = 0$, we obtain $w(t) = 0$ for all t, as required. $\square$

10.6.3. Theorem. *Let $\{T_t\}_{t \geqslant 0}$ be a strongly continuous semigroup of selfadjoint operators on a Hilbert space H such that $\|T_t\| \leqslant 1$. Then its generator L is selfadjoint and $L \leqslant 0$.*

Conversely, if an operator L is selfadjoint and $L \leqslant 0$, then the operators $T_t = \exp(tL)$ form a strongly continuous semigroup of selfadjoint operators with generator L and $\|T_t\| \leqslant 1$.

PROOF. The operator L is closed and symmetric. For every $x \in \mathfrak{D}(L)$ we have

$$\frac{d}{dt}(T_tx, T_tx) = 2(T_tLx, T_tx).$$

In addition, $(T_tx, T_tx) \leqslant (x, x)$ and $(T_0x, T_0x) = (x, x)$. Therefore, $(Lx, x) \leqslant 0$, since otherwise for sufficiently small $t \geqslant 0$ we would have $2(T_tLx, T_tx) > 0$ and then $(T_tx, T_tx) > (x, x)$. As shown in §10.7(ii), the operator L possesses a nonpositive selfadjoint extension G. This extension actually coincides with L, since both operators $L - I$ and $G - I$ have bounded inverse operators.

The converse assertion is obvious from the fact that in the separable case L can be represented as multiplication by a nonpositive function, which enables us to verify our assertion directly. The nonseparable case reduces to the separable one by decomposing H into a direct sum of separable subspaces invariant with respect to L. $\square$

We now prove the general Hille–Yosida theorem about generators of contracting semigroups, i.e., semigroups $\{T_t\}_{t\geqslant 0}$ with $\|T_t\| \leqslant 1$.

10.6.4. Theorem. (i) *Let L be the generator of a strongly continuous contracting semigroup $\{T_t\}_{t\geqslant 0}$ on a Banach space X. Then every real number $\lambda > 0$ belongs to the resolvent set of L and*

$$\|R_\lambda\| = \|(\lambda I - L)^{-1}\| \leqslant \frac{1}{\lambda}. \tag{10.6.1}$$

(ii) *Conversely, let L be a linear operator with a dense domain of definition $\mathfrak{D}(L)$ such that for every real number $\lambda > 0$ the operator $\lambda I - L$ has a bounded inverse $R_\lambda\colon X \to \mathfrak{D}(X)$ satisfying condition (10.6.1). Then L is the generator of a strongly continuous contracting semigroup $\{T_t\}_{t\geqslant 0}$ on X.*

PROOF. Corollary 10.5.7 and equality (10.5.5) give (i), where (10.6.1) follows from the fact that the integral of $e^{-\lambda t}$ over $(0, +\infty)$ is λ^{-1}. For the proof of (ii) we set

$$L_n := n^2 R_n - nI, \quad n \in \mathbb{N}.$$

Since $LR_n = nR_n - I$, we have $L_n = nLR_n$. The idea of the proof is to approximate L by the operators L_n. Since the operator L_n is bounded, with the aid of the usual exponent it generates the semigroup $T_t^{(n)} := \exp(tL_n)$. We show that

$$\lim_{n\to\infty} nR_n x = x, \quad x \in X. \tag{10.6.2}$$

If $x \in \mathfrak{D}(L)$, then this equality is true by the relations $nR_n x - x = R_n Lx$ and $\|R_n Lx\| \leqslant n^{-1}\|Lx\|$. Since $\mathfrak{D}(L)$ is dense in X and $\|nR_n\| \leqslant 1$, relation (10.6.2) is true for all $x \in X$. Using that for all $x \in \mathfrak{D}(L)$ we have $L_n x = nLR_n x = nR_n Lx$, from (10.6.2) we obtain the equality

$$\lim_{n\to\infty} L_n x = Lx, \quad x \in \mathfrak{D}(L). \tag{10.6.3}$$

By the definition of L_n we obtain

$$\|T_t^{(n)}\| = \|\exp(tL_n)\| = e^{-nt}\|\exp(n^2 t R_n)\| \leqslant e^{-nt} \sum_{k=0}^{\infty} \frac{1}{k!}(n^2 t)^k \|R_n^k\|$$

$$\leqslant e^{-nt} \sum_{k=0}^{\infty} \frac{1}{k!}(n^2 t)^k n^{-k} = e^{-nt} e^{nt} = 1.$$

Thus, the constructed semigroups are contracting.

We now show that these semigroups converge. To this end we estimate the quantity $\|T_t^{(n)} x - T_t^{(k)} x\|$ for $x \in \mathfrak{D}(L)$. It is readily seen that the operators L_n commute with $T_t^{(k)}$. This yields the relation

$$\frac{d}{dt} T_{s-t}^{(n)} T_t^{(k)} x = T_{s-t}^{(n)} T_t^{(k)}[L_k - L_n]x.$$

By the estimate $\|T_t^{(n)}\| \leqslant 1$ the norm of the right-hand side does not exceed $\|L_k x - L_n x\|$. Therefore,

$$\|T_t^{(n)} x - T_t^{(k)} x\| \leqslant t\|L_n x - L_k x\|.$$

Along with (10.6.3) this shows the existence of the limit

$$T_t x := \lim_{n \to \infty} T_t^{(n)} x, \quad x \in \mathfrak{D}(L),$$

which is uniform on every compact interval. The uniform boundedness of $T_t^{(n)}$ yields the existence of the limit for all $x \in X$. It is clear that $\{T_t\}$ is a continuous semigroup and $\|T_t\| \leqslant 1$.

It remains to show that L coincides with the generator of $\{T_t\}$. Passing to the limit in the equality

$$T_t^{(n)} x - x = \int_0^t T_s^{(n)} L_n x \, ds$$

as $n \to \infty$, for every $x \in \mathfrak{D}(L)$, on account of (10.6.3) we find

$$T_t x - x = \int_0^t T_s L x \, ds.$$

Let us denote the generator of $\{T_t\}$ by G. The previous equality shows (dividing by t and letting t go to zero) that $\mathfrak{D}(L)$ belongs to $\mathfrak{D}(G)$ and $Lx = Gx$ for all $x \in \mathfrak{D}(L)$. Thus, G extends L. However, this extension cannot be proper, since the operators $L - I$ and $G - I$ have bounded inverses (the first one by our assumption and the second one as the generator of a contracting semigroup). Hence $L = G$. $\qquad\square$

10.6.5. Corollary. *A closed densely defined operator L on a Banach space X is the generator of a strongly continuous operator semigroup precisely when there exist numbers $C \geqslant 0$ and $\beta \in \mathbb{R}^1$ such that every real number $\lambda > \beta$ belongs to the resolvent set of L and for all $n \in \mathbb{N}$ we have*

$$\|(\lambda I - L)^{-n}\| \leqslant \frac{C}{(\lambda - \beta)^n}. \tag{10.6.4}$$

PROOF. If L is the generator of a strongly continuous semigroup, then one has estimate (10.5.1). Suppose first that $\beta = 0$. Then we can introduce a new norm

$$\|x\|_0 := \sup_{t \geqslant 0} \|T_t x\|,$$

which is equivalent to the old one: $\|x\| \leqslant \|x\|_0 \leqslant C\|x\|$. The norms of operators on X corresponding to the new norm will be also denoted by $\|\cdot\|_0$. With respect to the new norm the operators T_t satisfy the estimate $\|T_t\|_0 \leqslant 1$. Hence all points λ with a positive real part are regular and $\|(\lambda I - L)^{-1}\|_0 \leqslant |\lambda|^{-1}$. Hence $\|(\lambda I - L)^{-n}\|_0 \leqslant |\lambda|^{-n}$, whence the desired estimate follows, since $\|A\|_0 \leqslant C\|A\|$ for all $A \in \mathcal{L}(X)$. In the general case we consider the semigroup of operators $S_t := e^{-t\beta} T_t$ with the generator $L - \beta I$. We have $\|S_t\| \leqslant C$, so by the previous step we obtain the regularity of all numbers λ with $\mathrm{Re}\,\lambda > \beta$ and estimate (10.6.4).

Conversely, suppose that L satisfies the indicated conditions. Again we start with the case $\beta = 0$. For any $\mu > 0$, we set

$$\|x\|_\mu := \sup\{\|\mu^n(\mu I - L)^{-n}\|: \ n = 0, 1, 2, \ldots\}.$$

This norm is equivalent to the original one, since $\|x\| \leqslant \|x\|_\mu \leqslant C\|x\|$. For the corresponding operator norm we have $\|\mu(\mu I - L)^{-1}\|_\mu \leqslant 1$. Moreover, the following inequality holds:

$$\|(\lambda I - L)^{-1}\|_\mu \leqslant \frac{1}{\lambda}, \quad 0 < \lambda \leqslant \mu. \tag{10.6.5}$$

For the proof we use the identity $R_\lambda - R_\mu = (\mu - \lambda)R_\lambda R_\mu$ for resolvents to obtain

$$\|R_\lambda\|_\mu \leqslant \|R_\mu\|_\mu + \|(\mu - \lambda)R_\lambda R_\mu\|_\mu \leqslant \frac{1}{\mu} + \frac{\mu - \lambda}{\mu}\|R_\lambda\|_\mu,$$

which gives $\lambda\|R_\lambda\|_\mu \leqslant 1$, as required. Now, whenever $n \in \mathbb{N}$ and $0 < \lambda \leqslant \mu$, by (10.6.5) we obtain

$$\|\lambda^n(\lambda I - L)^{-n}x\| \leqslant \|\lambda^n(\lambda I - L)^{-n}x\|_\mu \leqslant \|\lambda(\lambda I - L)^{-1}\|_\mu^n\|x\|_\mu \leqslant \|x\|_\mu.$$

Therefore, $\|x\|_\lambda \leqslant \|x\|_\mu$ if $0 < \lambda \leqslant \mu$. Let us introduce the norm

$$\|x\|_0 := \sup_{\mu > 0} \|x\|_\mu = \lim_{\mu \to \infty} \|x\|_\mu.$$

It is clear that $\|x\| \leqslant \|x\|_0 \leqslant C\|x\|$. Since $\|(\lambda I - L)^{-1}x\|_\mu \leqslant \lambda^{-1}\|x\|_\mu$ whenever $0 < \lambda \leqslant \mu$ (by (10.6.5)), letting $\mu \to \infty$ we obtain $\|(\lambda I - L)^{-1}x\|_0 \leqslant \lambda^{-1}\|x\|_0$, hence $\|(\lambda I - L)^{-1}\|_0 \leqslant \lambda^{-1}$. By the Hille–Yosida theorem the operator L is the generator of some continuous semigroup $\{T_t\}$ with $\|T_t\|_0 \leqslant 1$. Then we have the inequalities $\|T_t x\| \leqslant \|T_t x\|_0 \leqslant \|x\|_0 \leqslant C\|x\|$, i.e., $\|T_t\| \leqslant C$.

In the case of arbitrary β we apply the previous step to the shifted operator $L_0 = L - \beta I$. It is the generator of some continuous semigroup $\{S_t\}_{t \geqslant 0}$. The semigroup of operators $T_t := e^{\beta t}S_t$ has generator L. $\qquad\square$

There is yet another useful characterization of generators of contracting semigroups. A densely defined operator $(L, \mathfrak{D}(L))$ on a Banach space X is called *dissipative* if, for every $u \in \mathfrak{D}(L)$ with $\|u\| = 1$, there exists $l \in X^*$ with $\|l\| = 1$ such that $l(u) = 1$ and $\operatorname{Re} l(Lu) \leqslant 0$. If X is a Hilbert space, then this means that $\operatorname{Re}(u, Lu) \leqslant 0$. Dissipativity is equivalent to the property that for all $\lambda > 0$ and $u \in \mathfrak{D}(L)$ we have $\|\lambda u - Lu\| \geqslant \lambda\|u\|$ (see Exercise 10.7.41). Hence the generator of a contracting semigroup is dissipative. On account of the Hille–Yosida theorem this gives the following result due to Lumer and Phillips.

10.6.6. Theorem. *Let $(L, \mathfrak{D}(L))$ be a dissipative operator on a Banach space. Its closure is the generator of a strongly continuous contracting semigroup precisely when the set $(\lambda I - L)(\mathfrak{D}(L))$ is dense for some (and then for every) number $\lambda > 0$.*

10.7. Complements and Exercises

(i) Extensions of symmetric operators (461). (ii) Semibounded forms and operators (466). (iii) The Chernoff and Trotter theorems (470). (iv) The mathematical model of quantum mechanics (472). (v) Sturm–Liouville operators (478). Exercises (480).

10.7(i). Extensions of symmetric operators

In this subsection we discuss in greater detail the question about symmetric and selfadjoint extensions of symmetric operators.

We have seen in Example 10.2.3 that a closed symmetric operator can fail to be selfadjoint, but at the same time possess selfadjoint extensions. Let us consider an example of a closed symmetric operator that has no selfadjoint extensions.

10.7.1. Example. Let $H = L^2[0, +\infty)$. Set

$$\mathfrak{D}(A_0) = C_0^\infty(0, +\infty), \quad A_0 u = iu'.$$

The operator A_0 is symmetric, its closure A is a closed and symmetric operator, but has no selfadjoint extensions. Indeed, similarly to Example 10.2.3 we can prove that the set $\mathfrak{D}(A_0^*) = \mathfrak{D}(A^*)$ consists of functions $u \in L^2[0, +\infty)$ such that u is absolutely continuous on bounded intervals and $u' \in L^2[0, +\infty)$. In addition, $A_0^* u = iu'$. Note that the continuous version of any function $u \in \mathfrak{D}(A_0^*)$ must have the zero limit at infinity. Indeed,

$$|u(t)|^2 = |u(0)|^2 + \int_0^t [u'(s)\overline{u(s)} + u(s)\overline{u'(s)}]\, ds.$$

The right-hand side has a limit at infinity by the quadratic integrability of u and u'. It is clear that this limit must be zero.

The inclusion $A_0 \subset A_0^*$ yields that $A_0^{**} \subset A_0^*$. Then for all $u \in \mathfrak{D}(A_0^{**})$ and $v \in \mathfrak{D}(A_0^*)$ we have

$$\int_0^\infty iu'(t)\overline{v(t)}\, dt = \int_0^\infty u(t)\overline{iv'(t)}\, dt,$$

which by the integration by parts formula and the fact that u and v tend to zero at infinity gives the relation $u(0)\overline{v(0)} = 0$. Therefore, $u(0) = 0$. Conversely, if $u \in \mathfrak{D}(A_0^*)$ and $u(0) = 0$, then we have $u \in \mathfrak{D}(A_0^{**})$ and $A_0^{**} u = iu$. Thus, $\mathfrak{D}(A_0^{**})$ consists of functions $u \in \mathfrak{D}(A_0^*)$ such that $u(0) = 0$. We observe that $A_0^{**} = A = \overline{A_0}$, since the operator A_0^{**} is closed and its graph contains the closure of the graph of A_0.

Let B be a selfadjoint operator with $A \subset B$. Then $A \subset B = B^* \subset A^*$. However, $\mathfrak{D}(A)$ is of codimension 1 in $\mathfrak{D}(A^*)$, hence the set $\mathfrak{D}(B)$ must coincide with $\mathfrak{D}(A)$ or with $\mathfrak{D}(A^*)$, which is impossible, since A is not selfadjoint.

What is the reason that symmetric operators very similar by appearance are so different with respect to existence of selfadjoint extensions? It turns out that this is due to different defect numbers, which arise in case of bounded and unbounded intervals.

10.7.2. Definition. *Let T be a closed symmetric operator on a Hilbert space H. The dimensions $n_-(T)$ and $n_+(T)$ (possibly, infinite) of the orthogonal complements to the subspaces*

$$\operatorname{Ran}(T + iI) = \operatorname{Ker}(T^* - iI) \quad and \quad \operatorname{Ran}(T - iI) = \operatorname{Ker}(T^* + iI)$$

are called the defect numbers (defect indices) of the operator T.

In the terminology introduced at the end of §10.1 the defect $d_T(\lambda)$ can be written as $n_-(T) = d_T(-i)$, $n_+(T) = d_T(i)$.

Let us introduce the Caley transform of a densely defined symmetric operator T. As in the case of a selfadjoint operator, the operators $T + iI$ and $T - iI$ are injective. However, now their ranges need not be everywhere dense.

10.7.3. Proposition. *Let T be a densely defined symmetric operator on a Hilbert space H. Then the following assertions are true.*

(i) *One has the identity*

$$\|Tx + ix\|^2 = \|Tx\|^2 + \|x\|^2 = \|Tx - ix\|^2, \quad x \in \mathfrak{D}(T).$$

(ii) *The operator T is closed precisely when the subspaces $\operatorname{Ran}(T + iI)$ and $\operatorname{Ran}(T - iI)$ are closed.*

(iii) *There is a linear isometry U between the linear subspaces $\operatorname{Ran}(T + iI)$ and $\operatorname{Ran}(T - iI)$ defined by the formula*

$$U(Tx + ix) = Tx - ix, \quad x \in \mathfrak{D}(T).$$

The mapping U is called the Caley transform of the operator T.

PROOF. Assertion (i) is verified directly.

(ii) Let T be closed. Let $x_n \in \mathfrak{D}(T)$ and $Tx_n + ix_n \to y$. It follows from (i) that $\{x_n\}$ is a Cauchy sequence. Hence $x_n \to x \in H$. Then sequence $\{Tx_n\}$ converges to some vector $z \in X$. Since T is closed, we have $x \in \mathfrak{D}(T)$ and $z = Tx$, whence $y = Tx + ix$. Similarly we obtain that $\operatorname{Ran}(T - iI)$ is closed.

Conversely, suppose that the subspace $\operatorname{Ran}(T+iI)$ is closed. Let $x_n \in \mathfrak{D}(T)$, $x_n \to x$ and $Tx_n \to y$. Then $Tx_n + ix_n \to y + ix$. Since $\operatorname{Ran}(T + iI)$ is closed, we have $u \in \mathfrak{D}(T)$ and $y + ix = Tu + iu$. Hence $T(x_n - u) + i(x_n - u) \to 0$, which by (i) gives convergence $x_n - u \to 0$. Thus, $x = u \in \mathfrak{D}(T)$ and $y = Tx$. Assertion (iii) follows from (i). □

10.7.4. Corollary. *Any densely defined symmetric operator T possesses a closed symmetric extension.*

PROOF. By Proposition 10.1.9 the operator T^* is closed. Since it extends T by the symmetry of T, the operator T is closable. Its closure $\overline{T}$ is symmetric, because for all $x, y \in \mathfrak{D}(\overline{T})$ by definition there exist two sequences of vectors $x_n, y_n \in \mathfrak{D}(T)$ such that $Tx_n \to \overline{T}x$ and $Ty_n \to \overline{T}y$. Hence

$$(\overline{T}x, y) = \lim_{n\to\infty} (Tx_n, y_n) = \lim_{n\to\infty} (x_n, Ty_n) = (x, \overline{T}y).$$

By construction the operator $\overline{T}$ is closed (see Proposition 10.1.9). □

10.7.5. Corollary. *Let T be a closed symmetric operator. For every $\lambda \in \mathbb{C}$ with $\operatorname{Im} \lambda \neq 0$ one has the orthogonal decomposition*

$$H = \operatorname{Ran}(T - \lambda I) \oplus \operatorname{Ker}(T^* - \overline{\lambda}I).$$

PROOF. Since $\operatorname{Ran}(T - \lambda I)$ is closed, we can apply Proposition 10.1.13. □

10.7.6. Theorem. *The Caley transform establishes a one-to-one correspondence between symmetric operators T on H and isometric operators V such that $\mathfrak{D}(V)$ is a linear subspace in H and $\overline{\operatorname{Ran}(V - I)} = H$. Here T is closed if and only if $\mathfrak{D}(V)$ is a closed subspace.*

PROOF. We already know that if the operator T is symmetric, then its Caley transform V maps isometrically the linear subspace $\operatorname{Ran}(T+iI)$ onto the subspace $\operatorname{Ran}(T - iI)$. Moreover, T is closed if and only if this subspace is closed. Let us show that the range of $V - I$ is dense in H. Indeed, let $y \in H$ be an element such that

$$(Vz - z, y) = 0 \quad \forall z = (T + iI)x, \quad x \in \mathfrak{D}(T).$$

Since $V(T + iI)x = (T - iI)x$, this means that $(x, y) = 0$ for all $x \in \mathfrak{D}(T)$, i.e., $y = 0$. Conversely, suppose that the operator V possesses the indicated properties. We observe that the mapping $V - I$ is injective. Indeed, if $Vx - x = 0$ for some $x \in \mathfrak{D}(V)$, then for all $y \in \mathfrak{D}(V)$ we have

$$\big(Vx, (I - V)y\big) = (Vx, y) - (Vx, Vy) = (Vx, y) - (x, y) = 0,$$

which gives the equality $x = 0$, since the range of $V - I$ is dense. Now we set

$$\mathfrak{D}(T) := \operatorname{Ran}(V - I), \quad Ty = -i(V + I)x \quad \text{if } y = (V - I)x.$$

The operator T is densely defined. In addition,

$$(Ty, y) = -i(Vx + x, Vx - x)$$
$$= -i[(Vx, Vx) - (Vx, x) + (x, Vx) - (x, x)] = -2\operatorname{Re}(Vx, x),$$

which shows that the form (Tx, x) is real. Thus, the operator T is symmetric. Since

$$Ty + iy = -iVx - ix + iVx - ix = -2ix,$$

we have $\operatorname{Ran}(T + iI) = \mathfrak{D}(V)$. If this subspace is closed, then the operator T is closed. Finally, the Caley transform U of the operator T coincides with V, since $\operatorname{Ran}(T + iI) = \mathfrak{D}(V)$, i.e., $\mathfrak{D}(U) = \mathfrak{D}(V)$, moreover,

$$U(Ty + iy) = Ty - iy = -2iVx, \quad U(Ty + iy) = U(-2ix) = -2iUx$$

for all $x \in \mathfrak{D}(V)$.   □

The Caley transform reduces the problem of extending symmetric operators to the problem of extending linear isometries, which is easily solved.

10.7.7. Proposition. *If a symmetric operator $\widetilde{T}$ is an extension of the symmetric operator T, then the Caley transform of the operator $\widetilde{T}$ is an extension of the Caley transform of the operator T.*

Conversely, if T is a symmetric operator with the Caley transform V and $\widetilde{V}$ is an isometry operator extending V, then $\widetilde{V}$ is the Caley transform of some symmetric operator $\widetilde{T}$ that extends T.

PROOF. The inclusion $\mathrm{Ran}\,(T+iI) \subset \mathrm{Ran}\,(\widetilde{T}+iI)$ yields the first assertion. The second assertion follows from the previous theorem, which gives a symmetric operator $\widetilde{T}$ with the Caley transform $\widetilde{V}$. Moreover,

$$\mathfrak{D}(T) = \mathrm{Ran}\,(V-I) \subset \mathrm{Ran}\,(\widetilde{V}-I) = \mathfrak{D}(\widetilde{T})$$

and on $\mathfrak{D}(T)$ the operators T and $\widetilde{T}$ coincide, which is clear from the formula defining T by V (see the previous proof). $\qquad\square$

10.7.8. Lemma. *Let H_1 and H_2 be closed linear subspaces in a Hilbert space $H \neq 0$ and let $U\colon H_1 \to H_2$ be a linear isometry with $U(H_1) = H_2$. The mapping U can be extended to an isometry on all of H with values in H precisely when $\dim H_1^\perp \leqslant \dim H_2^\perp$. In the case of the equality of these two dimensions U extends to a unitary operator.*

PROOF. If the indicated condition is fulfilled, we can take a linear isometric embedding $V\colon H_1^\perp \to H_2^\perp$ (which in case of equal dimensions of these subspaces can be made surjective) and set

$$\widetilde{U}(x+u) = Ux + Vu, \quad x \in H_1, u \in H_1^\perp.$$

Conversely, if U has an isometric extension $\widetilde{U}$ to all of H, then we obtain $\widetilde{U}(H_1^\perp) \subset H_2^\perp$, since $\widetilde{U}$ preserves the orthogonality. $\qquad\square$

10.7.9. Theorem. *Let T be a closed symmetric operator and let $n_-(T)$, $n_+(T)$ be its defect numbers. Then*

(i) *the operator T is selfadjoint if and only if $n_-(T) = n_+(T) = 0$;*

(ii) *the existence of selfadjoint extensions of T is equivalent to the equality $n_-(T) = n_+(T)$;*

(iii) *the absence of proper symmetric extensions of T is equivalent to the equality to zero of at least one of its defect numbers;*

(iv) *if $n_-(T) = n_+(T) = \infty$, then T has symmetric extensions with an a priori given pair of defect numbers.*

PROOF. All assertions of this theorem follow from the established correspondence between symmetric operators T and their Caley transforms V taking into account the fact that the defect numbers $n_-(T)$ and $n_+(T)$ coincide with the dimensions $n_e(V)$ and $n_i(V)$ of the orthogonal complements to the subspaces $H_0 = \mathfrak{D}(V)$ and $H_1 = \mathrm{Ran}\,V$. If $n_e(V) \neq n_i(V)$, then V cannot be extended to a unitary operator. If $n_e(V) = n_i(V)$, then this is possible. If $n_e(V) = n_i(V) = \infty$, then for any pair of numbers n, m from $\{0, 1, \ldots, \infty\}$ we can take in $H_0^\perp$ and $H_1^\perp$ closed infinite-dimensional subspaces E_0 and E_1 having in $H_0^\perp$ and $H_1^\perp$ the codimensions n and m, respectively. Now we can extend the operator V with the aid of an isometry mapping from E_1 onto E_2. $\qquad\square$

10.7.10. Corollary. *All symmetric extensions of a closed symmetric operator T with defect numbers $n_+ = n_- = 1$ are selfadjoint.*

These extensions are described by a complex parameter θ with $|\theta| = 1$ in the following way: a unit vector u of the one-dimensional space $\mathfrak{D}(V)^{\perp}$ is mapped by the isometric extension V to the vector θv, where v is a unit vector of the one-dimensional space $(\operatorname{Ran} V)^{\perp}$.

The established facts yield the following criterion for the closure of a symmetric operator A to be a selfadjoint operator. Such an operator A is called *essentially selfadjoint*.

10.7.11. Corollary. *A densely defined symmetric operator A is essentially selfadjoint, i.e., its closure is selfadjoint, precisely when the equations $A^*u = iu$ and $A^*u = -iu$ have only zero solutions in $\mathfrak{D}(A^*)$.*

Let us give a more precise description of symmetric extensions.

10.7.12. Theorem. *Let T be a closed symmetric operator.*
(i) *For every λ with $\operatorname{Im} \lambda \neq 0$ one has the equality*

$$\mathfrak{D}(T^*) = \mathfrak{D}(T) + \operatorname{Ker}(T^* - \lambda I) + \operatorname{Ker}(T^* - \overline{\lambda} I),$$

where we have a direct algebraic sum.
(ii) *Let $D_0 \subset \operatorname{Ker}(T^* - \lambda I)$ and $R_0 \subset \operatorname{Ker}(T^* - \overline{\lambda} I)$ be closed linear subspaces of the same dimension n_0 (possibly, both infinite-dimensional), and let V_0 be some isometry from D_0 onto R_0. Then the formula*

$$\mathfrak{D}(\widetilde{T}) := \mathfrak{D}(T) + (V_0 - I)(D_0) \tag{10.7.1}$$

defines the domain of definition of some closed symmetric extension $\widetilde{T}$ of the operator T, where $\widetilde{T}(x + V_0 x_0 - x_0) = Tx - ix_0 - iV_0 x_0$ for all $x \in \mathfrak{D}(T)$, $x_0 \in D_0$. Conversely, every closed symmetric extension of T has the indicated form.

PROOF. (i) By the symmetry of the operator T the right-hand side of the equality to be proved is contained in the left-hand side. Let $y \in \mathfrak{D}(T^*)$. Since

$$H = \operatorname{Ran}(T - \lambda I) \oplus \operatorname{Ker}(T^* - \overline{\lambda} I)$$

according to Corollary 10.7.5, we obtain the representation

$$T^*y - \lambda y = (T - \lambda I)x + (\overline{\lambda} - \lambda)u, \quad \text{where } x \in \mathfrak{D}(T), \ u \in \operatorname{Ker}(T^* - \overline{\lambda} I).$$

On account of the equality $T^*x = Tx$ this gives $(T^* - \lambda I)(y - x - u) = 0$, i.e., $y - x - u \in \operatorname{Ker}(T^* - \lambda I)$. Let us verify that the sum of the three indicated spaces is direct. Let $x + u + v = 0$, where $u \in \mathfrak{D}(T)$, $(T^* - \lambda I)u = 0$ and $(T^* - \overline{\lambda} I)v = 0$. Applying $T^* - \lambda I$ and taking into account that $T^*x = Tx$, we obtain $(T^* - \lambda I)x + (\overline{\lambda} - \lambda)v = 0$. Using the orthogonality of $\operatorname{Ran}(T - \lambda I)$ and $\operatorname{Ker}(T^* - \overline{\lambda} I)$ once again, we obtain $x = 0$ and $v = 0$.

(ii) Assertion (i) yields that $(V_0 - I)(D_0) \cap \mathfrak{D}(T) = \{0\}$, since

$$V_0(D_0) \subset R_0 \subset \operatorname{Ker}(T^* - \overline{\lambda} I), \quad quad D_0 \subset \operatorname{Ker}(T^* - \lambda I).$$

Without loss of generality we can assume that $\lambda = i$. Let V be the Caley transform of T; H is the orthogonal sum of the closed subspaces $\operatorname{Ran}(T + iI) = \mathfrak{D}(V)$ and $\operatorname{Ker}(T^* - iI)$. Hence D_0 is contained in the orthogonal complement of $\mathfrak{D}(V)$.

Similarly, R_0 belongs to the orthogonal complement of $\mathrm{Ran}\,(T - iI) = \mathrm{Ran}\,V$. Thus, by means of V_0 we obtain an isometric extension $\widetilde{V}$ of the operator V. It has been shown that $\mathfrak{D}(T) = \mathrm{Ran}\,(V - I)$. Hence $\widetilde{V}$ corresponds to a symmetric extension $\widetilde{T}$ of the operator T satisfying (10.7.1). Moreover, $\widetilde{T} \subset \widetilde{T}^* \subset T^*$, hence $\widetilde{T}x = ix$ and $\widetilde{T}V_0 x = -iV_0 x$ for all $x \in D_0$. Since the symmetric extensions of T are obtained in the indicated way from the isometric extensions of V, we have described all closed symmetric extensions of T. $\qquad\square$

Considering extensions it is useful to include defect numbers in the parametric family $d_T(\lambda)$ (see the end of §10.1). As we know from Proposition 10.1.15, the function d_T is locally constant on the set of regular points. When applied to symmetric operators this leads to the following conclusions.

10.7.13. Proposition. *Let T be a closed symmetric operator. If $\mathrm{Im}\,\lambda \neq 0$, then the operator $T - \lambda I$ has a bounded inverse on the domain $\mathrm{Ran}\,(T - \lambda I)$, i.e., λ is a regular point. In addition, the number $d_T(\lambda)$ is constant in every half-plane $\mathrm{Im}\,\lambda > 0$ and $\mathrm{Im}\,\lambda < 0$.*

PROOF. Set $\lambda = \alpha + i\beta$. Since (Tx, x) is real for all $x \in \mathfrak{D}(T)$, we obtain

$$\|(T - \lambda I)x\|^2 = \|(T - \alpha I)x\|^2 + \beta^2\|x\|^2 + 2\mathrm{Re}\,((A - \alpha I)x, i\beta x)$$
$$= \|(T - \alpha I)x\|^2 + \beta\|x\|^2 \geqslant \beta^2\|x\|^2,$$

which proves the first assertion. The second assertion is obvious from Proposition 10.1.15. $\qquad\square$

10.7.14. Corollary. *If a closed symmetric operator T has a real regular point, then its defect numbers are equal. Hence T has selfadjoint extensions.*

10.7.15. Corollary. *Let T be a closed symmetric operator such that for some $m \in \mathbb{R}^1$ we have*

$$(Tx, x) \geqslant m(x, x) \quad \forall x \in \mathfrak{D}(T).$$

Then the ray $(-\infty, m)$ is contained in the set of regular points and the defect numbers of T are equal. Hence T has selfadjoint extensions.

In the next section we discuss in greater detail the situation of this corollary and see that there is an extension with the same bound.

10.7(ii). Semibounded forms and operators

In this subsection we study symmetric operators the quadratic forms of which are estimated from below by $\mathrm{const}\|x\|^2$. Such operators arise frequently in applications (of particular importance are operators with nonnegative forms). They possess certain special selfadjoint extensions bounded from below: the so-called Friedrichs extensions constructed through closures of forms.

Suppose that in a complex Hilbert space H we are given a dense linear subspace $\mathfrak{D}(Q)$ and a function

$$Q \colon \mathfrak{D}(Q) \times \mathfrak{D}(Q) \to \mathbb{C}$$

which is linear in the first argument and conjugate-linear in the second argument. It generates the quadratic form $x \mapsto Q(x, x)$ on $\mathfrak{D}(Q)$. This form is called *semibounded from below* if there exists a real number m such that

$$Q(x, x) \geqslant m(x, x), \quad x \in \mathfrak{D}(Q).$$

If m can be taken nonnegative, then the form is called *nonnegative*, and if m can be taken strictly positive, then the form is called *positive definite* or *positive*. The largest possible m is called the *precise lower bound of the form Q*.

We shall consider forms generated by symmetric operators by the formula

$$Q(x, x) := (Ax, x), \quad x \in \mathfrak{D}(A).$$

The terminology introduced above extends from forms to operators. In particular, a symmetric operator is called *semibounded from below* if so is its quadratic form; if the form is nonnegative or positive, then the operator is called nonnegative (respectively, positive). It is quite often in applications that the primary object is the quadratic form, which is used to construct the generating operator; next, its domains of definition and selfadjointness is studied. Here is a typical example: given a nonnegative Borel measure μ on $\mathbb{R}^n$, the so-called Dirichlet form on the space $L^2(\mu)$ is introduced by the formula

$$\mathcal{E}(\varphi, \varphi) = \int_{\mathbb{R}^n} \big(\nabla\varphi(x), \nabla\varphi(x)\big)\, \mu(dx), \quad \varphi \in C_0^\infty(\mathbb{R}^n).$$

The question arises: can we define this form by a selfadjoint operator? Say, if μ is Lebesgue measure, then the form $\mathcal{E}$ is generated by the Laplace operator Δ. Already in this example it is clear that the form generated by an operator can have a natural domain of definition larger than the domain of definition of the generating operator. For example, the natural domain of definition of the Laplace operator Δ on $L^2(\mathbb{R}^n)$ is the Sobolev class $W^{2,2}(\mathbb{R}^n)$, but the generated gradient Dirichlet form is naturally defined on the larger Sobolev class $W^{2,1}(\mathbb{R}^n)$. We shall see below that this phenomenon is very typical.

Suppose first that the form Q is positive-definite. Then $\mathfrak{D}(Q)$ can be equipped with the inner product

$$(x, y)_Q := Q(x, y), \quad x, y \in \mathfrak{D}(Q).$$

If $\mathfrak{D}(Q)$ is Hilbert with respect to $(\,\cdot\,, \cdot\,)_Q$, then the form Q is called *closed*.

10.7.16. Theorem. (i) *Let A be a positive selfadjoint operator. Then it generates a closed form*

$$Q_A(x, x) := (\sqrt{A}x, \sqrt{A}x), \quad \mathfrak{D}(Q_A) := \mathfrak{D}(\sqrt{A}).$$

Moreover, Q_A is the unique closed positive form Q satisfying the conditions

$$\mathfrak{D}(A) \subset \mathfrak{D}(Q), \quad (Ax, y) = Q(x, y) \text{ for all } x \in \mathfrak{D}(A), y \in \mathfrak{D}(Q). \quad (10.7.2)$$

(ii) *Conversely, every densely defined closed positive form Q can be obtained from some positive selfadjoint operator in this way and the corresponding operator is unique.*

PROOF. (i) Clearly, the form Q_A is closed, since the selfadjoint operator $\sqrt{A}$ is closed. Next, suppose that a positive form Q satisfies conditions (10.7.2). Then $Q_A(x,y) = Q(x,y)$ for all $x, y \in \mathfrak{D}(A)$. For the proof of the equality of the forms Q_A and Q we shall verify that $\mathfrak{D}(A)$ is dense in the Hilbert spaces $\mathfrak{D}(Q_A)$ and $\mathfrak{D}(Q)$ with their norms $\| \cdot \|_{Q_A}$ and $\| \cdot \|_Q$. If $y \in \mathfrak{D}(Q)$ and $Q(x,y) = 0$ for all $x \in \mathfrak{D}(A)$, then by (10.7.2) the vector y is orthogonal in H to the set $\mathrm{Ran}\, A$, which is dense, since A is positive. Hence $y = 0$. This proves that $\mathfrak{D}(A)$ is dense in $\mathfrak{D}(Q)$. The same reasoning applies to $\mathfrak{D}(Q_A)$.

(ii) For any $u \in H$, let us consider the functional $l_u(y) := (u,y)$ on the Hilbert space $\mathfrak{D}(Q)$ equipped with the inner product $(\,\cdot\,,\,\cdot\,)_Q$. Then

$$|l_u(y)| \leqslant \|u\| \|y\| \leqslant m^{-1/2}\|u\| \|y\|_Q.$$

By the Riesz representation theorem there exists a unique element $v \in \mathfrak{D}(Q)$ such that $l_u(y) = Q(v,y)$ for all $y \in \mathfrak{D}(Q)$. It is clear that the element v depends linearly on u. Set $Bu := v$. Then

$$Q(Bu, y) = (u,y), \quad y \in \mathfrak{D}(Q).$$

Since $\|Bu\|_Q^2 = (u, Bu) \leqslant \|u\| \|Bu\| \leqslant m^{-1/2}\|u\| \|Bu\|_Q$, we obtain the bound $\|Bu\|_Q \leqslant m^{-1/2}\|u\|$, i.e., B is a continuous operator from H to $\mathfrak{D}(Q)$ with the norm $\| \cdot \|_Q$. Hence B is continuous from H to H. Since for $y = Bu$ we have $(u, Bu) = Q(Bu, Bu)$, the operator B is selfadjoint. This operator is injective. Indeed, if $Bu = 0$, then $(y, u) = 0$ for all $y \in \mathfrak{D}(Q)$, whence $u = 0$. Hence the operator $A := B^{-1}$ on the domain $\mathfrak{D}(A) := \mathrm{Ran}\, B \subset \mathfrak{D}(Q)$ is selfadjoint. We have obtained the required operator. Indeed, for $u = Ax$ with $x \in \mathfrak{D}(A)$ we have

$$(Ax, y) = Q(x, y), \quad x \in \mathfrak{D}(A),\, y \in \mathfrak{D}(Q).$$

According to (i) the form Q_A coincides with Q. Let us verify that A is unique. If A_1 is one more selfadjoint operator generating the form Q, then

$$\mathfrak{D}(Q) = \mathfrak{D}(\sqrt{A}) = \mathfrak{D}(\sqrt{A_1}).$$

Hence for every $y \in \mathfrak{D}(A_1)$ we have

$$(x, A_1 y) = Q(x,y) = (\sqrt{A}x, \sqrt{A}y), \quad x \in \mathfrak{D}(Q).$$

This equality means that

$$\sqrt{A}y \in \mathfrak{D}\big((\sqrt{A})^*\big) = \mathfrak{D}(\sqrt{A}) \quad \text{and} \quad (\sqrt{A})^*\sqrt{A}y = A_1 y,$$

i.e., $y \in \mathfrak{D}(A)$ and $Ay = A_1 y$. Thus, $A_1 \subset A$. Similarly, $A \subset A_1$. $\qquad\square$

Suppose now that T is a symmetric densely defined operator semibounded from below. By Corollary 10.7.4 it has the closure that is a closed symmetric operator. This closure is also semibounded from below, because for every $x \in \mathfrak{D}(\overline{T})$ there is a sequence $\{x_n\} \subset \mathfrak{D}(T)$ such that $x_n \to x$ and $Tx_n \to \overline{T}x$. Hence the estimate $(Tx_n, x_n) \geqslant m(x_n, x_n)$ yields the estimate $(\overline{T}x, x) \geqslant m(x, x)$. Hence the operator T has selfadjoint extensions. The connection with quadratic forms described above enables us to construct a selfadjoint extension semibounded from below. To this end we shall deal with the closure of T and replace T by the positive operator $T_0 := T + \lambda I$ for some $\lambda > m$. Let us consider the positive definite form $Q(x) = (T_0 x, x)$.

10.7.17. Lemma. *Let Q be a positive definite quadratic form on a dense linear subspace $\mathfrak{D}(Q)$ in H. Suppose that the following condition is fulfilled:*

(C) *if a sequence $\{x_n\} \subset \mathfrak{D}(Q)$ is Cauchy with respect to the norm $\|\cdot\|_Q$ and $\|x_n\| \to 0$, then $Q(x_n, y) \to 0$ for all $y \in \mathfrak{D}(Q)$.*

Then the form Q is closable in the following sense: there is a closed quadratic form $\overline{Q}$ with a domain of definition $\mathfrak{D}(\overline{Q})$ in H such that $\mathfrak{D}(\overline{Q})$ is the completion of $\mathfrak{D}(Q)$ with respect to the norm $\|\cdot\|_Q$ and $\overline{Q}|_{\mathfrak{D}(Q)} = Q$.

PROOF. Let $\{x_n\}$ be a sequence in $\mathfrak{D}(Q)$ fundamental with respect to the norm $\|\cdot\|_Q$. Then it is fundamental in H and hence converges in H to some element $x \in H$. Let $\mathfrak{D}(\overline{Q})$ consist of all elements in H obtained in this way. Set $\overline{Q}(x, x) = \lim\limits_{n\to\infty} Q(x_n, x_n)$. We show that $\mathfrak{D}(\overline{Q})$ can be identified with the completion of $\mathfrak{D}(Q)$. For this we have to verify the following: if two Cauchy sequences $\{x_n\}$ and $\{y_n\}$ in $\mathfrak{D}(Q)$ converge in the completion to different elements, then their limits in H are also different. If they have a common limit x in H, then we obtain the sequence $z_n := x_n - y_n$ that is Cauchy with respect to the norm $\|\cdot\|_Q$ and converges to zero in H. By condition (C) we have $Q(z_n, y) \to 0$ for all $y \in \mathfrak{D}(Q)$. Since $\mathfrak{D}(Q)$ is dense in the completion, this means that $\{z_n\}$ converges in the completion to zero, contrary to our supposition that $\{x_n\}$ and $\{y_n\}$ converge to different elements of the completion. It is easy to see that we have obtained the desired closure. $\square$

Thus, the form $Q(x) = (Tx, x)$ has the closure, moreover, the closure is also positive definite. According to the established facts there exists a positive selfadjoint operator T_0 generating the closure of our form. This operator is called the *Friedrichs extension* of the operator $T + \lambda I$, and the operator $T_0 - \lambda I$ is called the Friedrichs extension of the operator T. It is not always the closure of T.

10.7.18. Example. Let the operator $A = -d^2/dt^2$ be defined on the domain $\mathfrak{D}(A) = C_0^\infty(0, 1)$. Then

$$(A\varphi, \varphi) = \int_0^1 |\varphi'(t)|^2 \, dt \geqslant 0.$$

The completion of $\mathfrak{D}(A)$ with respect to the norm $\big((A\varphi, \varphi) + (\varphi, \varphi)\big)^{1/2}$ coincides with the Sobolev space $W_0^{2,1}[0, 1]$ of all absolutely continuous functions φ such that $\varphi(0) = \varphi(1) = 0$ and $\varphi' \in L^2[0, 1]$. The Friedrichs extension of the operator A is the nonnegative selfadjoint operator $\widetilde{A} = -d^2/dt^2$ on the domain $\mathfrak{D}_0 = \{\varphi \in W^{2,2}[0, 1]\colon \varphi(0) = \varphi(1) = 0\}$. However, A has other nonnegative selfadjoint extensions: for example, $A_1 = -d^2/dt^2$ on the domain $\mathfrak{D}_1 = \{\varphi \in W^{2,2}[0, 1]\colon \varphi'(0) = \varphi'(1) = 0\}$. In addition, there are selfadjoint extensions of the operator A that are not nonnegative: for example, $A_2 = -d^2/dt^2$ on the domain $\mathfrak{D}_1 = \{\varphi \in W^{2,2}[0, 1]\colon \varphi'(0) = \varphi(0), \varphi'(1) = \varphi(1)\}$. Note that the operator A is not essentially selfadjoint (its closure is defined on functions $\varphi \in W^{2,2}[0, 1]$ with $\varphi(0) = \varphi(1) = \varphi'(0) = \varphi'(1) = 0$), which yields different selfadjoint extensions bounded from below (Exercise 10.7.33).

The proof of the next assertion can be found in [67, Chapter 10, §3].

10.7.19. Theorem. *Suppose that A is a densely defined symmetric operator bounded from below. Let $\widetilde{A}$ be its Friedrichs extension and let $\mathfrak{D}_Q$ be the domain of definition of the closure of the form generated by the operator A.*

(i) *Let A_1 be a selfadjoint extension of A such that $\mathfrak{D}(A_1) \subset \mathfrak{D}_Q$. Then we have $A_1 = \widetilde{A}$.*

(ii) *Let A_1 be a bounded from below selfadjoint extension of A and Q_1 the corresponding form. Then $\mathfrak{D}_Q \subset \mathfrak{D}_{Q_1}$ and $Q_1 = Q$ on $\mathfrak{D}_Q$.*

The previous example shows that the Friedrichs extension need not be a unique selfadjoint extension bounded from below. A description of all extensions bounded from below can be found in [**6**, §109].

10.7(iii). The Chernoff and Trotter theorems

Here we prove two interesting results connected with approximation of operator semigroups. The first result — Chernoff's theorem — was actually found later than Trotter's theorem, which we obtain below as a corollary. First we establish an auxiliary result.

10.7.20. Lemma. *Let L and L_n for every $n \in \mathbb{N}$ be the generators of strongly continuous contracting semigroups on a complex Banach space X and let $D \subset \mathfrak{D}(L) \bigcap (\bigcap_{n \geqslant 1} \mathfrak{D}(L_n))$ be a linear subspace such that $(\lambda I - L)(D)$ is dense in X. Suppose that $L_n x \to L x$ as $n \to \infty$ for every $x \in D$. Then, for every element $x \in X$, whenever $\operatorname{Re} \lambda > 0$ we have $(\lambda I - L_n)^{-1} x \to (\lambda I - L)^{-1} x$.*

PROOF. Let $\psi = (\lambda I - L)\varphi$, $\varphi \in D$. We recall that $\|(\lambda I - L_n)^{-1}\| \leqslant |\lambda|^{-1}$. Hence

$$\|(\lambda I - L_n)^{-1}\psi - (\lambda I - L)^{-1}\psi\|$$
$$= \|(\lambda I - L_n)^{-1}(\lambda I - L_n)\varphi - (\lambda I - L_n)^{-1}(L_n - L)\varphi - \varphi\|$$
$$= \|(\lambda I - L_n)^{-1}(L_n - L)\varphi\| \leqslant |\lambda|^{-1}\|(L_n - L)\varphi\|.$$

Thus, the lemma is true for all ψ from a dense set. On account of the bounds $\|(\lambda I - L_n)^{-1}\| \leqslant 1$ and $\|(\lambda I - L)^{-1}\| \leqslant 1$ this gives the desired assertion. $\square$

10.7.21. Theorem. *Let X be a Banach space and let $F \colon [0, +\infty) \to \mathcal{L}(X)$ be a mapping continuous on every vector and satisfying the following conditions: $F(0) = I$, $\|F(t)\| \leqslant e^{at}$ with some constant a and there is a dense linear subspace $D \subset X$ such that for all $x \in D$ there exists a limit $F'(0)x := \lim_{t \to 0} t^{-1}[F(t)x - x]$. Suppose that $F'(0)$ on D has the closure C that is the generator of a strongly continuous semigroup $\{T_t\}_{t \geqslant 0}$. Then, for every $x \in X$, as $n \to \infty$ we have $F(t/n)^n x \to T_t x$ uniformly in t from every compact interval.*

PROOF. It suffices to consider the case where $a = 0$ and $\{T_t\}_{t \geqslant 0}$ is a contracting semigroup (multiplying $F(t)$ by e^{-tC} with a sufficiently large $C > 0$). Let us note the following fact: if $T \in \mathcal{L}(X)$ and $\|T\| \leqslant 1$, then the operators $\exp(t(T - I))$ form a contracting semigroup and

$$\left\| \left[\exp(n(T - I)) - T^n\right]x \right\| \leqslant n^{1/2}\|(T - I)x\|.$$

Indeed, $\left\|\exp\big(t(T-I)\big)\right\| \leqslant e^{-t}\sum_{k=0}^{\infty}\frac{t^k}{k!}\|T\|^k \leqslant 1$. In addition,

$$\left\|\big[\exp\big(n(T-I)\big)-T^n\big]x\right\| = \left\|e^{-n}\sum_{k=0}^{\infty}\frac{n^k}{k!}(T^k-T^n)x\right\|$$

$$\leqslant e^{-n}\sum_{k=0}^{\infty}\frac{n^k}{k!}\|(T^k-T^n)x\|$$

$$\leqslant e^{-n}\sum_{k=0}^{\infty}\frac{n^k}{k!}\|(T^{|k-n|}-I)x\| \leqslant e^{-n}\sum_{k=0}^{\infty}\frac{|k-n|n^k}{k!}\|(T-I)x\|,$$

which is estimated by $n^{1/2}\|(T-I)x\|$, since

$$\sum_{k=0}^{\infty}|k-n|n^k/k! \leqslant \left(\sum_{k=0}^{\infty}|k-n|^2 n^k/k!\right)^{1/2}\left(\sum_{k=0}^{\infty}n^k/k!\right)^{1/2} \leqslant (ne^n)^{1/2}e^{n/2}$$

by the easily verified equality $\sum_{k=0}^{\infty}|k-n|^2 n^k/k! = ne^n$. Let us fix $\tau > 0$ and set $T := F(\tau/n)$, $L_n := n\tau^{-1}\big(F(\tau/n)-I\big)$. According to the assertion proved above, the operator $T-I$ generates a contracting semigroup $\exp[t(T-I)]$. Hence the operator L_n also generates a contracting semigroup $\exp(tL_n)$. By our conditions and Theorem 10.5.8 one has the pointwise convergence of the sequence of operators $\exp\big[n\big(F(\tau/n)-I\big)\big] = \exp(\tau L_n)$ to T_τ. Let $x \in D$. Applying the aforementioned assertion once again, as $n \to \infty$ we obtain

$$\|\exp(\tau L_n) - F(\tau/n)^n x\| \leqslant n^{1/2}\|(F(\tau/n)-I)x\|$$

$$= \frac{\tau}{n^{1/2}}\|n\tau^{-1}\big(F(\tau/n)-I\big)x\| \to 0.$$

Thus, for any $x \in D$ we have $F(\tau/n)^n x \to T_\tau x$. Then the same remains in force for all $x \in X$, since D is dense and $\|F(\tau/n)\| \leqslant 1$ in the considered case. A closer look at the proof shows that convergence is uniform in τ from every compact interval. $\qquad\square$

10.7.22. Corollary. *Let A and B be the generators of two strongly continuous contracting semigroups $\exp(tA)$ and $\exp(tB)$ on X. Suppose that $\mathfrak{D}(A)\cap\mathfrak{D}(B)$ is dense and the operator $A+B$ on this domain possesses the closure C that is the generator of a strongly continuous contracting semigroup $\exp(tC)$. Then $\exp(tC)x = \lim\limits_{n\to\infty}\big(\exp(tA/n)\exp(tB/n)\big)^n x$ for every $x \in X$ uniformly in t from every compact interval.*

PROOF. It suffices to set $F(t) = \exp(tA)\exp(tB)$. Then

$$\frac{F(t)-I}{t}x = t^{-1}[\exp(tA)(\exp(tB)x - x) + (\exp(tA)x - x)] \to Bx + Ax$$

for all $x \in \mathfrak{D}(A)\cap\mathfrak{D}(B)$, hence we can apply the theorem. $\qquad\square$

10.7.23. Example. Let $\{T_t\}_{t\geqslant 0}$ be a strongly continuous contracting semi-group with generator L on a Banach space X. Then

$$\lim_{n\to\infty}(I - tn^{-1}L)^{-n}x = T_t x, \quad x \in X.$$

Indeed, let us set

$$F(t) := (I - tL)^{-1} = \int_0^\infty e^{-s} T_{st} \, ds$$

and observe that $F'(0) = L$ on $\mathfrak{D}(L)$.

Note that in Chernoff's theorem the operator $F'(0)$ is automatically closable if it is densely defined. To see this it suffices to consider the case $a = 0$ in which the operator $F'(0)$ is obviously dissipative (since $l(F(t)x) \leqslant 1$ if $\|l\| = \|x\| = l$), hence is closable, see Exercise 10.7.42. Hence the main condition is that the set $(I - F'(0))(D)$ is dense. It does not follow from the dissipativity: it suffices to take a nonpositive symmetric operator that is not essentially selfadjoint.

10.7(iv). The mathematical model of quantum mechanics

The mathematical model of quantum mechanics used at present is described by a set of axioms formulated in the language of the theory of operators on a Hilbert space and introduced in the book by von Neumann [455] published in German in 1932. Actually, this model of von Neumann is a formalization of the model introduced in Dirac's book [153], the first edition of which was published in 1930. It is worth noting that one year after the publication of von Neumann's book, Kolmogorov's monograph "Foundations of probability theory" was published (also first in German). These two books by von Neumann and Kolmogorov give together a solution to the 6th Hilbert problem (one of the 23 problems posed in his lecture at the 2nd International Congress of Mathematicians in Paris in 1900) concerned with an axiomatic construction of probability theory and mechanics. The principal mathematical objects used for constructing the described model are complex separable Hilbert spaces and selfadjoint operators on such spaces. The principal physical objects are a quantum system (the investigated physical object, for example, an elementary particle, a family of such particles or a quantum computer), its observables (physical quantities that can be measured) and states. States can be pure and mixed. A state of the quantum system is called pure if it is not a probability mixture of other states. Every mixed state coincides with a probability mixture of pure states.

We now list the axioms with certain comments.

Axioms. 1. To every quantum system we associate a complex separable Hilbert space H, called its state space. To observables we associate selfadjoint operators on H (which in the framework of the considered model are often called observables). To pure states we associate orthogonal projections onto one-dimensional subspaces in H. Since the projection operator can be identified with its range, we can say that pure states are defined by one-dimensional subspaces in H (i.e., by elements of the projective Hilbert spaces). In addition, pure states can be represented by vectors generating the corresponding one-dimensional subspaces (they can be taken normalized). These vectors are called state vectors of the quantum system; for this reason H is called the state space. If every vector from the space H defines some (pure) state, then we say that the quantum system has no superselection rules (superselection rules are discussed in Comment 8).

Axioms. 2. A time evolution of a state of the quantum system is described by a one-parameter strongly continuous group of unitary operators $U(t)$. This means that if $\psi(t_1)$ is a state vector of the quantum system at time t_1, then its state vector at time t_2 is $\psi(t_2) = U(t_2 - t_1)\psi(t_1)$. According to Stone's theorem, we have $U(t) = \exp\left(-i\frac{t}{\hbar}\widehat{H}\right)$, where $\hbar$ is the Planck constant and $\widehat{H}$ is a selfadjoint operator, called the Hamiltonian of the system; it corresponds to (= is) an observable, called the energy. It is usually assumed that the operator $\widehat{H}$ is nonnegative definite or semibounded (from below), because it is believed that the energy of a physical system cannot decrease unboundedly.

Comments. 1. We have the equality $i\hbar\psi'(t) = \widehat{H}\psi(t)$, called the Schrödinger equation.

Axioms. 3. If $A_1, A_2, \ldots, A_n$ are pairwise commuting observables, then there exists an experiment that enables us to measure these observables simultaneously for any state of the system. Moreover, if ψ is the vector defining the state of the quantum system and $\Pi_j, j = 1, 2, \ldots, n$, are the projection-valued measures for A_j (see §7.9, §7.10(ii)), then the probability that the measured values of the observables A_j are contained in Borel sets $B_j \subset \mathbb{R}$ with $j = 1, \ldots, n$, equals the quantity $\|\Pi_1(B_1) \cdots \Pi_n(B_n)\psi\|^2 / \|\psi\|^2$.

Comments. 2. The expectation of the result of the measurement of the observable A in the pure state given by a vector ψ equals $(A\psi, \psi)/\|\psi\|^2$ (if the right-hand side is defined). Axiom 3 in turn follows (under certain assumptions) from this assertion.

Axioms. 4. An experiment of measurement of an observable A with a discrete spectrum can be performed in such a way that if the original state was pure and represented by a vector φ and the result of measurement gave a number λ, then immediately after measurement the system will be in a pure state that is the projection of the vector φ onto the subspace corresponding to the eigenvalue λ. Passage from the vector φ to its projection is called reduction of a state vector.

Comments. 3. This axiom is a reinforcement of the original axiom of von Neumann (according to which the dimension of the corresponding subspace equals one), called sometimes the Lüders postulate; it is still arguable in the literature whether this postulate can be accepted. Note that in the well-known textbook by Landau and Lifshits the original axiom of von Neumann is also rejected.

Comments. 4. In order to extend the axiom about reduction of state vectors to the case of observables with continuous spectra it is necessary to enlarge the considered model by using the theory of distributions. Actually, this was already done by Dirac himself (before von Neumann's axiomatic), but at a heuristic level.

Comments. 5. There are two types of evolution of quantum systems: the one described by Axiom 2 (unitary or Hamiltonian) and the one described by Axiom 4 concerning reduction of state vectors. In the first case the future of the state of the system is uniquely determined by its initial state; in the second case the future of the state depends on the initial state only in a probabilistic way. The question in which degree the second type of evolution can be reduced to the first type is still under active discussion. Evolution of the first type takes place in an isolated system (such systems are called closed); evolution of the second type is the result of interaction of the system with another system — the measurement device.

Axioms. 5. If a quantum system with a Hilbert space H consists of two subsystems with Hilbert spaces H_1 and H_2, respectively, then it is assumed that H can be identified with the Hilbert tensor product of the spaces H_1 and H_2 (see §7.10(vi); here the Hilbert tensor product will be denoted by $H_1 \otimes H_2$) or with one of its subspaces: the closure of the subspace of symmetric tensors or the closure of the space of antisymmetric tensors (the first option is used for describing particles called bosons and the second one for describing particles called fermions). It is supposed that if the quantum systems H_1 and H_2 are in pure states represented by vectors x_1 and x_2, respectively, then the state of the joint system is also pure and represented by the vector $x_1 \otimes x_2$. In addition, if A_1 and A_2 are observables of the corresponding subsystems, then their measurement is the same as the measurement of the observables $A_1 \otimes I_2$ and $I_1 \otimes A_2$ in the joint system; here and below I_1, I_2 are the identity mappings of the corresponding spaces. Quantum systems with Hilbert spaces H_1 and H_2 regarded as subsystems of their union are called open.

Comments. 6. The following theorem holds. *Suppose that the joint system is in the pure state represented by a vector $\varphi \in H = H_1 \otimes H_2$. Then there exists a nonnegative nuclear operator T_φ^1 on H_1 with $\mathrm{tr}\, T_\varphi^1 = 1$ such that for every bounded selfadjoint operator A on H_1 we have $\mathrm{tr}\,(AT_\varphi^1) = (A \otimes I_2 \varphi, \varphi)_H$.*

Indeed, let $\{e_n^2\}$ be an orthonormal basis in H_2; then the relation

$$(T_\varphi^1 x, z)_{H_1} = \|\varphi\|^{-2} \sum_{n=1}^{\infty} (x \otimes e_n^2, \varphi)_H \overline{(z \otimes e_n^2, \varphi)}_H, \quad x, z \in H_1,$$

uniquely defines T_φ^1. The operator T_φ^1 is called a density operator (generated by the state φ of the joint system); we shall say that it defines a mixed state of the system with the Hilbert space H_1 (generated by the pure state of the joint system). Every nonnegative nuclear operator with unit trace can be obtained in this way.

Comments. 7. In the framework of the considered model a density operator on a Hilbert space K is defined as an arbitrary nonnegative nuclear operator with trace 1. Let $L_1^+(K)$ be the set of all such operators. We shall say that an operator $T \in L_1^+(K)$ defines a mixed state; if the system is in this state, then the mean value of results of measurement of the observable A equals $\mathrm{tr}\, AT$ (if the expression on the right is defined; for example, this is true for bounded A). Pure states are a particular case of mixed states; namely, a pure state given by a normalized vector φ is also defined by the density operator $\varphi \otimes \varphi$ (which acts by the formula $\varphi \otimes \varphi \colon x \mapsto \varphi(x, \varphi)$). If a state of a joint system with a Hilbert space $H = H_1 \otimes H_2$ is itself mixed and given by a density operator S, then there exists a density operator T_S^1 on H_1 and a density operator T_S^2 on H_2 such that for all bounded observables A_1 on H_1 and A_2 on H_2 we have $\mathrm{tr}\,(A_1 T_S^1) = \mathrm{tr}\,[(A_1 \otimes I_2)S]$ and $\mathrm{tr}\,(A_2 T_S^2) = \mathrm{tr}\,[(I_1 \otimes A_2)S]$. The states T_S^1 and T_S^2 are called reductions of the state S; it is also customary to say that T_S^1 is the partial trace of the operator S with respect to H_2 and T_S^2 is the partial trace of the operator S with respect to H_1;

these partial traces are denoted by the symbols $\mathrm{tr}_{H_2} S$ and $\mathrm{tr}_{H_1} S$, respectively, the same symbols are used to denote the usual traces in the same spaces. Thus, $\mathrm{tr}_{H_1}(\mathrm{tr}_{H_2} S) = \mathrm{tr}\, S = \mathrm{tr}_{H_2}(\mathrm{tr}_{H_1} S)$.

We emphasize that by means of the states T_S^1 and T_S^2 of the subsystems that are reductions of the state S of the joint system this state S can be reconstructed only in the case when both states T_S^1 and T_S^2 are pure (then $S = T_S^1 {\otimes} T_S^2$). However, in the general case this equality can fail. This will be the case if the state S is pure and T_S^1 and T_S^2 are mixed. Thus, in quantum mechanics, unlike the classical mechanics, the states of subsystems can fail to determine the states of the joint system (but, as we have seen, the state of the joint system determines the states of the subsystems).

The one-parameter group $U(t)$ of unitary operators from Axiom 2 defines one-parameter strongly continuous groups (also called dynamical) $F(t)$ and $G(t)$ of continuous mappings of the Banach space of nuclear operators into itself and of the Banach space of bounded observables into itself in following way:

$$F(t)T = U(t)TU(t)^*, \qquad G(t)A = U(t)^*AU(t).$$

For every t, the mapping $F(t)$ is the extension by continuity of the mapping on finite linear combinations of pure states that is in turn the extension by linearity of the mapping $U(t)$. Every mapping $G(t)$ is adjoint to the corresponding mapping $F(t)$; in the definition of adjoint mappings we use Theorem 7.10.40, according to which the space of all continuous linear operators on a Hilbert space can be identified with the dual to the space of all nuclear operators (any continuous linear operator A is identified with the functional $T \mapsto \mathrm{tr}\, AT$). In this case the functions $t \mapsto F(t)T$ and $t \mapsto G(t)A$ are solutions to the Cauchy problems for the equations

$$f'(t) = i[\widehat{H}, f(t)]/\hbar, \qquad g'(t) = i[g(t), \widehat{H}]/\hbar$$

with initial conditions T and A, respectively. Usually these equations are called the Heisenberg equations, although the first of them is called sometimes the Schrödinger equation and the second one is called the Liouville equation. It is customary to say that the groups $U(t)$, $F(t)$ and $G(t)$, and also the Heisenberg and Schrödinger equations, define the Hamiltonian dynamics of the quantum system. If $U_H(t)$ is a one-parameter group of unitary transformations of H defining the dynamics of the quantum system, then, for any state T of the joint system, the function $\mathcal{F}$ given by the equality $\mathcal{F}(t) = \mathrm{tr}_{H_2} U(t)TU(t)^*$ defines the dynamics of the first subsystem generated by the dynamics of this joint system. Of course, the function $\mathcal{F}$ need not be a one-parameter semigroup. The equation it satisfies is called the master equation. Nevertheless, for describing the so-called Markov approximation of the dynamics of quantum systems interacting with quantum fields it becomes useful to employ limits of some families of functions like the function $\mathcal{F}$, depending on a parameter characterizing interaction (see Accardi, Lu, Volovich [3]). These limit functions are one-parameter semigroups and satisfy certain equations analogous to the backward Kolmogorov equation in the theory of diffusion processes.

Comments. 8. Superselection rules (if they are introduced) are defined by means of some projection-valued measure Π. It is supposed that physically admissible are only observables commuting with all operators $\Pi(B)$. If the measure Π is concentrated on a finite or countable set, then in a similar way physically admissible are only states representable by density operators commuting with all operators $\Pi(B)$. The set of vectors corresponding to physically admissible pure states coincides with the union of ranges of all projections $\Pi(\{q_j\})$, $j = 1, 2, \ldots$ If there are no superselection rules in a quantum system, then every linear combination of vectors representing (pure) states is again a vector representing some pure state; this fact, called the principle of superposition, reflects the so-called wave properties of quantum systems. The principle of superposition yields a fundamental difference between quantum and classical systems.

Comments. 9. Every mixed state can be obtained as a probability mixture of pure states. Namely, let ν be a probability measure on K such that $\nu(\{0\}) = 0$. Then the expectation values of the results of measurement of a bounded observable A in supposition that this measure determines the probability distribution of pure states equals $\int (Az, z)_K \|z\|^{-2} \nu(dz)$. It follows from Theorem 7.10.40 that there exists a nuclear nonnegative operator S on K such that for every bounded observable A this integral equals $\operatorname{tr}(AS)$ (the theorem gives the equality first on compact operators, but then it extends to all operators by pointwise approximation). This implies that for a measure ν (defined by the equality not uniquely) one can take the measure that is the product of the function $z \mapsto \|z\|^2$ and another probability measure μ. This means that $\operatorname{tr}(AS)$ is the integral of $(Az, z)_K$ against the measure μ, so that for μ we can take any probability measure with zero mean and the correlation operator S (among such measures there exists precisely one Gaussian measure). Thus, every mixed state can be obtained by two principally different procedures: as a probability mixture of pure states and as a state generated by a pure state of some larger system.

Comments. 10. The density operator characterizes the membership of the quantum system in some statistical ensemble (by the way, it is still under discussion whether the element of the Hilbert space defining a pure state of the quantum system is a characteristic of the individual system or only of the statistical ensemble to which it belongs). It should be emphasized that by means of an experiment it is impossible to distinguish mixed states obtained by the two procedures described at the end of the previous comment. However, if it is known a priori that the mixed state of the quantum system is a probability mixture of pure states, then this means that each copy of the corresponding statistical ensemble is in some pure state (if the system is in a mixed state generated by a pure state of some larger system, then the assumption that actually every copy of the corresponding statistical ensemble of copies of this system is in some pure state and the assumption about finite speed of propagation of interactions lead together to a contradiction with Bell's inequality from the classical probability theory). One can raise the question about estimating this state by means of a result of an individual experiment.

If, for example, the probability measure characterizing a mixed state is concentrated on the set of two orthogonal vectors $\{h, k\}$ (say, of unit norm), then for determining in which of the two pure states the considered system is, it suffices to accomplish a measurement of every observable that is a selfadjoint operator with eigenvectors h and k.

If some mixed state is given by a probability measure concentrated on the set of three vectors h, $\cos\frac{\pi}{3}h + \sin\frac{\pi}{3}k$, $\cos\frac{\pi}{3}h - \sin\frac{\pi}{3}k$ and its values do not vanish on singletons in this set, then there is no experiment enabling one to determine precisely in which of the three pure states the considered system is.

Nevertheless, there exists an experiment enabling one to give the best (in some sense) estimate of this state. It consists of a measurement of some observable A belonging to an enlarged quantum system obtained by adding to the investigated system (with a Hilbert space H) an auxiliary quantum system (with a Hilbert space K) that is in an especially selected pure state S_0. If Π_A is the projection-valued measure generated by the operator A, then the operator-valued set function $E(\cdot)$ defined by the equality $E(B) = \text{tr}_K\left[(I_H \otimes S_0)\Pi_A(B)\right]$ is a measure with values in the space of bounded nonnegative operators on H (called the partition of unity). In this case, if the state of the original quantum system is represented by a density operator S, then the probability that the measurement just described gives a value from a Borel set B equals $\text{tr}\left[SE(B)\right]$. More details can be found in the book Holevo [279]. The corresponding proofs are based on Naimark's theorem, according to which every symmetric operator on a Hilbert space possesses a selfadjoint extension in some larger Hilbert space.

Comments. 11. The presented system of axioms of quantum mechanics is not complete, because there exist non-isomorphic quantum systems. A large class of systems is obtained by the procedure of quantization of Hamiltonian systems of classical mechanics.

A symplectic locally convex space is a pair (E, I), where E is a locally convex space over $\mathbb{R}$ and $I\colon E \to E^*$ is a linear mapping (E^* is equipped with a topology that agrees with duality between E^* and E) such that $I^* = -I$. A Poisson bracket of complex or real functions f and g on E is the function $\{f, g\}$ on E defined by $\{f, g\}(x) = f'(x)(I(g'(x)))$. Now let $E = Q \times P(= Q \oplus P)$, where Q is a locally convex space and $P = Q^*$ is equipped with a topology consistent with duality between P and Q. The space E is identified with a space of linear functionals on it by means of the mapping associating to the element $(q, p) \in E$ the linear functional on E denoted by the symbol (p, q) and given by the formula $\langle (p, q), (q', p') \rangle = q(p') + q'(p)$. The mapping I is defined as follows: $I(p, q) = (q, -p)$. The linear space of smooth functions on E with the operation of multiplication defined by the Poisson bracket forms a Lie algebra, which is denoted by P_E; its subalgebra P_C generated by linear functions is a central extension of the commutative Lie algebra $E^* = E$ with the trivial multiplication (P_C is called the Heisenberg algebra generated by E). A mapping from the Lie algebra P_C to the set of selfadjoint operators on a Hilbert space H such that for all $k, g \in P_C$, $a, b \in \mathbb{R}$ we have $F(ak + bg) = aF(k) + bF(g)$ and $iF(\{k, g\}) = [F(k), F(g)] = F(k)F(g) - F(g)F(k)$ is called a representation in H of the canonical commutation relations (CCR) in the Heisenberg form; a unitary representation of the central extension E_C of the commutative group that is the additive group of $E = E^*$ (this extension is called the Heisenberg group or the Weyl group) is called a representation of the CCR in Weyl's form.

A Hamiltonian system is a collection $(E, I, \mathcal{H})$, where (E, I) is a symplectic locally convex space, called the phase space, and $\mathcal{H}$ is a real function on E, called the Hamiltonian. The Hamilton equation is the following equation with respect to the function f of a real argument, interpreted as time, taking values in E:

$$f'(t) = I\big(\mathcal{H}'(f(t))\big).$$

If the space Q is finite-dimensional, then this equation coincides with the standard system of Hamilton equations. An example of an infinite-dimensional Hamilton equation is the Schrödinger equation (in case of an infinite-dimensional space H of the quantum system). Indeed, let E be the realification of H, I the operator on E generated by multiplication by the imaginary unit on H and $\mathcal{H}(x) = -\frac{i}{2\hbar}(\widehat{H}x, x)$. Then $(E, I, \mathcal{H})$ is the Hamiltonian system and the corresponding Hamilton equation coincides with the Schrödinger equation.

A quantization of a Hamiltonian system is a procedure of defining an operator $\mathcal{H}(F(p), F(q))$ on H, where F is the representation of the CCR in the Heisenberg form. Since the operators $F(q)$ and $F(p)$ do not commute, there is no unique "natural" definition of the operator $\mathcal{H}(F(q), F(p))$; moreover, in the infinite-dimensional case also a representation of the CCR is not unique. If the space Q is finite-dimensional, then a privileged role is played by the so-called Schrödinger representation. For every $p \in P$, let $F(p)$ denote the operator of multiplication by the linear functional $q \mapsto p(q)$ acting on the space $L_2(Q)$; for every $q \in Q$, let $F(q)$ denote the operator $g \mapsto -i\hbar g'q$ on the same space; $F(q)$ is the impulse operator in the direction q. If $\mathcal{H}(q, p) = V(q) + \frac{1}{2}(Bp, p)$, where B is a linear operator on Q, then as the result of application of the described method of quantization we obtain the standard Schrödinger equation (with a potential V and an "anisotropic mass", characterized by the operator B). This is the form of the Hamiltonian of the Hamiltonian system introduced above, for which for the Hamiltonian equation we obtain the Schrödinger equation. The quantization of this Hamiltonian system is called the second quantization; the name is explained by the fact that usually the original Schrödinger equation is also obtained by means of a quantization. If E is infinite-dimensional, then it has no reasonable analog of Lebesgue measure (there is no nonzero translation-invariant σ-finite locally finite countably additive Borel measure); however, in place of it one can use a suitable Gaussian measure (the space of functions on Q quadratically integrable with respect to this measure is called the Wiener–Segal–Fock space).

10.7(v). Sturm–Liouville operators

We know that the operator $D = i d/dt$ is selfadjoint on $L^2[0, 1]$ on the domain consisting of absolutely continuous functions x with a derivative in $L^2[0, 1]$ satisfying the boundary condition $x(0) = x(1)$. Theorem 10.3.3 yields the selfadjointness of the operator $D^2 x = -x''$ of taking the second derivative on the

domain consisting of functions x for which the derivative x' is absolutely continuous, $x'' \in L^2[0,1]$, and $x(0) = x(1)$, $x'(0) = x'(1)$.

We now study a more general Sturm–Liouville operator

$$Lx(t) = -\big(p(t)x'(t)\big)' + q(t)x(t),$$

where the function p is continuously differentiable and strictly positive on $[0,1]$
and the function q is continuous and real. Let us take for the domain of definition
$\mathfrak{D}(L)$ the class of all absolutely continuous functions x on $[0,1]$ for which the
function x' is absolutely continuous, $x'' \in L^2[0,1]$, and the boundary conditions
$x(0) = x(1) = 0$ hold. On this domain the operator L is symmetric, since

$$\int_0^1 Lx(t)\overline{y(t)}\,dt = \int_0^1 \left[p(t)x'(t)\overline{y'(t)} + q(t)x(t)\overline{y(t)}\right] dt$$

and the same holds for $x(t)\overline{Ly(t)}$. But we have not yet proved that L is selfadjoint!
It follows from the symmetry that if L has eigenvectors, then the corresponding
eigenvalues are real and the eigenvectors corresponding to different eigenvalues
are orthogonal. Hence the set of all eigenvalues is at most countable.

Suppose that $\operatorname{Ker} L = 0$; to this case we can pass by replacing L with the
operator of the same form $L - \lambda I$, where $\lambda \in \mathbb{R}$ is not an eigenvalue (which
means that we replace q with $q - \lambda$). The selfadjointness of L can be seen
from the following reasoning. The operator $L_0 x = -x''$ is also injective on
$\mathfrak{D}(L)$ and maps $\mathfrak{D}(L)$ onto $L^2[0,1]$. The space $\mathfrak{D}(L)$ is a closed subspace in the
Sobolev space $H = W^{2,2}((0,1))$, which is a Hilbert space with the inner product
$(u,v)_H := (u,v)_{L^2} + (u',v')_{L^2} + (u'',v'')_{L^2}$. Since the operator $L_0 \colon H \to L^2[0,1]$
is continuous, the operator $L_0 \colon \mathfrak{D}(L) \to L^2[0,1]$ is invertible. Then the operator $L_1 x = -px''$ from $\mathfrak{D}(L)$ to $L^2[0,1]$ is also invertible, since multiplication
by the positive continuous function p is an invertible operator on $L^2[0,1]$. Our
operator $L \colon \mathfrak{D}(L) \to L^2[0,1]$ has the form $L = L_1 + S$, where the operator
$Sx = -p'x' + qx$, as is easily verified, is compact from H with the indicated
Hilbert norm to $L^2[0,1]$; hence it is also compact on $\mathfrak{D}(L)$. The injectivity of L
gives the surjectivity, since $L = L_1(I + L_1^{-1}S)$, where $L_1^{-1}S$ is compact on $\mathfrak{D}(L)$.

The operator $K = L^{-1} \colon L^2[0,1] \to \mathfrak{D}(L)$ is symmetric. Our discussion
shows that it takes the unit ball from $L^2[0,1]$ to a set bounded in H. The ball of H
is compact in $L^2[0,1]$ (even in $C[0,1]$). Hence L is the inverse to the selfadjoint
compact operator K and is also selfadjoint on $\mathfrak{D}(L) = K(L^2[0,1])$. By the
Hilbert–Schmidt theorem K has an orthonormal eigenbasis $\{e_n\}$ with eigenvalues
$k_n \to 0$ of finite multiplicity. Hence L has the same eigenbasis and eigenvalues
$\lambda_n = k_n^{-1}$, where $|\lambda_n| \to \infty$. We recall that all this has been done assuming the
injectivity of L. In the general case, as noted above, we have the injective operator
$L - \lambda_0 I$ for some $\lambda_0 \in \mathbb{R}$, so it has an orthonormal eigenbasis $\{e_n\}$ with real
eigenvalues μ_n; in this basis L is also diagonal and has eigenvalues $\lambda_n = \mu_n + \lambda_0$.

Conclusion: *the equation $Lx = y$ with our boundary conditions is solvable
for those and only those functions $y \in L^2[0,1]$ which are orthogonal in $L^2[0,1]$ to
the subspace $\operatorname{Ker} L$, consisting of all solutions to the equation $Lu = 0$ with given
boundary conditions.*

In case $\operatorname{Ker} L = 0$, the operator $K = L^{-1}$ is defined by a continuous real integral kernel $\mathcal{G}(t,s) = \mathcal{G}(s,t)$, called the Green's function of the operator L. To see this, we take two solutions u_1 and u_2 to the equation $Lu = 0$ with the boundary conditions $u_1(0) = 0$, $u_1'(0) = 1$ and $u_2(1) = 0$, $u_2'(1) = 1$. Since L is injective, the functions u_1 and u_2 are linearly independent. Then the Wronskian $\Delta := p(u_1'u_2 - u_1u_2')$ is a nonzero constant (its derivative equals zero). Finally, set $\mathcal{G}(t,s) = \Delta^{-1}u_1(s)u_2(t)$ if $0 \leqslant s \leqslant t \leqslant 1$, $\mathcal{G}(t,s) = \Delta^{-1}u_1(t)u_2(s)$ if $0 \leqslant t \leqslant s \leqslant 1$. It is straightforward to verify that K coincides with the integral operator $K_\mathcal{G}$ defined by $\mathcal{G}$, i.e., $LK_\mathcal{G}f = f$.

One can take more general selfadjoint boundary conditions; for example, for $p = 1$ and $q = 0$ one can obtain the aforementioned operator $(id/dt)^2$, and the boundary conditions we considered give a different selfadjoint extension of $-d^2/dt^2$ on $C_0^\infty(0,1)$.

In a similar way one can investigate multidimensional boundary value problems for second order elliptic operators (for example, for the Laplace operator Δ), although the Green's function cannot be constructed in such a simple manner.

Exercises

10.7.24.° Let T be a densely defined operator on a Hilbert space H such that T^* is densely defined and continuous. Prove that T is also continuous and extends to a bounded operator on H.

HINT: show that the adjoint operator for the extension of T^* by continuity extends T.

10.7.25.° Let S and T be densely defined operators on a Hilbert space H such that $S \subset T$. Prove that $T^* \subset S^*$.

10.7.26.° Let S and T be densely defined operators on a Hilbert space H such that $S \subset T$. Prove that if the operator T is closable, then S is closable as well and $\overline{S} \subset \overline{T}$.

10.7.27. Let S and T be densely defined closed operators on a Hilbert space H and suppose that the set $D := \mathfrak{D}(S) \cap \mathfrak{D}(T)$ is dense.

(i) Is it true that the operator $S + T$ on D is closable?

(ii) If $S + T$ is closable on D, then is it true that $(S + T)^* = S^* + T^*$ on the domain $\mathfrak{D}(S^*) \cap \mathfrak{D}(T^*)$? Is it true that $\mathfrak{D}(\overline{S^* + T^*}) = \mathfrak{D}(S^*) \cap \mathfrak{D}(T^*)$, where on the left we take the domain of the closure?

HINT: (i) Take $H = L^2[0,1] \oplus \mathbb{C}$, $S(u,z) = (u', u(1/2))$, $Tu = (-u', 0)$ for absolutely continuous functions u on $[0,1]$ with $u' \in L^2[0,1]$. Show that the operator $u \mapsto (0, u(1/2))$ is not closable. (iii) Consider the case where $T = -S + I$.

10.7.28. Let T_0 be a closed operator on a Hilbert space H and let T be an operator such that $T_0 \subset T$ and $\mathfrak{D}(T_0)$ has a finite codimension in $\mathfrak{D}(T)$. Prove that T is closed.

HINT: observe that the Hilbert space D_{T_0} is a closed subspace of finite codimension in D_T. Hence D_T is also complete.

10.7.29. Investigate selfadjoint extensions of the operator $\varphi \mapsto i\varphi'$ on the domain of definition consisting of smooth compactly supported functions in the spaces $L^2(\mathbb{R}^1)$ and $L^2[0, +\infty)$.

10.7.30. Investigate selfadjoint extensions of the operator $\varphi \mapsto \varphi''$ on the domain of definition consisting of smooth compactly supported functions in the spaces $L^2(\mathbb{R}^1)$ and $L^2[0, +\infty)$.

10.7.31.° Let A be a bounded selfadjoint injective operator on a Hilbert space H. Show that the set $A(H)$ is dense in H, and the inverse operator $T = A^{-1}$ on the domain $\mathfrak{D}(T) := A(H)$ is selfadjoint (verify this directly by definition).

10.7.32. (i) Let A be a symmetric operator bounded from below and having a finite defect number. Prove that every selfadjoint extension of A is bounded from below (possibly, by a number less than the original bound). (ii) Construct an example of a symmetric operator bounded from below that has a selfadjoint extension not bounded from below.

HINT: see [**503**, Chapter X, Section 3].

10.7.33. Suppose that a symmetric operator A bounded from below is not essentially selfadjoint. Prove that A has selfadjoint extensions bounded from below that are different from its Friedrichs extension.

HINT: we can assume that A is closed. Let B be the Friedrichs extension of A. Then $A \subset B = B^* \subset A^*$ and $\mathfrak{D}(B)$ is the direct sum of $\mathfrak{D}(A)$, $\mathrm{Ker}\,(A^* - iI)$, $\mathrm{Ker}\,(A^* + iI)$. Let $D_0 \subset \mathrm{Ker}\,(A^* - iI)$, $R_0 \subset \mathrm{Ker}\,(A^* + iI)$ and $V_0 \colon D_0 \to R_0$ be the subspaces and the isometry corresponding to the Friedrichs extension, i.e., $\mathfrak{D}(B) = \mathfrak{D}(A) \oplus (V_0 - I)(D_0)$. Let us write D_0 is a direct sum of the linear span of a nonzero vector e and a subspace D_1. Let $R_1 = V_0(D_1)$, $e_1 = V_0(e)$. Define V_1 as follows: $V_1|_{D_1} = V_0$, $V_1(e) = -e_1$. Then $V_1 \neq V_0$. The operator V_1 corresponds to some selfadjoint extension A_1 of A different from B. Observe that $(V - I)(D_0)$ and $(V_1 - I)(D_0)$ differ by a finite-dimensional subspace and on the intersection A_1 acts by the formula $A_1(x + V_0 x_0 - x_0) = V_0 x - i x_0 - i V_0 x_0$. This yields that A_1 is bounded from below.

10.7.34.* Let A be a closed densely defined operator and

$$\mathfrak{D}(A^*A) := \{x \in \mathfrak{D}(A) \colon Ax \in \mathfrak{D}(A^*)\}, \quad A^*Ax := A^*(Ax).$$

Prove that A^*A is selfadjoint.

HINT: consider the form $Q(x, y) = (Ax, Ay) + (x, y)$, which is positive and closed (since A is closed). Hence there is a selfadjoint operator B such that $Q(x, y) = (Bx, y)$ for all $x \in \mathfrak{D}(B)$ and $y \in \mathfrak{D}(Q)$. It follows from the proof of Theorem 10.7.16 that $\mathfrak{D}(B)$ is closed in the Hilbert space $\mathfrak{D}(Q)$ (with the corresponding norm). We show that $\mathfrak{D}(B) \subset \mathfrak{D}(A^*A)$. For all $v \in \mathfrak{D}(B)$ and $x \in \mathfrak{D}(A)$ we have $(Sv, x) = (Av, Ax)$, where $S := B - I$. Hence $x \mapsto (Ax, Av)$ is a continuous functional and $Av \in \mathfrak{D}(A^*)$, $A^*Av = Sv$. Thus, S is selfadjoint and $S \subset A^*A$. It is obvious that A^*A is symmetric, whence it is easy to derive that $A^*A = S$.

10.7.35. Let A be a bounded selfadjoint operator on a separable Hilbert space H, $x \in H$ and H_x the closure of the linear span of the vectors $\Pi(B)x$, where $B \in \mathcal{B}(\mathbb{R}^1)$. Prove that H_x equals the closure of the linear span of the sequence $\{x, Ax, A^2 x, \ldots\}$.

HINT: for $A = A_\varphi$ on $L^2(\mu)$, the closed subspaces generated by the functions $I_B \circ \varphi$ and by the functions $f \circ \varphi$, where f is a polynomial, coincide.

10.7.36.° Construct an example of a selfadjoint operator on a separable Hilbert space that cannot be the generator of a strongly continuous semigroup.

HINT: consider the operator of multiplication by the argument on $L^2(\mathbb{R})$.

10.7.37.° Construct an example of two different strongly continuous semigroups on a separable Hilbert space the generators of which coincide on a dense subspace.

10.7.38. Let A be a closed densely defined symmetric operator on a Hilbert space and let B be a bounded selfadjoint operator. Prove that the defect numbers of the operators A and $A + B$ coincide.

HINT: see [**6**, p. 352].

10.7.39. Suppose that A is the generator of a strongly continuous contracting semi-group $\{T_t\}_{t\geqslant 0}$ on a Hilbert space H and $D \subset \mathfrak{D}(A)$ is a linear subspace dense in H such that $T_t(D) \subset D$ for all t. Prove that the operator A on D is essentially selfadjoint, that is, A is the closure of $A|_D$.

HINT: see [**503**, Theorem X.49].

10.7.40. Let $T_t\varphi(x) = \varphi(x + t)$, $\varphi \in L^p(\mathbb{R}^1)$, $1 \leqslant p < \infty$. (i) Prove that we have obtained a strongly continuous semigroup. (ii) Prove that for $p = 1$ the semigroup of operators T_t^* on $L^\infty(\mathbb{R}^1) = \left(L^1(\mathbb{R}^1)\right)^*$ is not strongly continuous.

HINT: (i) first verify the continuity on compactly supported continuous functions. (ii) Consider T_t^* on $I_{[0,1]}$.

10.7.41. (i) Prove that an operator A on a domain $\mathfrak{D}(A)$ is dissipative precisely when $\|\lambda x - Ax\| \geqslant \lambda\|x\|$ for all $x \in \mathfrak{D}(A)$ and $\lambda > 0$.

(ii) Let A be a dissipative operator on a Banach space X. Prove that $\operatorname{Re} l(Ax) \leqslant 0$ for all $x \in \mathfrak{D}(A)$ and all $l \in X^*$ such that $l(x) = \|x\|^2 = \|l\|^2$ (not only for some l with this property as required by the definition).

HINT: (i) if A is dissipative, $x \in \mathfrak{D}(A)$, $\|x\| = 1$ and l is the corresponding functional, then $\|\lambda x - Ax\| \geqslant \operatorname{Re} l(\lambda x - Ax) \geqslant \lambda$ for all $\lambda > 0$. Conversely, let (ii) be fulfilled, $\|x\| = 1$, $l_\lambda \in X^*$, $l_\lambda(\lambda x - Ax) = \|l_\lambda\|^2 = \|\lambda x - Ax\|^2$. Let $f_\lambda = l_\lambda/\|l_\lambda\|$. Then $\operatorname{Re} f_\lambda(Ax) \leqslant 0$, since

$$\lambda\|x\| \leqslant \|\lambda x - Ax\| = f_\lambda(\lambda x - Ax) = \lambda\operatorname{Re} f_\lambda(x) - \operatorname{Re} f_\lambda(Ax) \leqslant \lambda\|x\| - \operatorname{Re} f_\lambda(Ax).$$

In addition, $\operatorname{Re} f_\lambda(x) \geqslant \|x\| + \lambda^{-1}\operatorname{Re} f_n(Ax)$. By the the weak-$*$ compactness of the balls in X^* the sequence $\{f_n\}$ has a limit point f. Clearly, $\operatorname{Re} f(Ax) \leqslant 0$ and $\operatorname{Re} f(x) \geqslant \|x\|$, whence $f(x) = \|x\| = 1$. (ii) Using the contracting semigroup $\{T_t\}_{t\geqslant 0}$ generated by A we fund that $|l(T_t x)| \leqslant \|l\|\,\|x\| = \|x\|^2$ and $\operatorname{Re} l(T_t x - x) = \operatorname{Re} l(T_t x) - \|x\|^2 \leqslant 0$; next, divide by t and let t go to zero.

10.7.42. Let A be a densely defined dissipative operator on a Banach space. Prove that A is closable and its closure is also dissipative.

HINT: let $x_n \in \mathfrak{D}(A)$, $x_n \to 0$ and $Ax_n \to y$. For every vector $z \in \mathfrak{D}(A)$ and all $\lambda > 0$ we have $\|x_n + \lambda z\| \leqslant \|x_n + \lambda z - \lambda A(x_n + \lambda z)\|$, which gives the estimate $\|\lambda z\| \leqslant \|\lambda(z - y) - \lambda^2 Az\|$, whence $\|z\| \leqslant \|z - y - \lambda Az\|$. As $\lambda \to 0$ we obtain $\|z\| \leqslant \|z - y\|$. Taking $z_n \in \mathfrak{D}(A)$ with $z_n \to 2y$, we obtain $y = 0$, so the operator A is closable. The dissipativity of $\overline{A}$ is clear from Exercise 10.7.41.

10.7.43. Let A be a dissipative operator on a Banach space X. (i) Prove that the closedness of A is equivalent to the closedness of the range of $\lambda I - A$ for some $\lambda > 0$ (then for all $\lambda > 0$). (ii) Prove that if the range of $\lambda I - A$ is X for some $\lambda > 0$, then this is true for all $\lambda > 0$.

HINT: see [**123**, §3.3].

10.7.44. Let A be a densely defined dissipative operator on a Banach space X. Prove the equivalence of the following properties. (i) The operator A^* on X^* is dissipative; (ii) the operator A^* on X^* is m-dissipative (i.e., is dissipative and for all $\lambda > 0$ the operators $A^* - \lambda I$ are surjective); (iii) the operator $\overline{A}$ is m-dissipative; (iv) the range of $\lambda I - A$ is dense for some $\lambda > 0$ (then for all $\lambda > 0$).

HINT: see [**123**, §3.3].

10.7.45. Let A be a densely defined dissipative operator on a Banach space X. Prove that $\overline{A}$ is the generator of a continuous contracting operator semigroup precisely when $\operatorname{Ker}(\lambda I - A^*) = 0$ for some $\lambda > 0$ (then for all $\lambda > 0$).

CHAPTER 11

Banach Algebras

In this chapter we give a brief introduction to the theory of Banach algebras. As well as operator theory, this theory has important applications in physics. One of the most important Banach algebras is the algebra of operators. Considerations of this chapter shed a new light on some already encountered objects and also complement some results obtained above.

11.1. Basic Definitions

First we consider purely algebraic concepts connected with Banach algebras. In this chapter we shall deal with *complex spaces*.

11.1.1. Definition. *A linear space $\mathcal{A} \neq 0$ is called an algebra if it is equipped with an associative multiplication that agrees with the linear structure, i.e., with every pair of elements $(a, b) \in \mathcal{A} \times \mathcal{A}$ their product $ab \in \mathcal{A}$ is associated such that $(ab)c = a(bc)$, $(a + b)c = ac + bc$, $a(b + c) = ab + ac$, $(\lambda a)b = \lambda(ab) = a(\lambda b)$ for all $a, b, c \in \mathcal{A}$ and all scalars λ.*

An algebra $\mathcal{A}$ is called unital if it possesses a unit, i.e., an element 1 such that $1a = a1 = a$ for all elements $a \in \mathcal{A}$.

An algebra $\mathcal{A}$ is called commutative if $ab = ba$ for all elements $a, b \in \mathcal{A}$.

It is readily seen that an algebra can have only one unit. A simple and important example of an algebra is the algebra of polynomials $\mathcal{P}[t]$ with complex coefficients with respect to a formal variable t. For functional analysis are particularly interesting and important the algebra $C_b(T)$ of bounded continuous complex functions on a topological space T, equipped with the natural operation of multiplication $fg(t) = f(t)g(t)$, and also the algebra $\mathcal{L}(X)$ of all bounded linear operators on a Banach space X (traditionally also denoted by the symbol $B(X)$) with the operation of the usual multiplication (composition) of operators. All these algebras are unital. A typical example of an algebra without a unit is the algebra of continuous functions on the real line tending to zero at infinity. In Example 11.1.8 below we list typical Banach algebras.

It is clear that $\mathcal{P}[t]$ and $C_b(T)$ are commutative, but $\mathcal{L}(X)$ is not if X is a Banach space with $\dim X > 1$.

Elements of a unital algebra can be substituted in polynomials with complex coefficients: for arbitrary $a \in \mathcal{A}$ and $p(z) = c_n z^n + c_{n-1} z^{n-1} + \cdots + c_1 z + c_0$

© Springer Nature Switzerland AG 2020 483
V. I. Bogachev, O. G. Smolyanov, *Real and Functional Analysis*,
Moscow Lectures 4, https://doi.org/10.1007/978-3-030-38219-3_11

we set naturally (as we did in previous chapters for functions and operators)

$$p(a) = c_n a^n + c_{n-1} a^{n-1} + \cdots + c_1 a + c_0 1$$

and call $p(a)$ a polynomial of a. Banach algebras introduced below enable one to substitute elements into many other functions.

For functional analysis are particularly important algebras equipped with topologies that agree with the algebraic structures. An important particular case: Banach algebras.

11.1.2. Definition. *A Banach algebra is an algebra $\mathcal{A}$ equipped with a norm with respect to which it is complete and satisfies the inequality*

$$\|ab\| \leqslant \|a\| \, \|b\|, \quad a, b \in \mathcal{A}.$$

Important examples of Banach algebras are the commutative algebras $C(K)$ and $L^1(\mathbb{R}^n)$ (with the operation of convolution) and the noncommutative algebra $\mathcal{L}(X)$ of bounded operators on a Banach space X and the algebra $\mathcal{K}(X)$ of compact operators on X equipped with the operator norm (the latter is not unital).

Since algebras in addition to the linear structure possesses a multiplication, natural mappings between them are multiplicative operators.

11.1.3. Definition. *A linear operator $\varphi\colon \mathcal{A} \to \mathcal{B}$ between two algebras is called a homomorphism if*

$$\varphi(ab) = \varphi(a)\varphi(b) \quad \text{for all } a, b \in \mathcal{A}.$$

If both algebras are unital and $\varphi(1_{\mathcal{A}}) = 1_{\mathcal{B}}$, then φ is called a unital homomorphism.

A homomorphism $\varphi\colon \mathcal{A} \to \mathbb{C}$ is called a character of the algebra $\mathcal{A}$.

A one-to-one continuous homomorphism of Banach algebras $\mathcal{A}$ and $\mathcal{B}$ is called an isomorphism.

If the algebra $\mathcal{A}$ has a unit 1, then for every nonzero character χ we have $\chi(1) = 1$, since for $a \in \mathcal{A}$ with $\chi(a) \neq 0$ we obtain $\chi(a) = \chi(1)\chi(a)$.

The reader already accustomed to the fact that in the infinite-dimensional case one should always take care of the continuity of linear functionals will be pleasantly surprised by the property that the characters of any Banach algebra are automatically continuous.

11.1.4. Proposition. *Every character χ of a Banach algebra $\mathcal{A}$ is continuous and $\|\chi\| \leqslant 1$. If the algebra $\mathcal{A}$ has a unit 1 of unit norm and $\|\chi\| > 0$, then we have $\|\chi\| = 1$.*

PROOF. Otherwise there exists an element $a \in \mathcal{A}$ such that $\chi(a) = 1$ and $\|a\| < 1$. By the estimate $\|a^n\| \leqslant \|a\|^n$ and the completeness of $\mathcal{A}$ the series $\sum_{n=1}^{\infty} a^n$ converges to some $b \in \mathcal{A}$. It follows from the continuity of multiplication that $a + ab = a + \lim_{n\to\infty} a \sum_{k=1}^{n} a^k = b$. Then $1 + \chi(b) = \chi(a + ab) = \chi(b)$, which is impossible.

If $\chi \neq 0$ and there is a unit of the algebra $\mathcal{A}$, then $\chi(1) = 1$, whence we obtain $\|\chi\| \geqslant 1$. Hence $\|\chi\| = 1$. $\qquad\square$

There is a standard construction of an extension of an algebra $\mathcal{A}$ without a unit to an algebra $\mathcal{A}_1$ with a unit. For $\mathcal{A}_1$ one can take the linear space $\mathcal{A} \otimes \mathbb{C}$ with the operation of multiplication

$$(a, \alpha)(b, \beta) := (ab + \alpha b + \beta a, \alpha \beta).$$

It is straightforward to verify that we obtain an algebra with the unit $(0, 1)$ (this algebra is commutative if so is $\mathcal{A}$), moreover, the mapping $a \mapsto (a, 0)$ is an isomorphism between $\mathcal{A}$ and a subalgebra in $\mathcal{A}_1$ of codimension 1. If $\mathcal{A}$ is a Banach algebra, then $\mathcal{A}_1$ is also a Banach algebra with the natural norm

$$(a, \alpha) \mapsto \|a\| + |\alpha|,$$

moreover, the image of $\mathcal{A}$ under the indicated isomorphism is closed in $\mathcal{A}_1$ and the unit in $\mathcal{A}_1$ has unit norm.

The multiplicative inequality for the norm on a Banach algebra is introduced for convenience and can be obtained by a suitable renorming, as we now prove.

11.1.5. Proposition. *Let $\mathcal{A}$ be a Banach space that is an algebra with a unit different from zero and let the mappings $x \mapsto ax$ and $x \mapsto xa$ be continuous. Then there exists an equivalent norm on $\mathcal{A}$ with respect to which $\mathcal{A}$ is a Banach algebra and the unit has unit norm.*

PROOF. Let us consider the representation of the algebra $\mathcal{A}$ by continuous linear operators on $\mathcal{A}$ by the formula

$$T_a(x) = ax, \quad a \in \mathcal{A}, x \in \mathcal{A}.$$

It is clear that the mapping $a \mapsto T_a$ is linear and $T_{ab} = T_a T_b$. Let us introduce a new norm by the formula $\|a\|_0 := \|T_a\|$, where on the right the operator norm is meant. For the original norm $\| \cdot \|$ we have

$$\|a\| = \|a1\| \leqslant \|T_a\| \, \|1\| = \|a\|_0 \|1\|.$$

On the other hand, the mapping $(a, x) \mapsto ax$ is separately continuous. Hence it is jointly continuous and we have the estimate $\|ax\| \leqslant M\|a\| \, \|x\|$ with some number M (see §12.5; we could also apply to $a \mapsto T_a$ the closed graph theorem). Hence $\|T_a\| \leqslant M\|a\|$, i.e., a new norm is equivalent to the old one. It is clear that $\|ab\|_0 = \|T_{ab}\| = \|T_a T_b\| \leqslant \|a\|_0 \|b\|_0$ and T_1 is the unit operator, so $\|1\|_0 = 1$ (the condition $\mathcal{A} \neq 0$ is needed to ensure this equality).   □

This proposition enables us to pass to the case of a unit with unit norm (which is sometimes included in the definition).

Some Banach algebras (called *involutive*) possess one more very important operation — the *involution*. This is an operation $a \mapsto a^*$ with the following properties:

(i) $(a + b)^* = a^* + b^*$ for all $a, b \in \mathcal{A}$,
(ii) $(\lambda a)^* = \overline{\lambda} a^*$ for all $\lambda \in \mathbb{C}$ and all $a \in \mathcal{A}$,
(iii) $(ab)^* = b^* a^*$ for all $a, b \in \mathcal{A}$,
(iv) $(a^*)^* = a$ for all $a \in \mathcal{A}$.

An element a is called *Hermitian* or *selfadjoint* if $a^* = a$. If $a^* a = aa^*$, then the element a is called *normal*.

11.1.6. Proposition. *Let $\mathcal{A}$ be an involutive algebra and $a \in \mathcal{A}$. Then*
(i) *the elements a^*a, $a + a^*$ and $i(a^* - a)$ are Hermitian;*
(ii) *there exist unique Hermitian elements $u, v \in \mathcal{A}$ such that $a = u + iv$;*
(iii) *if $\mathcal{A}$ is unital, then its unit is Hermitian.*

PROOF. Assertion (i) is obvious and gives the desired elements u and v by the formulas $u = (a + a^*)/2$, $v = i(a^* - a)/2$. If u' and v' are Hermitian and $a = u' + iv'$, then we set $w = v' - v$. Then w and iw are Hermitian, whence $iw = (iw)^* = -iw^* = -iw$. Hence $w = 0$, i.e., $v = v'$ and $u = u'$. So the decomposition in (ii) is unique. Assertion (iii) follows from (i), since $1^* = 1^*1$ is a Hermitian element and hence $1^* = (1^*)^* = 1$. $\qquad\square$

11.1.7. Definition. (i) *A Banach algebra $\mathcal{A}$ with an involution is called a star-algebra or $*$-algebra if $\|a^*\| = \|a\|$ for all $a \in \mathcal{A}$.*
(ii) *A Banach $*$-algebra $\mathcal{A}$ is called a C^*-algebra if $\|a^*a\| = \|a\|^2$ for all elements $a \in \mathcal{A}$.*

The equality in (ii) yields the equality in (i), since $\|a^*a\| \leqslant \|a^*\|\,\|a\|$, whence $\|a\| \leqslant \|a^*\|$ and $\|a^*\| \leqslant \|(a^*)^*\| = \|a\|$.

The most important examples of C^*-algebras are the commutative algebra $C(K)$ of continuous complex functions on a compact space K with the natural operation of conjugation and the noncommutative algebra $\mathrm{B}(H)$ of continuous linear operators on a Hilbert space H with the involution $A \mapsto A^*$. As we shall see below, these two examples are universal in a sense.

Let us give a number of model examples.

11.1.8. Example. (i) The algebra $C_b(T)$ of bounded continuous complex functions on a topological space T equipped with the usual sup-norm and the natural operation of pointwise multiplication $fg(t) = f(t)g(t)$ is a commutative Banach C^*-algebra with a unit.

(ii) The algebra $\mathcal{L}(X) = \mathrm{B}(X)$ of all bounded linear operators on a Banach space X (of dimension at least 2) with the operator norm and the usual operation of multiplication of operators is a noncommutative Banach algebra with a unit. If X is Hilbert, then the conjugation $A \mapsto A^*$ makes it a C^*-algebra.

(iii) An important example of an algebra without a unit is the complex space $L^1(\mathbb{R}^n)$ of equivalence classes of integrable functions, in which the product of f and g is their convolution $f * g$. This algebra is commutative and the usual complex conjugation serves as a natural involution.

(iv) A discrete analog of $L^1(\mathbb{R}^1)$ is the algebra $l^1(\mathbb{Z})$ of two-sided sequences $z = (z_n)_{n \in \mathbb{Z}}$ with its natural l^1-norm $\|z\| = \sum_{n \in \mathbb{Z}} |z_n|$ and the product

$$(x * y)_n := \sum_{k \in \mathbb{Z}} x_k y_{n-k}, \quad x = (x_n), y = (y_n).$$

It is straightforward to verify that $l^1(\mathbb{Z})$ is a commutative Banach algebra. It has a unit: the element $(\dots, 0, 0, 1, 0, 0, \dots)$, where 1 is at the place with number 0. An involution can be defined by $(x^*)_n := \overline{x_{-n}}$ or by $(x^*)_n := \overline{x_n}$.

(v) The algebra of sequences $l^1(\mathbb{Z})$ can be identified with the algebra W of all 2π-periodic continuous complex functions with absolutely convergent Fourier series, i.e., W consists of functions f of the form

$$f(s) = \sum_{n \in \mathbb{Z}} c_n(f)e^{ins}, \quad \|f\|_W := \sum_{n \in \mathbb{Z}} |c_n(f)| < \infty,$$

where

$$c_n(f) = \frac{1}{2\pi} \int_0^{2\pi} f(t)e^{-int} \, dt.$$

A multiplication in W is the usual pointwise multiplication of functions. Of course, here we have to verify the inequality $\|fg\|_W \leqslant \|f\|_W \|g\|_W$. By the absolute convergence of the Fourier series for f and g we have

$$
\begin{aligned}
c_n(fg) &= \frac{1}{2\pi} \int_0^{2\pi} f(t)g(t)e^{-int} \, dt \\
&= \frac{1}{2\pi} \int_0^{2\pi} \left(\sum_{k \in \mathbb{Z}} c_k(f)e^{ikt} \right) \left(\sum_{l \in \mathbb{Z}} c_l(g)e^{ilt} \right) e^{-int} \, dt \\
&= \sum_{k,l \in \mathbb{Z}} c_k(f)c_l(g) \frac{1}{2\pi} \int_0^{2\pi} e^{i(k+l-n)t} \, dt \\
&= \sum_{k,l \in \mathbb{Z}:\ k+l=n} c_k(f)c_l(g) = \sum_{k \in \mathbb{Z}} c_k(f)c_{n-k}(g).
\end{aligned}
$$

Thus,

$$
\begin{aligned}
\sum_{n \in \mathbb{Z}} |c_n(fg)| &\leqslant \sum_{n \in \mathbb{Z}} \sum_{k \in \mathbb{Z}} |c_k(f)| \, |c_{n-k}(g)| \\
&= \sum_{k \in \mathbb{Z}} \sum_{n \in \mathbb{Z}} |c_k(f)| \, |c_{n-k}(g)| = \sum_{k \in \mathbb{Z}} |c_k(f)| \sum_{n \in \mathbb{Z}} |c_{n-k}(g)|,
\end{aligned}
$$

as required. It is clear that the mapping $f \mapsto \left(c_n(f) \right)_{n=1}^{\infty}$ is a linear isometry between W and $l^1(\mathbb{Z})$ preserving products, so it is an isomorphism of these two algebras.

Not every Banach $*$-algebra is a C^*-algebra. For example, the algebra $C^1[0,1]$ of complex continuously differentiable functions on $[0,1]$ with the pointwise multiplication, the norm

$$\|x\|_{1,\infty} := \max_t |x(t)| + \max_t |x'(t)|$$

and the involution $x^* = \bar{x}$ is a Banach $*$-algebra (which is easily verified). However, it cannot be topologically isomorphic to a C^*-algebra, since it possesses a sequence of elements $x_n = x_n^*$ with $\|x_n\| \geqslant 1$ and $\|x_n^2\| \to 0$ (it suffices to take x_n such that $0 \leqslant x_n(t) \leqslant 1/n$, $x_n'(0) = 1$, $|x_n'(t)| \leqslant C$).

11.2. Ideals

In the theory of algebras an important role is played by ideals.

11.2.1. Definition. *Let $\mathcal{A}$ be an algebra. A linear subspace $\mathcal{B}$ in $\mathcal{A}$ is called a subalgebra if $ab \in \mathcal{B}$ for all $a, b \in \mathcal{B}$.*

A subalgebra $\mathcal{I}$ in $\mathcal{A}$ is called an ideal (more precisely, a two-sided ideal) if $ab \in \mathcal{I}$ and $ba \in \mathcal{I}$ for all $a \in \mathcal{A}$ and $b \in \mathcal{I}$.

11.2.2. Example. (i) In the algebra $\mathcal{L}(\mathbb{C}^n)$ of linear operators on $\mathbb{C}^n$ the class of all operators of the form λI is a subalgebra, but is not an ideal if $n > 1$ (here there are no nontrivial ideals, see Exercise 11.7.18).

(ii) In the algebra $C_b(T)$ the subspace of all functions vanishing on a given set $T_0 \subset T$ is an ideal.

(iii) Let X be a Banach space. The class $\mathcal{K}(X)$ of all compact operators on X is an ideal in $\mathcal{L}(X)$. In addition, the class $\mathcal{F}$ of all operators of the form $\lambda I + K$, where $K \in \mathcal{K}(X)$, is a subalgebra (but not necessarily an ideal). As we have already noted, S. Argyros and R. Haydon [**666**] constructed an example of an infinite-dimensional space X for which $\mathcal{F}$ coincides with $\mathcal{L}(X)$.

(iv) If $\mathcal{A}$ is a commutative algebra with a unit 1 and $a_0 \in \mathcal{A}$ is a fixed element, then $a_0 \cdot \mathcal{A} := \{a_0 a \colon a \in \mathcal{A}\}$ is an ideal.

Every ideal $\mathcal{I}$ in an algebra $\mathcal{A}$ generates the factor-space $\mathcal{A}/\mathcal{I}$ consisting of equivalence classes $a + \mathcal{I}$, $a \in \mathcal{I}$. Two elements belong to the same class if their difference belongs to $\mathcal{I}$. This factor-space can be made an algebra (the linear structure already exists: it is a factor-space of a linear space with respect to its subspace), by letting $(a + \mathcal{I})(b + \mathcal{I}) := ab + \mathcal{I}$. This operation is well-defined, since $\mathcal{I}$ is an ideal. If the algebra $\mathcal{I}$ has a unit 1, then $1 + \mathcal{I}$ is a unit in $\mathcal{A}/\mathcal{I}$.

Let us give an example of a classical Banach algebra that has very few closed ideals.

11.2.3. Theorem. *Let H be a separable Hilbert space. In the algebra of bounded operators $\mathcal{L}(H)$ the only closed ideal different from zero and $\mathcal{L}(H)$ is the algebra of compact operators.*

PROOF. Let $\mathcal{I}$ be a nontrivial ideal in $\mathcal{L}(H)$ and let $A \in \mathcal{I}$ be a nonzero element. We show that $\mathcal{I}$ contains all one-dimensional operators of the form $P_{a,b}x = (x, a)b$. For this we take a vector $u \in H$ such that $v := Au \neq 0$. We can assume that $(v, v) = 1$. Then

$$P_{v,b}AP_{a,u}x = (x, a)P_{v,b}Au = (x, a)P_{v,b}v = (x, a)b = P_{a,b}x.$$

Since $\mathcal{I}$ is a two-sided ideal, we have $P_{a,b} \in \mathcal{I}$. Hence $\mathcal{I}$ contains all bounded finite-dimensional operator, but then by the closedness of $\mathcal{I}$ also all compact operators.

Suppose that $\mathcal{I}$ contains a noncompact operator A. Let $A = U|A|$ be the polar decomposition of A. Then $|A| = U^*A \in \mathcal{I}$. The operator $|A|$ is selfadjoint, nonnegative and noncompact. Theorem 7.8.6 yields that there exists a closed infinite-dimensional subspace H_0 in H such that we have $|A|(H_0) = H_0$ and $\| |A|h\| \geqslant \varepsilon \|h\|$ for some $\varepsilon > 0$ and all $h \in H_0$. For example, if the operator

$|A|$ is written in the form of multiplication by a bounded nonnegative function φ on $L^2(\mu)$, then for H_0 one can take the subspace of functions vanishing outside the set $\{t\colon \varphi(t) \geqslant \varepsilon\}$ for sufficiently small ε. There is a linear isometry V of the whole space H and H_0 (since both subspaces are infinite-dimensional separable Hilbert spaces). By Proposition 7.6.7 the operator V^* isometrically maps H_0 onto H and is zero on the orthogonal complement of H_0. Hence

$$V^*|A|V(H) = V^*|A|(H_0) = V^*(H_0) = H.$$

In addition,

$$\|V^*|A|Vx\| = \| |A|Vx\| \geqslant \varepsilon\|Vx\| = \varepsilon\|x\|, \quad x \in H.$$

Thus, the operator $V^*|A|V \in \mathcal{I}$ is invertible. Hence $I \in \mathcal{I}$, whence the equality $\mathcal{I} = \mathcal{L}(H)$ follows. $\qquad\square$

Let us consider an example of a Banach algebra which has many closed ideals.

11.2.4. Example. Let K be a nonempty compact space and let $C(K)$ be the algebra of continuous complex functions on K. Then for every set $S \subset K$ (which can be assumed to be closed) the class $\mathcal{I}_S$ of all functions in $C(K)$ vanishing on S is a closed ideal. Conversely, every closed ideal in $C(K)$ has the form $\mathcal{I}_S$ with some closed set S. Every closed ideal of codimension 1 has the form $\{x\colon x(k) = 0\}$ for some point $k \in K$.

PROOF. The first assertion is obvious. Suppose that $\mathcal{I}$ is a closed ideal. The set

$$S := \{k \in K\colon x(k) = 0 \; \forall x \in \mathcal{I}\}$$

is closed and $\mathcal{I} \subset \mathcal{I}_S$. Let $x \in \mathcal{I}_S$. We show that $x \in \mathcal{I}$. By the closedness of $\mathcal{I}$ it suffices to verify that for every $\varepsilon > 0$ there exists $x_\varepsilon \in \mathcal{I}$ with $\|x - x_\varepsilon\| \leqslant \varepsilon$. We observe that for every $k \in K$ there exists $x_k \in \mathcal{I}$ with $x_k(k) = x(k)$. Indeed, if $k \in S$, then we take $x_k = 0$. If $k \notin S$, then there exists $u_k \in \mathcal{I}$ with $u_k(k) \neq 0$, so there is a required element x_k. Set $U_k := \{t\colon |x(t) - x_k(t)| < \varepsilon\}$. This gives an open cover of the compact space K. We choose a finite subcover $U_{k_1}, \ldots, U_{k_n}$. Let us take a subordinated partition of unity (see Theorem 1.9.16), i.e., functions $f_i \in C(K)$, $i = 1, \ldots, n$, for which $0 \leqslant f_i \leqslant 1$, $\sum_{i=1}^n f_i = 1$ and $f_i(t) = 0$ if $t \notin U_{k_i}$. Then $x_\varepsilon := \sum_{i=1}^n f_i x_{k_i} \in \mathcal{I}$, since $\mathcal{I}$ is an ideal. It remains to observe that

$$|x_\varepsilon(t) - x(t)| \leqslant \sum_{i=1}^n f_i(t)|x_{k_i}(t) - x(t)| \leqslant \sum_{i=1}^n f_i(t)\varepsilon = \varepsilon,$$

since the number $f_i(t)|x_{k_i}(t) - x(t)|$ equals 0 if $t \notin U_{k_i}$ and does not exceed $\varepsilon f_i(t)$ if $t \in U_{k_i}$.

If the closed ideal $\mathcal{I}$ is of codimension 1, then the corresponding set S cannot contain different points. $\qquad\square$

We shall see below that in a commutative Banach algebra there is a one-to-one correspondence between nonzero characters and ideals of codimension 1, and on C^*-algebras there is a rich supply of characters. In the noncommutative case characters do not play such a role; they can be absent at all (for example, for $\mathcal{L}(\mathbb{C}^n)$ and $\mathcal{L}(l^2)$, see Exercise 11.7.19).

11.3. Spectra

A particular role is played by invertible elements of unital algebras, which are analogs of invertible operators. With their aid one defines spectra of all elements, which is important for constructing a functional calculus.

We shall say that an element a of a unital algebra $\mathcal{A}$ has a *left inverse* $a_l^{-1} \in \mathcal{A}$ if $a_l^{-1}a = 1$. Similarly, an element a_r^{-1} is called a *right inverse* for a if $aa_r^{-1} = 1$.

If a has a left inverse a_l^{-1} and a right inverse a_r^{-1}, then they coincide, since

$$a_l^{-1} = a_l^{-1}(aa_r^{-1}) = (a_l^{-1}a)a_r^{-1} = a_r^{-1}.$$

In this case the element $a_l^{-1} = a_r^{-1}$ is called the *inverse* to a and denoted by the symbol a^{-1}, and the element a is called *invertible*.

For example, in the algebra $C_b(T)$ the unit is the function identically equal to 1 and invertible elements are functions f with $\inf_{t \in T} |f(t)| > 0$. In the algebra of operators $\mathcal{L}(X)$ the unit is the identity operator and invertible elements are precisely invertible operator. We observe that if X is infinite-dimensional, then an operator A can have a left inverse (or even many left inverses), but can fail to have a right inverse and hence need not be invertible. In the algebra $\mathcal{P}[t]$ only nonzero scalars are invertible.

The product of two invertible elements is invertible: it is obvious that $b^{-1}a^{-1}$ is inverse to ab for all invertible elements a and b. Therefore, the product of invertible elements $a_1, \ldots, a_n$ is always invertible. As we have seen by example of operators, the product of noninvertible elements can be the unit, i.e., an invertible element. However, if the elements $a_1, \ldots, a_n$ commute and their product is invertible, then each element $a_1, \ldots, a_n$ is invertible. Indeed, letting $b = (a_1 \cdots a_n)^{-1}$, we obtain that the element a_i has a left inverse $ba_1 \cdots a_{i-1}a_{i+1} \cdots a_n$ and a right inverse $a_1 \cdots a_{i-1}a_{i+1} \cdots a_n b$, which is obvious from the commutativity of $a_1, \ldots, a_n$.

We observe that if $\mathcal{A}$ has a unit, then no proper ideal $\mathcal{I}$ in $\mathcal{A}$ contains the unit (otherwise $\mathcal{I} = \mathcal{A}$) and hence $\mathcal{I}$ has no invertible elements.

11.3.1. Definition. *Let $\mathcal{A}$ be a unital algebra. A complex number λ is called a regular point of the element $a \in \mathcal{A}$ if the element $a - \lambda 1$ is invertible. In the opposite case λ is called a singular point of a.*

The set of all singular points of the element a is called its spectrum and denoted by the symbol $\sigma(a)$.

In the case $\mathcal{A} = \mathcal{L}(X)$ these notions reduce to the ones previously introduced for operators.

11.3.2. Example. (i) The spectrum of an element $f \in C_b(T)$ coincides with the closure of $f(T)$ in $\mathbb{C}$. In particular, all points λ for which $|\lambda| > \sup_{t \in T} |f(t)|$ are regular.

(ii) Let us consider the algebra $\mathbb{C}^X$ of all complex functions on a nonempty set X with the pointwise multiplication. Then the spectrum of an element f is the set of values of f.

The following properties of spectra have already been encountered in the operator case.

11.3.3. Theorem. *Let $\mathcal{A}$ be a unital algebra, let $a, b \in \mathcal{A}$, and let p be a polynomial with complex coefficients.*

(i) *If $\sigma(a) \neq \varnothing$, then*

$$\sigma(p(a)) = p(\sigma(a)) := \{p(z) : z \in \sigma(a)\}.$$

(ii) *If the element a invertible, then*

$$\sigma(a^{-1}) = \{\lambda \in \mathbb{C} : \lambda^{-1} \in \sigma(a)\}.$$

In addition, for every $b \in \mathcal{A}$ we have

$$\sigma(ab) = \sigma(ba).$$

PROOF. (i) We can apply the same reasoning as in the case of operators. We can assume that p is not a constant, since otherwise our assertion is trivial (but here it is important that the spectrum of a is not empty). Let $\lambda \in \mathbb{C}$. The complex polynomial $p - \lambda$ has the roots $\lambda_1, \ldots, \lambda_n$ and can be written in the form $p(z) - \lambda = c(z - \lambda_1) \cdots (z - \lambda_n)$, where $c \neq 0$. It is readily seen that

$$p(a) - \lambda 1 = c(a - \lambda_1 1) \cdots (a - \lambda_n 1).$$

As noted above, by the commutativity of $a - \lambda_i 1$ the element $p(a) - \lambda 1$ is invertible precisely when all $a - \lambda_i 1$ are invertible. Hence the inclusion $\lambda \in \sigma(p(a))$ is equivalent to the property that $\lambda_i \in \sigma(a)$ for at least one λ_i. This is fulfilled if and only if $\lambda \in p(\sigma(a))$, since $p(\lambda_i) = \lambda$ and there are no other roots.

(ii) It is obvious that the spectra of a and a^{-1} do not contain zero. We observe that for all $\lambda \neq 0$ we have

$$a^{-1} - \lambda 1 = -\lambda a^{-1}(a - \lambda^{-1} 1),$$

from which the first assertion follows. For the proof of the last assertion we observe that $ba - \lambda 1 = a^{-1}(ab - \lambda 1)a$. $\qquad \square$

Let us note a number of important properties connected with invertibility in Banach algebras. These properties have been also encountered in our study of operators.

11.3.4. Theorem. *Let $\mathcal{A}$ be a unital Banach algebra. Then the following assertions hold.*

(i) *For every $a \in \mathcal{A}$ with $\|a\| < 1$ the element $1 - a$ is invertible and*

$$(1 - a)^{-1} = 1 + \sum_{n=1}^{\infty} a^n.$$

(ii) *The set Inv $\mathcal{A}$ of all invertible elements $\mathcal{A}$ is open an the mapping $a \mapsto a^{-1}$ is its homeomorphism.*

(iii) *The spectrum of every element $a \in \mathcal{A}$ is a nonempty compact set in $\mathbb{C}$ belonging to the disc $\{z: |z| \leqslant \|a\|\}$.*

PROOF. (i) The series in the right-hand side of the indicated equality converges with respect to the norm, since $\|a^n\| \leqslant \|a\|^n$. We denote its sum by b. It is straightforward to verify that

$$(1 - a)b = b(1 - a) = 1,$$

which proves our assertion.

(ii) Let $a \in \text{Inv}\,\mathcal{A}$, $b \in \mathcal{A}$ and $\|b\| < \|a^{-1}\|^{-1}$. Then $\|a^{-1}b\| < 1$. According to (i) the element $1 + a^{-1}b$ is invertible. This yields the invertibility of $a + b = a(1 + a^{-1}b)$. Thus, $\text{Inv}\,\mathcal{A}$ is an open set. Let us verify the continuity of the mapping $h: a \mapsto a^{-1}$ on $\text{Inv}\,\mathcal{A}$. The continuity at the unit follows from assertion (i), because whenever $\|a\| < 1$ we have

$$\|1 - (1 - a)^{-1}\| \leqslant \sum_{n=1}^{\infty} \|a\|^n = \frac{\|a\|}{1 - \|a\|}.$$

This yields the continuity at every point $a \in \text{Inv}\,\mathcal{A}$, since

$$(a - b)^{-1} - a^{-1} = (1 - a^{-1}b)^{-1}a^{-1} - a^{-1}$$

for all $b \in \mathcal{A}$ with a sufficiently small norm.

(iii) To show that $\sigma(a)$ is closed we use that for every regular number all sufficiently close numbers are also regular because $\text{Inv}\,\mathcal{A}$ is open. If $|\lambda| > \|a\|$, then the element $a - \lambda 1 = \lambda(\lambda^{-1}a - 1)$ is invertible by the estimate $\|\lambda^{-1}a\| < 1$. However, it is less obvious that the spectrum is not empty. But also here the reasoning does not differ from the operator case. Namely, for regular values λ we let $R(\lambda) := (a - \lambda 1)^{-1}$. Then for all regular values λ and μ the following Hilbert identity holds:

$$R(\lambda) - R(\mu) = (\lambda - \mu)R(\mu)R(\lambda).$$

For the proof it suffices to observe that after multiplication of both sides from the left by $a - \mu 1$ and from the right by $a - \lambda 1$ we obtain $(\lambda - \mu)1$. The Hilbert identity yields that for every bounded linear functional $\varphi: \mathcal{A} \to \mathbb{C}$ the function $f_\varphi: \lambda \mapsto \varphi(R(\lambda))$ is holomorphic on the set of regular values. Indeed, for all $\lambda \neq \mu$ we have

$$\frac{f_\varphi(\lambda) - f_\varphi(\mu)}{\lambda - \mu} = \varphi(R(\mu)R(\lambda)),$$

which tends to $\varphi(R(\lambda)^2)$ as $\mu \to \lambda$. If the spectrum of a were empty, then all functions f_φ would be entire. Since the function f_φ is bounded by the estimate

$$\|(a - \lambda 1)^{-1}\| \leqslant |\lambda|^{-1}\|(\lambda^{-1}a - 1)^{-1}\|$$

and $\|(\lambda^{-1}a - 1)^{-1}\|$ tends to $\|1\|$ as $|\lambda| \to \infty$, all these functions are identically zero by the Liouville theorem. Hence $R(\lambda) = 0$, which is impossible, since $1 \neq 0$ due to the condition $\mathcal{A} \neq 0$. The obtained contradiction proves that the spectrum is not empty. $\qquad\square$

As in the operator case, the last assertion can be sharpened by introducing the spectral radius of an element a of a Banach unital algebra by the formula

$$r(a) := \inf\{\|a^n\|^{1/n}:\ n \in \mathbb{N}\}.$$

The same reasoning as in Proposition 7.1.7 leads to the following conclusion.

11.3.5. Proposition. *The equality*

$$r(a) = \lim_{n\to\infty} \|a^n\|^{1/n}$$

holds. In addition,

$$r(a) = \max\{|z|:\ z \in \sigma(a)\}.$$

From the facts proved above we obtain the following Gelfand–Mazur theorem.

11.3.6. Theorem. *Any unital Banach algebra in which all nonzero elements are invertible is isomorphic as an algebra to the field* $\mathbb{C}$.

PROOF. By the previous theorem for every element $a \in \mathcal{A}$ there exists a scalar $\lambda \in \mathbb{C}$ such that the element $a - \lambda 1$ is not invertible. Then $a - \lambda 1 = 0$, i.e., $a = \lambda 1$. Hence $\mathcal{A} = \{\lambda 1:\ \lambda \in \mathbb{C}\}$, whence the claim follows. $\square$

For involutive algebras there is a useful connection between the spectra and the involution.

11.3.7. Proposition. *Let $\mathcal{A}$ be an involutive Banach algebra with a unit. Then*

$$\sigma(a^*) = \overline{\sigma(a)}, \quad a \in \mathcal{A}.$$

In addition, if $\mathcal{A}$ is a C^*-algebra and $a = a^*$, then $\sigma(a) \subset \mathbb{R}^1$.

PROOF. We recall that $1^* = 1$. Hence the element $b = a - \lambda 1$ is invertible if and only if the element $b^* = a^* - \bar{\lambda}1$ is invertible. The inverse to b^* is $(b^{-1})^*$.

Let $a^* = a$ and $\lambda = \alpha + i\beta \in \sigma(a)$, where $\alpha, \beta \in \mathbb{R}^1$. We show that $\beta = 0$. For every $\gamma \in \mathbb{R}^1$ we have $\alpha + i(\beta + \gamma) \in \sigma(a + i\gamma 1)$. Hence there holds the inequality $|\alpha + i(\beta + \gamma)| \leqslant \|a + i\gamma 1\|$. Therefore,

$$\alpha^2 + (\beta + \gamma)^2 \leqslant \|a + i\gamma 1\|^2 = \|(a + i\gamma 1)^*(a + i\gamma 1)\| = \|a^2 + \gamma^2 1\|,$$

which does not exceed $\|a^2\| + \gamma^2$, whence $\alpha^2 + \beta^2 + 2\beta\gamma \leqslant \|a^2\|$ for all $\gamma \in \mathbb{R}^1$. This is only possible if $\beta = 0$. $\square$

Let us give one more result useful for the sequel and connected with spectra. A subset M of an involutive algebra is called *normal* if

$$a^* \in M \quad \text{and} \quad ab = ba \quad \text{for all } a, b \in M.$$

11.3.8. Proposition. *Let $\mathcal{A}$ be an involutive Banach algebra with a unit and let $\mathcal{B}$ be a normal subset not contained in any other normal sets different from $\mathcal{A}$. Then $\mathcal{B}$ is a closed commutative subalgebra of $\mathcal{A}$ and the spectrum of every element $b \in \mathcal{B}$ does not depend on whether b is considered as an element of $\mathcal{A}$ or of $\mathcal{B}$.*

PROOF. We observe that $\mathcal{B}$ contains all elements $x \in \mathcal{A}$ such that $xx^* = x^*x$ and $xy = yx$ for all $y \in \mathcal{B}$. Indeed, in this case by the normality of $\mathcal{B}$ we have $xy^* = y^*x$ and $x^*y = yx^*$ for all $y \in \mathcal{B}$. Hence $\mathcal{B} \cup \{x, x^*\}$ is also normal, which by the maximality of $\mathcal{B}$ proves our claim.

It follows that the sums and products of elements of $\mathcal{B}$ belong to $\mathcal{B}$, i.e., $\mathcal{B}$ is a commutative subalgebra. Let $b_n \in \mathcal{B}$ and $b_n \to b$. Since $b_n x = x b_n$ for all $x \in \mathcal{B}$ and the multiplication is continuous, we have $bx = xb$. Hence $b^*x = (x^*b)^* = (bx^*)^* = xb^*$ for all $x \in \mathcal{B}$. In particular, $b^*b_n = b_n b^*$, whence $b^*b = bb^*$. Hence $b \in \mathcal{B}$, which proves that $\mathcal{B}$ is closed. Note that we have not assumed the continuity of the involution.

It also follows that $\mathcal{B}$ contains the unit. We show that if an element $b \in \mathcal{B}$ is invertible in $\mathcal{A}$, then it is invertible also in $\mathcal{B}$. The element b^* is invertible in $\mathcal{A}$ and $(b^*)^{-1} = (b^{-1})^*$. This yields the normality of b^{-1}. In addition, $b^{-1}x = xb^{-1}$ for all $x \in \mathcal{B}$, since $xb = bx$. Therefore, $b^{-1} \in \mathcal{B}$, i.e., the element b is invertible in $\mathcal{B}$.   □

In the general case the spectrum of an element in a subalgebra is not always equal to its spectrum in the whole algebra (Exercise 11.7.13).

11.4. Functional Calculus

It has already been noted that elements of a Banach algebra with a unit can be substituted in polynomials. More generally, if Q is a rational function on $\mathbb{C}$ such that its poles do not belong to the spectrum $\sigma(a)$ of an element a of an unital Banach algebra $\mathcal{A}$, then we can write

$$Q(\lambda) = P(\lambda) + c_1(\lambda - \alpha_1)^{-k_1} + \cdots + c_n(\lambda - \alpha_n)^{-k_n},$$

where P is a polynomial and $\alpha_i \notin \sigma(a)$, which enables us to define the element

$$Q(a) := P(a) + c_1 R_{\alpha_1}^{k_1} + \cdots + c_n R_{\alpha_n}^{k_n} \in \mathcal{A},$$

where $R_\lambda = (a - \lambda 1)^{-1}$. Here we go further and define the action on a of functions analytic in a neighborhood of $\sigma(a)$. This will be done with the aid of the Cauchy theorem. If γ is a curve in $\mathbb{C}$ parametrized by points of an interval with the aid of a piece-wise mapping and F is a continuous mapping on γ with values in $\mathcal{A}$, then the integral

$$\int_\gamma F(z)\, dz$$

is defined in the same way as for complex functions. It coincides with the Bochner integral briefly discussed in §6.10(vi), but for the purposes of this section it suffices to deal with the Riemann integral; if γ is given by a piece-wise smooth mapping $\alpha\colon [0, 1] \to \mathbb{C}$, then the indicated integral can be defined as the Riemann integral of $F(\alpha(t))\alpha'(t)$ over $[0, 1]$, which coincides with the integral of F with respect to the measure that equals the image of the measure $\alpha'(t)\, dt$ under the mapping α. For every continuous linear functional l on $\mathcal{A}$ the result of application of l to the integral of F is the integral of the continuous complex function $l \circ F$ over the contour γ. The idea of constructing a functional calculus is seen from the following lemma.

We shall say that a piece-wise smooth contour γ in a domain Ω is embracing a compact set K if in every connected component of Ω the contour γ bounds a connected domain containing the closure of the intersection of K with the regarded component.

We recall that the mapping $z \mapsto (z\mathbf{1} - a)^{-1}$ is continuous on the complement of the spectrum of a. Hence we can integrate such mappings multiplied by continuous functions.

11.4.1. Lemma. *Let Q be a rational function holomorphic in a domain Ω containing $\sigma(a)$ and let $\gamma \subset \Omega$ be a piece-wise contour embracing $\sigma(a)$ in the indicated sense. Then*

$$Q(a) = \frac{1}{2\pi i} \int_\gamma Q(z)(z\mathbf{1} - a)^{-1}\, dz.$$

In particular, for all $\alpha \notin \sigma(a)$ we have

$$(\alpha\mathbf{1} - a)^n = \frac{1}{2\pi i} \int_\gamma (\alpha - z)^n (z\mathbf{1} - a)^{-1}\, dz, \quad n \in \mathbb{Z}$$

if the contour γ in $\mathbb{C}\backslash\{\alpha\}$ embraces $\sigma(a)$.

PROOF. It is clear that it suffices to prove the last assertion. Let $n = 0$. We show that

$$\frac{1}{2\pi i} \int_\gamma (z\mathbf{1} - a)^{-1}\, dz = 1.$$

Let γ_0 be the circle of some radius $r > \|a\|$. Then the series

$$(z\mathbf{1} - a)^{-1} = \sum_{k=0}^{\infty} z^{-k-1} a^k$$

converges on γ_0 with respect to the norm. Integrating this series in z over γ_0, we obtain the integral of $z^{-1}\mathbf{1}$, i.e., $2\pi i\mathbf{1}$. By the Cauchy theorem the integral does not change if we replace γ_0 by γ (here it suffices to have the Cauchy theorem for scalar functions, since we can apply arbitrary continuous linear functionals on $\mathcal{A}$ to the considered integrals).

If $n \neq 0$, then the integral in the right-hand side of the desired equality will be denoted by J_n. Let us show that

$$J_{n+1} = (\alpha\mathbf{1} - a)J_n.$$

Then the general case will reduce to the case $n = 0$. We observe that

$$(z\mathbf{1} - a)^{-1} = (\alpha\mathbf{1} - a)^{-1} + (\alpha - z)(\alpha\mathbf{1} - a)^{-1}(z\mathbf{1} - a)^{-1}.$$

Multiplying by $(\alpha - z)^n$ and integrating over γ, we obtain the relation indicated above, since the integral of $(\alpha - z)^n$ over γ is zero due to the fact that α is outside the domain bounded by the contour γ. $\square$

Now for a given domain $\Omega \subset \mathbb{C}$ we consider the algebra $H(\Omega)$ of all functions holomorphic in Ω and the family of elements

$$\mathcal{A}_\Omega := \{a \in \mathcal{A}: \ \sigma(a) \subset \Omega\}.$$

Let $f \in H(\Omega)$. For every $a \in \mathcal{A}_\Omega$ we take a piece-wise smooth contour $\gamma \subset \Omega$ embracing $\sigma(a)$ and set

$$f(a) := \frac{1}{2\pi i} \int_\gamma f(z)(z\mathbf{1} - a)^{-1} \, dz.$$

The integral exists by the continuity on γ of the mapping $z \mapsto f(z)(z\mathbf{1} - a)^{-1}$. This definition does not depend on the choice of the contour with the indicated properties. Let $\widetilde{H}(\Omega)$ denote the collection of all $\mathcal{A}$-valued mappings $\widetilde{f}$ on $\mathcal{A}_\Omega$ of the form $\widetilde{f}: a \mapsto f(a)$, where $a \in \mathcal{A}_\Omega$ and $f \in H(\Omega)$.

11.4.2. Theorem. *The mapping $f \mapsto \widetilde{f}$ is an isomorphism from the algebra $H(\Omega)$ onto the algebra $\widetilde{H}(\Omega)$. If a sequence of functions $f_n \in H(\Omega)$ converges to a function f uniformly on compact sets Ω, then*

$$\lim_{n \to \infty} f_n(a) = f(a), \quad a \in \mathcal{A}_\Omega.$$

In addition, if $f(z) = z$ and $g(z) = 1$ in Ω, then $f(a) = a$ and $g(a) = \mathbf{1}$ for all elements $a \in \mathcal{A}_\Omega$.

PROOF. The last assertion follows from the lemma. It is clear that the mapping $f \mapsto \widetilde{f}$ is linear. If a sequence $\{f_n\} \subset H(\Omega)$ converges to f uniformly on compact sets in Ω, then for every $a \in \mathcal{A}_\Omega$ we can take a contour γ common for all f_n and f and embracing $\sigma(a)$ in Ω. By the boundedness of the function $z \mapsto \|(z\mathbf{1} - a)^{-1}\|$ on γ we obtain convergence of $f_n(a)$ to $f(a)$.

We now show that if $f, g \in H(\Omega)$ and $h = fg$, then for all $a \in \mathcal{A}_\Omega$ we have $h(a) = f(a)g(a)$. For rational functions this follows from the lemma. In the general case we can apply the Runge theorem from complex analysis, according to which there exist two sequences of rational functions f_n and g_n holomorphic in Ω and converging, respectively, to f and g uniformly on compact sets in Ω. Then the rational functions $f_n g_n$ converge to h uniformly on compact sets in Ω, which completes the proof due to the continuity of the correspondence $f \mapsto \widetilde{f}$ established above. $\square$

11.4.3. Proposition. *Let $\mathcal{A}$ be a unital Banach algebra with a continuous involution and let $a \in \mathcal{A}$ be a Hermitian element such that its spectrum does not intersect $(-\infty, 0]$. Then there exists a Hermitian element b such that $b^2 = a$.*

PROOF. Let Ω be the complement in $\mathbb{C}$ to the ray $(-\infty, 0]$ of the real axis and let f be the branch of the square root in Ω with $f(1) = 1$. There is a sequence of polynomials P_n in z converging to f uniformly on compact sets in Ω. Such polynomials can be chosen with real coefficients: taking any polynomials Q_n converging to f uniformly on compact sets in Ω (which is possible by the Runge theorem) and observing that $\overline{f(z)} = f(\overline{z})$, we pass to the complex polynomials $P_n(z) = [Q_n(z) + \overline{Q_n(\overline{z})}]/2$. Set now $b_n := P_n(a)$. The elements b_n are Hermitian. By the theorem above they converge in norm to some element $b = f(a)$. Since we assume the continuity of the involution, the limit element is Hermitian. Finally, $b_n^2 = P_n(a)P_n(a) = P_n^2(a) \to a$, since $P_n(z) \to z$ uniformly on compact sets in Ω. Hence $b^2 = a$. $\square$

Note that one can omit the continuity of the involution, but then the proof becomes longer (see Theorem 11.20 and Theorem 11.26 in [**522**]). Below in Proposition 11.6.5 this result will be sharpened in the case of C^*-algebras.

11.5. Commutative Banach Algebras

Here we discuss the Gelfand transform that enables one to identify commutative C^*-algebras with algebras of continuous complex functions on compact spaces.

Let Δ be the set of all nonzero characters of a commutative Banach algebra $\mathcal{A}$ (we recall that all characters are continuous). Below we shall see that $\Delta \neq \varnothing$. The formula

$$\widehat{a}(\chi) := \chi(a), \quad \chi \in \Delta,$$

associates with every element $a \in \mathcal{A}$ the function $\widehat{a} \colon \Delta \to \mathbb{C}$. This function is called the *Gelfand transform* of the element a. Let $\widehat{A}$ be the set of all such functions $\widehat{a}$ with $a \in \mathcal{A}$. The Gelfand topology on Δ is defined as the weakest topology with respect to which all functions from $\widehat{A}$ are continuous.

An ideal of an algebra is called *proper* if it differs from the whole algebra.

11.5.1. Proposition. *Let $\mathcal{A}$ be a commutative Banach algebra with a unit and let $\mathcal{I}$ be a closed proper ideal in it. Then the Banach factor-space $\mathcal{A}/\mathcal{I}$ has a natural structure of a Banach algebra with a unit, and the factor-mapping $\pi \colon \mathcal{A} \to \mathcal{A}/\mathcal{I}$ is a continuous homomorphism of algebras.*

PROOF. A multiplication on $\mathcal{A}/\mathcal{I}$ is defined by the formula

$$\pi(a)\pi(b) := \pi(ab), \quad a, b \in \mathcal{A}.$$

This operation is well-defined, since for $a - a' \in \mathcal{I}$ and $b - b' \in \mathcal{I}$ we have $a'b' - ab = (a' - a)b' + a(b' - b)$, whence $a'b' - ab \in \mathcal{I}$. It is readily seen that $\mathcal{A}/\mathcal{I}$ is an algebra and π is a continuous homomorphism (observe that $\|\pi(a)\| \leqslant \|a\|$). Let us show that

$$\|\pi(a)\pi(b)\| \leqslant \|\pi(a)\| \, \|\pi(b)\|.$$

Let $\varepsilon > 0$. By the definition of the factor-norm in the Banach space $\mathcal{A}/\mathcal{I}$ (see §5.1) there exist elements $u, v \in \mathcal{I}$ such that

$$\|a + u\| \leqslant \|\pi(a)\| + \varepsilon, \quad \|b + v\| \leqslant \|\pi(b)\| + \varepsilon.$$

Since $(a + u)(b + v) - ab \in \mathcal{I}$, we have

$$\|\pi(ab)\| \leqslant \|(a + u)(b + v)\| \leqslant \|a + u\| \, \|b + v\| \leqslant (\|\pi(a)\| + \varepsilon)(\|\pi(b)\| + \varepsilon),$$

whence the desired inequality follows, because ε was arbitrary. It is clear that $\pi(1)$ serves as a unit in the factor-algebra. $\square$

A *maximal ideal* is a proper ideal not contained in a larger proper ideal. An important role is played by the following theorem connecting characters with maximal ideals.

11.5.2. Theorem. *Let $\mathcal{A}$ be a commutative Banach algebra with a unit and let Δ be the set of all nonzero characters of $\mathcal{A}$. Then the following assertions are valid.*

(i) *Every ideal in $\mathcal{A}$ is contained in some maximal ideal.*

(ii) *Every maximal ideal is closed.*

(iii) *Every maximal ideal is the kernel of some character from Δ and the kernel of every character from Δ is a maximal ideal.*

PROOF. (i) The family $\mathfrak{M}$ of all proper ideals containing a given ideal is partially ordered by inclusion. Every chain in it has a majorant: the union of all elements of this chain (which, as is easily seen, also belongs to $\mathfrak{M}$, since the unit cannot belong to a proper ideal). By Zorn's lemma $\mathfrak{M}$ has a maximal element.

(ii) Let $\mathcal{M}$ be a maximal ideal and let $\overline{\mathcal{M}}$ be its closure. It is straightforward to verify that $\overline{\mathcal{M}}$ is an ideal. Since $\mathcal{M} \subset \overline{\mathcal{M}}$ and $\mathcal{M}$ is a maximal ideal, it suffices to verify that $\overline{\mathcal{M}}$ does not coincide with $\mathcal{A}$. Since $\mathcal{M}$ contains no invertible elements and the set $\mathcal{U}$ of invertible elements is open, the closure of $\mathcal{M}$ is contained in the complement to $\mathcal{U}$, which completes the proof.

(iii) Let $\mathcal{M}$ be a maximal ideal. We have proved that it is closed. Hence the factor-algebra $\mathcal{A}/\mathcal{M}$ is Banach. Pick an element $a \notin \mathcal{M}$. Then

$$\mathcal{J} := \{ab + c\colon\ b \in \mathcal{A}, c \in \mathcal{M}\}$$

is an ideal strictly containing $\mathcal{M}$ (since it contains a). Hence we have $\mathcal{J} = \mathcal{A}$. Hence there exist elements $b \in \mathcal{A}$ and $c \in \mathcal{M}$ such that $ab + c = 1$. For the factor-mapping $\pi\colon \mathcal{A} \to \mathcal{A}/\mathcal{M}$ we obtain $\pi(a)\pi(b) = \pi(1)$. This means that all nonzero elements of the algebra $\mathcal{A}/\mathcal{M}$ are invertible. By the Gelfand–Mazur theorem there exists an isomorphism $h\colon \mathcal{A}/\mathcal{M} \to \mathbb{C}$. Let $\chi := h \circ \pi$. Then $\chi \in \Delta$ and $\mathcal{M}$ is the kernel of χ. Conversely, if $\chi \in \Delta$, then $\chi^{-1}(0)$ is an ideal in $\mathcal{A}$ of codimension 1, i.e., a maximal ideal. $\qquad\square$

11.5.3. Example. (i) Let $\mathcal{A} = C(K)$, where K is a compact space. It follows from Example 11.2.4 and the theorem proved above that all nonzero characters of $\mathcal{A}$ have the form $x \mapsto x(k)$, where $k \in K$.

(ii) Let $\mathcal{A} = l^1(\mathbb{Z})$ (see Example 11.1.8(iv)). For every $z \in \mathbb{C}$ with $|z| = 1$ let $\varphi_z(x) := \sum_{n \in \mathbb{Z}} x_n z^n$. This gives a homomorphism, since

$$\varphi_z(x * y) = \sum_{n \in \mathbb{Z}} (x * y)_n z^n = \sum_{n \in \mathbb{Z}} \sum_{k \in \mathbb{Z}} x_k y_{n-k} z^n$$

$$= \sum_{k \in \mathbb{Z}} x_k z^k \sum_{n \in \mathbb{Z}} y_{n-k} z^{n-k} = \varphi_z(x)\varphi_z(y).$$

On the other hand, a multiplicative functional φ on $l^1(\mathbb{Z})$ is given by some bounded sequence of numbers $\theta_n = \varphi(e_n)$, where e_n is the element with 1 on the nth place and 0 on all other places. Set $z := \theta_1$. If $\varphi \neq 0$, then $\|\varphi\| = 1$, as shown above. Since $e_n * e_k = e_{n+k}$, we obtain $\theta_n = z^n$. In addition,

$$|z| = |\varphi(e_1)| \leqslant \|e_1\| = 1, \quad |z^{-1}| = |\varphi(e_{-1})| \leqslant \|e_{-1}\| = 1,$$

which gives $|z| = 1$, i.e., $\varphi = \varphi_z$.

(iii) The previous assertion and Example 11.1.8(v) imply that nonzero characters of the algebra W of absolutely convergent Fourier series have the form $f \mapsto f(t)$, $t \in [0, 2\pi]$.

(iv) One can show (Exercise 11.7.17) that nonzero characters of the convolution algebra $L^1(\mathbb{R}^1)$ have the form

$$f \mapsto \int_{\mathbb{R}^1} e^{-its} f(s)\, ds, \quad t \in \mathbb{R}^1.$$

(v) If $\mathcal{A}$ is a commutative Banach algebra and $a_0 \neq 0$ is a noninvertible element (by Theorem 11.3.6 such elements exist if $\dim \mathcal{A} > 1$) and there exists $a \in \mathcal{A}$ with $a_0 a \neq 0$, then $a_0 \cdot \mathcal{A}$ is a nontrivial ideal. Hence this ideal is contained in some maximal ideal, i.e., there exists a nonzero character vanishing on $a_0 \cdot \mathcal{A}$. For a general Banach algebra it can happen that there are no nonzero ideals: for example, let $\mathcal{A} = \mathcal{L}(\mathbb{C}^n)$. This cannot happen for commutative C^*-algebras.

The spectrum of an element can be connected with its Gelfand transform.

11.5.4. Proposition. *Let $\mathcal{A}$ be a commutative Banach algebra with a unit and let Δ be the space of its maximal ideals. Then the following assertions are true.*

(i) *An element $a \in \mathcal{A}$ is invertible precisely when the function $\widehat{a}$ has no zeros. This is also equivalent to the property that a is not contained in a proper ideal.*

(ii) *A point λ belongs to $\sigma(a)$ precisely when exists a character $\chi \in \Delta$ such that $\chi(a) = \lambda$, i.e., $\sigma(a) = \widehat{a}(\Delta)$. In this case*

$$\max_{\chi \in \Delta} |\widehat{a}(\chi)| = r(a) \leqslant \|a\|,$$

where $r(a)$ is the spectral radius of a.

PROOF. (i) If the element a invertible, then

$$\chi(a)\chi(a^{-1}) = \chi(1) = 1.$$

Hence $\chi(a) \neq 0$. If a is not invertible, then the set $\mathcal{I} := \{ab \colon b \in \mathcal{A}\}$ is an ideal not containing the unit. Hence $\mathcal{I}$ is contained in some maximal ideal and belongs to the kernel of some nonzero character χ, i.e., $\widehat{a}(\chi) = \chi(a) = 0$. It remains to recall that a proper ideal cannot contain invertible elements.

(ii) It suffices to apply (i) to $a - \lambda 1$ and use Proposition 11.3.5. $\qquad\square$

11.5.5. Theorem. *Let $\mathcal{A}$ be a commutative Banach algebra with a unit and let Δ be the space of its maximal ideals. Then the following assertions hold.*

(i) *The space Δ is Hausdorff and compact.*

(ii) *The Gelfand transform is a homomorphism from $\mathcal{A}$ onto a subalgebra $\widehat{\mathcal{A}}$ of the algebra $C(\Delta)$ and its kernel coincides with the intersection of all maximal ideals of $\mathcal{A}$.*

PROOF. (i) Let K be the closed unit ball of the space $\mathcal{A}^*$ dual to the Banach space $\mathcal{A}$. By the Banach–Alaoglu–Bourbaki theorem the set K is compact in the weak-$*$ topology. Since $\|\chi\| \leqslant 1$, we have $\Delta \subset K$. The Gelfand topology is the restriction to Δ of the topology $\sigma(\mathcal{A}^*, \mathcal{A})$. Hence it suffices to verify the closedness of Δ in K in the weak-$*$ topology. Suppose that an element $\chi_0 \in K$

belongs to the closure of K in the indicated topology. We have to show that χ_0 is a nonzero multiplicative functional, i.e.,

$$\chi_0(ab) = \chi_0(a)\chi_0(b) \quad \text{if } a, b \in \mathcal{A} \text{ and } \chi_0(1) = 1.$$

There is a net of elements $\chi_\alpha \in \Delta$ converging to χ_0. Since they satisfy the relations above, these relations remain valid for χ_0. Assertion (ii) is verified directly.   □

11.5.6. Remark. If $\|a^2\| = \|a\|^2$ for all $a \in \mathcal{A}$, then the Gelfand transform is an isometry. This is seen from the inductively verified relation $\|a^{2^k}\| = \|a\|^{2^k}$, giving the equality $\|\widehat{a}\|_\infty = r(a) = \|a\|$.

Now we can prove the following important result.

11.5.7. Theorem. (THE GELFAND–NAIMARK THEOREM) *Let $\mathcal{A}$ be a commutative C^*-algebra with a unit 1 of unit norm and let Δ be the space of its maximal ideals. Then the Gelfand transform is an isometric isomorphism between $\mathcal{A}$ and $C(\Delta)$, moreover, $\widehat{a^*} = \overline{\widehat{a}}$.*

PROOF. Let $a \in \mathcal{A}$ be an element such that $a^* = a$. We show that the function $\widehat{a}$ is real, i.e., $\chi(a) \in \mathbb{R}^1$ for every element $\chi \in \Delta$. For any real t let $z = a + it1$. Writing $\chi(a) = \alpha + i\beta$, where $\alpha, \beta \in \mathbb{R}^1$, we obtain

$$\chi(z) = \alpha + i(\beta + t), \quad zz^* = a^2 + t^2 1,$$

whence we find that

$$\alpha^2 + (\beta + t)^2 = |\chi(z)|^2 \leqslant \|z\|^2 = \|zz^*\| \leqslant \|a\|^2 + t^2,$$

which is only possible for $\beta = 0$.

Every element $a \in \mathcal{A}$ can be represented in the form $a = u + iv$, where $u = u^*$, $v = v^*$, by setting $u := (a + a^*)/2$. We have $a^* = u - iv$. As shown above, the functions $\widehat{u}$ and $\widehat{v}$ are real, which gives the relation $\widehat{a^*} = \overline{\widehat{a}}$.

Thus, the algebra $\widehat{\mathcal{A}}$ is closed with respect to the complex conjugation. By the Stone–Weierstrass theorem it is dense in $C(\Delta)$. It remains to show that it is closed and that the Gelfand transform is an isometry. It suffices to verify the latter. Let $a \in \mathcal{A}$ and $b = aa^*$. Then $b^* = b$, whence $\|b^2\| = \|b\|^2$. As noted in Remark 11.5.6, this gives $\|\widehat{b}\|_{C(\Delta)} = r(b) = \|b\|$. It follows from what we have proved that $\widehat{b} = |\widehat{a}|^2$, whence

$$\|\widehat{a}\|_{C(\Delta)}^2 = \|\widehat{b}\|_{C(\Delta)} = \|b\| = \|aa^*\| = \|a\|^2.$$

Thus, we have obtained an isometry.   □

We consider three simple cases in which the Gelfand transform is described explicitly.

11.5.8. Example. (i) Let K be a compact space and let $\mathcal{A} = C(K)$. We know that nonzero characters of $\mathcal{A}$ have the form $x \mapsto x(k)$, where $k \in K$. Hence the Gelfand space Δ is naturally identified with K.

(ii) All nonzero characters of the algebra $\mathcal{A} = l^1(\mathbb{Z})$ have the following form: $\varphi_z\colon (x_n) \mapsto \sum_{n\in\mathbb{Z}} x_n z^n$, where $z \in \mathbb{C}$ and $|z| = 1$. Hence here Δ can be identified with the unit circumference in $\mathbb{C}$: the mapping $z \mapsto \varphi_z$ is a homeomorphism, since it is continuous and one-to-one.

(iii) On the convolution algebra $L^1(\mathbb{R}^1)$ without a unit the characters have the form $f \mapsto \sqrt{2\pi}\widehat{f}(t)$, where $\widehat{f}$ is the Fourier transform of the function f and $t \in \mathbb{R}^1$ (Exercise 11.7.17). Hence the Gelfand transform of the element f can be identified with $\sqrt{2\pi}\widehat{f}$.

As we have seen, a Banach algebra without a unit can be represented as a closed subalgebra of codimension 1 (that is a two-sided ideal) in a Banach algebra with a unit, and if the original algebra was commutative or a C^*-algebra, then the extension inherits these properties (in the case of a C^*-algebra it is also important to choose the right norm on the extension, see Exercise 11.7.22). Hence every commutative C^*-algebra can be realized as a closed two-sided ideal of codimension 1 in the algebra $C(\Delta)$, where Δ is a compact space. As shown in Example 11.2.4, such ideal has the form $\{x\colon x(\delta) = 0\}$ for some point $\delta \in \Delta$.

It follows from the established facts that in the form of an algebra of continuous functions we can obtain the commutative C^*-algebra $L^\infty(\mu)$, where μ is a probability measure on a measurable space $(\Omega, \mathcal{B})$. This example seems especially surprising, because on Ω there is no topology at all. Even in the case of Lebesgue measure λ on an interval the corresponding space Δ is rather complicated; for example, it is nonmetrizable, since the space $L^\infty(\lambda)$ is nonseparable. Under the isomorphism of the algebras $L^\infty(\mu)$ and $C(\Delta)$ the equivalence class of the indicator function of a measurable set E is taken to some continuous function φ_E satisfying the identity $\varphi_E^2 = \varphi_E$. Hence this function assumes only two values 0 and 1, i.e., is the indicator function of some set $\widetilde{E}$. By the continuity of φ both sets $\varphi^{-1}(0)$ and $\varphi^{-1}(1)$ are simultaneously open and closed. Hence the set $\widetilde{E}$ is open. Since linear combinations of simple functions are dense in $L^\infty(\mu)$, we see that linear combinations of functions with finite sets of values are dense in $C(\Delta)$. It follows that Ω has a topology base consisting of sets that are simultaneously open and closed (the so-called clopen sets). See also §11.7(i).

11.6. The Structure of C^*-Algebras

The main result of this section characterizes unital C^*-algebras as subalgebras in the algebras of bounded operators on Hilbert spaces. An important role in the proof is played by positive functionals. A linear functional f on a Banach algebra $\mathcal{A}$ with an involution is called positive if

$$f(xx^*) \geqslant 0 \quad \forall x \in \mathcal{A}.$$

11.6.1. Example. Let Ω be a nonempty compact space. A linear functional f on the complex or real Banach algebra $C(\Omega)$ is positive in the sense of Banach algebras precisely when it is defined by a nonnegative Radon measure μ on Ω by the formula

$$f(x) = \int_K x(\omega)\,\mu(d\omega).$$

PROOF. Since $x^*x(\omega) = |x(\omega)|^2 \geqslant 0$, the measure $\mu \geqslant 0$ generates a positive functional. Conversely, if the functional f is positive in the sense of Banach algebras, then, as shown below, it is continuous. By the Riesz theorem this functional is given by a Radon measure μ, which is nonnegative, since it assumes nonnegative values on all nonnegative functions in $\varphi \in C(\Omega)$, which is clear from the equality $\varphi = \varphi^{1/2}\varphi^{1/2}$, where $\varphi^{1/2} \in C(\Omega)$ and $(\varphi^{1/2})^* = \varphi^{1/2}$. □

We also observe that for every topological space T the positivity of a linear functional on the real space $C_b(T)$ in the sense of positivity on all squares of elements is equivalent to its positivity in the sense of ordered vector spaces, since nonnegative functions in $C_b(T)$ are squares of functions in $C_b(T)$.

We note a number of simple properties of positive functionals.

11.6.2. Proposition. *Let f be a positive functional on a unital Banach algebra with a continuous involution. Then*
  (i) $f(x^*) = \overline{f(x)}$ *for all* $x \in \mathcal{A}$;
  (ii) $|f(xy^*)|^2 \leqslant f(xx^*)f(yy^*)$ *for all* $x, y \in \mathcal{A}$;
  (iii) $|f(x)|^2 \leqslant f(1)f(xx^*) \leqslant f(1)^2 r(xx^*)$ *for all* $x \in \mathcal{A}$;
  (iv) *the functional f is continuous.*

PROOF. For any $x, y \in \mathcal{A}$ let

$$\alpha = f(xx^*), \quad \beta = f(yy^*), \quad \gamma = f(xy^*), \quad \delta = f(yx^*).$$

Since $f\big((x + \lambda y)(x^* + \overline{\lambda}y^*)\big) \geqslant 0$ for all $\lambda \in \mathbb{C}$, we have

$$\alpha + \overline{\lambda}\gamma + \lambda\delta + |\lambda|^2\beta \geqslant 0.$$

This relation with $\lambda = 1$ and $\lambda = i$ yields that both numbers $\gamma + \delta$ and $i(\delta - \gamma)$ are real. Hence $\delta = \overline{\gamma}$. For $y = 1$ we obtain assertion (i). Let us take $\theta \in \mathbb{R}^1$ such that $e^{i\theta}\gamma = |\gamma|$. Let $\lambda = te^{-i\theta}$, where $t \in \mathbb{R}^1$. Then $t^2\beta + 2|\gamma|t + \alpha \geqslant 0$, which gives (ii). The first inequality in (iii) is (ii) with $y = 1$. For the proof of the second one we fix $t > r(xx^*)$. Then $\sigma(t1 - xx^*)$ is contained in the open half-plane. By Proposition 11.4.3 there exists a Hermitian element $b \in \mathcal{A}$ such that $b^2 = t1 - xx^*$. Hence

$$tf(1) - f(xx^*) = f(b^2) \geqslant 0,$$

whence we obtain $f(xx^*) \leqslant f(1)r(xx^*)$. Since we assume that the involution is continuous, there exists a constant C such that $\|x^*\| \leqslant C\|x\|$. Therefore, we have $r(xx^*) \leqslant \|xx^*\| \leqslant C\|x\|^2$, which shows the continuity of f. □

Note that this proposition is valid without the assumption about the continuity of the involution (see Theorem 11.31 in [522]), but the proof is longer.

11.6.3. Theorem. *For every element z in a C^*-algebra $\mathcal{A}$ with a unit of unit norm there exists a positive functional f on $\mathcal{A}$ such that*

$$f(zz^*) = \|z\|^2, \quad f(1) = 1.$$

PROOF. Let $a = zz^*$. The element a is Hermitian. Let us consider the closed subalgebra $\mathcal{A}_0 \subset \mathcal{A}$ generated by the elements a and $\mathbf{1}$. It is clear that this subalgebra is commutative and closed with respect to the involution. By the Gelfand–Naimark theorem the Gelfand transform establishes an isometric isomorphism between the algebra $\mathcal{A}_0$ and the algebra $C(\Delta)$ of continuous complex functions on the compact space Δ of maximal ideals on $\mathcal{A}_0$. Let $\xi \in \Delta$ be a point such that $|\widehat{a}(\xi)| = \|\widehat{a}\|_{C(\Delta)} = \|a\|$. It is clear that the functional $f_0 \colon x \mapsto \widehat{x}(\xi)$ is positive, takes $\mathbf{1}$ to 1 and equals $\|a\|$ at a. We observe that $\|f_0\| = 1$. By the Hahn–Banach theorem this functional extends with the same norm to the whole algebra $\mathcal{A}$. We denote the extension by f and show that $f(xx^*) \geqslant 0$ for every $x \in \mathcal{A}$. Let us fix $x \in \mathcal{A}$ and set $b := xx^*$. As above, there is a closed subalgebra $\mathcal{A}_1$ in $\mathcal{A}$ generated by the elements b and $\mathbf{1}$ and closed with respect to the involution. By the Gelfand–Naimark theorem it is isometrically isomorphic to the algebra $C(\Delta_1)$ of continuous complex functions on the compact space Δ_1, the space of maximal ideals of $\mathcal{A}_1$, and to positive functionals there correspond positive functionals (by the equality $\widehat{a^*} = \overline{\widehat{a}}$ in Theorem 11.5.7). Hence it suffices to verify that every functional g on $C(\Delta_1)$ with the unit norm and $g(1) = 1$ is positive. As noted in Example 11.6.1, we have to verify that g is nonnegative on nonnegative functions. Let $\varphi \in C(\Delta_1)$ and $0 \leqslant \varphi \leqslant 1$. Set $g(\varphi) = \alpha + i\beta$. Then $|g(\varphi)| \leqslant 1$ and $|1 - g(\varphi)| = |g(1 - \varphi)| \leqslant 1$, which gives $\alpha \geqslant 0$. In addition, for all real t we have $|g(e^{it\varphi})| \leqslant 1$. This is only possible for $\beta = 0$, since it is easy to see that $g(e^{it\varphi}) = 1 + itg(\varphi) + o(t)$. $\qquad\square$

11.6.4. Theorem. (THE GELFAND–NAIMARK THEOREM) *Suppose that $\mathcal{A}$ is a C^*-algebra with a unit of unit norm. Then there exists a Hilbert space H along with an isometric isomorphism j from the algebra $\mathcal{A}$ onto a closed subalgebra in $\mathcal{L}(H)$ for which $j(a^*) = j(a)^*$.*

PROOF. First we show that for every nonzero element $h \in \mathcal{A}$ there exists a Hilbert space H_h along with a homomorphism $T_h \colon \mathcal{A} \to \mathcal{L}(H_h)$ such that for all $x \in \mathcal{A}$ we have

$$T_h(1) = I, \quad \|T_h(h)\| = \|h\|, \quad T_h(x^*) = \big(T_h(x)\big)^*, \quad \|T_h(x)\| \leqslant \|x\|.$$

It was shown in the previous theorem that there exists a positive functional f on $\mathcal{A}$ such that $f(1) = 1$ and $f(hh^*) = \|h\|^2$. By means of this functional we define an inner product. To this end we consider the linear space

$$L := \big\{ y \in \mathcal{A} \colon f(xy) = 0 \quad \text{for all } x \in \mathcal{A} \big\}.$$

The continuity of f yields that L is closed. On the factor-space $\mathcal{A}/L$ we define an inner product by the formula

$$(\widetilde{a}, \widetilde{b}) := f(b^*a),$$

where a and b are arbitrary representatives of the equivalence classes $\widetilde{a}$ and $\widetilde{b}$. This number is well-defined, i.e., does not depend on the choice of representatives of classes, since $f(b^*a) = 0$ if at least one of the elements a or b belongs to L. If $a \in L$, then this is clear from the definition of L, and if $b \in L$, then by assertion (i) of Proposition 11.6.2 we have $f(b^*a) = \overline{f(a^*b)} = 0$. The linearity of $(\widetilde{a}, \widetilde{b})$ in the first argument is obvious (with respect to the second argument this function is conjugate-linear). Finally, $f(a^*a) \geqslant 0$, moreover, by assertion (ii) of the aforementioned proposition the equality $f(a^*a) = 0$ yields that $a \in L$. For H we take the completion of the space $\mathcal{A}/L$ with the introduced inner product. The required operator is defined on $\mathcal{A}/L$ by the equality

$$T(x)\widetilde{y} := \widetilde{xy}.$$

The vector h is fixed, so the dependence of T on h is not indicated. It is easy to verify that this definition does not depend on the choice of a representative in the class. It is also clear that the mapping T is linear, takes the unit to the unit operator and $T(x_1x_2) = T(x_1)T(x_2)$. Let us show that

$$\|T(x)\| \leqslant \|x\|, \quad x \in \mathcal{A}.$$

This estimate will imply that T uniquely extends to a continuous operator on H with the required properties. We have

$$\|T(x)\widetilde{y}\|^2 = (\widetilde{xy}, \widetilde{xy}) = f(y^*x^*xy).$$

For any fixed y the functional $\varphi \colon x \mapsto f(y^*xy)$ is positive. According to assertion (iii) of Proposition 11.6.2 we have

$$|\varphi(x^*x)| \leqslant \varphi(1)\|x\|^2 = \|\widetilde{y}\|^2\|x\|^2,$$

i.e., $\|T(x)\widetilde{y}\|^2 \leqslant \|\widetilde{y}\|^2\|x\|^2$. Thus, $\|T(x)\| \leqslant \|x\|$. Since

$$\|\widetilde{1}\|^2 = f(1^*1) = f(1) = 1,$$

we have

$$\|h\|^2 = f(h^*h) = \|T(h)\widetilde{1}\|^2 \leqslant \|T(h)\|^2,$$

which gives the equality $\|T(h)\| = \|h\|$. Finally, the relations

$$(T(x^*)\widetilde{a}, \widetilde{b}) = (\widetilde{x^*a}, \widetilde{b}) = f(b^*x^*a) = f((xb)^*a) = (\widetilde{a}, \widetilde{xb})$$
$$= (\widetilde{a}, T(x)\widetilde{b}) = (T(x)^*\widetilde{a}, \widetilde{b})$$

show that $T(x^*)\widetilde{a} = T(x)^*\widetilde{a}$ for all $\widetilde{a} \in \mathcal{A}/L$. Since $\mathcal{A}/L$ is dense in H, the established equality extends to all elements in H.

Now for every nonzero element $h \in \mathcal{A}$ we take the corresponding Hilbert space H_h and operator T_h on H_h. For H we take the direct sum of all such spaces H_h, i.e., the space of all collections $v = \{v_h\}_{h \in \mathcal{A} \setminus \{0\}}$ with $v_h \in H_h$ such that at most countably many elements v_h differ from zero and

$$\|v\|^2 := \sum_h \|v_h\|_{H_h}^2 < \infty.$$

The linear structure is introduced component-wise, the inner product is given by the formula

$$(u, v)_H := \sum_h (u_h, v_h)_{H_h}.$$

To every element $a \in \mathcal{A}$ we put in correspondence an operator $B(a) \in \mathcal{L}(H)$ by the formula

$$(B(a)v)_h := T_h(a)v_h,$$

where T_h is the operator on H_h constructed above. The described construction is called the GNS-*construction* after the names of Gelfand, Naimark and Segal. Since $\|T_h(a)\| \leqslant \|a\| = \|T_a(a)\|$, we have

$$\|B(a)\| = \sup_h \|T_h(a)\| = \|a\|,$$

which completes the proof. □

An embedding of a C^*-algebra into the algebra of operators is not only of important theoretical significance, but is also useful for proofs of many technical assertions, which can be proved directly, but longer. Let us give an example complementing Proposition 11.4.3.

11.6.5. Proposition. *Let $\mathcal{A}$ be a C^*-algebra with a unit. Let $a \in \mathcal{A}$ be a Hermitian element with the spectrum in $[0, +\infty)$. Then there exists a Hermitian element b with the spectrum in $[0, +\infty)$ such that $b^2 = a$. In addition, for every $u \in \mathcal{A}$ the element u^*u is Hermitian with the spectrum in $[0, +\infty)$.*

Finally, an element $u \in \mathcal{A}$ is Hermitian with a nonnegative spectrum precisely when $l(u) \geqslant 0$ for every nonnegative functional l.

PROOF. In the case of the algebra of operators on a Hilbert space H these assertions are obvious, since for b one can take $\sqrt{a}$ and the operator u^*u is nonnegative. The last assertion characterizing nonnegative operators follows from the fact that for every vector $h \in H$ the functional $A \mapsto (Ah, h)$ is nonnegative and if $(Th, h) \geqslant 0$ for all h, then the operator T is nonnegative selfadjoint. It is clear that the considered properties are preserved by involutive isomorphisms. □

11.6.6. Remark. The element b constructed above can be taken as a definition of $\sqrt{a}$. Next for every $a \in \mathcal{A}$ one can set $|a| := \sqrt{a^*a}$ by analogy with the case of operators. Then the isomorphism j from the previous theorem preserves absolute values, i.e., $j(|a|) = |j(a)|$, where $|j(a)| = \sqrt{j(a)^*j(a)}$. Indeed, $j(a^*a) = j(a)^*j(a)$ by the proved properties of j. By our definition $|a|$ is exactly $j^{-1}\big(\sqrt{j(a)^*j(a)}\big)$. In the case of the algebras $L^\infty(\mu)$ and $C_b(T)$ this absolute value coincides with the pointwise defined absolute value of the function.

Note that any separable C^*-algebra can be embedded into the operator algebra $\mathcal{L}(l^2)$ (Exercise 11.7.21). However, the factor-algebra $\mathcal{C}(l^2) = \mathcal{L}(l^2)/\mathcal{K}(l^2)$, called the *Calkin algebra*, does not possesses this property (Exercise 11.7.26).

11.7. Complements and Exercises

(i) The algebras $C(K)$ and L^∞ (505). (ii) Von Neumann algebras (507). (iii) Haar measures and representations of groups (508). Exercises (509).

11.7(i). The algebras $C(K)$ and L^∞

Let $(\Omega, \mathcal{A}, \mu)$ be a probability space and let $L^\infty(\mu)$ be the complex Banach space of equivalence classes of bounded μ-measurable functions. This space with its natural multiplication corresponding to the pointwise multiplication of representatives of classes is a commutative C^*-algebra with the unit the role of which is played by the function 1. By the Gelfand–Naimark theorem the Gelfand transform is an isometric isomorphism between $L^\infty(\mu)$ and $C(\Delta)$, where Δ is the space of maximal ideals, compact in the Gelfand topology. The Gelfand transform $f \mapsto \widehat{f}$ will be denoted by j. Since $j(\overline{f}) = \overline{j(f)}$, for all $f \geqslant 0$ we have $j(f) \geqslant 0$ in the sense that in the equivalence class of $j(f)$ there is a pointwise nonnegative representative. This is seen from the fact that $j(f)$ has a real representative and $j(f) = j(f^{1/2})j(f^{1/2})$. In addition, if $\varphi \geqslant 0$, then $j^{-1}(\varphi) \geqslant 0$. The isomorphism j enables us to transport the measure μ to Δ.

11.7.1. Proposition. *On the σ-algebra in Δ generated by $C(\Delta)$ there exists a unique probability measure ν for which*

$$\int_\Omega f(\omega)\overline{g(\omega)}\,\mu(d\omega) = \int_\Delta \widehat{f}(\omega)\overline{\widehat{g}(\omega)}\,\nu(d\omega) \quad \forall\, f,g \in \mathcal{L}^\infty(\mu). \tag{11.7.1}$$

In addition, the spaces $L^2(\mu)$ and $L^2(\nu)$ are isometric.

PROOF. Let us consider the functional

$$\Lambda\colon \varphi \mapsto \int_\Omega j^{-1}(\varphi)(\omega)\,\mu(d\omega),$$

where in the integral we use a version of the element $j^{-1}(\varphi)$. It is clear that this functional is continuous and by the Riesz theorem (see Theorem 6.10.76) is represented as the integral of φ with respect to some nonnegative measure ν on the σ-algebra in Δ generated by $C(\Delta)$. We observe that $\Lambda(1) = 1$ and $\Lambda(\varphi) \geqslant 0$ for all $\varphi \geqslant 0$. Hence ν is a probability measure. It is clear that equality (11.7.1) is fulfilled if $f = g$. By the linearity of j it is fulfilled for all $f, g \in \mathcal{L}^\infty(\mu)$. Since $L^\infty(\mu)$ is dense in $L^2(\mu)$ and $C(\Delta)$ is dense in $L^2(\nu)$, the mapping j uniquely extends by the continuity to an isometry of the given L^2-spaces. The uniqueness of ν follows from the fact that the Gelfand transform maps $L^\infty(\mu)$ onto $C(\Delta)$ and any measure on the σ-algebra generated by $C(\Delta)$ is uniquely determined by the integrals of the elements of $C(\Delta)$. □

11.7.2. Remark. The measure ν can be uniquely extended to a regular Borel measure, i.e., a measure satisfying the equality

$$\nu(B) = \sup\{\nu(K)\colon K \text{ is compact in } B\}$$

for all Borel sets B. In this case $C(\Delta)$ is also dense in the corresponding L^2 (Exercise 3.12.44).

For ν-measurable functions the following stronger version of Luzin's theorem holds that fails for most of Borel measures on metric spaces, in particular, for Lebesgue measure on $[0, 1]$.

11.7.3. Corollary. *Let ν be the measure from the previous theorem. Then every bounded ν-measurable function ν-a.e. equals some continuous function.*

PROOF. It suffices to consider real functions. Let $f \in \mathcal{L}^\infty(\nu)$ take values in $[-M, M]$. There is a sequence $\{f_n\} \subset C(\Delta)$ converging to f in $L^2(\nu)$. Passing to a subsequence, we can assume that $f_n(s) \to f(s)$ for ν-a.e. s. Let us take a function $\theta\colon \mathbb{R}^1 \to [-M, M]$ for which $\theta(t) = t$ if $|t| \leqslant M$, $\theta(t) = M$ if $t > M$, and $\theta(t) = -M$ if $t < -M$. Then $\theta \circ f_n \to f$ ν-a.e. By the Lebesgue dominated convergence theorem we obtain convergence in $L^2(\nu)$. Hence $j^{-1}(\theta \circ f_n) \to j^{-1}(f)$ in $L^2(\mu)$. The sequence of functions $j^{-1}(\theta \circ f_n)$ is bounded in the norm of $L^\infty(\mu)$ by the number M, since $|\theta \circ f_n(s)| \leqslant M$. Therefore, $g := j^{-1}(f) \in L^\infty(\mu)$, which gives the equality $j(g) = f$ in $L^2(\mu)$, i.e., $j(g) = f$ ν-a.e. It remains to recall that $j(g) \in C(\Delta)$. $\square$

Let us give an example of using the isomorphism $j\colon L^\infty(\mu) \to C(\Delta)$.

11.7.4. Example. Suppose that a sequence $\{f_n\}$ converges weakly to zero in $L^\infty(\mu)$. Then $\{|f_n|\}$ also converges weakly to zero.

PROOF. In the notation from the previous reasoning we have $j(|f|) = |j(f)|$ for all $f \in L^\infty(\mu)$ (see Remark 11.6.6). Since j is a linear isometry, weak convergence of elements φ_n in $L^\infty(\mu)$ is equivalent to weak convergence of elements $j(\varphi_n)$ in $C(\Delta)$. Therefore, it suffices to verify that weak convergence of elements $\psi_n \in C(\Delta)$ to zero yields weak convergence of $|\psi_n|$ to zero. The desired implication is obvious from the fact that weak convergence in $C(\Delta)$ is equivalent to the pointwise convergence along with the uniform boundedness (see Example 6.6.6, which remains valid for general compact spaces). $\square$

11.7(ii). Von Neumann algebras

Let $\mathcal{A}$ be an involutive algebra and let M be its subset. The set

$$M' := \{a \in \mathcal{A}\colon ab = ba \ \forall b \in M\}$$

is called the *commutant* of M. It is clear that M' is a subalgebra in $\mathcal{A}$. If M is closed with respect to the involution, then M' is an involutive algebra (and if $\mathcal{A}$ is a C^*-algebra, then so is the algebra M'). The algebra $M'' = (M')'$ is called the *bicommutant* of M. An involutive subalgebra M of the algebra $\mathcal{L}(H)$ of all bounded operators on a Hilbert space H is called a *von Neumann algebra* if we have $M = M''$. For example, the whole algebra $\mathcal{L}(H)$ is a von Neumann algebra, since its commutant consists of the operators of the form λI. The ideal $\mathcal{K}(H)$ of compact operators is not a von Neumann algebra in the case of an infinite-dimensional space H, because $\mathcal{K}(H)' = \mathcal{L}(H)'$.

Let us mention the following theorem due to Sakai.

11.7.5. Theorem. *A C^*-algebra $\mathcal{A}$ is $*$-isomorphic to some von Neumann algebra precisely when the Banach space $\mathcal{A}$ is the dual to some Banach space.*

Commutative von Neumann algebras are described by the following theorem due to von Neumann.

11.7.6. Theorem. *Every commutative von Neumann algebra is isometrically $*$-isomorphic to $L^\infty(\mu)$ for some measure μ.*

A von Neumann algebra M is called a factor if $M \cap M'$ consists of the operators λI. Factors play an important role in the study of von Neumann algebras.

11.7(iii). Haar measures and representations of groups

The results of this chapter show in particular that reasonable algebras can be realized as operator algebras. Here we briefly discuss representations of groups by operators. A representation of a group G is its homomorphism into a group of linear operators on a linear space. A unitary (or orthogonal in the real case) representation is a homomorphism into the group $\mathcal{U}(H)$ of unitary operators on a Hilbert space H. This algebraic picture is complemented by certain topological and analytical conditions. For example, the group G is supposed to be topological, which means that it is equipped with a Hausdorff topology such that the operations $(g_1, g_2) \mapsto g_1 g_2$ and $g \mapsto g^{-1}$ of multiplication and inversion are continuous. In addition, one considers continuous or measurable representations. A classical example is the case where G is a compact topological group. Then a new important object appears: Haar measures. The Borel σ-algebra of a topological space T is the smallest σ-algebra containing all open sets. If T is a metric space, then this is the smallest σ-algebra with respect to which all continuous functions are measurable (the latter is called the Baire σ-algebra in the general case). Measures on Borel σ-algebras are called Borel measures.

11.7.7. Definition. *A left-invariant Haar measure on G is a nonzero bounded nonnegative Borel measure m on G that is inner compact regular in the sense that*

$$m(B) = \sup\{m(K): K \subset B, \ K \text{ is compact}\}$$

for all Borel sets B and $m(g \cdot B) = m(B)$ for all $g \in G$.

A right-invariant Haar measure is defined similarly with $m(B) = m(B \cdot g)$.

11.7.8. Theorem. *A left-invariant Haar measure exists on any compact group and is unique up to a positive factor. The same is true for right-invariant Haar measures.*

PROOF. To simplify some technical details we consider the case of a metric compact group (G, d) and show only the existence of a Haar probability measure. By the Riesz theorem (see Theorem 6.10.76) the space of Borel measures on G can be identified with the dual to $C(G)$. The subset P of $C(G)^*$ consisting of nonnegative functionals equal to 1 on 1 corresponds to probability measures. The elements of G generate continuous operators on $C(G)^*$ with the weak-$*$ topology by the formula $T_g l(f) = l(f_g)$, where $f_g(h) = f(gh)$ for $h, g \in G$. These operators are equicontinuous on P with the weak-$*$ topology and take P into P. Indeed, given $f_1, \ldots, f_n \in C(G)$ and $\varepsilon > 0$, we can find $\delta > 0$ such that $|f_i(g_1 h) - f_i(g_2 h)| < \varepsilon$ for all $i \leqslant n$, $h \in G$ whenever $d(g_1, g_2) < \delta$. Then $|T_{g_1} l(f_i) - T_{g_2} l(f_i)| < \varepsilon$ for all $l \in P$. By Kakutani's theorem (see Theorem 8.6.36) there is a common fixed point $l_0 \in P$ for all T_g. The functional l_0 corresponds to a left-invariant probability measure for G. Note that Rudin [522, Chapter 5] applies Kakutani's theorem in a different way to certain norm-compact sets. $\qquad\square$

Moreover, the theorem extends to locally compact groups if we admit Haar measures with values in $[0, +\infty]$; such measures are required to be finite on compact sets and not identically zero and also inner compact regular. This import result goes back to Haar with later developments due to Banach, Cartan, von Neumann, and Weil. Its proof (which is more involved) can be found in many textbooks, see [73, Chapter 9], [177], and [517]. Some authors require the inner compact regularity only for open sets, but add the condition that $m(B)$ be the infimum of $m(U)$ over all open sets $U \supset B$. The difference occurs for some non-σ-compact groups, but for separable metrizable locally compact groups both definitions coincide. For nonmetrizable groups, it is also possible to define the Haar measure only on the Baire σ-algebra. An in-depth study of Haar measures is given by Fremlin [197].

By using the Haar measure m we can construct the following unitary representation of G in the space $L^2(m)$:

$$U(g)\varphi(h) = \varphi(gh), \quad \varphi \in L^2(m), \ g, h \in G.$$

Obviously, $U(g)$ is a unitary operator by the left invariance of m. For example, if $G = S^1 = \{z \colon |z| = 1\}$, then m is the usual Lebesgue measure on S^1 and $U(g)$ is the shift by g. Similarly one defines the unitary operators $U(g)$ in the case of a locally compact group. For example, if $G = \mathbb{R}^n$, then the standard Lebesgue measures serves as the Haar measure and $U(x)\varphi(y) = \varphi(y + x)$.

Exercises

11.7.9° Let $\mathcal{A}$ be a Banach algebra with a unit. Prove the equality

$$\sigma(ab)\backslash\{0\} = \sigma(ba)\backslash\{0\}, \quad a, b \in \mathcal{A}.$$

Prove also that ab and ba have equal spectral radii.

11.7.10° The complex space l^p with $1 \leqslant p < \infty$ can be equipped with the coordinate-wise multiplication $(xy)_n = x_n y_n$. Show that we have obtained a Banach algebra.

11.7.11° Let $\mathcal{A}$ be a C^*-algebra. Show that $\|a\| = \|a^*\|$ and $\|aa^*\| = \|a\|^2$ for all elements $a \in \mathcal{A}$.

11.7.12° Let $\mathcal{M}(\mathbb{R}^n)$ be the space of all complex Borel measures of bounded variation on $\mathbb{R}^n$ with the norm $\|\mu\| = |\mu|(\mathbb{R}^n)$. The convolution will serve as a multiplication:

$$\mu * \nu(B) := \int_{\mathbb{R}^n} \int_{\mathbb{R}^n} I_B(x + y) \, \mu(dx) \, \mu(dy), \quad B \in \mathcal{B}(\mathbb{R}^n).$$

Show that $\mathcal{M}(\mathbb{R}^n)$ is a commutative Banach algebra with a unit, the role of which is played by the Dirac probability measure concentrated at the origin.

11.7.13. Let us consider the convolution algebra $\mathcal{A} = l^1(\mathbb{Z})$ and let us take its subalgebra $\mathcal{B} := \{(x_n) \in \mathcal{A} \colon x_n = 0 \text{ for all } n < 0\}$. Show that the element e_1 has different spectra in $\mathcal{A}$ and $\mathcal{B}$.

11.7.14° Let $\mathcal{A} = C(U)$ be the algebra of continuous complex functions on the unit disc U in $\mathbb{C}$ with the norm $f \mapsto \max_{z \in U} |f(z)|$, the pointwise multiplication and the involution $f^*(z) := \overline{f(\bar{z})}$. Show that $\|f^*\| = \|f\|$, but $\mathcal{A}$ is not a C^*-algebra.

11.7.15. Show that the convolution algebras $L^1(\mathbb{R}^n)$, $\mathcal{M}(\mathbb{R}^n)$ and $l^1(\mathbb{Z})$ are not C^*-algebras with respect to the involution defined either by $x^*(t) := \overline{x(-t)}$ or by $x^* := \bar{x}$, although $\|x^*\| = \|x\|$.

11.7.16. Prove that the algebra $L^1(\mathbb{R}^n)$ with the convolution operation has no unit. HINT: use the Fourier transform.

11.7.17. Prove that all nonzero characters of the convolution algebra $L^1(\mathbb{R}^1)$ have the form

$$f \mapsto \int_{\mathbb{R}^1} e^{-its} f(s)\,ds, \quad t \in \mathbb{R}^1.$$

HINT: see [**519**, § 9.22].

11.7.18.° Prove that $\mathcal{L}(\mathbb{C}^n)$ has no nontrivial ideals. More generally, every nonzero ideal in $\mathcal{L}(X)$ contains all finite-dimensional bounded operators for every Banach space X.

HINT: observe that any nonzero ideal contains all one-dimensional bounded operators.

11.7.19.° (i) Prove that there are no nonzero characters on $\mathcal{L}(\mathbb{C}^n)$, $n > 1$. (ii) Prove that there are no nonzero characters on $\mathcal{L}(l^2)$. (iii) Prove that there are nonzero characters on the algebra of operators of the form $\lambda I + K$, where $K \in \mathcal{K}(l^2)$.

HINT: consider the case $n = 2$.

11.7.20.° Let $\mathcal{A}$ be a C^*-algebra and let $h\colon \mathcal{A} \to \mathbb{C}$ be a homomorphism. Show that (i) $h(a) \in \mathbb{R}^1$ if $a = a^*$, (ii) $h(a^*) = \overline{h(a)}$ and $h(a^*a) \geqslant 0$ for all $a \in \mathcal{A}$.

11.7.21.° Let $\mathcal{A}$ be a separable C^*-algebra with a unit. Show that there exists an isometric $*$-homomorphism from $\mathcal{A}$ to $\mathcal{L}(l^2)$.

11.7.22. Let $\mathcal{A}$ be a C^*-algebra without a unit and let $\widetilde{\mathcal{A}}$ be its standard extension to a unital algebra consisting of all pairs of the form (a, α), where $a \in \mathcal{A}$ and $\alpha \in \mathbb{C}$, with the multiplication $(a, \alpha)(b, \beta) := (ab + \alpha b + \beta a, \alpha\beta)$. Show that $\widetilde{\mathcal{A}}$ has a unique norm making $\widetilde{\mathcal{A}}$ a C^*-algebra and coinciding with the original norm on the algebra $\mathcal{A}$ regarded as a subspace in $\widetilde{\mathcal{A}}$.

HINT: for $z = (a, \alpha)$ set $\|z\| := \|L_z\|$, where the operator L_z on $\mathcal{A}$ is defined by the formula $L_z x := ax + \alpha x$, $x \in \mathcal{A}$.

11.7.23. Prove that any nonzero character of the algebra $\mathcal{A}(U)$ of all functions holomorphic in the unit disc $U \subset \mathbb{C}$ and continuous on $\overline{U}$ (with the sup-norm) has the form $\varphi \mapsto \varphi(z)$, where $z \in \overline{U}$. Moreover, Δ with the Gelfand topology is homeomorphic to $\overline{U}$.

11.7.24. Prove that the Sobolev space $W^{p,1}(\mathbb{R}^n)$ with $p > n$ is a Banach algebra with respect to the pointwise multiplication.

HINT: use the embedding $W^{p,1}(\mathbb{R}^n) \subset L^\infty(\mathbb{R}^n)$.

11.7.25. Let X be an infinite-dimensional Banach space. Let us make it a Banach algebra by setting $xy \equiv 0$. Let $\{v_\alpha\}$ be a Hamel basis in X and $\{v_n\}$ its countable part with $\|v_n\| = 1$. We define an involution j as follows: $j(v_{2n-1}) = n^{-1}v_{2n}$, $j(v_{2n}) = nv_{2n-1}$, $j(v_\alpha) = v_\alpha$ for the remaining indices and then extend j by linearity. Show that we have obtained a discontinuous involution.

11.7.26. Let us consider the Calkin algebra $\mathcal{C}(l^2) = \mathcal{L}(l^2)/\mathcal{K}(l^2)$. Let E be a Hilbert space and let $\pi\colon \mathcal{C}(l^2) \to \mathcal{L}(E)$ be an isometric $*$-homomorphism. Prove that E is nonseparable.

11.7.27. Let X be a Banach space. Prove that the unit operator is an extreme point of the unit ball in $\mathcal{L}(X)$. Deduce from this that the unit of a Banach algebra $\mathcal{A}$ is an extreme point of the unit ball in $\mathcal{A}$.

HINT: let $\|I - T\| \leqslant 1$ and $\|I + T\| \leqslant 1$, where $T \in \mathcal{L}(X)$. Then $\|I - T^*\| \leqslant 1$ and $\|I + T^*\| \leqslant 1$, whence $\|l - T^*l\| \leqslant 1$, $\|l + T^*l\| \leqslant 1$ for all $l \in X^*$, $\|l\| \leqslant 1$, which gives $T^* = l$ if l is an extreme point of the unit ball in X^*. From the weak-$*$ compactness of the ball in X^* and the Krein–Milman theorem we obtain $T^* = 0$, so $T = 0$.

CHAPTER 12

Infinite-Dimensional Analysis

In this chapter we discuss foundations of differential calculus in infinite-dimensional spaces and some related questions. In the finite-dimensional case there are two different types of differentiability: differentiability at a point based on the consideration of increments of the function and also a global differentiability based on the consideration of the derivative as some independent object (as this is done in the theory of distributions and in the theory of Sobolev spaces). A similar, but more complicated, picture is observed in the infinite-dimensional case. Here we consider only the first type of differentiability, although now the second one plays an increasingly notable role in research and applications. At present foundations of differentiable calculus in infinite-dimensional spaces are usually not included in courses of functional analysis and are studied in courses of optimization or calculus of variations. However, we have decided to include this short chapter for a more complete representation of main directions of functional analysis as well as due to its conceptual connections with the linear theory.

12.1. Differentiability and Derivatives

Similarly to the case of functions of two real variables one can consider partial derivatives of functions on a linear space and one can also define derivatives in the spirit of the classical "main linear part of the increment of the function". We start with differentiability along directions. Suppose we are given two real linear spaces X and Y and a mapping $F\colon X \to Y$. Suppose that Y is equipped with some convergence (for example, is a locally convex space, as it will be the case everywhere below). Let $x_0 \in X$ and $h \in X$. We shall say that F has a *partial derivative* along h (or in the direction of h) if in Y there exists a limit

$$\partial_h F(x_0) := \lim_{t \to 0} \frac{F(x_0 + th) - F(x_0)}{t}.$$

Even if for every $h \in X$ the partial derivative $\partial_h F(x_0)$ exists, the mapping $h \mapsto \partial_h F(x_0)$ can fail to be linear (see Example 12.1.3). It is often useful to have not only the linearity of this mapping, but also the possibility to approximate F by it "up to an infinitely small quantity of higher order".

Various types of differentiability can be described by the following simple scheme of differentiability with respect to a class of sets $\mathcal{M}$. Let X, Y be locally convex spaces and let $\mathcal{M}$ be some class of nonempty subsets of X.

12.1.1. Definition. *The mapping* $F\colon X \to Y$ *is called differentiable with respect to* $\mathcal{M}$ *at the point* x *if there exists a sequentially continuous linear mapping from* X *to* Y, *denoted by* $DF(x)$ *or* $F'(x)$ *and called the derivative of* F *at the point* x, *such that uniformly in* h *from every fixed set* $M \in \mathcal{M}$ *we have*

$$\lim_{t \to 0} \frac{F(x + th) - F(x)}{t} = DF(x)h. \qquad (12.1.1)$$

Taking for $\mathcal{M}$ the collection of all finite sets we obtain the *Gateaux differentiability* (according to Mazliak [**696**], the correct spelling of Gateaux is without the circumflex, unlike the French word "gâteaux"). Obviously, the Gateaux derivative is unique if it exists. Thus, the Gateaux differentiability differs from the existence of derivatives $\partial_h F(x)$ by the property that we require in addition the linearity of the mapping $h \mapsto \partial_h F(x)$ and also its sequential continuity.

If $\mathcal{M}$ is the class of all compact subsets, we arrive at the *differentiability with respect to the system of compact sets*, which for normed spaces is called the *Hadamard differentiability*. One can also use the *differentiability with respect to the system of sequentially compact sets* (for normed spaces it coincides with the Hadamard differentiability); it is equivalent to the property that, as $t_n \to 0$, $t_n \neq 0$ and $h_n \to h$ in X, we have $[F(x + t_n h_n) - F(x)]/t_n \to DF(x)h$ in Y.

Finally, if X, Y are normed spaces and $\mathcal{M}$ consists of all bounded sets, then we obtain the definition of the *Fréchet differentiability* (of course, such definition can be also considered for locally convex spaces; then this differentiability is called the *bounded differentiability*).

The main idea of differentiability is a local approximation of the mapping F by a linear mapping, i.e., the representation

$$F(x + h) = F(x) + DF(x)h + r(x, h),$$

where the mapping $h \mapsto r(x, h)$ is in a sense an "infinitely small quantity of higher order" as compared to h. In the case of normed spaces the Fréchet differentiability gives the following meaning to this concept of smallness:

$$\lim_{\|h\| \to 0} \frac{\|r(x, h)\|}{\|h\|} = 0.$$

Symbolically this is denoted by $r(x, h) = o(h)$. In the more general case of differentiability with respect to $\mathcal{M}$ the smallness means the uniform (in h from every fixed $M \in \mathcal{M}$) relation

$$\lim_{t \to 0} \frac{r(x, th)}{t} = 0. \qquad (12.1.2)$$

It is easy to observe that this condition is equivalent to equality (12.1.1) if we set $r(x, h) := F(x + h) - F(x) - DF(x)h$. Of course, the concept of smallness can be given another sense, which will lead to another type of differentiability. Thus, as for functions on $\mathbb{R}$, the derivative plays the role of some tangent mapping.

12.1.2. Example. Let X be a Hilbert space and let $f(x) = (x, x)$. Then we have $f'(x) = 2x$. Indeed,

$$(x + h, x + h) - (x, x) = 2(x, h) + (h, h) \quad \text{and} \quad (h, h) = o(h).$$

Clearly, for mappings on the real line the differentiabilities of Gateaux, Hadamard and Fréchet coincide. In the space $\mathbb{R}^n$ with $n > 1$ the Hadamard definition is equivalent to the Fréchet definition and is strictly stronger than the Gateaux definition.

12.1.3. Example. (i) Let a function $f\colon \mathbb{R}^2 \to \mathbb{R}^1$ be defined by the formula

$$f(x) = r\cos 3\varphi, \quad x = (r\cos\varphi, r\sin\varphi), \quad f(0) = 0$$

in polar coordinates. At the point $x_0 = 0$ the partial derivatives

$$\partial_h f(x_0) = \lim_{t\to 0} t^{-1} f(th) = \lambda\cos 3\alpha$$

exist for all $h = (\lambda\cos\alpha, \lambda\sin\alpha) \in \mathbb{R}^2$, but the mapping $h \mapsto \partial_h f(x_0)$ is not linear. To see this, it suffices to take the vectors $(1,0)$ and $(0,1)$.

(ii) Let us define a function $f\colon \mathbb{R}^2 \to \mathbb{R}^1$ as follows:

$$f(x) = \begin{cases} 1 & \text{if } x = (x_1, x_2), \text{ where } x_2 = x_1^2 \text{ and } x_1 > 0, \\ 0 & \text{else.} \end{cases}$$

At the point $x = 0$ the Gateaux derivative exists and equals zero, since for every $h \in \mathbb{R}^2$ we have $\lim\limits_{t\to 0} t^{-1} f(th) = 0$ because $f(th) = 0$ if $|t| \leqslant \delta(h)$, where $\delta(h) > 0$. There is no Fréchet differentiability at zero, because $f(h) = 1$ if we take $h = (t, t^2)$.

For a locally Lipschitz (i.e., Lipschitz in a neighborhood of every point) mapping of normed spaces the Gateaux and Hadamard differentiabilities coincide.

12.1.4. Theorem. *Let X and Y be normed spaces and let $F\colon X \to Y$ be a locally Lipschitz mapping. If F is Gateaux differentiable at the point x, then at this point F is also Hadamard differentiable and the corresponding derivatives are equal.*

PROOF. Let K be compact in X and $\varepsilon > 0$. Let F satisfy the Lipschitz condition with constant L on the ball $B(x, r)$ with $r > 0$, let K be contained in the ball $B(0, R)$, and let $M := \max(L, R, \|DF(x)\|)$. We find a finite ε-net $h_1, \ldots, h_m$ in K. There is a number $\delta \in (0, r/R)$ such that, whenever $|t| < \delta$, for every $i = 1, \ldots, m$ we have

$$\|F(x + th_i) - F(x) - tDF(x)(h_i)\| \leqslant \varepsilon|t|.$$

Then, as $|t| < \delta$, for every $h \in K$ we obtain

$$\|F(x + th) - F(x) - tDF(x)(h)\| \leqslant \varepsilon|t| + 2M\varepsilon|t|,$$

since there exists h_i with $\|h - h_i\| \leqslant \varepsilon$, whence

$$\|F(x + th) - F(x + th_i)\| \leqslant M\|th - th_i\| \leqslant M\varepsilon|t|$$

and $\|tDF(x)h - tDF(x)h_i\| \leqslant M\varepsilon|t|$. Thus, F is Hadamard differentiable at x. It is clear that the Hadamard derivative serves as the Gateaux derivative, since the latter is unique. $\square$

In infinite-dimensional Banach spaces the Fréchet differentiability is strictly stronger than the Hadamard differentiability.

12.1.5. Example. The function

$$f\colon L^1[0,1] \to \mathbb{R}^1, \quad f(x) = \int_0^1 \sin x(s)\, ds \qquad (12.1.3)$$

is everywhere Hadamard differentiable, but nowhere Fréchet differentiable. The same is true for the mapping

$$F\colon L^2[0,1] \to L^2[0,1], \quad F(x)(s) = \sin x(s). \qquad (12.1.4)$$

PROOF. For the proof of differentiability of a mapping it is often useful to find a candidate for the derivative, which is done by calculating partial derivatives. For the function f we have the following equality:

$$f(x + th) = \int_0^1 \sin[x(s) + th(s)]\, ds.$$

Differentiating in t by the Lebesgue dominated convergence theorem we obtain

$$\partial_h f(x) = \int_0^1 h(s) \cos x(s)\, ds.$$

It is clear that the Gateaux derivative exists and is given by the functional

$$Df(x)h = \int_0^1 h(s) \cos x(s)\, ds.$$

Since $\|Df(x)\| \leqslant 1$, with the aid of the mean value theorem for functions on the real line we conclude that the function f is Lipschitzian (of course, this can be verified directly). By Theorem 12.1.4 we obtain the Hadamard differentiability.

For the mapping F the reasoning is similar. Here we have the operators $DF(x)$ on $L^2[0,1]$ and

$$(DF(x)h)(s) = (\cos x(s))h(s).$$

Let us see whether f and F are Fréchet differentiable. Let $x = 0$. Then $f(x) = 0$. We have to check whether the relation $f(h) - Df(0)h = o(\|h\|)$ holds. The left-hand side equals

$$\int_0^1 [\sin h(s) - h(s)]\, ds.$$

Since the Taylor expansion of $\sin h(s) - h(s)$ begins with h^3 and our space is L^1, we may suspect that the Fréchet differentiability fails here. In order to make sure that this is true, we take for h elements of the unit ball for which $f(th) - tDf(0)h$ will not be uniformly $o(t)$. Namely, let $h_k(s) = k$ if $0 \leqslant s \leqslant 1/k$ and $h_k(s) = 0$ if $s > 1/k$. Then

$$f(th_k) - tDf(0)h_k = k^{-1} \sin kt - t.$$

This quantity is not $o(t)$ uniformly in k: it suffices to take $t = k^{-1}$, which gives $t(\sin 1 - 1)$. For an arbitrary point x the reasoning is similar. Let us fix a version of x. We consider the expression

$$\int_0^1 \Big(\sin\big(x(s) + th(s)\big) - \sin x(s) - th(s)\cos x(s)\Big)\,ds.$$

The functions $\cos x(s)$ and $\sin x(s)$ have a common Lebesgue point $s_0 \in (0,1)$. For every ε this point is a Lebesgue point for the function

$$\sin\big(x(s) + \varepsilon\big) - \sin x(s) - \varepsilon\cos x(s).$$

Pick $\varepsilon \in (0,1)$ such that $\sin\big(x(s_0) + \varepsilon\big) - \sin x(s_0) - \varepsilon\cos x(s_0) \neq 0$. Set $h_k = kI_{E_k}$, $E_k = (s_0 - k^{-1}, s_0 + k^{-1})$. For $t = \varepsilon k^{-1}$ we obtain the quantity

$$\int_{E_k} \Big(\sin\big(x(s) + \varepsilon\big) - \sin x(s) - \varepsilon\cos x(s)\Big)\,ds$$

of order of smallness $Lk^{-1} = L\varepsilon^{-1}t$, where $L \neq 0$ is some number, since the limit of this quantity multiplied by $k/2$ is $\sin\big(x(s_0) + \varepsilon\big) - \sin x(s_0) - \varepsilon\cos x(s_0) \neq 0$ as $k \to \infty$. Similar estimates work in the case of F. $\qquad\square$

It is worth noting that if the function f is considered not on L^1, but on L^2, then it becomes Fréchet differentiable.

12.1.6. Example. The function f given by formula (12.1.3) on the space $L^2[0,1]$ is everywhere Fréchet differentiable. The mapping F given by (12.1.4) on $C[0,1]$ is everywhere Fréchet differentiable.

PROOF. A nuance making a difference in the properties of f on L^1 and L^2 is that the quantity $|f(x + h) - f(x) - Df(x)h|$ with the aid of the inequality $|\sin(x + h) - \sin x - h\cos x| \leqslant h^2$ is estimated by the integral of h^2, which is the square of the L^2-norm (infinite for some h in L^1). A similar reasoning applies to the mapping F on the space $C[0,1]$. Here $\|F(x + h) - F(x) - DF(x)h\|$ is estimated by $\|h\|^2$ in the case of the sup-norm, but not in the case of L^2-norm, when the indicated estimate leads to the integral of h^4. $\qquad\square$

Let us consider one more instructive infinite-dimensional example. It employs the function that is frequently encountered in applications: the distance to a set.

12.1.7. Example. Let X be an infinite-dimensional normed space and let K be a compact set. Set

$$f(x) = \text{dist}(x, K) = \inf\{\|x - y\|\colon y \in K\}.$$

Then the function f satisfies the Lipschitz condition, but is not Fréchet differentiable at points of K. If K is such that $\alpha K \subset K$ whenever $|\alpha| \leqslant 1$ and the set $\bigcup_{n=1}^{\infty} nK$ is everywhere dense in X, then f has the zero Gateaux derivative at the point $0 \in K$. For example, one can take for K the compact ellipsoid

$$K = \left\{(x_n) \in l^2\colon \sum_{n=1}^{\infty} n^2 x_n^2 \leqslant 1\right\}$$

in the Hilbert space l^2.

PROOF. Let $x \in K$. Then $f(x) = 0$. Suppose that at x there exists the Fréchet derivative $f'(x)$. This derivative can be only zero, since for every nonzero vector h the function $t \mapsto f(x + th)$ attains its minimum for $t = 0$. We shall arrive at a contradiction if we show that $f(x + h) - f(x) - f'(x)h = f(x + h)$ is not $o(\|h\|)$. For every $n \in \mathbb{N}$ we find a vector h_n such that $\|h_n\| \leqslant 1/n$ and the ball of radius $\|h_n\|/4$ centered at $x + h_n$ does not intersect K. This will give the estimate $f(x + h_n) \geqslant \|h_n\|/4$. The compact set K is covered by finitely many balls of radius $(4n)^{-1}$ centered at points $a_1, \ldots, a_k$. Let L be the finite-dimensional linear space generated by these centers. There is a vector h_n with $\|h_n\| = 1/n$ and $\operatorname{dist}(h_n, L) = 1/n$. This vector is what we need. Indeed, if there exists a vector $y \in K \cap B(x + h_n, \|h_n\|/4)$, then we obtain the following decomposition: $x = u + l_1$, $y = v + l_2$, where $l_1, l_2 \in L$, $\|u\| \leqslant (4n)^{-1}$, $\|v\| \leqslant (4n)^{-1}$ and $\|x + h_n - y\| \leqslant \|h_n\|/4$. Therefore,

$$\|h_n - (l_2 - l_1) + u - v\| \leqslant (4n)^{-1},$$

and hence $\|h_n - (l_2 - l_1)\| \leqslant 3(4n)^{-1}$ contrary to our choice of h_n, because we have $l_2 - l_1 \in L$.

Suppose now that K satisfies the indicated additional conditions. We show that at the point $0 \in K$ the Gateaux derivative exists and equals zero. For this we have to verify that for each fixed $h \in X$ we have $\lim_{t \to 0} t^{-1} f(th) = 0$. Let $\varepsilon > 0$. By our condition there exists a vector $v \in nK$ such that $\|h - v\| \leqslant \varepsilon$. Since $tv \in K$ whenever $|t| \leqslant n^{-1}$ by our condition, we have $f(tv) = 0$ for such t. Hence $|t^{-1} f(th)| \leqslant \varepsilon$ by the estimate $|f(th) - f(tv)| \leqslant \|th - tv\| \leqslant |t|\varepsilon$, which holds by the Lipschitzness of f. $\qquad\square$

In Exercise 12.5.20 it is suggested to verify that if, in addition, the set K is convex, then f has the zero Gateaux derivative at all points of $\bigcup_{0 \leqslant t < 1} tK$.

For normed spaces one can consider the strict differentiability, which is even stronger than the Fréchet differentiability.

12.1.8. Definition. *Let X and Y be normed spaces, let U be a neighborhood of a point $x_0 \in X$, and let $f \colon U \to Y$ be a mapping Fréchet differentiable at x_0. If for every $\varepsilon > 0$ there exists $\delta > 0$ such that, whenever $\|x_1 - x_0\|_X \leqslant \delta$ and $\|x_2 - x_0\|_X \leqslant \delta$, we have*

$$\|f(x_1) - f(x_2) - f'(x_0)(x_1 - x_2)\|_Y \leqslant \varepsilon\|x_1 - x_2\|_X,$$

then f is called strictly differentiable at the point x_0.

We observe that a mapping strictly differentiable at x_0 is continuous not only at the point x_0, but also in some ball centered at x_0, since the definition and the triangle inequality yield that in the ball of radius δ centered at x_0 the mapping f satisfies the Lipschitz condition with constant $\|f'(x_0)\| + \varepsilon$. Hence even for scalar functions on the real line the strict differentiability does not reduce to Fréchet differentiability.

If E is a linear subspace in X equipped with some stronger locally convex topology, then one can define the *differentiability along E* (in the respective sense) at the point x as the differentiability at $h = 0$ of the mapping $h \mapsto F(x+h)$ from E to Y in the corresponding sense. The derivative along the subspace E is denoted by the symbol $D_E F$. When E is one-dimensional, this gives the usual partial derivative $\partial_h F$.

12.2. Properties of Differentiable Mappings

The most important properties of differentiable mappings include the mean value theorem and the chain rule, i.e., the rule of differentiating the composition. The main role in obtaining multidimensional or infinite-dimensional versions of classical results is played by the corresponding assertions for the real line. However, here there are some subtleties, especially in the infinite-dimensional case, requiring some precautions. First we discuss the differentiability of compositions. Suppose that X, Y and Z are locally convex spaces and mappings

$$F\colon X \to Y \quad \text{and} \quad G\colon Y \to Z$$

are differentiable in a certain sense. Is the mapping $G{\circ}F\colon X \to Z$ differentiable in the same sense? The answer depends on the type of differentiability. For example, the composition of Gateaux differentiable mappings need not be Gateaux differentiable.

12.2.1. Example. Let us define a mapping $g\colon \mathrm{IR}^2 \to \mathrm{IR}^2$ by the formula $g\colon (x_1, x_2) \mapsto (x_1, x_2^2)$ and take the function $f\colon \mathrm{IR}^2 \to \mathrm{IR}^1$ from Example 12.1.3(ii). Then the composition $f{\circ}g\colon \mathrm{IR}^2 \to \mathrm{IR}^1$ is not Gateaux differentiable at the point $x = 0$. Moreover, this composition has no partial derivatives at zero along the vectors $(1, 1)$ and $(1, -1)$. Indeed,

$$f\big(g(x)\big) = \begin{cases} 1 & \text{if } x_1 = |x_2| > 0, \\ 0 & \text{in all other cases.} \end{cases}$$

Note that in this example the inner function is even Fréchet differentiable. It turns out that if the outer function is Fréchet (or Hadamard) differentiable, then the situation becomes better.

12.2.2. Theorem. *Let X, Y and Z be normed spaces, let $\Psi\colon X \to Z$ be the composition of two mappings $F\colon X \to Y$ and $G\colon Y \to Z$, and let $x_0 \in X$ and $y_0 = F(x_0)$. Suppose that the mapping G is Hadamard differentiable at the point y_0. If the mapping F is differentiable at the point x_0 either in the sense of Gateaux or in the sense of Hadamard, then the mapping Ψ is differentiable at x_0 in the same sense and*

$$\Psi'(x_0) = G'(y_0)F'(x_0). \tag{12.2.1}$$

If at the point x_0 the mapping F is Fréchet differentiable and G is differentiable at y_0 also in the Fréchet sense, then Ψ is differentiable at x_0 in the Fréchet sense and (12.2.1) is fulfilled.

PROOF. First we observe that if F has the partial derivative $\partial_h F$ at x_0, then there exists the partial derivative

$$\partial_h \Psi(x_0) = G'(y_0)\partial_h F(x_0).$$

Indeed, $F(x_0 + th) = F(x_0) + t\partial_h F(x_0) + r(t) = y_0 + t\partial_h F(x_0) + r(t)$, where $\|r(t)/t\| \to 0$ as $t \to 0$. In addition,

$$G(y_0 + u) - G(y_0) = G'(y_0)u + s(u),$$

where $\lim\limits_{t\to 0} t^{-1}s(tu) = 0$ uniformly in u in every fixed compact set. Hence

$$\Psi(x_0 + th) - \Psi(x_0) = G'(y_0)[t\partial_h F(x_0) + r(t)] + s\big(t\partial_h F(x_0) + r(t)\big),$$

where

$$\lim\limits_{t\to 0} t^{-1}\Big[G'(y_0)r(t) + s\big(t\partial_h F(x_0) + r(t)\big)\Big] = 0$$

in the sense of convergence with respect to the norm in Z.

This yields Gateaux differentiability of the mapping Ψ in the case where the mapping F is Gateaux differentiable.

Suppose that the mapping F is Hadamard differentiable at x_0 and the mapping G is Hadamard differentiable at y_0. For every fixed compact set $K \subset X$ we have

$$F(x_0 + h) = F(x_0) + F'(x_0)h + r(h),$$

where $\lim\limits_{t\to 0} \sup\limits_{h\in K} \|t^{-1}r(th)\| = 0$. Therefore,

$$\Psi(x_0 + h) - \Psi(x_0) - G'(y_0)F'(x_0)h$$
$$= G'(y_0)[F'(x_0)h + r(h)] - G'(y_0)F'(x_0)h + s\big(F'(x_0)h + r(h)\big)$$
$$= G'(y_0)r(h) + s\big(F'(x_0)h + r(h)\big),$$

where $\lim\limits_{t\to 0} \sup\limits_{h\in K} \|t^{-1}G'(y_0)r(th)\| = 0$ and

$$\lim\limits_{t\to 0} \sup\limits_{h\in K} \big\|t^{-1}s\big(tF'(x_0)h + r(th)\big)\big\| = 0.$$

The last equality follows from the fact that for every sequence $t_n \to 0$ and every sequence $\{h_n\} \subset K$, the sequence of vectors $F'(x_0)h_n + t_n^{-1}r(t_nh_n)$ is contained in the compact set $F'(x_0)(K) + \big(\{t_n^{-1}r(t_nh_n)\} \cup \{0\}\big)$, because $t_n^{-1}r(t_nh_n) \to 0$ by the definition of Hadamard differentiability.

Finally, in the case of Fréchet differentiability the same reasoning applies with balls in place of compacta. $\square$

A closer look at the proof enables us to modify it for mappings of locally convex spaces.

12.2.3. Theorem. *Let X, Y and Z be locally convex spaces, let a mapping $\Psi\colon X \to Z$ be the composition of two mappings $F\colon X \to Y$ and $G\colon Y \to Z$, and let $x_0 \in X$ and $y_0 = F(x_0)$. Suppose that the mappings F and G are differentiable with respect to the system of compact sets at points x_0 and y_0, respectively, and assume also that the operator $F'(x_0)$ takes compact sets to compact sets. Then Ψ is differentiable at x_0 with respect to the system of compact sets and*

$$\Psi'(x_0) = G'(y_0)F'(x_0).$$

An analogous assertion is true if both mappings are differentiable with respect to the system of bounded sets and the operator $F'(x_0)$ takes bounded sets to bounded sets.

Finally, if the mapping F is Gateaux differentiable at x_0 and the mapping G is differentiable at y_0 with respect to the system of compact sets, then Ψ is Gateaux differentiable at x_0 and the chain rule for $\Psi'(x_0)$ indicated above is true.

PROOF. The reasoning is analogous to the one given above. We only explain the role of additional restrictions on the mapping $F'(x_0)$ that are automatically fulfilled in the case of normed spaces. We have the following representation:

$$\Psi(x_0 + h) - \Psi(x_0) - G'(y_0)F'(x_0)h = G'(y_0)r(h) + s\big(F'(x_0)h + r(h)\big),$$

and we have to show that, for every fixed set K from the class with respect to which F is differentiable, for every sequence $\{h_n\} \subset K$ and every sequence numbers $t_n \to 0$, we have

$$\lim_{n\to\infty} t_n^{-1}G'(y_0)r(t_n h_n) + t_n^{-1}s\big(t_n F'(x_0)h_n + t_n t_n^{-1}r(t_n h_n)\big) = 0.$$

Hence we have to ensure the inclusion of the sequence $F'(x_0)h_n + t_n^{-1}r(t_n h_n)$ to some set from the class $\mathcal{K}$ with respect to which the outer mapping G is differentiable. For this the sequence of vectors $F'(x_0)h_n$ must belong to the class $\mathcal{K}$, since the sequence $t_n^{-1}r(t_n h_n)$ converges to zero and its addition does not influence containment in the classes of bounded or compact sets. □

An analogous assertion is true in the case of differentiability with respect to the system of sequentially compact sets, since $F'(x_0)$ is sequentially continuous (here it is not required that the operator $F'(x_0)$ take compact sets to compact sets).

12.2.4. Example. Suppose that a mapping $F: X \to Y$ between normed spaces is differentiable at a point x_0 in the sense of Gateaux, Hadamard or Fréchet and that $G: Y \to Z$ is a continuous linear operator with values in a normed space Z. Then the composition $G \circ F$ is differentiable at the point x_0 in the same sense as F.

The situation is similar for differentiable mappings of locally convex spaces in the sense of Gateaux or with respect to the systems of compact sets, sequentially compact sets or bounded sets if the operator G is linear and sequentially continuous.

The next theorem characterizing Hadamard differentiability is also connected with the theorem on the derivative of the composition.

12.2.5. Theorem. *A mapping $F: X \to Y$ between normed spaces is Hadamard differentiable at $x_0 \in X$ precisely when there exists a continuous linear mapping $L: X \to Y$ such that for every mapping $\varphi: \mathbb{R}^1 \to X$ differentiable at zero with $\varphi(0) = x_0$ the composition $F \circ \varphi: \mathbb{R}^1 \to Y$ is differentiable at the point 0 and we have $(F \circ \varphi)'(0) = L\varphi'(0)$.*

PROOF. If F is differentiable, then the composition is differentiable too as shown above. Suppose that we have numbers $t_n \to 0$ and vectors $h_n \to h$. A mapping $\varphi: \mathbb{R}^1 \to X$ will be defined as follows: $\varphi(t_n) = x_0 + h_n t_n$ and $\varphi(t) = x_0 + th$ if $t \notin \{t_n\}$. Then $\varphi(0) = x_0$ and $t^{-1}[\varphi(t) - \varphi(0)] \to h$ as $t \to 0$, since this difference quotient equals h_n if $t = t_n$ and equals h for all other t. We have

$$\frac{F(x_0 + t_n h_n) - F(x_0)}{t_n} = \frac{F\big(\varphi(t_n)\big) - F\big(\varphi(0)\big)}{t_n} \to L\varphi'(0) = Lh$$

as $n \to \infty$, which proves Hadamard differentiability of F at the point x_0. □

Note that if F and G are Gateaux, Hadamard or Fréchet differentiable at x, then, as is easily seen, $F + G$ is differentiable at x in the same sense and the equality $(F + G)'(x) = F'(x) + G'(x)$ holds.

Let us proceed to the mean value theorem. We recall that if a function f is differentiable in a neighborhood of the interval $[a, b]$, then

$$f(b) - f(a) = f'(c)(b - a)$$

for some point $c \in (a, b)$. This assertion does not extend in the same form to multidimensional mappings (even to mappings from $\mathbb{R}^1$ to $\mathbb{R}^2$). For example, let

$$f(x) = (\sin x, \cos x), \quad x \in \mathbb{R}^1.$$

Then $f(2\pi) = f(0)$, although the derivative of f does not vanish at any point. The right multidimensional analog of the mean value theorem deals with either inequalities or convex envelopes of the sets of values. We recall that the symbols $\operatorname{conv} A$ and $\overline{\operatorname{conv}} A$ denote, respectively, the convex envelope and the closed convex envelope of the set A in a locally convex space. The symbol $[a, b]$ denotes the line segment (closed interval) with the endpoints a and b of a linear space, i.e., the set of all vectors of the form $a + t(b - a)$, $t \in [0, 1]$. Similarly we define (a, b).

12.2.6. Theorem. *Let X and Y be locally convex spaces, let U be an open convex set in X, and let a mapping $F \colon U \to Y$ be Gateaux differentiable at every point in U. Then for every $a, b \in U$ one has the inclusion*

$$F(b) - F(a) \in \overline{\operatorname{conv}} \{F'(c)(a - b) \colon c \in (a, b)\}. \tag{12.2.2}$$

PROOF. Let E denote the set in the right-hand side. Let $l \in Y^*$. The function $\varphi \colon t \mapsto l\big(F(a + tb - ta)\big)$ is defined in a neighborhood of $[0, 1]$ by our assumption. In addition, it is differentiable in a neighborhood of $[0, 1]$. By the classical mean value theorem there exists a point $t \in (0, 1)$ such that

$$l\big(F(b)\big) - l\big(F(a)\big) = \varphi(1) - \varphi(0) = \varphi'(t) = l\big(F'(c)(b - a)\big) \leqslant \sup_{y \in E} |l(y)|,$$

where $c = a + t(b - a) \in U$. By a corollary of the Hahn–Banach theorem we conclude that $F(b) - F(a) \in E$. $\qquad\square$

Let us give a number of important corollaries.

12.2.7. Corollary. *Suppose that in the situation of the previous theorem we are given a sequentially continuous linear mapping $\Lambda \colon X \to Y$ (for example, $\Lambda = F'(a)$). Then for every $a, b \in U$ one has the inclusion*

$$F(b) - F(a) - \Lambda(b - a) \in \overline{\operatorname{conv}} \{[F'(c) - \Lambda](a - b) \colon c \in (a, b)\}. \tag{12.2.3}$$

PROOF. It suffices to apply the theorem above to the mapping $F - \Lambda$. $\qquad\square$

12.2.8. Corollary. *Let X and Y be normed space, let U be an open convex set in X, and let a mapping $F \colon U \to Y$ be Gateaux differentiable at every point in U. Then, for all $a, b \in U$, we have*

$$\|F(b) - F(a)\| \leqslant \sup_{c \in (a,b)} \|F'(c)\| \, \|a - b\|. \tag{12.2.4}$$

PROOF. For every $c \in U$ we have $\|F'(c)(b-a)\| \leqslant \|F'(c)\| \|b-a\|$, which by (12.2.2) gives (12.2.4).   □

12.2.9. Corollary. *Suppose that in the previous corollary we are given a continuous linear mapping* $\Lambda \colon X \to Y$. *Then, for all* $a, b \in U$, *we have*

$$\|F(b) - F(a) - \Lambda(b-a)\| \leqslant \sup_{c \in (a,b)} \|F'(c) - \Lambda\| \|a-b\|. \qquad (12.2.5)$$

12.2.10. Corollary. *Let X and Y be normed spaces, let U be an open convex set in X, and let a mapping $F \colon U \to Y$ be Gateaux differentiable at every point in U. Suppose that the mapping $x \mapsto F'(x)$ from U to the space of operators $\mathcal{L}(X, Y)$ with the operator norm is continuous at some point $x_0 \in U$. Then the mapping F is Fréchet differentiable at the point x_0. Moreover, F is strictly differentiable at x_0.*

PROOF. Let $\varepsilon > 0$ be such that the ball $B(x_0, \varepsilon) = \{x \colon \|x - x_0\| < \varepsilon\}$ is contained in U. As shown above, for all h with $\|h\| < \varepsilon$ we have

$$\|F(x_0 + h) - F(x_0) - F'(x_0)(h)\| \leqslant \sup_{c \in B(x_0, \varepsilon)} \|F'(c) - F'(x_0)\| \|h\|.$$

By the continuity of F' at x_0 we have

$$\lim_{\varepsilon \to 0} \sup_{c \in B(x_0, \varepsilon)} \|F'(c) - F'(x_0)\| = 0,$$

which gives Fréchet differentiability. The strict differentiability follows easily from estimate (12.2.5).   □

If the derivative of the mapping F is continuous with respect to the operator norm in a domain Ω, then F is called a C^1-mapping in Ω.

12.3. Inverse and Implicit Functions

In this section we consider the local invertibility of nonlinear mappings and existence of a functional dependence $y = y(x)$ between solutions of equations of the type $F(x, y) = 0$. Results of this kind are called, respectively, inverse function theorems and implicit function theorems. At present there exists a well developed theory covering such problems, but here we confine ourselves to some simplest infinite-dimensional analogs of the facts known from finite-dimensional calculus. Even these simplest theorems have many interesting and important applications. The results presented here do not employ refined tools and techniques and are applications of the contracting mapping theorem or some analogous reasoning. However, the method of application is elegant and instructive. The idea is that a given mapping is locally approximated by a simpler (in some sense) mapping. For such a simpler mapping in this section we take the identity mapping and linear invertible operators.

We start with the consideration of Lipschitzian homeomorphisms.

12.3.1. Theorem. *Let $U = B(a, r)$ be an open ball of radius $r > 0$ centered at a point a in a Banach space X and let $F \colon U \to X$ be a mapping such that*

$$\|F(x) - F(y)\| \leqslant \lambda \|x - y\| \quad \forall x, y \in U,$$

where $\lambda \in [0,1)$ *is a constant. Then there exists an open neighborhood* V *of* a *such that the mapping* $\Psi\colon x \mapsto x + F(x)$ *is a homeomorphism of* V *and the open ball* $W := B\big(\Psi(a), r(1 - \lambda)\big)$. *The inverse mapping* $\Psi^{-1}\colon W \to V$ *satisfies the Lipschitz condition with the constant* $(1 - \lambda)^{-1}$.

If, in addition, $\|F(a)\| < r(1 - \lambda)/2$, *then* Ψ *is a homeomorphism of some neighborhood of the point* a *and the open ball* $B\big(a, r(1 - \lambda)/2\big)$.

PROOF. Replacing F by $F - F(a)$, we can assume that $F(a) = 0$. Let us apply Theorem 1.4.4 to the space $T = W$ and the mapping $(w, x) \mapsto w - F(x)$ from $W \times U$ to X. All necessary conditions are fulfilled. In particular, the inequality $\|w - F(a) - a\| = \|w - a\| < r(1 - \lambda)$ holds for all $w \in W$. Hence there exists a continuous mapping $f\colon W \to U$ such that we have $f(w) = w - F\big(f(w)\big)$ for all $w \in W$, i.e., $\Psi\big(f(w)\big) = w$.

Let us show that f is a homeomorphism from W onto $f(W)$. It is clear that f is an injective mapping. The mapping Ψ on U is also injective, since the equality $\Psi(u_1) = \Psi(u_2)$ yields that

$$\|u_1 - u_2\| = \|F(u_1) - F(u_2)\| \leqslant \lambda \|u_1 - u_2\|.$$

This is only possible if $u_1 = u_2$. Thus, the mapping Ψ is a homeomorphism from the set $V := f(W)$ onto W, inverse to f. The set $V = \Psi^{-1}(W)$ is open by the continuity of Ψ. In addition, $a \in V$, since $\Psi(a)$ is the center of W.

The fact that f is Lipschitz with constant $(1 - \lambda)^{-1}$ is clear from the equality

$$f(w) - f(w') = w - w' + F\big(f(w')\big) - F\big(f(w)\big),$$

which gives the estimate $\|f(w) - f(w')\| \leqslant \|w - w'\| + \lambda \|f(w) - f(w')\|$, i.e., $\|f(w) - f(w')\| \leqslant (1 - \lambda)^{-1}\|w - w'\|$. If, in addition, $\|F(a)\| < r(1 - \lambda)/2$, then the ball $B\big(a, r(1 - \lambda)/2\big)$ belongs to W. $\square$

Now we apply the facts established above to differentiable mappings. First we clarify the condition for the differentiability of the inverse mapping for a differentiable homeomorphism.

12.3.2. Proposition. *Let* X *and* Y *be Banach spaces, let* $U \subset X$ *and* $V \subset Y$ *be open sets, and let* $F\colon U \to V$ *be a homeomorphism that is Fréchet differentiable at some point* $a \in U$. *For Fréchet differentiability of the mapping* $G = F^{-1}\colon V \to U$ *at the point* $b = F(a)$ *it is necessary and sufficient that the operator* $F'(a)$ *map* X *one-to-one onto* Y.

PROOF. If the mapping G is Fréchet differentiable at the point b, then by the chain rule we have $G'(b)F'(a) = I_X$ and $F'(a)G'(b) = I_Y$.

Suppose now that the operator $\Lambda := F'(a)$ is invertible. Passing to the mapping $\Lambda^{-1}F$, we reduce our assertion to the case $X = Y$ and $F'(a) = I$. In addition, we can assume that $a = 0$ and $F(a) = 0$. We have to show that $G'(0) = I$, i.e., that

$$G(y) - y = o(\|y\|) \quad \text{as } \|y\| \to 0.$$

We have

$$F(x) = x + \|x\|\varphi(x), \quad \lim_{\|x\| \to 0} \|\varphi(x)\| = 0. \tag{12.3.1}$$

Since F is a homeomorphism, we have $y = F(x)$, where $x = G(y)$, for all y in some neighborhood of zero, and for the corresponding x one has (12.3.1). Thus,

$$G(y) = x = y - \|x\|\varphi(x).$$

We have to show that $\|x\|\varphi(x) = o(\|y\|)$. Let us find a neighborhood of zero U_0 such that $\|\varphi(x)\| < 1/2$ for all $x \in U_0$. Then $\|x\| \leqslant 2\|y\|$ for such x and $y = F(x)$, i.e., $\|x\| \, \|\varphi(x)\| \leqslant 2\|y\| \, \|\varphi(x)\|$. It remains to observe that $\|x\| \to 0$ as $\|y\| \to 0$ by the continuity of G and hence $\varphi(x) \to 0$. $\qquad\square$

12.3.3. Theorem. *Let F be a continuously differentiable mapping from the open ball $U = B(x_0, r)$ in a Banach space X to a Banach space Y. Suppose that the operator $\Lambda := F'(x_0)$ maps X one-to-one onto Y. Then F maps some neighborhood V of the point x_0 one-to-one onto some neighborhood W of $F(x_0)$, moreover, the mapping $G := F^{-1}: W \to V$ is continuously differentiable and*

$$G'(y) = \big[F'\big(F^{-1}(y)\big)\big]^{-1}, \quad y \in W.$$

PROOF. The assertion reduces to the case $Y = X$ and $F'(x_0) = I$ if we pass to the mapping $\Lambda^{-1}F$. In addition, we can assume that $x_0 = 0$ and $F(x_0) = 0$. Having in mind to apply Theorem 12.3.1, we write F as $F(x) = x + F_0(x)$, where $F_0(x) = F(x) - x$. By the continuity of the derivative at x_0, there exists a ball $B(0, r)$ such that $\|F'(x) - I\|_{\mathcal{L}(X)} \leqslant 1/2$ for all $x \in B(0, r)$. By the mean value theorem the mapping F_0 is Lipschitz on $B(0, r)$ with constant $1/2$ (any constant less than 1 would suit for us). In addition, $F_0(0) = 0$. By the theorem cited, in the ball $W = B(0, r/2)$ we have a continuous inverse mapping G for F that takes this ball to some neighborhood of zero V. Proposition 12.3.2 ensures Fréchet differentiability of G at zero and the equality $G'(0) = I$. Note that so far we have used only the continuity of the derivative at zero. Since we assumed the continuity of the derivative at all points of U, in a neighborhood of x_0 this derivative is invertible, which shows the differentiability of the inverse mapping in a neighborhood of the point $F(x_0)$. Finally, the continuity of the mapping $y \mapsto G'(y)$ in this neighborhood follows from the formula $G'(y) = \big[F'(G(y))\big]^{-1}$ on account of the continuity of G and the continuity of the mapping $A \mapsto A^{-1}$ on the set of invertible operators. $\qquad\square$

We now obtain a criterion for F to be a diffeomorprhism.

12.3.4. Corollary. *Let F be a C^1-mapping from an open set U in a Banach space X to a Banach space Y. In order F be a C^1-diffeomorphism of the set U onto an open set in Y, it is necessary and sufficient that F be injective and the derivative $F'(x)$ be an invertible operator from X onto Y for all $x \in U$.*

PROOF. The necessity of these conditions is obvious. Let us prove their sufficiency. We already know that every point x in U possesses a neighborhood $U_x \subset U$ such that F is a C^1-diffeomorphism of U_x onto some open ball V_x centered at $F(x)$. Hence the set $F(U) = \bigcup_{x \in U} V_x$ is open. It follows from the results above that the inverse mapping $F^{-1}: V \to U$ is continuously differentiable. $\qquad\square$

We have already noted that in Theorem 12.3.3 the existence and differentiability of G at zero has been actually obtained under weaker conditions. Namely, we only needed (under the assumption that $x_0 = 0$ and $F(x_0) = 0$) that the auxiliary mapping $F_0(x) = F'(0)^{-1}F(x) - x$ be contracting in a neighborhood of zero. This condition is certainly weaker than the continuity of the derivative at zero and obviously follows from the strict differentiability of F at zero, which gives

$$F'(0)^{-1}F(x) - F'(0)^{-1}F(y) - x + y = \|x - y\|\psi(x, y)$$

with $\lim\limits_{x,y\to 0} \psi(x, y) = 0$. Hence F_0 is Lipschitz with an arbitrarily small constant in a suitable neighborhood of zero. This gives the following result.

12.3.5. Theorem. *Suppose that a mapping F from an open ball with center x_0 in a Banach space X to a Banach space Y is strictly differentiable at the point x_0 and the operator $D := F'(x_0)$ maps X one-to-one onto Y. Then F homeomorphically maps some neighborhood V of the point x_0 onto a neighborhood W of the point $F(x_0)$ and the mapping $G := F^{-1} \colon W \to V$ is Fréchet differentiable at the point $y_0 = F(x_0)$ and $G'(y_0) = D^{-1}$.*

12.3.6. Example. Let X be a Hilbert space and let $F(x) = Ax + B(x, x)$, where A is an invertible linear operator on X and $B \colon X \times X \to X$ is a continuous mapping linear in every argument. Then $F'(x)h = Ah + B(x, h) + B(h, x)$, since $\|B(h, h)\| \leqslant C|h|^2$, which is easily derived from the boundedness of B on some ball centered at zero. Hence $F'(0) = A$. Therefore, F maps diffeomorphically some neighborhood of zero onto a neighborhood of zero.

Let us proceed to the implicit function theorem. Suppose we are given three Banach spaces X, Y and Z and a continuously differentiable mapping F from an open set $U \subset X \times Y$ to Z. Suppose that for some point $(a, b) \in U$ we have $F(a, b) = 0$. We are interested in other solutions (x, y) to the equation $F(x, y) = 0$, sufficiently close to (a, b). Moreover, we would like to represent (locally) the set of solutions as a surface $y = f(x)$. It turns out that for this we need only one condition.

12.3.7. Theorem. *Suppose that in the situation described above the derivative $F'_Y(a, b)$ of the mapping F along Y at the point (a, b) is a linear isomorphism between Y and Z. Then there exists a neighborhood V_a of the point a in X, a neighborhood W_b of the point b in Y and a continuously differentiable mapping $f \colon V_a \to Y$ with the following properties: $V_a \times W_b \subset U$ and the conditions*

$$(x, y) \in V_a \times W_b \quad and \quad F(x, y) = 0$$

are equivalent to the conditions

$$x \in V_a \quad and \quad y = f(x).$$

In particular, $f(a) = b$. Thus, in the domain $V_a \times W_b$ all solutions to the equation $F(x, y) = 0$ are given by the formula $y = f(x)$.

PROOF. We shall reduce this theorem to the inverse mapping theorem. For this we consider an auxiliary mapping

$$F_1\colon U \to X \times Z, \quad (x,y) \mapsto (x, F(x,y)).$$

It is readily seen that this mapping is continuously differentiable and

$$F_1'(x,y)(u,v) = (u, F_X'(x,y)u + F_Y'(x,y)v), \quad u \in X, v \in Y.$$

Since $F_Y'(a,b)$ is a linear isomorphism between Y and Z, the mapping $F_1'(a,b)$ is an isomorphism between $X \times Y$ and $X \times Z$. Indeed, for any given $h \in X$ and $k \in Z$, there is a uniquely determined $u = h$ and then we take

$$v = [F_Y'(a,b)]^{-1}(k - F_X'(a,b)h)$$

with $F_1'(a,b)(u,v) = (h,k)$. As shown above, there exists a neighborhood U_0 of the point (a,b) that is mapped by F_1 diffeomorphically onto some neighborhood W of the point $(a, F(a,b)) = (a,0)$. In U_0 we can take a smaller neighborhood of the form $V_a \times W_b$, where V_a is a neighborhood of a and W_b is a neighborhood of b. Let $W_0 = F_1(V_a \times W_b)$. The diffeomorphism inverse to F_1 can be written in the form $(x,z) \mapsto (x, \varphi(x,z))$, where the mapping $(x,z) \mapsto \varphi(x,z)$ with values in Y is defined in a neighborhood of the point $(a,0)$ in $X \times Z$. We let $f(x) = \varphi(x,0)$ and show that we have obtained a desired mapping. Indeed, this mapping is continuously differentiable. In addition, the conditions $(x,y) \in V_a \times W_b$ and $F(x,y) = 0$ are equivalent to the conditions $(x,0) \in W_0$ and $\varphi(x,0) = y$ by the injectivity of the mapping F_1 on $V_a \times W_b$. The set $V_a' := \{x\colon (x,0) \in W_0\}$ is an open neighborhood of the point a. Replacing V_a by a smaller neighborhood $V_a \cap V_a'$, we obtain the desired properties for f. We have actually proved even more: for all z in some neighborhood of zero the solutions (x,y) to the equation $F(x,y) = z$ belonging to a sufficiently small neighborhood of (a,b) have a parametric representation $y = \varphi(x,z)$. $\square$

Differentiating the identity $F(x, f(x)) = 0$, we find that

$$f'(x) = -[F_Y'(x, f(x))]^{-1} F_X'(x, f(x))$$

in a neighborhood of a. In particular,

$$f'(a) = -[F_Y'(a,b)]^{-1} F_X'(a,b).$$

It is clear from the proof that the obtained representation gives all solutions (x,y) from a sufficiently small neighborhood of (a,b). But this assertion should not be understood in the sense that in a sufficiently small neighborhood of a there are no other differentiable mappings f with $F(x, f(x)) = 0$: we are talking only about f such that $y = f(x)$ belongs to a small neighborhood of b. For example, for the equation $x^2 + y^2 = 1$ on the plane in every neighborhood of the point $a = 0$ on the first coordinate line there are two differentiable functions $f_1(x) = \sqrt{1 - x^2}$ and $f_2(x) = -\sqrt{1 - x^2}$ for which $x^2 + f_1(x)^2 = 1$ and $x^2 + f_2(x)^2 = 1$. Here f_1 uniquely defines solutions from a neighborhood of the point $(0,1)$ and f_2 uniquely describes solutions from a neighborhood of the point $(0,-1)$.

It follows from our discussion that the mapping f is unique in the following sense: if a mapping f_0 from a neighborhood of the point a to the set W_b satisfies the equality $F\big(x, f_0(x)\big) = 0$, then f_0 coincides with f in some neighborhood of the point a.

If we compare the theorem on the implicit function with the theorem on the inverse function, then it is easy to observe that it corresponds to Theorem 12.3.3, but not to Theorem 12.3.5. It is also possible to obtain an analog of the latter.

12.3.8. Theorem. *Let F be a mapping from a neighborhood $U \subset X \times Y$ of the point (a, b) to the space Z such that $F(a, b) = 0$ and $\lim\limits_{x \to a} F(x, b) = 0$. Suppose that F is differentiable along the space Y at (a, b) and for every $\varepsilon > 0$ there is $\delta > 0$ such that*

$$\|F(x, y_1) - F(x, y_2) - F_Y'(a, b)(y_1 - y_2)\| \leqslant \varepsilon \|y_1 - y_2\|$$

if $\|x - a\| \leqslant \delta$, $\|y_1 - b\| \leqslant \delta$, $\|y_2 - b\| \leqslant \delta$ (for example, the derivative F_Y' exists and is continuous in U). Suppose that the operator $F_Y'(a, b)$ from Y onto Z is invertible. Then there exists a mapping f with values in Z defined in some neighborhood V_a of the point a and possessing the following properties: $f(a) = b$ and $F\big(x, f(x)\big) = 0$ for all $x \in V_a$.

PROOF. Without loss of generality we can assume that $a = 0$ and $b = 0$. Passing to the mapping $F_Y'(a, b)^{-1}F$, we can also assume that $Z = Y$ and $F_Y'(a, b) = I$. Let us find $r_0 > 0$ such that U contains the ball of radius r_0 centered at zero. For every $x \in B(0, r_0)$ we consider the mapping $\Psi_x \colon y \mapsto y - F(x, y)$. By assumption there exists $r \in (0, r_0)$ such that for all $y_1, y_2 \in B(0, r)$, $x \in B(0, r)$ we have

$$\|F(x, y_1) - F(x, y_2) - (y_1 - y_2)\| \leqslant \tfrac{1}{2}\|y_1 - y_2\|.$$

Hence for all $x \in B(0, r)$ the mapping Ψ_x on the ball $B(0, r)$ satisfies the Lipschitz condition with constant $1/2$. By assumption there exists $\delta \in (0, r)$ such that $\|F(x, 0)\| < r/2$ if $\|x\| < \delta$. For such x we have $\|\Psi_x(0)\| = \|F(x, 0)\| < r/2$. Thus, we are under the assumptions of Theorem 1.4.4, which gives a mapping $f \colon B(0, \delta) \to Y$ with $f(x) = f(x) - F\big(x, f(x)\big)$ for all $x \in B(0, \delta)$. Hence for such points x we have $F\big(x, f(x)\big) = 0$. $\qquad\square$

12.3.9. Remark. If in the situation of this theorem the mapping F is Fréchet differentiable and its derivative is continuous at the point (a, b), then the mapping f constructed above is differentiable at the point a and

$$f'(a) = -[F_Y'(a, b)]^{-1}F_X'(a, b).$$

This is proved in the same way as in Theorem 12.3.7.

12.4. Higher Order Derivatives

If a mapping $F \colon X \to Y$ between locally convex spaces is differentiable, then a new mapping $F' \colon x \mapsto F'(x)$ arises with values in the space of linear mappings from X to Y. We can equipped this space with some locally convex topology and study the differentiability of the mapping F', denoting its derivative (in a suitable sense) by F''.

In the case of Fréchet differentiability in normed spaces this leads to considering the derivative $F'\colon X \to \mathcal{L}(X, Y)$, $x \mapsto F'(x)$ of a Fréchet differentiable mapping $F\colon X \to Y$ of normed spaces. The space of operators $\mathcal{L}(X, Y)$ is equipped with the operator norm, so F' becomes a mapping with values in a normed space and we can speak of its Fréchet differentiability at the point x. If F' has a Fréchet derivative at x, then it is denoted by $F''(x)$ or $D^2 F(x)$. In this case $F''(x) \in \mathcal{L}(X, \mathcal{L}(X, Y))$.

We observe (see §12.5) that the space of operators $\mathcal{L}(X, \mathcal{L}(X, Y))$ can be canonically identified with the space $\mathcal{L}_2(X, X, Y)$ of bilinear continuous mappings from $X \times X$ to Y. To this end, to every operator $\Lambda \in \mathcal{L}(X, \mathcal{L}(X, Y))$ we set in correspondence a bilinear continuous mapping $\widehat{\Lambda}$ by the formula

$$\widehat{\Lambda}(u, v) = [\Lambda(u)](v).$$

If $\Lambda = F''(x)$, then the value of $\widehat{\Lambda}$ on the pair (u, v) is evaluated by the formula

$$\widehat{\Lambda}(u, v) = \partial_v \partial_u F(x).$$

Conversely, to every continuous bilinear mapping B from $X \times X$ to Y we associate a linear mapping Λ from X to the space of linear mappings from X to Y by the formula

$$[\Lambda(u)](v) := B(u, v).$$

We have $\Lambda \in \mathcal{L}(X, \mathcal{L}(X, Y))$ and $\widehat{\Lambda} = B$.

12.4.1. Example. Let X be a Hilbert space. Set $f(x) = (x, x)$. Then $f'(x) = 2x$ and $f''(x) = 2I$.

The higher order differentiability is defined by induction: a mapping F at a point x_0 has a derivative of order $k > 1$ if in some neighborhood of x_0 there exist derivatives $DF(x), \ldots, D^{k-1}F(x)$ and the mapping $x \mapsto D^{k-1}F(x)$ is differentiable at x_0.

The Fréchet derivative $D^k F(x_0)$ of order k is often identified with a continuous k-linear mapping from the space X^k to Y (see §12.5(ii)), i.e., the convention is that $D^k F(x_0) \in \mathcal{L}_k(X^k, Y)$.

12.4.2. Example. Let X and Y be normed space and let $\psi\colon X \to Y$ be the continuous homogeneous polynomial of degree k generated by a continuous symmetric k-linear mapping Ψ_k. Then the mapping ψ is infinitely Fréchet differentiable and

$$D\psi(x)h = k\Psi_k(x, \ldots, x, h).$$

For the proof we write $\Psi_k(x+h, \ldots, x+h) - \Psi_k(x, \ldots, x)$ using the additivity of Ψ_k in every argument. We obtain the term $k\Psi_k(x, \ldots, x, h)$ and a sum of terms in which h enters as an argument of Ψ_k at least twice. So this sum is $o(\|h\|)$, which proves our claim.

12.4.3. Theorem. *Let a mapping F between normed spaces X and Y have a derivative of order k at a point x_0. Then the multilinear mapping $D^k F(x_0)$ is symmetric. In particular, for $k = 2$ we have*

$$D^2 F(x_0)(u, v) = D^2 F(x_0)(v, u).$$

PROOF. It suffices to verify this assertion for scalar functions considering compositions $l \circ F$, where l is an element of Y^*. Then everything reduces to the case $X = \mathbb{R}^k$ known from calculus. For example, for $k = 2$ we have to take the function $f(t_1, t_2) := F(x_0 + t_1 u + t_2 v)$ on $\mathbb{R}^2$. Then

$$\frac{\partial}{\partial t_1} f(t_1, t_2) = DF(x_0 + t_1 u + t_2 v)(u),$$

whence $D^2 F(x_0)(u, v) = \frac{\partial}{\partial t_2} \frac{\partial}{\partial t_1} f|_{t_1 = t_2 = 0}$. A similar equality with the interchanged partial derivatives in t_1 and t_2 is true for $D^2 F(x_0)(v, u)$. $\square$

12.4.4. Theorem. (TAYLOR'S FORMULA) *Let U be an open set in a normed space X and let a mapping $F \colon X \to Y$ with values in a normed space Y have Fréchet derivatives up to order $n - 1$ in U.*

(i) *If at a point $x_0 \in U$ there exists the Fréchet derivative $D^n F(x_0)$, then, as $\|h\| \to 0$, we have*

$$\left\| F(x_0 + h) - F(x_0) - DF(x_0)h - \cdots - \frac{1}{n!} D^n F(x_0)(h, \ldots, h) \right\| = o(\|h\|^n).$$

(ii) *If in U there exists the Fréchet derivative $D^n F(x)$ and $\|D^n F(x)\| \leqslant M$, then, for every $x \in U$ and every $h \in X$ such that the interval with endpoints in x and $x + h$ belongs to U, we have*

$$F(x + h) = F(x) + DF(x)h + \cdots + \frac{1}{(n-1)!} D^{n-1} F(x)(h, \ldots, h)$$

$$+ \frac{1}{(n-1)!} \int_0^1 (1 - t)^{n-1} D^n F(x + th)(h, \ldots, h) \, dt. \quad (12.4.1)$$

In addition, the following estimate holds:

$$\| F(x + h) - F(x) - DF(x)h - \cdots$$

$$- \frac{1}{(n-1)!} D^{n-1} F(x)(h, \ldots, h) \| \leqslant M \frac{\|h\|^n}{n!}.$$

PROOF. (i) Let us apply induction on n. For $n = 1$ our assertion is true by the definition of the Fréchet derivative. Suppose it is true for some $n = k$. If the mapping F has the derivative $D^{k+1} F(x_0)$ at x_0, then the mapping

$$G \colon h \mapsto F(x_0 + h) - F(x_0) - DF(x_0)h - \cdots$$

$$- \frac{1}{(k+1)!} D^{k+1} F(x_0)(h, \ldots, h)$$

is differentiable in a neighborhood of zero. According to Example 12.4.2 we have

$$DG(h)v = DF(x_0 + h)v - DF(x_0)v - \cdots - \frac{1}{k!} D^{k+1} F(x_0)(h, \ldots, h, v),$$

which can be written as

$$DG(h) = \Psi(x_0 + h) - \Psi(x_0) - D\Psi(x_0)v - \cdots - \frac{1}{k!} D^k \Psi(x_0)(h \ldots, h),$$

where $\Psi(x) := DF(x)$. Applying the inductive assumption to the mapping Ψ, we obtain the relation $\|DG(h)\| = o(\|h\|^k)$. Since $G(0) = 0$, we have the estimate $\|G(h)\| \leqslant \sup_{0\leqslant t\leqslant 1} \|DG(th)\| \cdot \|h\| = o(\|h\|^{k+1})$ by the mean value theorem, which proves our assertion for $n = k + 1$.

(ii) In the proof of (12.4.1) we can assume that the space Y is separable, since on the interval with endpoints at x and $x + h$ the mapping F is continuous and hence takes values in a separable subspace of Y. In addition, we can pass to the completion of the space Y and assume that it is Banach (then the integral in (12.4.1) will be an element of the original space). The Y-valued mapping $t \mapsto (1 - t)^{n-1} D^n F(x + th)(h, \ldots, h)$ is measurable and bounded by our condition, hence it is Lebesgue integrable (see Example 6.10.66). It suffices to verify the equality of both parts of the regarded equality under the action of elements of Y^*. This reduces our assertion to scalar functions on the interval if we pass to the mapping $\varphi(t) = F(x + th)$. The second assertion in (ii) is a simple corollary of the first one. $\qquad\square$

As in the case of real functions on $\mathbb{R}$, in applications the first two derivatives turn out to be the most useful and frequently used. These derivatives describe the behavior of the function in a neighborhood of a point of local extremum.

12.4.5. Theorem. *Let U be an open set in a normed space X and let $f\colon X \to \mathbb{R}$ be a Gateaux differentiable function.*

(i) *If f has a minimum at the point $x_0 \in U$, then $Df(x_0) = 0$.*

(ii) *If the function f is twice Fréchet differentiable at the point $x_0 \in U$ and has a minimum at this point, then $D^2 f(x_0)(h, h) \geqslant 0$ for all $h \in X$.*

(iii) *Let the function f be twice Fréchet differentiable in U and let $x_0 \in U$ be a point such that $Df(x_0) = 0$ and for some $\lambda > 0$ we have $D^2 f(x_0)(h, h) \geqslant \lambda\|h\|^2$ for all $h \in X$. Then x_0 is a point of a strict local minimum.*

PROOF. Assertions (i) and (ii) follow from the one-dimensional case applied to the function $t \mapsto f(x + th)$, and (iii) follows from assertion (i) of the previous theorem, giving the estimate $f(x_0 + h) - f(x_0) \geqslant \lambda\|h\|^2/2$ for small $\|h\|$. $\qquad\square$

The condition $D^2 f(x_0)(h, h) > 0 \; \forall h \neq 0$ is not sufficient here. For example, for the function $f(x) = \sum_{n=1}^{\infty}(n^{-3}x_n^2 - x_n^4)$ on the space l^2 it holds for $x_0 = 0$, but x_0 is not a point of local minimum.

12.5. Complements and Exercises

(i) Newton's method (529). (ii) Multilinear mappings (530). (iii) Subdifferentials and monotone mappings (534). (iv) Approximations in Banach spaces (535). (v) Covering mappings (536). Exercises (537).

12.5(i). Newton's method

The known Newton method of solving the equation $f(x) = 0$ with a smooth function f on the interval $[a, b]$ employs recurrent approximations

$$x_{n+1} := x_n - f(x_n)/f'(x_n), \quad x_0 = a.$$

One can show that if $f'(x) \neq 0$ on $[a, b]$ and x^* is the unique root of this equation, then $x_n \to x^*$. In the infinite-dimensional case there is a modification of Newton's method, the so-called Newton–Kantorovich method, worked out by L.V. Kantorovich. We shall give a typical result, referring for a more detailed discussion to [312, Chapter XVIII]. Let X and Y be Banach spaces and let $F\colon X \to Y$ be a Fréchet differentiable mapping in the ball $B = B(x_0, r)$ such that its derivative F' satisfies the Lipschitz condition with constant L, i.e., $\|F'(x) - F'(y)\| \leqslant L\|x - y\|$. Suppose that the operator $F'(x_0)$ is invertible . Then we can set

$$x_{n+1} := x_n - [F'(x_0)]^{-1}[F(x_n)].$$

These modified approximations possess worse convergence properties, but use the derivative only at the point x_0. Let $M = \big\|[F'(x_0)]^{-1}\big\|$, $k = \big\|[F'(x_0)]^{-1}(F(x_0))\big\|$, $h = MkL$, let t_0 be the least root of the equation $ht^2 - t + 1 = 0$, and let $r \geqslant kt_0$.

12.5.1. Theorem. *If $h < 1/4$, then in the ball $\{x\colon \|x - x_0\| \leqslant kt_0\}$ the equation $F(x) = 0$ has a unique solution x^*, $x^* = \lim\limits_{n\to\infty} x_n$, moreover, the bound $\|x^* - x_n\| \leqslant kq^n/(1 - q)$ holds, where $q = ht_0 < 1/2$.*

PROOF. Passing to the mapping $[F'(x_0)]^{-1}F$, we can assume that $X = Y$, $F'(x_0) = I$, $M = 1$. In addition, we can assume that $x_0 = 0$. Let us set $\Psi(x) = x - F(x)$. Since the derivative of the mapping $x \mapsto x - F(x) + F(0)$ equals $I - F'(x)$, by the mean value theorem

$$\|\Psi(x)\| = \|x - F(x) + F(0) - F(0)\| \leqslant \|x - F(x) + F(0)\| + k$$

$$\leqslant \sup_{\|y\|\leqslant\|x\|} \|I - F'(y)\|\,\|x\| + k \leqslant L\|x\|^2 + k \leqslant Lk^2t_0^2 + k = kt_0$$

whenever $\|x\| \leqslant kt_0$, so Ψ takes the ball $\{x\colon \|x\| \leqslant kt_0\}$ into the same ball. In this ball Ψ is a contraction, since

$$\|\Psi'(x)\| = \|I - F'(x)\| \leqslant L\|x\| \leqslant Lkt_0 = (1 - \sqrt{1 - 4h})/2 = q < 1/2$$

by the equalities $h = Lk$ and $t_0 = (2h)^{-1}(1 - \sqrt{1 - 4h})$ in our case. It remains to observe that $x_{n+1} = \Psi(x_n)$ and apply Theorem 1.4.1 with the estimate of the rate of convergence obtained in the cited theorem. $\qquad\square$

12.5(ii). Multilinear mappings

We recall that a mapping

$$\Psi\colon X_1 \times \cdots \times X_k \to Y$$

from the product of linear spaces X_i to a linear space Y is called k-*linear* if for any fixed $k - 1$ arguments Ψ is linear in the remaining argument. A mapping that is k-linear for some k is called multilinear. If $X_1 = \cdots = X_k = X$, then such a mapping generates the mapping $\psi_k(x) := \Psi(x, \ldots, x)$ from X to Y, called homogeneous of order k. It is clear that ψ_k is also generated by any k-linear mapping obtained from Ψ by an arbitrary permutation of variables. Hence different k-linear mappings can generate the same homogeneous mapping. However, if Ψ is symmetric, i.e., invariant with respect to all permutations of arguments, then it is uniquely recovered by ψ_k, which is shown below. The symmetrization of the mapping Ψ on X^k is defined by

$$\widetilde{\Psi}: \; x \mapsto \frac{1}{k!} \sum_{\sigma \in S_k} \Psi(x_{\sigma(1)}, \ldots, x_{\sigma(k)}),$$

where summation is taken over all permutations σ of the collection $\{1, \ldots, k\}$. It is clear that the symmetrization of a k-linear mapping Ψ on X^k generates the same homogeneous of order k mapping as Ψ itself. Thus, a homogeneous mapping can be always generated by a symmetric multilinear mapping.

A mapping $\psi: X \to Y$ is called *polynomial* or a *polynom* if

$$\psi(x) = \psi_n(x) + \cdots + \psi_1(x) + \psi_0,$$

where ψ_0 is a constant element of Y and ψ_k is a homogeneous of order k mapping from X to Y, $1 \leqslant k \leqslant n$. In other words, there exist k-linear mappings Ψ_k such that $\psi(x) = \Psi_n(x, \ldots, x) + \cdots + \Psi_1(x) + \psi_0$. We say that ψ has a degree at most n; the exact degree of ψ is the minimal k for which ψ has a degree at most k. The homogeneous polynomials ψ_k are called homogeneous components of ψ.

The next result shows that the homogeneous components are uniquely determined by the polynomial and any homogeneous polynomial uniquely determines the symmetric multilinear mapping generating it. For an arbitrary mapping $\varphi: X \to Y$ and every $h \in X$ we define a mapping $\Delta_h \varphi: X \to Y$ by the formula

$$(\Delta_h \varphi)(x) := \varphi(x + h) - \varphi(x).$$

If $h_1, h_2 \in X$, then we set $\Delta_{h_2} \Delta_{h_1} \varphi := \Delta_{h_2}(\Delta_{h_1} \varphi)$. Since

$$\Delta_{h_2} \Delta_{h_1} \varphi(x) = \varphi(x + h_1 + h_2) - \varphi(x + h_1) - \varphi(x + h_2) + \varphi(x),$$

we have $\Delta_{h_2} \Delta_{h_1} \varphi = \Delta_{h_1} \Delta_{h_2} \varphi$. By induction let

$$\Delta_{h_n} \Delta_{h_{n-1}} \cdots \Delta_{h_1} \varphi := \Delta_{h_n}(\Delta_{h_{n-1}} \cdots \Delta_{h_1} \varphi).$$

One can observe that $\Delta_{h_n} \Delta_{h_{n-1}} \cdots \Delta_{h_1} \varphi$ is the sum of 2^n functions of the form $x \mapsto (-1)^{n-p} \varphi(x + h_{i_1} + \cdots + h_{i_p})$, where $i_1 < \cdots < i_p$ are indices from $\{0, \ldots, n\}$, and for $i_p = 0$ we set $h_{i_p} = 0$. It is easy to verify by induction that $\Delta_{h_n} \Delta_{h_{n-1}} \cdots \Delta_{h_1} \varphi(x)$ is a symmetric function of the arguments $h_1, \ldots, h_n$.

12.5.2. Theorem. *Let $\psi = \psi_n + \cdots + \psi_1 + \psi_0: X \to Y$ be a polynomial of degree at most $n \geqslant 1$. Let Ψ_n be an n-linear symmetric mapping such that we have $\Psi_n(x, \ldots, x) = \psi_n(x)$. Then*
   (i) *for every $h \in X$ the mapping $\Delta_h \psi$ is a polynomial of degree at most $n-1$;*
   (ii) *for any fixed $h_1, \ldots, h_n \in X$, the mapping $\Delta_{h_n} \cdots \Delta_{h_1} \psi$ is constant and equals $n! \Psi_n(h_1, \ldots, h_n)$.*

The proof is delegated to Exercise 12.5.25.

12.5.3. Corollary. *The homogeneous components of any polynomial ψ are unique. If ψ is a homogeneous polynomial of degree k, then the generating symmetric k-linear mapping is uniquely determined by the relation*

$$\Psi_k(x_1, \ldots, x_k) = \frac{1}{k!} \Delta_{x_1} \cdots \Delta_{x_k} \psi(a),$$

where the right-hand side is constant in a.

We now consider polynomial mappings of normed spaces. As in the case of linear mappings, of particular importance are continuous polynomials. The next assertion shows the equivalence of several natural conditions for the continuity. Below the product $X_1 \times \cdots \times X_k$ of normed spaces is equipped with the norm $(x_1, \ldots, x_k) \mapsto \|x_1\|_{X_1} + \cdots + \|x_k\|_{X_k}$.

The proofs of the following two theorems are delegated to Exercise 12.5.26.

12.5.4. Theorem. *Let X and Y be normed spaces and let $\psi = \psi_n + \cdots + \psi_0$ be a polynomial mapping, where each homogeneous component ψ_k is generated by a symmetric k-linear mapping Ψ_k. The following conditions are equivalent:*

(i) *all Ψ_k are continuous;*

(ii) *all homogeneous components ψ_k are continuous;*

(iii) *ψ is continuous;*

(iv) *ψ is continuous at some point;*

(v) *ψ is bounded on some ball of positive radius;*

(vi) *ψ is bounded on every ball.*

12.5.5. Theorem. *Let $X_1, \ldots, X_k, Y$ be normed spaces, where $k \in \mathbb{N}$, and let $\Psi \colon X_1 \times \cdots \times X_k \to Y$ be a k-linear mapping. The following conditions are equivalent:*

(i) *Ψ is continuous;*

(ii) *Ψ is continuous at some point;*

(iii) *Ψ is bounded on a neighborhood of some point;*

(iv) *Ψ is bounded on every bounded set in $X_1 \times \cdots \times X_k$;*

(v) *there exists $M \geqslant 0$ such that $\|\Psi(x_1, \ldots, x_k)\| \leqslant M \|x_1\|_{X_1} \cdots \|x_k\|_{X_k}$.*

Estimate (v) suggests the definition of the norm of a continuous multilinear mapping Ψ by the formula

$$\|\Psi\| := \sup\{\|\Psi(x_1, \ldots, x_k)\| \colon \|x_i\|_{X_i} \leqslant 1\}.$$

The linear space $\mathcal{L}_k(X_1, \ldots, X_k, Y)$ of all continuous k-linear mappings from $X_1 \times \cdots \times X_k$ to Y is a normed space with this norm (and is Banach if so is Y).

Continuous multilinear mappings can be identified with operators in the following way. If Ψ is a bilinear continuous mapping from $X_1 \times X_2$ to Y, then we associate to it the operator

$$\Lambda_\Psi \in \mathcal{L}\big(X_2, \mathcal{L}(X_1, Y)\big), \quad \Lambda_\Psi(x_2)(x_1) := \Psi(x_1, x_2).$$

It is not hard to verify that $\|\Lambda_\Psi\| = \|\Psi\|$. Conversely, every operator Λ_Ψ of class $\mathcal{L}\big(X_2, \mathcal{L}(X_1, Y)\big)$ generates an element $\mathcal{L}_2(X_1, X_2, Y)$ by the indicated formula. The described correspondence is one-to-one and is a linear isometry. Similarly, to every element $\Psi \in \mathcal{L}_k(X_1, \ldots, X_k, Y)$ we associate the operator

$$\Lambda_\Psi \in \mathcal{L}^k := \mathcal{L}\big(X_k, \mathcal{L}^{k-1})\big), \quad \Big(\big(\Lambda_\Psi(x_k)\big) \cdots \Big)(x_1) := \Psi(x_1, \ldots, x_k),$$

where $\mathcal{L}^1 := \mathcal{L}(X_1, Y)$. This correspondence is a linear isometry, i.e., one has

$$\|\Lambda_\Psi\| = \sup_{\|x_1\| \leqslant 1, \ldots, \|x_k\| \leqslant 1} \|\Psi(x_1, \ldots, x_k)\|.$$

This connection was already used in §12.4 when we discussed k-fold differentiable mappings. Polynomial mappings of Banach spaces possess additional properties.

12.5.6. Theorem. *If the spaces $X_1, \ldots, X_k$ are Banach, then the continuity of the multilinear mapping $\Psi\colon X_1 \times \cdots \times X_k \to Y$ is equivalent to its separate continuity in every argument.*

PROOF. Let $x_1 \in X_1$ and $\|x_1\|_{X_1} \leqslant 1$. We consider the multilinear mapping $\Psi_{x_1}\colon X_2 \times \cdots \times X_k \to Y$, $\Psi_{x_1}(x_2, \ldots, x_k) = \Psi(x_1, x_2, \ldots, x_k)$. If $k \geqslant 2$, then by induction we can assume that the continuity of every mapping Ψ_{x_1} is known. The family of these mappings is pointwise bounded: if $x_2, \ldots, x_k$ are fixed, then the quantity $\sup\{\|\Psi_{x_1}(x_2, \ldots, x_k)\|\colon \|x_1\| \leqslant 1\}$ is finite by the continuity and linearity in x_1. Hence it suffices to have the following analog of the Banach–Steinhaus theorem for multilinear mappings: if $X_1, \ldots, X_k, Y$ are Banach spaces and a family of continuous multilinear mappings $T_\alpha\colon X_1 \times \cdots \times X_k \to Y$ is pointwise bounded, then it is uniformly bounded on the product of the unit balls. This analog follows at once from the isometric identification of multilinear mappings and operators described above (and also is easily proved by the same reasoning as in the case of operators). $\qquad\square$

For incomplete spaces this theorem can fail. For example, if the space X of all polynomials on $[0, 1]$ is equipped with the norm from $L^1[0, 1]$, then the bilinear function

$$\Psi(x, y) = \int_0^1 x(t)y(t)\, dt$$

is discontinuous, although it is continuous in every argument separately.

We have defined above polynomials with the aid of homogeneous components. It is clear that the restriction of such a polynomial to every straight line is a usual polynomial, i.e., for every $a, b \in X$ the function $\psi_{a,b}\colon t \mapsto \psi(a + tb)$ is a polynomial in t of the form $c_n t^n + \cdots + c_0$, where $c_k \in Y$. This gives one more option to define polynomials.

12.5.7. Theorem. (i) *Let $\psi\colon X \to Y$ be a mapping between linear spaces such that for every $a, b \in X$ the mapping $\psi_{a,b}$ of a real variable is a polynomial of degree at most n. Then ψ is a polynomial of degree at most n.*

(ii) *Let X and Y be Banach spaces and let $\psi\colon X \to Y$ be a continuous mapping such that for every $a, b \in X$ the mapping $\psi_{a,b}$ is a polynomial. Then the degrees of such polynomials are uniformly bounded, hence ψ is a polynomial.*

The proof is delegated to Exercise 12.5.27 (assertion (ii) is nontrivial!).

12.5.8. Corollary. *A mapping $\psi\colon X \to Y$ between linear spaces is a polynomial of degree at most n precisely when for every linear function l on Y the composition $l \circ \psi$ is a polynomial of degree at most n. If Y is a normed space, then it suffices to take continuous linear functions l.*

12.5.9. Corollary. *Let X and Y be Banach spaces and let $\psi\colon X \to Y$ be a continuous mapping such that for all elements $a, b \in X$ and every continuous linear functional l on Y the function $l \circ \psi_{a,b}$ is a polynomial. Then ψ is a polynomial.*

PROOF. It is clear from the previous theorem that the real function on the space on $X \times Y^*$ defined by $(x, l) \mapsto l(\psi(x))$ is a polynomial, so the degrees of the polynomials $l \circ \psi$ are uniformly bounded. $\square$

It is readily verified that without the assumption of continuity of ψ the theorem can fail: it suffices to take a Hamel basis $\{v_\alpha\}$ in X, pick a countable part $\{v_{\alpha_n}\}$ and set $\psi(x) = \sum_{n=1}^{\infty} x_n^n$, where $x = \sum_n x_n v_{\alpha_n} + \sum_{\alpha \notin \{\alpha_n\}} x_\alpha v_\alpha$. It can also fail in the case of an incomplete space, although, as one can show, in place of completeness it suffices to require that $X \times X$ be a Baire space. On the incomplete space of all finite sequences $x = (x_1, \ldots, x_n, 0, 0, \ldots)$ with the norm from l^2 the function $\psi(x) = \sum_{n=1}^{\infty} \frac{1}{n!} x_n^n$ is continuous, is polynomial on every finite-dimensional subspace, but is not a polynomial on the whole space.

12.5(iii). Subdifferentials and monotone mappings

The geometric interpretation of the derivative of a function as a tangent line to the graph leads to a very important concept of subdifferential of a convex function. Let E be a real locally convex space, $U \subset E$ a nonempty open convex set, and let $f \colon U \to \mathbb{R}^1$ be a continuous convex function. A linear functional $l \in X^*$ is called a *subgradient* of the function f at the point $x_0 \in U$ if

$$f(x) - f(x_0) \geqslant l(x - x_0)$$

for all $x \in U$. The *subdifferential* of f at x_0 is defined to be the set $\partial f(x_0)$ of all subgradients of f at x_0.

It is clear that the set $\partial f(x_0)$ is convex. We show that, in addition, it is nonempty. Indeed, since the supergraph $G_f := \{(x, t) \colon t > f(x)\}$ is convex and open, the point $(x_0, f(x_0))$ can be separated from G_f by a closed hyperplane in $E \times \mathbb{R}^1$ of the form

$$\{(x, t) \colon l(x) + \alpha t = c\}, \quad l \in E^*, \alpha, c \in \mathbb{R}^1.$$

Note that $\alpha \neq 0$, so we can assume that $\alpha = 1$ and that $l(x) + c < t$ for all points $(x, t) \in G_f$, i.e., $l(x_0) + t > l(x) + f(x_0)$ for all $t > f(x)$, whence we obtain $l(x_0) + f(x) \geqslant l(x) + f(x_0)$, as required. If the function f is Gateaux differentiable at x_0, then the Gateaux derivative will be the unique element of the set $\partial f(x_0)$.

Among the most important results about subdifferentials we should mention in the first place the Moro–Rockafellar theorem and the Dubovitskii–Milyutin theorem. For continuous convex functions on locally convex spaces, the former gives the equality $\partial(f + g)(x) = \partial f(x) + \partial g(x)$, and the latter gives the equality $\partial \max(f, g)(x) = \mathrm{conv}\,(\partial f(x) \cup \partial g(x))$. For simplification of technical details we formulate these definitions and results not in their full generality.

The subdifferential of a convex function delivers the most important example of a multivalued monotone mapping. In order not to leave the framework of single-valued mappings, we mention a particular case of the general definition. Let X be a locally convex space. A mapping $F \colon X \to X^*$ is called *monotone* if

$$\langle F(x) - F(y), x - y \rangle \geqslant 0 \quad \forall x, y \in X.$$

For example, if the convex function f is Gateaux differentiable, then the mapping $x \mapsto f'(x)$ is monotone. Monotone mappings play an important role in applications. Mathematical models of many real world phenomena lead to differential and integral equations with monotone mappings. Among the best known results about monotone mappings we mention only the following two (see [93]).

12.5.10. Theorem. *If H is a Hilbert space, $F: H \to H$ is monotone and continuous on finite-dimensional subspaces and* $\lim\limits_{\|x\| \to \infty} \|F(x)\| = \infty$, *then* $F(H) = H$.

We observe that for every monotone mapping $F: H \to H$ that is continuous on finite-dimensional subspaces, the mapping $F + \lambda I$ with $\lambda > 0$ satisfies the conditions of the stated theorem and hence is surjective. Indeed,

$$\|F(x) + \lambda x\| \geqslant \|F(x) - F(0) + \lambda x\| - \|F(0)\|$$
$$\geqslant \langle F(x) - F(0) + \lambda x, x\|x\|^{-1}\rangle - \|F(0)\| \geqslant \lambda\|x\| - \|F(0)\|.$$

The mapping $F + \lambda I$ is injective by the inequality

$$\langle F(x) - F(y) + \lambda x - \lambda y, x - y\rangle \geqslant \lambda\|x - y\|^2.$$

Therefore, we obtain everywhere defined inverse mappings

$$J_\lambda := (\lambda F + I)^{-1}, \quad \lambda > 0.$$

The mappings $F_\lambda := \lambda^{-1}(I - J_\lambda)$ are called the *Yosida approximations* of F.

12.5.11. Theorem. *Let $F: H \to H$ be a monotone mapping continuous on finite-dimensional subspaces. The mappings F_λ possess the following properties*:
   (i) F_λ *is monotone and Lipschitz with constant λ^{-1}*;
   (ii) *for all $\lambda, \mu > 0$ we have $(F_\lambda)_\mu = F_{\lambda+\mu}$*;
   (iii) *for all x, as $\lambda \to 0$ we have*

$$\|F_\lambda(x)\| \uparrow \|F(x)\| \quad and \quad \|F_\lambda(x) - F(x)\| \to 0.$$

The proof of a more general theorem can be found in [39, §3.5.3], [93, Chapter 2]. Let us also note the following fact.

12.5.12. Theorem. *Let a mapping $F: H \to H$ of a Hilbert space H be continuous on finite-dimensional subspaces. The following conditions are equivalent*:
   (i) F *is monotone*,
   (ii) F *is monotone and $(F + I)(H) = H$*,
   (iii) *for every number $\lambda > 0$, the mapping $I + \lambda^{-1}F$ is a bijection of H and* $(I + \lambda^{-1}F)^{-1}$ *is a contraction*.

12.5(iv). Approximations in Banach spaces

Let us make several remarks about approximations by differentiable mappings in infinite-dimensional spaces. Let X be a separable Banach space with the closed unit ball U. The proofs of the following facts can be found in the literature cited below.

Every uniformly continuous real function f on U is uniformly approximated by Lipschitz functions that are Hadamard differentiable. However, on the space

$C[0, 1]$ even the norm is not approximated uniformly on U by Fréchet differentiable functions. On a Hilbert space, uniformly continuous functions are uniformly approximated by functions with bounded and continuous second Fréchet derivatives, however, on l^2 there is a Lipschitz function that is not approximated uniformly on U by functions with uniformly continuous second derivatives. Thus, even in the case of a Hilbert space the border between positive and negative results is passing between the continuity and uniform continuity of bounded second derivatives of approximating functions.

The situation with approximation of infinite-dimensional mappings is even more complicated. There exist uniformly continuous mappings from separable Banach spaces to l^2 which cannot be uniformly approximated by Lipschitz mappings. However, uniformly continuous mappings between Hilbert spaces possess uniform Lipschitz approximations with bounded Fréchet derivatives. The problems of constructing smooth approximations are also connected with the existence of smooth functions with bounded support on Banach spaces. Let us mention some interesting facts.

12.5.13. Theorem. (i) *On $C[0, 1]$ there are no nonzero Fréchet differentiable functions with bounded support.*

(ii) *If on a Banach space X and on its dual there are nonzero functions with bounded support and locally Lipschitz derivatives, then X is a Hilbert space (up to a renorming).*

(iii) *The existence of nonzero functions with bounded support and Lipschitz derivatives is equivalent to the existence of an equivalent norm with a Lipschitz derivative on the unit sphere.*

(iv) *On c_0 there is a nonzero C^∞-function with bounded support (on c_0 there is an equivalent norm that is real-analytic outside the origin).*

(v) *If X possesses a nonzero C^k-function with bounded support, then X contains an isomorphic copy of either c_0 or l^k.*

For a deeper acquaintance with this direction, see the books Benyamini, Lindenstrauss [56], Deville, Godefroy, Zizler [144] and the papers Nemirovskii, Semenov [699], Tsar'kov [710], and Bogachev [673].

12.5(v). Covering mappings

Here we discuss a number of interesting concepts and results connected with elementary objects such as Lipschitz mappings and fixed points, but discovered very recently. These concepts and results could be discussed already in the first sections of the first chapter, but we have kept them for completing our considerations, believing that for the reader, as for ourselves, it will be pleasant to learn in the end that until the present discoveries of something simple and bright still occur in the very basic things. The main object will be the concept of an α-covering mapping introduced by A. A. Milyutin. A number of fundamental results have been obtained in the papers Levitin, Milyutin, Osmolovskii [694], Dmitruk, Milyutin, Osmolovskii [685] and effectively reinforced in the recent paper Arutyunov [668], which we follow in our presentation. Closed balls in metric spaces X and Y will be denoted by the symbols $B_X(x,r)$ and $B_Y(x,r)$; in the case where $X = Y$ we omit indication of the space.

12.5.14. Definition. *Let X and Y be metric spaces. A mapping $\Psi \colon X \to Y$ is called covering with constant $\alpha > 0$, or α-covering, if for every ball $B_X(x,r)$ in X we have $B_Y\big(\Psi(x), \alpha r\big) \subset \Psi\big(B_X(x,r)\big)$.*

Note that any α-covering mapping is surjective, since the union of balls $B_Y\big(\Psi(x_0), \alpha n\big)$ with a fixed center x_0 is the whole space Y.

The main result is the following remarkable theorem from [**668**].

12.5.15. Theorem. *Let $\Psi \colon X \to Y$ be continuous and α-covering with some $\alpha > 0$ and let $\Phi \colon X \to Y$ satisfy the Lipschitz condition with constant $\beta < \alpha$. Suppose that X is complete. Then there exists a point $\xi \in X$ such that $\Psi(\xi) = \Phi(\xi)$. Moreover, for every $x_0 \in X$, there exists a point $\xi = \xi(x_0)$ such that $\Psi(\xi) = \Phi(\xi)$ and $d_X(x_0, \xi) \leqslant (\alpha - \beta)^{-1} d_Y\big(\Psi(x_0), \Phi(x_0)\big)$.*

PROOF. Replacing d_Y by d_Y/α, we can pass to the case $\beta < \alpha = 1$. There is a point $x_1 \in X$ with $\Psi(x_1) = \Phi(x_0)$ and
$$d_X(x_0, x_1) \leqslant d_Y\big(\Psi(x_0), \Phi(x_0)\big).$$
By induction we can construct points x_n such that
$$\Psi(x_n) = \Phi(x_{n-1}), \quad d_X(x_n, x_{n-1}) \leqslant \beta d_X(x_{n-1}, x_{n-2}). \tag{12.5.1}$$
Indeed, if points $x_0, \ldots, x_n$ are already found, there is a point x_{n+1} for which $\Psi(x_{n+1}) = \Phi(x_{n-1})$ and
$$d_X(x_{n+1}, x_n) \leqslant d_Y\big(\Psi(x_n), \Phi(x_n)\big) = d_Y\big(\Phi(x_{n-1}), \Phi(x_n)\big) \leqslant \beta d_X(x_{n-1}, x_n).$$
Using (12.5.1) and the formula for the sum of a progression it is easy to obtain the estimate $d_X(x_n, x_0) \leqslant (1-\beta)^{-1} d_Y\big(\Psi(x_0), \Phi(x_0)\big)$. The completeness of X gives a limit ξ of the sequence $\{x_n\}$ and this limit satisfies the desired conditions. $\square$

Let us show how from this theorem one can obtain some known results.

12.5.16. Example. (i) Let $X = Y$ be complete, $f(x) = x$, and let g be a contracting mapping. Then we obtain a fixed point of g, taking $\alpha = 1$ and $\beta < 1$ equal to the Lipschitz constant for g.

(ii) (Milyutin's theorem) Let X be a complete metric space, Y a normed space, $\Psi \colon X \to Y$ an α-covering continuous mapping, and let $\Phi \colon X \to Y$ satisfy the Lipschitz condition with constant $\beta < \alpha$. Then $\Psi + \Phi$ is $(\alpha - \beta)$-covering. Indeed, we have to show that if $x_0 \in X$ and $y_0 \in Y$, then there exists $\xi \in X$ with $d_X(x_0, \xi) \leqslant (\alpha - \beta)^{-1} \|y_0\|$ and $\Psi(\xi) + \Phi(\xi) = \Psi(x_0) + \Phi(x_0) + y_0$. To this end, let $\Phi_0(x) := \Psi(x_0) + \Phi(x_0) + y_0 - \Phi(x)$. It is clear that Φ_0 is Lipschitz with constant β. The theorem above gives a point ξ with $\Psi(\xi) = \Phi_0(\xi)$, as required.

Exercises

12.5.17. Let $\mathcal{A}$ be a Banach algebra (for example, the algebra of all bounded operators on a Banach space). Prove that the mapping $a \mapsto e^a$, $\mathcal{A} \to \mathcal{A}$ is Fréchet differentiable and find its derivative at zero.

12.5.18.° Prove that the norm on l^1 is nowhere differentiable.
HINT: given $x \in l^1$, consider the maximal $|x_i|$.

12.5.19.° Prove that the following mapping from $L^2[0,1]$ to $C[0,1]$ is Fréchet differentiable and evaluate its derivative:

$$F(x)(t) = \int_0^t \Psi(x(s), s, t)\, ds, \quad \text{where } \Psi \in C_b^1(\mathbb{R}^3).$$

12.5.20. Suppose that in the situation of Example 12.1.7 the set K is compact, convex and balanced and that $\bigcup_{n=1}^\infty nK$ is dense in X. Show that f has the zero Gateaux derivative at all points of $\bigcup_{0 \leqslant t < 1} tK$.

12.5.21. Let μ be a nonnegative measure and $p \in (1, +\infty)$. Prove that the mapping $F \colon x \mapsto |x|^p$ from $L^p(\mu)$ to $L^1(\mu)$ is Fréchet differentiable and $DF(x)h = p|x|^{p-1}h$.

12.5.22.° For $z \in \mathbb{C}$ let $f(z) = \exp(-z^{-4})$ if $z \neq 0$ and $f(0) = 0$. Show that f as a function on $\mathbb{R}^2$ has all partial derivatives $\partial_x^n \partial_y^k f$ on the whole plane, but is discontinuous at the origin, hence is not differentiable.

12.5.23.° Let H be the Hilbert space l^2 of two-sided sequences $x = (x_n)$, $n \in \mathbb{Z}$, let $\{e_n\}$ be its natural basis, and let T be the linear isometry such that $Te_n = e_{n-1}$. Let us consider the second order polynomial mapping

$$f(x) = T(x + \varepsilon(1 - (x,x))e_0), \quad \varepsilon \in (0, 1/2).$$

Show that f takes the closed unit ball U to U and is a diffeomorphism of some neighborhoods of U and a homeomorphism of U, but has no fixed points.

12.5.24. Prove that a function ψ on $\mathbb{R}^d$ is a polynomial if ψ is a polynomial in every argument.
HINT: apply the induction on d. For this assume that $d > 1$ and that for $d - 1$ the assertion is true. The elements of $\mathbb{R}^d$ will be written in the form (x, y), where $x \in \mathbb{R}^{d-1}$ and $y \in \mathbb{R}^1$. For every $x \in \mathbb{R}^{d-1}$ the function $y \mapsto \psi(x, y)$ is a polynomial of some degree. By the Baire theorem there is $n \in \mathbb{N}$ and a set $M \subset \mathbb{R}^{d-1}$ dense in some ball B such that for every $x \in M$ and all y we have $\psi(x, y) = a_n(x)y^n + \cdots + a_1(x)y + a_0(x)$, where a_j is some function on $\mathbb{R}^{d-1}$. Substituting $y = 0, 1, 2, \ldots, n$ we obtain $n + 1$ polynomials $\psi(x, k)$, $k = 0, 1, \ldots, n$, on $\mathbb{R}^{d-1}$. For every $x \in M$ we have $n + 1$ equalities $\psi(x, k) = a_n(x)k^n + \cdots + a_1(x)k + a_0(x)$. There are numbers c_{ij}, where $i, j = 0, 1, \ldots, n$, independent of x, for which

$$a_k(x) = c_{k0}\psi(x, 0) + c_{k1}\psi(x, 1) + \cdots + c_{kn}\psi(x, n),$$

since the determinant of this system with respect to $a_k(x)$ (the Vandermonde determinant) is not zero. Let us consider the polynomials $p_k(x) := \sum_{j=0}^n c_{kj}\psi(x, j)$ and the polynomial $\varphi(x, y) := \sum_{k=0}^n p_k(x)y^k$. For any $x \in M$ the functions $y \mapsto \psi(x, y)$ and $y \mapsto \varphi(x, y)$ are polynomials of degree n and coincide at $n + 1$ points $y = 0, 1, \ldots, n$. Hence they coincide for all y. Thus, the function ψ coincides with a polynomial on the set $M \times \mathbb{R}^1$, i.e., $\psi - \varphi = 0$ on this set. By the inductive assumption for every fixed y the function $x \mapsto \psi(x, y) - \varphi(x, y)$ is a polynomial. Since this polynomial vanishes on the set M dense in the ball, it is identically zero. Hence $\psi(x, y) = \varphi(x, y)$ for all (x, y).

12.5.25. Prove Theorem 12.5.2 by induction on n.
HINT: for $n = 1$ the assertion is verified directly. Suppose that it is true for $n - 1 \geqslant 1$. We have $\Delta_h \psi = \Delta_h \psi_n + \Delta_h(\psi_{n-1} + \cdots + \psi_0) = \Delta_h \psi_n + \varphi$, where φ is a polynomial of

degree at most $n - 2$ by the inductive assumption. In addition, by using the multilinearity and symmetry of Ψ_n we obtain

$$\Delta_h \psi_n = \Psi_n(x + h, \ldots, x + h) - \Psi_n(x, \ldots, x) = n\Psi_n(x, \ldots, x, h) + \eta(x),$$

where η is a polynomial of degree at most $n - 2$ and the argument x enters $\Psi_n(x, \ldots, x, h)$ $n - 1$ times. Thus, $\Delta_h \psi$ is the sum of the homogeneous polynomial $n\Psi_n(x, \ldots, x, h)$ of degree $n - 1$ and the polynomial $\varphi + \eta$ of degree at most $n - 2$, which completes the proof of assertion (i). We now make an inductive step to justify (ii). As already shown, we have $\Delta_{h_n} \psi = \varphi_{n-1} + \cdots + \varphi_0$, where φ_k are homogeneous polynomials of degree k and

$$\varphi_{n-1}(x) = g_{n-1}(x, \ldots, x), \quad g_{n-1}(x_1, \ldots, x_{n-1}) = n\Psi_n(x_1, \ldots, x_{n-1}, h_n).$$

It remains to use the inductive assumption for the polynomial $\Delta_{h_n} \psi$ of degree at most $n-1$, which shows that $\Delta_{h_1} \cdots \Delta_{h_{n-1}} (\Delta_{h_n} \psi)$ is the number $(n - 1)! g_{n-1}(h_1, \ldots, h_{n-1})$.

12.5.26. Prove Theorems 12.5.4 and 12.5.5.

HINT: each of the assertions (i)–(iv) implies the next one and (vi) implies (v). Show that (v) implies (i) (then also (vi)) by induction on n. For $n = 1$ this is true, since $\psi_1 = \Psi_1$ is a linear mapping. Suppose that the implication in question is true for some $n - 1 \geqslant 1$. By assumption there is a ball $B(a, r)$ with $r > 0$ on which $\|\psi(x)\|$ does not exceed some M. We know that $n!\Psi_n(h_1, \ldots, h_n) = \Delta_{h_1} \cdots \Delta_{h_n} \psi(a)$ is the sum of 2^n terms of the form $(-1)^{n-p}\psi(a + h_{i_1} + \cdots + h_{i_p})$, where $i_1 < \cdots < i_p$ are indices from $\{1, \ldots, n\}$ and $h_{i_p} = 0$ for $p = 0$. If we take vectors h_i with $\|h_i\| \leqslant r/n$, then $a + h_{i_1} + \cdots + h_{i_p} \in B(a, r)$. Hence $\|\Psi_n(h_1, \ldots, h_n)\| \leqslant 2^n M$ whenever $\|h_i\| \leqslant r/n$. This yields the estimate $\|\Psi_n(x_1, \ldots, x_n)\| \leqslant 2^n M(nR/r)^n$ whenever $\|x_i\| \leqslant R$, which gives the continuity of Ψ_n at zero. By the additivity of Ψ_n in every argument we obtain the continuity of Ψ_n everywhere. The proof of Theorem 12.5.5 is similar.

12.5.27. Prove Theorem 12.5.7 and Corollary 12.5.8.

HINT: (i) If $X = \mathbb{R}^k$, then the assertion is easily verified by induction on k with the aid of the fact that ψ is a polynomial in every variable. Hence the restriction of ψ to every finite-dimensional space is a polynomial of degree at most n. It follows that the mapping $\Psi_n \colon (x_1, \ldots, x_n) \mapsto \Delta_{x_1} \cdots \Delta_{x_n} \psi(0)$ is multilinear and the restriction of the mapping $\psi(x) - \Psi_n(x, \ldots, x)$ to every straight line is a polynomial of degree at most $n - 1$. Using induction on n, we obtain our claim. (ii) Show that the degrees of the polynomials $\psi_{a,b}$ are uniformly bounded. In the case of $\mathbb{R}^d$ this is true even without the assumption about the continuity of ψ, which can be derived from Baire's theorem (Exercise 12.5.24). In the infinite-dimensional case we also use Baire's theorem, but in another way. For every $n \in \mathbb{N}$ let us consider the set

$$M_n := \big\{ (a, b) \in X \times X \colon \psi_{a,b} \text{ has a degree at most } n \big\}.$$

By our condition, the union of all M_n is $X \times X$. By Baire's theorem some M_{n_1} is dense in some ball U of radius $r > 0$ centered at (u_0, v_0). Then $U \subset M_{n_1}$. Indeed, let $(u_n, v_n) \to (u, v)$ in U and $(u_n, v_n) \in M_{n_1}$. By the continuity of ψ for all $t \in \mathbb{R}^1$ we have $\psi(u + tv) = \lim_{n \to \infty} \psi(u_n + tv_n)$. It is not hard to verify that the pointwise limit of polynomials of degree at most n_1 on the real line is also a polynomial of degree at most n_1. Thus, $(u, v) \in M_{n_1}$. Passing to the function $x \mapsto \psi(x - a)$, we can assume that $a = 0$. Thus, whenever $\|u\| \leqslant r$ and $\|v - v_0\| \leqslant r$, all functions $\psi_{u,v}$ are polynomials of degree at most d. It follows that all functions $\psi_{u,v}$ possess this property. Indeed, let us fix u_1, v_1. Let us consider the restriction of ψ to the three-dimensional space E generated by u_1, v_1 and v_0. We can assume that $E = \mathbb{R}^3$, $v_1 = e_1$, $u_1 = e_2$, $v_0 = e_3$, where e_i are the vectors of the standard basis (if the vector v_0 belongs to the two-dimensional space generated by u_1 and v_1, then the problem becomes even simpler). We are given the function ψ that is a polynomial on every straight line and for all u from some ball centered at zero and all v from a ball centered at e_3 the polynomials $\psi_{u,v}$ have a degree at most n. We already know that such a function is a polynomial. Hence it is clear that its degree does not exceed n. For the proof of Corollary 12.5.8 observe that if for every linear function l the composition $l \circ \psi$ is a polynomial of degree at most n, then for every $a, b \in X$ we have $l\big(\psi(a+tb)\big) = c_n(a,b,l)t^n + \cdots + c_0(a,b,l)$, where the functions $l \mapsto c_k(a,b,l)$ are linear. Hence there exist elements $v_k(a,b)$ in the second algebraic dual to Y such that $\psi(a+tb) = v_n(a,b)t^n + \cdots + v_0(a,b)$ for all $t \in \mathbb{R}^1$. Substituting $t = 0, 1, \ldots, n$, we obtain that the elements $v_k(a,b)$ are linear combinations of the elements $\psi(a+kb) \in Y$, since the determinant of the corresponding linear system is nonzero (this is the so-called Vandermonde determinant). Thus, $v_k(a,b) \in Y$, so Theorem 12.5.7 applies. In the case of a normed space Y the same reasoning works with continuous functionals on Y.

12.5.28. Let H be a Hilbert space and $F\colon H \to H$. Prove that F is monotone precisely when $\|x - y\| \leqslant \|x - y + \lambda F(x) - \lambda F(y)\|$ for all $\lambda > 0$ and $x, y \in H$.

12.5.29. Let H be a Hilbert space and let $F\colon H \to H$ be a monotone mapping continuous on finite-dimensional subspaces. Prove that for every $v \in H$ the function $x \mapsto \big(F(x), v\big)$ is continuous. Give an example where F itself is not continuous.

12.5.30. Let X be a Hilbert space and let $\{T_t\}$ be a one-parameter group of linear operators on X with $\|T_t\| = 1$ such that for some $h \in X$ the bound $\|T_t h - h\| \leqslant C|t|$ holds whenever $t \in [-1, 1]$. Prove that the mapping $F\colon t \mapsto T_t h$ is continuously differentiable on $\mathbb{R}^1$ and Lipschitz with the constant $\|F'(0)\| \leqslant C$.

12.5.31. Let X, Y be metric spaces, let X be complete, and let $F\colon X \to Y$ be continuous and satisfy the following condition: for every $\varepsilon > 0$ there is $\delta > 0$ such that the closure of $F\big(U(x, \varepsilon)\big)$ contains $U\big(F(x), \delta\big)$ for all $x \in X$, where $U(a, r)$ is the open ball of radius r centered at a. Prove that F takes open sets to open sets.
HINT: see [**418**, p. 17, Lemma 3.9].

12.5.32. (J. Vanderwerff) Let X be an infinite-dimensional Banach space. Show that on X there is a continuous convex function that is not bounded on the closed unit ball.
HINT: take a sequence of functionals $f_n \in X^*$ with $\|f_n\| = 1$ that is weak-$*$ convergent to zero and set $f := \sum_{n=1}^{\infty} \varphi_n(f_n)$, where φ_n are even continuous convex functions on the real line with $\varphi_n(t) = 0$ for all $t \in [0, 1/2]$ and $\varphi_n(1) = n$.

12.5.33.* (Shkarin [**705**]) Prove that on an infinite-dimensional separable Hilbert space there is an infinitely Fréchet differentiable function f such that $f(x) = 0$ if $\|x\| \geqslant 1$, but $f'(x) \neq 0$ if $\|x\| < 1$. Moreover, there is a polynomial of degree 4 with this property.
HINT: consider $H = L^2[0, 1]$, $f(x) = \big(1 - (x, x)\big)\big((Ax, x) + 2(\varphi, x) + 4/27\big)$, where $Ax(t) = tx(t)$, $\varphi(t) = t(1 - t)$.

12.5.34.* Prove that there exists an infinitely Fréchet differentiable function f on l^2 that is bounded on the closed unit ball, but does not attain its maximum on it.
HINT: use the previous exercise.

Comments

Functional analysis is a relatively young mathematical course in the university curriculum. As an independent field functional analysis was formed shortly before its inclusion in the university programme: the formation of functional analysis as an area of mathematics is usually connected with the publication of Stefan Banach's monograph [**46**], which appeared at the beginning of the 1930s in the Polish language, next in French, and soon afterwards in English. Even a quick glance at the table of contents of this outstanding treatise shows how close it is to modern textbooks, more precisely, how large portions of modern courses are borrowed from Banach's book (about Banach's life, see Kałuża [**308**]). In spite of this, it should be noted that actually functional analysis as analysis in spaces of functions, curves, etc., (moreover, nonlinear analysis!) emerged much earlier in problems of calculus of variations: Newton, Euler, Bernoulli, and later other classics dealt with curves as independent objects and studied functions on sets of curves in connection with very nontrivial applied problems. This nonlinear functional analysis, that appeared even earlier than the linear one, became one of the most important sources of modern functional analysis, although many of its components belong now to other areas, such as calculus of variations, optimal control. However, elements of calculus of variations until present times are frequently included in university courses of functional analysis. Among the researchers who contributed considerably to the development of nonlinear functional analysis we should mention H. Poincaré, D. Hilbert, J. Hadamard, V. Volterra, M. Fréchet. The second important source for the creation of functional analysis was the theory of integral equations and the theory of partial differential equations. One should particularly single out the first quarter of the XX century, when simultaneously with the development of the general (and already abstract to a large degree) theory there was the process of accumulation of important partial examples and concrete results originating in mathematical physics and pushing forward the theory. Here one can mention again the names of H. Poincaré and D. Hilbert and also F. Riesz, H. Weyl, E. Schmidt, H. Lebesgue, J. Radon, T. Carleman, E. I. Fredholm. The end of this period was marked by the creation of quantum mechanics and attempts to work out an adequate mathematical apparatus for it on the basis of the theory of unbounded linear operators. A particular role was played by the works of J. von Neumann, M. Stone and F. Riesz. Simultaneously with progress in the theory of operators, starting from the 1920s a development of the "linear theory" of spaces

© Springer Nature Switzerland AG 2020
V. I. Bogachev, O. G. Smolyanov, *Real and Functional Analysis*,
Moscow Lectures 4, https://doi.org/10.1007/978-3-030-38219-3

in which these operators act was very active, involving also abstract spaces. In this connection, in the first place one should mention S. Banach, but great contributions of other researchers must be also noted (H. Hahn, E. Helly, S. Mazur, J. Schauder, N. Wiener). An important role was played by the famous Lvov mathematical school (see Duda [160]). Certainly, a more thorough look reveals numerous important subdirections in this powerful flow of research, such as linear and convex analysis (H. Minkowski, H. Hahn, E. Helly, S. Banach), general measure and integration theory (founded by Lebesgue at the beginning of the century and developed by J. Radon, F. Riesz, M. Fréchet, C. Carathéodory, N. N. Luzin, P. J. Daniell, N. Wiener, O. Nikodym, A. N. Kolmogorov, J. von Neumann, A. D. Aleksandrov, and Yu. V. Prohorov). At the same time a new stage of the development of nonlinear functional analysis began (N. N. Bogolioubov, N. M. Krylov, J. Schauder, A. N. Tychonoff, J. Leray). One should also not forget about close areas such as general topology (F. Hausdorff, P. S. Alexandroff, P. S. Urysohn, A. N. Tychonoff) and extremal problems (L. V. Kantorovich, L. S. Pontryagin and others). Modern functional analysis became the product of deep interactions between all these directions. The ideas generated in the first years of its formation led to completely new directions already in the earliest years. For example, the theory of Sobolev spaces initiated by S. L. Sobolev in the 1930s and the theory of distributions created by L. Schwartz at the end of the 1940s and the beginning of the 1950s became the most powerful tools in the modern theory of partial differential equations and in mathematical and theoretical physics. Another example is the theory of Banach algebras founded by I. M. Gelfand and developed by many researchers (to mention just a few: M. A. Naimark, G. E. Shilov and I. Segal). Let us also mention the theory of ordered spaces and vector lattices (F. Riesz, L. V. Kantorovich, M. G. Krein). Many motives of the first years of rapid development of functional analysis in the 1920–1930s inspired, and still continue to inspire, generations of researchers. One of these motives is geometry and topology of Banach spaces. Many fundamental discoveries in this direction were already made by Banach and other representatives of the Lwów school in the 1930s. In subsequent years considerable contributions were made by M. G. Krein, D. P. Milman, V. L. Shmulian, B. J. Pettis, N. Dunford, R. S. Phillips, W. F. Eberlein, A. A. Milyutin, A. Grothendieck, M. I. Kadets, A. Pełczyński, J. Lindenstrauss, P. Enflo, W. Gowers. Certainly, this list can be continued by mentioning a number of authors of solutions of longstanding problems and particularly known examples and counter-examples (many of which we mentioned in the text). Under the influence of the theory of Banach spaces and the theory of distributions the theory of locally convex spaces has gained considerable popularity. Some important results were obtained in the 1930s by A. N. Kolmogorov, A. N. Tychonoff, J. von Neumann, J. Leray, but as a large independent area this direction was formed in the 1950s. Among those who considerably contributed to this theory one should mention L. Schwartz, J. Dieudonné, G. Mackey, R. F. Arens, A. Grothdieck, G. Choquet. For some time the theory of locally convex spaces was one of the most fashionable directions in functional analysis; large chapters on this theory were included in textbooks (and we also pay tribute to this theory in the present book). Fashions pass away, but this theory had grounds to be popular, and today its main principles are still of value for general functional analysis as well as for diverse applications. The theory of Banach spaces and the theory of locally convex spaces include the study of linear operators on such spaces, but by the theory of linear operators it is customary to mean another great component of the modern functional analysis that grew from the aforementioned investigations in integral equations and differential operators. Characteristic features of this area that distinguish it from the geometry and topology of abstract Banach or locally convex spaces are an interest in concrete operators and concrete equations and the particular role of unbounded operators. Certainly, there is no formal division here, since the various listed areas actively interact with one another. Finally, it should be noted that many of these "linear" directions have "nonlinear" branches. Nonlinear spaces, nonlinear operators, and nonlinear equations are so important and ubiquitous objects in natural sciences that any separation inside general functional analysis of "nonlinear analysis" from its "linear" directions, which can be regarded as special parts of nonlinear analysis, can be doubtful. One can talk a lot about the importance of functional analysis for applications. We confine ourselves to only one example. When in the 1940s in attempts to create a nuclear weapon it became necessary to solve new and very difficult problems connected with modelling complicated processes, in addition to physicists a number of mathematicians, experts in nonlinear analysis, were recruited. In particular, in the USA these were J. von Neumann and S. Ulam and in the Soviet Union these were A. N. Tychonoff, S. L. Sobolev, I. G. Petrovskii, I. M. Gelfand, L. V. Kantorovich, N. N. Bogolioubov, M. V. Keldysh and others (see, for example, Tikhonova, Tikhonov [595]).

As it has already been noted, Banach's book [46] became the first textbook on functional analysis and a model for many subsequent texts until today. However, some elements of the modern course connected with Hilbert spaces were presented already in Vitali's book [618]. Approximately the same time Stone's monograph [575] appeared, which also influenced very much the formation of the modern appearance of functional analysis and its place in the university curriculum. Among the first university courses of functional analysis we should mention the classical text of Riesz, Sz.-Nagy [510] (which had many editions in several languages), Russian texts (later translated) by Smirnov [563] and Kolmogorov and Fomin [331] (which grew from Kolmogorov's lectures [330] at the Department of Mechanics and Mathematics and Fomin's lectures at the Department of Physics at Moscow University and had many editions in many languages). The course of "Analysis-III" (now "Functional analysis") introduced in the curriculum at Moscow University in 1945 was later lectured by such distinguished mathematicians as I. M. Gelfand and G. E. Shilov and other known experts. In our opinion, the textbook [331] is one of the best courses of functional analysis for university students (in spite of the absence of a systematic presentation of spectral theory).

Before we mention other texts, let us note three books of encyclopedic nature and large size: Dunford, Schwartz [164], Edwards [171], and Kantorovich, Akilov [312] (a survey of classical results is also given in Krein [617]). Returning to more textbook-style books, let us mention known treatises for in-depth study Reed, Simon [502] and Rudin [522]. There are hundreds of courses in different languages, oriented towards different audiences and varying greatly by the depth of presentation, size, style and methodic principles. We give a rather extensive list (which still does not pretend to be complete) of those books in English, French, German and Russian (and some selected titles in Spanish and Italian) which we were able to look through in the libraries of more than one hundred universities and mathematics institutes all over the world. Not repeating the *already mentioned books*, we start with more complete presentations:

Alt [17], Amann, Escher [20], Antonevich, Radyno [27], Appell, Väth [30], Arsen'ev [33], Avanissian [41], Bachman, Narici [42], Baggett [44], Bass [50], Berberian [57], Berezansky, Sheftel, Us [58], Bertrandias [62], Brézis [94], Brown, Page [97], Brown, Pearcy [98], Bühler, Salamon [100], Cheney [111], Choudhary, Nanda [115], Conway [127], Cotlar, Cignoli [131], DeVito [145], DiBenedetto [146], Eidelman, Milman, Tsolomitis [176], Einsiedler, Ward [177], Epstein [183], Farenick [187], Fedorov [188], Folland [195], Fuchssteiner, Laugwitz [201], Garnir, De Wilde, Schmets [203], Georgiev, Zennir [213], Goffman, Pedrick [221], Griffel [238], Grossmann [240], Ha [246], Heine [260], Helemskii [262], Heuser [269], Hille [273], Hirsch, Lacombe [275], Hirzebruch, Scharlau [276], Jain, Ahuja, Ahmad [288], Kaballo [300], [301], Kadets [304], Kantorovitz [314], Knyazev [328], Komornik [332], Korevaar [335], Kreyszig [356], Krishnan [357], Kutateladze [367], Lahiri [371], Lang [373], Larsen [374], Lavrent'ev, Savel'ev [376], Lax [378], Lelong [383], Limaye [390], Lösch [396], Maddox [405], Malkowsky, Rakočević [407], Mathieu [411], Maurin [413], Meise, Vogt [418], Miranda [428], Mukherjea, Pothoven [441], Nair [450], Naylor, Sell [454], Oden, Demkowicz [460], Pedersen [472], Pflaumann, Unger [478], Poroshkin [494], Pryce [497], Pugachev, Sinitsyn [498], Rao [499], Reddy [501], Roman [516], Ruckle [518], Rynne, Youngson [524], Sadovnichiĭ [526], Samuélidès, Touzillier [529], Schechter [535], Schröder [538], Sen [544], Siddiqi [552], Simon [555], Somasundaram [568], Stein, Shakarchi [573], Storch, Wiebe [576], Şuhubi [581], Sunder [582], Swartz [584], Taylor [589], Trénoguine [602], Triebel [609], Werner [628], Wilansky [631], Wloka [635], Wouk [638], Yosida [640].

Separately we mention texts containing elements of functional analysis (for example, as a brief course or part of advanced analysis):

Amerio [22], Artémiadis [34], Bhatia [65], Bobrowski [70], Boccara [71], Bollobás [77], Bonic [78], Bowers, Kalton [89], Bressan [92], Bridges [96], Burg, Haf, Wille [102], Cannarsa, D'Aprile [103], Caumel [108], Cerdà [109], Chacón, Rafeiro, Vallejo [110], Christensen [117], Davis [137], Dieudonné [150], Efimov, Zolotarev, Terpigoreva [174], Egle [175], Fichera [191], Giles [215], Gill, Zachary [216], Göpfert, Riedrich [233], Gostiaux [234], Groetsch [239], Haase [247], Hansen [253], Hengartner, Lambert, Reischer [266], Humpherys, Jarvis, Evans [285], Hunter, Nachtergaele [286], Jantscher [290], Jost [299], Katzourakis, Vărvărucă [319], Koliha [329], Krantz [342], Kubrusly [358], Kühner, Lesky [362], Limaye [391], Lindstrøm [392], Lusternik, Sobolev [399], [400],

MacCluer [403], Marle, Pilibossian [408], Michel, Herget [420], Mlak [433], Maurin [414], Montesinos, Zizler, Zizler [435], Moore [436], Muscat [445], Nachbin [447], Orlicz [464], Ovchinnikov [467], Packel [468], Panzone [469], Phillips [481], Ponnusamy [493], Promislow [495], Ray [500], Royden [517], Sacks [525], Sasane [531], Saxe [532], Segal, Kunze [542], Shilov [547], [549], Simmons [553], Skandalis [562], Sohrab [567], Sonntag [569], Swartz [585], Torchinsky [599], Tricomi [606], Vainberg [613], Wagschal [623], Weaver [625], Willem [634], Wong [637], Wulich [639], Young [641], Zaanen [642], Zaidman [643], Zamansky [645], Zimmer [658].

We have deliberately avoided trying to provide any more detailed classifications of the aforementioned texts, since this would lead to too broad a spectrum with unclear boundaries of transitions. Even with respect to their size, some of these books differ by more than a factor of ten. It would also not be easy to divide the list into groups according to their coverage of the university curriculum: some short texts cover a very broad programme, although without details, while others with a much greater size and a deeper presentation omit some parts of the standard programme (if we admit that there is some "standard programme"). In a number of texts only Hilbert spaces and operators on them are considered. However, it seems that for every lecturer creating a programme it is very useful to consult a large number of very different texts. Even weak points that are obviously present in many of them can be very instructive. In addition, it is important to bear in mind orientation to specific categories of students, for example, physicists, engineers, economists, applied mathematicians or students in computational mathematics. This is why we present here this list collected over several decades. It is worth noting that the modern electronic data bases and sophisticated search systems do not trivialize the task of completing such a bibliography, even if they do help enormously. Many books on our list have non-typical and unexpected titles and are not shown by an elementary search; some books, apparently are not included in available data bases.

Many of the books listed contain large collections of problems and exercises, but there are also separate problem books (for the entire course or its parts):

Abramovich, Aliprantis [2], Aliprantis, Burkinshaw [15], Alpay [16], Antonevich, Knjazev, Radyno [26], Benoist, Salinier [55], Borodin, Savchuk, Sheipak [82], Bouyssel [88], Costara, Popa [130], Dorogovtsev [157], Gelbaum [206], Gelbaum, Olmsted [207], George [212], Glazman, Ljubič [218], Gonnord, Tosel [232], Gorodetskii, Nagnibida, Nastasiev [231], Halmos [251], Kaczor, Nowak [302], Kirillov, Gvishiani [326], Kubrusly [359], Kudryavtsev, Kutasov, Chehlov, Shabunin [360], Lacombe, Massat [370], Leont'eva, Panferov, Serov [385], Letac [386], Makarov, Goluzina, Lodkin, Podkorytov [406], Samuélidès, Touzillier [530], Shiryaev [550], Sonntag [569], Telyakovskii [591], Torchinsky [600], Trenogin, Pisarevskii, Soboleva [601], Ulyanov et al. [610], Vladimirov et al. [621], Wagschal [623], Zuily [660].

We do not give a bibliography in measure theory and integration (a course of real analysis), since it is also very extensive and is well-presented in [73] (where only the collection of books counts more than 300 titles). Note only the following well known texts: Bruckner, Bruckner, Thomson [99], Dudley [161], Halmos [250], Hewitt, Stromberg [271], Natanson [453], Royden [517], Saks [528], and Stromberg [578].

Various additional information on the *geometry of Banach spaces* is presented in the following books: Albiac, Kalton [9], Beauzamy [51], Benyamini, Lindenstrauss [56], Bessaga, Pełczyński [64], Bourgin [87], Carothers [105], Casazza, Shura [107], Day [138], Defant, Floret [140], Deville, Godefroy, Zizler [144], Diestel [147], [148], Diestel, Uhl [149], van Dulst [163], Effros, Ruan [173], Fabian et al. [185], Fabian et al. [186], Fleming, Jamison [193], Guerre-Delabrière [243], Hájek, Montesinos Santalucía, Vanderwerff, Zizler [248], Johnson, Lindenstrauss [293], Kadets, Kadets [303], Lacey [369], Li, Queffélec [389], Lindenstrauss, Tzafriri [393], Marti [409], Megginson [417] (a very well written modern introduction), Milman, Schechtman [427], Morrison [439], Mushtari [446], Odinets, Yakubson [461], Pietsch [483], [485], Pietsch, Wenzel [486], Pisier [489], [490], Ryan [523], Semadeni [543], Singer [558], Tomczak-Jaegermann [597], Vakhania, Tarieladze, Chobanyan [614], and Wojtaszczyk [636]. Various special spaces and operators on them are considered in Duren, Schuster [168], Fetter, Gamboa de Buen [190], Hedenmalm, Korenblum, Zhu [259], Hoffman [278], Koosis [334], Krasnosel'skiĭ, Rutickiĭ [346], and Nikolski [456]. A survey of open problems in the theory of Banach spaces is given in Guirao, Montesinos, Zizler [244].

There is a very rich literature on *operator theory* and *spectral theory*, which can be found through electronic data bases and the following books: Abramovich, Aliprantis [1], Akhiezer, Glazman [6], Arveson [35], Aupetit [40], Birman, Solomjak [67], Bonsall, Duncan [79], [80], Brown, Pearcy [98], Carl, Stephani [104], Conway [128], [129], Davies [136], Dowson [159], Dunford, Schwartz [164]–[166], Gohberg, Goldberg [222], Gohberg, Goldberg, Kaashoek [223], [224], Gohberg, Krein [226], [227], Gohberg, Goldberg, Krupnik [225], Goldberg [228], Friedrichs [200], Helmberg [264], Istrățescu [287], Kato [317], Kreĭn [352], Levitan, Sargsjan [388], Lord, Sukochev, Zanin [395], Maslov [410], Moretti [438], Müller [442], Naĭmark [449], Pietsch [483], Pirkovskii [488], Plesner [491], Reed, Simon [503]–[505], Retherford [507], Ringrose [511], Schmüdgen [537], Simon [557], Stone [575], Vasilescu [616], Weidmann [626], Zhu [656].

Let us give some orienting bibliography on various other areas.

Integral operators and integral equations: Bukhvalov et al. [101], Edmunds, Kokilashvili, Meskhi [170], Fenyö, Stolle [189], Jörgens [296], Korotkov [338], Krasnov [350], Krasnov, Makarenko, Kiselev [351], Krasnosel'skiĭ, Zabreiko, Pustyl'nik, Sobolevskiĭ [349], Kress [355], Mihlin [421], [422], Mikhlin, Morozov, Paukshto [424], Mikhlin, Prössdorf [425], Okikiolu [462], and Prössdorf [496].

Operator semigroups: Brezis [93], Clément et al. [123], Davies [135], Engel, Nagel [180], [181], Goldstein [229], Hille [272], Hille, Phillips [274], Pazy [471], Reed, Simon [503], and Sinha, Srivastava [561].

Banach and operator algebras: Blackadar [**68**], Blecher, Le Merdy [**69**], Bonsall, Duncan [**81**], Bratteli, Robinson [**90**], Connes [**125**], Dales [**133**], Dales, Aiena, Eschmeier, Laursen, Willis [**134**], Dixmier [**155**], Douglas [**158**], Effros, Ruan [**173**], Fillmore [**192**], Gamelin [**202**], Gelfand, Raikov, Shilov [**209**], Gillman, Jerison [**217**], Graham [**236**], Helemskii [**261**], [**263**], Jorgensen, Tian [**297**], Kadison, Ringrose [**305**], [**306**], Kaniuth [**310**], Landsman [**372**], Larsen [**375**], Loomis [**394**], Murphy [**443**], Naïmark [**448**], Rickart [**509**], Sakai [**527**], Takesaki [**588**], and Żelazko [**653**].

Classical and abstract harmonic analysis and related problems of representation theory: Bachman, Narici, Beckenstein [**43**], Benedetto [**53**], Constantin [**126**], Deitmar, Echterhoff [**142**], van Dijk [**152**], Folland [**196**], Helson [**265**], Hewitt, Ross [**270**], Howell [**284**], Kirillov [**325**], Körner [**336**], [**337**], Krantz [**341**], Loomis [**394**], Muscalu, Schlag [**444**], Osilenker [**466**], Pereyra, Ward [**475**], Pinsky [**487**], Reiter, Stegeman [**506**], Rudin [**521**], Serov [**545**], Simon [**556**], Stein, Shakarchi [**572**], Sugiura [**580**], Székelyhidi [**586**], Terras [**592**], Varadarajan [**615**], Weisz [**627**], Willem [**633**], and also the references at the end of §9.2.

Topological vector spaces: Bogachev, Smolyanov [**75**], Bourbaki [**85**], Choquet [**113**], Edwards [**171**], Garsoux [**205**], Grothendieck [**241**], Horvath [**282**], Jarchow [**291**], Kalton, Peck, Roberts [**307**], Kantorovich, Akilov [**311**], Kelley, Namioka [**321**], Köte [**340**], Narici, Beckenstein [**451**], Pérez Carreras, Bonet [**476**], Pietsch [**482**], Robertson, Robertson [**512**], Rolewicz [**515**], Schaefer [**533**], and Trèves [**604**].

Distributions (generalized functions: Antosik, Mikusinski, Sikorski [**28**], Barros-Neto [**48**], Bhattacharyya [**66**], Bremermann [**91**], Choquet-Bruhat [**114**], Donoghue [**156**], Duistermaat, Kolk [**162**], El Kinani, Oudadess [**179**], Friedlander [**198**], Friedman [**199**], Garsoux [**205**], Gel'fand, Shilov [**210**], Gel'fand, Vilenkin [**211**], Grubb [**242**], Hervé [**268**], Hoskins, Sousa Pinto [**283**], Jantscher [**289**], Jones [**294**], Kanwal [**315**], Mikusinski, Sikorski [**426**], Mitrea [**430**], Mitrović, Žubrinić [**431**], Ortner, Wagner [**465**], Schwartz [**540**], [**541**], Shilov [**548**], Strichartz [**577**], Vladimirov [**619**], [**620**], Walter [**624**], and Zemanian [**654**], [**655**].

Sobolev spaces: Adams [**5**], Besov, Il'in, Nikolskiĭ [**63**], Brezis [**95**], Evans, Gariepy [**184**], Gol'dshteĭn, Reshetnyak [**230**], Haroske, Triebel [**256**], Leoni [**384**], Maz'ja [**415**], Maz'ya, Shaposhnikova [**416**], Mikhlin [**423**], Mitrović, Žubrinić [**431**], Nikol'skii [**457**], Sobolev [**565**], [**566**], Stein [**570**], Stein, Weiss [**574**], Triebel [**607**], [**608**], and Ziemer [**657**].

Some material on distributions and Sobolev spaces is also included in many books on partial differential equations and pseudodifferential operators, see, e.g., Hörmander [**281**], Mizohata [**432**], Petersen [**477**], Shubin [**551**], Taylor [**590**], and Trèves [**603**], [**605**].

Ordered vector spaces, vector lattices, positive linear and nonlinear operators: Akilov, Kutateladze [**8**], Aliprantis, Burkinshaw [**13**], Bukhvalov et al. [**101**], Kantorovič, Vulih, Pinsker [**313**], Kusraev [**365**], Luxemburg, Zaanen [**402**], Meyer-Nieberg [**419**], Schaefer [**533**], [**534**], and Vulikh [**622**].

Nonlinear analysis, convex analysis and functional aspects of the theory of extremal problems: Akhmerov et al. [7], Alekseev, Tikhomirov, Fomin [10], Aliprantis, Border [12], Ambrosetti, Arcoya [21], Aubin, Ekeland [38], Aubin, Frankowska [39], Averbuch, Smolyanov [669], [670], Barbu, Precupanu [47], Berger [60], Bogachev [74], Borwein, Zhu [83], Botelho [84], Brezis [93], Cartan [106], Cioranescu [119], Clarke [120], Deimling [141], Denkowski, Migórski, Papageorgiou [143], Deville, Godefroy, Zizler [144], Ekeland, Temam [178], Goebel [219], Goebel, Kirk [220], Granas, Dugundji [237], Holmes [280], Jeribi, Krichen [292], Joshi, Bose [298], Kantorovich, Akilov [312], Kesavan [322], Khatskevich, Shoiykhet [324] Kirk, Sims [327], Krasnosel'skiĭ [343], [344], Krasnosel'skiĭ et al. [347], Krasnosel'skij, Lifshits, Sobolev [345], Krasnosel'skiĭ, Zabreiko [348], Kusraev, Kutateladze [366], Kutateladze, Rubinov [368], Levin [387], Misjurkeev [429], Mordukhovich [437], Nirenberg [458], Papageorgiou, Winkert [470], Penot [474], Phelps [480], Polovinkin, Balashov [492], Rockafellar [513], Schirotzek [536], Schwartz [539], Singh, Watson, Srivastava [560], Smolyanov [564], Sveshnikov, Al'shin, Korpusov [583], Takahashi [587], Vainberg [611], [612], Zălinescu [644], Zeidler [646]–[650]. There are many texts on the classical calculus of variations, see, e.g., Gelfand, Fomin [208] and Giaquinta, Hildebrandt [214]. Functional analysis in *approximation theory*: Achieser [4], Altomare, Campiti [18], Altomare, Cappelletti Montano, Leonessa, Raşa [19], Apostol, Fialkow, Herrero, Voiculescu [29], Korovkin [339], Anastassiou [23], [24], Anastassiou, Gal [25], Gupta, Agarwal [245], Herrero [267], Singer [558], [559], and Tikhomirov [593].

Various applications (of course, applications are also considered in many books cited above): Atkinson, Han [36], Aubin [37], Balakrishnan [45], Ciarlet [118], Collatz [124], Cryer [132], Debnath, Mikusiński [139], Hutson, Pym, Cloud [258], Kesavan [323], Lebedev, Cloud [379], Lebedev, Vorovich, Cloud [380], Lebedev, Vorovich, Gladwell [381], Lebedev [382], Mordukhovich [437], Nowinski [459], Rolewicz [514], Tikhonov, Arsenin [594], and Zeidler [646]–[652].

Ideas and methods of functional analysis have become an indispensable part of mathematical physics, see Berezin, Shubin [59], Bogolubov, Logunov, Oksak, Todorov [76], Mackey [404], Mikhlin [423], von Neumann [455], Reed, Simon [505], Richtmyer [508], Schwartz [541], and Vladimirov [619], [620].

General topology: Arkhangel'skiĭ, Ponomarev [32], Bourbaki [86], Engelking [182], Kelley [320], Kuratowski [363], Simmons [553], and Wilansky [632].

On the *history* of the development of functional analysis, see Birkhoff, Kreyszig [671], Dieudonné [151], Duda [160], Dunford, Schwartz [164], Lützen [401], Mauldin [412], Monna [434], Narici [698], and Pietsch [485]. In addition, historical, biographical and bibliographic information is presented in many books listed above, in particular, in Edwards [171], Reed, Simon [503]–[505], Saxe [532], Simon [555], and Werner [628].

Note that this book does not contain references to original papers, the results of which have already been presented in books; we only mention some selected papers used in supplementary sections or exercises.

References

[1] Abramovich, Y. A., Aliprantis, C. D. An invitation to operator theory. Amer. Math. Soc., Providence, Rhode Island, 2002; xiv+530 pp.

[2] Abramovich, Y. A., Aliprantis, C. D. Problems in operator theory. Amer. Math. Soc., Providence, Rhode Island, 2002; xii+386 pp.

[3] Accardi, L., Lu, Y. G., Volovich, I. V. Quantum theory and its stochastic limit. Springer, New York, 2002; 493 pp.

[4] Achieser (Akhiezer), N. I. Theory of approximation. Translated from the Russian. Frederick Ungar Publ., New York, 1956; x+307 pp. (2nd Russian ed.: Moscow, 1965).

[5] Adams, R. A. Sobolev spaces. Academic Press, New York, 1975; 268 pp. (2nd ed.: Adams, R. A., Fournier, J. J. F. Academic Press, New York, 2003; xiii+305 pp.)

[6] Akhiezer, N. I., Glazman, I. M. Theory of linear operators in Hilbert space. Translated from the Russian. Dover Publ., New York, 1993; xiv+147+iv+218 pp.

[7] Akhmerov, R. R., Kamenskii, M. I., Potapov, A. S., Rodkina, A. E., Sadovskii, B. N. Measures of noncompactness and condensing operators. Translated from the Russian. Birkhäuser, Basel, 1992; viii+249 pp.

[8] Akilov, G. P.; Kutateladze, S. S. Ordered vector spaces. Nauka, Novosibirsk, 1978; 368 pp. (in Russian).

[9] Albiac, F., Kalton, N. J. Topics in Banach space theory. Springer, New York, 2006; xii+373 pp.

[10] Alekseev, V. M., Tikhomirov, V. M., Fomin, S. V. Optimal control. Translated from the Russian. Consultants Bureau, New York, 1987; xiv+309 pp.

[11] Alfsen, E. M. Compact convex sets and boundary integrals. Springer-Verlag, Berlin – New York, 1971; 210 pp.

[12] Aliprantis, C. D., Border, K. C. Infinite-dimensional analysis. A hitchhiker's guide. 2nd ed. Springer-Verlag, Berlin, 1999; xx+672 pp.

[13] Aliprantis, C. D., Burkinshaw, O. Locally solid Riesz spaces with applications to economics. 2nd ed. Amer. Math. Soc., Providence, Rhode Island, 2003; xii+344 pp.

[14] Aliprantis, C. D., Burkinshaw, O. Positive operators. Academic Press, Orlando, Florida, 1985; xvi+367 pp.

[15] Aliprantis, C. D., Burkinshaw, O. Problems in real analysis. A workbook with solutions. 2nd ed. Academic Press, San Diego, California, 1999; viii+403 pp.

[16] Alpay, D. An advanced complex analysis problem book. Topological vector spaces, functional analysis, and Hilbert spaces of analytic functions. Birkhäuser/Springer, Cham, 2015; ix+520 pp.

[17] Alt, H. W. Linear functional analysis. An application-oriented introduction. Springer-Verlag, London, 2016; xii+435 pp. (German ed.: Berlin, 2006)

[18] Altomare, F., Campiti, M. Korovkin-type approximation theory and its applications. De Gruyter, Berlin, 1994; xii+627 pp.

[19] Altomare, F., Cappelletti Montano, M., Leonessa, V., Raşa, I. Markov operators, positive semigroups and approximation processes. De Gruyter, Berlin, 2014; xii+313 pp.

[20] Amann, H., Escher, J. Analysis III. Birkhäuser, Basel – Boston – Berlin, 2001; 480 S.

[21] Ambrosetti, A., Arcoya, D. An introduction to nonlinear functional analysis and elliptic problems. Birkhäuser, Boston, 2011; xii+199 pp.

© Springer Nature Switzerland AG 2020
V. I. Bogachev, O. G. Smolyanov, *Real and Functional Analysis*,
Moscow Lectures 4, https://doi.org/10.1007/978-3-030-38219-3

[22] Amerio, L. Analisi matematica con elementi di analisi funzionale. V. III. Methodi matematici e applicazioni. Parte I. Unione Tipografico – Editrice Torinese, Turin, 1981; viii+418 pp.

[23] Anastassiou, G. A. Moments in probability and approximation theory. Longman Scientific & Technical, Harlow; John Wiley & Sons, New York, 1993; x+411 pp.

[24] Anastassiou, G. Quantitative approximations. Chapman & Hall/CRC, Boca Raton, Florida, 2001; xiv+607 pp.

[25] Anastassiou, G. A., Gal, S. G. Approximation theory. Moduli of continuity and global smoothness preservation. Birkhäuser, Boston, 2000; xiv+525 pp.

[26] Antonevich, A. B., Knjazev, P. N., Radyno, Ya. V. Problems and exercises in functional analysis. Vysh. Shkola, Minsk, 1978; 205 pp. (in Russian).

[27] Antonevich, A. B., Radyno, Ya. V. Functional analysis and integral equations. Belorus. Gos. Univ., Minsk, 2003; 431 pp. (in Russian).

[28] Antosik, P., Mikusinski, J., Sikorski, R. Theory of distributions. The sequential approach. Elsevier, Amsterdam; Polish Sci. Publ., Warsaw, 1973; xiv+273 pp.

[29] Apostol, C., Fialkow, L. A., Herrero, D. A., Voiculescu, D. Approximation of Hilbert space operators. V. II. Pitman, Boston, 1984; xi+524 pp.

[30] Appell, J., Väth M. Elemente der Funktionalanalysis. Vektorräume, Operatoren und Fixpunktsätze. Vieweg, Wiesbaden, 2005; xvi+349 S.

[31] Arias de Reyna, J. Pointwise convergence of Fourier series. Lecture Notes in Math. V. 1785. Springer-Verlag, Berlin, 2002; xviii+175 pp.

[32] Arkhangel'skiĭ, A. V., Ponomarev, V. I. Fundamentals of general topology. Problems and exercises. Translated from the Russian. Reidel, Dordrecht, 1984; xvi+415 pp.

[33] Arsen'ev, A. A. Lectures in functional analysis for beginners in mathematical physics. 2nd ed. Regular and Chaotic Dynamics, Moscow – Izhevsk, 2011; 524 pp. (in Russian).

[34] Artémiadis, N. K. Real analysis. Southern Illinois University Press, Carbondale; Feffer & Simons, London – Amsterdam, 1976; xii+581 pp.

[35] Arveson, W. A short course on spectral theory. Springer-Verlag, New York, 2002; x+135 pp.

[36] Atkinson, K., Han, W. Theoretical numerical analysis. A functional analysis framework. 3d ed. Springer, Dordrecht, 2009; xvi+625 pp.

[37] Aubin, J.-P. Applied functional analysis. With exercises by Bernard Cornet and Jean-Michel Lasry. 2nd ed. Wiley-Interscience, New York, 2000; xviii+495 pp.

[38] Aubin, J.-P., Ekeland, I. Applied nonlinear analysis. John Wiley & Sons, New York, 1984; xi+518 pp.

[39] Aubin, J.-P., Frankowska, H. Set-valued analysis. Birkhäuser Boston, Boston, 1990; xx+461 pp.

[40] Aupetit, B. A primer on spectral theory. Springer-Verlag, New York, 1991; xii+193 pp.

[41] Avanissian, V. Initiation à l'analyse fonctionnelle. Presses Universitaires de France, Paris, 1996; xiv+546 pp.

[42] Bachman, G., Narici, L. Functional analysis. Academic Press, New York – London, 1966; xiv+530 pp.

[43] Bachman, G., Narici, L., Beckenstein, E. Fourier and wavelet analysis. Springer-Verlag, New York, 2000; x+505 pp.

[44] Baggett, L. W. Functional analysis. A primer. Marcel Dekker, New York, 1992; xii+267 pp.

[45] Balakrishnan, A. V. Applied functional analysis. 2nd ed. Springer-Verlag, New York – Berlin, 1981; xiii+373 pp.

[46] Banach, S. Theory of linear operations. North-Holland, Amsterdam, 1987; x+237 pp. (French ed.: Warszawa, 1932).

[47] Barbu, V., Precupanu, T. Convexity and optimization in Banach spaces. 4th ed. Springer, Dordrecht, 2012; xii+368 pp.

[48] Barros-Neto, J. An introduction to the theory of distributions. Marcel Dekker, New York, 1973; ix+221 pp.

[49] Bary, N. K. A treatise on trigonometric series. V. I, II. Translated from the Russian. Pergamon Press, Macmillan, New York, 1964; xxiii+553 pp., xix+508 pp.

[50] Bass, J. Cours de mathématiques. T. III: Topologie, intégration, distributions, équations intégrales, analyse harmonique. Masson et Cie, Paris, 1971; 405 pp.

[51] Beauzamy, B. Introduction to Banach spaces and their geometry. 2nd ed. North-Holland, Amsterdam, 1985; xv+338 pp.

[52] Beauzamy, B. Introduction to operator theory and invariant subspaces. North-Holland, Amsterdam, 1988; xiv+358 pp.

[53] Benedetto, J. J. Harmonic analysis and applications. CRC Press, Boca Raton, Florida, 1997; xx+336 pp.

[54] Bennett, C., Sharpley, R. Interpolation of operators. Academic Press, Boston, 1988; xiv+469 pp.

[55] Benoist, J., Salinier, A. Exercices de calcul intégral: avec rappels de cours. Masson, Paris, 1997; ix+212 pp.

[56] Benyamini, Y., Lindenstrauss, J. Geometric nonlinear functional analysis. Amer. Math. Soc., Providence, Rhode Island, 2000; 488 pp.

[57] Berberian, S. K. Lectures in functional analysis and operator theory. Springer-Verlag, New York – Heidelberg, 1974; ix+345 pp.

[58] Berezansky, Yu. M., Sheftel, Z. G., Us, G. F. Functional analysis. V. 1. Translated from the Russian. Birkhäuser, Basel – Boston – Berlin, 1996; xvi+423 pp.

[59] Berezin, F. A., Shubin, M. A. The Schrödinger equation. Translated from the Russian. Kluwer, Dordrecht, 1991; xviii+555 pp.

[60] Berger, M. S. Nonlinearity and functional analysis. Lectures on nonlinear problems in mathematical analysis. Academic Press, New York – San Francisco – London, 1977; xix+417 pp.

[61] Bergh, J., Löfstrom, J. Interpolation spaces. An introduction. Springer-Verlag, Berlin – New York, 1976; x+207 pp.

[62] Bertrandias, J.-P. Mathématiques pour l'informatique. 1: Analyse fonctionnelle. Librairie Armand Colin, Paris, 1970; 230 pp.

[63] Besov, O. V., Il'in, V. P., Nikolskiĭ, S. M. Integral representations of functions and imbedding theorems. V. I, II. Translated from the Russian. Winston & Sons, Washington; Halsted Press, New York – Toronto – London, 1978, 1979; viii+345 pp., viii+311 pp.

[64] Bessaga, C., Pełczyński A. Selected topics in infinite-dimensional topology. Polish Scientific Publ., Warszawa, 1975; 353 pp.

[65] Bhatia, R. Notes on functional analysis. Hindustan Book Agency, New Delhi, 2009; x+237 pp.

[66] Bhattacharyya, P. K. Distributions. Generalized functions with applications in Sobolev spaces. De Gruyter, Berlin, 2012; xxxviii+833 pp.

[67] Birman, M. Sh., Solomjak, M. Z. Spectral theory of selfadjoint operators in Hilbert space. Translated from the Russian. Reidel, Dordrecht, 1987; xv+301 pp.

[68] Blackadar, B. Operator algebras. Theory of C^*-algebras and von Neumann algebras. Springer-Verlag, Berlin, 2006; xx+517 pp.

[69] Blecher, D. P., Le Merdy, C. Operator algebras and their modules — an operator space approach. The Clarendon Press, Oxford University Press, Oxford, 2004; x+387 pp.

[70] Bobrowski, A. Functional analysis for probability and stochastic processes. An introduction. Cambridge University Press, Cambridge, 2005; xii+393 pp.

[71] Boccara, N. Functional analysis. An introduction for physicists. Academic Press, Boston, 1990; xiv+327 pp.

[72] Bochner, S. Lectures on Fourier integrals: with an author's supplement on monotonic functions, Stieltjes integrals, and harmonic analysis. Princeton University Press, Princeton, New Jersey, 1959; 333 pp.

[73] Bogachev, V. I. Measure theory. V. 1, 2. Springer, Berlin, 2007; xvii+500 pp., xiii+575 pp.

[74] Bogachev, V. I. Differentiable measures and the Malliavin calculus. Amer. Math. Soc., Rhode Island, Providence, 2010; xvi+488 pp.

[75] Bogachev, V. I., Smolyanov, O. G. Topological vector spaces and their applications. Springer, Cham, 2017; x+456 pp.

[76] Bogolubov, N. N., Logunov, A. A., Oksak, A. I., Todorov, I. T. General principles of quantum field theory. Translated from the Russian. Kluwer, Dordrecht, 1990; xx+694 pp.

[77] Bollobás, B. Linear analysis. An introductory course. 2nd ed. Cambridge University Press, Cambridge, 1999; xii+240 pp.

[78] Bonic, R. A. Linear functional analysis. Gordon and Breach Science Publ., New York – London – Paris, 1969; xiv+124 pp.

[79] Bonsall, F. F., Duncan, J. Numerical ranges of operators on normed spaces and of elements of normed algebras. Cambridge University Press, London – New York, 1971; iv+142 pp.

[80] Bonsall, F. F., Duncan, J. Numerical ranges. II. Cambridge University Press, New York – London, 1973; vii+179 pp.

[81] Bonsall, F. F., Duncan J. Complete normed algebras. Springer-Verlag, New York – Heidelberg, 1973; x+301 pp.

[82] Borodin, P. A., Savchuk, A. M., Sheipak, I. A. Exercises in functional analysis. Parts I, II. 3d ed. MCCME, Moscow, 2017; 337 pp. (in Russian).

[83] Borwein, J. M., Zhu, Q. J. Techniques of variational analysis. Springer-Verlag, New York, 2005; vi+362 pp.

[84] Botelho, F. Functional analysis and applied optimization in Banach spaces. Applications to non-convex variational models. Springer, Cham, 2014; xviii+560 pp.

[85] Bourbaki, N. Topological vector spaces. Chapters 1–5. Springer-Verlag, Berlin, 1987; viii+364 pp.

[86] Bourbaki, N. General topology. Chapters 1-4. Chapters 5-10. Elements of Mathematics. Springer-Verlag, Berlin, 1998; vii+437 pp., iv+363 pp.

[87] Bourgin, R. D. Geometric aspects of convex sets with the Radon-Nikodym property. Lecture Notes in Math. V. 993. Springer-Verlag, Berlin, 1983; xii+474 pp.

[88] Bouyssel, M. Intégrale de Lebesgue. Mesure et intégration. Exercices avec solutions et rappels de cours. Cépaduès–éditions, Toulouse, 1997; 383 pp.

[89] Bowers, A., Kalton, N. J. An introductory course in functional analysis. Springer, New York, 2014; xvi+232 pp.

[90] Bratteli, O., Robinson, D. W. Operator algebras and quantum statistical mechanics. V. 1. C^*- and W^*-algebras, algebras, symmetry groups, decomposition of states. V. 2. Equilibrium states. Models in quantum statistical mechanics. Springer-Verlag, New York – Berlin, 1979, 1997; xii+500 pp., xiv+519 pp.

[91] Bremermann, H. Distributions, complex variables, and Fourier transforms. Addison-Wesley, Reading, 1965; xii+186 pp.

[92] Bressan, A. Lecture notes on functional analysis. With applications to linear partial differential equations. Amer. Math. Soc., Providence, Rhode Island, 2013; xii+250 pp.

[93] Brézis, H. Opérateurs maximaux monotones et semi-groupes de contractions dans les espaces de Hilbert. North-Holland, Amsterdam-London; American Elsevier, New York, 1973; vi+183 pp.

[94] Brézis, H. Analyse fonctionnelle. Théorie et applications. Masson, Paris, 1983; xiv+234 pp.

[95] Brezis, H. Functional analysis, Sobolev spaces and partial differential equations. Springer, New York, 2011; xiii+599 pp.

[96] Bridges, D. S. Foundations of real and abstract analysis. Springer-Verlag, New York, 1998; xiv+322 pp.

[97] Brown, A. L., Page, A. Elements of functional analysis. Van Nostrand Reinhold, London – New York – Toronto, 1970; xi+394 pp.

[98] Brown, A., Pearcy, C. Introduction to operator theory. I. Elements of functional analysis. Springer-Verlag, New York – Berlin – Heidelberg, 1977; xiv+474 pp.

[99] Bruckner, A. M., Bruckner, J. B., Thomson, B. S. Real analysis. Prentice-Hall, 1997; 713 pp.

[100] Bühler, T., Salamon, D. A. Functional analysis. Amer. Math. Soc., Providence, Rhode Island, 2018; xiv+466 pp.

[101] Bukhvalov, A. V., Gutman, A. E., Korotkov, V. B., Kusraev, A. G., Kutateladze, S. S., Makarov, B. M. Vector lattices and integral operators. Translated from the Russian. Kluwer, Dordrecht, 1996; x+462 pp.

[102] Burg, K., Haf, H., Wille, F. Höhere Mathematik für Ingenieure. B. V: Funktionalanalysis und partielle Differentialgleichungen. Teubner, Stuttgart, 1991; xviii+446 S.

[103] Cannarsa, P., D'Aprile, T. Introduction to measure theory and functional analysis. Springer, Cham, 2015; xiv+314 pp.

[104] Carl, B., Stephani, I. Entropy, compactness and the approximation of operators. Cambridge University Press, Cambridge, 1990; x+277 p.

[105] Carothers, N. L. A short course on Banach space theory. Cambridge University Press, Cambridge, 2005; xii+184 p.

[106] Cartan, H. Differential calculus. Differential forms. Houghton Mifflin, Boston, Massachusetts, 1970; 167 pp., 160 pp. (French ed.: Paris, 1967)

[107] Casazza, P. G., Shura, T. J. Tsirel'son's space. Lecture Notes in Math. V. 1363. Springer-Verlag, Berlin, 1989; viii+204 pp.

[108] Caumel, Y. Cours d'analyse fonctionnelle et complexe. 2nd ed. Cépaduès É'ditions, Toulouse, 2009; 238 pp.

[109] Cerdà, J. Linear functional analysis. Amer. Math. Soc., Providence, Rhode Island; Real Soc. Matem. Espan., Madrid, 2010; xiii+330 pp.

[110] Chacón, G. R., Rafeiro, H., Vallejo J. C. Functional analysis – a terse introduction. De Gruyter, Berlin, 2017; xiii+227 pp.

[111] Cheney, W. Analysis for applied mathematics. Springer-Verlag, New York, 2001; viii+444 pp.

[112] Chernoff, P. R. Product formulas, nonlinear semigroups, and addition of unbounded operators. Mem. Amer. Math. Soc., N 140. Amer. Math. Soc., Providence, Rhode Island, 1974; v+121 pp.

[113] Choquet, G. Lectures on analysis. V. I: Integration and topological vector spaces. W.A. Benjamin, New York – Amsterdam, 1969; xx+360+xxi pp.

[114] Choquet-Bruhat, Y. Distributions. Masson et Cie, Paris, 1973; x+232 pp.

[115] Choudhary, B., Nanda, S. Functional analysis with applications. John Wiley & Sons, New York, 1989; xii+344 pp.

[116] Christensen, J. P. R. Topology and Borel structure. North-Holland, Amsterdam – London, Amer. Elsevier, New York, 1974; 133 pp.

[117] Christensen, O. Functions, spaces, and expansions. Birkhäuser, Boston, 2010; xix+266 pp.

[118] Ciarlet, P. G. Linear and nonlinear functional analysis with applications. SIAM, Philadelphia, 2013; xiv+832 pp.

[119] Cioranescu, I. Geometry of Banach spaces, duality mappings and nonlinear problems. Kluwer, Dordrecht, 1990; xiv+260 pp.

[120] Clarke, F. H. Optimization and nonsmooth analysis. 2nd ed. SIAM, Philadelphia, 1990; xii+308 pp.

[121] Clarke, F. H. Functional analysis, calculus of variations and optimal control. Springer, London, 2013; xiv+591 pp.

[122] Clarke, F.H., Ledyaev, Yu. S., Stern, R. J., Wolenski, P. R. Nonsmooth analysis and control theory. Springer-Verlag, New York, 1998; xiv+276 pp.

[123] Clément, Ph., Heijmans, H. J. A. M., Angenent, S., van Duijn, C. J., de Pagter, B. One-parameter semigroups. North-Holland, Amsterdam, 1987; x+312 pp.

[124] Collatz, L. Functional analysis and numerical mathematics. Academic Press, New York – London, 1966; xx+473 pp.

[125] Connes, A. Noncommutative geometry. Academic Press, London, 1994; xiv+661 pp.

[126] Constantin, A. Fourier analysis. Part I. Theory. Cambridge University Press, Cambridge, 2016; xiv+353 pp.

[127] Conway, J. B. A course in functional analysis. 2nd ed. Springer-Verlag, New York, 1990; xvi+399 pp.

[128] Conway, J. B. The theory of subnormal operators. Amer. Math. Soc., Providence, Rhode Island, 1991; xvi+436 pp.

[129] Conway, J. B. A course in operator theory. Amer. Math. Soc., Providence, Rhode Island, 2000; xvi+372 pp.

[130] Costara, C., Popa, D. Exercises in functional analysis. Kluwer, Dordrecht, 2003; x+451 pp.

[131] Cotlar, M., Cignoli, R. An introduction to functional analysis. North-Holland, Amsterdam – London; 1974; xiv+585 pp.

[132] Cryer, C. W. Numerical functional analysis. The Clarendon Press, Oxford University Press, New York, 1982; iv+417+151 pp.

[133] Dales, H. G. Banach algebras and automatic continuity. The Clarendon Press, Oxford University Press, New York, 2000; xviii+907 pp.

[134] Dales, H. G., Aiena, P., Eschmeier, J., Laursen, K., Willis, G. A. Introduction to Banach algebras, operators, and harmonic analysis. Cambridge University Press, Cambridge, 2003; xii+324 pp.

[135] Davies, E. B. One-parameter semigroups. Academic Press, London – New York, 1980; viii+230 pp.

[136] Davies, E. B. Linear operators and their spectra. Cambridge University Press, Cambridge, 2007; xii+451 pp.

[137] Davis, M. A first course in functional analysis. Gordon and Breach, New York – London – Paris, 1966; xii+110 pp.

[138] Day, M. M. Normed linear spaces. 3d ed. Springer-Verlag, New York – Heidelberg, 1973; viii+211 pp.

[139] Debnath, L., Mikusiński, P. Introduction to Hilbert spaces with applications. 2nd ed. Academic Press, San Diego, California, 1999; xx+551 pp.

[140] Defant, A., Floret, K. Tensor norms and operator ideals. North-Holland, Amsterdam, 1993; xii+566 pp.

[141] Deimling, K. Nonlinear functional analysis. Springer-Verlag, Berlin, 1985; xiv+450 pp.

[142] Deitmar, A., Echterhoff, S. Principles of harmonic analysis. 2nd ed. Springer, Cham, 2014; xiv+332 pp.

[143] Denkowski, Z., Migórski, S., Papageorgiou, N. S. An introduction to nonlinear analysis: theory. Kluwer, Boston, 2003; xvi+689 pp.

[144] Deville, R., Godefroy, G., Zizler, V. Smoothness and renormings in Banach spaces. Longman Scientific & Technical, Harlow; John Wiley & Sons, New York, 1993; xii+376 pp.

[145] DeVito, C. L. Functional analysis and linear operator theory. Addison-Wesley, Redwood City, California, 1990; x+358 pp.

[146] DiBenedetto, E. Real analysis. Birkhäuser, Boston, 2002; xxiv+485 pp.

[147] Diestel, J. Geometry of Banach spaces – selected topics. Lecture Notes in Math. V. 485. Springer-Verlag, Berlin – New York, 1975; xi+282 pp.

[148] Diestel, J. Sequences and series in Banach spaces. Springer, Berlin – New York, 1984; xi+261 pp.

[149] Diestel, J., Uhl, J. J. Vector measures. Amer. Math. Soc., Providence, Rhode Island, 1977; xiii+322 pp.

[150] Dieudonné, J. Foundations of modern analysis. Academic Press, New York – London, 1960; xiv+361 pp.

[151] Dieudonné, J. History of functional analysis. North-Holland, Amsterdam – New York, 1981; vi+312 pp.

[152] van Dijk, G. Introduction to harmonic analysis and generalized Gelfand pairs. De Gruyter, Berlin, 2009; x+223 pp.

[153] Dirac, P. A. M. The principles of quantum mechanics. 3d ed. Oxford, Clarendon Press, 1947; xii+311 pp.

[154] Ditkin, V. A., Prudnikov, A. P. Integral transforms and operational calculus. Pergamon Press, Oxford – Edinburgh – New York, 1965; xi+529 pp. (2nd Russian ed.: Moscow, 1975).

[155] Dixmier, J. C^*-algebras. North-Holland, Amsterdam – New York – Oxford, 1977; xiii+492 pp.

[156] Donoghue, W. F., Jr. Distributions and Fourier transforms. Academic Press, New York, 1969; viii+315 pp.

[157] Dorogovtsev, A. Ya. Mathematical analysis. Collection of exercises. Vischa Shkola, Kiev, 1987; 408 pp. (in Russian).

[158] Douglas, R. G. Banach algebra techniques in operator theory. 2nd ed. Springer-Verlag, New York, 1998; xvi+194 pp.

[159] Dowson, H. R. Spectral theory of linear operators. Academic Press, London – New York, 1978; xii+422 pp.

[160] Duda, R. Pearls from a lost city. The Lvov school of mathematics. Amer. Math. Soc., Providence, Rhode Island, 2014; xii+231 pp.

[161] Dudley, R. M. Real analysis and probability. Wadsworth & Brooks, Pacific Grove, California, 1989; xii+436 pp.

[162] Duistermaat, J. J., Kolk, J. A. C. Distributions: theory and applications. Springer, New York, 2010; xvi+445 pp.

[163] van Dulst, D. Reflexive and superreflexive Banach spaces. Mathematisch Centrum, Amsterdam, 1978; v+273 pp.

[164] Dunford, N., Schwartz, J. T. Linear operators, I. General Theory. Interscience, New York, 1958; xiv+858 pp.

[165] Dunford, N., Schwartz, J. T. Linear operators. Part II: Spectral theory. Self adjoint operators in Hilbert space. John Wiley & Sons, New York – London, 1963; ix+859 pp.

[166] Dunford, N., Schwartz, J. T. Linear operators. Part III: Spectral operators. John Wiley & Sons, New York – London – Sydney, 1971; xx+668 pp.

[167] Duoandikoetxea, J. Fourier analysis. American Math. Soc., Providence, Rhode Island, 2001; xviii+222 pp.

[168] Duren, P., Schuster, A. Bergman spaces. Amer. Math. Soc., Providence, Rhode Island, 2004; x+318 pp.

[169] Dym, H., McKean, H. P. Fourier series and integrals. Academic Press, New York – London, 1972; x+295 pp.

[170] Edmunds, D. E., Kokilashvili, V., Meskhi, A. Bounded and compact integral operators. Kluwer, Dordrecht, 2002; xvi+643 pp.

[171] Edwards, R. E. Functional analysis. Theory and applications. Cor. repr. of the 1965 original. Dover Publ., New York, 1995; xvi+783 pp.

[172] Edwards, R. E. Fourier series. A modern introduction. V. 1, 2. 2nd ed. Springer-Verlag, New York – Berlin, 1979, 1982; 224 pp., xi+369 pp.

[173] Effros, E. G., Ruan, Z.-J. Operator spaces. Clarendon Press, Oxford, 2000; xvi+363 pp.

[174] Efimov, A. V., Zolotarev, Yu. G., Terpigoreva, V. M. Mathematical analysis. Advanced topics. Part 2. Application of some methods of mathematical and functional analysis. Mir, Moscow, 1985; 371 pp.

[175] Egle, K. Elemente der Funktionalanalysis. Verlag Anton Hain, Königstein, 1980; 187 S.

[176] Eidelman, Y., Milman, V., Tsolomitis, A. Functional analysis. An introduction. Amer. Math. Soc., Providence, Rhode Island, 2004; xvi+323 pp.

[177] Einsiedler, M., Ward, T. Functional analysis, spectral theory, and applications. Springer, Cham, 2017; xiv+614 pp.

[178] Ekeland, I., Temam, R. Convex analysis and variational problems. North-Holland, Amsterdam – Oxford; Amer. Elsevier, New York, 1976; ix+402 pp.

[179] El Kinani, A., Oudadess, M. Distribution theory and applications. World Sci., Hackensack, New Jersey, 2010; xvi+202 pp.

[180] Engel, K.-J., Nagel, R. A short course on operator semigroups. Springer, New York, 2006; x+247 pp.

[181] Engel, K.-J., Nagel, R. One-parameter semigroups for linear evolution equations. Springer-Verlag, New York, 2000; xxii+586 pp.

[182] Engelking, P. General topology. Polish Sci. Publ., Warszawa, 1977; 626 pp.

[183] Epstein, B. Linear functional analysis. Introduction to Lebesgue integration and infinite-dimensional problems. W.B. Saunders, Philadelphia – London – Toronto, 1970; x+229 pp.

[184] Evans, C., Gariepy, R. F. Measure theory and fine properties of functions. CRC Press, Boca Raton – London, 1992; viii+268 pp.

[185] Fabian, M., Habala, P., Hájek, P., Montesinos Santalucía, V., Pelant J., Zizler V. Functional analysis and infinite-dimensional geometry. Springer-Verlag, New York, 2001; x+451 pp.

[186] Fabian, M., Habala, P., Hájek, P., Montesinos, V., Zizler, V. Banach space theory. The basis for linear and nonlinear analysis. Springer, New York, 2011; xiv+820 pp.

[187] Farenick, D. Fundamentals of functional analysis. Springer, Cham, 2016; xiv+451 pp.

[188] Fedorov, V. M. A course in functional analysis. Lan', S.-Petersburg, 2005; 352 pp. (in Russian).

[189] Fenyö, S., Stolle, H. W. Theorie und Praxis der linearen Integralgleichungen. B. 1–4. Birkhäuser, Basel, 1982–1984; 328 S., 376 S., 548 S., 708 S.

[190] Fetter, H., Gamboa de Buen, B. The James forest. With a foreword by Robert C. James and a prologue by Bernard Beauzamy. Cambridge University Press, Cambridge, 1997; xii+255 pp.

[191] Fichera, G. Lezioni sulle trasformazioni lineari. V. I. Introduzione all'analisi lineare. Istituto Matematico, Università, Trieste, 1954; xvii+502+iv pp.

[192] Fillmore, P. A. A user's guide to operator algebras. John Wiley & Sons, New York, 1996; xiv+223 pp.

[193] Fleming, R. J., Jamison, J. E. Isometries on Banach spaces: function spaces. Chapman & Hall/CRC, Boca Raton, Florida, 2003; x+197 pp.

[194] Folland, G. B. Fourier analysis and its applications. Wadsworth & Brooks/Cole Advanced Books & Software, Pacific Grove, California, 1992; x+433 pp.

[195] Folland, G. B. Real analysis: modern techniques and their applications. 2nd ed. Wiley, New York, 1999; xiv+386 p.

[196] Folland, G. B. A course in abstract harmonic analysis. 2nd ed. CRC Press, Boca Raton, Florida, 2016; xiii+305 pp.

[197] Fremlin, D. Measure theory. V. 1–5. University of Essex, Colchester, 2003.

[198] Friedlander, F. G. Introduction to the theory of distributions. Cambridge University Press, Cambridge, 1982; vii+157 pp.

[199] Friedman, A. Generalized functions and partial differential equations. Prentice-Hall, Englewood Cliffs, New Jersey, 1963; xii+340 pp.

[200] Friedrichs, K. O. Perturbation of spectra in Hilbert space. Amer. Math. Soc., Providence, Rhode Island, 1965; xiii+178 pp.

[201] Fuchssteiner, B., Laugwitz, D. Funktionalanalysis. Bibliographisches Institut, Mannheim – Vienna – Zürich, 1974; 219 S.

[202] Gamelin, T. W. Uniform algebras. Prentice-Hall, Englewood Cliffs, New Jersey, 1969; xiii+257 pp.

[203] Garnir, H. G., De Wilde, M., Schmets, J. Analyse fonctionnelle. T. I–III. Birkhäuser Verlag, Basel–Stuttgart, 1968, 1972, 1973; 562 pp., 287 pp., 375 pp.

[204] Garsia, A. M. Topics in almost everywhere convergence. Markham Publ., Chicago, 1970; x+154 pp.

[205] Garsoux, J. Espaces vectoriels topologiques et distributions. Dunod, Paris, 1963; xiii+324 pp.

[206] Gelbaum, B. Problems in real and complex analysis. Springer, New York, 1992; x+488 pp.

[207] Gelbaum, B. R., Olmsted, J. M. H. Counterexamples in analysis. Corr. repr. of 2nd ed. Dover, Mineola, New York, 2003; xxiv+195 pp.

[208] Gelfand, I. M., Fomin, S. V. Calculus of variations. Translated from the Russian. Prentice-Hall, Englewood Cliffs, New Jersey, 1963; vii+232 pp.

[209] Gelfand, I., Raikov, D., Shilov, G. Commutative normed rings. Translated from the Russian. Chelsea, New York, 1964; 306 pp.

[210] Gel'fand, I. M., Shilov, G. E. Generalized functions. V. 1. Properties and operations. V. 2. Spaces of fundamental and generalized functions. Translated from the Russian. AMS Chelsea Publ., Providence, Rhode Island, 2016; xviii+423 pp., x+261 pp.

[211] Gel'fand, I. M., Vilenkin, N. Ya. Generalized functions. V. 4. Applications of harmonic analysis. Translated from the Russian. AMS Chelsea Publ., Providence, Rhode Island, 2016; xiv+384 pp.

[212] George, C. Exercises in integration. Springer-Verlag, Berlin – New York, 1984; 550 pp.

[213] Georgiev, S., Zennir, K. Functional analysis with applications. De Gruyter, Berlin, 2019; ix+393 pp.

[214] Giaquinta, M., Hildebrandt, S. Calculus of variations. I. The Lagrangian formalism. II. The Hamiltonian formalism. Springer-Verlag, Berlin, 1996; xxx+474 pp., xxx+652 pp.

[215] Giles, J. R. Introduction to the analysis of normed linear spaces. Cambridge University Press, Cambridge, 2000; xiv+280 pp.

[216] Gill, T. L., Zachary, W. W. Functional analysis and the Feynman operator calculus. Springer, Cham, 2016; xix+354 pp.

[217] Gillman, L., Jerison, M. Rings of continuous functions. Van Nostrand, Princeton – New York, 1960; ix+300 pp.

[218] Glazman, I. M., Ljubič, Ju. I. Finite-dimensional linear analysis. A systematic presentation in problem form. Translated from the Russian. Dover, Mineola, New York, 2006; xx+520 pp.

[219] Goebel, K. Concise course on fixed point theorems. Yokohama Publ., Yokohama, 2002; iv+182 pp.

[220] Goebel, K., Kirk, W. A. Topics in metric fixed point theory. Cambridge University Press, Cambridge, 1990; viii+244 pp.

[221] Goffman, C., Pedrick, G. First course in functional analysis. Prentice-Hall, Englewood Cliffs, 1965; xi+282 pp.

[222] Gohberg, I., Goldberg, S. Basic operator theory. Birkhäuser, Boston, 1981; xiii+285 pp.

[223] Gohberg, I., Goldberg, S., Kaashoek, M. A. Basic classes of linear operators. Birkhäuser, Basel, 2003; xviii+423 pp.

[224] Gohberg, I., Goldberg, S., Kaashoek, M. A. Classes of linear operators. V. I, II. Birkhäuser, Basel, 1990, 1993; xiv+468 pp., x+552 pp.

[225] Gohberg, I., Goldberg, S., Krupnik, N. Traces and determinants of linear operators. Birkhäuser, Basel, 2000; x+258 pp.

[226] Gohberg, I. C.; Krein, M. G. Introduction to the theory of linear nonselfadjoint operators. Translated from the Russian. Amer. Math. Soc., Providence, Rhode Island, 1969; xv+378 pp.

[227] Gohberg, I. C.; Krein, M. G. Theory and applications of Volterra operators in Hilbert space. Translated from the Russian. Amer. Math. Soc., Providence, Rhode Island, 1970; x+430 pp.

[228] Goldberg, S. Unbounded linear operators: Theory and applications. McGraw-Hill, New York – Toronto – London, 1966; viii+199 pp.

[229] Goldstein, J. A. Semigroups of linear operators and applications. The Clarendon Press, Oxford University Press, New York, 1985; x+245 pp.

[230] Gol'dshteĭn, V. M., Reshetnyak, Yu. G. Quasiconformal mappings and Sobolev spaces. Translated from the Russian. Kluwer, Dordrecht, 1990; xx+371 pp.

[231] Gorodetskii, V. V., Nagnibida, N. I., Nastasiev, P. P. Methods for solving problems in functional analysis. Vyshcha Shkola, Kiev, 1990; 480 pp. (in Russian).

[232] Gonnord, S., Tosel, N. Topologie et analyse fonctionnelle. Thèmes d'analyse pour l'agrégation. Ellipses, Paris, 1996; 160 pp.

[233] Göpfert, A., Riedrich, Th. Funktionalanalysis. 4e Aufl. Teubner, Stuttgart, 1994; 136 S.

[234] Gostiaux, B. Cours de mathématiques spéciales. Tome 3: Analyse fonctionnelle et calcul différentiel. Presses Universitaires de France, Paris, 1993; viii+443 p.

[235] Grafakos, L. Classical and modern Fourier analysis. Pearson/Prentice Hall, 2004; 859 pp.

[236] Graham, A. Introduction to Banach spaces and algebras. Oxford University Press, Oxford, 2010; 384 pp.

[237] Granas, A., Dugundji, J. Fixed point theory. Springer, New York, 2003; xv+690 pp.

[238] Griffel, D. H. Applied functional analysis. Dover, Mineola, New York, 2002; 390 pp.

[239] Groetsch, Ch. W. Elements of applicable functional analysis. Marcel Dekker, New York, 1980; x+300 pp.

[240] Grossmann, S. Funktionalanalysis. B. I,II: Im Hinblick auf Anwendungen in der Physik. Akademische Verlagsgesellschaft, Frankfurt am Main, 1970; xv+158 S., xi+157 S.

[241] Grothendieck, A. Topological vector spaces. Gordon and Breach, New York – London – Paris, 1973; x+245 pp.

[242] Grubb, G. Distributions and operators. Springer, New York, 2009; xii+461 pp.

[243] Guerre-Delabrière, S. Classical sequences in Banach spaces. Marcel Dekker, New York, 1992; xvi+207 pp.

[244] Guirao, A. J., Montesinos, V., Zizler, V. Open problems in the geometry and analysis of Banach spaces. Springer, Cham, 2016; xii+169 pp.

[245] Gupta, V., Agarwal, R. P. Convergence estimates in approximation theory. Springer, Cham, 2014; xiv+361 pp.

[246] Ha, D. M. Functional analysis. V. 1: a gentle introduction. Matrix Editions, Ithaca, New York, 2006; xvi+640 pp.

[247] Haase, M. Functional analysis. An elementary introduction. Amer. Math. Soc., Providence, Rhode Island, 2014; xviii+372 pp.

[248] Hájek, P., Montesinos Santalucía, V., Vanderwerff, J., Zizler, V. Biorthogonal systems in Banach spaces. Springer, New York, 2008; xviii+339 pp.

[249] Halilov, Z. I. Fundamentals of functional analysis. Izd. Azerb. Gos. Univ., Baku, 1949; 169 pp. (in Russian).

[250] Halmos, P. Measure theory. Van Nostrand, New York, 1950; xi+304 pp.

[251] Halmos, P. R. A Hilbert space problem book. 2nd ed. Springer-Verlag, New York – Berlin, 1982; xvii+369 pp.

[252] Halmos, P. R., Sunder, V. S. Bounded integral operators on L^2 spaces. Springer-Verlag, Berlin – New York, 1978; xv+132 pp.

[253] Hansen, V. L. Functional analysis. Entering Hilbert space. 2nd ed. World Sci., Hackensack, New Jersey, 2006; xvi+176 pp.

[254] Hardy, G.H., Littlewood, J.E., Pólya, G. Inequalities. Repr. of the 1952 ed. Cambridge University Press, Cambridge, 1988; xii+324 pp.

[255] Hardy, G. H., Rogosinski, W. W. Fourier series. Cambridge University Press, Cambridge, 1944; 100 pp.

[256] Haroske, D., Triebel, H. Distributions, Sobolev spaces, elliptic equations. European Math. Soc., Zürich, 2008; ix+294 pp.

[257] Hausdorff, F. Grundzüge der Mengenlehre. Leipzig, 1914; 476 S. English transl. of the 3rd ed.: Set theory. Chelsey, New York, 1962; 362 pp.

[258] Hutson, V., Pym, J. S., Cloud, M. J. Applications of functional analysis and operator theory. 2nd ed. Elsevier, Amsterdam, 2005; xiv+426 pp.

[259] Hedenmalm, H., Korenblum, B., Zhu, K. Theory of Bergman spaces. Springer-Verlag, New York, 2000; x+286 pp.

[260] Heine, J. Topologie und Funktionalanalysis. Grundlagen der abstrakten Analysis mit Anwendungen. Oldenbourg, München, 2002; x+745 S.

[261] Helemskii, A. Ya. Banach and locally convex algebras. Translated from the Russian. The Clarendon Press, Oxford University Press, New York, 1993; xvi+446 pp.

[262] Helemskii, A. Ya. Lectures and exercises on functional analysis. Translated from the Russian. Amer. Math. Soc., Providence, Rhode Island, 2006; xviii+468 pp.

[263] Helemskii, A. Ya. Quantum functional analysis. Non-coordinate approach. Translated from the Russian. Amer. Math. Soc., Providence, Rhode Island, 2010; xviii+241 pp.

[264] Helmberg, G. Introduction to spectral theory in Hilbert space. North-Holland, Amsterdam – London; John Wiley & Sons, New York, 1969; xiii+346 pp.

[265] Helson, H. Harmonic analysis. 2nd ed. Hindustan Book Agency, New Delhi, 2010; viii+227 pp.

[266] Hengartner, W., Lambert, M., Reischer, C. Introduction à l'analyse fonctionnelle. Presses de l'Université du Québec, Montreal, 1981; v+538 pp.

[267] Herrero, D. A. Approximation of Hilbert space operators. V. 1. 2nd ed. Longman Scientific & Technical, Harlow; John Wiley & Sons, New York, 1989; xii+332 pp.

[268] Hervé, M. Transformation de Fourier et distributions. Presses Universitaires de France, Paris, 1986; 182 pp.

[269] Heuser, H. Funktionalanalysis. Theorie und Anwendung. 4e Aufl., Teubner, Stuttgart, 2006; 696 S.; English transl.: Functional analysis. John Wiley & Sons, Chichester, 1982; xv+408 pp.

[270] Hewitt, E., Ross, K. A. Abstract harmonic analysis. V. I. Structure of topological groups, integration theory, group representations. V. II: Structure and analysis for compact groups. Analysis on locally compact Abelian groups. Springer-Verlag, Berlin – New York, 1979, 1970; ix+519 pp., ix+771 pp.

[271] Hewitt, E., Stromberg, K. Real and abstract analysis. Englewood Springer, Berlin – New York, 1975; x+476 pp.

[272] Hille, E. Functional analysis and semi-groups. Amer. Math. Soc., New York, 1948; xii+528 pp.

[273] Hille, E. Methods in classical and functional analysis. Addison-Wesley, Reading, 1972; ix+486 pp.

[274] Hille, E., Phillips, R. S. Functional analysis and semi-groups. Rev. ed. Amer. Math. Soc., Providence, Rhode Island, 1957; xii+808 pp.

[275] Hirsch, F., Lacombe, G. Elements of functional analysis. Springer-Verlag, New York, 1999; xiv+393 pp. (French ed.: Masson, Paris, 1997).

[276] Hirzebruch, F., Scharlau, W. Einführung in die Funktionalanalysis. Bibliographisches Institut, Mannheim, 1991; 178 S.

[277] Hodel, R. Cardinal functions. In: Handbook of set-theoretic topology, ed. by K. Kunen and J. E. Vaugha, pp. 1–62. North-Holland, Amsterdam – Oxford, 1984.

[278] Hoffman, K. Banach spaces of analytic functions. Dover, New York, 1988; viii+216 pp.

[279] Holevo, A. S. Statistical structure of quantum theory. Lecture Notes in Physics. Springer-Verlag, Berlin, 2001; x+159 pp.

[280] Holmes, R. B. Geometric functional analysis and its applications. Springer-Verlag, New York – Heidelberg, 1975; x+246 pp.

[281] Hörmander, L. The analysis of linear partial differential operators. I–IV. Springer-Verlag, Berlin – New York, 1983, 1985; ix+391 pp., viii+391 pp., viii+525 pp., vii+352 pp.

[282] Horvath, J. Topological vector spaces and distributions. V. I. Addison-Wesley, Reading – London – Don Mills, 1966; xii+449 pp.

[283] Hoskins, R. F., Sousa Pinto, J. Theories of generalised functions. Distributions, ultradistributions and other generalised functions. Horwood Publ., Chichester, 2005; xii+293 pp.

[284] Howell, K. B. Principles of Fourier analysis. 2nd ed. CRC Press, Boca Raton, Florida, 2017; xvi+788 pp.

[285] Humpherys, J., Jarvis, T. J., Evans, E. J. Foundations of applied mathematics. V. 1. Mathematical analysis. SIAM, Philadelphia, 2017; xx+689 pp.

[286] Hunter, J. K., Nachtergaele, B. Applied analysis. World Sci., River Edge, New Jersey, 2001; xiv+439 pp.

[287] Istrățescu, V. I. Introduction to linear operator theory. Marcel Dekker, New York, 1981; xii+579 pp.

[288] Jain, P. K., Ahuja, O. P., Ahmad, Kh. Functional analysis. New Age International, New Delhi, 1995; x+326 pp.

[289] Jantscher, L. Distributionen. De Gruyter, Berlin – New York, 1971; 367 S.

[290] Jantscher, L. Hilberträume. Akademische Verlagsgesellschaft, Wiesbaden, 1977; 294 S.

[291] Jarchow, H. Locally convex spaces. Teubner, Stuttgart, 1981; 548 pp.

[292] Jeribi, A., Krichen, B. Nonlinear functional analysis in Banach spaces and Banach algebras. Fixed point theory under weak topology for nonlinear operators and block operator matrices with applications. CRC Press, Boca Raton, Florida, 2016; xvi+355 pp.

[293] Johnson, W. B., Lindenstrauss, J. (ed.) Handbook of the geometry of Banach spaces. V. I, II. North-Holland, Amsterdam, 2001, 2003; x+1005 pp., xii+1007–1866 pp.

[294] Jones, D. S. The theory of generalised functions. 2nd ed. Cambridge University Press, Cambridge – New York, 1982; xiii+539 pp.

[295] Jørboe, O. G., Mejlbro, L. The Carleson–Hunt theorem on Fourier series. Lecture Notes in Math. V. 911. Springer, Berlin – New York, 1982; 123 pp.

[296] Jörgens, K. Linear integral operators. Pitman, Boston – London, 1982; x+379 pp.

[297] Jorgensen, P., Tian, F. Non-commutative analysis. World Sci., Hackensack, New Jersey, 2017; xxviii+533 pp.

[298] Joshi, M. C., Bose, R. K. Some topics in nonlinear functional analysis. Wiley Eastern, New Delhi, 1985; 311 pp.

[299] Jost, J. Postmodern analysis. Springer-Verlag, Berlin, 1998; xviii+353 pp.

[300] Kaballo, W. Grundkurs Funktionalanalysis. Spektrum Akademischer Verlag, Heidelberg, 2011; xii+345 S.

[301] Kaballo, W. Aufbaukurs Funktionalanalysis und Operatortheorie. Distributionen - lokalkonvexe Methoden - Spektraltheorie. Springer Spektrum, Berlin, 2014; xi+493 S.

[302] Kaczor, W. J., Nowak, M. T. Problems in mathematical analysis III: integration. Amer. Math. Soc., Providence, Rhode Island, 2003; 356 pp.

[303] Kadets, M. I., Kadets, V. M. Series in Banach spaces. Conditional and unconditional convergence. Translated from the Russian. Birkhäuser, Basel, 1997; viii+156 pp.

[304] Kadets, V. M. A course in functional analysis and measure theory. Translated from the Russian. Springer, Cham, 2018; xxii+539 pp.

[305] Kadison, R. V., Ringrose, J. R. Fundamentals of the theory of operator algebras. V. I. Elementary theory. V. II. Advanced theory. Amer. Math. Soc., Providence, Rhode Island, 1997; xvi+398 pp., i-xxii+676 pp.

[306] Kadison, R. V., Ringrose, J. R. Fundamentals of the theory of operator algebras. V. III. Special topics. Elementary theory – an exercise approach. V. IV. Special topics. Advanced theory – an exercise approach. Birkhäuser, Boston, 1991, 1992; xiv+273 pp., i-xv+586 pp.

[307] Kalton, N. J., Peck, N. T., Roberts, J. W. An F-space sampler. Cambridge University Press, Cambridge, 1984; xii+240 pp.

[308] Kałuża, R. The life of Stefan Banach. Through a reporter's eyes. Birkhäuser, Boston, 1996; x+137 pp.

[309] Kammler, D. W. A first course in Fourier analysis. 2nd ed. Cambridge University Press, Cambridge, 2007. xviii+843 pp.

[310] Kaniuth, E. A course in commutative Banach algebras. Springer, New York, 2009; xii+353 pp.

[311] Kantorovich, L. V., Akilov, G. P. Functional analysis in normed spaces. Translated from the Russian. Macmillan, New York, 1964; xiii+771 pp. (Rusian ed.: Moscow, 1959).

[312] Kantorovich, L. V., Akilov, G. P. Functional analysis. Translated from the Russian. 2nd edition. Pergamon Press, Oxford– Elmsford, New York, 1982; xiv+589 pp.

[313] Kantorovič, L. V., Vulih, B. Z., Pinsker, A. G. Functional analysis in partially ordered spaces. GITTL, Moscow, 1950; 548 pp. (in Russian).

[314] Kantorovitz, S. Introduction to modern analysis. Oxford University Press, New York, 2003; xii+434 pp.

[315] Kanwal, R. P. Generalized functions. Theory and applications. 3d ed. Birkhäuser, Boston, 2004; xviii+476 pp.

[316] Kashin, B. S., Saakyan, A. A. Orthogonal series. Translated from the Russian. Amer. Math. Soc., Providence, Rhode Island, 1989; xii+451 pp.

[317] Kato, T. Perturbation theory for linear operators. 2nd ed. Springer-Verlag, Berlin – New York, 1976; xxi+619 pp.

[318] Katznelson, Y. An introduction to harmonic analysis. 2nd ed. Dover Publ., New York, 1976; xiv+264 pp.

[319] Katzourakis, N., Vărvărucă, E. An illustrative introduction to modern analysis. CRC Press, Boca Raton, Florida, 2018; xv+541 pp.

[320] Kelley, J. L. General topology. Van Nostrand, Toronto –New York – London, 1955; xiv+298 pp.

[321] Kelley, J. L., Namioka, I. Linear topological spaces. 2nd ed. Springer-Verlag, New York – Heidelberg, 1976; xv+256 pp.

[322] Kesavan, S. Nonlinear functional analysis. A first course. Hindustan Book Agency, New Delhi, 2004; x+176 pp.

[323] Kesavan, S. Functional analysis. 2nd corr. repr. Hindustan Book Agency, New Delhi, 2017; xii+269 pp.

[324] Khatskevich, V., Shoiykhet, D. Differentiable operators and nonlinear equations. Birkhäuser, Basel, 1994; x+280 pp.

[325] Kirillov, A. A. Elements of the theory of representations. Translated from the Russian. Springer-Verlag, Berlin – New York, 1976; xi+315 pp.

[326] Kirillov, A. A., Gvishiani, A. D. Theorems and problems in functional analysis. Translated from the Russian. Springer-Verlag, New York – Berlin, 1982; ix+347 pp.

[327] Kirk, W. A., Sims, B. (eds.) Handbook of metric fixed point theory. Kluwer, Dordrecht, 2001; xiv+703 pp.

[328] Knyazev, P. N. Functional analysis. 2nd ed. Editorial URSS, Moscow, 2003; 208 pp. (in Russian).

[329] Koliha, J. J. Metrics, norms and integrals. An introduction to contemporary analysis. World Sci., Hackensack, New Jersey, 2008; xviii+408 pp.

[330] Kolmogorov, A. N. Lectures on a course of Analysis–III. Moscow Univ., Moscow, 1946–1947 (in Russian).

[331] Kolmogorov, A. N., Fomin, S. V. Introductory real analysis. V. 1. Metric and normed spaces. V. 2. Measure. The Lebesgue integral. Hilbert space. Transl. from the 2nd Russian ed. Corr. repr. Dover, New York, 1975; xii+403 pp. (Russian ed.: Elements of the theory of functions and functional analysis, Moscow, 1954, 1960).

[332] Komornik, V. Lectures on functional analysis and the Lebesgue integral. Springer-Verlag, London, 2016; xx+403 pp.

[333] König, H. Eigenvalue distribution of compact operators. Birkhäuser, Basel, 1986; 262 pp.

[334] Koosis, P. Introduction to H_p spaces. 2nd ed. Cambridge University Press, Cambridge, 1998; xiv+289 pp.

[335] Korevaar, J. Mathematical methods. V. I: linear algebra, normed spaces, distributions, integration. Academic Press, London, 1968; x+505 pp.

[336] Körner, T. W. Fourier analysis. Second edition. Cambridge University Press, Cambridge, 1989; xii+591 pp.

[337] Körner, T. W. Exercises for Fourier analysis. Cambridge University Press, Cambridge, 1993; x+385 pp.

[338] Korotkov, V. B. Integral operators. Nauka, Novosibirsk, 1983; 224 pp. (in Russian).

[339] Korovkin, P. P. Linear operators and approximation theory. Translated from the Russian. Gordon and Breach, New York; Hindustan Publ., Delhi, 1960; vii+222 pp.

[340] Köthe, G., Topological vector spaces. V. I, II. Springer-Verlag, New York, 1969, 1979; xv+456 pp., xii+331 pp.

[341] Krantz, S. G. Explorations in harmonic analysis. With applications to complex function theory and the Heisenberg group. Birkhäuser, Boston, 2009; xiv+360 pp.

[342] Krantz, S. G. A guide to functional analysis. Mathematical Association of America, Washington, 2013; xii+137 pp.

[343] Krasnosel'skiĭ, M. A. Topological methods in the theory of nonlinear integral equations. Translated from the Russian. Macmillan, New York, 1964; xi+395 pp.

[344] Krasnosel'skiĭ, M. A. Positive solutions of operator equations. Translated from the Russian. Noordhoff, Groningen, 1964; 381 pp.

[345] Krasnosel'skij, M. A., Lifshits, Je. A., Sobolev, A. V. Positive linear systems. The method of positive operators. Translated from the Russian. Heldermann, Berlin, 1989; viii+354 pp.

[346] Krasnosel'skiĭ, M. A., Rutickiĭ, Ja. B. Convex functions and Orlicz spaces. Translated from the Russian. Noordhoff, Groningen, 1961; xi+249 pp.

[347] Krasnosel'skiĭ, M. A., Vainikko, G. M., Zabreiko, P. P., Rutitskii, Ya. B., Stetsenko, V. Ya. Approximate solution of operator equations. Translated from the Russian. Wolters-Noordhoff, Groningen, 1972; xii+484 pp.

[348] Krasnosel'skiĭ, M. A., Zabreiko, P. P. Geometrical methods of nonlinear analysis. Translated from the Russian. Springer-Verlag, Berlin, 1984; xix+409 pp.

[349] Krasnosel'skiĭ, M. A., Zabreiko, P. P., Pustyl'nik, E. I., Sobolevskiĭ, P. E. Integral operators in spaces of summable functions. Translated from the Russian. Noordhoff, Leiden, 1976; xv+520 pp.

[350] Krasnov, M. L. Integral equations. Introduction to the theory. 2nd ed. Komkniga, Moscow, 2006; 304 pp. (in Russian).

[351] Krasnov, M. L., Makarenko, G. I., Kiselev, A. I. Problems and exercises in the calculus of variations. Mir Publ., Moscow, 1975; 222 pp.

[352] Kreĭn, S. G. Linear equations in Banach spaces. Translated from the Russian. Birkhäuser, Boston, 1982; xii+102 pp.

[353] Kreĭn, S. G., Petunin, Yu. I., Semenov, E. M. Interpolation of linear operators. Translated from the Russian. Amer. Math. Soc., Providence, Rhode Island, 1982; xii+375 pp.

[354] Krengel, U. Ergodic theorems. De Gruyter, Berlin – New York, 1985; 357 pp.

[355] Kress, R. Linear integral equations. 2nd ed. Springer, New York, 1999; xiv+365 pp.

[356] Kreyszig, E. Introductory functional analysis with applications. John Wiley & Sons, New York, 1989; xvi+688 pp.

[357] Krishnan, V. K. Textbook of functional analysis. A problem-oriented approach. Prentice-Hall of India, New Delhi, 2001; x+394 pp.

[358] Kubrusly, C. S. Elements of operator theory. Birkhäuser, Boston, 2001; xiv+527 pp.

[359] Kubrusly, C. S. Hilbert space operators. A problem solving approach. Birkhäuser, Boston, 2003; xvi+149 pp.

[360] Kudryavtsev, L. D., Kutasov, A. D., Chehlov, V. I., Shabunin, M. I. A collection of problems in mathematical analysis. Functions of several variables. Nauka, Fizmatlit, Moscow, 1995; 496 pp. (in Russian).

[361] Kufner, A., Kadlec, J. Fourier series. Iliffe Books, London, 1971; 13+358 pp.

[362] Kühner, E., Lesky, P. Grundlagen der Funktionalanalysis und Approximationstheorie. Vandenhoeck & Ruprecht, Göttingen, 1977; 216 S.

[363] Kuratowski, K. Topology. V. 1. Academic Press, New York – London, Polish Sci. Publ., Warsaw, 1966; xx+560 pp.

[364] Kurosh, A. G. Lectures in general algebra. Translated from the Russian. Pergamon Press, Oxford– Edinburgh – New York, 1965; x+364 pp.

[365] Kusraev, A. G. Dominated operators. Kluwer, Dordrecht, 2000; xiv+446 pp.

[366] Kusraev, A. G., Kutateladze, S. S. Subdifferentials: theory and applications. Kluwer, Dordrecht, 1995; x+398 pp.

[367] Kutateladze, S. S. Fundamentals of functional analysis. Kluwer, Dordrecht, 1996; xiv+276 pp. (5th Russian ed.: Novosibirsk, 2006).

[368] Kutateladze, S. S., Rubinov, A. M. Minkowski duality and its applications. Nauka, Novosibirsk, 1976; 254 pp. (in Russian).

[369] Lacey, H. E. The isometric theory of classical Banach spaces. Springer-Verlag, Berlin – New York, 1974; x+270 pp.

[370] Lacombe, G., Massat, P. Analyse fonctionnelle. Exercices corrigés. Dunod, Paris, 1999; 356 p.

[371] Lahiri, B. K. Elements of functional analysis. 6th ed. World Press, Calcutta, 2005; xvi+559 pp.

[372] Landsman, K. Foundations of quantum theory. From classical concepts to operator algebras. Springer, Cham, 2017; xv+881 pp.

[373] Lang, S. Real and functional analysis. 3d ed. Springer, New York, 1993; xiv+580 pp.

[374] Larsen, R. Functional analysis: an introduction. Marcel Dekker, New York, 1973; xii+497 pp.

[375] Larsen, R. Banach algebras. An introduction. Marcel Dekker, New York, 1973; xi+345 pp.

[376] Lavrent'ev, M. M., Savel'ev, L. Ya. Operator theory and ill-posed problems. Translated from the Russian. VSP, Leiden, 2006; xvi+680 pp.

[377] Lawrentjew, M. A., Schabat, B. W. Methoden der komplexen Funktionentheorie. VEB Deutscher Verlag der Wissenschaften, Berlin, 1967; x+846 S. (5th Russian ed.: Moscow, 1987).

[378] Lax, P. D. Functional analysis. John Wiley & Sons, New York, 2002; xx+580 pp.

[379] Lebedev, L. P., Cloud, M. J. The calculus of variations and functional analysis. With optimal control and applications in mechanics. World Sci., River Edge, New Jersey, 2003; xiv+420 pp.

[380] Lebedev, L. P., Vorovich, I. I., Cloud, M. J. Functional analysis in mechanics. 2nd ed. Springer, New York, 2013; x+308 pp.

[381] Lebedev, L. P., Vorovich, I. I., Gladwell, G. M. L. Functional analysis. Applications in mechanics and inverse problems. Kluwer, Dordrecht, 1996; viii+239 pp.

[382] Lebedev, V. I. An introduction to functional analysis in computational mathematics. Birkhäuser, Boston, 1997; x+255 pp.

[383] Lelong, P. Introduction à l'analyse fonctionnelle. I: Espaces vectoriels topologiques. Les cours de Sorbonne. Centre de Documentation Universitaire, Paris, 1971; ii+230 pp.

[384] Leoni, G. A first course in Sobolev spaces. Amer. Math. Soc., Providence, Rhode Island, 2009; xvi+607 pp.

[385] Leont'eva, T. A., Panferov, V. S., Serov, V. S. Exercises in the theory of functions of a real variable. Moscow State Univ., Moscow, 1997; 208 pp. (in Russian).

[386] Letac, G. Exercises and solutions manual for "Integration and probability" by Paul Malliavin. Springer-Verlag, New York, 1995; viii+142 pp.

[387] Levin, V. L. Convex analysis in spaces of measurable functions and its applications in mathematics and economics. Nauka, Moscow, 1985; 352 pp. (in Russian).

[388] Levitan, B. M., Sargsjan, I. S. Introduction to spectral theory: selfadjoint ordinary differential operators. Translated from the Russian. Amer. Math. Soc., Providence, Rhode Island, 1975; xi+525 pp.

[389] Li, D., Queffélec, H. Introduction à l'étude des espaces de Banach. Soc. Math. de France, Paris, 2004; xxiv+627 pp.

[390] Limaye, B. V. Functional analysis. 2nd ed. New Age International Publ., New Delhi, 1996; x+612 pp.

[391] Limaye, B. V. Linear functional analysis for scientists and engineers. Springer, Singapore, 2016; xiv+254 pp.

[392] Lindstrøm, T. L. Spaces: an introduction to real analysis. Amer. Math. Soc., Rhode Island, Providence; 369 pp.

[393] Lindenstrauss, J., Tzafriri, L. Classical Banach spaces. V. I,II. Springer, Berlin – New York, 1977, 1979; xiii+190 pp., x+243 pp.

[394] Loomis, L. H. An introduction to abstract harmonic analysis. Van Nostrand, Toronto – New York – London, 1953; x+190 pp.

[395] Lord, S., Sukochev, F., Zanin, D. Singular traces. Theory and applications. De Gruyter, Berlin, 2013; xvi+452 pp.

[396] Lösch, F. Höhere Mathematik: eine Einführung für Studierende und zum Selbststudium (von H. Mangoldt, K. Knopp). B. 4: Mengenlehre, Lebesguesches Maß und Integral, Topologische Räume, Vektorräume, Funktionalanalysis, Integralgleichungen. 4e Aufl. S. Hirzel Verlag, Stuttgart, 1990; xv+612 S.

[397] Lukashenko, T. P., Skvortsov, V. A., Solodov, A. P. Generalized integrals. Librokom, Moscow, 2010; 280 pp. (in Russian).

[398] Lunardi, A. Interpolation theory. 3d ed. Edizioni della Normale, Pisa, 2018; xiv+199 pp.

[399] Lusternik, L. A., Sobolev, V. J. Elements of functional analysis. Hindustan Publishing Corp., Delhi; Halsted Press, John Wiley& Sons, New York, 1974; 360 pp. (1st Russian ed.: Moscow, 1951).

[400] Lusternik, L. A., Sobolev, V. I. A brief course of functional analysis. Vyssh. Shkola, Moscow, 1982; 272 pp. (in Russian).

[401] Lützen, J. The prehistory of the theory of distributions. Springer-Verlag, New York – Berlin, 1982; viii+232 p.

[402] Luxemburg, W. A. J., Zaanen, A. C. Riesz spaces. V. I, II. North-Holland, Amsterdam – London; American Elsevier, New York, 1971, 1983; xi+514 pp., xi+720 pp.

[403] MacCluer, B. D. Elementary functional analysis. Springer, New York, 2009; x+207 pp.

[404] Mackey, G. W. Mathematical foundations of quantum mechanics. Reprint of the 1963 original. Dover Publ., Mineola, New York, 2004; xii+137 pp.

[405] Maddox, I. J. Elements of functional analysis. 2nd ed. Cambridge University Press, Cambridge, 1988; xii+242 pp.

[406] Makarov, B. M., Goluzina, M. G., Lodkin, A. A., Podkorytov, A. N. Selected problems in real analysis. Translated from the Russian. Amer. Math. Soc., Rhode Island, Providence, 1992; 370 pp.

[407] Malkowsky, E., Rakočević, V. Advanced functional analysis. CRC Press, Boca Raton, Florida, 2019; xix+446 pp.

[408] Marle, Ch.-M., Pilibossian, Ph. Analyse fonctionnelle. Ellipses, Paris, 2004; 134 pp.

[409] Marti, J. T. Introduction to the theory of bases. Springer-Verlag, New York, 1969; xii+149 pp.

[410] Maslov, V. P. Operational methods. Translated from the Russian. Mir, Moscow, 1976; 559 pp.

[411] Mathieu, M. Funktionalanalysis: ein Arbeitsbuch. Spektrum, Akad. Verl., Heidelberg, 1998; vii+393 S.

[412] Mauldin, R. D. (ed.) The Scottish Book. Mathematics from the Scottish Cafe. Birkhäuser, Boston, 1981; xiii+268 pp.

[413] Maurin, K. Methods of Hilbert spaces. 2nd ed. Polish Scientific Publ., Warsaw, 1972; 553 pp.

[414] Maurin, K. Analysis. Part II. Integration, distributions, holomorphic functions, tensor and harmonic analysis. Reidel, Dordrecht, 1980; xvii+829 pp.

[415] Maz'ja, V. G. Sobolev spaces. Springer-Verlag, Berlin, 1985; xix+486 pp.

[416] Maz'ya, V. G., Shaposhnikova, T. O. Theory of Sobolev multipliers. With applications to differential and integral operators. Springer-Verlag, Berlin, 2009; xiv+609 pp.

[417] Megginson, R. E. An introduction to Banach space theory. Springer-Verlag, New York, 1998; xx+596 pp.

[418] Meise, R., Vogt, D. Introduction to functional analysis. The Clarendon Press, Oxford University Press, New York, 1997; x+437 pp.

[419] Meyer-Nieberg, P. Banach lattices. Springer-Verlag, Berlin, 1991; xvi+395 pp.

[420] Michel, A. N., Herget, Ch. J. Applied algebra and functional analysis. Corr. repr. of the 1981 original. Dover Publ., New York, 1993; x+484 pp.

[421] Mihlin, S. G. Linear integral equations. Translated from the Russian. Gordon and Breach, New York; Hindustan Publ. Corp. (India), Delhi, 1960; vii+223 pp.

[422] Mikhlin, S. G. Multidimensional singular integrals and integral equations. Translated from the Russian. Pergamon Press, Oxford – New York – Paris, 1965; xi+255 pp.

[423] Mikhlin, S. G. Mathematical physics, an advanced course. Translated from the Russian. North-Holland, Amsterdam – London; American Elsevier, New York, 1970; xv+561 pp.

[424] Mikhlin, S. G., Morozov, N. F., Paukshto, M. V. The integral equations of the theory of elasticity. Translated from the Russian. Teubner, Stuttgart, 1995; 375 pp.

[425] Mikhlin, S. G., Prössdorf, S. Singular integral operators. Springer-Verlag, Berlin, 1986; 528 pp.

[426] Mikusinski, J.; Sikorski, R. Théorie élémentaire des distributions. Gauthier-Villars, Paris, 1964; vi+108 pp.

[427] Milman, V. D., Schechtman, G. Asymptotic theory of finite dimensional normed spaces. With an appendix by M. Gromov: Isoperimetric inequalities in Riemannian manifolds. Lecture Notes in Math. V. 1200. Springer-Verlag, Berlin – New York, 1986; viii+156 pp.

[428] Miranda, C. Istituzioni di analisi funzionale lineare. V. I,II. Pitagora Editrice, Bologna, 1978, 1979; iii+596 pp., 748 pp.

[429] Misjurkeev, I. V. Introduction to nonlinear functional analysis. Permsk. Gos. Univ., Perm, 1968; 308 pp. (in Russian).

[430] Mitrea, D. Distributions, partial differential equations, and harmonic analysis. Springer, New York, 2013; xxii+460 pp.

[431] Mitrović, D., Žubrinić, D. Fundamentals of applied functional analysis. Distributions – Sobolev spaces – nonlinear elliptic equations. Longman, Harlow, 1998; x+399 p.

[432] Mizohata, S. The theory of partial differential equations. Cambridge University Press, New York, 1973; xii+490 pp.

[433] Mlak, W. Hilbert spaces and operator theory. Kluwer, Dordrecht; Polish Sci. Publ., Warsaw, 1991; xii+290 pp.

[434] Monna, A. F. Functional analysis in historical perspective. John Wiley & Sons, New York – Toronto, 1973; viii+167 pp.

[435] Montesinos, V., Zizler, P., Zizler, V. An introduction to modern analysis. Springer, Cham, 2015; xxxii+863 pp.

[436] Moore, R. E. Computational functional analysis. Ellis Horwood, Chichester; Halsted Press [John Wiley & Sons], New York, 1985; 156 pp.

[437] Mordukhovich, B. S. Variational analysis and generalized differentiation. I. Basic theory. II. Applications. Springer-Verlag, Berlin, 2006; xxii+579 pp., xxii+610 pp.

[438] Moretti, V. Spectral theory and quantum mechanics. Mathematical foundations of quantum theories, symmetries and introduction to the algebraic formulation. 2nd ed. Springer, Cham, 2017; xxii+950 pp.

[439] Morrison, T. J. Functional analysis. An introduction to Banach space theory. John Wiley & Sons, New York, 2001; xiv+359 pp.

[440] Mozzochi, C. J. On the pointwise convergence of Fourier series. Lecture Notes in Math. V. 199. Springer-Verlag, Berlin, 1971; vii+87 pp.

[441] Mukherjea, A., Pothoven, K. Real and functional analysis. Plenum Press, New York, 1978; x+529 pp.

[442] Müller, V. Spectral theory of linear operators and spectral systems in Banach algebras. Birkhäuser, Basel, 2003; x+381 pp.

[443] Murphy, G. J. C^*-algebras and operator theory. Academic Press, Boston, 1990; x+286 pp.

[444] Muscalu, C., Schlag, W. Classical and multilinear harmonic analysis. V. I, II. Cambridge University Press, Cambridge, 2013; xviii+370 pp., xvi+324 pp.

[445] Muscat, J. Functional analysis. An introduction to metric spaces, Hilbert spaces, and Banach algebras. Springer, Cham, 2014; xii+420 pp.

[446] Mushtari, D. Kh. Probabilities and topologies on linear spaces. Kazan Mathematics Foundation, Kazan', 1996; xiv+233 pp.

[447] Nachbin, L. Introduction to functional analysis: Banach spaces and differential calculus. Marcel Dekker, New York, 1981; ix+166 pp.

[448] Naĭmark, M. A. Normed algebras. Translated from the Russian. Wolters-Noordhoff Publ., Groningen, 1972; xvi+598 pp.

[449] Naĭmark, M. A. Linear differential operators. Part I: Elementary theory of linear differential operators. Part II: Linear differential operators in Hilbert space. Frederick Ungar Publ., New York, 1967, 1968; xiii+144 pp., xv+352 pp.

[450] Nair, M. T. Functional analysis. A first course. Prentice-Hall of India, New Delhi, 2005; 448 pp.

[451] Narici, L., Beckenstein, E. Topological vector spaces. Marcel Dekker, New York, 1985; xii+408 pp.

[452] Narasimhan, R. Analysis on real and complex manifolds. North-Holland, Amsterdam, 1985; xiv+246 pp.

[453] Natanson, I. P. Theory of functions of a real variable. Translated from the Russian. V. 1, 2. Frederick Ungar Publ., New York, 1955, 1961; 277 pp., 265 pp.

[454] Naylor, A., Sell, G. Linear operator theory in engineering and science. 2nd ed. Springer-Verlag, New York – Berlin, 1982; xv+624 pp.

[455] von Neumann, J. Mathematical foundations of quantum mechanics. Princeton University Press, Princeton, New Jersey, 1996; xii+445 pp. (German ed.: Berlin, 1932).

[456] Nikolski, N. K. Operators, functions, and systems: an easy reading. V. 1, 2. Amer. Math. Soc., Providence, Rhode Island, 2002; xiv+461 pp., xiv+439 pp.

[457] Nikol'skii, S. M. Approximation of functions of several variables and imbedding theorems. Translated from the Russian. Springer-Verlag, New York – Heidelberg, 1975; viii+418 pp.

[458] Nirenberg, L. Topics in nonlinear functional analysis. New York University, Courant Inst. Math. Sci., New York; Amer. Math. Soc., Providence, Rhode Island, 2001; xii+145 pp.

[459] Nowinski, J. L. Applications of functional analysis in engineering. Plenum Press, New York, 1981; xv+304 pp.

[460] Oden, J. T., Demkowicz, L. F. Applied functional analyis. 3d ed. CRC Press, Boca Raton, Florida, 2018; xxi+609 pp.

[461] Odinets, V. P., Yakubson, M. Ya. Projections and bases in normed spaces. 2nd ed., URSS, Moscow, 2004; 152 pp. (in Russian).

[462] Okikiolu, G. O. Aspects of the theory of bounded integral operators in L^p-spaces. Academic Press, London – New York, 1971; ix+522 pp.

[463] Olevskiĭ, A. M. Fourier series with respect to general orthogonal systems. Springer-Verlag, New York – Heidelberg, 1975; viii+136 pp.

[464] Orlicz, W. Linear functional analysis. World Sci., River Edge, New Jersey, 1992; xvi+246 pp.

[465] Ortner, N., Wagner, P. Fundamental solutions of linear partial differential operators. Theory and practice. Springer, Cham, 2015; xii+398 pp.

[466] Osilenker, B. Fourier series in orthogonal polynomials. World Sci., River Edge, New Jersey, 1999; vi+287 pp.

[467] Ovchinnikov, S. Functional analysis. An introductory course. Springer, Cham, 2018; xii+205 pp.

[468] Packel, E. W. Functional analysis. A short course. Corrected reprint of the 1974 original. Robert E. Krieger Publ., Huntington, New York, 1980; xvii+172 pp.

[469] Panzone, R. Lecciones preliminares de análisis funcional. Universidad Nacional del Sur, Instituto de Matemática, Bahía Blanca, 1983; v+196 pp.

[470] Papageorgiou, N. S., Winkert, P. Applied nonlinear functional analysis. An introduction. De Gruyter, Berlin, 2018; x+612 pp.

[471] Pazy, A. Semigroups of linear operators and applications to partial differential equations. Springer-Verlag, New York, 1983; viii+279 pp.

[472] Pedersen, M. Functional analysis in applied mathematics and engineering. Chapman & Hall/CRC, Boca Raton, Florida, 2000; x+298 pp.

[473] Pełczyński, A. Linear extensions, linear averagings, and their applications to linear topological classification of spaces of continuous functions. Dissert. Math. (Rozprawy Mat.) 1968. V. 58. 92 pp.

[474] Penot, J.-P. Analysis – from concepts to applications. Springer, Cham, 2016; xxiii+669 pp.

[475] Pereyra, M. C., Ward, L. A. Harmonic analysis. From Fourier to wavelets. Amer. Math. Soc., Providence, Rhode Island; Institute for Advanced Study, Princeton, New Jersey, 2012; xxiv+410 pp.

[476] Pérez Carreras, P., Bonet, J. Barrelled locally convex spaces. North-Holland, Amsterdam, 1987; xvi+512 pp.

[477] Petersen, B. E. Introduction to the Fourier transform & pseudodifferential operators. Pitman, Boston, 1983; xi+356 pp.

[478] Pflaumann E., Unger H. Funktionalanalysis. I, II. Bibliographisches Institut, Mannheim – Vienna – Zürich, 1968, 1974; 240 S., 338 S.

[479] Phelps, R. R. Lectures on Choquet's theorem. 2nd ed. Lecture Notes in Math., V. 1757. Springer-Verlag, Berlin, 2001; viii+124 pp.

[480] Phelps, R. R. Convex functions, monotone operators and differentiability. 2nd ed. Lecture Notes in Math. V. 1364. Springer-Verlag, Berlin, 1993; xii+117 pp.

[481] Phillips, E. R. An introduction to analysis and integration theory. Dover Publications, New York, 1984; xxviii+452 pp.

[482] Pietsch, A. Nuclear locally convex spaces. Springer-Verlag, New York – Heidelberg, 1972; ix+193 pp.

[483] Pietsch, A. Operator ideals. North-Holland, Amsterdam – New York, 1980; 451 pp.

[484] Pietsch, A. Eigenvalues and s-numbers. Cambridge University Press, Cambridge, 1987; 360 pp.

[485] Pietsch, A. History of Banach spaces and linear operators. Birkhäuser, Boston, 2007; xxiv+855 pp.

[486] Pietsch, A., Wenzel, J. Orthonormal systems and Banach space geometry. Cambridge University Press, Cambridge, 1998; x+553 pp.

[487] Pinsky, M. A. Introduction to Fourier analysis and wavelets. Amer. Math. Soc., Providence, Rhode Island, 2009; xx+376 pp.

[488] Pirkovskii, A. Yu. Spectral theory and functional calculus for linear opeartors. MCCME, Moscow, 2010; 176 pp. (in Russian).

[489] Pisier, G. The volume of convex bodies and Banach space geometry. Cambridge University Press, Cambridge, 1999; xv+250 pp.

[490] Pisier, G. Introduction to operator space theory. Cambridge University Press, Cambridge, 2003; vii+478 pp.

[491] Plesner, A. I. Spectral theory of linear operators. V. I, II. Translated from the Russian. Frederick Ungar Publ., New York, 1969; xii+235 pp., viii+271 pp.

[492] Polovinkin, E. S., Balashov, M. V. Elements of convex and strongly convex analysis. Fizmatlit, Moscow, 2004; 416 pp. (in Russian).

[493] Ponnusamy, S. Foundations of functional analysis. Alpha Science International, Pangbourne, 2002; xvi+457 pp.

[494] Poroshkin, A. G. Functional analysis. 2nd ed., Vuz. Kniga, Moscow, 2007; 432 pp. (in Russian).

[495] Promislow, S. D. A first course in functional analysis. Wiley-Interscience, Hoboken, New Jersey, 2008; xiv+307 pp.

[496] Prössdorf, S. Some classes of singular equations. North-Holland, Amsterdam – New York, 1978; xiv+417 pp.

[497] Pryce, J. D. Basic methods of linear functional analysis. Hutchinson, London, 1973; 320 pp.

[498] Pugachev, V. S., Sinitsyn, I. N. Lectures on functional analysis and applications. World Sci., River Edge, New Jersey, 1999; xx+730 pp.

[499] Rao, K.-C. Functional analysis. 2nd ed. Narosa, New Delhi, 2006; 282 pp.

[500] Ray, W. O. Real analysis. Prentice Hall, Englewood Cliffs, New Jersey, 1988; xii+307 pp.

[501] Reddy, B. D. Introductory functional analysis. With applications to boundary value problems and finite elements. Springer-Verlag, New York, 1998; xiv+471 pp.

[502] Reed, M., Simon, B. Methods of modern mathematical physics. I. Functional analysis. 2nd ed. Academic Press, New York, 1980; xv+400 pp.

[503] Reed, M., Simon, B. Methods of modern mathematical physics. II. Fourier analysis, self-adjointness. Academic Press, New York – London, 1975; xv+361 pp.

[504] Reed, M., Simon, B. Methods of modern mathematical physics. III. Scattering theory. Academic Press, New York – London, 1979; xv+463 pp.

[505] Reed, M., Simon, B. Methods of modern mathematical physics. IV. Analysis of operators. Academic Press, New York – London, 1978; xv+396 pp.

[506] Reiter, H., Stegeman, J. D. Classical harmonic analysis and locally compact groups. 2nd ed. The Clarendon Press, Oxford University Press, New York, 2000. xiv+327 pp.

[507] Retherford, J. R. Hilbert space: compact operators and the trace theorem. Cambridge University Press, Cambridge, 1993; xii+131 pp.

[508] Richtmyer, R. D. Principles of advanced mathematical physics. V. I. Springer-Verlag, New York – Heidelberg, 1978; xv+422 pp.

[509] Rickart, C. E. General theory of Banach algebras. Van Nostrand, Princeton – New York, 1960; xi+394 pp.

[510] Riesz, F., Sz.-Nagy, B. Functional analysis. Reprint of the 1955 original. Dover Publ., New York, 1990; xii+504 pp.

[511] Ringrose, J. R. Compact non-self-adjoint operators. Van Nostrand Reinhold, London, 1971; vi+238 pp.

[512] Robertson, A. P., Robertson, W. Topological vector spaces. 2nd ed. Cambridge University Press, London – New York, 1973; viii+172 pp.

[513] Rockafellar, R. T. Convex analysis. Princeton University Press, Princeton, New Jersey, 1970; xviii+451 pp.

[514] Rolewicz, S. Functional analysis and control theory. Linear systems. Reidel, Dordrecht; Polish Sci. Publ., Warsaw, 1987; xvi+524 pp.

[515] Rolewicz, S. Metric linear spaces. 2nd ed. Reidel, Dordrecht; Polish Sci. Publ., Warsaw, 1985; xii+459 pp.

[516] Roman, P. Some modern mathematics for physicists and other outsiders. An introduction to algebra, topology, and functional analysis. V. 2: Functional analysis with applications. Pergamon Press, New York, 1975; 288 pp.

[517] Royden, H. L. Real analysis. 3d ed., Prentice Hall, Englewood Cliffs, New Jersey, 1988; 444 pp. (1st ed.: Macmillan, 1963).

[518] Ruckle, W. H. Modern analysis. Measure theory and functional analysis with applications. PWS-Kent Publ., Boston, 1991; xv+265 pp.

[519] Rudin, W. Real and complex analysis. 2nd ed., McGraw-Hill, New York, 1974; xii+452 pp.

[520] Rudin, W. Principles of mathematical analysis. 3d ed. McGraw-Hill, New York – Auckland – Düsseldorf, 1976; x+342 pp.

[521] Rudin, W. Fourier analysis on groups. Reprint of 1962 original. John Wiley & Sons, New York, 1990; x+285 pp.

[522] Rudin, W. Functional analysis. 2nd ed., McGraw-Hill, New York, 1991; xviii+424 pp.

[523] Ryan, R. A. Introduction to tensor products of Banach spaces. Springer-Verlag London, London, 2002; xiv+225 pp.

[524] Rynne, B. P., Youngson M. A. Linear functional analysis. 2nd ed. Springer-Verlag London, London, 2008; x+324 pp.

[525] Sacks, P. Techniques of functional analysis for differential and integral equations. Elsevier/Academic Press, London, 2017; x+310 pp.

[526] Sadovnichiĭ, V. A. Theory of operators. Translation from the Russian. Consultants Bureau, New York, 1991; xii+396 pp.

[527] Sakai, S. C^*-algebras and W^*-algebras. Springer, Berlin, 1971; xii+253 pp.

[528] Saks, S. Theory of the integral. Warszawa, 1937; xv+343 pp.

[529] Samuélidès, M., Touzillier, L. Analyse fonctionnelle. Collection La Chevêche, Cépaduès Éditions, Toulouse, 1989; iv+289 pp.

[530] Samuélidès, M., Touzillier, L. Problèmes d'analyse fonctionnelle et d'analyse harmonique. Collection La Chevêche, Cépaduès Éditions, Toulouse, 1993; vi+392 pp.

[531] Sasane, A. A friendly approach to functional analysis. World Sci., Hackensack, New Jersey, 2017; xiv+379 pp.

[532] Saxe, K. Beginning functional analysis. Springer-Verlag, New York, 2002; xii+197 pp.

[533] Schaefer, H. H. Topological vector spaces. Springer-Verlag, Berlin – New York, 1971; xi+294 pp.

[534] Schaefer, H. H. Banach lattices and positive operators. Springer-Verlag, New York – Heidelberg, 1974; xi+376 pp.

[535] Schechter, M. Principles of functional analysis. 2nd ed. Amer. Math. Soc., Providence, Rhode Island, 2002; xxii+425 pp.

[536] Schirotzek, W. Nonsmooth analysis. Springer, Berlin, 2007; xii+373 pp.

[537] Schmüdgen, K. Unbounded self-adjoint operators on Hilbert space. Springer, Dordrecht, 2012; xx+432 pp.

[538] Schröder, H. Funktionalanalysis. 2e Aufl., Verlag Harri Deutsch, Thun, 2000; viii+384 S.

[539] Schwartz, J. T. Nonlinear functional analysis. Notes by H. Fattorini, R. Nirenberg and H. Porta. With an additional chapter by Hermann Karcher. Gordon and Breach, New York – London – Paris, 1969; 236 pp.

[540] Schwartz, L. Théorie des distributions. T. I,II. Hermann & Cie, Paris, 1950, 1951; 148 pp., 169 pp. (2e ed.: Hermann, Paris, 1966; xiii+420 pp.)

[541] Schwartz, L. Méthodes mathématiques pour les sciences physiques. Hermann, Paris, 1961; 392 pp.

[542] Segal, I., Kunze, R. A. Integrals and operators. 2nd ed. Springer, Berlin – New York, 1978; xiv+371 pp.

[543] Semadeni, Z. Banach spaces of continuous functions. V. I. Polish Sci. Publ., Warszawa, 1971; 584 pp.

[544] Sen, R. A first course in functional analysis: theory and applications. Anthem Press, London, 2013; xviii+468 pp.

[545] Serov, V. Fourier series, Fourier transform and their applications to mathematical physics. Springer, Cham, 2017; xi+534 pp.

[546] Shen, A., Vereshchagin, N. K. Basic set theory. Translated from the Russian. Amer. Math. Soc., Providence, Rhode Island, 2002; viii+116 pp.

[547] Shilov, G. Ye. Mathematical analysis. A special course. Translated from the Russian. Pergamon Press, Oxford – New York – Paris, 1965; xii+481 pp.

[548] Shilov, G. E. Generalized functions and partial differential equations. Translated from the Russian. Gordon and Breach, New York – London – Paris, 1968; xii+345 pp.

[549] Shilov, G. E. Elementary functional analysis. Translated from the Russian. The M.I.T. Press, Cambridge, 1974; vii+334 pp.

[550] Shiryaev, A. N. Problems in probability. Translated from the Russian. Springer, New York, 2012; xii+427 pp.

[551] Shubin, M. A. Pseudodifferential operators and spectral theory. Translated from the Russian. 2nd ed. Springer-Verlag, Berlin, 2001; xii+288 pp.

[552] Siddiqi, A. H. Functional analysis and applications. Springer, Singapore, 2018; xvii+562 pp.

[553] Simmons, G. F. Introduction to topology and modern analysis. McGraw-Hill, New York – London, 1963; xv+372 pp.

[554] Simon, B. Trace ideals and their applications. 2nd ed. Amer. Math. Soc., Providence, Rhode Island, 2005; viii+150 pp.

[555] Simon, B. Real analysis. A comprehensive course in analysis, Part 1. Amer. Math. Soc., Providence, Rhode Island, 2015; xx+789 pp.

[556] Simon, B. Harmonic analysis. A comprehensive course in analysis, Part 3. Amer. Math. Soc., Providence, Rhode Island, 2015; xviii+759 pp.

[557] Simon, B. Operator theory. A comprehensive course in analysis, Part 4. Amer. Math. Soc., Providence, Rhode Island, 2015; xviii+749 pp.

[558] Singer, I. Bases in Banach spaces. I, II. Springer-Verlag, New York – Berlin, 1970, 1981; viii+668 pp., viii+880 pp.

[559] Singer, I. Abstract convex analysis. John Wiley & Sons, New York, 1997; xxii+491 pp.

[560] Singh, S., Watson, B., Srivastava, P. Fixed point theory and best approximation: The KKM-map principle. Kluwer, Dordrecht, 1997; x+220 pp.

[561] Sinha, K. B., Srivastava, S. Theory of semigroups and applications. Hindustan Book Agency, New Delhi, 2017; x+167 pp.

[562] Skandalis, G. Topologie et analyse fonctionnelle: mathématiques pour la licence. Cours et exercices avec solutions. Dunod, Paris, 2000; xi+323 pp.

[563] Smirnov, V. I. A course of higher mathematics. V. 5. Translated from the Russian. Pergamon Press, Oxford – New York; Addison-Wesley, , Reading, London, 1964; xiv+635 pp. (1st Russian ed.: Moscow, 1947).

[564] Smolyanov, O. G. Analysis on topological linear spaces and its applications. Moskov. Gos. Univ., Moscow, 1979; 86 pp. (in Russian).

[565] Sobolev, S. L. Some applications of functional analysis in mathematical physics. Translated from the Russian. Amer. Math. Soc., Providence, Rhode Island, 1991; viii+286 pp. (1st Russian ed.: Novosibirsk, 1962).

[566] Sobolev, S. L. Cubature formulas and modern analysis. Translated from the Russian. An introduction. Gordon and Breach, Montreux, 1992; xvi+379 pp.

[567] Sohrab, H. H. Basic real analysis. Birkhäuser, Boston, 2003; xiv+559 pp.

[568] Somasundaram, D. First course in functional analysis. Narosa, New Delhi, 2006; ix+399 pp.

[569] Sonntag, Y. Topologie et analyse fonctionnelle. Cours de licence avec 240 exercices et 30 problèmes corrigés. Ellipses, Paris, 1998; 512 pp.

[570] Stein, E. M. Singular integrals and differentiability properties of functions. Princeton University Press, Princeton, New Jersey, 1970; xiv+290 pp.

[571] Stein, E. M. Harmonic analysis: real-variable methods, orthogonality, and oscillatory integrals. Princeton University Press, Princeton, New Jersey, 1993; xiii+695 pp.

[572] Stein, E. M., Shakarchi, R. Fourier analysis. An introduction. Princeton University Press, Princeton, New Jersey, 2003; xvi+311 pp.

[573] Stein, E. M., Shakarchi, R. Functional analysis. Introduction to further topics in analysis. Princeton University Press, Princeton, New Jersey, 2011; xviii+423 pp.

[574] Stein, E. M., Weiss, G. Introduction to Fourier analysis on Euclidean spaces. Princeton University Press, Princeton, New Jersey, 1971; x+297 pp.

[575] Stone, M. H. Linear transformations in Hilbert space and their applications to analysis. Amer. Math. Soc. Colloq. Publ., V. 15, New York, 1932; viii+622 pp.; Reprint: Amer. Math. Soc., Providence, Rhode Island, 1990.

[576] Storch, U., Wiebe, H. Lehrbuch der Mathematik. B. 4: Analysis auf Mannigfaltigkeiten, Funktionentheorie, Funktionalanalysis. Spektrum Akademischer Verlag, Heidelberg, 2001; 889 S.

[577] Strichartz, R. S. A guide to distribution theory and Fourier transforms. CRC Press, Boca Raton, Florida, 1994; x+213 pp.

[578] Stromberg, K. An introduction to classical real analysis. Wadsworth, Belmont, 1981; ix+575 pp.

[579] Suetin, P. K. Classical orthogonal polynomials. 2nd ed., Nauka, Moscow, 1979; 415 pp. (in Russian).

[580] Sugiura, M. Unitary representations and harmonic analysis. An introduction. 2nd ed. North-Holland, Amsterdam; Kodansha, Tokyo, 1990; xvi+452 pp.

[581] Şuhubi, E. S. Functional analysis. Kluwer, Dordrecht, 2003; xii+691 pp.

[582] Sunder, V. S. Functional analysis. Spectral theory. Birkhäuser, Basel, 1997; x+241 pp.

[583] Sveshnikov, A. G., Al'shin, A. B., Korpusov, M. O. Nonlinear functional analysis and its applications to partial differential equations. Nauch. Mir, Moscow, 2008; 400 pp. (in Russian).

[584] Swartz, Ch. An introduction to functional analysis. Marcel Dekker, New York, 1992; xiv+600 pp.

[585] Swartz, Ch. Elementary functional analysis. World Sci., Hackensack, New Jersey, 2009; x+181 pp.

[586] Székelyhidi, L. Harmonic and spectral analysis. World Sci., Hackensack, New Jersey, 2014; xvi+230 pp.

[587] Takahashi, W. Nonlinear functional analysis. Fixed point theory and its applications. Yokohama Publ., Yokohama, 2000; iv+276 pp.

[588] Takesaki, M. Theory of operator algebras. V. I–III. Springer-Verlag, Berlin, 1979, 2003; vii+415 pp., xxii+518 pp., xxii+548 pp.

[589] Taylor, A. E. Introduction to functional analysis. John Wiley & Sons, New York; Chapman & Hall, London, 1958; xvi+423 pp.

[590] Taylor, M. E. Pseudodifferential operators. Princeton University Press, Princeton, New Jersey, 1981; xi+452 pp.

[591] Telyakovskii, S. A. A collection of problems in the theory of functions of a real variable. Nauka, Moscow, 1980; 112 pp. (in Russian).

[592] Terras, A. Harmonic analysis on symmetric spaces – Euclidean space, the sphere, and the Poincaré upper half-plane. 2nd ed. Springer, New York, 2013; xviii+413 pp.

[593] Tikhomirov, V. M. Some questions in approximation theory. Izdat. Moskov. Univ., Moscow, 1976; 304 pp. (in Russian).

[594] Tikhonov, A. N., Arsenin, V. Y. Solutions of ill-posed problems. Translated from the Russian. Winston & Sons, Washington, John Wiley & Sons, New York – Toronto – London, 1977; xiii+258 pp.

[595] Tikhonova, A. A., Tikhonov, N. A. Andrei Nikolaevich Tikhonov. Sobranie, Moscow, 2006; 240 pp. (in Russian).

[596] Titchmarsh, E. C. Introduction to the theory of Fourier integrals. 3d ed. Chelsea, New York, 1986; x+394 pp.

[597] Tomczak-Jaegermann, N. Banach-Mazur distances and finite-dimensional operator ideals. Longman Scientific & Technical, Harlow; John Wiley & Sons, New York, 1989; xii+395 pp.

[598] Torchinsky, A. Real variable methods in harmonic analysis. Academic Press, New York, 1986; 462 pp.

[599] Torchinsky, A. Real variables. Addison-Wesley, New York, 1988; 403 pp.

[600] Torchinsky, A. Problems in real and functional analysis. Amer. Math. Soc., Providence, Rhode Island, 2015; x+467 pp.

[601] Trenogin, V. A., Pisarevskii, B. M., Soboleva, T. S. Problems and exercises in functional analysis. Nauka, Moscow, 1984; 256 pp. (in Russian).

[602] Trénoguine, V. Analyse fonctionnelle. Mir, Moscow, 1985; 528 pp.

[603] Trèves, J. F. Linear partial differential equations with constant coefficients: existence, approximation and regularity of solutions. Gordon and Breach, New York, 1966; x+534 pp.

[604] Trèves, F. Topological vector spaces, distributions and kernels. Academic Press, New York – London, 1967; xvi+624 pp.

[605] Trèves, F. Introduction to pseudodifferential and Fourier integral operators. V. 1. Pseudodifferential operators. V. 2. Fourier integral operators. Plenum Press, New York – London, 1980; xxvii+299+xi pp., xiv+349+xi pp.

[606] Tricomi, F. G. Istituzioni di analisi superiore (metodi matematici della fisica). 2 ed. CEDAM, Padua, 1970; ix+472 pp.

[607] Triebel, H. Interpolation theory, function spaces, differential operators. North-Holland, Amsterdam – New York, 1978; 528 pp. I

[608] Triebel, H. Theory of function spaces. Birkhäuser, Basel, 1983; 284 pp.

[609] Triebel, H. Higher analysis. Johann Ambrosius Barth Verlag, Leipzig, 1992; 473 pp.

[610] Ulyanov, P. L., Bakhvalov, A. N., D'yachenko, M. I., Kazaryan, K. S., Cifuentes, P. Real analysis in exercises. Fizmatlit, Moscow, 2005; 416 pp. (in Russian).

[611] Vainberg, M. M. Variational methods for the study of nonlinear operators. With a chapter on Newton's method by L. V. Kantorovich and G. P. Akilov. Translated from the Russian. Holden-Day, San Francisco – London – Amsterdam, 1964; x+323 pp.

[612] Vainberg, M. M. Variational method and method of monotone operators in the theory of non-linear equations. Translated from the Russian. Halsted Press, New York – Toronto, 1973; xi+356 pp.

[613] Vainberg, M. M. Functional analysis. Prosveshchenie, Moscow, 1979; 128 pp. (in Russian).

[614] Vakhania, N. N., Tarieladze, V. I., Chobanyan, S. A. Probability distributions in Banach spaces. Translated from the Russian. Kluwer, 1991; xxvi+482 pp.

[615] Varadarajan, V. S. An introduction to harmonic analysis on semisimple Lie groups. Cambridge University Press, Cambridge, 1999; x+316 pp.

[616] Vasilescu, F. H. Analytic functional calculus and spectral decompositions. Reidel, Dordrecht, 1982; xiv+378 pp.

[617] Vilenkin, N. Ya., Gorin, E. A., Kostyuchenko, A. G.; Krasnosel'skiĭ, S. G., Kreĭn, S. G., Maslov, V. P., Mityagin, B. S., Petunin, Yu. I., Rutitskii, Ya. B., Sobolev, V. I., Stetsenko, V. Ya., Faddeev, L. D., Tsitlanadze, E. S. Functional analysis. Translated from the Russian. Wolters-Noordhoff Publ., Groningen, 1972; xv+380 pp.

[618] Vitali, G. Geometria nello spazio Hilbertiano. N. Zanichelli, Bologna, 1929; 283 pp.

[619] Vladimirov, V. S. Equations of mathematical physics. Translated from the Russian. Mir, Moscow, 1984; 464 pp.

[620] Vladimirov, V. S. Methods of the theory of generalized functions. Taylor & Francis, London, 2002; xiv+311 pp.

[621] Vladimirov, V. S., Mikhailov, V. P., Vasharin, A. A., Karimova, Kh. Kh., Sidorov, Yu. V., Shabunin, M. I. A collection of problems on the equations of mathematical physics. Translated from the Russian. Springer-Verlag, Berlin, 1986; 288 pp.

[622] Vulikh, B. Z. Introduction to the theory of partially ordered spaces. Translated from the Russian. Wolters-Noordhoff, Groningen, 1967; xv+387 pp.

[623] Wagschal, C. Topologie et analyse fonctionnelle. Exercices corrigés. Hermann, Paris, 2003; iv+526 pp.

[624] Walter, W. Einführung in die Theorie der Distributionen. 3e Aufl. Bibliogr. Inst., Mannheim, 1994; xiv+240 S.

[625] Weaver, N. Measure theory and functional analysis. World Sci., Hackensack, New Jersey, 2013; viii+202 pp.

[626] Weidmann, J. Linear operators in Hilbert spaces. Springer, New York, 1980; xiii+402 pp.

[627] Weisz, F. Convergence and summability of Fourier transforms and Hardy spaces. Birkhäuser/Springer, Cham, 2017; xxii+435 pp.

[628] Werner, D. Funktionalanalysis. 7e Aufl. Springer-Verlag, Berlin, 2011; xiii+552 S.

[629] Widder, D. V. The Laplace transform. Princeton University Press, Princeton, New Jersey, 1946; x+406 pp.

[630] Wiener, N. The Fourier integral and certain of its applications. Reprint of the 1933 ed. Cambridge University Press, Cambridge, 1988; xviii+201 pp.

[631] Wilansky, A. Functional analysis. Blaisdell – Ginn, New York – Toronto – London, 1964; xvi+293 pp.

[632] Wilansky, A. Topology for analysis. Ginn, Waltham – Toronto – London, 1970; xiii+383 pp.

[633] Willem, M. Analyse harmonique réelle. Hermann, Paris, 1995; ii+236 pp.

[634] Willem, M. Functional analysis. Fundamentals and applications. Cornerstones. Birkhäuser/-Springer, New York, 2013; xiv+213 pp.

[635] Wloka, J. Funktionalanalysis und Anwendungen. De Gruyter, Berlin – New York, 1971; 291 pp.

[636] Wojtaszczyk, P. Banach spaces for analysts. Cambridge University Press, Cambridge, 1991; xiv+382 pp.

[637] Wong, Y.-C. Some topics in functional analysis and operator theory. Science Press, Beijing, 1993; vi+327 pp.

[638] Wouk, A. A course of applied functional analysis. John Wiley & Sons, New York, 1979; xvii+443 pp.

[639] Wulich, B. S. Einführung in die Funktionalanalysis. Teil 1, 2. Teubner, Leipzig, 1961; vii+202 S., iv+171 S. (2nd Russian ed.: Moscow, 1967).

[640] Yosida, K. Functional analysis. 6th ed. Springer-Verlag, Berlin, 1980; xii+501 pp.

[641] Young, N. An introduction to Hilbert space. Cambridge University Press, Cambridge, 1988; x+239 pp.

[642] Zaanen, A. C. Linear analysis. Measure and integral, Banach and Hilbert space, linear integral equations. Interscience, New York; North-Holland, Amsterdam; Noordhoff, Groningen, 1953; vii+601 pp.

[643] Zaidman, S. Functional analysis and differential equations in abstract spaces. Chapman & Hall/CRC, Boca Raton, Florida, 1999; vi+226 pp.

[644] Zălinescu, C. Convex analysis in general vector spaces. World Sci., River Edge, New Jersey, 2002; xx+367 pp.

[645] Zamansky, M. Introduction à l'algèbre et l'analyse modernes. 3me éd. Dunod, Paris, 1967; xviii+435 pp.

[646] Zeidler, E. Nonlinear functional analysis and its applications. I. Fixed-point theorems. Springer-Verlag, New York, 1986; xxi+897 pp.

[647] Zeidler, E. Nonlinear functional analysis and its applications. II/A. Linear monotone operators. Springer-Verlag, New York, 1990; xviii+467 pp.

[648] Zeidler, E. Nonlinear functional analysis and its applications. II/B. Nonlinear monotone operators. Springer-Verlag, New York, 1990; xvi+734 pp.

[649] Zeidler, E. Nonlinear functional analysis and its applications. III. Variational methods and optimization. Springer-Verlag, New York, 1985; xxii+662 pp.

[650] Zeidler, E. Nonlinear functional analysis and its applications. IV. Applications to mathematical physics. Springer-Verlag, New York, 1988; xxiv+975 pp.

[651] Zeidler, E. Applied functional analysis. Main principles and their applications. Springer-Verlag, New York, 1995; xvi+404 pp.

[652] Zeidler, E. Applied functional analysis. Applications to mathematical physics. Springer-Verlag, New York, 1995; xxx+479 pp.

[653] Żelazko, W. Banach algebras. Elsevier, Amsterdam – London – New York; PWN–Polish Scientific Publ., Warsaw, 1973; xi+182 pp.

[654] Zemanian, A. H. Generalized integral transformations. 2nd ed. Dover Publ., New York, 1987; xvi+300 pp.

[655] Zemanian, A. H. Distribution theory and transform analysis. An introduction to generalized functions, with applications. 2nd ed. Dover, New York, 1987; xii+371 p.

[656] Zhu, K. Operator theory in function spaces. 2nd ed. Amer. Math. Soc., Providence, Rhode Island, 2007; xvi+348 pp.

[657] Ziemer, W. Weakly differentiable functions. Springer-Verlag, New York – Berlin, 1989; xvi+308 pp.

[658] Zimmer, R. J. Essential results of functional analysis. University of Chicago Press, Chicago, 1990; x+157 pp.

[659] Zorich, V. A. Mathematical analysis. V. I, II. Translated from the Russian. Springer, Berlin – New York, 2004; xviii+574 pp., xv+681 pp.

[660] Zuily, C. Distributions et équations aux dérivées partielles. Exercices corrigés. Hermann, Paris, 1986; 245 pp.

[661] Zygmund, A. Trigonometric series. V. I, II. 3d ed. Cambridge University Press, Cambridge, 2002; xiv+383 pp., viii+364 pp.

PAPERS:

[662] Abramovich, Y. A., Aliprantis, C. D., Burkinshow, O. *The invariant subspace problem: some recent advances.* Rend. Istit. Mat. Univ. Trieste. 1998. №29. P. 3–79.

[663] Alekhno, E. A. *Some properties of the weak topology in the space L_∞.* Vestsi Nats. Akad. Navuk Belarusi Ser. Fiz.-Mat. Navuk. 2006. №3. P. 31–37 (in Russian).

[664] Alekhno, E. A., Zabreiko, P. P. *On the weak continuity of the superposition operator in the space L_∞.* Vestsi Nats. Akad. Navuk Belarusi Ser. Fiz.-Mat. Navuk. 2005. №2. P. 17–23 (in Russian).

[665] Altman, M. *A general maximum principle for optimization problems.* Studia Math. 1968. V. 31. P. 319–329.

[666] Argyros, S. A., Haydon, R. G. *A hereditarily indecomposable* $\mathcal{L}_\infty$*-space that solves the scalar-plus-compact problem.* Acta Math. 2011. V. 206, №1. P. 1–54.

[667] Arias de Reyna, J. *Dense hyperplanes of first category.* Math. Ann. 1980. B. 249, №2. S. 111–114.

[668] Arutyunov, A. V. *Covering mappings in metric spaces, and fixed points.* Dokl. Akad. Nauk. 2007. V. 416, №2. P. 151–155 (in Russian); English transl.: Dokl. Math. 2007. V. 76, №2. P. 665–668.

[669] Averbuch, V. I., Smolyanov, O. G. *Theory of differentiation in linear topological spaces.* Uspekhi Mat. Nauk. 1967. V. 22, №6. P. 201–260 (in Russian); English transl.: Russian Math. Surveys. 1967. V. 22, №6. P. 201–258.

[670] Averbuch, V. I., Smolyanov, O. G. *The various definitions of the derivative in linear topological spaces.* Uspekhi Mat. Nauk. 1968. V. 23, №4. P. 67–116 (in Russian); English transl.: Russian Math. Surveys. 1968. V. 23, №4. P. 67–113; An addendum: ibid., №5. P. 223–224.

[671] Birkhoff, G., Kreyszig, E. *The establishment of functional analysis.* Historia Math. 1984. V. 11, №3. P. 258–321.

[672] Bittner, L. *A remark concerning Hahn–Banach's extension theorem and the quasilinearization of convex functionals.* Math. Nachr. 1971. B. 51. S. 357–362.

[673] Bogachev, V. I. *Smooth measures, the Malliavin calculus and approximations in infinite dimensional spaces.* Acta Math. Univ. Carolinae, Math. et Phys. 1990. V. 31, №2. P. 9–23.

[674] Bogachev, V. I., Kirchheim, B., Schachermayer, W. *On continuous restrictions of linear mappings between Banach spaces.* Acta Math. Univ. Carolinae, Math. et Phys. 1989. V. 30, №2. P. 5–9.

[675] Bogachev, V. I., Smolyanov, O. G. *Generalized functions obtained by the regularization of nonintegrable functions.* Dokl. Akad. Nauk 2008. V. 419, №6. P. 731–734 (in Russian); English transl.: Dokl. Math. 2008. V. 77, №2. P. 302–305.

[676] Bogachev, V. I., Smolyanov, O. G., Schachermayer, W. *Continuous restrictions of linear mappings.* In: "Mathematics Today'92", pp. 115–126. Kiev, Vyscha Shkola, 1992 (in Russian).

[677] Brown, A., Pearcy, C. *Structure of commutators of operators.* Ann. Math. (2). 1965. V. 82. P. 112–127.

[678] Brown, A., Pearcy, C. *Spectra of tensor products of operators.* Proc. Amer. Math. Soc. 1966. V. 17. P. 162–166.

[679] Brown, A., Pearcy, C. *Operators of the form* $PAQ - QAP$. Canad. J. Math. 1968. V. 20. P. 1353–1361.

[680] Brown, A., Pearcy, C., Salinas, N. *Perturbations by nilpotent operators on Hilbert space.* Proc. Amer. Math. Soc. 1973. V. 41. P. 530–534.

[681] Chernoff, P. R. *Note on product formulas for operator semigroups.* J. Funct. Anal. 1968. V. 2, №2. P. 238–242.

[682] Dixmier, J. *Sur les variétés J d'un espace de Hilbert.* J. Math. Pures Appl. (9). 1949. V. 28. P. 321–358.

[683] Dixmier, J. *Étude sur les variétés et les opérateurs de Julia, avec quelques applications.* Bull. Soc. Math. France. 1949. V. 77. P. 11–101.

[684] Dixmier, J., Foias, C. *Sur le spectre ponctuel d'un opérateur.* Hilbert Space Operators Operator Algebras (Proc. Internat. Conf., Tihany, 1970). Colloq. Math. Soc. Janos Bolyai 5, pp. 127–133. North-Holland, Amsterdam, 1972.

[685] Dmitruk, A. V., Milyutin, A. A., Osmolovskii, N. P. *Lyusternik's theorem and the theory of extrema.* Uspekhi Mat. Nauk. 1980. T. 35, №6. P. 11–46 (in Russian); English transl.: Russian Math. Surveys. 1980. V. 35, №6. P. 11–51.

[686] Douglas, R. G. *On majorization, factorization, and range inclusion of operators on Hilbert space.* Proc. Amer. Math. Soc. 1966. V. 17. P. 413–415.

[687] Fillmore, P. A., Williams, J. P. *On operator ranges.* Advances in Math. 1971. V. 7. P. 254–281.

[688] Fonf, V. P., Johnson, W. B., Pisier, G., Preiss, D. *Stochastic approximation properties in Banach spaces.* Studia Math. 2003. V. 159, №1. P. 103–119.

[689] Gowers, W. T. *A solution to Banach's hyperplane problem.* Bull. London Math. Soc. 1994. V. 26. P. 523–530.

[690] Gurarii, V. I. *Countably linearly independent sequences in Banach spaces.* Uspekhi Mat. Nauk. 1981. V. 36, №5. P. 171–172 (in Russian); English transl.: Russian Math. Surveys. 1981. V. 36, №5. P. 151–152.

[691] Kalisch, G. K. *On operators on separable Banach spaces with arbitrary prescribed point spectrum.* Proc. Amer. Math. 1972. V. 34, №1. P. 207–208.

[692] Kaufman, R. *Representation of Souslin sets by operators.* Integral Equat. Oper. Theory. 1984. V. 7, №6. P. 808–814.

[693] Lacey, M. T. *Carleson's theorem: proof, complements, variations.* Publ. Mat. 2004. V. 48, №2. P. 251–307.

[694] Levitin, E. S., Milyutin, A. A., Osmolovskii, N. P. *Conditions of high order for a local minimum in problems with constraints.* Uspekhi Mat. Nauk. 1978. V. 33, №6. P. 85–148 (in Russian); English transl.: Russian Math. Surveys. 1978. V. 33, №6. P. 97–168.

[695] Lifshits, E. A. *Ideally convex sets.* Funkt. Anal. Pril. 1970. V. 4, №4. P. 76–77 (in Russian); English transl.: Funct. Anal. Appl. 1970. V. 4, №4. P. 330–331.

[696] Mazliak, L. *The ghosts of the École Normale. Life, death and destiny of Réne Gateaux.* Statist. Sci. 2015. V. 30, №3. P. 391–412.

[697] Mazur, S., Orlicz, W. *Sur les espaces métriques linéaires. II.* Studia Math. 1953. V. 13. P. 137–179.

[698] Narici, L. *On the Hahn–Banach theorem.* Advanced courses of mathematical analysis. II, pp. 87–122, World Sci., Hackensack, New Jersey, 2007 (http://at.yorku.ca/p/a/a/o/58.htm).

[699] Nemirovskiĭ, A. S., Semenov, S. M. *On polynomial approximation of functions on Hilbert space.* Mat. Sb. (N.S.) 1973. V. 92, №2. P. 257–281 (in Russian); English transl.: Math. USSR-Sbornik. 1973. V. 21, №2. P. 255–277.

[700] Nikol'skaya, L. N. *Structure of the point spectrum of a linear operator.* Matem. Zametki. 1974. V. 15, №1. P. 149–158 (in Russian); English transl.: Math. Notes. 1974. V. 15, №1. P. 83–87.

[701] von Neumann, J. *Zur Theorie des Unbeschränkten Matrizen.* J. Reine Angew. Math. 1929. B. 161. S. 208–236.

[702] Odell, E., Schlumprecht, Th. *A universal reflexive space for the class of uniformly convex Banach spaces.* Math. Ann. 2006. B. 335, №4. S. 901–916.

[703] Pearcy, C. *On commutators of operators on Hilbert space.* Proc. Amer. Math. Soc. 1965. V. 16. P. 53–59.

[704] Shevchik, V. V. *On subspaces of a Banach space that coincide with the ranges of continuous linear operators.* Rev. Roumaine Math. Pures Appl. 1986. V. 31, №1. P. 65–71.

[705] Shkarin, S. A. *On the Rolle theorem in infinite-dimensional Banach spaces.* Mat. Zametki. 1992. V. 51, №3. P. 128–136 (in Russian); English transl.: Math. Notes. V. 51, №3-4. P. 311–317.

[706] Shkarin, S. A. *Continuous restrictions of linear operators.* Infin. Dimens. Anal. Quantum Probab. Relat. Top. 2001. V. 4, №1. P. 121–136.

[707] Smolyanov, O. G. *Almost closed linear subspaces of strict inductive limits of sequences of Fréchet spaces.* Matem. Sb. (N.S.) 1969. V. 80, №4. P. 513–520 (in Russian); English transl.: Math. USSR-Sbornik. 1969. V. 9, №4. P. 479–485.

[708] Smolyanov, O. G., Shkarin, S. A. *Structure of spectra of linear operators in Banach spaces.* Matem. Sbornik. 2001. V. 192, №4. P. 99–114 (in Russian); English transl.: Sbornik Math. 2001. V. 192, №4. P. 577–591.

[709] Sudakov, V. N. *Criteria of compactness in function spaces.* Uspehi Mat. Nauk. 1957. V. 12, №3. P. 221–224 (in Russian).

[710] Tsar'kov, I. G. *Smoothing of Hilbert-valued uniformly continuous mappings.* Izv. Ross. Akad. Nauk Ser. Mat. 2005. V. 69, №4. P. 149–160 (in Russian); English transl.: Izv. Math. 2005. V. 69, №4. P. 791–803.

[711] Werner, D. *Recent progress on the Daugavet property.* Irish Math. Soc. Bulletin. 2001. V. 46. P. 77–97.

[712] Zabreiko, P. P. *A theorem for semiadditive functionals.* Funk. Anal. Pril. 1969. V. 3, №1. P. 86–88 (in Russian); English transl.: Funct. Anal. Appl. 1969. V. 3, №1. P. 70–72.

Index

Printed in the United States
By Bookmasters